EVOLUTION AND THE AQUATIC ECOSYSTEM
Defining Unique Units in Population Conservation

D1284021

Support for the Symposium on Evolution and the Aquatic Ecosystem:
Defining Unique Units in Population Conservation
and publication of this proceedings
was provided by

American Fisheries Society
Genetics Section
Western Division

Edison Electric Institute

Environmental Protection Agency

Rayma and Norman Sharber

Turner Foundation Inc.

U.S. Department of Agriculture
Forest Service

U.S. Department of Commerce
National Marine Fisheries Service

U.S. Department of the Interior
Bureau of Land Management
Bureau of Reclamation
Fish and Wildlife Service
National Biological Service
National Park Service

Evolution and the Aquatic Ecosystem: Defining Unique Units in Population Conservation

Jennifer L. Nielsen
Editor

Dennis A. Powers
Consulting Editor

American Fisheries Society Symposium 17

*Proceedings of the Symposium on Evolution and the Aquatic Ecosystem:
Defining Unique Units in Population Conservation*

Held at Monterey, California, USA
23–25 May 1994

American Fisheries Society
Bethesda, Maryland
1995

The American Fisheries Society Symposium series is a registered serial.
Suggested citation formats follow.

Entire book

Nielsen, J. L., editor. 1995. Evolution and the aquatic ecosystem: defining unique units in population conservation. American Fisheries Society Symposium 17, Bethesda, Maryland.

Article within the book

Healey, M. C., and A. Prince. 1995. Scales of variation in life history tactics of Pacific salmon and the conservation of phenotype and genotype. Pages 176–184 in J. L. Nielsen, editor. Evolution and the aquatic ecosystem: defining unique units in population conservation. American Fisheries Society Symposium 17, Bethesda, Maryland.

Library of Congress Catalog Card Number: 95-083247

ISBN 0-913235-94-6   ISSN 0892-2284

Printed in the United States of America on recycled, acid-free paper.

American Fisheries Society
5410 Grosvenor Lane, Suite 110
Bethesda, Maryland 20814-2199, USA

CONTENTS

PREFACE

There is little doubt that the next wave of scientific discoveries and medical breakthroughs will be deeply vested in our growing knowledge of genetic architecture, gene function, and evolutionary processes. Since the time of Charles Darwin, the picture of evolution has portrayed a world in which biodiversity results from modification by natural selection through descent. The underlying genetic mechanism is presumed to be gene mutation—random change in nucleic acid sequence—that confers an adaptive advantage or disadvantage on an individual; natural selection then determines whether that mutation will be passed to subsequent generations or not. The Darwinian model supplanted Lamarck's earlier view that organisms can acquire adaptations by volition and pass them to their offspring. By either of these theories, the evolutionary forces leading to biodiversity act most effectively at the level of individuals, not at the levels of groups, populations, or species. Yet it is these higher levels of biological organization that must be addressed by public policies intended to preserve biodiversity. Linking the evolutionary origins of diversity to the practical conservation of that diversity is one of the major scientific challenges of our day.

New genetic concepts threaten to blur the distinction between Darwinian and Lamarckian modes of evolution. One of these is genetic mutualism, which involves cooperation between individuals of different species and between different components of an individual's genetic material in ways that lead to evolutionary shifts. Another is epigenetic inheritance, in which genes produce biochemical effects (epimutations) that alter the tempo and diversity of evolutionary change. Still others are directed mutation (the ability of cells to undergo spontaneous mutation while in a resting state) and adaptive mutation (the controversial idea that some cells are able to sense their environment and to use this information to produce specific beneficial mutations). One of the deep schisms in evolutionary theory today involves the proposition that subgenetic modes of cellular heredity involved with the mechanisms of differentiation and development (i.e., epigenesis) can be translated and transmitted across generations.

The creative and controversial climate surrounding molecular and cellular genetics and evolution today, however, must not distract us from developing more practical, effective ways to measure and define the life histories and evolutionary biologies of the organisms that make up natural biological diversity. Molecular approaches to diagnosing units of populations and species for conservation purposes are improving, but we cannot turn our backs on the studies of morphology, ecology, and behavior that have shaped our concepts of the natural world for the last century. As John Avise stated in his 1994 volume *Molecular Markers, Natural History and Evolution* (Chapman and Hall, New York), molecular geneticists must be careful not to stray too far from the "real stuff" of adaptive evolution. I feel that the next 10–20 years will bring important interactions among presently divergent biological disciplines as we probe the underlying mechanisms of evolution—mechanisms that are probably enormously diverse and interactive among themselves.

The need for disciplinary interface led to the framework I developed for an American Fisheries Society (AFS) conference on "Evolution and the Aquatic Ecosystem," which was held May 23–25, 1994, in Monterey, California. Because my own scientific training had been broad, I was able to recruit a wide array of formidable talent to address the evolutionary basis of aquatic biodiversity in a symposium framework. The readiness of such distinguished scientists to openly discuss empirical, theoretical, and philosophical insights gave the conference high levels of energy and enthusiasm. Their perspectives were further broadened by a panel of resource agency administrators who opened a dialog on the roles of science and government in developing policies to conserve aquatic diversity. The protracted discussions that followed each conference session reflected enthusiasm for this interdisciplinary approach from all who attended.

I extended the approach by giving conference papers the benefit of cross-disciplinary peer reviews. This led to controversy and even discomfort in some cases, but it added to the overall value of the perspectives included in this volume. The proceedings contain all but four of the papers given in Monterey; time limitations and other obligations prevented Oliver Rider, Eldridge Bermingham, Linda Park, and Paul Bentzen from submitting their papers for publication. Their contributions to the symposium itself were highly valued nonetheless.

The Monterey meeting led to new and better dialogue among individual scientists and to clearer goals of communication and cooperation among agencies, the public, and academia. I feel the results presented here represent a model of the effects achievable when large, complex environmental problems are approached through sound interdisciplinary science.

I am indebted to AFS, especially to Pamela McClelland and the Society's Fisheries Action Network, for providing the opportunity to create and implement this forum. I also thank my consulting editor Dennis Powers for providing a comfortable working environment in which to accomplish my editing chores and the U.S. Forest Service, Pacific Southwest Research Station, for salary and support throughout the evolution of this project.

Thanks are also due to Christine Gan and Cindy Carpanzano who attended to a myriad of details before, during, and following the symposium and who contributed significantly to the success of the project. Dale Burkett is thanked for his skillful facilitation of the interchange of ideas between federal agency administrators and the audience. I am grateful to the California–Nevada Chapter of the AFS and its members for their logistical support at the conference. Special thanks are due to Susan Ellis, Ramona Swenson, Lesa Meng, Dave Stride, and Marie Piché for their generous support and assistance during the conference. I'd like to thank Julie Packard of the Monterey Bay Aquarium who provided an educational but fun respite from our work by donating an evening at the aquarium. Many people were involved in raising funds in support of the symposium and its products; they all worked hard in the belief that this project would make an important contribution to the future of conservation biology.

This symposium resulted in two products that will carry forward the information presented and ideas shared during the conference. Jud Monroe is recognized for taking on the enormous task of synthesizing the symposium presentations into a succinct Executive Summary for distribution to symposium participants as well as members of Congress and their staffs. The real heroes are the more that 85 individuals who reviewed the manuscripts presented in this proceedings. These reviewers represent a broad range of disciplines and a diversity of geographic locales. The attention they lavished on these papers resulted in a book of the highest scientific merit. Finally, editorial and production quality was upheld by Bob Kendall, Beth Staehle, Eva Silverfine, and Janet Harry of the American Fisheries Society Editorial Office.

I dedicate this book to three advisors from my graduate days at the University of California at Berkeley: George Barlow, Howard Bern, and Kelley Thomas. They set the interdisciplinary stage for me by being the best at their chosen disciplines but confident and open to the influence of divergent minds. I continue to draw inspiration from their example.

JENNIFER L. NIELSEN
Editor

INTRODUCTION

J. L. NIELSEN

U.S. Forest Service, Pacific Southwest Research Station and Hopkins Marine Station
Stanford University, Pacific Grove, California 93950–3094, USA

D. A. POWERS

Hopkins Marine Station, Stanford University

As an evolutionist, I am uncomfortably aware that there is always more than one way of achieving an end.
(M. Ruse, *Struggle for the Soul of Science*, 1994)

This volume contains the proceedings of the May 1994 symposium "Evolution and the Aquatic Ecosystem: Defining Unique Units in Population Conservation," which was held in Monterey, California. It presents the findings and opinions of many important scientists and administrators on the value and feasibility of defining unique units of aquatic populations for the purpose of conservation. The outline of this book generally follows the symposium's structure. Three perspectives—on conservation philosophy (Callicott), the "evolutionarily significant unit" (Waples), and ecosystem management (Mundy et al.)—open the proceedings. Scientific papers are organized in four categories—morphology and systematics, behavior, genetics, and ecosystems—each introduced by the Monterey session chair for that category. The volume concludes with highlights of a facilitated discussion between a panel of U.S. natural resource agency administrators and the audience over the scientific basis of conservation policy, and with the individual remarks of the five federal panelists themselves.

At the time this project was conceived, controversy surrounded the use of evolutionarily significant units (ESUs) to define populations of aquatic organisms. Robin Waples, who wrote the National Marine Fisheries Service's (NMFS) 1991 working definition of ESUs for Pacific salmon, had laid out two criteria for such units: "an ESU is a population (or group of populations) that (1) is substantially reproductively isolated from other conspecific population units, and (2) represents an important component in the evolutionary legacy of the species." These criteria were not universally accepted or clearly understood within the scientific, regulatory, and public arenas. Both were challenged as subjective because of nonspecific terms such as "substantially" and "important." These concerns carried into the Monterey conference, where many questions were voiced over how such criteria could be objectively and consistently applied to real-world situations. In this volume, Waples addresses these concerns at considerable length. Despite the initial controversy, the ESU concept pervades the accompanying papers as both a positive and a contrary model by which to judge population structure in aquatic ecosystems.

We feel that the ESU concept is a direct intellectual descendent of both the "biological species concept," which requires absolute reproductive isolation for the definition of a species, and the "evolutionary species concept," which replaces reproductive isolation with spatial or temporal isolation within a population. As such, the idea has its roots in an early debate between two divergent branches of systematics over the spatial organization and temporal continuity of species and population structure (original papers from both sides of this issue are cited by Mayden and Wood). This ongoing conflict between a focus on reproductive isolation and one on evolutionary directionality in defining units of natural diversity provides a context for investigating ESUs in aquatic ecosystems.

In the National Research Council's *Science and the Endangered Species Act* (Washington, DC, 1995), the National Committee on Scientific Issues in the Endangered Species Act, the National Board on Environmental Studies and Toxicology, and the National Commission on Life Sciences all agreed that the "distinct population segment" language of the Endangered Species Act (ESA) is consistent with our contemporary understanding of "evolutionary units." They defended the practice of listing subspecies and distinct population segments (which presently constitute about 20% of listed taxa: T. Eisner, J. Lubchenco, E. O. Wilson, D. S. Wilcov, and M. J. Bean, *Science* 268:1231, 1995) if these subunits represent a distinct evolutionary pathway or "evolutionary unit" (EU). The report takes issue with the "vague" use of "significant" in ESUs, and urges that reproductive isolation remain the fundamental criterion for listing taxa under the ESA. (The scientific literature, however, gives little evidence of absolute reproductive isolation between subspecies, so EUs defined solely by this criterion necessarily would be species.) The conceptual difference between EUs as defined by the National Research Council's publication and ESUs as defined by the NMFS policy on Pacific salmon harkens back to the old argument between scientists who supported the biological species concept (which requires reproductive isolation), and those who were dissatisfied with the nondimensionality of such a requirement for organisms without sexual reproduction.

In any case, the concepts of evolutionary unit and evolutionarily significant unit both refer to populations that maintain their identity at appropriate temporal and spatial scales within a unique evolutionary lineage. The argument over "significance" as a decision criterion for listings under the ESA relates not to evolutionary mechanisms themselves but to our currently poor understanding of those mechanisms. Thus, we believe, the argument is over policy and public liability, not over scientific viability, and we expect that a debate over evolutionary "significance" will follow us into the next century. The debate must not distract us from the important conservation issues immediately at hand in every aquatic ecosystem we care to survey.

A second recurring argument within this volume concerns the role of molecular genetics in the definition of an "important component in the evolutionary legacy of the species." Waples (herein) argues that the evolutionary legacy of a species is "the genetic variability that is a product of past evolutionary events and that represents the reservoir upon which future evolutionary potential depends." Others suggest alternative means of judging evolutionary history at the population level in aquatic ecosystems. Among these are *measurable* genetic variability (Phillips and Ehlinger, Utter et al., Allendorf, Moritz et al., Stepien, Dizon et al., Thorgaard et al.), life history characteristics (Healey and Prince, Wood, Baylis), and heritable character traits associated with morphology or behavior (Barlow, Hard, Noakes et al., Baylis, Stauffer et al.). The general flavor of these papers supports an effort to gain agreement from as many scientific disciplines as possible before ESUs or EUs are defined.

The need for interaction and dialog among disciplines was well summarized by Robert Behnke when he pointed out that defining ESUs is not simple; "one size will not fit all." Organisms have ecological contexts that must be understood if aquatic biodiversity is to be adequately protected. Ecological diversity is a template for biological diversity, and environmental variability through space and time must be diagnosed at scales ranging from

stream segments (Grossman et al.) and small wetlands (Sada et al.) to entire landscapes (Bisson, Reeves et al.). Populations of a species typically are distributed among patchy habitats; whether individuals can move between populations, the extent to which they do so, and the historical relationships among the populations themselves strongly influence definitions of unit biodiversity (Fausch and Young, Li et al., Schlosser and Angermeier). Some scientists, recognizing that species do not evolve in biological isolation, urge that species assemblages be considered the units of diversity for conservation purposes (Mina and Golubtsov, Angermeier and Schlosser). Clearly, ecological and biogeographic knowledge must be used to broaden the efficiency of ESU definitions for aquatic ecosystems.

In practical terms, however, agreement among disciplines depends on each discipline's ability to relate its information to a concept, logic, or model shared by all. Such common ground may be difficult to find. For example, homology (similarity of structure or function traceable to common ancestry) underlies several calls for phylogenetic techniques in this book (Bernatchez, Smith et al., Mayden and Wood). Other authors, though, find it difficult to apply this or any model when spatial and temporal scales cannot be defined for the evolution of particular populations or when different parts of a genome within a population give discordant information (Healey and Prince, Smith et al., Mundy et al.). Conversely, patterns of variation in genetic markers may not match patterns deduced by other disciplines.

We are still a long way from understanding the genetic variation that underlies most phenotypic traits despite intensive studies of fruit flies, mice, worms, and humans. We are in a new era in technology and resolution that will improve our understanding of molecular genetics. But using this understanding in isolation from natural history, ecology, ethology, and systematics divorces the power of new genetic techniques from population-scale issues in conservation and resource management. It is only by integrating multiple biological disciplines that we can bring our best science to bear effectively on the environmental problems of our era.

The loss of genetically distinct populations within species is, at the moment, at least as important as the loss of entire species. Once a species is reduced to a remnant, its ability to benefit humanity ordinarily declines greatly, and its total extinction in the relatively near future becomes much more likely. By the time an organism is recognized as endangered, it is often too late to save it. (P. R. Ehrlich, *Biodiversity*, 1988)

PART ONE

PERSPECTIVES

American Fisheries Society Symposium 17:3–7, 1995

Conservation Ethics at the Crossroads

J. Baird Callicott

Department of Philosophy, University of North Texas and University of Wisconsin at Stevens Point
Stevens Point, Wisconsin 54481, USA

Abstract.—For most of the twentieth century no consensus on conservation values existed, but, in North America, two simple conflicting philosophies of conservation—the wise use of natural resources and wilderness preservation—prevailed. The former philosophy dominated state and federal conservation agencies. Hence fisheries were managed for maximum production of sport and meat. The latter philosophy dominated private conservation organizations, such as the Sierra Club. But because native fishes were out of sight (under scenic waters), they were mostly out of preservationists minds and rarely benefited from protectionist efforts. Today both philosophies of conservation are obsolete. They seem to be giving way to a new conservation cleavage reflecting a long-standing chasm in ecology between a focus on species populations and communities, on the one hand, and a focus on ecosystems, on the other. From the former point of view, preserving species (and subspecies) or, aggregatively, preserving "biodiversity," is the overriding goal of conservation ethics. From the latter point of view, preserving ecosystem functions, now called ecosystem health, is the overriding goal. Fortunately, these recent conceptions of conservation ethics are not as mutually exclusive as were resource management versus wilderness preservation. Preserving biodiversity, however, is a more stringent conservation norm, because ecosystem health may not necessarily be compromised if a rare subspecies is replaced by a more commonplace cousin or even if a species proper is replaced by another that fills the same niche. Both of these recent conservation ethics, but especially biodiversity preservation, have been complicated by the current emphasis on dynamism in ecology. Biodiversity preservation has thus become the preservation of evolutionarily significant units; meanwhile proponents of ecosystem health struggle to define the health of ecosystems in circumstances of constant perturbation and shifting patch mosaics.

To an outside observer, like me, concern for conserving species, subspecies, and varieties of aquatic organisms that constitute evolutionarily significant units (ESUs) appears to be an astonishing alteration of course in fisheries management—a field in which a devil-may-care policy of stocking any and every pond and lake, stream and river with nonnative game fish was recently so dominant. For example, E. P. Pister (Desert Fishes Council, personal communication) reports that in the eastern Sierra recreation area alone "hatcheries reared 2.5 million catchable-size trout weighing 576 tons for planting in roadside waters" and that put-and-take "programs of similar magnitude are conducted in other areas of the state [of California] and the West."

In addition to the scientific uncertainties surrounding the ESU concept, there are also axiological uncertainties. Why *should* conserving ESUs be a primary goal of fisheries management? When any new concept in any area of policy comes along, it's a good idea to step back and look at it in historical perspective.

Preservationism and Resourcism

The myth of superabundance—the belief that the natural bounty of the Americas was inexhaustible—prevailed until the late nineteenth century when the slaughter of the bison to near extinction and the closing of the frontier called it into question. But the need for conservation dawned on a few far-sighted thinkers even earlier. Henry David Thoreau (1863), for one, saw a mixed blessing at best in the march of Euro-American settlement and civilization across the land. George Perkins Marsh (1864), for another, was alarmed by extensive deforestation and its effects on soils, waters, and microclimates. By the end of the nineteenth century two simple, distinct, and conflicting conservation philosophies began to take definite shape.

The first derived from the work of Thoreau and his literary and artistic contemporaries who expressed—through the medium of painting as well as prose and poetry—a newly discovered beauty in the wild American landscape. Its most able turn-of-the-century spokesperson was John Muir (1894). The basic idea was to protect pristine natural scenery from despoliation as a consequence of economic exploitation—mining, logging, grazing, and plowing. The prime candidates for preservation were places, such as the Yosemite and Hetch Hetchy valleys in the California Sierra and the Yellowstone Plateau in Wyoming, that were extraordinarily scenic and picturesque and that offered the opportunity to enjoy a suite of solitary pursuits—everything from hunting and fishing to alpine camping and hiking to the cultivation of religious experience and philosophical insight. Elsewhere, I have called this philosophy of conservation the "romantic-transcendental preservation ethic" (Callicott 1991). One might call it preservationism for short.

The second turn-of-the-century conservation philosophy derived from the work of Marsh. Its most able spokesperson was Gifford Pinchot (1947). Pinchot did not think of conservation as protecting Nature from economic exploitation. Quite the contrary. He understood conservation to be the "wise use" of "natural resources." The foregoing generation of pioneers erred, in Pinchot's opinion, not in their preoccupation with extracting material wealth from the land, but in the monstrous waste resulting from the way they went about it, the inequitable distribution of the booty, and the ruination that they left in their wake. Elsewhere, I have called this the "resource conservation ethic" (Callicott 1991). One might call it resourcism for short. By whatever name, my colleague, Curt Meine, has summed it up quite concisely.

> The guiding principle of utilitarian conservation was to manage resources so as to produce commodities and services "for the greatest good of the greatest number for the longest time." To this end, wild nature was not to be preserved, but actively manipulated by scientifically informed experts to improve and sustain yields. Those yields were to be harvested and processed efficiently, and the economic gains allocated equitably. How . . . ? By strengthening the oversight role of government, enacting science-based regulations and resource management practices, developing the resources with a minimum of waste, and distributing the benefits of development fairly among all users. (Meine 1995:12.)

Resourcism became institutionalized in federal and state agencies (Fox 1981). First, the Forest Service was created in 1905, and Pinchot was appointed Chief. Eventually there followed the federal Bureau of Land Management and Fish and Wildlife Service and state departments of fish and game or natural resources—all guided by the precepts of resourcism. The wilderness preservation philosophy inspired the creation of a host of private conservation organizations (Fox 1981). John Muir himself founded the Sierra Club, which was followed by the Wilderness Society, the Nature Conservancy, Defenders of Wildlife, and so on. There are exceptions to the correlation of government agencies with resourcism and private conservation organizations with preservationism. The National Park Service, for example, is increasingly oriented toward preservation, whereas such private hook and bullet groups as Trout Unlimited and Ducks Unlimited clearly remain oriented toward commodity production.

A three-way struggle over American natural resources has, thus, characterized most of the twentieth century. Entrepreneurs invoking the preconservation era seventeenth-century philosophy of private property, articulated by the English philosopher John Locke (1961), think that they have a God-given right to own real estate and all the organisms on, over, or under their lands and waters and to do with them whatever they please. Federal and state resource managers believe that they have a statutory mandate to regulate entrepreneurial exploitation of natural resources in the interest of equity and maximum sustainable yield, especially on public lands and waters. And preservationists of various stripes plead for restraint on the human exploitation of Nature, hoping to protect places of great beauty from all extractive uses.

Aquatic resources, however, have been somewhat anomalous in that the Lockean philosophy of private property and the Muirean philosophy of wilderness preservation have played a diminished role, leaving them largely to the undisputed aegis of the Pinchovian resource conservation philosophy. Unlike terrestrial organisms (e.g., trees), fish are rarely claimed as private property—because the oceans are a commons and, in many jurisdictions, property rights end at streambanks and shorelines. And because fish are out of sight under scenic waters, they were mostly out of preservationists' minds, and, thus, unlike terrestrial fauna, fish have only recently (and incompletely) benefited from protectionist efforts, even in the national parks.

The Advent of Ecology

The development of ecology as a science and then as a worldview began to force a change in the bifurcated state of classical American conservation philosophy. In 1939, Aldo Leopold, who began his career in the Forest Service and thus in the resourcist school of thought, succinctly pointed out how ecology undermines Pinchot's basic assumptions:

> Ecology is a new fusion point for all the natural sciences. . . . The emergence of ecology has placed the economic biologist in a peculiar dilemma: with one hand he points out the accumulated findings of his search for utility, or lack of utility, in this or that species; with the other he lifts the veil from a biota so complex, so conditioned by interwoven cooperations and competitions, that no [one] can say where utility begins or ends. No species can be "rated" without the tongue in the cheek; the old categories of "useful" and "harmful"

have validity only as conditioned by time, place, and circumstance. The only sure conclusion is that the biota as a whole is useful, and biota includes not only plants and animals, but soils and waters as well. (Leopold 1939:727.)

Aldo Leopold, as everyone knows, was a prophet. And prophets, as everyone also knows, are ignored in their own time. So no one paid much attention to Leopold for another quarter century. Foresters went on clear-cutting "senescent" stands of timber (that is, old growth) and planting even-aged monocultures of fast-growing trees. Wildlife managers went on promoting deer, "controlling varmints" (that is, exterminating predators), releasing pen-raised pheasants, and ignoring nongame reptiles, birds, and mammals. And fish managers went on raising game fish in hatcheries, stocking them everywhere, and purging "trash fish" (Pister 1987).

But when an environmental crisis was acknowledged in the 1960s, *A Sand County Almanac* became the bible of the contemporaneous conservation movement. "A thing is right when it tends to preserve the integrity, stability, and beauty of the biotic community; it is wrong when it tends otherwise" became the eleventh commandment (Leopold 1949:224–225). This turn of events shifted the philosophical balance of power in favor of the wilderness preservation school of thought. Originally and publicly, preservationism was no less anthropocentric and utilitarian in its axiological foundations than was resourcism (Muir 1901). Outdoor recreationists, nature aesthetes, solitude seekers, landscape painters, nature poets, and transcendental philosophers put pristine natural areas to their own preferred "higher" uses without consuming them and without spoiling them—at least not to the extent that loggers, ranchers, miners, and other "developers" did. But, going all the way back to Muir, if not to Thoreau, there was a nonanthropocentric, non-utilitarian undercurrent running in the wilderness preservation philosophy. As early as the 1860s Muir (1916), in a journal he kept as he walked from Kentucky to Florida in 1867–1868, was privately broaching the idea that nature had rights. Thus a century later, it was easy to think that the best way to implement the nonanthropocentric Leopold land ethic was to preserve and restore as much wilderness as possible.

From the beginning, moreover, the concept of a natural balance or equilibrium had enjoyed a central place in ecology. It is implicit, for instance, early on in Frederick Clements's notion of succession to climax (Clements 1916). During the Great Depression, Arthur Tansley (1935) introduced the non-metaphorical concept of an ecosystem to replace the older organismic and community paradigms. Due to the influence of Eugene Odum (1959), by midcentury the ecosystem paradigm—which had originally been modeled on physics, more specifically on thermodynamics—reigned in ecology. Stability, balance, or equilibrium was conceived, accordingly, as working like a thermostat via negative feedback. The wilderness preservation philosophy fit this model to a tee. Conservation consists in preserving the balance of nature. When human beings meddle with poorly understood, devilishly complex ecological interactions we risk upsetting natural equilibria. Ideally, we should interfere not at all. But because that is impossible, we should preserve as much undisturbed pristine nature as we can—consistent with our own human needs, not wants—and modify those ecosystems that we must exploit as gently and cautiously as we can.

The New Dynamism

Since about 1975, however, the old "balance of nature" idea has been under unremitting and withering attack (Shrader-Frechette and McCoy 1993). In contemporary ecology, perturbation by wind, fire, flood, drought, disease, and so on is regarded as a normal, not abnormal, state of affairs. And ecosystems, if we can even still meaningfully speak of such "things," are believed to be constantly changing (Botkin 1990).

When the top-down, physics-like ecosystem paradigm—introduced by Tansley (1935), quantified by Raymond Lindeman (1942), and systematized by Odum (1959)—proved to be less predictive than promised, the bottom-up, biology-based community paradigm made a comeback in ecology (Golley 1994). This turn of events has given rise to a new bifurcation in conservation philosophy.

A biotic community is composed of species populations and their interactions. The new field of conservation biology has grown out of the old utilitarian wildlife (and fishery) management fields—the limitations of which Leopold (1939) pointed out—and community ecology (Primack 1993). In conservation biology, preserving populations of species and subspecies and genetic diversity within and interactions among populations—or, aggregatively, preserving "biodiversity"—is the overriding goal of conservation ethics (Primack 1993). But community compositions are dynamic, constantly changing (Botkin 1990). New species invade communities where formerly they were absent—with or without human assistance (Kornberg and Williamson 1988).

Extirpations occur, sometimes followed by recolonization, sometimes not—with or without human interference (MacArthur and Wilson 1967). And species extinctions happened long before human beings came on the scene and will go on happening long after we are gone (Raup and Sepkoski 1984). How can we conceive of conservation in a dynamic, ever-shifting matrix?

Because species extinction is a natural process—indeed 99% of all species that ever existed on Earth are now extinct—some anticonservationists argue that there can be nothing objectively wrong with the episode of anthropogenic species extinction that is now occurring (Shrader-Frechette 1989). There is an obvious and definitive retort to such an argument. There may be nothing objectively wrong, per se, with anthropogenic species extinction, but the temporal and spatial scales at which it is presently occurring are abnormal (Wilson 1988). We human beings are perhaps Earth's first biological agent of a rare, abrupt, mass extinction event (Raup 1991). In general, therefore, consideration of both spatial and temporal scale is crucial in taking account of change in a sound conservation philosophy. The changes that industrialized human beings are imposing on biotic communities are far more widely distributed and far more frequent than were historical changes. Wind or fire, for example, often leveled a patch of old growth forest (Norse 1990). The effect was not altogether unlike that of clear-cutting. But forestry in old growth has reversed historic patch dynamics. Stands of old growth are becoming islands in a sea of clear-cuts and second growth, rather than vice versa (Harris 1984). And silvicultural rotations are far more frequent than those resulting from preindustrial disturbance regimes. In aquatic ecosystems, very slow natural processes, such as the desiccation of the Great Basin after the last Ice Age, have caused extinctions and genetically isolated remnant species and subspecies populations (Smith 1981). Other slow natural processes, such as erosion, have resulted in the joining of watersheds, and that, in turn, may have caused extinctions and hybridizations between originally distinct species and subspecies (Smith 1981). So the sorting and mixing of species is also a natural process. Past fisheries management, however, has abnormally accelerated the rate of such processes and expanded the regions in which they occur (Miller et al. 1989). I suggest that the concept of evolutionarily significant units is a way of building change *at the appropriate spatial and temporal scales* into conservation biology's program of preserving biodiversity (Vogler and DeSalle 1994).

Ecosystem—as opposed to community—ecology is dealing with issues of appropriate spatial and temporal scale in a different way. In Lindeman's classic paper, "The Trophic-Dynamic Aspect of Ecology," the species of the study area, Cedar Bog Lake, were grouped according to trophic level (Lindeman 1942). What was important about a species, from the trophic-dynamic perspective, was its place in the food web and the niche it filled in the flow of energy from autotrophs to the top carnivores, not its species-specific characteristics (Golley 1994). From this point of view, conservation consists of preserving the niche, not the species that happens to fill it. Hence, if one species replaces another—say, brown trout *Salmo trutta* replace golden trout *Oncorhynchus aguabonita* in California waters—the ecosystem remains essentially unchanged. But ecosystems, no less than biotic communities, change—with or without human interference (Botkin 1990). Again, spatial and temporal scales are crucial in evaluating change. Some routine historic changes—such as the draining of a pond caused by the breaching of a beaver dam—are sudden, but local; and some—such as sedimentation—are ubiquitous, but gradual. From an ecosystem ecology point of view, only the imposition of orderly anthropogenic change is consistent with a conservation ethic. In other words, from a dynamic systems point of view, conservation consists of normalizing the extent and frequency of sudden drastic change and damping the rate of widespread change so as to preserve ecological functions—energy flows, nutrient cycling, soil stability, water quality, and so on—in the midst of change. In short, anthropogenic change is consistent with conservation ethics, from the ecosystem point of view, if "ecosystem health" is maintained (Costanza et al. 1992).

Whatever else the new policy of ecosystem management may mean, it certainly means that resource extraction is subordinate to conservation desiderata. The new bifurcation in conservation philosophy—the preservation of biodiversity versus the preservation of ecosystem health—is, thus, not so mutually exclusive as was the old resourcism versus preservationism. Preserving biodiversity (rendered dynamic by the concept of ESUs) is, from one point of view, a more stringent conservation norm than is ecosystem health, because ecosystem functions may remain unimpaired when one species replaces or hybridizes with another. However, preserving ecosystem health is, from another point of view, a more stringent conservation norm than is preserving ESUs, because such units might be preserved in small reserves or even ex situ, while disorderly and ecologically dysfunctional changes are imposed elsewhere. As these observations suggest, the preservation of biodiversity and of ecosystem health are, rather, complementary conservation norms.

Indeed, in combination they more finely articulate the conservation ideal envisioned by Aldo Leopold (1949): a human harmony with nature. We have an obligation to preserve ESUs. We cannot, however, make a nature reserve out of the whole planet. Complementing biodiversity sanctuaries, we also have an obligation to carry on our extractive economic activities in ways that do not compromise ecosystem health. That is, we should also strive, as Leopold (1991) wrote in 1935, to integrate beauty and utility.

Acknowledgments

Support was provided in part by a grant from the Great Lakes Fishery Commission.

References

Botkin, D. B. 1990. Discordant harmonies: a new ecology for the twenty-first century. Oxford University Press, New York.

Callicott, J. B. 1991. Conservation ethics and fishery management. Fisheries 16(2):22–28.

Clements, F. E. 1916. Plant succession: an analysis of the development of vegetation. Carnegie Institution of Washington Publication 290.

Costanza, R., B. G. Norton, and B. D. Haskell. 1992. Ecosystem health: new goals for environmental management. Island Press, Washington, DC.

Fox, S. 1981. John Muir and his legacy: the American conservation movement. Little, Brown, Boston.

Golley, F. B. 1994. A history of the ecosystem concept in ecology: more than the sum of the parts. Yale University Press, New Haven, Connecticut.

Kornberg, H., and M. H. Williamson. 1988. Quantitative aspects of the ecology of biological invasions. Cambridge University Press, Cambridge, UK.

Harris, L. D. 1984. The fragmented forest: island biogeography theory and the preservation of biotic diversity. University of Chicago Press, Chicago.

Leopold, A. 1939. A biotic view of land. Journal of Forestry 37:727–730.

Leopold, A. 1949. A Sand County almanac and sketches here and there. Oxford University Press, New York.

Leopold, A. 1991. Land pathology. Pages 212–217 in S. L. Flader and J. B. Callicott, editors. The river of the mother of God and other essays by Aldo Leopold. University of Wisconsin Press, Madison.

Lindeman, R. L. 1942. The trophic-dynamic aspect of ecology. Ecology 23:399–418.

Locke, J. 1961. An essay concerning human understanding, 2 volumes. E. P. Dutton & Co., New York.

MacArthur, R., and E. O. Wilson. 1967. The theory of island biogeography. Princeton University Press, Princeton, New Jersey.

Marsh, G. P. 1864. Man and nature; or physical geography as modified by human action. Charles Scribner's Sons, New York.

Meine, C. 1995. The oldest task in human history. Pages 7–35 in R. L. Knight and S. F. Bates, editors. A new century for natural resources management. Island Press, Washington, DC.

Miller, R. R., J. D. Williams, and J. E. Williams. 1989. Extinctions of North American fishes during the past century. Fisheries 14(6):22–38.

Muir, J. 1894. The mountains of California. Century, New York.

Muir, J. 1901. Our national parks. Houghton Mifflin, Boston.

Muir, J. 1916. A thousand mile walk to the gulf. Houghton Mifflin, Boston.

Norse, E. A. 1990. Ancient forests in the Pacific Northwest. Island Press, Washington, DC.

Odum, E. P. 1959. Fundamentals of ecology, 2nd edition. W. B. Saunders, Philadelphia.

Pinchot, G. 1947. Breaking new ground. Harcourt, Brace, New York.

Pister, E. P. 1987. A pilgrim's progress from group a to group b. Pages 221–232 in J. B. Callicott, editor. Companion to a Sand County almanac. University of Wisconsin Press, Madison.

Primack, R. B. 1993. Essentials of conservation biology. Sinauer, Sunderland, Massachusetts.

Raup, D. M. 1991. Extinction: bad genes or bad luck. W. W. Norton, New York.

Raup, D. M., and D. R. Sepkoski. 1984. Periodicity in extinctions in the geologic past. Proceedings of the National Academy of Sciences of the United States of America 81:801–805.

Shrader-Frechette, K. S. 1989. Ecological theories and ethical imperatives: can ecology provide a scientific justification for the ethics of environmental protection? Pages 73–104 in W. R. Shea and B. Sitter, editors. Scientists and their responsibility. Watson Publishing International, Canton, Massachusetts.

Shrader-Frechette, K. S., and E. D. McCoy. 1993. Method in ecology: strategies for conservation. Cambridge University Press, Cambridge, UK.

Smith, G. R. 1981. Late cenozoic freshwater fishes of North America. Annual Review of Ecology and Systematics 12:163–191.

Tansley, A. G. 1935. The use and abuse of vegetational concepts and terms. Ecology 16:284–307.

Thoreau, H. D. 1863. Excursions. Ticknor and Fields, Boston.

Wilson, E. O. 1988. The current state of biological diversity. Pages 3–18 in E. O. Wilson, editor. Biodiversity. National Academy Press, Washington, DC.

Vogler, A. P., and B. DeSalle. 1994. Diagnosing units of conservation management. Conservation Biology 8:354–363.

American Fisheries Society Symposium 17:8–27, 1995

Evolutionarily Significant Units and the Conservation of Biological Diversity under the Endangered Species Act

Robin S. Waples

National Marine Fisheries Service, Northwest Fisheries Science Center, Coastal Zone and Estuarine Studies Division, 2725 Montlake Boulevard East, Seattle, Washington 98112, USA

Abstract.—The U.S. Endangered Species Act (ESA) considers "distinct" populations of vertebrates to be "species" (and hence eligible for legal protection) but does not explain how distinctness should be evaluated. A review of the legislative and legal history of the ESA indicates that in implementing the ESA with respect to vertebrate populations, the Fish and Wildlife Service and the National Marine Fisheries Service (NMFS) should strive to conserve genetic diversity scientifically but sparingly. Based on these precepts, NMFS developed a species policy to guide ESA listing determinations for Pacific salmon *Oncorhynchus* spp. According to the policy, a population (or group of populations) will be considered distinct if it represents an evolutionarily significant unit (ESU) of the biological species. The unifying theme of the NMFS species policy is the desire to identify and conserve important genetic resources in nature, thus allowing the dynamic process of evolution to continue largely unaffected by human factors. A review of case histories in which the NMFS policy has been applied shows that it is flexible enough to provide guidance on many difficult issues for Pacific salmon, such as anadromy versus nonanadromy, variation in life history patterns, and the role of hatchery fish in regards to the ESA. Collectively, these case studies also provide insight into approaches for dealing with scientific uncertainty. Some criticisms of the ESU concept (e.g., that it is too subjective and relies too much—or not enough—on genetics) are discussed, as is its applicability to biological conservation outside the ESA.

As amended in 1978 (16 U.S.C. §§ 1532[16]), the U.S. Endangered Species Act (ESA) allows listing of "any subspecies of fish or wildlife or plants, and any distinct population segment of any species of vertebrate fish or wildlife which interbreeds when mature" [Section 3(15)]. This language indicates that the scope of the ESA extends beyond the traditional biological definition of species to include smaller biological units. Unfortunately, the ESA does not explain how population distinctness shall be evaluated or measured, and this omission has led to considerable confusion in application of the ESA to vertebrate populations. For example, the Fish and Wildlife Service (FWS) has used a variety of criteria for evaluating population distinctness in species such as grizzly bears, bald eagles, desert tortoises, spotted owls, and alligators.

The issue of vertebrate populations and the ESA is particularly challenging with respect to Pacific salmon *Oncorhynchus* spp. because their strong homing instinct leads to the formation of a large number of local spawning populations that might arguably be considered distinct. Furthermore, recent and widespread declines in Pacific salmon populations have raised the possibility that many might qualify as threatened or endangered "species" under the ESA (e.g., Nehlsen et al. 1991). Following receipt of petitions in 1990 for ESA listing of several Columbia and Snake river salmon populations,

the National Marine Fisheries Service (NMFS) published a technical paper (Waples 1991a) and an interim policy (*Federal Register* 56 [13 March 1991]: 10542) on defining distinct population segments of Pacific salmon under the ESA. After a public comment period, the technical paper (Waples 1991b) and the policy (*Federal Register* 56 [20 November 1991]:58612) were revised and published later the same year. The intention of the technical papers and the policy was to provide a biologically sound framework for considering populations under the ESA, with specific guidance for the complex issues involving Pacific salmon.

After briefly summarizing the important concepts in the NMFS species policy, I will explain how the policy has been applied in a number of listing determinations for Pacific salmon. Finally, I will discuss some criticisms of the salmon policy and its applicability to the broader question of defining units for biological conservation outside the ESA.

The National Marine Fisheries Service Salmon Policy

In spite of the failure of the ESA to provide explicit guidance on defining "distinct" populations, an examination of the legislative and legal history of the ESA revealed three guiding principles: (1) an important motivating factor behind the ESA was the desire to preserve genetic variability, both within and between species (e.g., 93rd Congress, 1st Session, 1973. House of Representatives Report 412); (2) the ESA [Section 4(b)(1)(A)] stipulates that listing decisions should be based "solely on the basis of the best scientific and commercial data available"; and (3) a congressional report in 1979 stated that "the committee is aware of the great potential for abuse of this authority and expects the FWS to use the ability to list populations sparingly and only when biological evidence indicates that such action is warranted" (96th Congress, 1st Session, 1979. Senate Report 151). Although not quite self-contradictory, the charge to conserve genetic resources scientifically but sparingly presents a delicate challenge to scientists and policymakers alike.

To balance these themes in a framework consistent with both the letter and intent of the ESA, I adopted as a unifying concept the evolutionarily significant unit (ESU). This term had already seen limited use in the literature (e.g., Ryder 1986), and its usefulness for ESA considerations was suggested by Andrew Dizon of NMFS, who has used an approach based on the ESU for identifying conservation units of marine mammals (Dizon et al. 1992). The framework I developed differs somewhat from that of Dizon et al. and relies on a simple, two-part test for determining whether a population is an ESU (Waples 1991b:12):

> A vertebrate population will be considered distinct (and hence a "species") for purposes of conservation under the Act if the population represents an evolutionarily significant unit (ESU) of the biological species. An ESU is a population (or group of populations) that (1) is substantially reproductively isolated from other conspecific population units, and (2) represents an important component in the evolutionary legacy of the species.

The evolutionary legacy of a species is the genetic variability that is a product of past evolutionary events and that represents the reservoir upon which future evolutionary potential depends. Some have interpreted this term to mean that NMFS will attempt to determine which populations will play an important role in future evolution of the species. This is not the case; such an attempt would be misguided and probably futile as well. Rather, the intention is to identify the important genetic building blocks of the species as a whole and (because we cannot tell which will be important in the future) conserve as many as possible so that the dynamic process of evolution will not be unduly constrained. In essence, then, the ESU policy of NMFS seeks to implement Aldo Leopold's (1953:147) sage advice:

"To keep every cog and wheel is the first precaution of intelligent tinkering."

The NMFS policy identifies a number of types of evidence that should be considered in evaluating each of the two ESU criteria. What follows is only a brief summary of the most important points; readers should consult Waples (1991b) for a more detailed discussion of this topic.

Isolation does not have to be absolute, but it must be strong enough to permit evolutionarily important differences to accrue in different population units. Important types of information to consider include movements of tagged fish, natural recolonization rates, measurements of genetic differences between populations, and evaluations of the efficacy of natural barriers. Each of these measures has its strengths and limitations. Data from protein electrophoresis or DNA analyses can be particularly useful for evaluation of isolation because they reflect levels of gene flow that have occurred over evolutionary time scales.

The key question with respect to a population's evolutionary legacy is, if the population became extinct, would this represent a significant loss to the ecological–genetic diversity of the species? An affirmative answer would lead to a strong presumption that the unit under consideration is an ESU. Again, a variety of types of information should be considered. Phenotypic and life history traits such as size, fecundity, migration patterns, and age and time of spawning may reflect local adaptations of evolutionary importance, but interpretation of these traits is complicated by their sensitivity to environmental conditions. Data from protein electrophoresis or DNA analyses provide valuable insight into the process of genetic differentiation among populations but little direct information regarding the extent of adaptive genetic differences. Habitat differences suggest the possibility for local adaptations but do not prove that such adaptations exist.

The ESU policy of NMFS provides a framework for addressing several issues of particular concern for Pacific salmon, including anadromous versus nonanadromous population segments, differences in run timing, groups of populations, introduced populations, and the role of hatchery fish. However, although the policy establishes a simple, two-part test for identifying ESUs, it by no means amounts to a simple formula. As illustrated below, application of the policy can be quite complex and often involves professional judgement. This result, however, can be attributed more to the complexities of biological processes than to the policy itself.

Application of the National Marine Fisheries Service Policy in Endangered Species Act Status Reviews

In response to ESA petitions for a number of Pacific salmon populations, NMFS has conducted a series of status reviews to determine whether listings as threatened or endangered species were warranted. By law, a listing determination must be made within 1 year of receipt of an ESA petition. Because the ESA stipulates that these listing determinations should be made on the basis of the best scientific information available, NMFS formed a team of scientists with a background in various aspects of salmon biology to conduct the status reviews. This biological review team (BRT) discussed and evaluated scientific information contained in an extensive public record developed for each of the status reviews. Conclusions of the BRT were used by NMFS in making the formal listing determinations announced in the *Federal Register*, and more extensive scientific reports were published for each of the status reviews as National Oceanic and Atmospheric Administration (NOAA) technical memoranda. The following summary identifies some key issues addressed in status reviews for Pacific salmon completed through mid-1994. Although the status reviews involve evaluation of two questions—is it a species as defined by the ESA? and, if so, is it threatened or endangered?—only the former question is addressed in detail here. For each case study discussed, citations are provided for *Federal Register* notices announcing listing determinations as well as for the NOAA technical memoranda. Geographic features mentioned in the text can be found on Figure 1.

Recent listing determinations by NMFS announced too late to allow discussion in this paper include mid-Columbia River summer chinook salmon *Oncorhynchus tshawytscha* (status review, Waknitz et al. 1995; listing, *Federal Register* 59 [23 September 1994]:48855) and Deer Creek (Puget Sound) summer steelhead *O. mykiss* (*Federal Register* 59 [21 November 1994]:59981).

Snake River Sockeye Salmon

The first of five petitions received by NMFS in 1990 was for Snake River sockeye salmon *O. nerka* (status review, Waples et al. 1991a; listing, *Federal Register* 56 [20 November 1991]:58619). Historically, sockeye salmon occurred in at least six to eight lake systems in the Snake River basin, but by 1990 the only population remaining was in Redfish Lake in Idaho. Extinction of this population also seemed imminent as only 4, 2, and 0 adults returned to spawn in 1988, 1989, and 1990, respectively. Because the nearest sockeye salmon population was over 900 river kilometers away in the upper Columbia River, strong reproductive isolation of Redfish Lake *O. nerka* was not in question. The lengthy freshwater migration and distinctive spawning habitat (almost 1,500 km from the ocean and 2,000 m in elevation, both unequalled by any other sockeye salmon population in the world) provided strong support for the second ESU criterion.

The status review, however, was complicated by uncertainty about the relationship between sockeye salmon and kokanee, the latter being a resident, freshwater form of *O. nerka* that was also native to Redfish Lake (Evermann 1896). In 1910, Sunbeam Dam was constructed about 30 kilometers downstream from Redfish Lake, and by all accounts the dam was a serious obstruction to passage of anadromous fish until its partial removal in 1934. According to one view (Chapman et al. 1990), the dam resulted in the extirpation of the original sockeye salmon population, and anadromous *O. nerka* returning since 1934 were derived from kokanee. Because kokanee were relatively abundant (total estimated abundance was approximately 25,000 fish; Bowler 1990), Chapman et al. (1990) argued that no entity was actually endangered: the original sockeye salmon population was extinct (and hence could not be listed under the ESA), and the kokanee presumably could continue to produce a modest number of anadromous fish indefinitely. An alternative view (Waples et al. 1991a) to that proposed by Chapman et al. was that the original sockeye salmon population had persisted, either by achieving limited passage through the dam or by spawning in areas below Sunbeam Dam and recolonizing Redfish Lake after 1934.

Early in the status review for Snake River sockeye salmon, there was considerable uncertainty about how to consider the kokanee population and the effects of Sunbeam Dam in the listing determination. Once the species policy was developed, it was possible to construct the flow diagram shown in Figure 2. From this diagram it is clear that the first key question to be addressed was, are Redfish Lake sockeye salmon and kokanee separate gene pools? A negative answer would lead to consideration of the sockeye salmon–kokanee gene pool as a single unit in ESA evaluations (right branch of flow diagram), whereas an affirmative answer would lead to

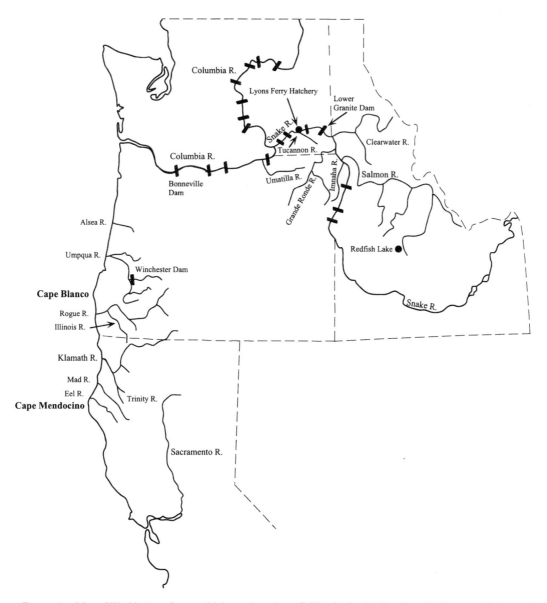

FIGURE 1.—Map of Washington, Oregon, Idaho, and northern California showing location of salmon populations and geographical features discussed in the text.

consideration of the sockeye salmon population as a separate entity (left branch of diagram). For reasons mentioned above, this latter option would presumably lead to recognition of Redfish Lake sockeye salmon as an ESU and, because of its extremely low abundance, eligible for listing as an endangered species under the ESA.

Empirical data could be found to support each of the proposed hypotheses about post-Sunbeam Dam sockeye salmon in Redfish Lake. There is no doubt

that the dam was a serious impediment to migration, and it may have completely blocked adult sockeye salmon for as many as 8–10 years (Chapman et al. 1990). Furthermore, it has been observed in several cases (e.g., Foerster 1947; Kaeriyama et al. 1992) that kokanee can produce anadromous offspring, and a study of Redfish Lake in the 1960s found more smolts of *O. nerka* emigrating from the lake in 1 year than could plausibly be explained by the number of anadromous adults spawning in the

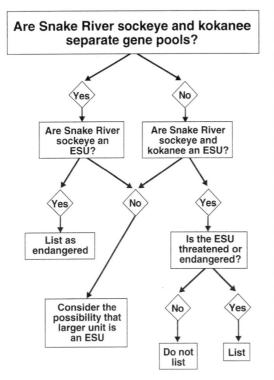

Are Snake River sockeye and kokanee separate gene pools?

FIGURE 2.—Flow diagram that results from applying the framework of the National Marine Fisheries Service evolutionarily significant unit (ESU) policy (Waples 1991b) to the status review for Snake River sockeye salmon (reproduced from Waples et al. 1991a).

lake the previous year (Bjornn et al. 1968). On the other hand, historical records indicate that anadromous fish passed over Sunbeam Dam in at least some years following the rebuilding of a fish ladder in 1920, and eyewitnesses reported seeing "big redfish" (presumably sockeye salmon and not the smaller kokanee) spawning in Redfish Lake in several years between 1927 and 1933. Therefore, there is no conclusive evidence that the original sockeye salmon population was extirpated. Furthermore, the kokanee population in Redfish Lake spawns in the inlet stream (Fishhook Creek) in August and September, whereas all recent records of sockeye salmon spawning have been on the·lake shore in October and November. Thus, substantial reproductive isolation of sockeye salmon and kokanee in Redfish Lake seemed possible, as had been demonstrated between the two forms in several lake systems in British Columbia (Foote et al. 1989).

At the time the listing determination had to be made, there was agreement among the BRT on one

issue: there was not enough scientific information to determine with any degree of certainty which of the two scenarios was true. Although a protein electrophoretic study conducted by NMFS during the status review demonstrated that kokanee from Redfish and nearby Alturas lakes were genetically distinct from other samples of *O. nerka* from the Pacific Northwest, no adult sockeye salmon returned in 1990, so a comparison of the two forms in Redfish Lake could not be made. Because there was a lack of consensus within the BRT (as well as within the broader community of fishery biologists in the Pacific Northwest) on how to interpret the limited information about the effects of Sunbeam Dam, the listing determination had to be made in the absence of conclusive scientific information. After considerable discussion, it was concluded that the most appropriate approach was to proceed under the assumption that a component of the native sockeye salmon gene pool still persisted in Redfish Lake and was distinct from the kokanee. A factor that weighed heavily in this consideration was the recognition that the consequences of taking the alternate course (i.e., assuming that recent anadromous *O. nerka* in Redfish Lake were derived from kokanee) and being wrong were irreversible, because the original sockeye salmon gene pool could easily become extinct before the mistake was realized.

Accordingly, a proposal to list Snake River sockeye salmon as an endangered species was published in April 1991, with the final rule coming in November of that year. Subsequently, samples taken from sockeye salmon adults that returned in 1991–1993 established that there are large allele frequency differences between sockeye salmon and kokanee within Redfish Lake (about 25–50% at several loci detected by protein electrophoresis [R. S. Waples, G. A. Winans, and P. B. Aebersold, NMFS, Seattle, unpublished data] and about 20% at a nuclear DNA locus [S. Cummings and G. Thorgaard, Washington State University, and E. Brannon, University of Idaho, personal communication]), thus providing strong support for the hypothesis that recent anadromous *O. nerka* returning to Redfish Lake are not merely the product of seaward drift of Fishhook Creek kokanee.

Snake River Chinook Salmon

In 1990, NMFS was petitioned to list three Snake River runs of chinook salmon *O. tshawytscha* as threatened or endangered species: spring-, summer-, and fall-run fish (status reviews, Matthews and Waples 1991; Waples et al. 1991b; listing, *Fed-*

eral Register 57 [22 April 1992]:14653). The different runs are defined by the season during which adults enter freshwater (or can be enumerated as they pass across a dam). According to the ESU policy of NMFS, the first step in evaluating the petitions was to determine whether the different runs were reproductively isolated. If so, they could be considered separately in the species determination; if not, they would be considered together as a single unit in determining whether the larger unit met the second criterion to be an ESU.

Several lines of evidence made it clear that Snake River fall chinook salmon were strongly isolated from spring- and summer-run fish. Whereas the latter two forms spawn in late summer in upper level tributaries (generally at 1,300–2,000 m elevation), Snake River fall chinook salmon spawn later (generally in October and November) in the main stem and its lower tributaries, at elevations of approximately 500 m or less. Genetic data showing substantial allele frequency differences (e.g., 25–50% at multiple loci) between fall chinook salmon and the other two forms in the Snake River supported the evidence for spatial and temporal isolation. Furthermore, juvenile Snake River fall chinook salmon migrate to sea as subyearlings in their first year of life (ocean-type life history; Healey 1991), whereas spring- and summer-run fish in the basin migrate as yearlings (stream-type life history).

The closest relatives to Snake River fall chinook salmon are fall chinook salmon from the Columbia River, which also have an ocean-type juvenile life history. Several years of genetic data from protein electrophoresis showed modest but consistent allele frequency differences between fall chinook salmon from the Columbia and Snake rivers, suggesting the possibility for long-term reproductive isolation (summarized by Waples et al. 1991b). Further evidence of reproductive isolation came from a tagging study of Columbia River fall chinook salmon conducted in the early 1980s that found high homing fidelity for fish from the mid-Columbia region and no evidence of straying into the Snake River (McIsaac and Quinn 1988).

Two lines of evidence were key in evaluating the second ESU criterion—contribution of Snake River fall chinook salmon to the ecological–genetic diversity of the species as a whole. First, the Snake River has greater turbidity, a higher pH and total alkalinity, and a higher and more variable temperature than does the Columbia River. During the summer months, when juvenile fall chinook salmon are rearing or migrating in the river, water temperatures in the Snake River can be as much as 6–8°C warmer

than in the Columbia River (Utter et al. 1982). Thus, the Snake River population may have developed physiological tolerance for elevated water temperatures, behavioral strategies to avoid warm water, or both. Second, several years of data from recovery of marked hatchery fish showed that fall chinook salmon from the two rivers differed in their ocean distribution; with Columbia River fish are more commonly taken in Alaskan waters and Snake River fish are more common in waters off California, Oregon, and Washington (Waples et al. 1991b).

As with the Snake River sockeye salmon, however, there was a difficult issue that remained to be resolved: evidence that isolation of the Snake River population began to break down in the 1980s as a result of straying by hatchery fish from the Columbia River. In particular, a large-scale program had been initiated in the early 1980s in an attempt to restore fall chinook salmon to the Umatilla River, the last major tributary of the Columbia River below the confluence with the Snake River. Apparently as a result of poor acclimation of juveniles and inadequate river flows for returning adults, in the late 1980s fish from the Umatilla program began to appear in the Snake River in alarming numbers. At Lyons Ferry Hatchery, a facility designed to provide a means of conserving the genetic diversity of Snake River fall chinook salmon, stray Columbia River hatchery fish constituted almost 40% of the broodstock spawned in 1989. Equally disturbing, a sample of tagged fish taken in 1990 at Lower Granite Dam (the uppermost dam on the Snake River) indicated that stray Columbia River hatchery fish had penetrated some distance into the Snake River, and the presence of strays on the spawning grounds was verified by recovery of carcasses of several tagged fish.

These data raised concerns that the distinctiveness of the Snake River population (i.e., the qualities that made it a species under the ESA) had been compromised by the stray fish from the Columbia River. However, given that (1) there was general agreement that an ESU was present until at least the early 1980s, (2) substantial straying of Columbia River hatchery fish had occurred only within the last generation, and (3) no direct evidence existed for genetic change to wild fall chinook salmon in the Snake River, the BRT felt it would be premature to conclude that the ESU no longer exists. Snake River fall chinook salmon were listed as a threatened species in 1992.

As was the case for fall chinook salmon, life history and genetic data indicated that the closest relatives to the spring and summer chinook salmon

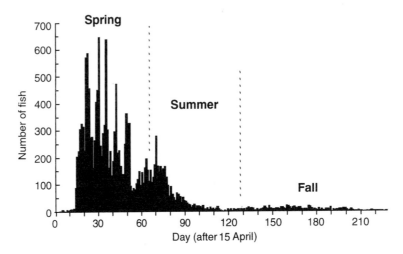

FIGURE 3.—Timing of adult chinook salmon passing Lower Granite Dam on the Snake River, Washington, on their upstream spawning migration in 1989. By convention, fish passing the dam prior to 16 June are considered spring run, those passing between 17 June and 16 August are considered summer run, and those passing from 17 August on are considered fall run.

from the Snake River are found in the Columbia River. However, the upper tributary spawning areas of the two rivers are well isolated from each other geographically. Distinctive habitat features of the upper Snake River basin (similar to those described above for Snake River sockeye salmon) also indicated that Snake River spring and summer chinook salmon were in a different ESU than were the Columbia River populations. Thus, the key issue that remained to be resolved was the relationship between spring- and summer-run fish in the Snake River.

Although two or more modes in adult run timing can be observed in any given year (Figure 3), it was clear that the inflexible dates for defining the runs were not sufficient to demonstrate that two independent units existed in the Snake River. Thus, the BRT focused on information for individual spawning populations. Some subbasins are considered to have only one run type; in others, both spring- and summer-run fish are believed to occur. Fish that spawn slightly earlier and at higher elevation are generally considered to be spring-run fish, and those that spawn later and at lower elevation are considered to be summer-run fish. Therefore, a key question was whether summer chinook salmon from a particular stream were more closely related to summer chinook salmon from other streams than they were to spring chinook salmon from the same stream. That is, the BRT looked for evidence to indicate whether the two run types were independent, monophyletic evolutionary units.

After reviewing available phenotypic and life history information, the BRT was unable to find any characters that consistently distinguished spring- and summer-run fish in the Snake River. Genetic data from protein electrophoresis showed a similar pattern. A hierarchical gene diversity analysis (Waples et al. 1993) of samples from both run types found that most (78%) of the total intersample diversity was attributable to geographic differences (between localities within drainages or between drainages within run times). In contrast, differences between Snake River spring- and summer-run fish as a whole accounted for little (8%) of the total intersample diversity (Figure 4). On the basis of the genetic, phenotypic, and life history information, the BRT concluded that Snake River spring/summer chinook salmon should be considered a single ESA species, and they were listed as threatened under the ESA.

In 1994, NMFS published an emergency rule (*Federal Register* 59 [18 August 1994]:42529) that temporarily revised the status of Snake River spring/summer and fall chinook salmon from threatened to endangered. Subsequently (*Federal Register* 59 [28 December 1994]:66784), NMFS initiated a process to finalize the change in status for Snake River chinook salmon.

Lower Columbia River Coho Salmon

Coho salmon *O. kisutch*, which have become extinct in most upper areas of the Columbia River

Gene diversity analysis

Snake River spring/summer chinook salmon

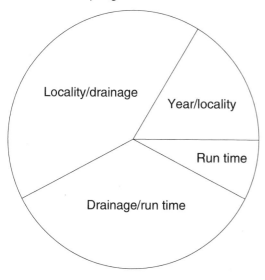

FIGURE 4.—Partitioning total intersample gene diversity (F_{ST} = 0.034) into various hierarchical components for Snake River spring/summer chinook salmon (modified from Waples et al. 1993). The geographic component to the intersample diversity (localities within drainages and drainages within run times) is much larger than that due to run-timing differences.

basin in recent decades, were historically abundant in the lower Columbia River (below Bonneville Dam), with adult returns averaging approximately 1 million fish per year early in this century. By the 1950s, overfishing and habitat degradation had dramatically reduced annual returns to perhaps 5% of their historic levels, and state fishery managers embarked on a large-scale hatchery program in an attempt to restore the runs. Although abundance has been quite variable over the last three decades, adult returns have averaged approximately 500,000 fish in many recent years. However, the number of naturally spawning fish continued to decline, prompting a petition in 1990 to list naturally spawning coho salmon in the lower Columbia River under the ESA (status review, Johnson et al. 1991; listing, *Federal Register* 56 [27 June 1991]:29553).

Although the BRT concluded that there was probably at least one ESU of coho salmon historically in the Columbia River, the difficulty was in determining what remained of this ESU in the 1990s. After declining in the 1950s to a small fraction of their historic abundance, natural popula-

tions suffered three decades of extremely high (85–95%) harvest rates directed at the more productive hatchery fish. Furthermore, extensive stock transfers among the hatcheries (Figure 5) and widespread releases of juvenile hatchery fish into virtually every major tributary in the lower Columbia River posed substantial threats to the genetic integrity of the remaining natural populations.

The status review for lower Columbia River coho salmon raised important questions about the role of hatchery fish in ESA evaluations. These issues have played a role in each of the other status reviews conducted to date, but that for Columbia River coho salmon was unusual because of the overwhelming influence of artificial propagation. Because of the emphasis in the ESA on conserving species and their ecosystems, the ESU policy of NMFS focuses on "natural" fish, which are defined as progeny of fish that spawn naturally, whether of wild or hatchery origin (Waples 1991b). This approach directs attention to fish that spend their entire life cycle in natural habitat. Implicit in this approach is the recognition that fish hatcheries are not a substitute for natural ecosystems. More details about NMFS policy regarding artificial propagation of Pacific salmon in relation to the ESA can be found elsewhere (Hard et al. 1992; *Federal Register* 58 [5 April 1993]:17573).

In the lower Columbia River, approximately 25,000 coho salmon spawn naturally each year (Johnson et al. 1991), but the vast majority of these were reared in hatcheries as juveniles and thus are not natural fish. The general consensus of fishery biologists in the region was that if any naturally spawning fish remained that represented the historic legacy of the indigenous populations, they would be found in streams in which spawning occurred much later (post-December) than that of most hatchery populations. However, little biological information was available for the late-spawning populations, and attempts to gather new information during the status review were largely unsuccessful. The NMFS conducted an extensive genetic survey of coho salmon from the lower Columbia River and compared new results with data from previously published studies. The genetic data provided some evidence for reproductive isolation of Columbia River coho salmon from those in other geographic regions but found "no apparent geographical or temporal structuring or separation between hatchery and wild stocks" (Johnson et al. 1991:28).

After considerable discussion, the BRT concluded that they were unable to identify any remaining natural populations of coho salmon in the lower

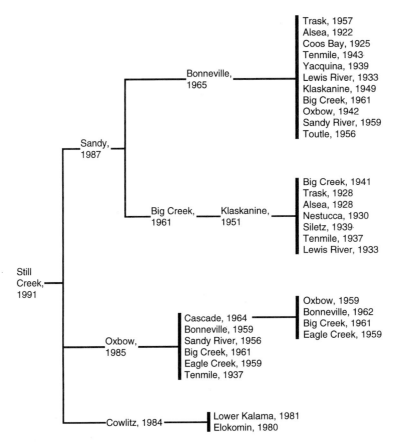

FIGURE 5.—Each of three stocks of coho salmon (Cowlitz, Oxbow, and Sandy) released into Still Creek between 1984 and 1987 are of mixed origin. For example, the Oxbow stock received transfers of eggs or fish from at least six other stocks between 1937 and 1964, and each of these six stocks has a mixed history as well (origins shown for only Cascade stock). Still Creek was identified as one of the sites in the lower Columbia River most likely to have a remnant run of native, naturally spawning coho salmon. Reproduced from Johnson et al. (1991).

Columbia River that warranted protection under the ESA. In 1991, NMFS found the petition for an ESA listing to be unwarranted.

Illinois River Winter Steelhead

In May 1992, NMFS received a petition for ESA listing of winter-run steelhead *O. mykiss* from the Illinois River, a tributary of the Rogue River in southern Oregon (status review, Busby et al. 1993; listing, *Federal Register* 58 [20 May 1993]:29390). In the Pacific Northwest, steelhead are generally considered to be either summer or winter run, based (as in chinook salmon) on the timing of adult entry into freshwater. The petitioners argued that Illinois River winter steelhead were an ESA species because of phenotypic and life history differences in comparison with other steelhead populations from the Rogue River.

Although Busby et al. (1993) found some evidence that Illinois River steelhead differed from other populations in the Rogue River in traits such as smolt age and age and size at spawning, the differences were small and the biological significance unclear. Furthermore, in none of these traits did Illinois River steelhead differ substantially from other coastal steelhead populations; if anything, the Rogue River populations were the ones that were somewhat distinctive. A protein electrophoretic survey of 15 steelhead populations from Oregon and northern California conducted by NMFS found evidence for genetic heterogeneity among populations, but no evidence for distinctiveness of steelhead from the Illinois River (Busby et al. 1993). In fact, in pairwise comparisons using genetic distance values, three of the four Illinois River samples were genetically more similar to a population from out of the basin than they were to any of the other Illinois River samples.

On the basis of this information, NMFS concluded that Illinois River winter steelhead did not by themselves constitute an ESA species, and the petition was found to be not warranted. However, NMFS also recognized that the petitioned population was part of a larger ESU whose boundaries remained to be determined. Accordingly, a broader status review was initiated to determine the boundaries of that ESU and whether it was threatened or endangered. That broader status review has recently been completed (Busby et al. 1994), and it provides a good example of how multiple lines of evidence can be used to define ESUs.

A prominent physical and ecological feature of the west coast of the United States is the Klamath Mountains Geological Province, which extends from the vicinity of Cape Blanco in southern Oregon to include the Klamath River basin in northern California. Geologically, the province is distinctive in that it includes northern extensions of formations typical of the California Coastal Ranges and the Sierra Nevada. Ecologically, the province includes areas that are warmer and drier than are coastal regions to the north and south; interior valleys receive less precipitation than does any other location in the Pacific Northwest west of the Cascade Range. The vegetation combines elements from California, the northern coast, and eastern Oregon, including a large number of endemic species (Whittaker 1960). Zoogeographic studies have also found similarities between the freshwater fish faunas of the Klamath and Rogue rivers, the two major rivers draining the Klamath Mountains Geological Province.

The "half-pounder" life history form of steelhead also appears to be restricted to southern Oregon and northern California, having been described from the Rogue, Klamath, Eel, and Mad rivers. Following their smolt migration, half-pounders spend only a few months at sea before returning to freshwater at a size that inspired their name. After overwintering in freshwater, half-pounders return to sea before their spawning migration the following fall or winter. Although the factors responsible for this life history strategy are poorly understood, expression of this trait is likely due to a combination of genetic and environmental factors.

The nearshore ocean environment in this region is strongly affected by seasonal upwelling. The strength and consistency of upwelling south of Cape Blanco yields highly productive waters. The area of increased upwelling extends, with some local variations, as far south as 33°N.

Patterns of ocean migration of salmon and steelhead may reflect reproductive isolation of spawning populations. Chinook salmon populations from south of Cape Blanco are generally considered south migrating (e.g., to ocean areas off southern Oregon and California), whereas most stocks from north of Cape Blanco are considered north migrating (Nicholas and Hankin 1988). Other studies (see Pearcy 1992) suggest that coho salmon and steelhead from south of Cape Blanco also may not be strongly migratory, remaining instead in the productive oceanic waters off southern Oregon and California.

Several lines of evidence (geology, ecology, life history, zoogeography, and genetics) thus suggest that Cape Blanco forms the northern boundary for the ESU that contains Illinois River winter steelhead. To the south, Cape Mendocino is a natural landmark associated with changes in ocean currents and represents the approximate southern limit of summer steelhead and the half-pounder life history strategy. However, the Klamath River basin forms the southern boundary of the Klamath Mountains Geological Province as well as the Klamath–Rogue freshwater zoogeographic zone. Furthermore, genetic data compiled by NMFS showed a sharp transition between steelhead populations from the Klamath River basin and those farther south (Busby et al. 1994). Therefore, the BRT concluded that the geographic boundaries of the ESU that contains Illinois River steelhead extend from Cape Blanco in the north to include the Klamath River basin in the south. In early 1995, NMFS proposed that Klamath Mountains Province steelhead (including summer-, fall-, and winter-run fish) be listed as a threatened species under the ESA (*Federal Register* 60 [16 March 1995]:14253).

Umpqua River Cutthroat Trout

In April 1993, NMFS was petitioned to list searun cutthroat trout *O. clarki* from the North and South Umpqua rivers under the ESA (status review, Johnson et al. 1994; listing, *Federal Register* 59 [8 July 1994]:35089). Because *O. clarki*, like *O. mykiss* and *O. nerka*, has both anadromous and resident life history types, the first issue to address was the relationship between the two forms in the Umpqua River. In this case, the situation was complicated by the possibility that a third (potamodromous) life-history type occurs in the Umpqua River, consisting of fish that migrate extensively in freshwater but do not enter the ocean.

This potamodromous life history type provides a possible link between anadromous and resident fish that may act to retard divergence between the life history forms. Sea-run cutthroat trout also have other distinctive life history traits that may have a similar effect. Unlike sockeye salmon, sea-run cutthroat trout do not necessarily die after spawning, and it is generally believed that the incidence of repeat spawning is higher than it is in steelhead. Sea-run cutthroat trout are also distinctive among anadromous salmonids in that they do not overwinter at sea; rather, it is believed that each year after only a few months at sea they return to freshwater and overwinter there. In addition, it is thought that a cutthroat trout that has gone to sea and returned may spend an entire year (or more) in freshwater before migrating to the sea again.

In combination, these traits suggest that opportunities for reproductive isolation between the different life history forms are not as great as in other *Oncorhynchus* species. Therefore, the BRT concluded that, at least until better information is developed, all life history forms of *O. clarki* in the Umpqua River should be considered part of the same ESU. The BRT also concluded that cutthroat trout from the Umpqua River probably were substantially reproductively isolated from other coastal cutthroat trout populations.

On three other key issues, however, the BRT concluded that there was insufficient information to allow a completely scientific determination. First, there remains some uncertainty about the geographic boundaries of the ESU. There are some distinctive features of the Umpqua River drainage, which originates in the Cascade Mountains rather than in the Coast Range, where most other coastal rivers originate. Anadromous cutthroat trout in the Umpqua River migrate farther inland than do cutthroat trout in most other coastal rivers, and the petitioners (ONRC et al. 1993) suggested that elevated water temperatures may have promoted adaptations in Umpqua River cutthroat trout. However, it is not clear whether water temperatures were historically elevated, and virtually no biological information was available during the status review on resident or potamodromous forms of Umpqua River cutthroat trout.

Second, the evolutionary lineage of present-day anadromous cutthroat trout in the Umpqua River is uncertain. Counts of adult (presumably sea-run) fish crossing Winchester Dam (river kilometer 190)

are probably the most reliable data for Umpqua River cutthroat trout, and these records show that the number of fish declined to very low levels in the mid-1950s, increased dramatically from about 1960 to 1975, and rapidly declined again after about 1976. The period of increase coincides almost exactly with releases into the Umpqua River of sea-run cutthroat trout from the Alsea River Hatchery on the Oregon coast. The most parsimonious explanation for the sudden increase in adults passing Winchester Dam in 1960–1975 is that they represent predominantly Alsea River Hatchery fish. Alsea River fish have a later run timing than the Umpqua River fish, and a shift toward later run timing in fish returning to Winchester Dam occurred after 1960 (Johnson et al. 1994). There is some evidence of a shift back toward the original run timing after cessation of the hatchery program. The unresolved issue is, what do the few remaining anadromous fish represent: remnants of the original Umpqua River gene pool, descendants of the Alsea River Hatchery fish, or a mixed lineage?

Finally, although the precarious status of the remaining sea-run fish in the Umpqua River was not in question, the status of the other two forms of *O. clarki* was essentially unknown. The existence of potamodromous fish in the Umpqua River was still largely a hypothesis, and the total information available to the BRT on abundance of resident *O. clarki* in the Umpqua River drainage amounted to a list of lakes believed to contain cutthroat trout.

This status review was thus distinctive in the lack of reliable information on virtually every key aspect of cutthroat trout biology, both for the species in general and for Umpqua River sea-run cutthroat trout in particular. Given this reality, and the legal obligation to decide whether to propose a listing within 1 year, NMFS was faced with two general options: (1) reject the petition because either (a) the petitioned entity is not an ESA species, or (b) the species is not threatened or endangered; or (2) propose a listing. Option 1b would require that an ESA species be defined and found not to be threatened or endangered. This, clearly, was not what the BRT concluded. Because the BRT concluded that all life history forms of cutthroat trout in the Umpqua River were part of the same ESU, option 1a was feasible but not particularly satisfactory and would not have represented a resolution of the issue. Option 2, however, also was problematical because it would presume that an ESU was identified and determined to be threatened or endangered, determinations the BRT concluded could not be made from a purely scientific standpoint. Option 2 would therefore require that NMFS adopt a conservative approach on each of the three unresolved issues: geographic extent of the ESU, the effects of introduced hatchery fish, and the abundance of resident and potamodromous fish. In the end, NMFS elected to take this approach and propose a listing of Umpqua River cutthroat trout as an endangered species.

Sacramento River Winter-Run Chinook Salmon

The winter run of chinook salmon from the Sacramento River was the first salmon population listed as a "distinct population segment" under the ESA (listing, *Federal Register* 55 [5 November 1990]: 46515). Although this action preceded the development of the ESU policy of NMFS, the listing determination did discuss biological characteristics of the population (e.g., unique run timing within the species and temporal isolation from other chinook salmon runs in the river) that are relevant to ESU evaluations.

Smith et al. (1995, this volume) express concern that although the winter run is listed, the other runs of chinook salmon in the Sacramento River (spring, fall, and late-fall) may not qualify for ESA protection under the NMFS ESU policy. Because the ESA status of these populations have not been formally considered by NMFS, it would be premature to speculate whether any or all of the other runs should be listed. Although NMFS has not yet determined how many ESUs include Sacramento River chinook salmon, each of these runs belongs to some ESU, and the ESU (or ESUs) as a whole could be listed if determined to be threatened or endangered. Even if it were concluded that an ESU contained more than one nominal run type, conservation of the diversity within the ESU could still be accomplished by following a procedure similar to that for Snake River spring/summer chinook salmon.

Atlantic Salmon

Following receipt of a petition to list all U.S. populations of Atlantic salmon as a threatened or endangered species under the ESA, NMFS and the FWS conducted a joint status review (Anonymous 1995) that used the ESU policy of NMFS as the basis for addressing the species issue. The status review concluded that, because Atlantic salmon had been extirpated from many rivers in southern New England as far back as the eighteenth century, populations of mixed, nonnative origin that had been reintroduced into these streams did not qualify for protection under the ESA (*Federal Register* 60 [17 March 1995]:14410). However, NMFS and the FWS also concluded that populations from seven rivers in Maine did represent an historic ESU of Atlantic salmon and announced an intention to issue a proposed rule regarding the status of this ESU in the near future.

Comprehensive Status Reviews

The status reviews discussed above were all initiated by ESA petitions. Ideally, of course, status reviews should be conducted proactively on a comprehensive basis, because ESUs can best be identified in the broader context of the range of variation found within the entire species. Toward this end, NMFS recently announced (*Federal Register* 59 [12 September 1994]:46808) that it was initiating comprehensive ESA status reviews for all Pacific salmon and anadromous trout (cutthroat trout and steelhead). The status reviews will cover all populations in California, Oregon, Idaho, and Washington and are scheduled for completion in 1996. Recently, NMFS announced results of the first of these comprehensive status reviews, that for coho salmon (status review, Weitkamp et al., in press; listing, *Federal Register* 60 [25 July 1995]:38011). The status review identified six ESUs of coho salmon from central California to southern British Columbia, and NMFS proposed listing the three southernmost ESUs as threatened species and declared that two of the remaining ESUs should be considered candidate species for future listing.

Discussion of Case Studies

The listing determinations described above illustrate how the ESU policy of NMFS has been applied in a variety of case studies. Two recurring themes are worth noting.

Life history diversity.—Two of the more common traits contributing to life history diversity in anadromous Pacific salmonids are run timing and anadromy versus nonanadromy. In applying the ESU policy of NMFS to these issues, the first question to address is whether the different forms are reproductively isolated. Results of the status reviews conducted to date indicate that there is no universal answer to that question. Substantial evidence for reproductive isolation led to separate ESA consideration for Snake River fall versus spring/summer chinook salmon and for Redfish Lake sockeye salmon versus kokanee, although conclusive evidence in the latter case was not obtained until after the final listing determination. On the other hand, inability to find consistent biological differences between spring and summer chinook salmon in the Snake River, summer, fall, and winter steelhead in the Klamath Mountains Geological Province, and resident, potamodromous, and anadromous cutthroat trout in the Umpqua River led to inclusion of multiple life history forms within these ESUs. These results show that ESA evaluations of life history diversity in Pacific salmonids will have to continue to be guided strictly by biological considerations and not by convention.

Uncertainty.—Collectively, the listing determinations described above provide considerable insight into how NMFS has addressed the issue of scientific uncertainty, which can be expected to arise to some extent in virtually every attempt to define conservation units. In general, NMFS has demonstrated a strong inclination to give the benefit of the doubt to the resource in these situations. Thus, NMFS proceeded with a listing of Snake River sockeye salmon in spite of uncertainty about the relationship between sockeye salmon and kokanee in Redfish Lake and has proposed a listing of Umpqua River cutthroat trout in spite of substantial uncertainty on several key issues. With Snake River fall chinook salmon, NMFS also elected to go forward with a listing in spite of evidence for a high level of straying by Columbia River hatchery fish in recent years and considerable uncertainty about the effects of this straying on the remaining natural population in the Snake River.

How far should the benefit of the doubt extend to the resource in cases of uncertainty? This is not primarily a scientific question, but some insight into how NMFS has dealt with this issue can be gained by considering the status review for Lower Columbia River coho salmon. Considerable uncertainty also existed about the status of naturally spawning coho salmon in the lower Columbia River, but NMFS did not propose a listing. In this case, the evidence for massive and long-term effects on natural coho salmon populations from overharvest, habitat degradation, and artificial propagation was overwhelming. Given that reality, the BRT looked for tangible evidence that coho salmon which retained the distinctive characteristics of the original population still existed in the lower Columbia River. Lacking convincing evidence for this, the BRT concluded that it could not identify a population or populations that warranted protection under the ESA.

Criticism of the Evolutionarily Significant Unit Concept and the National Marine Fisheries Service Species Policy

Because scientists often cannot even agree on how to define taxonomic or biological species, it is not surprising that a framework for formally recognizing smaller units should be controversial. This section briefly reviews some of the criticisms of the ESU concept and the ESU policy of NMFS that have been raised in the nearly 4 years since it was published. Many of the comments were made in public meetings or in submissions to the ESA administrative record, but formal literature references are provided as available.

Comment: Defining "distinct" populations under the ESA is primarily a legal and policy issue rather than a biological one (Rohlf 1994).

The ESA is concerned with avoiding extinctions. Because extinction is a biological process, it makes little sense to discuss extinction of units that do not have an underlying biological basis. Although legal and policy issues have a legitimate role in establishing the general context under which populations are considered (hence the admonition of the 1979 Senate report [96th Congress, 1st session, 1979. Senate Report 151] to use the ability to list vertebrate populations "sparingly"), it would be inconsistent and contrary to the goals of the ESA to abandon fundamental biological principles in defining units for conservation.

Comment: The ESA [Section 2(a)(3)] states that species are of "esthetic, ecological, educational, historical, recreational, and scientific value to the Nation and its people," and these qualities should be considered in evaluating population distinctness.

These are all good reasons why it is important to conserve biological diversity in general, but they are not necessarily good reasons for deciding which units to conserve. Because extinction is irreversible, effective conservation of biological diversity must be based on long-term considerations. In contrast, society's view of which species are of recreational or aesthetic value is subject to change, and many species that will be economically or scientifically important in the future may not be recognized as such today. Coggins (1991:64) described one of the remarkable features of the ESA.

In 1918, Congress acted to protect birds because they sang prettily; in 1940, Congress singled out the bald eagle for preservation as the living national symbol. By

1973, Congress had dispensed with species-by-species evaluations of good and bad in terms of value to *Homo sapiens*. It simply said that all fauna and flora species of whatever utility were entitled to continued existence.

Comment: The ESU policy of NMFS would allow fragmentation and loss through attrition of large, composite ESUs. It does not adequately consider population viability and ignores metapopulation structure.

The first concern is understandable: if NMFS were to make determinations of threatened and endangered status largely on the basis of absolute abundance levels, it might be possible to avoid listing a large ESU even if it faced pervasive declines throughout its geographic range. In practice, however, this is not how NMFS has approached the issue. The NMFS status reviews (e.g., Matthews and Waples 1991; Waples et al. 1991b; Busby et al. 1993) have identified a number of factors that should be considered in evaluating the level of risk faced by an ESU, including (1) absolute numbers of fish and their spatial and temporal distribution; (2) current abundance in relation to historical abundance and current carrying capacity of the habitat; (3) trends in abundance; (4) natural and human-influenced factors that cause variability in survival and abundance; (5) possible threats to genetic integrity; and (6) recent events that have predictable short-term consequences for the ESU.

An example of application of this approach to a large ESU is the listing determination for Snake River spring/summer chinook salmon. Although total abundance for the ESU averaged about 20,000–30,000 adults per year in the decade preceding the listing determination, this represented only a fraction of historical levels. Furthermore, a substantial proportion of the total run was hatchery fish, and the remaining natural spawners were thinly spread over a large geographic area. Finally, recent trends in abundance were uniformly downward throughout the ESU, and this pattern was not expected to improve in the near term because a series of drought years had adversely affected juveniles that would form the basis for subsequent years' adult returns. After considering all these factors, NMFS listed the ESU as a threatened species in 1992. However, NMFS also pointed out that considerable diversity exists within the ESU, in habitat as well as population characteristics, and indicated that conservation of this diversity was important to maintaining viability of the ESU (Matthews and Waples 1991). A draft recovery plan (Bevan et al. 1994) submitted to NMFS by an independent recovery team also recognized this diversity and identified

about 40 subpopulations within the ESU that are important to conserve as separate management units.

Recognition of multiple management units in no way limits the ability of those units to function as a metapopulation. Neither NMFS nor the recovery team has suggested that natural dispersal among subpopulations be restricted in any way. If efforts to address the root causes of the decline of Snake River spring/summer chinook salmon are effective, then natural processes should lead to reestablishment of a population structure approximating that which occurred historically.

The ESU policy of NMFS is inherently hierarchical and, far from ignoring metapopulation structure, is easily compatible with this concept. For example, Waples (1991b) discussed how the ESU concept can be applied to groups of populations as well as individual populations. The key is to focus on units that are largely independent from other population units over evolutionary time scales. Populations within such larger groups might exchange individuals on a regular basis, in which case the group might be considered a metapopulation in the sense that term is commonly used. Alternatively, populations within an ESU might experience gene flow only sporadically, with years or decades during which little or no exchange occurs. In any case, however, the ESU policy of NMFS does not focus on subpopulations recently isolated by human factors. Rather, ESU evaluations are based on inferences about historical levels of gene flow that have occurred over evolutionary time scales.

Comment: The ESU policy of NMFS relies too heavily on genetic information. It cannot be used in situations in which genetic data are not available.

It is true that genetics plays a central role in the NMFS ESU concept. However, this is a natural and inevitable consequence of focusing on the conservation of biological units. Extinction of a biological unit is irreversible because it involves the permanent loss of genetic resources capable of regenerating that unit; therefore, a program aimed at avoiding extinction must focus on conserving genetic resources. The ESU concept of NMFS is firmly rooted in genetic principles; it could not be otherwise and still accomplish the goals of the ESA.

The term genetics, however, is often used in a much more restrictive sense—that is, to refer to traits that can be detected by genetic procedures such as protein electrophoresis or DNA analyses. As noted above, genetic data of this type can be instrumental in making ESU determinations, but primarily by providing information about reproductive isolation. The second ESU criterion focuses on adaptive genetic differences that generally must be inferred from nongenetic information. This point is clearly articulated in the species definition paper (Waples 1991b), and much of the effort in the status reviews conducted to date has been directed toward compiling and evaluating phenotypic, life history, and habitat information relevant to the criterion of ecological–genetic diversity.

It is not clear what has spawned the apparently common misconception that ESU determinations cannot be made in the absence of biochemical genetic data; certainly this idea is not supported by an examination of the ESA record of NMFS. For example, when Snake River sockeye salmon were listed as an endangered species in 1991, no genetic data were available for the population because no anadromous adults returned to spawn in 1990. Similarly, no genetic data were available for cutthroat trout from the Umpqua River when they were proposed for listing as an endangered species in 1994. Furthermore, the species definition paper (Waples 1991b) and *Federal Register* notice announcing the final NMFS species policy (*Federal Register* 56 [20 November 1991]:58612) state clearly that such data are not required for an ESU determination. Notably, the vast majority of listing determinations made by the FWS since the ESA was adopted in 1973 have been made in the absence of genetic data. If genetic data are not available, however, evidence to support an ESU must be found elsewhere, which inevitably places a greater burden of proof on other characters. Because data for other characters are often open to multiple interpretations, lack of genetic data may add complexity and contribute uncertainty to ESU determinations.

As can be seen from the discussion that follows, others feel that genetic characteristics are the only traits that should be considered in defining ESUs.

Comment: The concept of evolutionary significance is too subjective to apply in practice.

Biological processes are complex and do not easily lend themselves to tidy categorizations. In recognition of this, the ESU policy of NMFS adopts a holistic, multidisciplinary approach to defining units for conservation under the ESA. Because this approach involves evaluations based on scientific judgement, it has been criticized as being too subjective. Subjectivity can come into play in ESU determinations in two principal ways: in determining

how to synthesize diverse types of information and in determining the level of differentiation required for evolutionary significance.

One approach to avoiding subjectivity would be to adopt an objective yardstick for defining units of conservation and apply it uniformly. This approach, however, has its problems as well. First, although it might be possible to identify a number of essentially objective standards for this purpose, the choice of which standard to adopt would necessarily be somewhat subjective (if not arbitrary). Second, even if there were consensus on an appropriate method to use, it would still be necessary to choose threshold values that would guide the identification of conservation units. As Moritz et al. (1995, this volume) point out, "there is no theoretically sound answer to the question 'How much difference is enough?'"; hence, this aspect of the "objective" approach would involve subjectivity as well.

In response to the latter difficulty, some have suggested that ESUs should be stringently diagnosable on the basis of reciprocal monophyly of mitochondrial DNA (Moritz 1994; Moritz et al. 1995) or any heritable trait (Vogler and DeSalle 1994). This approach focuses on qualitative rather than quantitative differences between conservation units, thus avoiding the problem of determining how much difference is enough. The goal of identifying ESUs that are monophyletic is reasonable, because otherwise the ESUs would be of doubtful biological value. However, this approach has several other drawbacks. First, one of the more lively topics in evolutionary biology involves arguments over the various methods that have been used for constructing phylogenies (e.g., Swofford and Olsen 1990; Hillis and Huelsenbeck 1992). One result seems clear: there is no single method that uniformly produces the "best" phylogeny in all situations. Thus, an approach that depends upon inferring phylogenetic relationships is not immune to subjectivity.

Second, the proposal to require that ESUs be uniquely defined by characteristics not found in other ESUs confuses the problem of identifiability with the goals of conservation; the result is a criterion that is overly restrictive. For example, Vogler and DeSalle (1994) would consider a biological unit an ESU only if all individuals in the unit shared at least one heritable trait not found in any individuals from any other unit. Most units that would meet this criterion would typically be recognized as species or subspecies already, which raises the question, what additional conservation benefits would be derived from identifying such units as ESUs? Moreover, Moritz et al. (1995) admit that if their criterion were followed, a number of widely accepted fish species (e.g., many African cichlids and desert pupfish) would not even be recognized as ESUs. They argue that this result does not really cause a conservation problem because such species would generally be recognized and protected through other means, perhaps through recognition of what they term management units. The whole point of the ESA, however, is to prevent extinction of taxa that have not been adequately conserved by other methods, and there is no shortage of these.

If Moritz et al.'s ESU criterion were applied to anadromous Pacific salmonids, the answer would be fairly clear and very simple: the only ESUs would be the recognized species of *Oncorhynchus*. Although Vogler and DeSalle's method allows characters other than mitochondrial DNA to be considered, I believe the result would be essentially the same. Regardless whether one thinks this is a reasonable outcome, it would result in the failure to recognize considerable ecological–genetic diversity within each of the salmon species, diversity which provides the raw material for evolution and which the ESA mandates be conserved.

In the final analysis, whether the concept of evolutionary significance is too subjective to apply in practice should be determined by examining case histories of its application. A number of such examples have been described in this paper, and I believe that collectively they demonstrate the ESU concept can be applied successfully to real biological problems.

Comment: The ESU concept may work for salmon, but it is not applicable to other organisms.

There are several levels at which the relevance of the ESU concept might be considered, all of which are discussed to some extent in papers in this proceedings. Unfortunately, the level at which the ESU concept is being considered is often not clearly articulated, which contributes to confusion in the discussion of its usefulness. Below, I identify four possible levels for approaching biological conservation and briefly discuss the relevance of the ESU concept to each.

The first level is identification of "distinct population segments" of salmon under the ESA. This is the specific purpose for which the ESU concept of the NMFS was developed, and much of this paper has been devoted to a discussion of its usefulness on this level.

On a second level, one can ask whether the ESU concept could be applied to all vertebrate popula- tions under the purview of the ESA. Some issues covered by the ESU policy of NMFS (e.g., anadromy versus nonanadromy) have little direct relevance for most other species. However, other issues that might appear to be esoteric to salmon have clear analogues in other organisms. For example, the biological consequences of straying in salmon are similar to the consequences of migration or dispersal in other species. Similarly, although run timing is not a concept commonly applied to organisms other than anadromous fish, differences in mating season and other life history traits may be equally important for other vertebrates. More generally, any attempt to define biologically meaningful units for conservation should consider the same types of information identified in the NMFS ESU policy: migration, gene flow, and factors affecting reproduction; physical and ecological features of the habitat; and genetic, phenotypic, and life history characteristics. The basic framework of the ESU concept of NMFS is thus in no way specific to salmon and could be applied to other vertebrates under the ESA.

On a third level, one can ask whether the ESU concept could be applied more broadly to the problem of biological conservation in general. Again, I believe the basic framework is flexible enough to provide guidance on this issue for most organisms. There are two caveats, however. First, the concept of reproductive isolation is not particularly useful with organisms that reproduce clonally or by parthenogenesis. Reproductive isolation is briefly discussed below under "Other Species Concepts." Second, the concept of evolutionary significance is truly meaningful only when placed in an appropriate context. The ESA provides a legislative and legal context for interpreting the term "evolutionary significance," the basic precepts of which are, in essence, "conserve genetic diversity, do it scientifically, but do it sparingly." Outside the ESA, conservation efforts might be guided by any of several alternative contexts for interpreting evolutionary significance. For example, at one extreme, every individual might be considered evolutionarily significant because each potentially contributes to the future evolutionary trajectory of the species. Alternatively, evolutionary significance might be interpreted in terms of much larger units (Figure 6). The key factor is how conservative one wants to be (or can afford to be) in attributing evolutionary significance to a biological unit.

Finally, a fourth level of consideration may be appropriate for cases in which the struggle to obtain basic human necessities (food, water, and shelter)

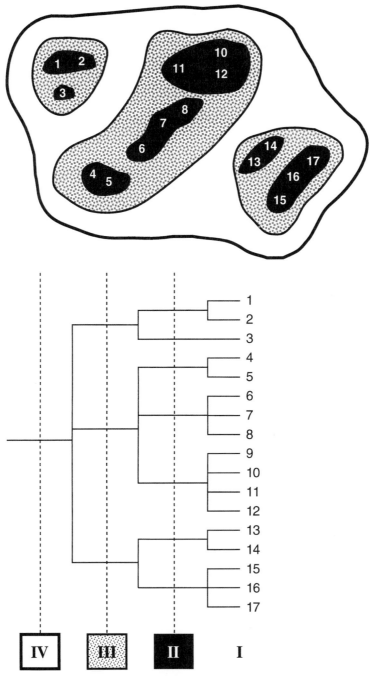

FIGURE 6.—Identifying units for conservation involves two essentially independent steps. First, one or more types of data are used to estimate evolutionary relationships among biological units (e.g., populations), as depicted here in a schematic diagram (top) and a phylogenetic tree (bottom). Additional data might resolve some of the uncertainties apparent in these diagrams (e.g., the relationships among populations 9, 10, 11, and 12). Second, a decision must be made regarding the appropriate hierarchical level on which to focus conservation efforts. In the example shown, approaches focusing on 1 all-inclusive unit, 3 units (stippled shading), 7 units (dark shading), or 17 individual populations all would be consistent with the biological data. Determining the appropriate hierarchical level may involve considering social, economic, and legal factors in addition to scientific ones.

precludes many conservation options (e.g., Mina and Golubtsov 1995, this volume; Stauffer et al. 1995, this volume). Societies in this situation often do not have the luxury of being concerned with the loss of biological populations, and even conservation of individual species may be beyond reach. Although the general concept of evolutionary significance might still be useful in guiding conservation efforts in such situations, the specific framework of the ESU concept as outlined here has little direct relevance to consideration of biological units above the level of species.

Other Species Concepts

In recent years, there have been a number of alternatives suggested to the traditional biological species concept of a group of actually (or potentially) interbreeding populations isolated from other such groups (e.g., Otte and Endler 1989; Avise 1994; see also discussions by Mayden and Wood 1995, this volume; Smith et al. 1995, this volume). In general, discussions of alternative species concepts have focused on a higher level of biological organization than that considered by the ESU concept of NMFS. That is, although the ESA considers distinct populations of vertebrates to be species, that legal definition does not change the biological reality that, in an evolutionary sense, populations are different entities than species. This is true whether species are defined according to biological, phylogenetic, evolutionary, recognition, or cohesion species concepts.

Nevertheless, the issue of reproductive isolation deserves a brief discussion here because it plays a central role in both the biological species concept and the ESU concept of NMFS. Critics of the biological species concept have pointed out that use of reproductive isolation to identify species has two limitations: (1) the test is difficult, if not impossible, to apply to strictly allopatric populations, and (2) many "good" species are not completely isolated reproductively from other species.

Neither of these difficulties represents a real problem for the NMFS ESU concept. First, consideration of allopatric populations as possible ESUs need only focus on the strength and duration of isolation that has actually occurred, not whether the allopatric units are hypothetically capable of reproducing successfully. Second, there is no requirement in the ESU concept of NMFS that reproductive isolation be absolute; rather, it need only be strong enough to allow important differences to develop in different population units. Factors af-

fecting the level of differentiation may include pre- and post-mating isolating mechanisms and selection for locally adapted genes or genotypes. Thus, the NMFS ESU concept takes a functional approach that focuses on the evolutionary consequences of reproductive isolation rather than on the isolation itself.

Joint Interagency Policy

As this paper was being finalized, the FWS and NMFS announced a draft, joint policy on recognition of distinct vertebrate population segments under the ESA (*Federal Register* 59 [21 December 1994]:65884). The joint policy is intended to be consistent with the NMFS salmon policy but will be applicable to all vertebrate species. Although the joint policy does not focus on evolutionary significance, it, like the NMFS salmon policy, proposes a two-part test for determining whether populations are distinct under the ESA. The first criterion (discreteness) is similar to the reproductive isolation criterion in the NMFS salmon policy, and the second (significance to the biological species) is roughly analogous to the contribution to ecological–genetic diversity criterion of the NMFS salmon policy. It is anticipated that NMFS will continue to use the salmon policy for Pacific salmon, and the joint policy will be used for other vertebrates.

Acknowledgments

Developing the ESU concept of NMFS involved integrating information and ideas from a large number of individuals. Particularly important resources included Utter's (1981) paper on distinct populations of salmon and the suggestions of Pat Montanio and Karl Gleaves, who provided background on the legislative and legal history of the ESA. The discussion of how the ESU concept has been applied in ESA status reviews for Pacific salmon relied heavily on information collected by a large number of individuals, both within and outside NMFS. I thank Paul Moran and Linda Park for discussion of molecular genetics issues, and Fred Allendorf, Jeff Hard, Dennis Hedgecock, Willa Nehlsen, and David Wilcove for helpful comments on earlier drafts of this paper.

References

Anonymous. 1995. Status review for anadromous Atlantic salmon in the United States. U.S. Fish and Wildlife Service and National Marine Fisheries Service, January 1995. (Available from U.S. Fish and Wildlife Service, 22 Bridge Street, Concord, New Hampshire 03301, USA.)

Avise, J. C. 1994. Molecular markers, natural history and evolution. Chapman and Hall, New York.

Bevan, D., and six coauthors. 1994. Snake River salmon recovery team: final recommendations to the National Marine Fisheries Service. NOAA (National Oceanic and Atmospheric Administration) NMFS (National Marine Fisheries Service) Environmental and Technical Services Division, Portland, Oregon.

Bjornn, T. C., D. R. Craddock, and D. R. Corley. 1968. Migration and survival of Redfish Lake, Idaho, sockeye salmon, *Oncorhynchus nerka*. Transactions of the American Fisheries Society 97:360–373.

Bowler, B. 1990. Additional information on the status of Snake River sockeye salmon. Report to Endangered Species Act administrative record for sockeye salmon. Idaho Department of Fish and Game, Boise.

Busby, P. J., O. W. Johnson, T. C. Wainwright, F. W. Waknitz, and R. S. Waples. 1993. Status review for Oregon's Illinois River winter steelhead. NOAA (National Oceanic and Atmospheric Administration) Technical Memorandum NMFS (National Marine Fisheries Service) NWFSC-10, Northwest Fisheries Science Center, Seattle.

Busby, P. J., T. C. Wainwright, and R. S. Waples. 1994. Status review for Klamath Mountains Province steelhead. NOAA (National Oceanic and Atmospheric Administration) Technical Memorandum NMFS (National Marine Fisheries Service) NWFSC-19, Northwest Fisheries Science Center, Seattle.

Chapman, D. W., W. S. Platts, D. Park, and M. Hill. 1990. Status of Snake River sockeye salmon. Final report for Pacific Northwest Utilities Conference Committee, Portland, Oregon.

Coggins, G. C. 1991. Snail darters and pork barrels revisited: reflections on endangered species and land use in America. Pages 62–74 *in* K. A. Kohm, editor. Balancing on the brink of extinction: the Endangered Species Act and lessons for the future. Island Press, Washington, DC.

Dizon, A. E., C. Lockyer, W. F. Perrin, D. P. DeMaster, and J. Sisson. 1992. Rethinking the stock concept: a phylogeographic approach. Conservation Biology 6:24–36.

Evermann, B. W. 1896. A report upon salmon investigations in the headwaters of the Columbia River in the state of Idaho, in 1895. U.S. Fish Commission Bulletin 16:151–202.

Foerster, R. E. 1947. Experiment to develop sea-run from land-locked sockeye salmon (*Oncorhynchus nerka kennerly*). Journal of the Fisheries Research Board of Canada 7(2):88–93.

Foote, C. J., C. C. Wood, and R. E. Withler. 1989. Biochemical genetic comparison of sockeye salmon and kokanee, the anadromous and nonanadromous forms of *Oncorhynchus nerka*. Canadian Journal of Fisheries and Aquatic Sciences 46:149–158.

Hard, J. J., R. P. Jones, Jr., M. R. Delarm, and R. S. Waples. 1992. Pacific salmon and artificial propagation under the Endangered Species Act. NOAA (National Oceanic and Atmospheric Administration) Technical Memorandum NMFS (National Marine Fisheries Service) NWFSC-2, Northwest Fisheries Science Center, Seattle.

Healey, M. C. 1991. Life history of chinook salmon (*Oncorhynchus tshawytscha*). Pages 311–393 *in* C. Groot and L. Margolis, editors. Pacific salmon life histories. University of British Columbia Press, Vancouver.

Hillis, D. M., and J. P. Huelsenbeck. 1992. Signal, noise, and reliability in molecular phylogenetic analyses. Journal of Heredity 83:119–195.

Johnson, O., T. A. Flagg, D. J. Maynard, G. B. Milner, and F. W. Waknitz. 1991. Status review for lower Columbia River coho salmon. NOAA (National Oceanic and Atmospheric Administration) Technical Memorandum NMFS (National Marine Fisheries Service) F/NWC-202, Northwest Fisheries Science Center, Seattle.

Johnson, O. W., R. S. Waples, T. C. Wainwright, K. G. Neely, F. W. Waknitz, and L. T. Parker. 1994. Status review for Oregon's Umpqua River sea-run cutthroat trout. NOAA (National Oceanic and Atmospheric Administration) Technical Memorandum NMFS (National Marine Fisheries Service) NWFSC-15, Northwest Fisheries Science Center, Seattle.

Kaeriyama, M., Urawa, S., and Suzuki, T. 1992. Anadromous sockeye salmon (*Oncorhynchus nerka*) derived from nonanadromous kokanees: life history in Lake Toro. Scientific Reports of the Hokkaido Salmon Hatchery 46:157–174.

Leopold, A. 1953. Round river: from the journals of Aldo Leopold. Oxford University Press, New York.

Matthews, G. M., and R. S. Waples. 1991. Status review for Snake River spring and summer chinook salmon. NOAA (National Oceanic and Atmospheric Administration) Technical Memorandum NMFS (National Marine Fisheries Service) F/NWC-200, Northwest Fisheries Science Center, Seattle.

Mayden, R. L., and R. M. Wood. 1995. Systematics, species concepts, and the evolutionarily significant unit in biodiversity and conservation biology. American Fisheries Society Symposium 17:58–113.

McIsaac, D. O., and T. P. Quinn. 1988. Evidence for a hereditary component in homing behavior of chinook salmon (*Oncorhynchus tshawytscha*). Canadian Journal of Fisheries and Aquatic Sciences 45:2201–2205.

Mina, M., and A. Golubtsov. 1995. Faunas of isolated regions as principal units in the conservation of freshwater fishes. American Fisheries Society Symposium 17:145–148.

Moritz, C. 1994. Defining 'evolutionarily significant units' for conservation. Trends in Ecology and Evolution 9:373–375.

Moritz, C., S. Lavery, and R. Slade. 1995. Using allele frequency and phylogeny to define units for conservation and management. American Fisheries Society Symposium 17:249–262.

Nehlsen, W., J. E. Williams, and J. A. Lichatowich. 1991. Pacific salmon at the crossroads: stocks at risk from California, Oregon, Idaho, and Washington. Fisheries 16(2):4–21.

Nicholas, J. W., and D. G. Hankin. 1988. Chinook salmon populations in Oregon coastal river basins: description of life histories and assessment of trends in run strengths. Oregon Department of Fish and Wildlife Information Report 88-1, 359p. (Available from Oregon Department of Fish and Wildlife, P.O. Box 59, Portland, Oregon 97207, USA.)

ONRC (Oregon Natural Resources Council), the Wilderness Society, and Umpqua Valley Audubon Society. 1993. Petition for a rule to list the north and south Umpqua River sea-run cutthroat trout as threatened or endangered under the Endangered Species Act and to designate critical habitat. (Document submitted to the U.S. Department of Commerce, National Oceanic and Atmospheric Administration, and National Marine Fisheries Service Northwest Division, Seattle, Washington, April 1993.) Oregon Natural Resources Council, 522 SW 5th, Suite 1050, Portland, OR, 97204.

Otte, D., and J. A. Endler, editors. 1989. Speciation and its consequences. Sinauer, Sunderland, Massachusetts.

Pearcy, W. G. 1992. Ocean ecology of North Pacific salmonids. University of Washington Press, Seattle.

Rohlf, D. J. 1994. There's something fishy going on here: a critique of the National Marine Fisheries Service's definition of species under the Endangered Species Act. Environmental Law 24:617–671.

Ryder, O. A. 1986. Species conservation and systematics: the dilemma of subspecies. Trends in Ecology and Evolution 1(1):9–10.

Smith, G., J. Rosenfield, and J. Porterfield. 1995. Processes of origin and criteria for preservation of fish species. American Fisheries Society Symposium 17:44–57.

Stauffer, J. R., Jr., N. J. Bowers, K. R. McKaye, and T. D. Kocher. 1995. Evolutionarily significant units among cichlid fishes: the role of behavioral studies. American Fisheries Society Symposium 17:227–244.

Swofford, D. L., and G. J. Olsen. 1990. Phylogeny reconstruction. Pages 411–501 in D. M. Hillis and C. Moritz, editors. Molecular systematics. Sinauer, Sunderland, Massachusetts.

Utter, F. 1981. Biological criteria for definition of species and distinct intraspecific populations of anadromous salmonids under the U.S. Endangered Species Act of 1973. Canadian Journal of Fisheries and Aquatic Sciences 38:1626–1635.

Utter, F. M., W. J. Ebel, G. B. Milner, and D. J. Teel. 1982. Population structures of fall chinook salmon, *Oncorhynchus tshawytscha*, of the mid-Columbia and Snake Rivers. NOAA (National Oceanic and Atmospheric Administration) NMFS (National Marine Fisheries Service) Northwest and Alaska Fisheries Center, Processed Report 82-10, Auke Bay, Alaska.

Vogler, A. P., and R. DeSalle. 1994. Diagnosing units of conservation management. Conservation Biology 8:354–363.

Waknitz, F. W., G. M. Matthews, T. Wainwright, and G. A. Winans. 1995. Status review for mid-Columbia River summer chinook salmon. NOAA (National Oceanic and Atmospheric Administration) Technical Memorandum NMFS (National Marine Fisheries Service) NWFSC-22, Northwest Fisheries Science Center, Seattle.

Waples, R. S. 1991a. Definition of "species" under the Endangered Species Act: application to Pacific salmon. NOAA (National Oceanic and Atmospheric Administration) Technical Memorandum NMFS (National Marine Fisheries Service) F/NWC-194, Northwest Fisheries Science Center, Seattle.

Waples, R. S. 1991b. Pacific salmon, *Oncorhynchus* spp., and the definition of "species" under the Endangered Species Act. U.S. National Marine Fisheries Service Marine Fisheries Review 53(3):11–22.

Waples, R. S., and six coauthors. 1993. A genetic monitoring and evaluation program for supplemented populations of salmon and steelhead in the Snake River basin. Annual Research Report to Bonneville Power Administration DE-A179-89BP00911, Portland, Oregon.

Waples, R. S., O. W. Johnson, and R. P. Jones, Jr. 1991a. Status review for Snake River sockeye salmon. NOAA (National Oceanic and Atmospheric Administration) Technical Memorandum NMFS (National Marine Fisheries Service) F/NWC-195, Northwest Fisheries Science Center, Seattle.

Waples, R. S., R. P. Jones, Jr., B. R. Beckman, and G. A. Swan. 1991b. Status review for Snake River fall chinook salmon. NOAA (National Oceanic and Atmospheric Administration) Technical Memorandum NMFS (National Marine Fisheries Service) F/NWC-201, Northwest Fisheries Science Center, Seattle.

Weitkamp, L. A., and six coauthors. In press. Status review of coho salmon from Washington, Oregon, and California. NOAA (National Oceanic and Atmospheric Administration) Technical Memorandum NMFS (National Marine Fisheries Service) NWFSC-XX, Northwest Fisheries Science Center, Seattle.

Whittaker, R. H. 1960. Vegetation of the Siskiyou Mountains, Oregon and California. Ecological Monographs 30(3):279–338.

American Fisheries Society Symposium 17:28–38, 1995

Selection of Conservation Units for Pacific Salmon: Lessons from the Columbia River

PHILLIP R. MUNDY

Fisheries and Aquatic Sciences, 1015 Sher Lane
Lake Oswego, Oregon 97034, USA

THOMAS W. H. BACKMAN

Columbia River Inter-Tribal Fish Commission
729 Northeast Oregon Street, Suite 200
Portland, Oregon 97232, USA

J. M. BERKSON

Columbia River Inter-Tribal Fish Commission and
Department of Biology, Montana State University
Bozeman, Montana 59717, USA

Abstract.—We present a working concept of a conservation unit for Pacific salmon *Oncorhynchus* spp. that we believe to be biologically effective, legally defensible, and consistent with a composite tribal and biological perspective on resource management. The two principal elements of the conservation unit definition are achieving viability for all life history types and maintaining sufficient geographic distribution. Conservation unit definition requires the ability to identify the biological species, its life history types, and the nature of the life history types' dependence on other components of the ecosystem, such as habitat. Identifying sufficient geographic range compels identification of the collection of habitats occupied by the life history types. Defining geographic sufficiency also requires determining the likelihood of persistence of a collection of salmon populations. We compare our working definition with the current salmon conservation unit, the evolutionarily significant unit (ESU), and find the ESU does not include the standards needed to enable recovery of damaged salmon populations. We explore the application of the current ESU to recovery of both spring chinook salmon *O. tshawytscha* and sockeye salmon *O. nerka*, and we suggest that the geographic range of the present ESU for both species is inadequate to effect recovery under the U.S. Endangered Species Act (ESA). In the legal context, the ESU is not materially different from its predecessor, the distinct population segment. Finally, we apply our conservation unit definition to identify four groups of salmon, some of which are not presently protected, that should qualify as species for listing under the ESA.

The impetus for a symposium to refine the definition of a conservation unit follows the listing of populations of Pacific salmon *Oncorhynchus* spp. in California, Oregon, Washington, and Idaho under the U.S. Endangered Species Act (ESA; 16 U.S.C. §§ 1531 to 1544). Defining a conservation unit means identifying the group of animals on which conservation actions are to be taken. The definition of a conservation unit for listed salmon species is biologically and politically important because it determines both the geographic area of interest and the number of populations that may eventually become federally listed species. As a matter of political concern, should the conservation unit definition focus on small, individual geographic areas, such as springs and small streams, a very large number of salmon populations may qualify for listing as threatened or endangered species. On the biological side, should the conservation unit definition encompass large geographic areas, such as one or more major river basins, including within the definition protection for the full diversity of salmon populations that any large area might contain would be challenging.

All salmon recovery actions that involve individuals from populations that are federally listed salmon species are now subject to federal oversight. Salmon recovery actions taken under federal oversight need to be consistent with the purposes of the ESA and appropriate to the biology of the species and its component populations. Hence federal salmon recovery oversight needs to be guided by a definition of the conservation unit that is legally defensible and can be tested for effectiveness during its application in specific recovery actions. We refer to such a working, testable explanation of the conservation unit as an operational definition.

In addition to the perspectives of the ESA and biological science, the salmon conservation unit definition needs to represent the perspectives of Native American tribes, many of which have treaty rights to salmon. Our view of the tribal perspective comes from our experiences working as biologists for tribal governments within Native American cultures in Alaska, Puget Sound, and the Columbia River basin over the past 18 years. The tribal perspectives relevant to the conservation unit definition concern sustainable use and geographic distribution.

The tribal perspective defines conservation as harvesting in a manner consistent with sustaining human uses of the salmon populations and their life history types, such as spring, summer, fall, and winter adult migrations (runs), for time periods equal to at least the next seven generations of humans. Thus, the tribal perspective on conservation includes the concept of indefinitely sustaining all species and life history types of salmon at levels of abundance sufficient to permit human uses. A second key concept from the tribal perspective that is relevant to the conservation unit definition is maintaining the historical geographic distribution of salmon populations and life history types. Tribal fishing rights are geographically defined either by, in the case of treaty fishing tribes, the extent of the areas ceded to the federal government, or, in the case of nontreaty tribes, the locations of traditional fishing sites.

A conservation unit definition that is consistent with the tribal perspective protects the salmon populations at all the localities where these fish have been traditionally harvested, it protects the salmon where they have the potential for sustaining themselves indefinitely at levels of abundance consistent with human use, and it protects geographic ranges of habitat that contain the full diversity of salmon life history types on which traditional cultural and economic uses are based.

As biologists, we see two purposes to be met by developing an operational definition of the conservation unit for the Pacific salmon. These purposes are essentially the same as those of the tribal perspective described above: achieving conservation and maintaining geographic distribution. The core of our concern, conservation, is to provide for the indefinite persistence of the historical assemblage of species, populations, and life history types. Maintaining the diversity of life history types is essential to the persistence of animal populations in general (Cole 1954; Schaffer 1974) and to salmon conservation in particular (Thompson 1951; Thorpe 1989). Conservation implies maintaining abundance, because providing for persistence requires maintaining the target populations at suitable levels of abundance based on their average productivity (Ricker

1954; Cushing 1983). Environmental stochasticity, random environmental effects (May 1973; Leigh 1981), demographic stochasticity, and random demographic effects (MacArthur and Wilson 1967; Richter-Dyn and Goel 1972) all act to increase a population's probability of extinction as population size decreases. In order to provide for the persistence of all life history types, the geographic range needs to include the appropriate number and quantity of habitat types (Frissell et al. 1986).

A Working Concept of the Salmon Conservation Unit

Our working concept of a salmon conservation unit is a collection of geographically based spawning aggregates that covers an area large enough to contain all of the biological species' life history types as well as sufficient quantity and quality of habitat to permit the conservation unit's long-term viability. The scientific basis for inclusion of these two elements, multiple spawning aggregates and sufficient geographic range, is drawn from the literature of population biology (Emlen 1984), including life history strategies (Cole 1954) and metapopulation dynamics (Hanski and Gilpin 1991).

The conservation unit needs to include multiple spawning aggregates, also called local populations (Andrewartha and Birch 1984), occupying multiple spawning localities. We define a spawning aggregate as a classic Mendelian population: a group of freely interbreeding individuals that produce viable offspring (Dobzhansky 1970). A conservation unit that has spatial structure will be far better protected against the risks posed by environmental and human-made uncertainty than would a conservation unit based on a single spawning aggregate that has no spatial structure (den Boer 1968; Vance 1980; Andrewartha and Birch 1984). Brown and Kodric-Brown (1977) termed the effect of geographic dispersal, speaking in reference to the metapopulation, the rescue effect.

In practice, we expect the number of salmon spawning aggregates that are combined to form a conservation unit will depend on a number of factors related to the physiography of the region and the biology of the populations. Such factors include the degree to which the spawning localities are connected by suitable habitat (see Stanford and Ward 1992), the degree to which populations exchange breeding individuals (Vance 1980), and the productivities (Ricker 1954) of the spawning aggregates. The study of metapopulations provides a body of literature that deals with the preceding issues of connectivity, rates of exchange, and rates of productivity among aggregates, and the effects of these factors on population viability (Hanski and Gilpin 1991).

The metapopulation, as applied to Pacific salmon, is a collection of spawning aggregates of a biological species that are genetically related and demographically connected to one another to varying degrees and that are viable (Gilpin and Soulé 1986; Thomas 1990) for arbitrarily long periods of time by virtue of these connections. The risk of extirpation (localized extinction of a portion of a biological species) would be defined for the Pacific salmon metapopulation in terms of the population dynamics of its component populations and their degree of demographic connectivity. In salmon, demographic connectivity refers to the extent to which the spawning aggregates are capable of providing reproductively viable individuals to one another in a process known as straying (Quinn 1984, 1985; Pascual and Quinn 1995).

Including the maintenance of sufficient geographic range in the definition for the conservation unit is based on requirements to include all extant life history types and to identify a collection of spawning aggregates that is expected to be viable over an arbitrarily long period of time. In the case of the Columbia River basin, we would expect the size of a geographic area that would contain the full range of life history types for a biological salmon species to be quite large, because there has been long-term, and geographically extensive, destruction of freshwater habitat and associated spawning aggregates (Rich 1941; NPPC 1986). Another reason the geographic area containing a salmon conservation unit in the Columbia River basin may be quite large is that population levels of all species are generally low, and declining, throughout the region (Norman and King 1992), and population viability is proportional to population size.

Critique of the Evolutionarily Significant Unit Concept

The present conservation unit for Pacific salmon, the evolutionarily significant unit (ESU), as defined by the National Marine Fisheries Service (NMFS; Waples 1991), is a biological definition of species for the purposes of the ESA. The ESU has been legally adopted as the standard conservation unit for recovery of endangered and threatened salmon populations by the NMFS (Waples 1991; Matthews and Waples 1991; Waples et al. 1991). The salmon ESU definition establishes two criteria for identify-

ing a conservation unit: (1) a state of substantial reproductive isolation and (2) possession of qualities that constitute a significant component of the evolutionary legacy of the species. These criteria for placing a population within an ESU are obviously open to a broad range of interpretations. For example, what degree of reproductive isolation is considered substantial, and what constitutes a significant component in the evolutionary legacy of a species are but two of the many key questions left unanswered by the present ESU definition.

The salmon ESU definition, as adopted and implemented by the NMFS, is not operational in our experience. In the short amount of time the ESU has been applied to damaged salmon populations, it has not established and promoted conditions conducive to long-term population viability because it sets no standards by which to design and implement salmon recovery actions. The standards missing from the ESU are those for achieving conservation and providing sufficient geographic range. Achieving conservation by including appropriate life history types is a standard that defines sufficient geographic range. The present ESU fails to define other standards that are also related to sufficient geographic range, such as population viability and the minimum number of animals needed—a threshold, or minimum viable population size (Thomas 1990)—to maintain a population's normal function within the ecosystem.

The following brief excerpts from the case histories of recovery operations for chinook salmon *O. tshawytscha* and sockeye salmon *O. nerka* in the Columbia River basin's Snake River illustrate the inadequacy of the ESU as a conservation unit definition. In the case of chinook salmon, the loss of 300 adult spawners from the ESU of the threatened species was narrowly averted by legal action. In the case of sockeye salmon, the recovery actions have been limited to what appears to be only a single spawning aggregate; such recovery actions appear to have increased the time and expense required for sockeye salmon recovery, and leave the viability of these sockeye salmon in serious doubt.

Snake River Chinook Salmon

Evidence of the nonoperational nature of the current salmon ESU definition was provided during its application in the 1993 management season for spring chinook salmon in the Imnaha River (Figure 1), a Snake River tributary in eastern Oregon (Berg 1993). The complexity of the Imnaha spring chinook salmon management situation exemplifies

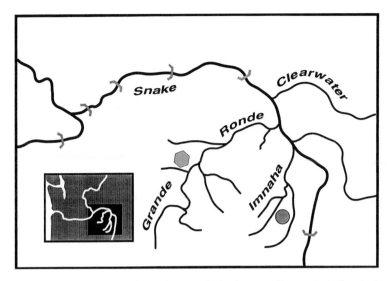

FIGURE 1.—Map of the Imnaha and Grande Ronde watersheds of eastern Oregon, including the locations (from left to right) of McNary Dam on the Columbia River; Ice Harbor, Lower Monumental, Little Goose, and Lower Granite dams on the Snake River; and Lookingglass Hatchery (hexagon) and the Imnaha satellite hatchery (circle).

the complexity of the Snake and Columbia rivers salmon management situation as a whole, so the reader is asked to bear with the detail of the following explanation. Imnaha River spring chinook salmon are part of the ESU identified by NMFS of Snake River basin spring–summer chinook salmon; however, not all Imnaha River spring chinook salmon are part of the federally listed threatened spring–summer chinook salmon species. How can this be? Members of the salmon ESU can fall outside the protection of the ESA because the salmon ESU is not actually a conservation unit in the sense that it can be used to define groups of salmon to which specific management actions apply. In the case of the Imnaha, as of 1993 NMFS had determined that those spring chinook salmon which reared for any amount of time in the Imnaha–Grande Ronde artificial propagation facilities (Figure 1) were not part of the listed species, although they were part of the ESU (M. H. Schiewe, National Marine Fisheries Service, personal communication). The artificially propagated spring chinook salmon were included as part of the ESU because there were no discernable differences in allelic or heritable phenotypic characters between the wild and artificially spawned and reared spring chinook salmon. Although data are unpublished, phenotypic and genotypic characteristics are examined annually by NMFS.

The lack of measurable genetic and phenotypic differences between Imnaha River hatchery and wild spring chinook salmon may be due to the fact that wild fish are systematically bred into the artificially propagated population each generation and that the artificial propagation program, which dates to 1982, has been intentionally operated to minimize, or eliminate, hatchery practices that might impose selective forces. The hatchery program was initiated under the federal Lower Snake River Compensation Program as mitigation for the effects of the lower Snake River dams and in response to clear and steady declines in annual abundance of wild Imnaha spring chinook salmon (Figure 2).

To this complex brew of salmon production methods and public policies is added the multiparty management structure imposed by the political history of the region. The Imnaha River salmon management program is conducted under the terms of the Federal District Court's (Portland, Oregon) Columbia River Fisheries Management Plan (Plan). The Plan resulted from the treaty Indian fishing rights litigation known as United States v. Oregon, 699 F. Supp. 1456 (D. Or. 1988). In the case of the Imnaha, the parties to the Plan are the Nez Perce Tribe of Idaho, the Confederated Tribes of the Umatilla Indian Reservation of Oregon, the Oregon Department of Fish and Wildlife (ODFW), and NMFS. The Imnaha management program is also bound to be consistent with the results of ESA consultations between NMFS and those responsible for Imnaha River harvest and hatchery operations—the U.S. Fish and Wildlife Service, the

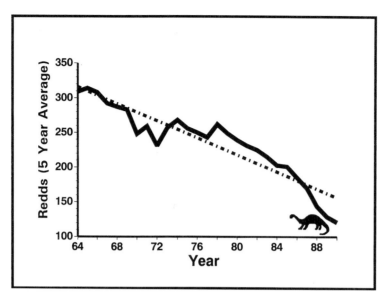

FIGURE 2.—Five-year moving average of the annual spring chinook salmon redd counts for the Imnaha River, eastern Oregon. The dashed line is a linear fit to these data.

ODFW, and the Bureau of Indian Affairs—on behalf of the tribes concerned.

Additional complexity is added to Imnaha salmon management because state of Oregon employees, who operate the Imnaha–Grande Ronde hatchery complex with federal funds and the advice and consent of the parties to the Plan, are also bound by the terms of the Oregon Wild Fish Policy (OWFP; OAR-635-07-501), although the other parties are not bound. As applied in the Imnaha, the OWFP requires that the ratio of hatchery to wild salmon spawners is not to exceed unity (1:1). Unfortunately, the OWFP application did not distinguish between hatchery salmon that are genetically and phenotypically indistinguishable from the wild salmon and hatchery fish that may be radically different from the wild salmon. Therein lies the tale, and the lesson, for designers of conservation units for salmon.

The OWFP was originally designed to serve the laudable purpose of preventing native salmon populations from being genetically eradicated through the process of spawning with large numbers of allocthonous salmon originating in hatcheries. But the OWFP, as the ODFW attempted to apply it in 1993, would have resulted in the prespawning termination of approximately 300 adult spring chinook salmon that were part of the ESU of a threatened species. The elimination of these hatchery returns was planned by ODFW in order to maintain the ratio of hatchery to wild chinook salmon near unity. The ODFW plan to dispose of approximately 300 spring chinook salmon spawners was supported by NMFS as consistent with recovery objectives and the ESU concept.

That an action so inimical to Imnaha spring chinook salmon recovery efforts could have been planned by the ODFW, with the approval of NMFS, is a stunning example of the damage that can be done in the attempt to apply a nonoperational definition, such as the current salmon ESU, to an actual management situation. The loss of 300 spawners in a river system that arguably has not been fully seeded with spring chinook salmon spawners in the past 36 years and from a population whose numbers have been steadily declining for the 25 years ending in 1989 (Figure 2) and whose members are within the ESU of a threatened species, would have been a catastrophe. By comparison, the Imnaha River is projected by the management entities to receive approximately only 400 spring chinook salmon spawners, both hatchery and wild combined, in all of 1994.

In the end, through the intervention of the treaty fishing tribes and the good offices of the federal court, ODFW agreed to allow a portion of the 300 condemned spring chinook salmon to spawn naturally and the balance to be artificially spawned. The ODFW plan was abandoned only after the Nez Perce Tribe and the Confederated Tribes of the

Umatilla Indian Reservation, with the support of the Columbia River Inter-Tribal Fish Commission and its two other member tribes, objected strenuously to the ODFW plan to sacrifice the spawners. Working through a court-supervised arbitration procedure provided under the Columbia River Fisheries Management Plan, the tribes questioned the scientific basis of the ODFW elimination plan. One of us (Mundy) participated in the hearings as a scientific advisor on behalf of the tribes. The tribes pointed out that the majority of the parents of the hatchery spawners had been wild, and most of the spawners condemned by ODFW were first-generation hatchery fish. Also disclosed during arbitration was that (1) the OWFP hatchery to wild fish ratio of 1:1 had no scientific basis, not being based on empirical evidence or published scientific literature; (2) there were no differences in allelic frequencies between hatchery-reared fish and wild fish; and (3) there were no measurable differences in heritable phenotypic traits of adaptive significance between the hatchery and wild spring chinook salmon. There were some differences in the age compositions of the two groups of chinook salmon; the differences were acknowledged by the parties to be the result of experimental rearing conditions imposed on some hatchery residents in an attempt to increase their smolt-to-adult survival.

The application of the current salmon ESU in the Imnaha in 1993 was a narrowly averted disaster because this ESU definition (and the OWFP as well) is blind to the demographic risks of extirpation. In pursuing the reasonable and essential objective of protecting the genetic diversity (Lande and Barrowclough 1987) of Pacific salmon populations under the ESA, the ESU definition has failed to take into account that genetic diversity becomes zero when population numbers reach zero. The action required under OWFP in the name of protecting genetic diversity in the Imnaha in 1993 would have contributed to increased risk of extirpation. On the other hand, protecting against demographic risks of extirpation with no regard for protection of genetic diversity would not be in the best interest of the long-term viability of either the Imnaha spring chinook salmon populations or the Snake River basin spring and summer chinook salmon of which the Imnaha population is a part. In formulating the definition of a unique conservation unit, a balance needs to be struck in defending populations against demographic and genetic risks because, taken to extremes, these two types of defensive actions can be mutually exclusive.

Snake River Sockeye Salmon

Although Columbia River basin sockeye salmon, which spawn and rear in association with freshwater lakes, have been excluded by dams from all but 2 of the up to 30 lakes they originally occupied (Figure 3), only one very small sockeye salmon population in a major tributary system, the Snake River basin, was listed in 1991 as endangered. The listed population was that of Redfish Lake in the Salmon River drainage of Idaho. Due to the small number of adult sockeye salmon returning to the Snake River to spawn each year at the time of the listing (Figure 4), the entire population was originally taken into a captive broodstock program to complete its life cycle in an artificial environment. Although a few sockeye salmon have been returned to the rearing lake from this program (P. Lumley, Columbia River Inter-Tribal Fish Commission, personal communication), the actual source of these sockeye salmon is not known and the present populations are probably not genetically identical to the originally native populations (Bjornn et al. 1968).

Clearly the Redfish Lake sockeye salmon recovery program is already a failure from the tribal perspective because a potentially productive salmon population has been entirely removed from the environment. Further, because the sockeye salmon broodstock selected for the recovery program is extremely limited in numbers, and therefore may also be limited in phenotypic traits of adaptive significance for the Snake River basin, the time by which a harvestable population of sockeye salmon will be established in the Snake River basin appears very far away.

Due to the application of the ESU to only the extremely small sockeye salmon population of Redfish Lake, the wider sockeye salmon populations of the Columbia River basin received no protection under the ESA, nor can they be used as broodstock for transplanting to accelerate the Snake River recovery program. The rationale for not including all Columbia River basin sockeye salmon in the ESU of the federally listed sockeye salmon species is not clear, given the loss of 96% of the Columbia sockeye salmon's freshwater habitat (Norman and King 1992; see Figure 3); the concomitant reduction in total numbers of spawning adult Columbia sockeye salmon returning to the Columbia River basin from several million a year (Craig and Hacker 1940) to less than 200,000 annually (NPPC 1986; Norman and King 1992); the nearly total reproductive isolation of these populations; and the significance of the evolutionary legacy accruing to populations at the

FIGURE 3.—Lake rearing sites of extant (black fish symbol), extirpated (dinosaur symbol), and endangered Redfish Lake (white fish symbol) sockeye salmon populations in the Columbia River basin.

current southeastern limit of the sockeye salmon's geographic range. In fact, meeting the last two criteria alone should have been sufficient for Columbia River basin sockeye salmon to qualify as an ESU under the existing definition. The terms of the ESA permit protecting populations that are in sharp decline in abundance and have had large losses in habitat, as is clearly the case with all of the Columbia basin sockeye salmon populations (Figure 3).

Thus, the present salmon ESU is silent on how to recognize significant geographic assemblages of spawning sockeye salmon populations and on how to determine threshold population levels that would require listing under the ESA. Our experience is that the issue of the threshold number warranting a listing of threatened or endangered under the ESA is not separable from the issue of what constitutes an ESA species. Redfish Lake clearly illustrates the connection between the standards of sufficiency of geographic range and abundance and viability.

Because of the narrow interpretation of the ESU

in the case of sockeye salmon, enormous financial resources of the Bonneville Power Administration have been allocated to the recovery of a population whose long-term viability is in serious doubt. By increasing the geographic scope of the listing, it is our opinion that populations that still have annual returns denominated in the thousands could be afforded some protection while they yet have the potential to recover.

The Legal Context

In addition to the flaws created by its biological discontinuities, the salmon ESU may be procedurally and legally superfluous. In attempting to provide objective biological criteria for defining species under the ESA (see Waples 1991), the salmon ESU may have succeeded only in offering its own arbitrary set of criteria in exchange for those of another ESA species definition, the distinct population segment. The intertwining of legal and biological problems within the present salmon ESU is eloquently

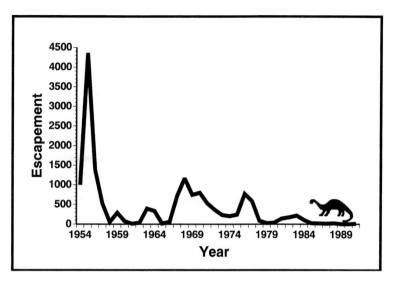

FIGURE 4.—Time series of abundance (escapement, number of fish) for sockeye salmon populations of the Salmon River, tributary to the Snake River basin. Dinosaur = extripations. (Bjornn et al. 1968; Norman and King 1992).

summarized by Rohlf (1994). Rohlf found the ESU to be inconsistent with the policies established by the U.S. Congress to implement the ESA and that the ESU definition missed biological factors which determine a species' persistence. Rohlf argued that Congress never intended to determine what constitutes a species under the ESA solely on the basis of scientific grounds, as NMFS does in the case of the ESU. He concludes that the technical shortcomings of the ESU do not permit the NMFS to protect salmon by means of the best scientific data available.

Conclusions

In practice, the definition of the conservation unit for Pacific salmon, as for any such classification, is necessarily arbitrary. Although tribal perspectives and biological principles can provide some guidance, ultimately there is no acceptable minimum group of animals on which to target conservation efforts, because the extirpation of even the smallest spawning aggregate is unacceptable. On the other hand, as the Imnaha River case study demonstrates, it is not prudent to assume, a priori, that the spawners of any given locality are evolutionarily significant. Such an assumption can, and did, lead to the conclusion that one collection of salmon spawning aggregates needed to be protected by sacrificing other, potentially evolutionarily significant, spawning aggregates. In view of the currently low salmon population levels, all Columbia River basin salmon

are evolutionarily significant enough to warrant protection.

We do not believe it is within the capabilities of science to judge the evolutionary significance of one salmon spawning aggregate against that of another. Based on evaluation of phenotypic traits, we do believe it is appropriate to restore to viability those spawning aggregates that are adapted to local environmental conditions (Ricker 1972; Riddell and Leggett 1981; Thorpe 1986). When such adaptive phenotypic traits have been lost from the spawners at a locality, it may be necessary to acquire phenotypically suitable individuals from other watersheds to replace these traits in order to promote the long-term viability of the population, subject to the usual precautions, such as pathology screening.

Based on our working definition of the conservation unit, four ESA species could be federally listed in the Columbia River basin under the ESA: chinook salmon, sockeye salmon, coho salmon *O. kisutch*, and chum salmon *O. keta*. Native populations of all four species have been in steep numerical decline within the Columbia River basin during most of this century (Craig and Hacker 1940; Norman and King 1992; see Figure 5), all four species have suffered extensive loss of, and damage to, critical freshwater habitat (NPPC 1986), all extant life history types are contained within this geographic region, the geographic region is sufficient to support viable populations of all four species, all four species are substantially reproductively iso-

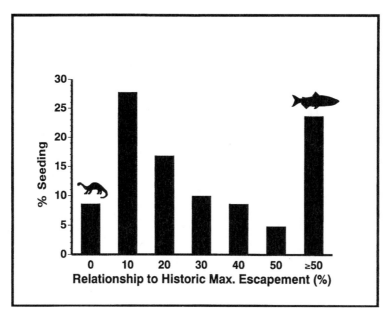

FIGURE 5.—Frequency distribution of seeding (number of redds) at the 292 spawning sites of wild chinook salmon surveyed in 1990 as a percentage of the site's historical maximum seeding, Columbia River basin (Mundy and Fryer 1995). Note that the number of redds at 28% of the survey sites did not exceed 10% of the site's historical maximum, and that less than 25% of the sites had seeding levels equal to or greater than 50% of their historical maximum in 1990. See Figure 3 for symbols.

lated by geographic separation from other populations of their biological species, and all four species contain a significant part of the evolutionary legacy of their biological species. For each species, the viability and genetic diversity of each spawning aggregation within each river basin (Cowlitz, Deschutes, Yakima, Imnaha, and others), needs to be secured by protecting and restoring spawning and rearing habitat and by implementing ecologically responsible stocking programs that strike a balance between the demographic risks of extirpation and the risks of losing genetic diversity (RASP 1992; Cuenco et al. 1993).

All salmon in the Columbia River basin need more protection than they have received to date, if they are to be available to future generations of humans. We recommend abandoning the current ESU approach, which protects some spawning aggregates to the detriment of others, in favor of a geographically comprehensive approach that seeks to establish and protect a diversity of life history types over the entire Columbia River basin.

Acknowledgments

We thank the Columbia River Inter-Tribal Fish Commission and the American Fisheries Society and its sponsors for making this effort possible. The views expressed are those of the authors, and they do not necessarily represent those of the Commission's four member tribes, the American Fisheries Society, or its sponsors. The written record of the hearings on the disposition of the Imnaha spring chinook spawners in 1993 (United States v. Oregon) was provided to us by Paul Lumley, Chair of the Technical Advisory Committee that serves the Federal District Court (Portland). The traditional Native American concept of managing salmon for the benefit of the next seven human generations was explained to us by Delbert Frank of the Confederated Tribes of the Warm Springs Indian Reservation, Warm Springs, Oregon. We thank Matthew Schwartzberg for his advice and help in the preparation of slides and illustrations. Jeffrey Fryer provided the information for the geographic distribution of extant and extirpated sockeye salmon populations from the draft of his doctoral dissertation.

References

Andrewartha, H. G., and L. C. Birch. 1984. The ecological web. More on the distribution and abundance of animals. University of Chicago Press, Chicago.

Berg, L. 1993. Oregon's wild fish policy hurts more than helps: the Imnaha spring chinook story. Wana Chinook Tymoo 3&4:8–13. Columbia River Inter-Tribal Fish Commission, Portland, Oregon.

Bjornn, T. C., D. R. Craddock, and D. R. Corley. 1968. Migration and survival of Redfish Lake, Idaho, sockeye salmon, *Oncorhynchus nerka*. Transactions of the American Fisheries Society 97:360–373.

Brown, J. H., and A. Kodric-Brown. 1977. Turnover rates in insular biogeography: effect of immigration on extinction. Ecology 58:445–449.

Cole, L. C. 1954. The population consequences of life history phenomena. Quarterly Reviews in Biology 29:103–137.

Craig, J. A., and R. L. Hacker. 1940. The history and development of the fisheries of the Columbia River. U.S. Bureau of Fisheries Bulletin 49:133–216.

Cuenco, M. L., T. W. H. Backman, and P. R. Mundy. 1993. The use of supplementation to aid in natural stock restoration. Pages 269–293 *in* J. G. Cloud and G. H. Thorgaard, editors. Genetic conservation of salmonid fishes. Plenum Press, New York.

Cushing, D. H. 1983. Key papers on fish populations. IRL Press, Washington, DC.

den Boer, P. J. 1968. Spreading of risk and stabilization of animal numbers. Acta Biotheoretica 18:165–194.

Dobzhansky, T. 1970. Genetics of the evolutionary process. Columbia University Press, New York.

Emlen, J. M. 1984. Population biology. The coevolution of population dynamics and behavior. Macmillan, New York.

Frissell, C. A., W. J. Liss, C. E. Warren, and M. D. Hurley. 1986. A hierarchical framework for stream habitat classification: viewing streams in a watershed context. Environmental Management 10:199–214.

Gilpin, M. E., and M. E. Soulé. 1986. Minimum viable populations: processes of species extinction. Pages 7–16 *in* M. E. Soulé, editor. Conservation biology: the science of scarcity and diversity. Sinauer, Sunderland, Massachusetts.

Hanski, I., and M. E. Gilpin. 1991. Metapopulation dynamics: brief history and conceptual domain. Pages 3–16 *in* M. E. Gilpin and I. Hanski, editors. Metapopulation dynamics. Academic Press, New York.

Lande, R., and G. F. Barrowclough. 1987. Effective population size, genetic variation, and their use in population management. Pages 87–124 *in* M. E. Soulé, editor. Viable populations for conservation. Cambridge University Press, New York.

Leigh, E. G., Jr. 1981. The average lifetime of a population in a varying environment. Journal of Theoretical Biology 90:213–239.

MacArthur, R. H., and E. O. Wilson. 1967. The theory of island biogeography. Princeton University Press, Princeton, New Jersey.

May, R. M. 1973. Stability and complexity in model ecosystems. Princeton University Press, Princeton, New Jersey.

Matthews, G. M., and R. S. Waples. 1991. Status review for Snake River spring and summer chinook salmon. NOAA (National Oceanic and Atmospheric Administration) Technical Memorandum NMFS (National Marine Fisheries Service) F/NWC-200, Northwest Fisheries Science Center, Seattle.

Mundy, P. R., and J. K. Fryer. 1995. Abundance based criteria for recognition of damaged salmon populations. Pages 202–212 *in* Salmon management in the 21st century: recovering stocks in decline. Proceedings, 1992 northwest Pacific chinook and coho salmon workshop. American Fisheries Society, Idaho Chapter, Boise, Idaho.

Norman, G., and S. King. 1992. Status report, Columbia River fish runs and fisheries, 1938–91. Washington Department of Fisheries and the Oregon Department of Fish and Wildlife, Portland.

NPPC (Northwest Power Planning Council). 1986. Appendix D of the 1987 Columbia River basin fish and wildlife program. Compilation of information on salmon and steelhead losses in the Columbia River basin, appendices March 1986. Northwest Power Planning Council, Portland, Oregon.

Pascual, M. A., and T. P. Quinn. 1995. Geographical patterns of straying of fall chinook salmon, *Oncorhynchus tshawytscha* (Walbaum), from Columbia River (USA) hatcheries. Aquaculture and Fisheries Management 25(Supplement 2):17–30.

Quinn, T. P. 1984. Homing and straying in Pacific salmon. Pages 357–362 *in* J. D. McCleave, editor. Mechanisms of migration in fishes. Plenum Press, New York.

Quinn, T. P. 1985. Homing and the evolution of sockeye salmon (*Oncorhynchus nerka*). Pages 353–366 *in* M. A. R. Rankin, editor. Migration: mechanisms and adaptive significance. Contributions to Marine Science 27.

RASP (Regional Assessment of Supplementation). 1992. Supplementation in the Columbia basin summary report series. Final report, project no. 85-62, contract number DE-AC06-75RL01830. Bonneville Power Administration, Portland, Oregon.

Richter-Dyn, N., and N. S. Goel. 1972. On the extinction of a colonizing species. Theoretical Population Biology 3:406–433.

Rich, W. H. 1940. The present state of the Columbia River salmon resources. Page 429 *in* Proceedings of the sixth Pacific science congress. University of California Press, Berkeley.

Ricker, W. E. 1954. Stock and recruitment. Journal of the Fisheries Research Board of Canada 11:559–623.

Ricker, W. E. 1972. Heredity and environmental factors affecting certain salmonid populations. Pages 19–60 *in* R.C. Simon and P. Larkin, editors. The stock concept in Pacific salmon. H. R. MacMillan Lectures in Fisheries, University of British Columbia, Vancouver.

Riddell, B. E., and W. C. Leggett. 1981. Evidence of an adaptive basis for geographic variation in body morphology and time of downstream migration of juvenile Atlantic salmon (*Salmo salar*). Canadian Journal of Fisheries and Aquatic Sciences 38:308–320.

Rohlf, D. J. 1994. There's something fishy going on here: a critique of the National Marine Fisheries Service's definition of species under the Endangered Species Act. Environmental Law 24:617–671.

Schaffer, W. M. 1974. Selection for optimal life histories: the effects of age structure. Ecology 55:291–303.

Stanford, J. A., and J. V. Ward. 1992. Management of aquatic resources in large catchments: recognizing interactions between ecosystem connectivity and environmental disturbance. Pages 91–124 in R. J. Naiman, editor. Watershed management. Springer-Verlag, New York.

Thomas, C. D. 1990. What do real population dynamics tell us about minimum viable population size? Conservation Biology 4:144–156.

Thompson, W. F. 1951. An outline for salmon research in Alaska. University of Washington, College of Fisheries, Fisheries Research Institute Circular 18, Seattle.

Thorpe, J. E. 1986. Age at first maturity in Atlantic salmon (Salmo salar): freshwater period influences and conflicts with smolting. Canadian Special Publication of Fisheries and Aquatic Sciences 89:39–52.

Thorpe, J. E. 1989. Developmental variation in salmonid populations. Journal of Fish Biology 35(Supplement A):295–303.

Vance, R. R. 1980. The effect of dispersal on population size in a temporally varying environment. Theoretical Population Biology 18:343–362.

Waples, R. S. 1991. Pacific salmon, Oncorhynchus spp., and the definition of "species" under the Endangered Species Act. U.S. National Marine Fisheries Service Marine Fisheries Review 53(3):11–22.

Waples, R. S., O. W. Johnson, and R. P. Jones, Jr. 1991. Status review for Snake River sockeye salmon. NOAA (National Oceanic and Atmospheric Administration) Technical Memorandum NMFS (National Marine Fisheries Service), F/NWC-195, Northwest Fisheries Science Center, Seattle.

PART TWO

MORPHOLOGY AND SYSTEMATICS

American Fisheries Society Symposium 17:41–43, 1995

SESSION OVERVIEW
Morphology and Systematics

ROBERT J. BEHNKE

Department of Fishery and Wildlife Biology
Colorado State University, Fort Collins, Colorado 80523, USA

I propose, only half in jest, that a license to practice conservation biology be required of all agency personnel involved with decision making in regards to biodiversity and endangered species issues. To be granted this license, some level of understanding would be required about Darwinian evolution by natural selection and all it implies regarding coevolution, coadaptation, and adaptive strategies involved with niche filling. I would also require an understanding of distinctions between evolutionary genetics (the result of natural selection) and controlled laboratory population genetics (the result of artificial selection) and the caution necessary to make extrapolations from the second to the first (i.e., and understanding of the dangers of inductive reasoning). Finally, I would also like to see an understanding of the pros, cons, and limitations of any method or technique in order to provide the most accurate answers to specific questions for defining evolutionary diversity and its significance—the need to use an integrated, eclectic approach. It is better to take the time to contemplate probable answers to questions of uncertainty than to seek precise answers to irrelevant questions.

The lure of deterministic methods or models offering instant, simplistic answers to complex questions can be overwhelming to agency people involved in the decision-making process. The danger is that data, rules, or quantitative indices are substituted for thinking and critical judgement.

My hypothetical licensed practitioner should realize that defining the evolutionarily significant unit (ESU) is not a simple matter. One size will not fit all. This truism was brought out during the conference when someone suggested that the ESU could be better called the ecologically significant unit. Although this matter was not pursued further at the conference, I will give my interpretation, based on evolutionary time scale, concerning the distinctions and implications between evolutionary and ecological in regards to defining the significance of biodiversity.

The U.S. Endangered Species Act (ESA; 16 U.S.C. §§ 1531 to 1544) defines a species (of vertebrate animals) to include intraspecific units—subspecies and "population segments which interbreed when mature." Thus, a local population or deme is the smallest population segment qualifying as a vertebrate species for protection under the ESA. The intraspecific diversity contained in a widely distributed anadromous species such as chinook salmon *Oncorhynchus tshawytscha* consists of numerous life history forms distinguished by different times of spawning runs from the ocean, different distances of spawning migration and times of spawning, different juvenile life histories, and different patterns of ocean migration. The evolutionary relationships of these intraspecific life history and ecological forms are "within basin." That is, all races, populations, and demes of chinook salmon within a large river basin are more closely related to each other than they are to analogous forms in other river basins (Behnke 1992). Life history adaptations in Pacific salmon and steelhead *O. mykiss* have independently arisen many times during the past 10,000 years (they are polyphyletic rather than shared synapomorphies).

A conservation strategy should aim to preserve the range of adaptiveness in a species in order to maintain its evolutionary options, and this is the basic issue for better definition of the ESU. The range of intraspecific, ecological life history adaptive capabilities has been evolutionarily programmed into the genome (the regulatory genome) by natural selection, but because of the relatively short evolutionary time span involved, and probable limited gene flow among populations, we should not expect that these adaptive properties can be detected or understood from molecular genetic data (of the structural genome). These adaptive properties, however, are the most important attributes for defining the evolutionarily (or ecologically) significant unit if our goal is to preserve the range of adaptiveness within a species.

A more common perception of evolutionary significance that the ESA is designed to protect I would equate with phylogenetic or taxonomic significance. A taxonomic hierarchy inclusively groups assemblages into higher categories: species, genera, families, orders, etc. The coelacanth *Latimeria cha-*

lumnae and the bowfin *Amia calva* are the sole existing species of ancient lineages and have been reproductively isolated from all other phylogenetic lineages for about 350 million and 150 million years, respectively. Their evolutionary significance is essentially self-defined by their phylogenetic history.

The methods, evidence, concepts, and pertinent questions used for defining phylogenetic or taxonomic evolutionary significance can be quite different from those necessary for defining the ecological and adaptive significance of intraspecific evolution. It's a matter of evolutionary time scale. The extinction of *Latimeria chalumnae* would irreversibly complete the extinction of crossopterygian fishes. The extinction of the winter run of chinook salmon of the Sacramento River or the fall run of the Snake River would reduce the range of adaptiveness in the species, but this could be considered as potentially reversible. That is, if historical environmental conditions and selective pressures were restored, other races of the species could give rise to forms duplicating the life histories of the extinct forms, but it would likely take hundreds of generations. Thus, with the evolutionarily significant ecological-adaptive unit, we might have as a goal the preservation of the range of intraspecific adaptiveness for the next 100 to 1,000 years, in hope that conditions for survival will improve.

Certainly, molecular genetics can play an important role in better defining intraspecific diversity in relation to ascertaining the significance of evolutionary units in certain situations. For example, how is a species structured? Bernatchez and Dodson (1991; personal communication) used mitochondrial DNA analysis to resolve the phylogeny of diversity in North American lake whitefish *Coregonus clupeaformis* and rainbow smelt *Osmerus mordax* after all other methods had failed to resolve the question of sympatric, reproductively isolated populations. Do these sympatric pairs represent ancient monophyletic lineages (as with the lake whitefish) or more recent (late Pleistocene), independent, polyphyletic origins (as with rainbow smelt)? Understanding the phylogenetic structure of intraspecific diversity is important for planning strategies for the conservation of diversity in relation to the irreversibility or irreplacibility of extinctions.

A question concerning the evolutionary significance of the southernmost populations of steelhead (do they represent the native genotype or have they been thoroughly homogenized by stocking of nonnative hatchery steelhead?) could only be answered with confidence by modern molecular techniques. Nielsen et al. (1994) found unique DNA sequences

in steelhead south of San Francisco Bay. These declining populations of southern steelhead do represent native populations and are worthy of protection and restoration.

My point is that phrasing the right questions in need of answers is of critical importance and should come before the methods of analysis are chosen. Also, we should promote the advantages of an eclectic, integrated approach to develop various lines of evidence. We should avoid bickering over what methods, concepts, and personal agendas are superior for defining the ESU and seek a common ground for furthering the cause of the preservation of biodiversity.

In relation to the definition of species in the ESA and its possible changes during reauthorization of the ESA, Stelle (1994) wrote, "I personally hope this question is not answered by the legislature. One of my worst nightmares envisions a congressional floor debate regarding the definition of 'subspecies' or 'distinct population.' This is an inherently scientific issue with no real place in the legislative process, and it should be resolved by scientists." If scientists can't agree on what is an ESU, it will likely be defined by the legislature.

This brings up my final, but most important, point. We should attempt to communicate the knowledge underlying our conservation ethic beyond our own peer group (Behnke 1994). Brussard (1994) raised the question, "Why do we want to conserve biodiversity anyway?" He pointed out that we haven't been highly successful in communications and influence at various levels of society. Our failure to communicate effectively the positive aspects of biodiversity preservation and the need for an ESA is illustrated by an article, "Better Red than Dead" in *Newsweek* 12 December 1994. The article tells about the "endangered salmon bake" held in Stanley, Idaho (headwaters of the Salmon River drainage, which has two endangered races of chinook salmon and the endangered population of sockeye salmon *O. nerka* of Redfish Lake). Helen Chenoweth, newly elected congresswoman from Idaho, spoke at the event. Congresswoman Chenoweth's environmental platform for the election was essentially that of the Wise Use Movement. Among her remarks to the audience was, "How can I take the salmon's endangered status seriously when you can buy a can at Albertson's?"

Evidently, the outrageously fallacious notions on evolution, extinctions, and values of biodiversity propagandized by groups such as the Wise Use Movement were more effective in forming the opinions of Congresswoman Chenoweth and most of the voters in Idaho who elected her than were any of the pro-environmental positions attempting to explain why we want to preserve biodiversity. Can we do a better job of communications by the next election? If not, the ESU may become extinct.

References

Behnke, R. J. 1992. Native trout of western America. American Fisheries Society Monograph 6.

Behnke, R. J. 1994. Writing for lay people. Fisheries 19(9):30.

Bernatchez, L., and J. J. Dodson. 1991. Phylogeographic structure in mitochondrial DNA of the lake whitefish (*Coregonus clupeaformis*) in North America and its relationship to Pleistocene glaciations. Evolution 45:1016–1035.

Brussard, P. F. 1994. Why do we want to conserve biodiversity anyway? Society for Conservation Biology Newsletter 1(4):1, Reno, Nevada.

Nielsen, J. L., C. A. Gan, J. M. Wright, D. B. Morris, and W. K. Thomas. 1994. Biogeographic distributions of mitochondrial and nuclear markers for southern steelhead. Molecular Marine Biology and Biotechnology 3:281–293.

Stelle, W. W. 1994. Major issues in reauthorization of the Endangered Species Act. Environmental Law 24:321–328.

American Fisheries Society Symposium 17:44–57, 1995

Processes of Origin and Criteria for Preservation of Fish Species

GERALD SMITH, JONATHAN ROSENFIELD, AND JEAN PORTERFIELD

Museum of Zoology, University of Michigan
Ann Arbor, Michigan 48109, USA

Abstract.—The origin of fish species is largely controlled by time, geography, and ecology. In this context, species are lineages diagnosed by unique characters, including ecological traits. Diagnoses of species lineages and evolutionarily significant units may require diverse evidence because they are not uniform, reducible, taxonomic units—they are historical entities that have arisen through diverse processes. Ecological and sexual selection, especially, promote different kinds of species attributes. The results reflect the relative timing of acquisition of three kinds of changes associated with the process of lineage divergence: reproductive, morphological, and ecological differentiation. In fishes, these three changes are not necessarily correlated with each other or with genetic indices of divergence because each arises as an interaction of genetics and environment, and each may be reversed. Species are theoretically delimited when reproductive isolation reaches irreversibility; however, many potentially introgressible lineages possess sufficient differentiation to be irreplaceable in nature. Fish species are defined here as reproductively independent lineages, diagnosed by different genetically based character states and ecological roles. The distinction between independence and isolation depends on the number and fitness of introgressed progeny. We recommend that individuality of lineages be assessed by broad comparative study of morphology, behavior, ecology, heritable continuity, karyology, proteins, and DNA.

What species really are, biologically, depends on what causes their integration, their continuity, and their distinction from each other. Probably this is not always the same in different cases, at least for the predominant cause. These three attributes themselves are represented to different degrees in different species. There are just the same sorts of problems with taxa at other levels. We should therefore refrain from definitions that define away what are empirically real possibilities in the diversity of life. (From Van Valen 1988:61.)

In this paper we contrast examples of morphological, ecological, and reproductive lineages of salmon *Oncorhynchus* spp., whitefish (Coregoninae), suckers (Catostomidae), and darters (Percidae) that illustrate different kinds of species and other evolutionarily significant units (ESUs) for purposes of conservation. Different definitions lead to different estimations of significant biodiversity (Cracraft 1987) and different justification for action (Rojas 1992). These problems lead to the present focus: How can we identify and prioritize the significant units in fish biology and conservation? We will suggest that monothetic (single criterion) concepts concerning the classification levels to be designated species or ESUs are not useful (Sokal and Crovello 1970; Van Valen 1988; deQueiroz and Donoghue 1988) because species lineages are not homogeneous units in a reductionist system. We recommend that individual lineages be considered for species recognition and ESU action and that these lineages be evaluated in terms of criteria derived from four species concepts that have particu-

lar relevance to the irreplaceability and ecology of threatened populations. These concepts are (1) genetic isolation (Mayr 1942, 1963), (2) evolutionary individuality (Simpson 1951; Ghiselin 1974; Wiley 1981), (3) mate-recognition systems (Paterson 1978, 1993), and (4) ecological interactions (Van Valen 1976, 1988).

Species Concepts and Definitions

Mayr (1942, 1963, 1970, 1988) thoroughly explicated the concept and application of the biological species concept. Mayr (1970) wrote that "species are groups of interbreeding natural populations that are reproductively isolated from other such groups." Mayr's familiar definition is widely accepted, but it has been criticized because it defines species in terms of a questionable speciation process—evolution of mechanisms for avoidance of outbreeding—and by relation to entities other than the species itself (Paterson 1985). Ghiselin (1974) avoided these problems by identifying species as historical lineages, not classes, and by defining the species category as "the most extensive [unit] in the natural economy such that reproductive competition occurs among [its] parts." The recognition of lineages, not classes, is crucial to irreplaceability because lineages (and species) can go extinct, arbitrary classes do not; individual lineages are real in nature, classes are defined. Paterson (1985, 1993: 147) echoed Ghiselin's definition when he defined (bisexual) species as "that most inclusive population of individual biparental organisms which share a common fertilization system." These definitions place emphasis on processes by which members of a lineage of sexual organisms interact with each other in their ecological context to achieve fertilization. Because of the nature of fish reproduction, and its environmental dependence, this emphasis has special power for fish taxonomy and conservation. Ghiselin and Paterson also criticized the isolation definition because it defines species by a relational property—with whom it does not breed. But in practice, relational diagnoses of individual lineages is necessitated by our dependence on the comparative method.

Paterson's concept implies that a new species arises by a selected shift of adaptations for mating in new habitats. This idea complements Van Valen's ecological species concept.

> A species is a lineage (or a closely related set of lineages) which occupies an adaptive zone minimally different from that of any other lineage in its range and which evolves separately from all lineages outside its range. (From Van Valen 1976:233.)

Van Valen thus defined ecological species and offered the crucial insight that "species are maintained for the most part ecologically, not reproductively." If so, reproductive isolation is not central to the practical biology of species. The environmental context of a species—its adaptive zone—is fundamental (Van Valen 1976). A fragment of a population that responds to its environment in ways that are heritably different from related lineages is a new species. Ghiselin (1987) disagreed with this emphasis on ecology as confusion of individuals with their context (see also Eldredge 1985), but we are persuaded by Van Valen (1976, 1988) that species origins and distinctive histories are usually integrated with their ecology. Therefore, ecological traits have considerable potential to contribute to species descriptions and diagnoses. It is especially apparent that goals of conservation biology—preservation of ecosystem processes and species (U.S. Endangered Species Act [ESA] 16 U.S.C. §§ 1531 to 1544; Ehrlich and Ehrlich 1981; Soulé 1985; Wilson 1992; Society for Conservation Biology 1987)—would be enhanced by the ecological emphasis in Van Valen's concept of species.

Together, the above concepts direct our search to practical evidence for evolutionary individuality based on ecological and reproductive uniqueness. Such evidence is necessary for definition of significant evolutionary units.

Integration, Continuity, and Distinction

Integration

Van Valen (1988) identified the need to understand the causes of integration, continuity, and distinction of species. Members of an evolutionary lineage share systems for integrating their developmental processes and their ecological physiology (homeostasis). Adaptive integration involves three parts—information gathering, response, and integration of responses (Levins and Lewontin 1985). "What is different even between similar species is what constitutes stress, that is, how signals are interpreted" (Levins and Lewontin 1985). Because of descent with modification, we expect some fraction of a species' developmental and physiological integration to be unique, depending on the ecology and age of the species. Ontogenetic evidence for different integrative systems could have special value in diagnosing lineages (Zelditch and Fink 1995).

Continuity

Mechanisms for getting mates and fertilizing gametes (e.g., Paterson 1985, 1993) are foremost among causes of genetic continuity. Mating systems in freshwater fishes involve habitat, mate choice, migrations, site specificity, seasonality, and integrated behaviors. Species mate-recognition systems are dependent upon hormonal and behavioral responses to seasonality, spatial and temporal distribution of resources and predators, and demographics (Mann et al. 1984). Spawning time and place are naturally selected to place eggs where newly hatched larvae will find or be carried to abundant resources at the optimal time. These factors indicate the context dependence of continuity. Disruption of continuity occurs in two phenomena at opposite extremes in the species question: speciation and introgression. Traditional tools for evaluating continuity versus discontinuity of diverging populations (i.e., morphological, behavioral, biochemical, and molecular nonoverlap) also estimate reproductive independence. Primary and secondary introgression are identified by comparing a combination of genetically informative data sources (e.g., Dowling et al. 1989) in a cladistic context: morphological, protein, and mitochondrial DNA (mtDNA) evidence can assess the amount of introgression into a lineage. But presence of introgression does not preclude lineage individuality. Less-fit backcross progeny (and sterile hybrids) do not constitute a threat to the system of genetic coherence. In fish, the ease of age determination allows the fitness of backcrosses to be estimated by their age-specific survivorship (Dowling and Moore 1985). Evolutionarily significant units can be prioritized by evaluating these measures of distinctness in the framework of an ESU's system of continuity.

Continuity is also a geography-dependent effect of demography, the breeding system, and past gene flow (Wright 1946). Because fish breeding systems depend on time and place, seasonal light and temperature changes cue internal responses through hormonal cycles related to gonad maturity (Bye 1984; Stacey 1984). Thus, a change in this context can reorganize the systems of integration and continuity into different pathways (Paterson 1982, 1993: 86). Subtle changes in the interaction of these factors with geography (Mann et al. 1984) and mate choice (Houde and Endler 1990) can promote population differentiation or its reverse, fusion.

Distinction

Van Valen's third factor embodies morphological, ecological, and genetic aspects of evolution. It depends on the duration and degree of reproductive separation (Dobzhansky 1937; Mayr 1963). But distinction may evolve without total isolation, depending on the strength of selection on life history and mate recognition. Evolution of characters signifying distinction is correlated with, but not locked into, a progression by which context-dependent reproductive isolation becomes intrinsic and irreversible (Mayr 1963). Evolution of distinction requires four steps: (1) partial geographic, ecological, or temporal separation of the breeding members of subpopulations, (2) reduced gene exchange between subpopulations, (3) reduced fitness of intermediates, and (4) irreversible, intrinsic genetic isolation. At step 4 the lineage is unambiguously a species (e.g., Mayr 1982; Kluge 1990; Eldredge 1993). However, morphological, genetic, and molecular divergence may not be highly correlated with each other or with this sequence: different ecological contexts and breeding systems can create distinct lineages before gene flow stops. Introgression, being under ecological as well as genetic control, is not necessarily eliminated as other steps in divergence occur. Irreversible genetic isolation is not requisite to the production of ecologically significant units in freshwater fishes.

Distinctness in mtDNA is not a reliable criterion for reproductive independence, by itself, contrary to current assumptions: evolution of profound differences in mtDNA can occur within lineages (Gach 1993), and mtDNA variability may be completely shared between lineages (Burnham 1993). Mitochondrial DNA differences may be unmeasurable among some lineages in species flocks (Meyer et al. 1990; Moran et al. 1994). In addition, mtDNA data may be subject to introgressive loss of historical evidence (Smith 1992). Though "genetic," DNA data cannot substitute for inheritance experiments to demonstrate either hybrid inviability or a genetic basis of morphological, behavioral, and ecological characters. In summary, no criterion except hybrid inviability or strong behavioral isolation is necessary or sufficient to demonstrate reproductive distinctness by itself. Even propensity to breed or not breed with a certain phenotype can be context dependent in fish (Breder and Rosen 1966).

Quantifying distinctness is satisfactorily accomplished by some combination among three kinds of techniques: metric (Bookstein et al. 1985; Rohlf and Bookstein 1990), character cladistics (Cracraft 1983, 1987; Nixon and Wheeler 1990), and DNA sequence cladistics (Moritz et al. 1995, this volume). Nonoverlapping or unique apomorphies or reciprocal monophyly indicate lineage individuality if the characters are genetically determined, genetically independent, and geographically congruent.

Our goal is to define species and evolutionarily and ecologically significant units to be concordant with important levels of biological diversity, recognizing lineages generally identified as significant by fish biologists (e.g., Behnke 1970, 1993; Rosen 1978). We suggest that fish species are reproductively independent lineages which can be diagnosed by different genetically based character states and ecological roles. Reproductive independence, or individualization, is demonstrated by evidence of selection against hybrids and backcross progeny in nature (Dowling and Moore 1985; Dowling et al. 1989). Reproductive independence is also evaluated by genetic differences, phylogenetic character comparisons, and behavioral experiments with the species' closest sister group and with its closest sympatric relative. Character differences may occur in morphology, behavior, physiology, phenology, reproductive attributes, life history, ecology, proteins, karyotypes, or DNA, but a character difference in only one of these systems would not satisfactorily diagnose a species because it could be interpretable as polymorphism or plasticity. Congruence of independent evidence for parent–offspring continuity determines the kind of genetic individuality diagnosable as a species. Characters of species are least ambiguous when they are diagnostic apomorphies, but other differences are valid evidence because many species are, or will become, paraphyletic, as small peripheral populations differentiate from larger, polytypic populations (Mishler and Dono-ghue 1982; Kluge 1990).

Evolutionarily and ecologically significant units are most convincing when they meet the criteria given above for species. Our point is that a diversity of evidence is crucial for diagnoses of ESUs because ecological and social decisions are urgent.

The criterion of reciprocal nonoverlap in apo-morphic characters is powerful evidence for unique-ness (Moritz et al. 1995) of species as well as ESUs. But character overlap, in the form of rare individ-uals (up to about 5%, for example) of mixed phe-notype, does not refute independence of species or other significant lineages because separate lineages will often tolerate (or benefit from) some genetic introgression without such introgression leading to loss of individuality or to significant breakdown of mate-recognition systems (discussed below). A lin-eage's distinct ecological role is evaluated by com-paring life history and other ecological evidence to its closest sister lineage and its closest sympatric relative (Van Valen 1976). Ecological data (like other data) must be evaluated by independent evi-dence for geographic and phylogenetic congruence as well as genetic, rather than ecophenotypic, cau-sation. In the examples below, we will explore some relationships between ecological and evolutionary significance and reproductive independence.

Case Histories

The following four examples illustrate different genetic, morphological, behavioral, and ecological diversification in fishes. We indicate ways in which ecological selection and sexual selection might dif-ferentially influence fish divergence and lead to different justification and methods for conservation of threatened populations.

Pacific Salmon

Five forms of Pacific salmon *Oncorhynchus* spp. inhabit North America. In the southern part of their range, most populations of Pacific salmon are threatened (Nehlsen et al. 1991; Frissel 1993; P. B. Moyle and R. M. Yoshiyama, unpublished). Since 1991, several populations of Pacific salmon (e.g., Snake River chinook salmon *O. tshawytscha* and Snake River sockeye salmon *O. nerka*) have been added to the list of threatened and endangered species, but at least two listing petitions (Illinois River, Oregon, steelhead *O. mykiss* and lower Co-lumbia River coho salmon *O. kisutch*) have been denied. Three current petitions to list populations of coho salmon as threatened are under federal

review.[1] The diverse coho salmon petitions reflect petitioners' different opinions about ESU policy and what constitutes a species under the ESA. These differences exemplify the effect of the "spe-cies problem" in thwarting conservation efforts.

Four seasonal runs of chinook salmon of the Sacramento River typify many biological and man-agement aspects of Pacific salmon. Chinook salmon have a 2–5 year life cycle, are semelparous and anadromous, and return with high fidelity to their natal streams (Quinn and Fresh 1984; McIsaac and Quinn 1988; Healy 1991). As with Pacific salmon in many rivers, distinct groups of Sacramento chinook salmon divide their fluvial habitat spatially and tem-porally into distinct adaptive zones. Four Sacra-mento chinook runs, winter, spring, fall, and late fall, are named for the timing of their migrations into freshwater. Every life history stage—egg, lar-vae, juvenile, and spawning adult—is present year-round in the Sacramento River as a result of this temporal partitioning.

Most authors recognize the distinctiveness of chi-nook salmon populations in different drainages and among different runs in the same drainage. Regard-ing Sacramento chinook salmon, P. B. Moyle and R. M. Yoshiyama (unpublished) conclude

> The runs . . . are differentiated by the maturity of fish entering freshwater, time of spawning migrations, spawning areas, incubation times, incubation tempera-ture requirements, and migration of juveniles. Differ-ences in the life histories effectively isolate spring chi-nook salmon from other runs . . . the traits are undoubtedly inherited. Allozymic differences between inland populations of California chinook salmon have also been observed. . . . Therefore, each run . . . must be considered to be genetically distinct, in some cases from other runs in the same stream.

---

[1]Hope, D. 1993. Petition to the State of California Fish and Game Commission to list coho salmon *Oncorhynchus kisutch*, of Scott and Waddel creeks, California. Submitted for review to the National Marine Fisheries Service, Spring 1993.

Oregon Trout, Inc., et al. 1993. Consolidated petitions for listing sub-species of Pacific Coast *Oncorhynchus kisutch* pursuant to the Endangered Species Act of 1973 as amended. Submitted to the National Marine Fisheries Service, Summer 1993.

Pacific Rivers Council, et al. 1993. Petition for a rule to list, for designation of critical habitat and for a status review of coho salmon throughout its range under the Endangered Species Act. Submitted to the National Ma-rine Fisheries Service, 7 October 1993.

In July 1995, NMFS proposed a rule (for public com-ment) to list three coho salmon ESUs in Oregon and California as threatened (*Federal Register* 60 [25 July 1995]:38011–38030).

Hybridization is presumed to occur between some runs of Sacramento chinook salmon and between Sacramento chinook salmon and those from other rivers, but evidence compiled by Moyle and Yoshiyama (unpublished) makes it clear that natural gene flow, through spatial and temporal strays, is not breaking down distinctions among populations. Morphological, molecular, or protein evidence for the amount of differentiation is not available, but it should not be assumed that lineage individuality is not present. At face value, the known ways in which the four breeding populations differ indicate that they are well into an early stage of ecological speciation. Nielsen (1994) has identified one unique mtDNA marker in the winter run. Each run occupies a separate adaptive zone, and each constitutes a separate, temporally based fertilization system. Does it matter whether there is gene flow through strays if each run is maintaining its historical individuality? Does this amount of gene flow justify destruction of these reproductively independent and irreplaceable lineages? We contend that protection of these populations is justified because they play a long-term, irreplaceable role in the ecology of the Sacramento River.

Like chinook salmon in many other rivers, Sacramento River forms have suffered serious population declines over the past half century. Pre-dam population estimates (at the turn of the century) suggest that the four spawning runs cumulatively included millions of individuals (Clark 1929; Reynolds et al. 1990). Threats to the existence of Sacramento chinook salmon stocks are representative of threats to all Pacific salmon (for a review, see Fraidenburg and Lincoln 1985; Nehlsen et al. 1991; Frissel 1993). They are endangered by several impoundments; for example, Shasta Dam blocks access to the winter run's historic spawning grounds on the Pit and McCloud rivers and periodically releases lethally warm water onto spawning sites (Hallock and Rectenwald 1990)[2]. Thousands of migrating juvenile salmon are disoriented and killed by pumps that divert freshwater from the Sacramento–San Joaquin rivers and San Francisco Bay delta[3]. Logging, grazing, and small water diversions threaten to destroy spawning habitat of the only remaining pure populations of the spring run in tributaries of the Sacramento (P. B. Moyle and R. M. Yoshiyama, unpublished). Commercial fishing also contributes to population declines (Fraidenburg and Lincoln 1985). Finally, Feather River Hatchery contributed to destruction of different runs by creating hybrids (P. B. Moyle and R. M. Yoshiyama, unpublished) and by reducing effective population size (Bartley et al. 1992).

In 1990 the winter run was listed as a threatened species under the ESA, and most intense conservation efforts are directed toward this form (J. Smith, U.S. Fish and Wildlife Service, personal communication); however, the spring run, with estimated spawning populations below 500 genetically pure individuals in recent years (P. B. Moyle and R. M. Yoshiyama, unpublished), is at least as threatened as the winter run.

Sacramento River chinook salmon populations illustrate the effect of species definitions on conservation decisions. If the goal were to preserve some remnant of each species that is irreversibly isolated, then any subunit of chinook salmon of the North Pacific rim might serve to be perpetuated, ignoring subunits suspected to be incompletely isolated. Although the winter run of Sacramento chinook salmon is being protected under the ESA, other runs may garner less protection if they fail to meet strict interpretation of the reproductive isolation criterion. The ESA definition may not extend protection to populations including hybrids (O'Brien and Mayr 1991), and some spring-run chinook salmon in the Sacramento River are known to hybridize with fall-run chinook salmon due to hatchery practices and habitat restriction (Campbell and Moyle 1990; P. B. Moyle and R. M. Yoshiyama, unpublished). But introgression and its effects need to be evaluated, not assumed without evidence. Small, spring-run populations are suspected to exist in at least two tributaries to the Sacramento River (Campbell and Moyle 1990; P. B. Moyle and R. M. Yoshiyama, unpublished), but they may not meet ESU definition of separate units if populations have hybridized. (Sacramento chinook salmon are repre-

---

[2]Rosenfield, J. A. 1992. Summer 1992 water temperatures in the Sacramento River, California, during the spawning and incubation period of winter-run chinook salmon. Exhibit #11 *in* Pacific Coast Federation of Fishermen's Associations et al. *v.* Lujan et al., Civil Number s-92 1492 LKK JFM. Sacramento Federal Circuit Court, Judge Lawrence K. Carlton. Sierra Club Legal Defense Fund. San Francisco, California.

---

[3]Brown, R. L. 1992. Winter-run chinook salmon losses at CVP and SWP Delta intake during the period 16 January through 30 April 1992. Exhibit WRINT DWR-31. California Department of Water Resources Testimony in U.S. Environmental Protection Agency San Francisco Bay-Delta Hearings, Sacramento, California.

sentative of a general problem [Allendorf and Leary 1988; Bartley et al. 1990; Goodman 1990; Bartley et al. 1992; Meffe 1992] in that efforts to preserve them through hatchery propagation threatens their genetic existence because different stocks are mixed.)

The National Marine Fisheries Service's (NMFS) definition of the ESU would appear to be more restrictive than the species definition of the ESA because it requires both reproductive isolation and evolutionary significance (Waples 1991). Sacramento chinook salmon runs might not qualify for protection under the ESU policy of the NMFS if evidence for hybrids disqualifies them as (1) "substantially reproductively isolated" and (2) a "significant" portion of the ecological and genetic diversity of the species.

An ecological concept supports definition of each Sacramento chinook salmon run as a significant ecological and evolutionary unit. Each run occupies an adaptive zone minimally different from that of any other lineage in its range, and each is evolving separately from all lineages outside its range (Van Valen 1976).

Whitefishes and Ciscoes

The whitefishes and ciscoes of the subfamily Coregoninae (Figure 1) are a notorious species problem in ichthyology and a devastated group of North American fishes. Their relationships and classification have been confused by differentiation of separate, ecologically unique, sympatric feeding phenotypes in the absence of irreversible reproductive isolation (Smith and Todd 1984, 1992; Todd and Smith 1992).

The local units at one scale, for example, the species of ciscoes in the Laurentian Great Lakes (Koelz 1927) or the races in European lakes (Svärdson 1952, 1965, 1979), were morphologically and ecologically distinct. In the Great Lakes, seven sympatric species of *Coregonus* were reproductively isolated from each other by different times and places of spawning (Koelz 1927; Smith and Todd 1984). The widespread cisco, or lake herring, is subdivided outside the Great Lakes into dozens of local races displaying so much diversity (Koelz 1931) that the cisco is marginally undiagnosable relative to certain other species across the Holarctic region (McPhail and Lindsey 1970). Systematists have been divided about the level of species recognition in this system. Early workers like Koelz (1927) described many more species and subspecies than are recognized today (Robins et al. 1991); other workers believed that only one variable species of cisco existed in the

Great Lakes (R. M. Bailey, University of Michigan, personal communication).

The controversy is more fundamental than lumping versus splitting; it exists because morphological differentiation and reproductive isolation are decoupled and in conflict with traditional species criteria. The hierarchy in Figure 1 shows that branch points more ancient than recognized species have left several examples of character overlap. Cases of character overlap suggest past gene flow and compromise diagnoses of lineages. However, many subunits within species are sympatric, ecologically unique, morphologically differentiated, and evolutionarily independent (Smith and Todd 1984). The search for molecular and protein characters to clarify species has not found abundant corroborating evidence for genetic differentiation (Todd et al. 1981; Bernatchez and Dodson 1990; Shields et al. 1990), suggesting that these lineages are recent or introgressed. There is some evidence that introgression was caused when populations were fished to low abundance and stray individuals joined larger schools (Hubbs 1955; Smith 1972; Todd and Steadman 1988).

Twenty-five original populations of seven cisco species in the five principal Great Lakes are now reduced to 11, of which 6 are threatened or rare (Todd and Smith 1992). The 14 extinctions were caused by overfishing, habitat degradation, and competition with exotic species (Fleischer 1992). Spatially and temporally concentrated spawning aggregations brought these species into extreme jeopardy. State and federal agencies delayed their responses to decline of these species, partly because species differences were known only to specialists and partly because evidence of introgression indicated potential for recolonization, redifferentiation, and recovery (Smith 1972). Application of a concept that recognizes morphological and ecological evidence for independence, despite some introgression, would have encouraged conservation in the past and would dictate better protection now.

Suckers

The family Catostomidae includes about 65 mostly benthic species in North American rivers and lakes (Smith 1993). Among the six most diverse U.S. fish families, suckers rank first in percent of endangered species, with 12% endangered (Warren and Burr 1994); two species of suckers have become extinct. Suckers are ecologically important because they constitute significant proportions of the biomass and energy flow in temperate lakes and

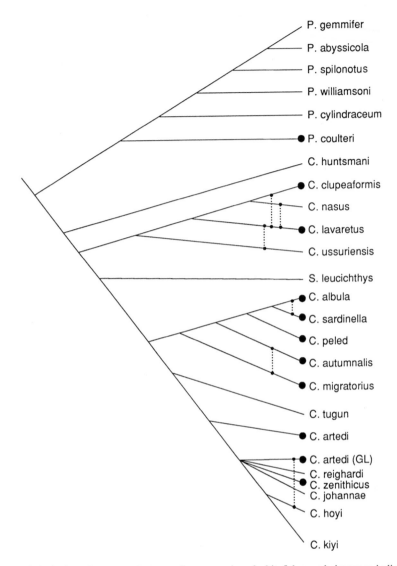

FIGURE 1.—Morphological overlap occurs between diverse species of whitefishes and ciscoes as indicated by dotted lines connecting branches: *Coregonus albula* and *C. sardinella*; *C. autumnalis* and *C. migratorius*; *C. lavaretus* and *C. nasus*; *C. lavaretus* and *C. clupeaformis*; and *C. lavaretus* and *C. ussuriensis*. Significant morphological, ecological, and reproductive differentiation occurs within species marked by dots at branch ends: *C. albula*; *C. sardinella*; *C. autumnalis*; *C. migratorius*; *C. peled*; *C. lavaretus*; *C. nasus*; *C. clupeaformis*; *C. artedi*; *C. zenithicus*; and *Prosopium coulteri* (McCart 1970). Data are from Todd and Smith (1980, 1992); abbreviations are *Prosopium* (P.), *Coregonus* (C.), *Stenodus* (S.), and Great Lakes (GL).

streams, feeding on benthos and providing forage for predators.

This family illustrates an important conclusion of this paper: lineages can be independent for millions of years without gaining irreversible reproductive isolation. Even ancient, well-diagnosed species in different genera are subject to context-dependent introgression (Anderson 1953; Hubbs 1955). The high incidence of introgression among suckers is probably caused by a mating system in which several males spawn simultaneously with each spawning female. Because of evidence for introgression, sucker species are currently classified into broad species. Many of the species are polytypic with morphologically, ecologically, and biochemically distinct subunits (Buth 1979, 1980; Smith and Koehn

1971) that have long histories documented by fossils (Smith 1993). Hybrids have been documented among many of the species (Hubbs et al. 1943). Twelve examples of significant introgression were documented by Smith (1992), of which six occur between lineages that diverged at least 3 million years ago (Figure 2). The most dramatic of these involves the introgressive replacement, following the drought of the mid-1930s, of the June sucker *Chasmistes liorus*, an endemic species in Utah Lake, by a hybrid population with many characteristics of the Utah sucker *Catostomus ardens* (Miller and Smith 1981). According to fossil evidence, *Chasmistes* and *Catostomus* were well-differentiated genera with considerable species diversity in western North America prior to 3 million years ago (Miller and Smith 1981; Smith 1993). *Chasmistes* and *Catostomus* remained distinct during the long history of Pliocene Lake Idaho, yet intermediates were discovered in Utah Lake (Miller and Smith 1981). Introgression is caused when different spawning assemblages are forced together during periods of low river discharge. Catostomids thus illustrate the special interaction of ecology and breeding systems: habitat destruction can lead to introgression. The response should be to correct the habitat damage, not to abandon conservation because of weakened reproductive isolation.

Darters

The genera *Etheostoma* and *Percina* of the family Percidae are the most species-rich genera of North American freshwater fishes. They include more than 135 recognized species, or 17% of the total freshwater fish species diversity in the United States. They are small, benthic, invertebrate feeders, often with brightly colored males and diverse nesting habits (Page 1985). As in other examples of exceptionally diverse fish clades (e.g., African rift-lake Cichlidae), species groups are morphologically and ecologically differentiated (Page and Swofford 1984), but diagnosis of members of closely related species pairs is usually based on colors. In darters, sister species are usually allopatric, whereas in cichlid flocks, sister species are often sympatric.

Many darter species are vicariant populations in small headwater streams and are isolated by long stretches of large-stream habitat. Morphological and ecological differentiation are causally connected (Page and Swofford 1984), but there is little evidence of ecological or morphological differentiation between sister species—primary differentiation is in male breeding coloration (Table 1). Our

hypothesis is that females choose mates based on color intensities that indicate a male's breeding readiness (Williams 1992). This depends on the assumption that sperm maturity and bright coloration are jointly correlated with androgen levels in the brief spring breeding season. If this is true, then local variations in male breeding color could eventually differentiate in different drainages, contributing to lineage distinctness (Dominey 1984). Tests of the female mate-choice hypothesis are being conducted by one of us (J. Porterfield, University of Michigan, unpublished).

Sixty percent of darter species have restricted distributions (Williams et al. 1989). Forty-eight darter species are listed as endangered, threatened, or of special concern (Williams et al. 1989). Several aspects of this conservation problem are unusual. First, it may be the original condition for many darter populations to be geographically restricted. Second, populations are not necessarily expected to be reproductively isolated when only male breeding colors have differentiated (Ryan and Wagner 1987). Sexually selected differences are not predictably rejected by females of related species (i.e., *Drosophila*, Kaneshiro 1989). Female darters might readily mate with more or less brightly colored stray males from sister species. Introgression would be difficult to recognize because the only distinctions are obscured by ontogenetic and seasonal variation. Species pairs of darters need more study to determine whether they are ecologically and genetically distinct. If those darter lineages that are distinct only in male colors are a special stage of speciation, perhaps newly formed, then they are of scientific interest in studies of the speciation process. Those darter species that have also acquired distinct ecological roles, distinctive morphological characters, or DNA and protein differences have diverged further and are more likely to be irreplaceable units in the functioning of stream ecosystems.

Discussion

A major contribution of the phylogenetic method is to enable an organized perception of hierarchical, successive lineage branching: the sequence of divergences and the characters that occurred in association with them. Lineages may acquire distinctive advancements in morphology, ecology, and mate-recognition systems in any order (Figure 3). Within chinook salmon, there are certain lineages that are ecologically, behaviorally, and reproductively independent units without obvious morphological markers. For our purposes, the seasonally separated chi-

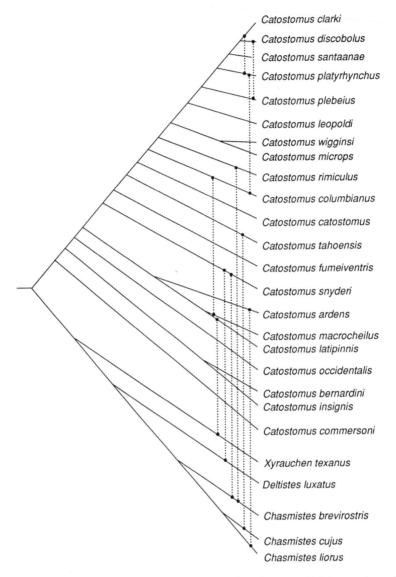

Catostomus clarki
Catostomus discobolus
Catostomus santaanae
Catostomus platyrhynchus
Catostomus plebeius
Catostomus leopoldi
Catostomus wigginsi
Catostomus microps
Catostomus rimiculus
Catostomus columbianus
Catostomus catostomus
Catostomus tahoensis
Catostomus fumeiventris
Catostomus snyderi
Catostomus ardens
Catostomus macrocheilus
Catostomus latipinnis
Catostomus occidentalis
Catostomus bernardini
Catostomus insignis
Catostomus commersoni
Xyrauchen texanus
Deltistes luxatus
Chasmistes brevirostris
Chasmistes cujus
Chasmistes liorus

FIGURE 2.—Introgression occurs between distantly related species of catostomid fishes as shown by dotted lines connecting branches (Miller and Smith 1981; Smith 1992; T. Dowling, Arizona State University, personal communication). Data are from Smith (1993).

nook salmon runs represent lineages with substantial genetic independence, ecological and economic significance, and great evolutionary interest.

Seven forms of ciscoes in the Laurentian Great Lakes represent ecologically and morphologically distinct lineages that achieved genetic independence as indicated by their reproductive allopatry and allochrony. Ecological evolution and morphological distinctness brought about reproductive in-

dependence, which persisted until breeding systems were destroyed by overfishing. Reproductive isolation was linked to ecological stability of breeding systems in ciscoes. Efforts to protect the populations were biologically and politically limited.

Endangered Catostomidae illustrate reversed sequences of stages of differentiation. Catostomid clades diverged substantially in ecology, morphology, and biochemistry as their lineages diversified. But some of these lineages that diverged more than

TABLE 1.—Distribution of kinds and combinations of characters of 91 species within a subgenera of *Etheostoma* (from Page 1983).

	External morphology	Meristics	General pigment	No other characters	All
Male breeding colors	14, 4[a], 1[b], 5[c]	5, 1[d]	4	44	
Morphometrics	1, 1[e]				
No other characters	1	1	2	6	
All					1[f]

[a]Diagnosed by male color + external morphology + meristics.
[b]Diagnosed by male color + external morphology + general pigment.
[c]Diagnosed by male color + external morphology + meristics + general pigment.
[d]Diagnosed by male color + meristics + general pigment.
[e]Diagnosed by morphometrics + male color + external morphology + meristics.
[f]Diagnosed by a combination of all kinds of characters.

3 million years ago may lack proper levels of reproductive isolation when their spawning habitat is disturbed.

Many of the darters of *Percina* and *Etheostoma* are reproductively independent, according to one

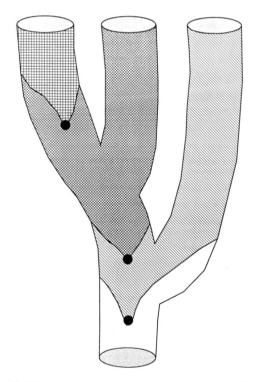

FIGURE 3.—Among lineages in the process of individuation, acquisition of new, unique advancements (three solid symbols and shading) for ecological differentiation, morphological differentiation, and mate recognition may occur as apomorphies in any sequence, resulting in different kinds of evolutionarily significant units. There is no fixed, necessary sequence of morphological, ecological, or reproductive apomorphies.

class of morphological evidence—male color differences—that bears directly on their mate-recognition systems. These are naturally vulnerable species, being small fish in restricted and frequently degraded habitats. They have scientific and cultural importance and evolutionary independence, though many species pairs lack ecological distinction.

The diversity of kinds of fish lineages illustrates the confusion caused by assuming that there is a real category at the species level and by emphasizing that level for practical purposes. Reproductive isolation, as an ultimate criterion, does not serve this issue well; it caused underestimation of biological diversity (Behnke 1970; Cracraft 1983). There is a need to use the concept of ESUs to protect lineages that show evidence that they are irreplaceable functional parts of aquatic ecosystems. This concept can be consistent with a species concept that adds ecology to reproductive independence—the populations to be conserved would be ecologically and evolutionarily significant units.

Lineage individuality (as an alternative to isolation) is a concept that connects the biological, morphological, genetic, and phylogenetic species definitions. Addition of an ecological component to the concept will empower conservation efforts aimed at preserving ecosystem functions and ESUs. Evidence for genetic and ecological independence qualifies a population for conservation. Three criteria (modified from Waples 1991) may help assign priority: (1) is the lineage independent as indicated by ecological, behavioral, or morphological uniqueness? (2) is the lineage independent, as indicated by genetic evidence? (3) is the population irreplaceable with respect to ecosystem function?

Agreement among different sources of evidence for uniqueness implies that gene flow is not destroying independence. Evidence can be supplied by studies of ecology and biology as a part of preparation for listing. Populations that were historically unique, but are introgressed because of anthropogenic destruction of their ecological context, are in special need of habitat restoration.

Summary

The species category is neither real nor filled with homogeneous units. The hierarchy of evolutionary history is not a reductionist system. In the past, species have been spoken of generally, as if they were created equal. But the species zone of the evolutionary continuum necessarily includes a whole spectrum of real lineages that represent early and late stages of different evolutionary processes. Position of potential ESUs along the continuum can be gauged by adding the bits of evidence attesting to their independence: reproductive, morphological, behavioral, ecological, and genetic. In addition, lineages are born into the evolutionary continuum by interaction of different processes in two arenas: sexual selection in different mating systems and ecological selection in different adaptive zones. In seasonal lineages of site-faithful stocks, lacustrine lineages differentiated by time and place of spawning, and temperate stream fishes with restricted breeding seasons, sexual selection is powerfully integrated with ecological selection, when females choose reproductive time and place in response to successful recruitment of young. The diversity of this kind of ecological and genetic adaptation is fundamental to the functions of aquatic ecosystems. Evidence of these complex ecological interactions should be considered among criteria for ESUs. Thus we have the opportunity to scale management effort to potential ecological consequences by acting on the strength of evidence for ecological interdependence of lineages as well as evidence for genetic independence of lineages.

Acknowledgments

Jennifer Nielsen prompted this discussion and, with the participants of the symposium organized by her, added to it substantially. James Albert constructed Figure 3. Tom Todd provided extensive information about *Coregonus* and read the manuscript. Catherine Badgley and Shane Webb reviewed the manuscript. Bonnie Miljour prepared Figures 1 and 2.

References

Allendorf, F. W., and R. F. Leary. 1988. Conservation and distribution of genetic variation in a polytypic species, the cutthroat trout. Conservation Biology 2:170–184.

Anderson, E. 1953. Introgressive hybridization. Biological Reviews 28:280–307.

Bartley, D. M., M. Bagely, G. A. Gall, and B. Bentley. 1992. Use of linkage disequilibrium data to estimate effective size of hatchery and natural fish populations. Conservation Biology 6:365–375.

Bartley, D. M., G. A. Gall, and B. Bentley. 1990. Biochemical genetic detection of natural and artificial hybridization of chinook and coho salmon in northern California. Transactions of the American Fisheries Society 119:431–437.

Behnke, R. J. 1970. The application of cytogenetic and biochemical systematics to phylogenetic problems in the family Salmonidae. Transactions of the American Fisheries Society 99:237–248.

Behnke, R. J. 1993. Status of biodiversity of taxa and nontaxa of salmonid fishes: contemporary problems of classification and conservation. Pages 43–48 *in* J. G. Cloud and G. H. Thorgaard, editors. Genetic conservation of salmonid fishes. Plenum, New York.

Bernatchez, L., and J. J. Dodson. 1990. Allopatric origin of sympatric populations of lake whitefish (*Coregonus clupeaformis*) as revealed by mtDNA restriction analysis. Evolution 44:1263–1271.

Bookstein, F. L., B. Chernoff, R. L. Elder, J. M. Humphries, G. R. Smith, and R. E. Strauss. 1985. Morphometrics in evolutionary biology. The geometry of size and shape change with examples from fishes. Academy of Natural Sciences of Philadelphia, Special Publication 15, Philadelphia.

Breder, C. M., Jr., and D. E. Rosen. 1966. Modes of reproduction in fishes. Natural History Press, Garden City, New York.

Burnham, M. 1993. Intralacustrine speciation of *Salvelinus namaycush* in Lake Superior: an investigation of genetic and morphological variation and evolution of lake trout in the Laurentian Great Lakes. Doctoral dissertation. University of Michigan, Ann Arbor.

Buth, D. G. 1979. Genetic relationships among the torrent suckers, genus *Thoburnia*. Biochemical Systematics and Ecology 7:311–316.

Buth, D. G. 1980. Evolutionary genetics and systematic relationships in the catostomid genus *Hypentelium*. Copeia 1980:280–290.

Bye, V. J. 1984. The role of environmental factors in the timing of reproductive cycles. Pages 187–206 *in* G. W. Potts and R. J. Wooton, editors. Fish reproduction: strategies and tactics. Academic Press, New York.

Campbell, E. A., and P. B. Moyle. 1990. Historical and recent population sizes of spring-run chinook salmon in California. Pages 155–216 *in* T. J. Hassler, editor. Proceedings, 1990 northeast Pacific chinook and coho salmon workshop. American Fisheries Society. Humboldt State University, Arcata, California.

Clark, G. H. 1929. Sacramento–San Joaquin salmon (*Oncorhynchus tshawytscha*) fishery of California. California Department of Fish and Game Fish Bulletin 17: 1–74.

Cracraft, J. 1983. Species concepts and speciation analysis. Current Ornithology 1:159–187.

Cracraft, J. 1987. Species concepts and the ontology of evolution. Biology & Philosophy 2:329–346.

deQueiroz, K., and M. J. Donoghue. 1988. Phylogenetic systematics and the species problem. Cladistics 4:317–338.

Dobzhansky, T. 1937. Genetics and the origin of species. Reprint ed., 1982. Columbia University Press, New York.

Dominey, W. J. 1984. Effects of sexual selection and life history on speciation: species flocks in African cichlids and Hawaiian *Drosophila*. Pages 231–249 *in* A. A. Echelle and I. Kornfield, editors. Evolution of fish species flocks. University of Maine Press, Orono.

Dowling, T., and W. S. Moore. 1985. Evidence for selection against hybrids in the family Cyprinidae (genus *Notropis*). Evolution 39:152–158.

Dowling, T., G. R. Smith, and W. Brown. 1989. Reproductive isolation and introgression between *Notropis cornutus* and *Notropis chrysocephalus* (family Cyprinidae): comparison of morphology, allozymes, and mitochondrial DNA. Evolution 43:620–634.

Ehrlich, P. R., and A. Ehrlich. 1981. Extinction: the causes and consequences of the disappearance of species. Random House, New York.

Eldredge, N. 1985. Unfinished synthesis. Biological hierarchies and modern evolutionary thought. Oxford University Press, New York.

Eldredge, N. 1993. What, if anything, is a species? Pages 3–20 *in* W. H. Kimbel and L. B. Martin, editors. Species, species concepts, and primate evolution. Plenum Press, New York.

Fleischer, G. W. 1992. Status of coregonine fishes in the Laurentian Great Lakes. Polskie Archiwum Hydrobiologii 39:247–259.

Fraidenburg, M. E., and R. H. Lincoln. 1985. Wild chinook salmon management: an international conservation challenge. North American Journal of Fisheries Management 5(3a):311–328.

Frissel, C. A. 1993. Topology of extinction and endangerment of native fishes in the Pacific Northwest and California (U.S.A.). Conservation Biology 7(2):342–354.

Gach, M. H. 1993. Evolution of mitochondrial DNA in the brook stickleback, *Culaea inconstans* (Teleostei: Gasterosteidae): description and transmission genetics of sequence and length variation. Doctoral dissertation. University of Michigan, Ann Arbor.

Ghiselin, M. T. 1974. A radical solution to the species problem. Systematic Zoology 23:536–544.

Ghiselin, M. T. 1987. Hierarchies and their components. Paleobiology 13:108–111.

Goodman, M. L. 1990. Preserving the genetic diversity of salmonid stocks: a call for federal regulation of hatchery programs. Environmental Law 20:11–166.

Hallock, R., and H. Rectenwald. 1990. Environmental factors contributing to the decline of the winter-run chinook salmon on the upper Sacramento River. Pages 141–145 *in* T. J. Hassler, editor. Proceedings of the 1990 northeast Pacific chinook and coho salmon workshop. American Fisheries Society. Humboldt State University, Arcata, California.

Healy, M. C. 1991. Life history of chinook salmon. Pages 313–393 *in* C. Groot and L. Margolis, editors. Pacific salmon life histories. University of British Columbia Press, Vancouver.

Houde, A. E., and J. A. Endler. 1990. Correlated evolution of female mating preference and male color pattern in *Poecilia reticulata*. Science 248:1405–1408.

Hubbs, C. L. 1955. Hybridization between fish species in nature. Systematic Zoology 4:1–20.

Hubbs, C. L., L. C. Hubbs, and R. E. Johnson. 1943. Hybridization in nature between species of catostomid fishes. Contributions from the Laboratory of Vertebrate Biology of the University of Michigan 22:1–76.

Kaneshiro, K. W. 1989. The dynamics of sexual selection and founder effects in species formation. Pages 279–296 *in* L. V. Giddings, K. Y. Kaneshiro, and W. W. Anderson, editors. Genetics, speciation, and the founder principle. Oxford University Press, New York.

Koelz, W. 1927. Coregonid fishes of the Great Lakes. U.S. Bureau of Fisheries Bulletin 43:297–643.

Koelz, W. 1931. The coregonid fishes of northeastern America. Papers of the Michigan Academy of Science, Arts, and Letters.

Kluge, A. G. 1990. Species as historical individuals. Biology & Philosophy 5:417–431.

Levins, R., and R. Lewontin. 1985. The dialectical biologists. Harvard University Press, Cambridge, Massachusetts.

Mann, R. H. K., C. A. Mills, and D. T. Crisp. 1984. Geographical variation in the life history tactics of some species of freshwater fish. Pages 171–186 *in* G. W. Potts and R. J. Wooton, editors. Fish reproduction: strategies and tactics. Academic Press, New York.

Mayr, E. 1942. Systematics and the origin of species. Columbia University Press, New York.

Mayr, E. 1963. Animal species and evolution. Harvard University Press, Cambridge, Massachusetts.

Mayr, E. 1970. Populations, species, and evolution. Harvard University Press, Cambridge, Massachusetts.

Mayr, E. 1982. The growth of biological thought. Harvard University Press, Cambridge, Massachusetts.

Mayr, E. 1988. Toward a new philosophy of biology: observations of an evolutionist. Harvard University Press, Cambridge, Massachusetts.

McCart, P. 1970. Evidence for the existence of sibling species of pygmy whitefish (*Prosopium coulteri*) in three Alaskan lakes. Pages 81–98 *in* C. C. Lindsey and C. S. Woods, editors. Biology of coregonid fishes. University of Manitoba Press, Winnepeg.

McIsaac, D. O., and T. P. Quinn. 1988. Evidence for a hereditary component in homing behavior of chinook salmon (*Oncorhynchus tshawytscha*). Canadian Journal of Fisheries and Aquatic Sciences 45:2201–2205.

McPhail, J. D., and C. C. Lindsey. 1970. Freshwater fishes of northwestern Canada and Alaska. Fisheries Research Board of Canada Bulletin 173:1–381.

Meffe, G. K. 1992. Techno-arrogance and halfway technologies: salmon hatcheries on the Pacific coast of North America. Conservation Biology 6(3):350–354.

Meyer, A. T., T. D. Kocher, P. Basasibwaki, and A. C. Wilson. 1990. Monophyletic origin of Lake Victoria cichlid fishes suggested by mitochondrial DNA sequences. Nature 347:550–553.

Miller, R. R., and G. R. Smith. 1981. Distribution and evolution of *Chasmistes* (Pisces: Catostomidae) in western North America. Occasional Papers of the Museum of Zoology University of Michigan 696:1–46.

Mishler, B. D., and M. J. Donoghue. 1982. Species concepts: a case for pluralism. Systematic Zoology 31(4): 503–511.

Moran, P., I. Kornfield, and P. N. Reinthall. 1994. Molecular systematics and radiation of the haplochromine cichlids (Teleostei: Perciformes) of Lake Malawi. Copeia 1994:274–288.

Moritz, C., S. Lavery, and R. Slade. 1995. Using allele frequency and phylogeny to define units for conservation and management. American Fisheries Society Symposium 17:249–262.

Nehlsen, W., J. E. Williams, and J. A. Lichatowich. 1991. Pacific salmon at the crossroads: stocks at risk from California, Oregon, Idaho and Washington. Fisheries 16(2):4–21.

Nielsen, J. L. 1994. Molecular genetics and stock identification in Pacific salmon (*Oncorhynchus* spp.). Doctoral dissertation. University of California, Berkeley.

Nixon, K. C., and Q. D. Wheeler. 1990. An amplification of the phylogenetic species concept. Cladistics 6:211–223.

O'Brien, S. J., and E. Mayr. 1991. Bureaucratic mischief: recognizing endangered species and subspecies. Science 25:1187–1188.

Page, L. M. 1983. Handbook of darters. TFH Publications, Neptune, New Jersey.

Page, L. M. 1985. Evolution of reproductive behaviors in percid fishes. Illinois Natural History Survey Bulletin 33:275–295.

Page, L. M., and D. L. Swofford. 1984. Morphological correlates of ecological specialization in darters. Environmental Biology of Fishes 11:139–159.

Paterson, H. E. H. 1978. More evidence against speciation by reinforcement. South African Journal of Aquatic Science 74:369–371.

Paterson, H. E. H. 1982. Perspective on speciation by reinforcement. South African Journal of Aquatic Science 78:53–57.

Paterson, H. E. H. 1985. The recognition concept of species. Pages 21–29 *in* E. S. Vrba, editor. Species and speciation. Transvaal Museum Monograph 4, Pretoria, South Africa.

Paterson, H. E. H. 1993. Evolution and the recognition concept of species. *In* Collected writings. S. F. McVey, editor. The Johns Hopkins University Press, Baltimore, Maryland.

Quinn, T. P., and K. Fresh. 1984. Homing and straying in chinook salmon (*Oncorhynchus tshawytscha*) from Cowlitz River hatchery, Washington. Canadian Journal of Fisheries and Aquatic Sciences 41:1078–1082.

Reynolds, F. L., R. L. Reavis, and J. Schuler. 1990. Sacramento valley salmon and steelhead restoration and enhancement. Pages 14–34 *in* T. J. Hassler, editor. Proceedings, 1990 northeast Pacific chinook and coho salmon workshop. American Fisheries Society. Humboldt State University, Arcata, California.

Robins, C. R., and six coauthors. 1991. Common and scientific names of fishes from the United States and Canada. American Fisheries Society Special Publication 20.

Rohlf, F. J., and F. L. Bookstein. 1990. Proceedings of the Michigan morphometrics workshop. University of Michigan Museum of Zoology Special Publication 2:1–380.

Rojas, M. 1992. The species problem and conservation: what are we protecting? Conservation Biology 6(2): 170–179.

Rosen, D. E. 1978. Vicariant patterns and historical explanation in biogeography. Systematic Zoology 27: 159–188.

Ryan, M. J., and W. E. Wagner, Jr. 1987. Asymmetries in mating preferences between species: female swordtails prefer heterospecific males. Science 236:595–597.

Shields, B. A., K. S. Guise, and J. C. Underhill. 1990. Chromosomal and mitochondrial DNA characterization of a population of dwarf cisco (*Coregonus artedii*) in Minnesota. Canadian Journal of Fisheries and Aquatic Sciences 47(8):1562–1569.

Simpson, G. G. 1951. The species concept. Evolution 5:285–298.

Smith, G. R. 1992. Introgression in fishes: significance for paleontology, cladistics, and evolutionary rates. Systematic Biology 41(1):41–57.

Smith, G. R. 1993. Cladistics and biogeography of the Catostomidae, freshwater fishes of North America and Asia. Pages 778–826 *in* R. L. Mayden, editor. Ecological biogeography of North American freshwater fishes. Stanford University Press, Stanford, California.

Smith, G. R., and R. K. Koehn. 1971. Phenetic and cladistic studies of biochemical and morphological characteristics of *Catostomus*. Systematic Zoology 20:282–297.

Smith, G. R., and T. N. Todd. 1984. Evolution of species flocks of fishes in north temperate lakes. Pages 45–68 *in* A. Echelle and I. Kornfield, editors. Evolution of fish species flocks. University of Maine Press, Orono.

Smith, G. R., and T. N. Todd. 1992. Morphological cladistic study of coregonine fishes. Polskie Archiwum Hydrobiologii 39:479–490.

Smith, S. H. 1972. Factors of ecological succession in oligotrophic fish communities of the Laurentian Great Lakes. Journal of the Fisheries Research Board of Canada 29:717–730.

Society for Conservation Biology. 1987. Conservation Biology 1. Society for Conservation Biology. Blackwell Scientific Publications, Boston.

Sokal, R. R., and T. J. Crovello. 1970. The biological species concept: a critical evaluation. American Naturalist 104:127–153.

Soulé, M. E. 1985. What is conservation biology? Bioscience 35(11):727–734.

Stacey, N. E. 1984. Control of the timing of ovulation by exogenous and endogenous factors. Pages 207–222 *in* G. W. Potts and R. J. Wooton, editors. Fish reproduction: strategies and tactics. Academic Press, New York.

Svärdson, G. 1952. The coregonid problem. IV. The significance of scales and gill rakers. Institute for Freshwater Research, Drottningholm 33:204–232.

Svärdson, G. 1965. The coregonid problem. VII. The isolating mechanisms in sympatric species. Institute for Freshwater Research, Drottningholm 46:95–123.

Svärdson, G. 1979. Speciation of Scandinavian *Coregonus*. Institute for Freshwater Research, Drottningholm 57:1–95.

Todd, T. N., and G. R. Smith. 1980. Differentiation in *Coregonus zenithicus* in Lake Superior. Canadian Journal of Fisheries and Aquatic Sciences 37:2228–2235.

Todd, T. N., and G. R. Smith. 1992. A review of differentiation of Great Lakes ciscoes. Polskie Archiwum Hydrobiologii 39:261–267.

Todd, T. N., G. R. Smith, and L. E. Cable. 1981. Environmental and genetic contributions to morphological differentiation in ciscoes (Coregoninae) of the Great Lakes. Canadian Journal of Fisheries and Aquatic Sciences 38:59–67.

Todd, T. N., and R. M. Steadman. 1988. Hybridization of ciscoes (*Coregonus* spp.) in Lake Huron. Canadian Journal of Zoology 67:1679–1685.

Van Valen, L. 1976. Ecological species, multi-species, and oaks. Taxon 25:233–239

Van Valen, L. 1988. Species, sets, and the derivative nature of philosophy. Biology & Philosophy 3:49–66.

Waples, R. S. 1991. Definition of "species" under the federal Endangered Species Act: application to Pacific salmon. NOAA (National Oceanic and Atmospheric Administration) Technical Memorandum NMFS (National Marine Fisheries Service) F/NWC-194, Northwest Fisheries Science Center, Seattle.

Warren, M. L., Jr., and B. M. Burr. 1994. Status of freshwater fishes of the United States: overview of an imperiled fauna. Fisheries 19:6–17.

Wiley, E. O. 1981. Phylogenetics: the theory and practice of phylogenetic systematics. Wiley, New York.

Williams, G. 1992. Natural selection: domains, levels, and challenges. Oxford University Press, New York.

Williams, J. E., and seven coauthors. 1989. Fishes of North America endangered, threatened, or of special concern: 1989. Fisheries 14:2–20.

Wilson, E. O. 1992. The diversity of life. Harvard University Press, Cambridge, Massachusetts.

Wright, S. 1946. Isolation by distance under diverse systems of mating. Genetics 31:125–156.

Zelditch, M. L., and W. L. Fink. 1995. Allometry and developmental integration of body growth in a piranha, *Pygocentrus nattereri* (Teleostei:ostariophysi). Journal of Morphology 223(3):341–355.

American Fisheries Society Symposium 17:58–113, 1995

Systematics, Species Concepts, and the Evolutionarily Significant Unit in Biodiversity and Conservation Biology

RICHARD L. MAYDEN

Department of Biological Sciences, University of Alabama
Box 870344, Tuscaloosa, Alabama 35487, USA

ROBERT M. WOOD

Department of Biology, Saint Louis University
3507 Laclede, St. Louis, Missouri 63103, USA

Abstract.—Biodiversity is the product of descent with modification. Descent is intrinsic to all organisms, and all types of their attributes are modified through a unique history. The temporal and spatial patterns of descent reflect processes responsible for both the origins and current maintenance of organisms, species, and lineages. The paradigm of phylogenetic systematics is the only demonstrably accurate method designed to recover these patterns, as well as to reconstruct and corroborate past evolutionary events of species, infraspecific entities, and their attributes. Species are fundamental in the evolution of biodiversity because they are viewed, since Darwinism, as the nuclear elements of evolution. Thus, understanding species and assimilating their evolution through a systematic method are the progressional links to understanding biological systems. All but one existing conceptualization of species (the evolutionary species concept) are fundamentally inconsistent with both the theoretical and empirical domains of evolutionary biology. Consequently, the use of any of these alternative concepts of species not only weakens one of the two fundamental links but is inherently damning to our abilities to comprehend universal biological systems. Considering our shared responsibilities for our shared resources, it is important to realize that these responsibilities—that is, the organisms and taxa upon which we depend and with which we work and their evolutionary histories—extend to the heart of our livelihoods. By using nonphylogenetic methods and all but one of the currently employed conceptualizations of species, we would be negligent in our responsibilities to these resources. Namely, many species would never be recognized, understood, utilized, or conserved. Only with input from phylogenetic systematics and the evolutionary species concept can all naturally occurring biodiversity, as presently understood, have the opportunity to be recognized and preserved in perpetuity.

The topics of species, species concepts, and speciation are of consequential importance to a diverse array of scientific and nonscientific areas involving biodiversity. One important reason for this general interest is that species as taxa serve as fundamental devices with which we measure diversity, general health and changes in environments, and many other properties of this planet. For biologists and nonbiologists alike, individual species serve as fundamental elements in natural biotic systems through which we assimilate varied notions and observations in numerous disciplines including agriculture, ecology, ethology, evolutionary biology, geography, geology, medicine, and systematic biology. Consequently, a knowledge base relevant to species (existence and identification) and the processes involved in their derivation (speciation) are of fundamental importance to many areas of our society. In any of the above fields, subject organisms possessing a proper scientific name are assumed to represent natural, singularly derived by-products of descent and not artificially contrived constructs that poorly reflect natural patterns and order.

Critical to estimating and conserving biodiversity is the accurate identification of the diversity of our natural world. If the ability to identify diversity correctly is impaired, then the remainder of our efforts to conserve and understand these entities further will be ineffective. Even ideal conservation protocols, if based on unnatural inventories of diversity, cannot be extrapolated to general systems and be effective. Conversely, general, previously effective conservation methods cannot be expected to be effective with unnatural classifications of diversity. Thus, those practicing conservation science, or any other disciplines dependent upon accurate estimates of historical descent, must clearly understand what constitutes diversity and must be able to differentiate theories and methods conducive to its accurate recognition from those that are inaccurate and known to exclude natural products of descent.

Herein we argue that not all methods used in enumerating diversity are equally effective. Rather, it is clear that the methods used to estimate biodiversity and effect its eventual conservation must incorporate a more pluralistic approach from comparative biology than is currently practiced. The most effective approach must incorporate both a conceptual view of species that permits the recognition of known types of diversity and general methodologies from systematic biology designed to better understand patterns and processes associated with genealogical diversity. Both of these elements work in concert to aid us in a more accurate understanding of naturally occurring diversity and can preclude the unintentional loss of diversity or its improper recovery or management. Only some systematic methods and species concepts are capable of accurately identifying natural diversity and genealogical relationships. Others are incapable of providing accurate estimates of natural systems. Where the phylogenetic relationships of a conservation unit have not been assessed or an inappropriate concept of species is applied then conservation efforts may be profitless.

We briefly illustrate the importance and interrelationships of systematic methods and concepts of species in effectively revealing natural diversity, either as species or evolutionarily significant units (ESUs), in conservation biology. We illustrate our philosophy in two separate parts, both relating to the recovery and understanding of diversity. First, we provide an elementary review of systematic theory and methods as is relevant to discussions relating to biodiversity and concepts of species. Second, we review and evaluate four contemporary concepts of species and the ESU concept for their effectiveness in revealing naturally occurring biological diversity. Although presented as two separate discussions, the topics of systematics and concepts of species are inseparable in achieving the mission of conserving unique evolutionary lineages.

What is Biodiversity?

Natural biological diversity (biodiversity) is envisioned in many ways, from unique sequences of nucleic acids through ecosystems to the entire Earth, depending upon the scale of focus. The Office of Technology Assessment (1987) defines biodiversity as: "the variety and variability among living organisms and the ecological complexes in which they occur."

Recently, Wilson (1992:37–38) discussed biodiversity.

Of what then is biodiversity composed? Since antiquity biologists have felt a compelling need to posit an atomic unit by which diversity can be broken apart, then described, measured, and reassembled. Let me put the matter as strongly as this important issue merits. Western science is built on the obsessive and hitherto successful search for atomic units, with which abstract laws and principles can be derived. Scientific knowledge is written in the vocabulary of atoms, subatomic particles, molecules, organisms, ecosystems, and many other units, including species. The metaconcept holding all of the units together is hierarchy, which presupposes levels of organization. . . . And so the search proceeds relentlessly for natural units until, like the true grail, they are found and all rejoice. Scientific fame awaits those who discover the lines of fracture and the processes by which lesser natural units are joined to create larger natural units.

So, the species concept is crucial to the study of biodiversity. It is the grail of systematic biology. Not to have a natural unit such as the species would be to abandon a large part of biology into free fall, all the way from the ecosystem down to the organism.

Regardless of the scale of focus, common to all of these different ideas is that ultimately each depends upon our success at understanding the final definitions of lineages because these are the fundamental elements of communities, ecosystems, and this planet. Through time, lineages (composed of individual organisms) diverge or may become extinct, they are the vehicles through which unique codes of nucleic acids are carried from generation to generation and through which these codes become modified over time. Lineages also participate in all known natural processes in communities, such as competition, death, extinction, food webs, parasitism, predation, and reproduction. Thus, the fundamental elements of biodiversity are the unique lineages produced through descent. In a simple form, at the scale of focus of the present paper, we may define biodiversity as the collection of unique biological lineages historically derived through descent.

Contributions from Systematic Biology

Elements of biological diversity are the primary subjects of a diverse array of academic programs as well as varied commercial and research interests in scientific and nonscientific areas alike. Despite this great breadth of disciplines, we view the fundamental foci of the vast majority of interest in biodiversity as the discovery, understanding, use, and conservation of natural biological entities.

Biological diversity is the natural by-product of descent with modification (Darwin 1859; Futuyma 1986). Fundamental elements of this descent are lineages, as individual organisms, populations, and species (e.g., cutthroat trout *Oncorhynchus clarki*) and as higher taxa (family Salmonidae). Some lineages are ancestral, some are descendants, some are extinct, and some are extant. We argue that through an accurate inventory and understanding of this natural diversity, we, as biologists, are in an appropriate position to perceive natural order in communities and ecosystems. From an accurate understanding of diversity and natural order follows more effective and successful efforts at protecting not only populations and species but communities and ecosystems inclusive of these lineages. The domain of systematic biology is fundamental to achieving this goal. This area of comparative biology has as its purview not only the conceptual basis for the recognition of species as taxa but the unique charge of recovering the descent of life. Thus, the effective development of programs to assess and manage biological diversity accurately must incorporate contributions from systematic biology.

For some biologists the title systematics or taxonomy may conjure up images of static and strictly observational practices neither to be envied nor expected to provide significant information to contemporary science. Traditionally, practitioners of systematics and taxonomy were viewed as merely describing species, faunas, or floras, building and maintaining museum collections, but contributing little else to the academic community. This attitude changed dramatically in recent years with the advent of phylogenetic systematics. For the first time, testable hypotheses of historical descent are achievable. These hypotheses serve as templates with which investigators are able to evaluate numerous historical and proximal questions pertinent to individual organisms, populations, species, and ecosystems (Mayden 1992a). This revolution also can be correlated with the academic community coming to the realization that history is as important as current and future environmental conditions in deliberations of and challenges to biodiversity.

Any effective program in the conservation and preservation of diversity, regardless of species-based or ecosystem-based approaches, must incorporate the theoretical and methodological aspects derived from systematic biology and associated subfields. Unfortunately, input from systematic biology traditionally was limited in identifying natural biological entities and in post hoc interpretations of patterns and processes, both instrumental in making management and conservation decisions (Mayden 1992b). Considerable interest has been expressed in (1) the need to identify equivalent types of biological units for conservation, (2) the various criteria to be used in identifying these entities, (3) a means to partition environmental and historical variation in populations and species, and (4) a gen-

eral, holistic approach to effect conservation procedures, but input from systematic biology had been largely missing from these discussions and operations. We do not pretend that systematics can or will solve all foreseeable problems. However, with accurate accounting of diversity much of the confusion over assessments of organismal characters, the purported conflicting information about "distinctive" units, and the lack of consensus in data and analyses will vaporize if interpretations of data, analyses, and diversity are rooted firmly in the theory and methods of systematics.

Contemporary systematic biology is charged with discovering natural diversity, revealing patterns of natural order, and elucidating origins and relationships among the products of evolutionary descent (Figure 1; Wiley 1981; Mayden and Wiley 1992). This field uniquely provides avenues for the elucidation of historical patterns of descent of lineages (populations or species) and of the attributes (characters) possessed by lineages and modified through descent. Within systematic biology, we specifically refer to and embrace phylogenetic systematics (Hennig 1966; Wiley 1981; Mayden and Wiley 1992). We specifically exclude other methods commonly referred to as phenetics, numerical taxonomy, or evolutionary systematics. As such, we reject any methods wherein similarity is the sole criterion used to formulate hypotheses of relationships and develop biological classifications. Only under specific evolutionary models are these latter methods capable of recovering natural order. Furthermore, we do not advocate character-type bias. All character sets are potentially informative to understanding diversity and genealogical relationships. Characters are only markers that we, *Homo sapiens*, attempt to use to unveil diversity and relationships. There is nothing inherently better nor more reliable about one character set (e.g., molecular) relative to others (e.g., morphology), so long as they are all heritable traits and reflect descent and genetic intercommunication.

Finally, some specific diversity issues and questions demand a pluralistic approach. In these instances information gained from systematic biology must also incorporate the theories and methods from other fields of specialization, such as population biology, ecology, and genetics. Although these latter fields undoubtedly contribute valuable information in conservation biology and other areas, these disciplines should not be considered equivalent to systematic biology in assessing species diversity, descent, or any historical questions. Unquestionably, the theories and methods of population

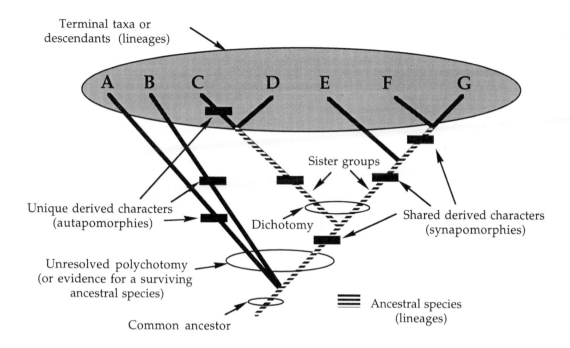

Natural Groups

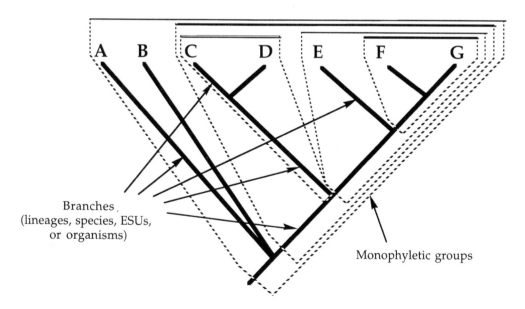

FIGURE 1.—General terminology used for phylogenetic trees.

biology are valuable in estimates of population and species parameters, their genetic structure, and evidence of population equilibrium. However, these estimates are not historical. They do not provide direct estimates of descent and, without input from systematics, cannot provide equivalent information concerning diversity and diversification. These fields have as their purview proximal explanations and observations of populations, species, communities, and ecosystems (Brooks 1985; Brooks and McLennan 1991, 1993; Mayden 1992a).

Phylogenetic Systematics versus Other Schools

The general principles of phylogenetic systematics are outlined by Wiley (1981), Wiley et al. (1991), and Mayden and Wiley (1992). The information provided here is a survey of the discipline to inform the reader of the general importance of this science to biodiversity and in preparation for subsequent discussions.

It is essentially impossible to observe descent with modification in real time (but see Hillis et al. 1992). Biotic diversity produced via descent but observed today evolved in the past over a landscape potentially quite different from today's. The sequence of events remains unrecorded except in rare instances documented by fossil history. Therefore, reconstructions of historical pattern and process must be inferred from observations and interpretations of existing data from organisms, regardless of whether this diversity is extant or extinct. Systematics is therefore a historical science, unique in that the focus of study, organisms and patterns of diversification, cannot be repeated. However, reconstructions of descent are repeatably testable using independent observations from the organisms in question.

Prior to Willi Hennig's (1966) development of phylogenetic systematics, an explicit methodology designed to reconstruct natural patterns of descent was unavailable. Phylogenetics revolutionized comparative biology and in doing so established renewed faith for systematics within the scientific community. Hennig's methods are unique from other systematic methodologies in that, for the first time, researchers can trace descent of organisms and address many types of evolutionary questions simply by searching for and identifying attributes of organisms that were modified (derived) through their descent. Once these modified attributes are identified, researchers can group organisms or determine sister-group relationships on the basis of shared and derived attributes by applying the general principle of parsimony in the evolution of attributes. This procedure results in a hierarchical branching diagram or phylogenetic tree wherein organisms or species possessing shared and derived attributes are hypothesized to be descendants from an ancestral species wherein the modified traits first evolved. Using multiple attributes and parsimony argumentation, genealogical histories are reconstructed to represent descent with modification. Importantly, these trees contain the supporting data for the hypothesized species relationships, and these relationships can be reevaluated and tested using additional attributes, taxa, or organisms. Finally, once organisms are grouped on the basis of attributes that were modified through their descent, then the hypothesis of descent is presented in an unequivocal biological classification of this information.

Because organisms, their attributes, and their temporal and spatial locations are the direct result of their descent, knowledge of descent is critical to understanding these, and other, fundamental elements of organisms. There are three important aspects to phylogenetic systematics that make this discipline an important tool to practitioners of comparative, conservation, and evolutionary biology. First, this discipline offers a repeatable method for reconstructing descent at various levels of complexity from individual organisms to species to natural higher taxa. Knowledge of descent permits not only our eventual understanding of the processes responsible for producing genealogical relationships, but these relationships further our understanding of the historical contributions to, or constraints on, natural systems (e.g., communities and ecosystems). Second, phylogenetic methods permit an explicit evaluation of the evolutionary origins and changes in attributes possessed by organisms and partition those useful in recovering descent from those that are not. Understanding of the origins of attributes is critical not only to our reconstructions of descent but in our interpretations of how we recognize biological diversity and how we explain the origin and importance of attributes that lineages possess. Third, phylogenetics provides a general and unequivocal method for the organization of the information currently known about diversity and genealogical relationships in biological classifications. A vast array of professions expect biological classifications to be accurate with respect to current understanding of diversity. The generation of inaccurate classifications is misleading to all those depending upon these information retrieval systems.

This philosophy of systematics and understanding of descent differs trenchantly from other schools in at least two very important areas: (1) the meaning of similarity and (2) the proliferation of information in genealogies. Similarity of two or more organisms can result from at least four very different processes, only one of which is informative for reconstructing phylogenetic relationships. First, a similar trait may have evolved in two or more species independently through convergent or parallel evolution (bird and butterfly wings, type 1; Figure 2A). Second, organisms can be similar to one another because they inherited the similar, compared attributes from a nonimmediate ancestor (the jaws of salmonid fishes relative to other gnathostome vertebrates, type 2; Figure 2B). Third, organisms can be similar because the trait shared was inherited from an immediate common ancestor, an ancestor in which the trait actually evolved (the jaws of all gnathostomes relative to agnathan vertebrates, type 3; Figure 2C). Fourth, similarity can result from the transfer of genetic information between distantly related taxa through gene exchange, either in Recent or ancestral populations (type 4). Importantly, in the fourth case, depending upon the phylogenetic relationships between the two taxa sharing genes, the end result may mimic either type 1 or type 3 similarities (Figure 2A or 2C). However, the origins of these four types of similarity are very different evolutionarily and can have important conservation implications, as demonstrated later. The types can only be differentiated, however, using other features of the organisms and the organisms' phylogenetic relationships. Thus, similarity of organisms is not necessarily clear cut, easily understood, nor always indicative of kinship or sister-group relationships (Figure 2).

The reason we reject the first and fourth types of similarity for determining sister-group relationships is easily understood. The reason only the third type of similarity is useful in reconstructing patterns of evolutionary relationships also is easily understood. The second type of similarity (Figure 2B) provides no unique information for the group of organisms in question (say salmonid fishes) that is not also found in other unrelated organisms or taxa (e.g., jaws are also present in percid fishes, bats, and hummingbirds). Relationships and the evolution of attributes of organisms or taxa should be based solely on the third type of similarity. Contrary to some opinion, this does not mean that other attributes possessed by organisms are discarded or excluded from evaluation or analysis. Rather, traits of the second type of similarity are those already used to reconstruct relationships at a higher level (e.g., grouping all gnathostome vertebrates).

The methods of phenetics, numerical taxonomy, and evolutionary systematics do not provide accurate depictions of reconstructions of descent of organisms and their attributes, nor the accurate conveyance of this information in classifications. In phenetics or numerical taxonomy, relationships are based on overall similarity without regard to the type of similarity. This has two important consequences when attempting to decipher historical patterns. First, groupings of organisms (taxa) will result from an agglomeration of attributes that fall under all four types of similarity. Consequently, when comparing the different attributes across different groups of organisms the patterns of covariation can, and will, appear confusing and conflicting. Both natural and artificial groupings of taxa result. Natural taxa (or monophyletic groups) are groupings that contain all the descendants from a common ancestor. Artificial taxa are those that share no common ancestors and are artifacts of an investigator's misunderstanding of the evolution of attributes of organisms or are groupings derived by the first and second types of similarity.

The second important consequence of applying nonphylogenetic methods to historical questions involves interpretations of the origins of the suites of attributes possessed by organisms or species. By not partitioning similarities of attributes into those shared by immediate common ancestors (or uniquely evolved in a species) from those evolved in nonimmediate ancestors, it is impossible to interpret accurately attribute origins and evolution. Whereas this may appear trivial, it is not for two reasons. First, just as artificial higher taxa may be recognized through this faulty method of character logic, naturally occurring populations, species, or species complexes may be misinterpreted as either hybrids or intergrades strictly because the researcher did not clearly differentiate the modes through which the traits evolved. Organisms, populations, or species can be misinterpreted as "intermediate" between two other such entities because the researcher is using and considering equivalent both the second and third type of similarity. For example, agnathan fishes appear intermediate between Cephalochordata chordates and Gnathostomata chordates because they possess shared, derived (type 3) Vertebrata traits with the gnathostomes and also share retained, primitive traits

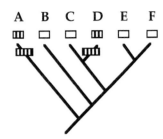

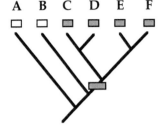

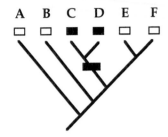

A. ⫿ Trait evolves in A and D through convergent or parallel evolution. Type 1 similarity in A and D.

B. Taxa C and D share ▦ trait because it is a retained, primitive attribute evolved in a nonimmediate ancestor. Type 2 similarity in C and D.

C. Taxa C and D share ■ trait because it is a shared derived attribute evolved in an immediate common ancestor. Type 3 similarity in C and D.

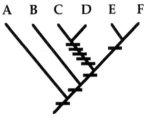

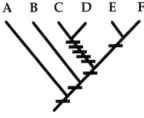

D. True relationships of taxa A–F and distribution of ▬ shared derived attributes in a phylogenetic analysis.

E. Possible phenetic reconstruction of relationships. Taxa C plus D are misplaced because they are so different from other taxa in sharing so many distinct traits relative to other traits.

F. Possible reconstruction using evolutionary systematics. Shared derived traits may be properly placed.

Classification
Genus *Aus*
  species *aus*
  species *bus*
  species *cus*
  species *dus*
  species *eus*
  species *fus*

Classification
Genus *Aus*
  species *aus*
  species *bus*
  species *eus*
  species *fus*

Genus *Cus*
  species *cus*
  species *dus*

Classification
Genus *Aus*
  species *aus*
  species *bus*
  species *eus*
  species *fus*

Genus *Cus*
  species *cus*
  species *dus*

G. Phylogenetic. Tree and classification are consistent, and sister-group relationships are maintained.

H. Phenetic. Tree and classification are consistent, and sister-group classification may reflect dendrogram.

I. Evolutionary systematic. Tree and classification are inconsistent. The latter attempts to show differences shared by taxa C and D in the former.

FIGURE 2.—(**A–C**) Different types of similarities shared by taxa. (**D–F**) Likely reconstructions for taxa A to F where relationships are known (**D**), by use of the methods of phylogenetic systematics (**D**), phenetics (**E**), and evolutionary systematics (**F**). (**G–I**) Logical classifications for the taxa A to F based on classification methods advocated by the schools of systematics and used to derive relationships in **D–F**.

(type 2) with Cephalochordata chordates. Second, if the different types of similarity are confused when comparing different taxa for different attributes the attributes appear conflicting and discordant. This is an artifact of perceiving all similarities as equivalent, when they are not!

For the purposes of elucidating and conserving biodiversity, the possible confusion of types 2 and 3 similarities is an extremely important consideration. Naturally occurring entities "behaving" like populations or species can be mistaken as entities of lesser conservation importance (e.g., hybrids or intergrades) because they possess a mixture of, for example, types 2 and 3 similarities. In actuality, however, because descent is largely a hierarchical process and different attributes of organisms are modified through this descent at different levels, all organisms, species, and higher taxa possess a mixture of these two types of similarity. Just as when reconstructing relationships of organisms, species, or higher taxa, consequential conclusions as to the status of populations or species regarding gene flow or the naturalness of these entities must be based solely on type 3 similarities. Populations and species sharing type 2 similarities should not be assumed to have acquired them as a result of interactions with other species. Rather, the existence of type 2 similarities can result solely from their retention from a common ancestor. Detailed examples of this differentiation are provided in following sections.

A second area in which phylogenetic systematics differs from other schools of systematic thought involves the philosophy in dissemination of genealogical information through biological classifications (Figure 2D–F and G–I). With classifications generated using methods of other schools (Figure 2H–I), it is impossible to know if the groupings contained within a classification are reflective of and consistent with the pattern of descent (Figure 2E–F) observed for the subject organisms. For example, in other schools it is possible for some sister-groups descended from a common ancestor to be placed in artificial taxonomic groups, that is, in different higher taxa or in taxa containing subtaxa that do not share a common ancestor. This practice is strictly an artifact of the alternative schools relying heavily upon the overall degree of similarity between populations, species, or higher taxa to derive classifications and not relying on genealogical connections shared between them (compare Figure 2D–F with G–I).

Fundamentals of Phylogenetic Systematics

There are seven general principles of phylogenetic systematics as the discipline is currently envisioned (Mayden and Wiley 1992).

The phylogenetic principle.—There is a single and historically unique phylogeny (genealogical history) relating all organisms. In other words, descent with modification is singular, and there is a single tree of life to be reconstructed. This is complicated in some instances by differences that may exist between taxon trees and character trees.

The relationship principle.—Relationship in the post-Darwinian world specifically refers to "blood" or genealogical relationship. The various concepts of the pre-Darwinian world of relationship or affinities based strictly on a concept of overall similarity or sharing of "essential" characters is explicitly rejected under this principle. Because overall similarity per se does not reflect direct ancestor–descendant relationship it must be excluded as a criterion.

The auxiliary principle.—Never assume that similar features of two organisms are not homologous and arose independently; always assume that they are homologous unless evidence exists to the contrary. When the attributes of two or more organisms are the same the attributes must be considered homologous a priori. Rejection of this hypothesis comes from evidence through parsimony argumentation of multiple characters for the same organisms. Where there are conflicting patterns of shared, derived attributes (type 3 similarities), some attributes are shared by taxa through type 1, 2, or 4 similarities.

The grouping principle.—Only certain homologous characters, apomorphies (or derived characters), can be used to group organisms into natural evolutionary groups (clades or monophyletic groups). Synapomorphies are the shared and derived attributes (type 3 similarity; Figure 2C) that are modified through the descent of the subject organisms and provide indicators of unique ancestor–descendant relationships. Other homologues, specifically plesiomorphies (or primitive attributes; shared, primitive attributes are symplesiomorphies; type 2 similarity; Figure 2B) and convergences and parallelisms (homoplasy; type 1 similarity; Figure 2A) are not indicative of unique descent and hence are not reliable in reconstructing relationships.

The character placement principle.—Because characters or the attributes of organisms are the features that have evolved, they have a place in the descent of the organisms. Their proper place in the phylogeny is where they arose during evolutionary history. Because during descent some of the attributes of organisms become modified, a proper phylogeny or evolutionary tree is one in which the taxa are placed in correct genealogical order and the characters are placed where they first arose, following the grouping principle (Figure 2A–F).

The inclusion–exclusion principle.—When using the grouping principle of characters to reconstruct phylogenies of organisms or taxa, the information from independent synapomorphies can be combined into a single hypothesis of relationship if that information allows for the complete inclusion or complete exclusion of groups formed by the independent synapomorphies. Under this principle descent and modifications of attributes are hierarchical and predict the inclusive–exclusive nesting of groupings of organisms or taxa (Figure 2D). Where evolution is nonhierarchical, or reticulate, or where similarities between organisms are derived from type 1, 2, or 4, there will be conflicting group membership based on hypothesized independent synapomorphies. Thus, overlap of group membership by independent synapomorphies leads to the generation of two or more hypotheses of relationship because the information cannot be combined into a single hypothesis. Because there is only one organismal or species phylogeny (the phylogenetic principle), one of the groupings (or possibly both) is false. This can become more complicated when differences exist between phylogenies generated for the organisms and the attributes or genes (e.g., particularly with reticulate evolution). In this case, both phylogenies may, in actuality, be reflective of a true pattern of descent, one tracking descent of the organisms or taxa and one tracking the descent of the attribute or gene.

The classification principle.—A truly evolutionary classification is one that is logically consistent with the relationships of the species classified. Of the many possible evolutionary classifications, the preferred classification is one that is fully informative about the common ancestry relationships as they have been reconstructed (Figure 2G).

The Routine Phylogenetic Analysis

Using the above principles as procedural guidelines one can infer historical patterns of descent as well as discover naturally occurring entities of biodiversity. An essential objective in research with biological diversity is that the entities discovered during the process should be natural groups (sensu Wiley 1981; species or subgroups and monophyletic

groups). Natural groups and individuals (philosophical term) are the by-products of nature and include those things that result from and participate in natural processes. In phylogenetics both species and natural supraspecific taxa (monophyletic groups) are natural groups. A monophyletic group is defined as a group inclusive of the ancestor and all descendants of that ancestor and are referred to as historical groups (philosophical term). These groups are discovered through the auxiliary, grouping, and inclusion–exclusion principles by means of the identification of shared and derived (synapomorphic) attributes of the subject organisms.

Because the process of discovering monophyletic groups involves searching for and using attributes of organisms that have been modified during descent to trace the true history of descent, synapomorphic attributes of organisms can include any and all heritable aspects. There is no inherent bias as to what types of attributes are informative for the discovery of descent and natural groups. Traits may include any of those detected in various types of data sets from morphology, physiology, ecology, genetics, behavior, etc. Because all monophyletic groups originate as species (e.g., ancestral species; Wiley 1981), this same logic applies not only to monophyletic groups but also to the discovery and descriptions of individuals, such as species or distinct entities within. Hence, all discoverable and heritable types of traits are equally informative towards the discovery, description, and justification of naturally occurring biological diversity from distinct entities within species to species and supraspecific groups.

The discovery of and distinction between synapomorphic and symplesiomorphic characters for phylogenetic reconstruction is best accomplished using the outgroup comparison method (Figure 3). Although there are at least four other methods for estimating character polarity in phylogenetic systematics, outgroup comparison involves the fewest number of inherent assumptions and is the most reliable (Mayden and Wiley 1992). Outgroups are taxa chosen to be used in character comparisons with the group of organisms being subjected to phylogenetic analysis, the ingroup (Figure 3). Such comparisons are more reliable if the taxa used are closely related to the ingroup and number more than one. The general function of the outgroup is to give the researcher an approximation of what the ancestral conditions would have been for the characters being examined in the ingroup (Figure 3). If these differ from the character conditions present in members of the ingroup, we know that character modification occurred during the ingroup's descent.

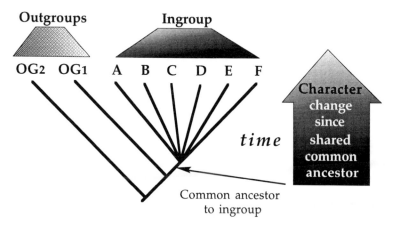

FIGURE 3.—Outgroups and ingroups in phylogenetic systematics, initial hypotheses of relationships in an analysis, and the general logic behind the use of outgroups to identify plesiomorphic (primitive) and apomorphic (derived) attributes to recover descent with modification.

For example, when two or more homologous characters occur among taxa within the ingroup, those homologues shared with the outgroups are symplesiomorphic homologues (type 2 similarity). They occur within the ingroup due strictly to the retention of ancestral traits in some ingroup members during the history of descent. Those homologues not found in the outgroups, but present in some ingroup members, are synapomorphic homologues (type 3 similarity). These homologous characters are the modified or derived characters that allow us to trace the descent of members of the ingroup and are used in the grouping, character placement, and inclusion–exclusion principles to infer phylogenetic relationships.

A final step in the analysis is the generation of a biological classification that is consistent with the evolutionary relationships inferred through character analysis. This classification serves as an information retrieval system about the common ancestry relationships for the ingroup and is fully consistent with these relationships. Phylogenetic classifications are based strictly on sister-group relationships. They are not influenced by the different classes of character changes (e.g., morphology and behavior) nor the amount of character change that may have occurred during a particular speciation event. In their rigor to represent sister-group relationships, phylogenetic classifications are inherently more stable than are classifications devised to represent amount and class of character evolution in the descent of a group (e.g., evolutionary systematics; Figure 2F, I). Given that a phylogenetic analysis of a single group of organisms is conducted with three

different types of data and the same sister-group relationships result for the group, the three phylogenetic classifications will be consistent. However, the three evolutionary systematic classifications will, in all likelihood, be inconsistent with one another because of the known variation in rates of evolutionary change for the different types of data and the ambiguous nature of viewing some attributes as being more "important" than others (Figure 4).

We examine these principles and methods using an example with four operational taxonomic units (OTUs; Figure 5). Theoretically, these OTUs could represent individual organisms (reproducing sexually, asexually, or parasexually), populations, subspecies, species, or natural supraspecific taxa. The a priori assumption in this analysis is that the group chosen (ingroup) is monophyletic and all of the terminal OTUs (extant or extinct) are descendants of a common ancestor (Figure 5A). No other information about the sister-group relationships of the four OTUs is provided. This tree only suggests that all four OTUs descended from a common ancestor. How the content of the ingroup is initially decided varies (Mayden and Wiley 1992). In most cases, ingroups are established based on a preexisting phylogenetic analysis or classification, such as a species, a subgenus, a genus, etc. It should be understood, however, that the initial content of the group is a hypothesis that is testable and is tested through further analysis.

Because the actual events leading to the production of the observed biodiversity within this ingroup cannot be observed, we must rely upon attributes possessed by these terminal OTUs to infer their

A–C. Phylogenies and varied data

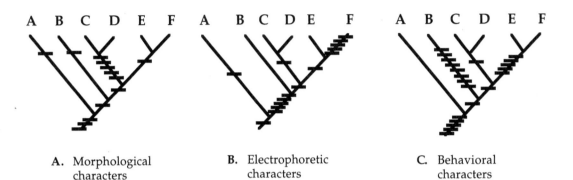

A. Morphological characters

B. Electrophoretic characters

C. Behavioral characters

D. Evolutionary systematic classification and varied data

Genus *Aus*
  species *aus*
  species *bus*
  species *eus*
  species *fus*
Genus *Cus*
  species *cus*
  species *dus*

Genus *Aus*
  species *aus*
Genus *Bus*
  species *bus*
  species *cus*
  species *dus*
  species *eus*
Genus *Fus*
  species *fus*

Genus *Aus*
  species *aus*
  species *cus*
  species *dus*
Genus *Bus*
  species *bus*
Genus *Eus*
  species *eus*
  species *fus*

E. Phylogenetic classification and varied data

Genus *Aus*
  species *aus*
Genus *Bus*
  species *bus*
Genus *Cus*
  species *cus*
  species *dus*
  species *eus*
  species *fus*

Genus *Aus*
  species *aus*
Genus *Bus*
  species *bus*
Genus *Cus*
  species *cus*
  species *dus*
  species *eus*
  species *fus*

Genus *Aus*
  species *aus*
Genus *Bus*
  species *bus*
Genus *Cus*
  species *cus*
  species *dus*
  species *eus*
  species *fus*

FIGURE 4.—Phylogenies and classifications. (**A–C**) Three phylogenetic reconstructions for the same group of taxa A–F based on three different data sets. Note the relationships are resolved exactly the same for each of the data sets. (**D**) Evolutionary classifications derived from the different data sets for the same organisms. Note that these classifications are different for each data set because of emphasis on the amount of divergence that occurred for different speciation events. (**E**) Phylogenetic classifications derived from the different data sets for the same organisms. Note that these classifications are identical, regardless of the data set.

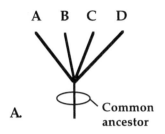

A.

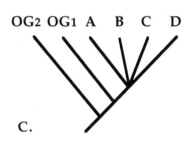

B.

	Characters									
Taxa	1	2	3	4	5	6	7	8	9	10
A	1	1	1	1	1	1	1	0	0	0
B	1	1	1	1	1	1	0	1	0	1
C	1	1	1	1	0	0	0	0	0	0
D	1	1	0	0	0	0	0	0	1	1
OG1	0	0	0	0	0	0	0	0	0	0
OG2	0	0	0	0	0	0	0	0	0	0

C.

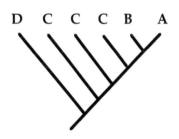

D.

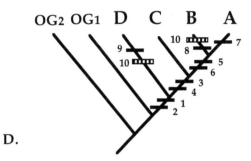

E.

F.

G.

H.

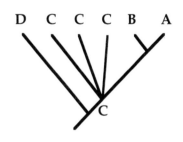

FIGURE 5.—(**A–D**) Phylogenetic analysis for taxa A–D and two outgroups (OG1, OG2). (**E–H**) Possible genealogical relationships that are compatible with the observation that taxon C possesses no known autapomorphic attributes (derived attribute for a single taxon).

ancestor–descendant relationships. To accurately infer sister-group relationships the investigator must search for characteristics possessed by any of the four OTUs that were modified through the OTUs' descent since they shared a common ancestor. If no characters available for our examination were modified through the OTUs descent or if each taxon diverged independently from their common ancestor, then we will be unable to reconstruct any hierarchical patterns of descent in the ingroup and our initial hypothesis stands (Figure 5A). However, in our heuristic example there is character variation informative for phylogeny reconstruction. We examine 10 different characteristics of these OTUs representing any variety of data types (e.g., morphology, behavior, genetics, etc.). For each character, we numerically code (e.g., 0, 1) the different homologous conditions of the character that is observed in the ingroup OTUs. Following the auxiliary principle, when the conditions of a character found in two or more OTUs appear similar we must assume that they are homologous; they are thus coded identically in this example (Figure 5B).

Employing the grouping principle we can then use the observed similarities and differences between the four terminal OTUs as shared and derived features to decipher sister-group relationships. However, at this stage we do not know which of the two homologous conditions (0 or 1) for any of the characters should be used to group these OTUs into natural taxa. To group only by derived and shared attributes, we need to know which of the homologues were modified during the descent of the group and reflect this history. For this, we need an estimate of conditions existing in the common ancestor for the OTUs A–D (Figure 5A). Because the common ancestor is likely not in existence, we employ close relatives to the ingroup yet OTUs that are not part of the ingroup. These OTUs represent the outgroups (Figure 5C) and serve to estimate "primitive" attributes for the ingroup for the 10 homologous characters. Homologous characters shared between the outgroups and members of the ingroup are thus shared and primitive, or symplesiomorphies (Figure 5B: coded as 0). These features exist in the ingroup only because the most recent common ancestor to A–D possessed the feature and passed it on to its descendants. Because both outgroups, and likely the common ancestor to the ingroup, possessed these plesiomorphic homologues, their shared occurrence between any two or more ingroup OTUs does not reflect a unique history of descent within the ingroup. Rather, the alternative homologues for each of the characters represents

the derived and shared homologue, or synapomorphy (Figure 5B: coded as 1). The occurrence of these homologues in any descendants is reflective of inherited modifications that occurred through their descent, since they shared a common ancestor.

Employing the grouping, character placement, and inclusion–exclusion principles, we infer the relationships of our ingroup OTUs based on the identified synapomorphic homologous characters (Figure 5D). Characters 1 to 6 and 10 are synapomorphic characters; characters 7 to 9 are autapomorphic characters, derived homologues possessed by only one taxon. Characters 1 to 6 can be combined into a single hypothesis of relationships because they are consistent with the inclusion–exclusion principle. Character 10, however, cannot be included in this grouping because the hypothesized relationship predicted by this character (taxon B and D sistergroups) is inconsistent with the hypothesis argued by characters 1 to 6, specifically characters 3 to 6 (Figure 5D). With parsimony argumentation, character 10 is hypothesized to have evolved independently in OTUs B and D through convergent or parallel evolution (or gene exchange), both referred to as homoplasious evolution. By means of the character placement principle, all of the synapomorphic, autapomorphic, and homoplasious characters can be placed onto the phylogenetic hypothesis (Figure 5D).

In the resulting hypothesis characters 1 and 2 became modified from homologous condition 0 to 1 in the common ancestor to OTUs A–D. This ancestral taxon underwent a divergence event to give rise to taxon D and the common ancestor to A–C. Characters 3 and 4 were modified in the ancestor to A–C, and this taxon subsequently diverged to produce the descendants taxon C and the common ancestor to A plus B. In the common ancestor to A plus B characters 5 and 6 became modified before this ancestor diverged to produce both A and B. The OTUs A, B, and D all possess autapomorphic characters; OTUs B and D also possess the homoplasious character 10, which should be considered an additional autapomorphic character for each of these OTUs because the condition present in taxon B is not considered homologous to that in taxon D (convergent or parallel evolution).

Interestingly, in this example, taxon C possesses no known autapomorphic characters. Rather, it possesses only symplesiomorphies inherited from former common ancestors and synapomorphies 3 and 4. Further analysis, using additional character data, may (or may not) reveal autapomorphic characters for C. Regardless, given the data currently available for this ingroup, all four of the OTUs can be diagnosed on the basis of the presence or absence of specific synapomorphic characters and autapomorphic characters, except taxon C. Without autapomorphic characters for taxon C it is impossible to state without qualification in a classification that C is a natural biological entity, as currently understood. It may eventually be found to be a natural group wherein all members of the taxon are closest relatives (Figure 5E). On the other hand, it could also represent a surviving ancestor or an unnatural grouping of organisms wherein some members of C are more closely related to A plus B than to other members of C (Figure 5F, G). In this latter case, it is possible then that not all members of taxon C will "behave" as a natural entity and respond similarly to the same physical, chemical, or behavioral stimuli in their environment. Finally, it is also possible that C is a surviving ancestor and will never be found to possess any autapomorphic characters (Figure 5H). Hence, in these latter instances our measure of biodiversity prior to a phylogenetic analysis was not accurate, and our conservation practices employed to manage and protect a species like C may be inadequate.

The most parsimonious phylogeny for this group A–D (Figure 5D) provides valuable information concerning the relative ages of members, their sister-group relationships, and the evolution of the traits that they possess. Using this phylogeny we can state that the speciation event leading to D predated the event leading to C, and both of these events predated the event leading to both A and B. Likewise, A is most closely related to B, C is most closely related to the common ancestor to A plus B, and so on. This tree provides an indispensable test of the homologies of the characters possessed by the species. We accept the hypothesis of homology for all of the characters, except character 10, which is interpreted to have been developed in the two species by convergent or parallel evolution. Finally, with all of the characters superimposed on the phylogeny the rate of evolution of the various characters used in the analysis is easily visualized. In this example there has been relatively uniform evolution in attributes with each cladogenetic event. However, if we were to compare the single phylogeny based on different data sets in our previous example (Figure 4A–C) it is apparent that rates of evolution are unequal both within and among data sets. Applications of these types of data are explored in detail in the following section.

Applications of Phylogenies

Phylogenetic relationships derived through systematic analysis are necessary in several important respects for assessing biodiversity and pursuing conservation biology. Recently, the diverse array of applications stemming from the theories, methods, and products of systematic analyses towards understanding historically derived questions have been explored in detail. Many examples of incorporating phylogenetic hypotheses into comparative investigations are provided by Brooks (1985), Wiley and Mayden (1985), Lauder (1986, 1990), Mayden (1987a, 1992a), Funk and Brooks (1990), Brooks and McLennan (1991), and works cited therein. Here, we provide a brief discussion of and examples relative to (1) phylogenetic relationships, (2) biological classifications, (3) inventories and diversity measures of biotas, (4) patterns of diversity in time and space, (5) patterns and origins in space and time of traits possessed by organisms, both ancestral and recent, and (6) the predictive value of phylogenies. Pertinent applications of phylogenies for modes of speciation, rates of evolution, origins of communities and life history traits, or conservation biology are covered in the previously referenced works and Mayden (1986, 1992b). Without the historical information provided in the phylogenetic hypothesis of the subject organisms, estimates of all of these and related questions are inherently indirect and may be grossly inaccurate.

Phylogenetic Relationships

The most important contribution of systematics to the biological community is delivery of genealogical hypotheses (Figure 1). Other contributions addressed here depend upon the successful determination of sister-group relationships. Phylogenies provide the important historical backdrop or template that allows better understanding of the origin and diversification of biological entities from a species or a group of species upwards to natural supraspecific taxa and communities (Brooks and McLennan 1991; Mayden 1992a). Consequently, understanding the history and processes of diversification is critical to understanding and developing effective programs of promoting biodiversity and its conservation.

Evaluation of biological entities from a historical perspective allows researchers to partition information about contemporary entities into attributes having historical origins and those derived proximally. Attributes include such general features of an organism as morphological structures, coloration, physiology, behavior, ecology, genetics, protein function and structure, mode of reproduction (asexual or sexual), and propensity for entities to alternate generations and, for sexually reproducing entities, to interbreed (Figure 6A). Which qualities of an organism, population, subspecies, species, supraspecific taxon, community, or ecosystem are historically constrained and do not owe their origins to or are not influenced by current environmental conditions? Conversely, which attributes are derived or autapomorphic, and which are influenced by current environmental conditions? Within a community or ecosystem, which taxon associations are "inherited" from ancestral communities where they coevolved, and which associations originated from recently derived invasions via dispersal or extinction? Finally, relative to estimating biodiversity and the application of some species concepts, which associations that involve interbreeding are relevant to falsifying species divergence, and which ones are not? Are the species comparisons or the comparisons of attributes possessed by the species appropriate for such deterministic comparisons?

Without sister-group relationships it is impossible to know if the species observed interbreeding within an ecosystem are (1) descendants from an immediate common ancestor (Figure 6C) or (2) more distantly related (Figure 6D). In the former, the existence of interbreeding is relevant to the diversity question and may suggest incomplete divergence. In the latter the interbreeding is not between closest genealogical relatives, and reproductive isolation, as a measure of completed divergence, is irrelevant. Rather, each of the participating species is more closely related to (and shared a more recent common ancestor with) some other species (ancestral or not) than the one with which it may be currently interbreeding. An effective approach to the discovery and conservation of natural biological entities is entirely dependent upon the successful separation of these types of differences in the temporal evolution of attributes and associations (Mayden 1992b). The eventual survival of a biological entity may be fundamentally linked to historically constrained attributes of their life cycle or some constrained species association (Figure 6A, e.g., taxa B and D). Attributes and species associations inherited from ancestral organisms or ecosystems may be inflexible to modification under current environmental changes at the scale necessary or within the time frame required to survive. Once these classes of features and associations are identified, then protocols used in conservation practices can be developed that will be more efficient and effective at achieving our objectives.

Biological Classifications

Classifications serve as fundamental information retrieval systems or devices that communicate ideas about the inanimate and animate worlds more effectively and efficiently. In biological systems, classifications should be natural (Wiley 1981) and reflect a summary of genealogical relationships of diversity (Figure 2D, G). They are relied upon by many scientists for communicating clearly, devising proper experimental procedures, and conducting informative comparisons. If classifications are inconsistent with the phylogenetic relationships of the organisms being classified, then all those depending upon them as devices reflecting the natural world will be woefully mislead and any resulting comparisons or communications will be flawed (compare naturalness of classifications in Figure 2).

Diversity of Biotas

In large part, inventories of floras and faunas are conducted without the aid of phylogenetic hypotheses of the organisms, populations, or species. With the enormous worldwide diversity in biological communities and the limited availability of phylogenies, this will likely be the status quo for some time. Traditionally, measures of species diversity are conducted without regard to the origins of traits possessed by the organisms as being either apomorphic or plesiomorphic. This is not a condemnation of these studies because it is often impossible to conduct such detailed work in regions where the flora and fauna are poorly known, or, for some taxa, when a phylogenetic perspective may be difficult if not impossible. However, for other regions and taxa, we are better able to differentiate between plesiomorphic and apomorphic attributes, and in these instances a new perspective of diversity is emerging.

The increasing application of phylogenetic perspectives in inventory and diversity studies has unveiled two important developments for biodiversity studies. These include a heightened awareness for critically evaluating (1) the origins of attributes possessed by and considered diagnostic of species or other natural entities, and (2) sister-group relationships of populations or other natural entities hypothesized to represent a species. First, the realization that species possess a combination of both plesiomorphic and apomorphic attributes permits researchers to evaluate and differentiate between

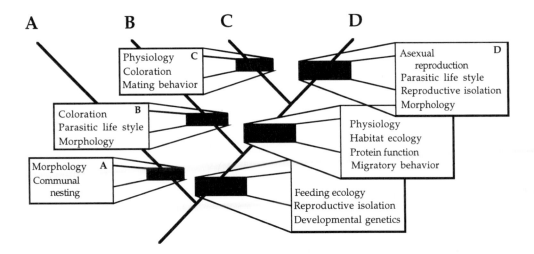

A. Evolution of attributes

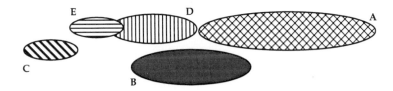

B. Geographic distributions

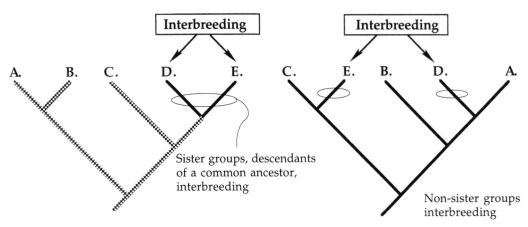

C. Assumed relationships

D. Actual relationships

FIGURE 6.—(**A**) The evolution of all classes of heritable attributes in the history of taxa A–D. (**B**) Geographic distributions of taxa A–E. Note that all species are allopatric except E and D, for which gene exchange has been hypothesized. (**C**) Possible phylogenetic relationship for taxa A–E based on standard a priori assumption of relationship of E and D under concepts enforcing an operational criterion of reproductive isolation of species. (**D**) Actual phylogenetic relationships of taxa A–E. Note that the two taxa hypothesized to be exchanging genes are not closest relatives.

the types of similarity shared by two or more populations. In this sense, populations may appear similar to one another for a particular attribute but may not be closest relatives because the similarities shared are plesiomorphic and are not indicative of a recently shared ancestral history. Likewise, one or more populations may appear intermediate between two purported parental species, subspecies, or ESUs only because the populations in question share plesiomorphic traits with one and apomorphic traits with another. In this case the intermediate populations represent the sister-group to those with which they share the apomorphic features. For example, in Figure 6A taxon C appears intermediate between B and D because it shares apomorphic features with D and plesiomorphic features with B. The origin and identification of these "intermediate" populations, and their eventual conservation, can be understood only within the context of a phylogeny depicting the evolution of the attributes in question (additional discussion follows). Excellent examples of phylogenetically valid, but intermediate-appearing, species represented by populations that were previously hypothesized as intergrades between two subspecies are provided by the minnow genera *Luxilus* (Mayden 1988a; Dowling et al. 1989) and *Notropis* (Wood and Mayden 1992).

The second important development for biodiversity studies is that with an increasing number of phylogenetic hypotheses available on relationships among populations of species once thought to represent natural groups, the documentation of sister-group relationships has exposed more diversity than was previously perceived. Species, subspecies, and ESUs as natural groups should, following verification through phylogenetic analysis of populations, be confirmed as natural groups. That is, all populations or demes of each of these natural entities should be more closely related to one another (or equally related, as in populations of ancestral species) than any of them are to other species, subspecies, or ESUs.

Following analyses of this nature, it is not uncommon for the traditional view of a species to change. Two excellent examples of such transformations involve the spotted darter *Etheostoma maculatum* and the southern studfish *Fundulus stellifer*. Prior to 1989, the spotted darter was thought to include three allopatrically distributed subspecies, *E. m. maculatum*, *E. m. sanguifluum*, and *E. m. vulneratum* (Figure 7A, upper), all sharing largely similar, plesiomorphic morphologies. When Etnier and Williams (1989) described the boulder darter *E.*

wapiti as a new and endangered species, their phylogenetic analysis revealed that the three subspecies of the spotted darter were not closest relatives (Figure 7A, middle). Rather, the wounded darter *E. vulneratum* was found to be most closely related to the boulder darter, and the spotted darter and the bloodfin darter *E. sanguifluum* were more closely related to the coppercheek darter *E. aquali*. Wood (1993) evaluated the relationships of this group by means of protein electrophoresis and resolved patterns of relationship consistent with those of morphology, except this tree provided better resolution and placed the smallscale darter *E. microlepidum* as the sister-group to the *E. maculatum–sanguifluum–aquali* clade (Figure 7A, lower).

Prior to an electrophoretic study by Rogers and Cashner (1987), the southern studfish was thought to consist of a single species (Figure 7B, upper). However, a phylogenetic analysis of this species and other members of the subgenus *Xenisma* revealed that populations largely from the Tallapoosa River drainage are more closely related to the northern studfish *F. catenatus* than to populations of southern studfish from the Coosa and Alabama rivers (Figure 7B, lower). After further evaluation of the complex, morphological characters were discovered to diagnose the Coosa and Alabama river populations of southern studfish from other populations, now referred to as the stippled studfish *F. bifax* (Cashner et al. 1988).

Examples also exist wherein taxonomic changes were made at the level of species without reference to a genealogical hypothesis validating sister-group relationships. The golden trout *Oncorhynchus aguabonita* represents an example of such a change. This species was described as a subspecies of cutthroat trout *O. clarki* and later as a subspecies of the rainbow trout *O. mykiss*. Since then it has variously been considered a subspecies of the rainbow trout or a distinct species (Figure 7C, upper). Without the aid of a phylogenetic hypothesis or phylogenetic interpretations of the shared characters, Behnke (1992:162) considered this taxon to represent a subspecies of the rainbow trout. This conclusion was reached on the basis of gross similarity and apparent discordant character variation, even though "subspecies of the Sacramento and Columbia river basins share cutthroatlike trout characters . . . [and] it is not known if they arose from a single ancestor or if they represent separate branches off a line leading to the more advanced coastal rainbow trout." The confusing status of the attributes of the various forms of the rainbow and cutthroat trouts owes its origin to the lack of a phylogenetic inter-

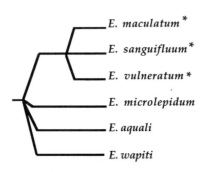

Prephylogenetic hypothesis

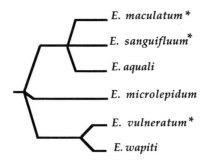

Phylogenetic hypothesis
(Etnier and Williams 1989)

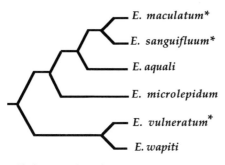

Phylogenetic hypothesis
(Wood 1993)

A. *Etheostoma maculatum* group

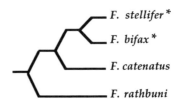

Prephylogenetic hypothesis

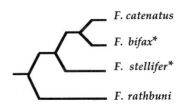

Phylogenetic hypothesis
(Cashner et al. 1988)

B. *Fundulus stellifer* group

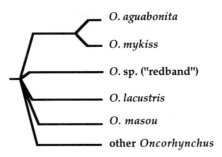

Prephylogenetic hypothesis
(Behnke 1992)

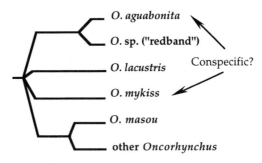

Phylogenetic hypothesis
(Stearley 1992)

C. *Oncorhynchus aguabonita*

FIGURE 7.—Classifications made for some groups of North American fishes with and without the use of genealogical information. (**A**) Relationships of the *Etheostoma maculatum* species group. (**B**) Relationships of some species in the Fundulus subgenus *Xenisma*. (**C**) Relationships of some species of *Oncorhynchus*. Included in the clade are the fossil *O. lacustris* and cherry salmon *O. masou*.

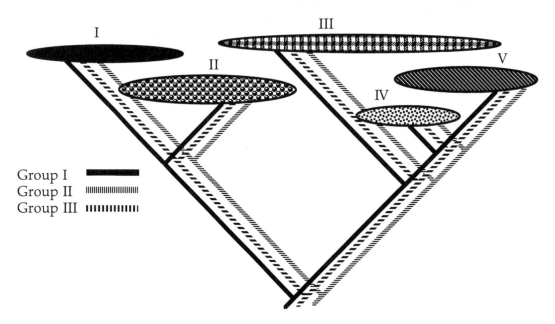

FIGURE 8.—General model of use of phylogenetic information in biogeographic studies of communities. Represented are five (I–V) different communities, ecosystems, or areas of endemism. Groups I–III are three different groups of organisms found in these areas.

pretation in not only the relationships of these entities but also in interpretations of their collective features. Populations of the golden trout possess distinctive qualities that may warrant its recognition as a distinct species. These qualities, however, were not embraced by Behnke (1992) as representative of the "necessary" (but subjectively determined) level of divergence for a species. Tragic in such thinking is that golden trout is not most closely related to the freshwater and anadromous rainbow trout but is most closely related to the freshwater "redband" trout *Oncorhynchus* sp. (Stearley 1992; Figure 7C, lower).

Patterns of Diversity

Phylogenies form the essential ingredients of historical biogeography (Figure 8). Likewise, phylogenetic and biogeographic hypotheses, together, form the central backdrop for investigations into the origins of communities and species associations, ages of speciation events and species associations, and the general evolution of biotas with Earth history (Brooks 1985; Wiley and Mayden 1985; Mayden 1985, 1987a, 1987b, 1988b, 1992a, 1992b; Brooks et al. 1992). Investigations into ecological biogeography evaluate and compare diversities of species,

subspecies, and ESUs from different areas and use these diversity statistics to derive similarities between faunas and floras. However, as with the problems observed when using a phenetic method in systematics and not differentiating between primitive and derived similarities of species, ecological biogeographic studies should not be interpreted as historical. Unlike phylogenetically based historical biogeographic studies, ecological-based biogeographic studies can not address directly the origins of species associations, ages of speciation events, or origins of communities. Through historically based biogeographic analysis the researcher is better able to partition proximal from historical associations within communities. Finally, only through the historically based approach to biogeographic analysis can researchers more efficiently identify species and geographic regions containing the largest proportions of our evolutionary legacy and in most critical need of conservation and protection (Humphries et al. 1991; Vane-Wright et al. 1991; Williams et al. 1991; Brooks et al. 1992). Should future conservation practices become more ecosystem based rather than species based, then the use of phylogenies of different groups of organisms in historical biogeography will be of fundamental importance to this endeavor.

Origins and Patterns of Traits

An equally fundamental application of phylogenies to the mission of conservation and management of biodiversity is understanding the origins of attributes of organisms and their distributions in time and space. Because attributes of organisms are the elements used in the discovery and justification of diversity, it is vital that we understand as much about these traits as possible. To understand the origin of traits in taxa means that we are better able to interpret the origins and significance of the diversity. Features of organisms can be either homologous or homoplasious. Homologous features are types 2 and 3 similarities in two or more species, occurring because the features are inherited through a shared common ancestor (Figure 2). Homoplasious features are types 1 and 4 similarities, occurring through convergent or parallel evolution or through interbreeding. Errors made in the identifications of type 1, 2, 3, or 4 similarities will result in not only misidentification of natural diversity but can lead to the eventual loss of valuable diversity. The confusing and largely unstable taxonomic history of trout systematics, especially those of the rainbow and cutthroat trout complexes, is rooted in this very misunderstanding of character evolution and the confounding of these four types of similarity. This problem and the eventual solution through the application of systematics is demonstrated easily through both simple and complex examples.

As a simple example, we investigate three allopatric metapopulations (Figure 9A, B), two of which (I, II) are separated easily from one another on the basis of the two available morphological features. The third metapopulation (III) is intermediate between metapopulations I and II for these characters, which complicates the picture. A standard interpretation of this pattern is that the mixed array of shared attributes in III represents either previous or current genetic interchange between I and II or clinal variation and that III represents a series of introgressed or intergrade populations. Given this conclusion, neither I nor II likely would be recognized in traditional approaches as distinct species. With the mixed genotypes in III, it would not be a candidate for recognition nor any type of protection and management. However, if the pattern of variation, homology, and evolution of these traits is considered in light of the phylogenetic relationships of the metapopulations, a quite different conclusion is available (Figure 9C, left). That is, given these relationships (Figure 9C, left) metapopulation III is intermediate between I and II because it retains some ancestral features (morphology 1) with I, yet shares some derived features (morphology 2) with II. At a larger scale, it is also possible that the three metapopulations examined are not even closest relatives, and other metapopulations (or species) are equally appropriate in evaluating the status of I, II, and III (Figure 9C, right). In either case, the origins and distributions of these homologues is consistent with the evolutionary history of at least two speciation events producing I, II, and III without any degree of "inter-metapopulation" gene flow.

Another simple example involves the origin of a single feature across all three metapopulations (Figure 9D). However, in this instance, metapopulation III is polymorphic for the conditions considered diagnostic for I and II. Again, when presented with this particular pattern of variation for a homologous character, the standard assumption is that I and II are closest relatives, have not diverged sufficiently, and have produced III through introgression or intergradation (Figure 9E). This assumption is without any appraisal of the phylogenetic relationships of the group. If these three metapopulations are closest relatives (Figure 9F, left), the origin of the polymorphism can be explained simply as the evolution of the "B" condition in the shared ancestor to II and III, followed by the fixation of the "B" condition in metapopulation II. Metapopulation III is polymorphic for "A" and "B" because it retains the type 2 similarity ("A") from the common ancestor yet diverged from an immediate common ancestor where the type 3 similarity ("B") first evolved. This ancestral polymorphism (retained from the common ancestor to II and III) may be maintained in metapopulation III because of environmental or evolutionary stasis. Another alternative is that the three metapopulations are not closest relatives (Figure 9F, right), and other entities that may be distinct species (e.g., VII) share the polymorphism. In both instances the intermediate genotype or phenotype was initially assumed to have a proximal explanation, that is, the intermediate populations result from ongoing gene exchange. However, their existence is a historical phenomenon of simple character evolution in entities that have not exchanged genes for millions of years!

In a more complicated example, we examine the status of four metapopulations for five character types (Figure 10A, B). In this case, metapopulations I and IV are quite distinctive from one another and are diagnosable. However, when one considers metapopulations II and III, the distinctiveness of I and IV is blurred by the sharing in these populations of the traits considered diagnostic for I and IV (Figure 10B). A standard, ahistorical explanation of this pattern of character variation (Figure 10C) is that II and III represent intergrade or hybrid metapopulations derived singly or independently from the interactions between I and IV. Alternatively, one might interpret this variation to mean that II and III inhabit an ecologically intermediate environment relative to I and IV and represent intermediates of a cline, a situation derived through either primary or secondary contact. Again, this seemingly convincing example of character variation involving incomplete speciation has a drastically different interpretation when viewed from a perspective cognizant of the different types of character similarities and their origins (Figure 10D). From a phylogenetic perspective, the intermediate status of metapopulations II and III is only an illusion, induced from a lack of familiarity with the four types of similarities and confusion of polymorphic conditions with type 4 similarities. The retention of a plesiomorphic condition (from ancestors or descendants of these ancestors) together with possession of a more recently evolved homologue in a polymorphic state could be confused easily with an intergrade or clinal origin for such populations when one operates in a world indifferent to phylogenetic history (Figure 10C). However, if the origins of traits are placed in their proper phylogenetic order by the use of outgroups, then the four types of similarities among attributes can be partitioned accordingly, and we are successful in accurately identifying natural biological entities for conservation and management (Figure 10D).

Predictive Value

What if the general biological habits of an endangered taxon are unknown and the taxon is so rare we are incapable of understanding its biology and offering corrective measures before it goes extinct? This is not an unusual or unexpected situation, especially because society lags in even accounting for biological diversity let alone understanding all of its biology. What if we are mandated to identify and prioritize species or communities for preservation? This, too, is not unlikely because worldwide population growth remains largely unchecked, and we are being forced to make such decisions. Whereas decisions can and are being made in an ahistorical context, informed decisions cannot be made in an ahistorical context. Before we can understand how to protect biological diversity successfully and make important decisions regarding maintenance of critical elements of biological diversity, we must have understanding of historical processes that produced the currently observed patterns. The predictive value of sister-group relationships of taxa and their geographic distributions offers the most effective avenue for the pursuit of these goals.

A phylogenetic hypothesis of the endangered taxon and close relatives provides strong predictive potential for the conservation biologist if life history aspects of close relatives are known or partially known (Mayden 1992b). Because many attributes of organisms are historically constrained and exist in species due to inheritance from common ancestors, the biological qualities of close relatives may be used to predict the qualities necessary for the survival of imperiled taxa. Though this may seem like pure guess work, it is not. Not all attributes of species exist in their present form because of functions finely honed to an existing environment. Most attributes exist in species because their ancestors possessed them. This general predictability is used effectively and efficiently in the worldwide propagation of some fish species. A testimony to the signif-

FIGURE 9.—Different interpretations, or misinterpretations, for the distributions of attributes across different metapopulations of species. (A) Two different morphological characters in three different metapopulations. Note that without a phylogenetic perspective the discordant variability of the two characters in metapopulation III is interpreted as intergradation between metapopulations I and II. (B) Distributions of the three metapopulations (I, II, and III) and close relatives (IV and V). (C) Two possible phylogenetic resolutions of the three metapopulations, and phylogenetic interpretations for the evolution of the morphological characters. Left tree involves only the three metapopulations. Right tree includes all five metapopulations. Note that these are two different resolutions and character interpretations. (D) A single morphological character observed for three metapopulations. Note that in this example metapopulation III possesses the attributes found in both I and II, and without a phylogenetic perspective this variation is interpreted as intergradation between metapopulations I and II. (E) Standard assumptions invoking processes to explain the origin of attributes occurring in III. Note that no ancestral conditions are known for this example, and III is interpreted to be the result of genetic exchange or its occurrence in an ecologically intermediate environment. (F) Two possible phylogenetic resolutions and the metapopulations and phylogenetic interpretations for the evolution of the morphological character. Left tree involves only the three metapopulations. Right tree includes all eight metapopulations. Note that these are two different resolutions and character interpretations.

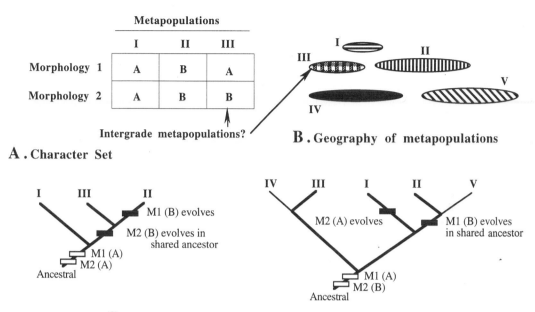

A. Character Set

B. Geography of metapopulations

C. Possible relationships among metapopulations

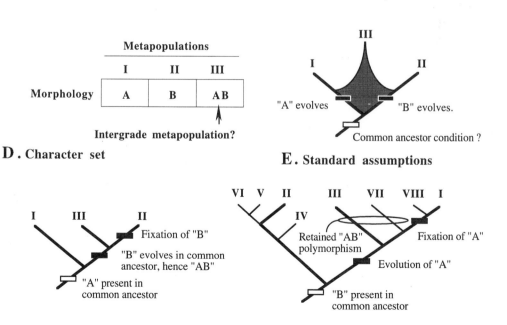

D. Character set

E. Standard assumptions

F. Possible relationships among metapopulations

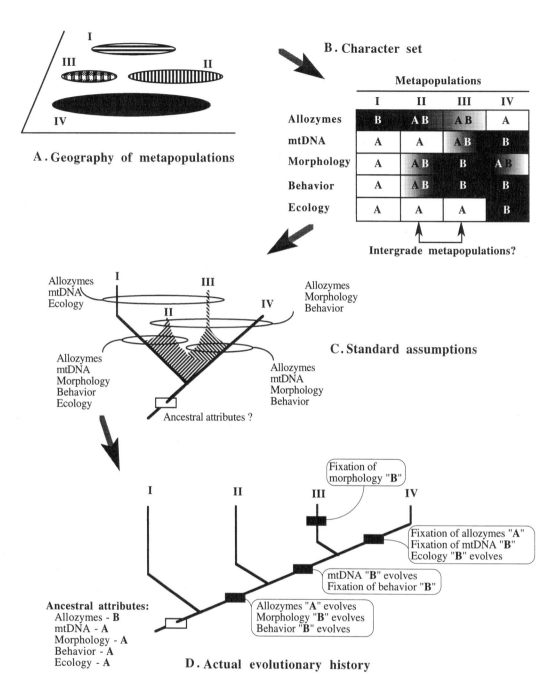

FIGURE 10.—Different interpretations, or misinterpretations, for the distributions of attributes across different metapopulations of species. (**A**) Distributions of metapopulations I–IV. (**B**) Attributes from an array of character types distributed across metapopulations (mtDNA is mitochondrial DNA). (**C**) Standard assumptions invoking process for the origin of the mixed attributes in metapopulations II and III relative to I and IV. Note that no ancestral conditions are known. Some attributes are shared between various combinations of I–IV, but in a nonhierarchical manner. These shared attributes are interpreted as evidence for genetic exchange or the occurrence of II and III in an ecologically intermediate environment. (**D**) Actual phylogenetic history of metapopulations I–IV based on optimizing the evolution of attributes. Note that the once discordant variation is absent when the evolution of the traits is evaluated with respect to the phylogeny of the organisms. No genetic exchange need be hypothesized.

icance of historically constrained and predictable biologies is the observation that many North American game species are imported worldwide as non-indigenous food or game sources, in lieu of costs and time associated with maximizing harvest of native species. To use hitherto unexploited species for management would necessarily require considerable primary research into details of their life cycles to maximize their harvest. In the same vein, recovery plans and protective measures of imperiled taxa should incorporate information gained from phylogenetic hypotheses about them. Biological information gained from closest relatives can be used to evaluate critical biological properties needed for the recovery of imperiled taxa (Brooks et al. 1992).

Knowledge of geographic distributions and the phylogenetic hypotheses of multiple species inhabiting expansive regions also is critically important to prioritizing species, communities, and ecosystems for preservation (Brooks et al. 1992). The most important elements provided through phylogenetic hypotheses in these decision-making processes involves the origins of historical constraints, keystone species, and species associations; the evolutionary potential of species or communities; and diversity indices as measures of importance of communities. Though genealogical information of taxa is critical to making such decisions, information from other areas of comparative biology about the organisms (e.g., ecology, physiology, and behavior) is equally essential to successful conservation.

A hypothetical example involving aquatic communities of the Pacific Northwest provides a worthy heuristic device that imparts the significance and efficiency of basing such concerns and decisions on phylogenetic hypotheses. It is likely that the future of this aquatic biota will involve prioritization for conserving aquatic communities in specific waterways. Should these priorities be based on estimates of species diversity; areas of highest endemism or genetic, ecological, and behavioral diversity; or unique combinations of species associations? Alternatively, do we preserve all communities because it is too difficult to decide which ones are most important? Making these types of decisions with any type of data will never be easy. However, it is less difficult in a more biologically informed and predictive manner if genealogical and biogeographic histories of the communities and species are used.

As an example, we use a series of aquatic communities inhabiting parallel rivers draining into the Pacific Ocean (Figure 11A). We know the number of species or ESUs occurring in these rivers and aspects of their autecologies and synecologies. With only this information, we are left with the argument to preserve communities on the basis of diversity (indices) and some unique biologies. In order to minimize error in prioritizing communities, we may have to argue for complete preservation. However, let us focus on just three hypothetical species occurring in these communities and superimpose a phylogenetic perspective. These results can be extrapolated to larger communities. As general background information, we already have reviewed the nature of parsimony argumentation and the evolution of homologous traits. Furthermore, it is known that most divergence (species or subspecific levels) occurs in allopatry and that this divergence is commonly replicated by numerous groups of organisms, producing replicated biogeographic patterns (Wiley and Mayden 1985; Lynch 1989). We know from biological investigations that two (I and II) of the three species from this example are anadromous and the third (III) is entirely freshwater, perhaps potandromous. The two anadromous species have three seasonal entities or run times, in ascending the rivers to spawn, much like many forms of *Oncorhynchus* spp. currently occupying these rivers. We know of other behavioral variables for these two species and of physiological and ecological variability in the freshwater species.

Our initial hypothesis is that for species I and II the different seasonal runs have evolved independently within each river community (Figure 11B). As hypothesized for species of *Oncorhynchus* (Waples 1991; Behnke 1992), this presumably occurred in order to maximize the use of available habitats in time and space and to serve as genetic reservoirs of each species should conditions for the individuals of one or more runs in a river be eliminated. This hypothesis suggests that each entity is equally important for each river community to ensure the future of the species. Furthermore, this hypothesis necessitates that the different run forms of each river are sister-groups and the different run times have evolved independently in each river due to similar environmental conditions (type 1 similarities). Does this hypothesis survive testing using phylogenetic methods, and how can phylogenetic methods assist in evaluating these communities for future preservation?

Phylogenetic analysis of populations of each species indicate that those of each run time form monophyletic groups and that the spring and summer runs form sister-groups, sister to those of the fall run (Figure 11C). Thus, this falsifies our initial hypothesis that the different run times evolved independently in each river system (Figure 11B).

Consequently, the similarities of different run times across communities are not type 1 similarities but are shared similarities of either type 2 or 3. Phylogenetic resolution for populations within each of the different run times for each species also provides valuable information regarding historical constraints within each run time (Figure 11D, E). Replicated patterns of relationships exist for both of these species. The different run-time populations from communities A, B, and C form monophyletic groups, sister to monophyletic groups inclusive of those from communities D–G (Figure 11D, E). The polychotomous resolution for all run-time populations of either species in communities A, B, and C suggest that there are no apparent historical constraints for these communities, and they are all biologically equivalent. However, for species I, fall-run populations from communities D and E form a monophyletic group and the spring-run populations from communities F and G form a monophyletic group. Remaining spring- and fall-run populations form polychotomous relationships within the D–G communities, and no phylogenetic resolution within these communities exists for any summer-run populations (Figure 11D, E). These populations are apparently biologically equivalent. For species II, the only apparent historical constraints exist for the spring-run populations in communities F and G, which form a monophyletic group. All other relationships of the different run times in species II for communities D–G are unresolved; here they are all assumed to be biologically equivalent, given existing data. For the freshwater species III, populations from communities A–C form a monophyletic group, replicating the pattern observed for species I and II (Figure 11F). A polychotomous phylogenetic history is found within A–C and between the A–C group relative to communities from D–F, suggesting that populations from A–C and from D–F are different but are biologically equivalent within each group.

A summary for the three species in shared communities provides extremely valuable information towards understanding and prioritizing communities and species for conservation (Figure 11G). These data dramatically alter conclusions that would have been derived based on our initial hypothesis for the origins of the diversity in these communities (Figure 11B). Using our initial hypothesis, we would conclude that because each of the different runs evolved independently in each community for each species, they are adapted independently to the ecosystems of particular river systems. Prioritization of communities for conservation, given this assumption, is impossible because each community has uniquely evolved biological entities and is equally valuable. The phylogenetic backdrop, however, allows us to identify directly communities with unique evolutionary entities through their possession of derived attributes and historical constraints for different run times, populations, and species (Figure 11G). From this we can hypothesize that communities A, B, and C (a), communities D and E (b), and communities F and G (c) are biologically unique from one another; however, within each of A, B, and C or D and E or F and G, the communities are biologically equivalent. An informed decision is possible and can be made about the future conservation of communities A–G. If we must choose between communities, we may choose within the biologically equivalent "a's," "b's," or "c's," but would not sacrifice any "a" over any "b" or "c," etc., because of their documented evolutionary uniqueness and likely irreplaceability.

The superimposition of additional biological information onto this example may or may not complicate the decision process. Many other aspects of communities should be included in these consider-

---

FIGURE 11.—Hypothetical example of employing phylogenetic methods in biodiversity and conservation. (**A**) Seven aquatic communities in parallel river systems along the Pacific Northwest. Numbers in elipses represents number of taxa native to each community. (**B**) A prior assumption of evolutionary relationships among the spring, summer, and fall run-time entities known for the anadromous species I and II. Note that in each of the seven communities the different run-time entities are assumed to be closest relatives (i.e., descended from a common ancestor within each community). (**C**) Actual phylogenetic analysis for the different run-time entities for species I and II. Note that these analyses are consistent with one another and argue that within each species the entities of each of the run times are more closely related to one another than any of them are to members of other run times. (**D, E**) Phylogenetic analysis of the different run-time populations in the seven communities for the anadromous species I and II, respectively. (**F**) Phylogenetic analysis of populations in the seven communities for freshwater species III. (**G**) Summary of the phylogenetic information for species I, II, and III regarding equivalance comparisons across these biological communities. Solid vertical lines are hypotheses of equivalent biological entities within each species. Dashed lines indicate historical differences between communities, indicating that they are not equivalent biological entities. Right column is summary of equivalent biological communties based the phylogenetic information recovered for species I, II, and III; heterogeneity of letters indicates distinct groupings of biological communities.

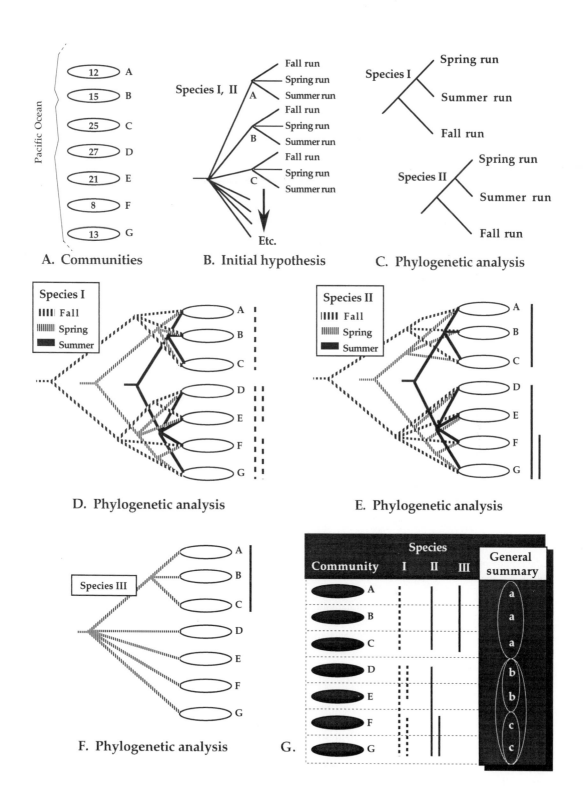

A. Communities

B. Initial hypothesis

C. Phylogenetic analysis

D. Phylogenetic analysis

E. Phylogenetic analysis

F. Phylogenetic analysis

G.

ations but only in conjunction with phylogenetic information. Information gained about species associations, food webs, and keystone species may be important. Within the "biologically equivalent" communities identified as either "a," "b," or "c," there may be unique associations of taxa that are consequential to the structure and function of a community. If such is the case, then these communities may appear equivalent and interchangeable at the phylogenetic level, but they are not at a functional ecological level. For example, if among the A–C communities a unique functional ecological relationship exists in A but not in B or C, then priorities would be shifted to A and between B and C.

Towards a Natural Species Concept

The species concept is crucial to the study of biodiversity. . . . Not to have a natural unit such as the species would be to abandon a large part of biology into free fall, all the way from the ecosystem down to the organism. (Wilson 1992:37–38.)

Biological concepts serve a fundamental purpose in science as bridges between patterns observed in nature and our discovery of processes responsible for their production and maintenance. Our conceptualization of "things" existing in nature involves not only our thoughts or ideas about the things but also words or terms composed to communicate effectively our understanding of the things in this world. Thus, a concept should be viewed as a general characterization or description of our thoughts about things in nature. With respect to the natural world, a concept designed to communicate ideas or thoughts can both adequately and accurately reflect natural things. Alternatively, a concept can be inadequate and inaccurate for two reasons. First, the idea about what forms a natural thing can be inadequate for the natural things, and a faulty concept ensues. Second, the ideas or perceptions about what forms a natural thing can be accurate and adequate, but the words or terms used in describing these things in a concept can be inadequate or inaccurate, and a faulty concept ensues. In both cases, our perception or description of species as things occurring in nature do not adequately reflect natural biological diversity and order.

Several concepts exist ostensibly delimiting the species as a taxon. Historically, two standard concepts were employed, (1) typological (or morphological, essentialist) and (2) nominalistic (Mayr and Ashlock 1991). These concepts were abandoned because they are theoretically unsound and do not accurately reflect biological reality. The typological concept relies essentially on the perception that there are a limited number of "universals," that species have limited variation in attributes and are constant through time, and that they differ from one another by sharp discontinuities. This concept is inconsistent with current knowledge of biological systems and thus is rejected as a conceptual bridge between pattern and process. Most importantly, this concept relies upon a particular (but inherently subjective) degree of difference as a criterion for the existence and recognition of biological diversity termed species. Given life cycle variations, sexual dimorphism, and existence of sibling species (species by and large appearing identical morphologically) in nature, this concept must be rejected. As stated by Mayr and Ashlock (1991:25), "Degree of difference thus cannot be considered a decisive criterion in the ranking of taxa as species." The nominalistic concept holds that in nature only individual organisms are produced and have reality, species are only human constructs for convenience of communication and have no reality. This concept is rejected because of overarching evidence that not only do parent–offspring connections exist in nature but so do phylogenetic connections between an ancestor and descendants.

There are four contemporary species concepts employed in research concerned with recovering biodiversity. These are the biological, phylogenetic, recognition, and evolutionary species concepts. These concepts vary in their age of origin and are derived from different sets of working hypotheses about the production and maintenance of biodiversity, some relying more than others upon scientific theory, logic, and current empirical observations of natural populations. Therefore, each concept may hold quite different consequences for our earlier listed objectives, "the discovery, understanding, use, and conservation of natural biological entities." More recently, the ESU was proposed as a conceptual basis to identify distinct populations within species for the preservation of their evolutionary legacy.

It is true that biodiversity exists, that nature has produced this diversity through descent with modification, and that *Homo sapiens* as a species is, among other things, morally obligated as a steward of the world to discover and maintain these natural systems. The implications of employing a particular species concept in biodiversity endeavors are clear. When we measure and study biotic diversity through species as taxa, the particular species concept employed, together with all of its theoretical implications, will have a profound effect on our ability to realize this obligation and achieve our objectives. If the species concept adopted is incapable of accurately identifying natural diversity then we will have failed in both our objectives and obligations by failing to recover accurately the essential products of descent. Thus, before discussing the latter four contemporary species concepts and briefly evaluating each, we will first examine theoretical and empirical observations about biodiversity.

Scale of Focus

The phrase "descent with modification" was popularized by Charles Darwin's 1859 thesis on evolution. This single phrase aptly communicates the concept through which the production of diversity is currently understood. Descent implies a unique ancestor–descendant pattern of history. Modification implies that attributes of organisms are altered during descent. Important in this phrase is that no specific mechanisms, timing, quantities, or any other particular aspects are specified as to what is being modified during this descent.

Homo sapiens is a vision-oriented species. By and large, our evaluations of biodiversity employ technologies making the best of this attribute. That is, we rely almost exclusively upon external morphological qualities that we visually perceive to document the reality of biodiversity. Interestingly, at one time our limits of scale were those the investigator could observe with a hand lens. Today, we use advanced technologies that permit not only easy recognition of morphology but the selection of scale for evaluating morphological variability and divergence from coarse to fine. Concomitant with the technological advancement in perception of morphological attributes is phenomenal development of technologies permitting investigators to explore, among others, behavioral, chemical, genetic, and physiological variability and divergences in nature. The question then becomes, because we possess these enhanced technologies for discovering natural variation, does this mean that the diversity identified by any of these latter classes of traits is of any less or more value in natural systems than that identified solely through morphological attributes which were once identified with only a hand lens? The answer is obviously a resounding no!

It is now abundantly clear that validly recognized species vary with respect to the types of traits and the amount of change in these traits that occur in anagenesis and speciation, or descent. In the descent of a species not all of its attributes are expected to be modified, and this logically includes attributes that may not be all that important in its unique life cycle, including its morphology. Sibling species serve as perfect falsifiers of a concept holding an amount of morphological divergence as the criterion of species or history of speciation; asexual and parasexual species serve as falsifiers of a concept holding sexual reproduction as the threshold. Thus, not all species should be expected to differ from one another for either allozymes or DNA sequences that we are currently able to visualize. While some features are modified in the origin of species others remain the same and exist as retained primitive features, or symplesiomorphies, in descendant taxa. Which features become apomorphies and which remain plesiomorphies in descendants is not constant, understood, or predictable. The randomness of evolution predicts it is an unpredictable process.

Thus, in our conceptualization of species and biodiversity, the adoption of an unnatural, limited worldview of acceptable, natural entities qualifying as the fundamental units of biodiversity is hopelessly damaging to our ability to achieve our earlier identified objectives. The problems directly associated with a faulty philosophy and inadequate conceptualization of natural biological systems can be illustrated easily in a simple heuristic example based on fishes and variable dimensions.

Dimensionality

To begin this example, imagine that we live in one-dimensional space. If we compare a sample of fishes from this world, they would all appear the same (Figure 12A; Table 1) (except for their positions in the figure to illustrate the point). In this world our estimate of diversity is one entity, say one species or ESU.

If we enhance our perception of the world and move into a two-dimensional world we add additional information, in this case the length of the fishes (Figure 12B). With this additional resolution we now see two different entities existing in the same world that we once thought included only one entity. If we add a third dimension (Figure 12C), we see that the shape of the fishes informs us of four different entities, estimated by four different body shapes. In a four-dimensional world (Figure 12D), where we can visualize color, we estimate at least eight different entities. Four new entities are added to our estimate of biodiversity that were formerly cryptically concealed within the other entities. In a five-dimensional world (Figure 12E), where we in-

MAYDEN AND WOOD

TABLE 1.—Common and scientific names of fishes referred to in Figures 12 and 14.

Common name	Scientific name
Bayou darter	*Etheostoma rubrum*
Bluebreast darter	*E. camurum*
Coppercheek darter	*E. aquali*
Greenbreast darter	*E. jordani*
Greenfin darter	*E. chlorobranchium*
Highback chub	*Notropis hypsinotus*
Longnose shiner	*N. longirostris*
Orangefin darter	*E. bellum*
Orangefin shiner	*N. ammophilus*
Redline darter	*E. rufilineatum*
Rough shiner	*N. baileyi*
Sabine shiner	*N. sabinae*
Sharphead darter	*E. acuticeps*
Smallscale darter	*E. microlepidum*
Spotfin chub	*Cyprinella monacha*
Spotted darter	*E. maculatum*
Tippecanoe darter	*E. tippecanoe*
Wounded darter	*E. vulneratum*
Yellowcheek darter	*E. moorei*
Yellowfin shiner	*N. lutipinnis*
Yoke darter	*E. juliae*

corporate and can perhaps visualize their genetics, we estimate at least nine different entities. In a six-dimensional world (Figure 12F), where we incorporate and visualize the ecologies of species, we estimate at least 10 different entities. Within this dimension, a new entity can be identified by the existence of a novel ecology, distinct from the primitive ecology shared by all of the other entities.

As our available resolution and our resources increase (dimensions) the cumulative number of entities (e.g., species or ESUs) may also increase (Figure 13A). However, it is not always true that with increasing resolution and additional resources there is concomitant increase in diversity. For some groups of organisms there may be stasis for attributes during and following (or between) speciation events (Figure 13B). Here modifications occur only within a few dimensions, presumably with respect to mate-recognition systems. At the other end of the spectrum, there are various magnitudes of anagenetic change for multiple suites of dimensions that can be modified during descent. However, this increase or rate of change need not be linear. In reality it is impossible to discover a mathematical function that would predict with any certainty the number of distinct entities given a set of available dimensions (attributes). That is, modification of any particular set of attributes is unpredictable during descent and uncorrelated (except for genetic and phenotypic linkages). We also cannot necessarily predict the level of differentiation that will exist for any given dimension (Figure 13C). However, it is

logically invalid to conclude that the lack of divergence at one dimension (set of attributes) necessarily negates the reality of diversity at another dimension (Figures 12, 13).

That divergence of biological entities occurs for multiple dimensions (attributes) in an unpredictable fashion during the descent of entities (species) is best understood and exemplified through the evaluation of both entities and their attributes within a phylogenetic perspective. Phylogenetic systematics, in combination with a logical and holistic philosophy of biological entities, provides a conducive methodology and philosophy to recover natural patterns of divergence at multiple levels. As exemplified in sections on application of systematic methods, comparative data taken from biological entities may appear conflicting and confusing and result in inaccurate accounts of diversity without a phylogenetic approach. Yet, when rooted in the history of the organisms and their attributes, and when accounting for different rates of change for attributes at multiple dimensions, these same data can be completely congruent. Now let us look at reality.

Darters of the Subgenus Nothonotus.—For an example we examine darters of the eastern North American genus *Etheostoma*, subgenus *Nothonotus* (Table 1). The subgenus was examined phylogenetically and is a monophyletic group (Wood 1993). This group of fishes is one of the best examples with which to examine diversity and conservation issues because considerable data are available documenting attributes of their morphology, genetics, ecology, and reproductive behavior. Currently, 19 species are recognized within the subgenus, all on the basis of morphological attributes (Figure 14A; Wood 1993; Wood and Mayden 1993). Wood (1993) examined species relationships within this group using protein electrophoresis. The ecology and reproductive behavior of these fishes were studied by numerous authors and much of this is summarized in Page (1983, 1985), Robison and Buchanan (1988), and Etnier and Starnes (1993).

We look at diversity within this group with the understanding that we can visualize variability in attributes associated with their ecologies, genetics, and reproductive behaviors, as with morphological attributes. If we used the general habitat of these fishes to measure biodiversity, we would recognize only one species. All of *Nothonotus* inhabit similar types of habitats in fast-flowing rivers and streams. Examining this group at only the dimension of genetic variability (no morphology, ecology, or reproductive behavior), we would recognize only five

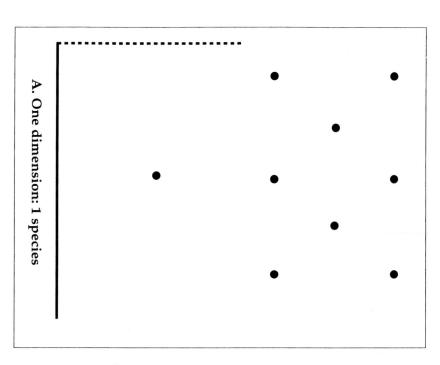

A. One dimension: 1 species

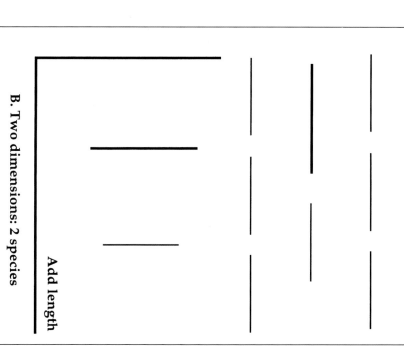

B. Two dimensions: 2 species

Add length

FIGURE 12.—Heuristic example demonstrating the importance of perceiving the reality of species diversity (numbers of species) as including all types of heritable attributes. (**A**) One-dimensional world where everything looks the same; diversity equals 1 species. (**B**) Two-dimensional world, adding length; diversity equals 2 species. (**C**) Three-dimensional world, adding depth; diversity equals 4 species. (**D**) Four-dimensional world, adding color; diversity equals 8 species. (**E**) Five-dimensional world, adding genetics; diversity equals 9 species. (**F**) Six-dimensional world, adding ecology; diversity equals 10 species. Different dimensions in this example are equivalent to different types of attributes. Drawings used by permission of Houghton-Mifflin Co. All rights reserved.

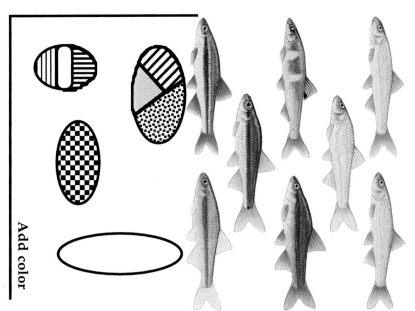

C. Three dimensions: 4 species

Add shape

D. Four dimensions: 8 species

Add color

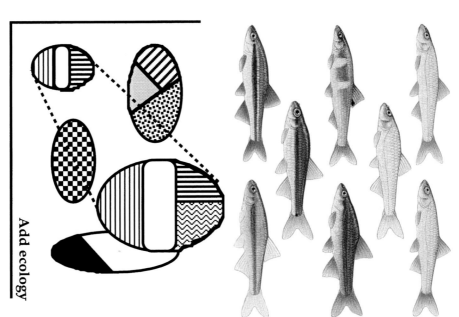

E. Five dimensions: 9 species   Add genetics

F. Six dimensions: 10 species   Add ecology

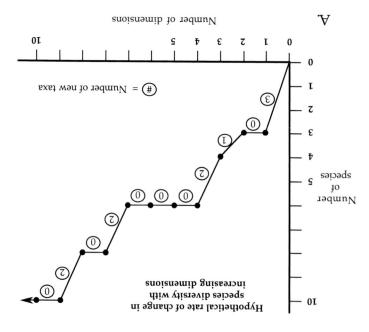

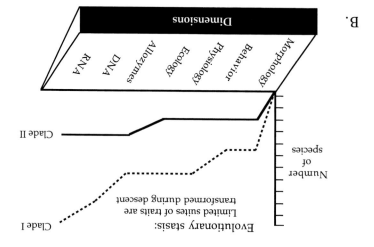

FIGURE 13.—Examples of different types of character evolution in the descent of a monophyletic group. (**A**) Increase in the number of dimensions (types of attributes) may or may not increase the cumulative number of species recognized. Numbers in circles represent the number of species within a monophyletic group added by incorporating a new dimension. (**B**) Example of general evolutionary stasis for multiple suites of attributes in the evolution of two monophyletic groups. Note that one cannot predict from one dimension to another, nor from one group to another for a given dimension, the amount of evolutionary change that will be observed. (**C**) Example of general anagenetic change for a monophyletic group. Note that in this example during the evolution of this group there has been change for nearly all of the different dimensions, but that the rate of change for any dimension within the group is variable. As in (**B**) one cannot predict from one group to another the amount of change for a particular dimension that may occur during descent.

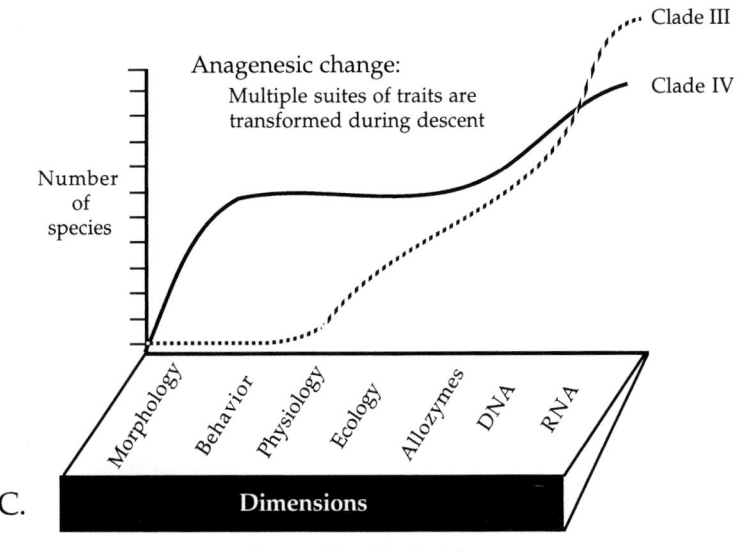

FIGURE 13.—Continued.

species on the basis of fixed allelic differences (Figure 14B). If unique and fixed alleles constituted a criterion for species biodiversity, then we would recognize only 10 species (Figure 14C). Finally, examining only the dimension of reproductive behavior we would recognize only two species, each representing one of two modes of reproduction in this group (Figure 14D; reproductive mode unknown for the greenfin darter). One group of species are substrate spawners, depositing eggs in the gravel substrate without parental care, and another group are egg clumpers, depositing eggs in clumps beneath large stones and tending their nests (Page 1985).

Obviously, none of these dimensions are solely capable of capturing and identifying the totality of the natural diversity within *Nothonotus*, as produced from descent with modification for the various dimensions examined. In fact, for each of these dimensions (or data sets) taken alone, including morphology, there are hidden, naturally occurring biological entities that are best represented as species. Without all of the dimensions taken together some species will necessarily be excluded artificially from our inventory and our efforts to conserve them. Furthermore, at first glance, comparison of the individual data sets for their ability to recover diversity within *Nothonotus* reveals an apparent conflicting and incongruous distribution of variability. The distribution of reproductive ecologies are incongruent with fixed genetic variability, and these are incongruent with morphological variability.

However, this is a twofold artifact. First, there have been variable and unpredictable rates of anagenesis across the different data sets during the descent of the group. Second, the observer does not possess a historical perspective for the group and insight into the evolution of the attributes themselves. With a historical perspective the variable rates of change in the different data types becomes obvious, and these different data sets are internally consistent with the descent of species (Wood 1993).

Thus, our sole reliance upon any one of the data sets in this example ultimately results in ignorance and nonrecognition and perhaps loss of biological diversity. A fixation on morphological divergence (or any single set of attributes) as a criterion for achieving the status of species is fundamentally flawed. This reflects poorly upon our abilities to perceive natural patterns and concomitantly our abilities to understand the processes responsible for them. Variable rates of anagenesis is not a phenomenon limited to a few obscure groups but is widespread. Among the 50 freshwater fish families recorded from North America, 15 contain geographically restricted organisms or organisms possessing notable evolutionary divergence but not necessarily on a morphological dimension (Table 2). Families containing this "hidden" diversity account for 871 of the 1,061 known species (Burr and Mayden 1992), or 82% of the fauna. The subgenus *Nothonotus* contains entities that are currently behaving as species but involve sibling or cryptic species. However, it should be clearly understood that

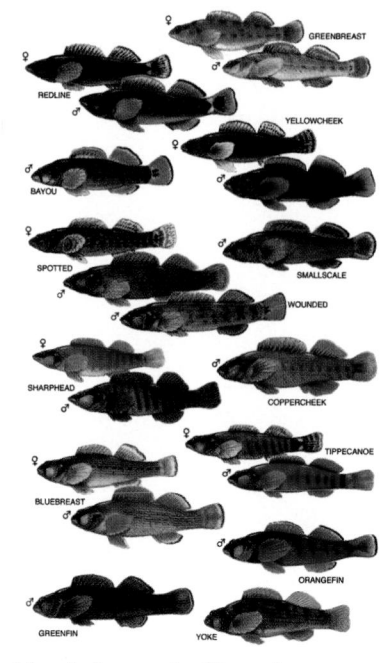

A. Morphology only: 19 species

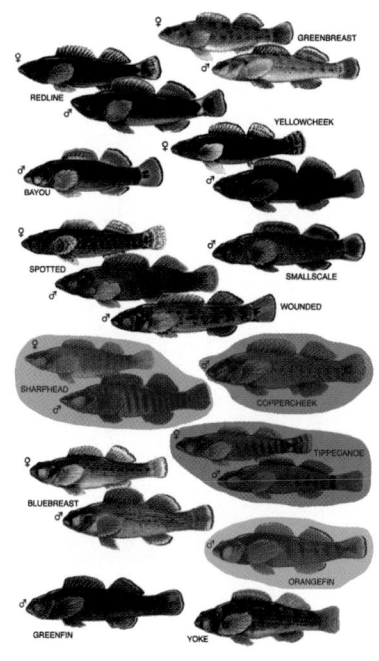

B. Fixed alleles only: 5 species

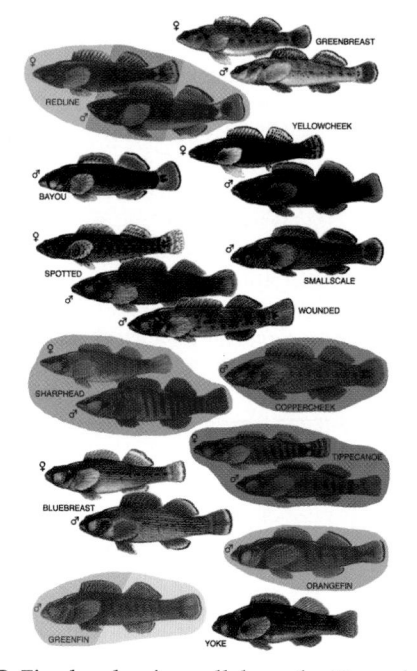

C. Fixed and unique alleles only: 10 species

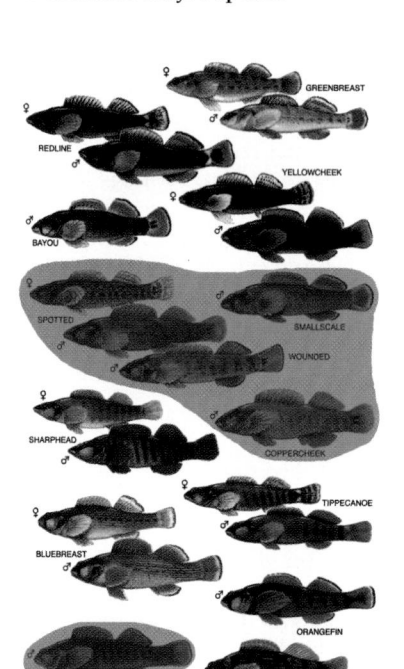

D. Reproductive behavior only: 2 species

FIGURE 14.—Diversity within the darter subgenus *Nothonotus*, genus *Etheostoma* (Percidae) based on four different data sets. (**A**) Morphological data reveals 19 validly recognized species (not all species illustrated here). (**B**) Fixed genetic differences between species reveals only five distinct taxa, some not recognized by morphological data presented in (**A**). (**C**) Unique or fixed genetic differences reveal 10 taxa, some not recognized by morphological data presented in (**A**). (**D**) Reproductive behavior variability reveals only two taxa. Drawings used by permission of Houghton-Mifflin Co. All rights reserved.

TABLE 2.—North American fish families that contain cryptic species. These 15 families contain cryptic species with minimal to no morphological divergence, and together their diversities account for 82% of the fauna.

Family	Species
Cyprinidae	Chubs, minnows, and shiners
Catostomidae	Suckers
Ictaluridae	Bullhead catfishes
Esocidae	Pikes
Salmonidae	Salmons, trouts, and whitefishes
Amblyopsidae	Cavefishes
Fundulidae	Topminnows[a]
Atherinidae	Silversides
Poeciliidae	Livebearers
Gasterosteidae	Sticklebacks
Cottidae	Sculpins
Centrarchidae	Sunfishes
Elassomatidae	Pygmy sunfishes[b]
Percidae	Darters and perches
Cichlidae	Cichlids

[a]Part of family Cyprinodontidae as delineated by Robins et al. (1991).
[b]Part of family Centrarchidae as delineated by Robins et al. (1991).

species are only cryptic if an observer's perception of reality is constrained. If the observer perceives reality as existing at multiple dimensions of attributes possessed by organisms, all of which can be subject to modification during descent, then there are no cryptic species in *Nothonotus*, no loss of information, and no diversity excluded from consideration. This scenario involves real data and organisms and demonstrates potential pitfalls in understanding biological diversity. This need not be the case for present and future generations concerned with biodiversity. Rather, we can accurately identify naturally occurring biological entities behaving as species if we use both a systematic perspective and a conceptualization of species capable of recovering the diversity produced through descent with modification. We explore the conceptualization of species in the following sections.

Qualities of Species Concepts

Given the universal issues in the recognition of natural biological diversity, what types of qualities should we look for in a conceptualization of species and natural biological entities that will enable us to meet our desired objectives? We envision at least five basic qualities that are undesirable (poor) and at least five qualities that are desirable (good) in a concept of species.

Poor qualities.—Any concept of species with any of the following qualities will handicap our efforts in studies of biodiversity and will neither reflect natural biological patterns nor elucidate process.

1. Concepts that knowingly exclude unique, historically derived biological entities. This is a basic issue in the conceptualization of species as entities, independent of other such entities participating in natural processes such as selection, speciation, and divergence. A concept that accepts only a portion of this natural diversity is not only ineffective and inconsistent but also is misleading to those attempting to evaluate natural patterns and processes.

2. Concepts that are dependent upon a particular threshold of divergence for a particular character source. Knowledge gained from systematic and evolutionary biology is consistent with the observation that the process of descent is not teleological or goal oriented. During a genealogical history many types of modifications occur in organisms, not just those in a particular class of attributes that are conveniently visualized and readily comprehended by humans. Requisite thresholds of divergence in a particular class of traits is necessarily subjective and, as such, is subject to the perception and worldview of the observer, not to any ontological status of the entity in question. Thus, if a concept focuses on an amount of divergence for a particular class of traits (e.g., morphology or reproductive isolation) that is an a priori requisite before an entity can exist in our world, then our understanding of the natural processes and patterns of this world will necessarily be highly constrained by such artificial and arbitrary definitions or boundaries.

3. Concepts wherein the reality of diversity is entirely dependent upon its relation to other diversity, or a relational logistic concept. "These organisms represent a species, say species X, because they are different or isolated from species Y." Based on this logic, species are defined relative to or by their relation to non-conspecific populations through isolation rather than by their relation to conspecifics through cohesion of the species. Thus, the reality of a "thing" in nature behaving like a species is totally dependent upon its relation to other "things," given a binomial and interpreted by us as real. Natural things that are individuals, like species, have reality irrespective of relational comparisons. Species X exists regardless of whether or not species Y exists. Species X would still exist even if species Y had gone extinct!

4. Concepts that are atemporal and nondimensional. As seen in previous examples, most often the attributes possessed by a species are a combination of primitive and derived traits that can

be understood only within the context of descent of a lineage. If a concept does not have dimensional components it is not logically consistent nor functionally effective in systematic and evolutionary biology, wherein processes are deduced from understanding patterns.

5. Concepts that are inconsistent with evolutionary models and inflexible to revision. Ineffective conceptualizations of species, like theories, must be either abandoned or revised in light of discoveries in the natural world. The preservation of a theory or concept for historical or authoritative reasons, or because of prevailing dogma and widespread application is detrimental to the ultimate objectives of biodiversity endeavors.

Good qualities.—We perceive that the eventual conceptualization of species and other natural biological entities must be sensitive to the listed areas of concern. As such, we offer the following properties as essential qualities of a meaningful species concept.

1. Concept that is consistent with known biological patterns derived through biological processes associated with descent with modification. Because species are individuals and a by-product of descent, our ability to discover and explore this diversity will be contingent upon our possession of a logically compatible conceptualization of this diversity.

2. Concept that is not dependent upon a particular threshold of divergence for a particular character source. With divergence occurring in potentially many attributes of organisms during their descent, it is illogical to emphasize a particular class of attributes to justify or deny the existence of species diversity. Whereas it may be an operational concern of convenience to the naturalist, it is ultimately detrimental as a primary factor for the conceptualization of naturally occurring diversity. Here, the operational concerns must be subservient to the realities of evolution.

3. Concept that is nonrelational. The existence of individuals in nature is not dependent upon the existence of other such individuals. Species exist in nature. As such, they can be discovered, described, and christened with names. They also may be evaluated for cohesiveness, irrespective of the existence of other such entities. The only exception to this generality includes direct ancestor–descendant relationships; that is, if an ancestral species did not exist then there would be no descendants.

4. Concepts that are multidimensional. Species evolve through different processes and possess a mixture of primitive and derived attributes. Without a perception of time or space, the variability and distribution of traits may be confusing and appear conflicting. Within the framework of species as lineages in time and space, this quality provides a theoretical basis for interpreting the phylogenetic branching diagrams as evolutionary trees denoting descent. Any concept employed in discovering entities and their attributes, understanding their origins, and incorporating time and space is theoretically superior and will enhance progress in studies of biodiversity.

5. Concepts that accommodate unique, historically derived entities and are flexible to revision given new information. The most important function of a species concept is to permit investigators to recognize taxa derived through descent and behaving as individuals. If there is anything that we should have learned about the discovery of natural patterns and processes over the last two centuries, it is that our understanding is not static. For a concept to continue to be effective, it must be capable of accommodating known, naturally occurring individuals as species.

Species Concepts, Definitions, and Operationalism

Before beginning comparisons and evaluations of the four concepts of species, we must consider conceptualism, definitions, operationalism, and their differences. Concepts of biological systems serve as fundamental links between pattern and process in nature, are employed in every discipline, and help guide our perception of nature and natural systems. They are derived through observation, study, and synthesis of both theory and empirical knowledge of nature. Concepts are universals existing only in the mind of the observer or as predicates that may be properly affirmed of reality. We may use a conceptual view of species, derived through theory and empiricism, to discover the reality of natural patterns and processes as well as species. In this knowledge acquisition system there is feedback between observed pattern–process and the concepts used to understand them, wherein the species concept corresponds to the universal existence of things in nature that behave like species.

One quality consistently argued in the conceptualization of species, either implicitly or explicitly, is that the concept should be operational. That is, anyone should be able to follow a prescribed set of identifiable and repeatable operations and at the end of the operations be able to tell (with a certain for the discovery, understanding, use, and conservation of natural biological entities?

Biological Species Concept

A species is a group of interbreeding natural populations that is reproductively isolated from other such groups. (Mayr and Ashlock 1991.)

The biological species concept (BSC) was reviewed in detail by its strongest proponent, Ernst Mayr, in several publications (1942, 1957, 1963, 1982a, 1982b, 1982c, 1988a, 1988b, in press). It also was reviewed by other authors, including Dobzhansky (1935, 1970), Wiley (1981), Sloan (1987), Coyne et al. (1988), Paterson (1993), and Avise (1994). As recently espoused by Mayr and Ashlock (1991) and Mayr (in press), species consist of reproductive and genetic unit. Individuals of a species seek and recognize one another for mating and thereby maintain an "intercommunicating" gene pool that, regardless of the individuals that constitute it, interacts as a unit with other species with which it shares its environment" (Mayr and Ashlock 1991: 26–27). For Mayr (in press) "each biological species is an assemblage of well balanced, harmonious genotypes and . . . indiscriminate interbreeding of individuals, no matter how different genetically, would lead to an immediate breakdown of these harmonious genotypes. . . . As a result, there was a high selective premium for the acquisition of mechanisms, now called isolating mechanisms, that would favor breeding with conspecific individuals and inhibit mating with non-conspecifics. This consideration provides the true meaning of species. The species is a device for the protection of harmonious, well integrated genotypes. It is this insight on which the biological species concept is based." Central to this concept, and the sole criterion for the reality of a species, is thus the idea of reproductive isolation of species from other such species. "A species is a protected gene pool" that is "shielded by its own devices (isolating mechanisms) against unsettling gene flow from other gene pools" (Mayr and Ashlock 1991). The word interbreeding in the definition above "indicates a propensity; a spatially or chronologically isolated population, of course, is not interbreeding with other populations but may have the propensity to do so when the extrinsic isolation is terminated" (Mayr, in press). Accordingly, speciation is the process of gaining or achieving reproductive isolation (Mayr 1963, 1970).

The BSC specifically excludes asexual or parasexual species that are known to exist in nature, and

level of confidence) if what he or she has is a species. The requirement of such an execution places certain and distinct limits on what can be recognizable as defined by the criteria of the operational "concept." The word and idea of concept as used here is placed in quotes because by forcing it to be operational by a set of what may be artificial criteria distinguishes it from the use of a universal concept in the preceding discussion. Whereas the operational "concept" is perhaps more convenient, convenience is not a criterion that should be optimized when attempting to discover and understand pattern and process in the natural world. Operationalism is a fundamental fault of any species concept adopting it. What is operational is determined strictly by the perceived reality of the viewer. If the viewer's senses perceive only a portion of reality and these are expressed in an operational definition of what reality consists of, then we will never know otherwise. If, however, the viewer is capable of perceiving or conceptualizing all of reality, then all of diversity will be discovered without placing limits on what can be recognized with an operational "concept." For instance, it is a mistake for someone who is red-green color blind to mandate a concept of species based on the operational criterion of color. Anyone discussing species diversity in hummingbirds, flowering plants, or darters, with this person would continually be frustrated with what reality is.

Evaluations of Concepts

In considering the four contemporary species concepts we attempt to identify both positive and negative aspects of each concept as they relate to achieving our earlier stated objectives for biodiversity. Positive aspects are those elements of a concept conducive to either recovering biodiversity or making the concept more user friendly. By "user friendly" we mean the perception that the concept is operational. Although an operational definition of species is not positive in its ability to recover natural diversity, a large element of the scientific community considers this aspect of a concept to be positive. Negative aspects are those elements of a concept interpreted as either adversely affecting efforts to recover biodiversity or making the concept less user friendly. Finally, given the existence of both positive and negative aspects, we provide a recommendation regarding the effectiveness of the concept in light of our previously specified objectives. That is, can we use the concept some have relegated diversity of this type to the class of pseudospecies (Dobzhansky 1970). The concept also is viewed as being an operational definition in that "taxa of the species category can be delimited against each other by operationally defined criteria, for example, interbreeding versus noninterbreeding of populations" (Mayr and Ashlock 1991:27). This concept is relational because "A is a species in relation to B and C because it is reproductively isolated from them." Finally, it is a nondimensional concept that "has its primary significance with respect to sympatric and synchronic populations . . . and these are precisely the situations where the application of the concept poses the fewest difficulties. The more distant two populations are in space and time, the more difficult it becomes to test their species status in relation to each other but the more biologically irrelevant this status becomes" (Mayr and Ashlock 1991). The obvious question here is why is this irrelevant?

Positive aspects.—Two aspects of this concept can be viewed as positive. First, if you, the observer, are a sexually reproducing thing in nature, this concept allows for the comfortable perception of the purity and integrity of species maintained through reproductive isolating mechanisms when in sympatry. Second, based on reproductive isolation as the criterion, this concept is an operational definition whereby the observer can perform a series of operations and expect an unequivocal result, if all of diversity exists in sympatry. Thus, although these are both considered positive by our criteria for evaluating a concept, they are both nonetheless poor qualities of a concept.

Negative aspects.—At least 10 elements of this concept should be viewed as counterproductive toward discovering and understanding biodiversity. The BSC has received substantial criticism in recent years including issues dealing with (1) the absence of a lineage perspective, (2) its nondimensionality, (3) erroneous operational qualities as a definition, (4) its exclusion of non-sexually reproducing organisms, (5) indiscriminate use of a reproductive isolation criterion, (6) confusion of isolating mechanisms with isolating effects, (7) implicit reliance upon group selection, (8) its relational nature, (9) its teleological overtones, and (10) its employment as a typological concept, no different from the frequently criticized morphological species concept. Each of these unfavorable attributes of the BSC is briefly addressed below. In some discussions we refer to Paterson (1993), which represents a collection of his writings on species.

Some of these problems (6–8) may be the result of the conflation of ideas about things in nature, the processes being attributed to these things and their attributes, and the terms and phrases used to describe the things, their origins, and their attributes (see criticisms by Paterson 1993 and responses by Mayr 1988b). However, conflation in this scientific context is not just an academic issue nor something that should be taken lightly. Rather, as Paterson (1993:116) stated in quoting Sir Charles Lyell, "The ordinary naturalist is not sufficiently aware that, when dogmatizing on what species are, he is grappling with the whole question of the organic world and its connections with a time past and with man." The use of convenient, yet imprecise, ideas and terms about such consequential topics as species can be confusing and fatal to our understanding their very nature. As Paterson (1993:159) has perceptively noted, "the stated properties [of "things" (e.g., species)] determine a number of logical consequences." Other problems with the BSC (1–5, 9–10), however, are not merely the result of imprecise conceptualizations and writings but are logical flaws of the concept or flaws resulting from inadequate knowledge of the phylogenetic histories of species and their attributes.

The absence of a lineage perspective (1) constrains adherents of this concept from a theoretical basis in interpreting phylogenetic trees as genealogical trees documenting descent of species as genealogical trees documenting descent of species and their attributes. This has several important consequences. Foremost of these is the absence of a historical perspective needed for discovering and correctly interpreting both sister-group relationships and the attributes of individual organisms and species, including the ability to reproduce. Documented sister-group relationships are critical to ensure that appropriate questions are being asked for appropriate taxa. For example, under the BSC, reproductive isolation is required to achieve species status when two or more sister species diverge from an immediate common ancestor. It is thus important that sister species are being compared, not species descended from different ancestral species. Unfortunately, this type of assurance is rarely, if ever, provided by adherents to the BSC. Rather, the criteria for such "reproductive" comparisons include only geographic sympatry or phenetic similarity.

As previously emphasized, species possess both primitive and derived traits, and the origins of these traits can be understood only within a phylogenetic context. When evaluating the existence of morphological, genetic, behavioral, or physiological traits of a particular species, it is critical to have a historical

As such, it is impossible for practitioners of the concept to employ it in discovering and accurately understanding the process of speciation. Second, allopatric speciation is, by far, the most prominent mode through which biodiversity is produced or speciation occurs. This observation is universally accepted among biologists. Recent studies from a variety of organisms, for which phylogenetic relationships are resolved, unequivocally support allopatric speciation as the primary mode, representing between 70 and 93% of speciation events (Lynch 1989; Grady and LeGrande 1992; Chesser and Zink 1994). If the vast majority of realized biodiversity occurs in allopatry then it is illogical to employ a nondimensional concept.

Given that the majority of speciation events occur in allopatry either through vicariance or peripheral isolation, it logically follows that the majority of sister species are allopatrically distributed, an observation noted even by the early naturalist Wallace (1855). Arguing that the BSC is, in fact, operational is completely erroneous (3). For the majority of the comparisons being performed to defend the existence of species in nature, the tests themselves (e.g., sympatry with isolation) are being conducted between species that descended from different, unrelated ancestors. In these instances the question being asked and the answer generated are essentially meaningless with respect to the process of speciation. Only for a minority of the case comparisons is it possible that the entities being compared are sister species that evolved via sympatric speciation and thus descended from an immediate common ancestor. Only in these cases is the measurement of a propensity to interbreed of meaning under the BSC. However, for the vast majority of cases operationalism is falsely asserted even if we know the sister species. The validity of species is not based on a reproductive isolation criterion but is inferred by geographic distance and degree of divergence. (Also see [5] below, isolation criterion.)

The exclusive application of the BSC to only sexually reproducing organisms (4) automatically eliminates from our inventories of biodiversity and conservation programs those entities in nature behaving as species but reproducing asexually or parasexually or possibly of hybrid origin. That asexual species do represent real entities with individual qualities is adequately demonstrated through three observations. First, some species reproduce asexually and sexually during different phases of their life cycles (e.g., plants and fungi). Does this mean that such organisms are species during some periods but not others? Second, many sexually reproducing spe-

perspective in mind before drawing conclusions as to the reality of the taxa in question. For instance, reproductive isolation is a derived trait modified in a species during or closely following the speciation event. Logically, the ancestral species that gave rise to the two or more descendant sister species had some degree of panmixis; reproductive isolation within this ancestor was not an issue. Any change from this condition would be a modification of the ancestral condition and, by definition, would be derived. Conversely, the ability to reproduce can be retained; it can be a primitive feature of organisms that was not modified during a speciation event. This does not mean that those attributes of the organisms important for species recognition were not modified during descent, only that the ability to produce offspring was not modified. This also does not mean that the two or more descendants of a common ancestor will not respond to different or similar selective regimes in the same way when in the same or different geographic regions. Rosen (1979) was the first to point out this flaw of the BSC in demonstrating experimentally that sister species in the poeciliid genus *Xiphophorus* were reproductively isolated from one another, but if one of these species was crossed with a more distant relative they were reproductively compatible.

For Mayr (in press) ''the word 'interbreeding' indicates a propensity; a spatially or chronologically isolated population, of course, is not interbreeding with other populations but may have the propensity to do so when the extrinsic isolation is terminated.'' It is not stated with whom this propensity to interbreed is to occur. Because of the limitation of this concept to sympatric populations, one can only assume that any sympatric species is considered a likely candidate. It seems to us that the only valid comparison for this criterion is the sister species. However, this too is a problematic aspect of this concept. Further, what is propensity? How does one measure propensity to interbreed when not in sympatry? Is there a threshold for a particular class of attributes (e.g., genetic or morphological) that can be surveyed in the populations that qualify a population as a species? These are significant problems for the BSC.

The nondimensionality (2) of the BSC is a principal defect of the concept for two significant reasons. First, similar to the difficulties of a concept without a lineage perspective, the theoretical basis is lacking for the divergence of sister species in geographic or temporal isolation and their acquisition of reproductive isolation or their independence due to the incidental product of adaptive evolution.

cies that we recognize as valid biological entities today owe their existence to the fact that their ancestors underwent a symbiotic speciation event that lead to multicellularity and higher forms of life that reproduce sexually (Margulis 1970). Third, many asexually reproducing species are descendants of, or are in clades derived through descent from, sexually reproducing ancestral species. Does this mean that the ancestor or some sister taxa should be considered a species but not its descendants or other sister taxa?

Much satisfaction and comfort is gained by many in knowing that the BSC relies upon an isolation criterion (5) for determinations of species as taxa. Unfortunately, most are satisfied in only knowing of this criterion and remain unenlightened or disinterested in its etymology and its actual application in the BSC to natural systems. Paterson (1993:142–143) and others suggest that some of this underlying bias towards the popularity of the isolation criterion has cultural origins, dating well before Darwin's separation from teleological interpretations, to "prehistoric attitudes of 'purity of stock' and 'purity of line,' engendered by the practices of ancient plant and animal breeders." . . . "In English, notice how approbative are words such as 'pure,' 'pure-bred,' and 'thoroughbred,' and how pejorative are those like 'mongrel,' 'bastard,' 'halfbreed,' and 'hybrid.' Such cultural biases, which act subtly, almost subliminally, through the vocabulary and imagery of languages, might well predispose the unwary to favor ideas like that of 'isolating mechanisms' with the role of 'protecting the integrity of species.'"

The BSC is presented by Mayr (1988b) and Mayr and Ashlock (1991) as a popular concept in the scientific community, but scrutiny of its implications on pattern and process ultimately may corroborate the cultural bias in this devotion. There are at least two areas of elementary concern with the strict application of a reproductive isolation criterion, both involving a historical perspective to diversity. First, how is the reproductive isolation criterion applied towards understanding diversity? In reality, the criterion for the recognition of species is relevant to sister species, species descended from an immediate common ancestor (Figure 6C). If the concern truly involves issues of descendants reaching a state of genetic independence through reproductive isolation, then the most important comparison involves sister entities. In practice, the implementation of the BSC rarely is founded on a phylogenetic hypothesis wherein the investigator has demonstrated that the entities being compared are, in fact, descendants of an immediate common

ancestor. Rather, the comparisons are between sympatric taxa or with allopatric taxa that are assumed to have some relevance to the question at hand. In many instances the implied close relationship is based solely on overall similarity because a phylogeny does not exist. Even Mayr (1988b:436), a strong proponent of systematic and evolutionary biology, a priori assumes that the occurrence of hybridization is part and parcel with being sister taxa in observing that "the enormous number of secondary hybrid belts . . . indicate how often temporarily isolated populations were unable to evolve through selection efficient isolating mechanisms, when they made secondary contact with sister or parental populations." Thus, in practice, because of the preponderance of allopatric speciation the comparison made may not involve sister species, especially where sympatric species are concerned (Figure 6D).

Second, the observation that mate-recognition systems and discontinuities exist in nature is undeniable. However, the reality of the unyielding and equivocal nature of the reproductive isolation criterion provides merely a false sense of security for many adherents to the BSC and ultimately results in our loss of understanding of natural systems. Combining the lack of a lineage perspective with the reproductive isolation criterion in this concept we necessarily have to reject many ancestral species that now are known to have temporarily lost this isolation and yet gone on to produce significant numbers of species that are currently recognized as valid. Our use of ancestral species here refers to a species that may no longer exist in a contemporaneous community but existed as a species that produced descendant species. Current studies have revealed data from some contemporary species that ancestral species of different phyletic clades were capable of interbreeding, yet continued to evolve independently and produce descendant species.

The nondimensionality of the BSC logically forces emphasis on extant diversity and processes of extant species' interactions, as if that is all that is important or has ever happened. This, in effect, deemphasizes historical patterns and processes that, combined again without a lineage perspective, necessarily results in our ignorance of, or ambivalence for, contemporary processes (gene exchange) operating in ancestral species. The breakdown of species isolation by ancestral species has been an important evolutionary mechanism for some groups of organisms and has resulted in their evolutionary success in diversity. For many plant groups speciation via hybridization is renowned. In North American fishes speciation via hybridization or the breakdown of isolating mechanisms in ancestral lineages is known to be important in the families Cyprinidae, Poeciliidae, and Atherinidae (Echelle et al. 1982; Meffe and Snelson 1989; DeMarais et al. 1992; Dowling and DeMarais 1993). Even in the face of this propensity to interbreed in ancestral communities (biotic communities that existed prior to the Recent, wherein species coexisted in communities and may or may not have interbred), these lineages have maintained their identities and yielded significant diversity. As more of these ancestral "hybrid" lineages are discovered with our enhanced technologies and methodologies, are we to exclude them and their subsequent diversity from our inventories? Given the purported popularity of the BSC and in light of its multiple inconsistencies with nature, there must be some truth in the cultural hypotheses advanced by Paterson (1993) and others.

Paterson (1993) criticized the BSC as logically flawed in confusing the origin of isolating mechanisms with the origin of isolating effects in allopatric speciation (6). Evolutionarily, through natural selection, attributes of organisms (and species) may be advantageous, disadvantageous, or neutral. Some attributes may spread through a population or species not because they are selected for as an advantageous trait, but instead because they may be pleiotropically related to another attribute(s) that is selected for in a given environment. Williams (1966) referred to the principal advantageous phenotypic effects of an allele or attribute as the "function." All other consequences of the pleiotropic selection on the "functional" attribute, whether negative, positive, or neutral, are referred to as "effects." A similar observation was made by Gould and Lewontin (1979) in that not all attributes possessed by a species need exist because of an adaptive advantage. Rather, attributes may be present in a species and have a selective advantage for the species at a later time. Traits of this latter class are referred to in phylogenetic systematics as retained primitive characters (or plesiomorphies).

A central premise to the BSC is that there is "a high selective premium for the acquisition of (isolating) mechanisms . . . that would favor breeding with conspecific individuals and inhibit mating with non-conspecifics" (Mayr, in press). Given that the selection regime required for the evolution of traits required to protect the integrity of a species' gene pool would necessarily develop in an environment where this integrity is being compromised, the character would have to evolve as a function (or mechanism) in sympatry with those species most threat-

ening to the immediate breakdown of harmonious genotypes. If isolating mechanisms are, in fact, the functional alleles or attributes, then they would evolve through selective pressures via sympatric speciation. Unfortunately, this is not only inconsistent with the mode of speciation (allopatric) espoused by those promulgating the BSC (Mayr 1963; Dobzhansky 1970), it is the least frequently observed mode of speciation in biotic systems (Lynch 1989) and is questioned by some as a valid mode of speciation (Paterson 1993). Thus, because speciation is a process that occurs largely in allopatry and many sister species may never occur in sympatry, it is only logical that if attributes resulting in the "preservation of genotypes" do evolve they should be thought of as incidental products of adaptive evolution (i.e., isolation effect) and not adaptational devices (Paterson 1993). Mayr (1988b) takes exception to the criticisms leveled by Paterson and continues to defend the BSC. As Darwin[1] once wrote in a letter to Huxley "Nature never made species mutually sterile by selection, nor will men."

The BSC is criticized as essentially relying on group selection (7), not individual selection, for the evolution of purported isolating mechanisms (Paterson, 1993). Mayr (1963) states that "it is the function of isolating mechanisms to prevent such a breakdown and to protect the integrity of the genetic system of species." Recently, Mayr and Ashlock (1991:26–27) summarize a species, within the context of the BSC, as "a protected gene pool. It is a Mendelian population shielded by its own devices (isolating mechanisms) against unsettling gene flow from other gene pools. Genes of the same gene pool form harmonious combinations because they have become coadapted by natural selection. Mixing the genes of two different species usually leads to a high frequency of disharmonious gene combinations; mechanisms that prevent this are therefore favored by selection." A similar perspective is implied in other writings by Mayr and many other adherents of the BSC. From such descriptions of the BSC, Paterson (1993:159) concludes that "If the isolating mechanisms are indeed ad hoc characters, they are adaptations sensu stricto, as the appellation 'mechanism' implies. This in turn means that they are the products of natural selection and, finally, that speciation must occur either in sympatry or in parapatry. Furthermore, if the function of the isolating mechanisms is really to protect the genetic integrity of a species, as Mayr has claimed, one should notice

---

[1] Letter from C. Darwin to T. H. Huxley, 7 January 1867.

that the species is a population, a group, and there-fore, a character which has been selected to protect this group character must have been subject to group selection, not individual selection."

Although group selection has been invoked in the evolution of social systems of some organisms, it is neither universally accepted nor demonstrated as a general phenomenon of species and speciation (Wilson 1983). One can conclude from the above statements regarding the BSC, as well as many others by Mayr and Dobzhansky, that the isolating mechanisms are adaptations of the group (demes, populations, or species) which have evolved for the good of the species, through group selection, rather than through individual selection. The question then becomes whether the evolution of such a property is for the good of the species or for the good of the individual; or would it be more clear to state that the property is an incidental effect of adaptive evolution? Mayr (1988b) did not respond to the group selection criticisms raised in the combined writings of Paterson (1993).

As discussed previously, the relational nature (8) of a concept is regarded as a disadvantage of a conception of reality of things possessing qualities of individuals. Paterson (1993) also identified this problem with the BSC. Although the statement that "A is a species in relation to B and C because it is reproductively isolated from them" may appear perfectly harmless and logical, the necessity of this statement for a concept has detrimental implications as previously discussed. On the more practical side, however, why focus on the A with B and C comparisons? Why not compare A with species X, Y, and Z? The implication of the BSC and this type of statement is that A was thought at one time to have shared a common ancestor with B, C, or both. Where are the data to substantiate that A, B, and C evolved from an immediate common ancestor, are sister species, and provide a meaningful comparison? Why else would one be interested in reproductive isolating mechanisms and these particular species? In the vast majority of the cases where the BSC is being employed or advocated the phylogenetic data are not available. This makes comparisons of A to B and C just as logical as A to X, Y, and Z with respect to isolating mechanisms and their evolution.

Repeating statements from an earlier paper, Mayr (1957:15, 1988b:435) responded to Paterson's criticisms of a relational concept, "Since the non-dimensional species concept is based on relationship [of populations toward each other] the word species is equivalent to words like, let us say, the word brother . . . an individual is a brother only with respect to someone else. Being a brother is not an inherent property, as hardness is a property of a stone. Describing a presence or absence of relationship makes this species concept non-arbitrary." This is an interesting comparison because Mayr's choice of words is particularly revealing as to his metaphysical understanding and philosophical position with regard to species. Brother is like the term sister-group. Both refer to genealogical, or blood, relationship (except in cases of adoption) strictly implicating a shared ancestral descent. Like brother, if there is only one species then there can be no sister-group. The word species, however, refers to a universal concept of diversity that is composed of individual organisms, demes, and populations and is argued to form the highest level on which natural selection operates. Even if there was only one species in the universe, species would still have meaning; brother would not. Thus, one word is relative, requires at least two entities, and refers strictly to genealogical relationship. The other word is a universal and has meaning regardless of relationships or the number of entities. It is surprising that Mayr (1988b) would choose a word with genealogical connotations to defend the relational (or referential) nature of the BSC, especially because it is a rare event that sister-group relationships are employed in biological decisions invoking the BSC.

A teleological explanation (9) is one that consists of specifying a goal or purpose towards the attainment of which an event or activity is a means. Paterson (1993) criticized the traditional BSC as having teleological overtones in that the qualities referred to as isolating mechanisms serve a purpose or allow species to achieve a goal. That is, there is a high selective premium for the acquisition of mechanisms, now called isolating mechanisms, that can be used as devices in the protection of well-balanced, harmonious, and coadapted genotypes from the indiscriminate interbreeding of non-conspecific individuals, no matter how different genetically, that would ultimately lead to an immediate breakdown of the harmonious genotypes. Even as late as 1976, Dobzhansky stated that "In general, the working hypothesis which seems to me fruitful is that species are not accidents but adaptive devices through which the living world has deployed itself to master a progressively greater range of environments and ways of living." Likewise, recently Mayr (in press) poses the questions "Why are there species? Why do we not find in nature simply an unbroken continuum of similar or more widely diverging individuals?" and responds with "The reason, of course, is that each biological species is an assemblage of well balanced, harmonious genotypes and that an indiscriminate interbreeding of individuals, no matter how different genetically, would lead to an immediate breakdown of these harmonious genotypes." Similarly, in the same paper Mayr argues that "two closely related sympatric species retain their distinction ... because they are genetically programmed not to mix." Alternatively, Dobzhansky (1950:405) viewed species as "the largest and most inclusive ... reproductive community of sexual and cross-fertilizing individuals which share in a common gene pool." Furthermore, "we are almost forced to conjecture that the isolating mechanisms are merely byproducts of some other differences between the organisms in question, these latter differences having some adaptive value and consequently being subject to natural selection" (Dobzhansky 1935:349). Likewise, for Mayr (1963:548), "Isolating mechanisms have no selective value as such until they are reasonably efficient and can prevent the breaking-up of gene complexes. They are ad hoc mechanisms. It is therefore somewhat difficult to comprehend how isolating mechanisms can evolve in isolated populations." Although these latter confessions are largely without teleological overtones (one may question the use of "reasonably efficient," "prevent the break-up," and the last sentence in Mayr's phrase), preceding interpretations definitely specify goals or purpose. While Paterson (1993) and Mayr (1988a, 1988b) disagree with respect to the teleological tendencies of the traditional BSC, these disagreements serve as excellent examples of the importance of the choice of words in describing a concept. As stated by Paterson (1993:36) "In science, inadequate concepts and inconsistent logic often lie at the root of disputation and disagreement ... Moreover, [Paterson 1993: 111] conflation [of concepts] makes the testing of hypotheses difficult."

Typological concepts of species (10) are rejected as appropriate characterizations of nature because they rely on the existence of universals fitting the concept. Here individuals of a species are members of a class. Typological species have limited variation in their attributes, are constant through time, and differ from one another by sharp discontinuities. Most importantly, this type of concept relies upon a set degree of difference as a criterion for the existence and recognition of biological diversity termed species (sensu Mayr and Ashlock 1991). With the typological morphological species concept a threshold of morphological difference was required for justifying the reality of a species. As stated by Mayr

and Ashlock (1991:25), "Degree of difference thus cannot be considered a decisive criterion in the ranking of taxa as species." A fundamental problem with the typological metaphysical perception of diversity is its relational nature. As discussed above the same problem exists with the BSC, except that the point of reference or relational quality here is reproductive isolation when in sympatry and a propensity to interbreed when in allopatry. Given actual employment of the BSC, the propensity to interbreed in the latter case is nearly universally determined by the degree of difference between the entities in question. For Mayr (in press) "It is the concept of reproductive isolation that provides the yardstick for delimitation of species taxa and this can be studied directly only [in] the nondimensional situation. However, since species taxa have extension in space and time, species status of non-contiguous populations must be determined by inference." Isolation is almost never shown a priori between sister species but is inferred on the basis of divergence that may restrict gene flow between species of unknown relationship to the species in question. Basically, this equates to substantial morphological divergence and no other data sets. Thus, degree of difference as a decisive criterion in the BSC is sexual reproduction and reproductive isolation or sexual reproduction and divergence in morphology. Just as cryptic or sibling species are excluded from our recognition within a morphological species concept, species that are asexually reproducing, products of hybrid origin, or unrelated species that retain the primitive ability to reproduce under some circumstances are logically excluded when using the BSC. In either case, known diversity is excluded from consideration because of both the typological and relational nature of the concepts.

Recommendation.—We recommend limited usage of this concept because it is known to exclude significant biodiversity and misrepresent natural biological systems.

For Mayr and Ashlock (1991:27) an ostensible justification for the BSC is popularity: "The importance of the biological species concept lies in the fact that it is the concept employed in the largest number of biological disciplines, particularly ecology, physiology, and behavioral biology." Curiously, as resolution and understanding of natural systems advanced through the refinement of theory and empirical methods in systematic biology, popularity of this concept has waned. Cracraft (1983) and Paterson (1993) argue that the BSC is not as popular as Mayr and other proponents have suggested, especially in disciplines charged with recovering the diversity of biotas. As discussed by these authors, the BSC has received substantive criticism, not only because it is not operational and there are particular problems with the classification of the biota, but also because those disciplines charged with studies of biodiversity have found "that it does not function well in helping us to understand the pattern and process of taxonomic diversification" (Cracraft 1983:162). When one considers the magnitude of unfavorable qualities of this concept, together with the various aspects currently known about biodiversity through systematic and evolutionary biology, it is almost inconceivable that the BSC has been popular in the general biological literature. Of all the concepts in general operation this single concept is the most artificially restrictive in the types of biological entities that can be recognized. Equally disturbing is the fact that this concept masquerades under the pretense that it is not only an important biological concept but is also operational. Paterson (1993:92–103) argues that much of the attraction to the BSC has stemmed from the retention of a teleological and idealistic philosophical view of species predating the Darwinian revolution. This philosophy is argued to be rooted in a deep-seated bias inherent in Western culture stemming from the biblical book of Genesis and shackles the minds of many biologists today (Masters et al. 1984; Paterson 1993).

Phylogenetic Species Concepts

At least three different species concepts are identified as phylogenetic. These concepts have developed out of phylogenetic systematics and a general need among some researchers of an operational lineage definition of species that is process free. Furthermore, with the growing popularity of phylogenetics, some felt it important to estimate the smallest biological units suitable for phylogenetic analysis. That is, for some, the species is the smallest biological unit appropriate for phylogenetic analysis, and infraspecific biological units are inappropriately evaluated in this context (Nixon and Wheeler 1990; Wheeler and Nixon 1990). This same perspective holds that species diversity must be understood before a phylogenetic analysis is performed. Others argue the converse and defend the position that hierarchical patterns exist within species and phylogenetic methods are appropriate from this level to supraspecific taxa (de Queiroz and Donoghue 1988, 1990; McKitrick and Zink 1988).

Common to all of these concepts is an attempt to identify the smallest biological entities (i.e., species) that are participating in natural processes and are diagnosable, monophyletic, or both. The species is thus the biological entity and the unit product of natural selection and descent. As such, the taxonomic category subspecies, fraught with ambiguities between a category of convenience and naturalness, is not an evolutionary unit and has no ontological status under the phylogenetic species concept (PSC; Cracraft 1983; McKitrick and Zink 1988; Warren 1992).

The different types of phylogenetic concepts may be grouped into three general classes. The first class includes those incorporating monophyly as the primary criterion for recognition of species; the second class includes those concepts using diagnosability as the primary criterion; and the third class includes those requiring both monophyly and diagnosability as criteria.

Monophyly and the Phylogenetic Species Concept

For Rosen (1978, 1979) and de Queiroz and Donoghue (1988, 1990) species have reality if they are monophyletic and supported by autapomorphic (unique and derived) attributes. As such, any biological entity possessing a uniquely derived character, of any type, magnitude, or quantity, qualifies as a species. Those not possessing autapomorphic attributes do not constitute a species, as traditionally viewed, but are referred to as "metaspecies" by some. The application of this concept necessitates a phylogenetic analysis. A lucid discussion is offered in the de Queiroz and Donoghue papers.

For Rosen (1978:176) "a geographically constrained group of individuals with some unique apomorphous character, is the unit of evolutionary significance." De Queiroz and Donoghue (1988) did not explicitly define species but viewed them as monophyletic entities supported by shared and derived attributes.

Diagnosability and the Phylogenetic Species Concept

Another class of concepts emphasizes the a priori diagnosability of species, irrespective of the criterion of monophyly. There are two purported benefits of this perspective. First, process is not invoked before pattern is observed. Second, phylogenetic methodologies are argued to be applicable to only genealogical relationships of species and supraspecific taxa, not to tokogenetic relationships of infraspecific entities (sensu Nixon and Wheeler 1990; Wheeler and Nixon 1990). Tokogenetic relationships are those genealogical connections existing between individual organisms. Unlike most phylogenetic relationships that are hierarchical and non-reticulate, tokogenetic relationships are reticulate. To conduct a phylogenetic analysis below the level of species would confuse the reticulate tokogenetic relationships with hierarchical phylogenetic relationships.

A species for Eldredge and Cracraft (1980:92) is "a diagnosable cluster of individuals within which there is a parental pattern of ancestry and descent, beyond which there is not, and which exhibits a pattern of phylogenetic ancestry and descent among units of like kind." Later, Cracraft (1983:170) considered the species to be "the smallest diagnosable cluster of individual organisms within which there is a parental pattern of ancestry and descent," deleting the reference to reproductive isolation. For Nelson and Platnick (1981:12) species are "simply the smallest detected samples of self perpetuating organisms that have unique sets of characters." Nixon and Wheeler (1990) consider species as "the smallest aggregation of populations (sexual) or lineages (asexual) diagnosable by a unique combination of character states in comparable individuals (semaphoronts)." Wheeler and Platnick (in press) define species as "the smallest aggregation of (sexual) populations or (asexual) lineages diagnosable by a unique combination of character states."

For proponents of this concept, monophyly, paraphyly, and polyphyly apply only to groups of species and above but not to any level of organization below. This class of PSCs does not rely upon a phylogenetic analysis for implementation. Rather, its application is argued to be like an analysis of supraspecific taxa. The operation of this concept involves the search for heritable variation that can be partitioned into either "characters" (fixed attributes for a specific entity) or "traits" (variable attributes for a specific entity). Nixon and Wheeler's (1990:217) perspective of a phylogenetic analysis places constraints on the types of attributes that can be used in an analysis to include only those that are deemed characters or "those that are found in all comparable individuals in a terminal lineage" (excluding instances of sex-linked attributes). In their argument relevant evolutionary transformation "is not limited to mutation, which is only the first step in evolutionary change, but is instead tied to fixation, the final step in evolutionary change." Thus, species are delimited by the distributions of fixed, diagnostic characters across populations. Where variability exists in an attribute within the taxon of question this attribute is considered inappropriate for that level of analysis. If this taxon happens to be an entity considered a species, then Nixon and Wheeler argue that there are only tokogenetic relationships and no phylogenetic relationships to be explored. Finally, the operation(s) necessary for the practical delineation of tokogenetic and phylogenetic relationships is not developed explicitly by those favoring this concept. Without knowing if you are dealing with one or more species ahead of time, one is not likely to know if phylogenetic methods are appropriate. Likewise, the difference is unclear between the theoretical inapplicability of phylogenetic methods in tokogenetic systems versus the applicability of the same methods for resolving relationships of species derived via hybrid origin; both contain reticulate patterns of history.

Monophyly, Diagnosability, and the Phylogenetic Species Concept

The PSC of McKitrick and Zink (1988) is a modification of the PSC provided by Cracraft (1983) but incorporates the criterion of monophyly for species. Although a definition was not provided by McKitrick and Zink (1988), they identified a species as the smallest diagnosable cluster of individual organisms forming a monophyletic group within which there is a parental pattern of ancestry and descent. Because in their conceptualization all recognized monophyletic taxa are diagnosable, this concept and its methods for the discovery of species are equivalent to the monophyletic PSC above.

Positive aspects.—There are several positive aspects to these concepts that make them particularly useful for the operation of discovering biodiversity. Many of these qualities solve some of the perceived problems with the BSC. In all classes the PSC is an operational definition, whether one uses diagnosability or monophyly. The set of operations necessary to discover diversity associated with species are clearly outlined. These concepts of diversity incorporate the notion of lineages, making them particularly appropriate for reconstructing histories of descent and interpretating the evolution of attributes. These concepts view the ability to interbreed as a shared primitive attribute and not necessarily of consequence in the recognition of species as taxa. Rather, they view using the ability to interbreed as an arbitrary criterion for the delineation of diversity. Because the ability to interbreed is ancestral, it is artificial to attempt to draw lines in the genealogical history of descent demarcating such changes, if they even occur. These concepts also have the ability to recognize both sexual and asexual species and possess no implied modes of selection nor speciation. Finally, in the execution of these concepts there is no inherently arbitrary divergence or distinction between species or subspecies in a polytypic species. Rather, subspecies are not considered evolutionary units and have no ontological status. Species, being by definition the evolutionary unit, are essentially rendered equivalent in their applications in comparative biological analyses (Cracraft 1983; Warren 1992).

Negative aspects.—Regardless of the positive aspects of the different classes of the PSC, there are some important problems with these concepts. Some of these negative aspects may be viewed as purely operational, whereas others can both preclude the recognition of some diversity behaving as species as well as overestimate the diversity of species. First, for two classes of these concepts, a phylogenetic analysis of either individual organisms or populations of organisms is required before species delineations are possible. This may be considered purely an operational problem, but for Nixon and Wheeler, this is both a theoretical and methodological problem. On the other hand, when relying exclusively on diagnosability, without the use of phylogenetic argumentation of character interpretation, it also is possible that species will not be comparable. In this case one may diagnose a species that evolved from a single speciation event. However, one may also use attributes to diagnose something sharing only primitive, convergent, or parallel attributes that exists as a geographically confined entity in nature only through multiple, possibly unrelated, speciation events. Only through a phylogenetic analysis can one effectively reveal the differences between these two similar and diagnosable types of diversity.

Second, use of the monophyletic criterion necessarily excludes all surviving or extinct ancestral species because the only traits possessed by an ancestral species are those of all the descendants (Figures 5H, 15). Thus, it will be impossible to recognize ancestral species. For de Queiroz and Donoghue (1988, 1990) any group of organisms or populations lacking autapomorphic characters and having unresolved relationships is termed a "metaspecies." Because both ancestral and descendant species are both individuals (Wiley 1981) and presumably continue to participate in the same natural processes, it seems unwise to draw artificial distinctions between these two equivalent naturally occurring biological entities. Furthermore, because an ancestral species necessarily possesses only the attributes of the

group to which it is ancestral, it is unclear how the definitions of the diagnosability class of the PSC are capable of recognizing ancestral species. Although ancestral species rarely have been discovered, they are known to survive speciation events and persist in contemporary communities (Echelle and Echelle 1992). However, if a concept is not capable of recognizing ancestral species when they do exist, they always will be difficult to discover.

Third, all three classes of concepts under the PSC are relational concepts, wherein the discovery and justification of a species is dependent upon the existence of and comparison with other species. Unlike the BSC, the relational nature of the PSC does not involve the inherently arbitrary nature of relative divergence among congeners or groupings of taxa based on an inferred propensity for interbreeding. Rather, the PSCs necessitate comparative measures with congeners to employ both the operational criterion of monophyly based on autapomorphic attributes as well as the operational criterion of diagnosability based on plesiomorphic and apomorphic traits. Although the relational component may appear to involve trivial argumentation, it remains true that species still represent fundamental units of evolution and exist independent of our abilities to perceive them. If our abilities to reveal true patterns in nature are inhibited by strict dependence upon operational criteria like these, then our abilities to discover and understand natural processes will likewise be endangered.

Finally, with the exclusive reliance upon the diagnosability criterion a researcher may overestimate the diversity of species. In paleontology a chronospecies or grade (Simpson 1961) concept is commonly employed to identify ancestral species. As Wiley (1981) demonstrated, chronospecies are artificial constructs and are not biologically equivalent to species. Although Nixon and Wheeler (1990: 219) do not advocate recognizing chronospecies or even ancestral species, "the PSC is character-based and will identify the same species as application of a chronospecies concept, when those species are based on character-state distributions." With such a perception of reality, the ambiguity and confusion as to the identification and recovery of natural biological entities is overshadowed by the necessary artifacts of this operational definition. The chronospecies problem is not an issue with the monophyly-based PSC.

Recommendation.—Although there are problems that can be perceived with the exclusive use of any of the classes of the PSC, there are also important positive aspects of these concepts over the BSC. We

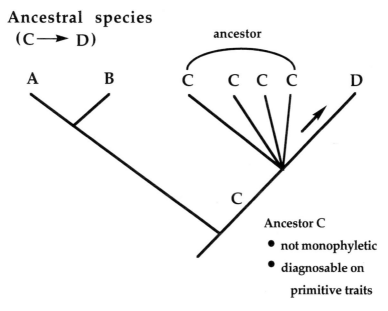

FIGURE 15.—Example of how ancestral species C will always lack any unique, derived features supporting their monophyly. Ancestral species possess only those features possessed by their descendants (C plus D), prior to any unique anagenesis in any descendants. Ancestral species C will be excluded from biotic inventories based on species concepts necessitating monophyly.

agree with the conclusions of Warren (1992) in that the PSC may serve as a good operational surrogate to a concept of species that is not implicated with as many problems in discovering and understanding biodiversity. Thus, we recommend cautious use of the various forms of the PSC, with the understanding that these concepts are capable of both excluding naturally occurring biodiversity known as species, overestimating biodiversity in the fossil record, and producing noncomparable biological entities thought to be species.

Recognition Species Concept

A species is that most inclusive population of individual, biparental organisms which share a common fertilization system. (Paterson 1993:105.)

The recognition species concept (RSC) was introduced in the writings of Hugh E. H. Paterson, collectively published in Paterson (1993). It was developed out of a general dissatisfaction with the BSC. Paterson's view is that the BSC neither adequately nor accurately represents the patterns and processes of natural diversity and its implementation inhibits progress towards related goals. For Paterson (1993:158), "One of the fundamental foundation stones of evolutionary and population biology is the species." Furthermore (Paterson

1993:40), "Just as a clear view of species is needed in order to understand speciation, so clarity on how speciation occurs is mandatory if we are to grapple successfully with ecological concepts such as the niche and species diversity."

For Paterson the biological limits for gene recombination (sensu Carson 1957) are determined by the mate-recognition system, more precisely, a specific mate-recognition system (SMRS). This recognition system consists of a series of specific coadapted signals and releasing properties exchanged between partners; successful functioning of the recognition system is dependent on the partners being able to receive information through complementary systems. The system is functional across a broad array of conceivable signal-reception methods from elaborate behaviors to chemicals and pheromones to cellular recognition with gametes. This coadapted complex is maintained by strong stabilizing selection as long as the species inhabits its natural habitat; this coadapted complex changes when the natural habitat for the species (perhaps ancestral) is changed through geographic or temporal disjunctions. At this point the coadapted complex of signals exchanged between partners may become altered via directional selection in the new habitats occupied by the descendant groups of daughter populations (or species). Paterson (1993:33) argues that "a new SMRS, derived in this way, determines a new gene pool and, hence, a new species. According to the recognition concept, species are populations of individual organisms which share a common specific mate-recognition system.... Species are, thus, incidental effects of adaptive evolution."

This model does not invoke a major role for selection in the evolution of positive assortative mating or the development of isolating mechanisms and does not require sympatry and "evolutionary reinforcement" to complete speciation. All of these assumptions are built into the BSC. Under the BSC "the use of the term 'isolating mechanisms' carries with it the necessity to accept speciation by reinforcement as the way in which [species] evolve" (Paterson 1993:3). Selection is ultimately believed to perfect adaptations that serve as isolating mechanisms for gene pools of species, the mechanisms then take on a teleological focus implying that the adaptations are "fashioned by selection for the goal attributed to it" [and] "If this is accepted, a further implication is that natural selection has a direct role in the production of species diversity" (Paterson 1993:3). The fallacy that selection is responsible for producing adaptations that, by design, are responsible for the isolation of gene pools is obvious from the observation that in large part the documented cases of speciation are the direct result of total allopatry (Lynch 1989), a speciation model that does not involve secondary contact or reinforcement of isolating mechanisms. Thus, if isolating mechanisms are products of descent they are the result of chance rather than design (Paterson 1993).

The general question regarding the origin of species within the RSC is not what are the characters and mechanisms that have evolved in the recognition or reproductive systems of a species which prevent successful matings and resulting ontogenetic development between sympatric species. Rather, the more appropriate question is what are the characters and mechanisms that have evolved in the recognition or reproductive systems of species that ensure effective syngamy, development, and future generations within a population occupying its preferred or natural habitat? (Paterson 1993).

Positive aspects.—Although this concept was criticized superficially by Mayr (1988b) as being nothing more than the BSC disguised under a new name, it is fundamentally different from the BSC. The RSC provides an important new and less constraining perspective to examining diversity that is not pervaded with process-directed overtones. The RSC provides sound argumentation for the use of

not only morphological but ecological, behavioral, physiological, and biochemical systems in the recognition of natural diversity. The concept is not constrained to a specific mode of selection or speciation and is not linked to selection for reproductive isolation. Species are considered incidental products of adaptive evolution, not adaptive devices as in the BSC. This perspective views the diversification process as one in which species and their SMRSs evolve as isolation effects, not as isolating mechanisms. The RSC is a nonrelational concept and thus does not require specific levels of divergence of particular attributes to exist in a taxon for it to be considered a species. Rather, species are considered independently, as things with qualities of individuals. The only exception is that species must have an SMRS. The occurrence of hybridization between species takes on a whole new context within the framework of the RSC. Because species are not adaptational devices and "There is no idealistic specific integrity conceived, hybridization may have good or bad consequences which are subject to natural selection. Hybrids are often preserved as allopolyploids, for example" (Paterson 1993:162). This concept also allows identification of ancestral species that survived speciation events and will accept species of hybrid origin as valid, except when they are asexual. The RSC is an operational concept in that when one identifies an SMRS, then one has discovered a species. Finally, the RSC is neither a typological nor a teleological concept of diversity.

Negative aspects.—Although there are important positive aspects to this concept that allow correct identification of species in a largely process-free environment, there are important problems with universal application of the RSC. We identify three areas of concern involving (1) strict reliance upon and knowledge of the SMRS, (2) lack of a lineage perspective, and (3) exclusion of some types of diversity, such as asexual species.

The RSC inherently assumes that all speciation events result in modifications of mate-recognition sequences. Although this eventually may be shown the case, there is little or no empirical evidence to suggest that the modification of an SMRS always occurs with speciation events. Perhaps part of the difficulties associated with using this as a universal criterion is that few investigations focus on such occurrences, especially from a phylogenetic perspective for which comparisons are conducted between sister-groups. Furthermore, there is difficulty associated with discovering, studying, and understanding SMRSs. This, of course, is no excuse for

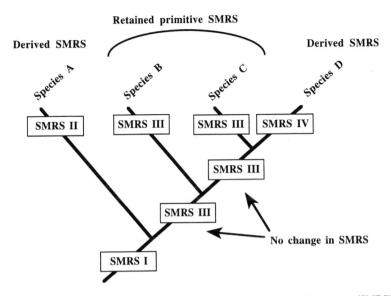

FIGURE 16.—Example of how the retention of an ancestral specific mate-recognition system (SMRS) can occur in the evolution of a monophyletic group of species. Because species B and C share the same SMRS, they would be considered the same species using the recognition species concept, even though C is genealogically more closely related to (and shares a unique common ancestor with) species D.

our lack of pursuit of such phenomena. However, it generally is not considered a user-friendly means of discovering diversity. In all fairness to the concept and Paterson (1993), he provides several examples in which the RSC and the search for SMRSs were employed in difficult (e.g., the mosquito genus *Anopheles*) diversity questions. The concept can be viewed as operational, but there may be substantial effort required on the part of the biological community for its universal implementation.

The RSC does not incorporate a lineage perspective in the evolution of species or attributes and is a nondimensional concept. As a result, it is impossible to interpret correctly the evolution of qualities possessed by species or sister-group relationships. Like other attributes of species, SMRSs can be either primitive or derived. If there ever exists the retention of a plesiomorphic SMRS in the evolution of a group, then we will underestimate diversity following from one or more speciation events (Figure 16). There is no more reason to suspect that divergence in the SMRS is coupled with cladogenesis than there is reason to argue that speciation occurs only if there is reproductive isolation, as in the BSC. It is unlikely that anagenesis will always result in new adaptations to the fertilization system, thus driving speciation as an "incidental effect-... from the adaptation of the characters of the fertilization system ... to a new habitat, or way of

life" (Paterson 1993:147–148). Rather, there may be mutations or accumulations of mutations in other premating or postmating systems of different species, none of which are related to determining the "new field for gene recombination" through SMRS. This actually is evidenced by the example used (Paterson 1993:146–147) dealing with orchid species. These species differ from one another only by their pollinators; there are apparently no physical or chemical differences among these orchid species that are used as SMRSs. Without a phylogenetic perspective of the system, the SMRS may be interpreted as derived or "adapted" to a new environment, when in fact it represents a retained, plesiomorphic state remaining unchanged through one or more speciation events (Figure 16).

Like the BSC, the RSC excludes some known types of diversity from being recognized, namely asexual species. Paterson (1993:160) realized the limitation of this concept but argued that "obligatory uniparental organisms are of course part of diversity of life, but they follow a different genetic strategy from sexual organisms and cannot be forced under one heading for human minds more concerned with tidiness than comprehension of complex phenomena."

Recommendation.—The many positive aspects of this concept over the BSC warrants its general appreciation for investigations of biological systems.

Although SMRSs are rather poorly investigated, Paterson (1993) provides several examples and logical arguments for their use in investigations of biological systems. However, given that the RSC can exclude known biological diversity—asexual species and species with retained primitive SMRSs—we recommend cautious reliance upon this concept.

Evolutionary Species Concept

The evolutionary species concept (ESC) was championed originally by Simpson (1951, 1961) out of a general dissatisfaction with the nondimensionality of the BSC. Wiley (1978, 1981) developed the concept further and argued for its general application to biological systems. Unlike aforementioned concepts, the ESC largely was ignored, until recently. Wiley and Mayden (in press) reviewed the ESC and developed it further. They also argue that the ESC is the only available concept with the capacity to accommodate all known types of biologically equivalent diversity. This concept of species as developed from Wiley (1978, modified from Simpson 1961) is characterized as

> a single lineage of ancestor–descendant populations which maintains its identity from other such lineages and which has its own evolutionary tendencies and historical fate.

This characterization was modified by Wiley and Mayden (in press) following criticism by Frost and Hillis (1990) regarding asexual species. In the revised concept the connotation of "ancestor–descendant populations" as a lineage was excluded to accommodate uniparental species. Following Wiley and Mayden (in press) a species is

> an entity composed of organisms which maintains its identity from other such entities through time and over space, and which has its own independent evolutionary fate and historical tendencies.

Positive aspects.—The ESC shares all of the positive aspects of the RSC except that it is not an operational concept. The ESC is a lineage concept that is nonrelational. Thus, the attributes and patterns of species can be correctly interpreted with respect to their unique descent. The ESC accommodates asexual species, species formed by hybridization events, and ancestral species. It does not require knowledge of, nor specific changes in, the SMRS. There is no threshold of a particular class of attributes needed for the existence of a species. Finally, reproductive isolation, as a plesiomorphic similarity, is viewed as an artificial criterion for the delineation of organismic diversity. Like the RSC,

the occurrence of hybridization does not a priori invalidate speciation.

Negative aspects.—The only perceived problem with this concept is that it has not been viewed as operational or user friendly.

Recommendation.—The ESC is the most flexible of the species concepts in that it accepts all types of proposed species discovered to date. It shares these positive aspects with the RSC but is not limited to only those entities with changes in SMRS. Because of these positive attributes, we recommend the use of the ESC as a conceptual basis for units of biodiversity.

Evolutionarily Significant Unit

Recognizing the importance of protecting elements of aquatic communities for their continued existence as functional communities, Waples (1991: 12) developed the concept of the evolutionarily significant unit (ESU) in an attempt to identify and conserve genetic resources in nature. The ESU was defined by Waples as

> a population (or group of populations) that (1) is substantially reproductively isolated from other conspecific population units, and (2) represents an important component in the evolutionary legacy of the species.

The U.S. Endangered Species Act (ESA; 16 U.S.C. §§ 1531 to 1544) recognizes "distinct" populations of vertebrates for protection as species, but, fortunately or unfortunately, does not provide guidance in identification of a distinct population. The anticipated significance of the ESU was to assist biologists in the identification of these distinct populations on the basis of scientific data. Once identified, these populations could then be considered for protection and conservation of genetic resources as distinct populations under the ESA. The history of the ESU, and its theoretical and empirical applications with salmonids, was reviewed by Waples (1995, this volume).

The intent of the ESU is critically important in conservation biology. However, some may wonder why the ESU is being considered in discussions of phylogenetic systematics and species. First, we are of the opinion that if the ESU is to be successful in its anticipated goals it must be consistent with theoretical and empirical foundations of evolutionary biology and with the most effective conceptualization of species. Second, it is unequivocally true that some of the entities satisfying criteria of the ESU also satisfy criteria of one or more contemporary concepts of species. Thus, discussions of species concepts do not necessarily focus "on a higher level of biological organization than that considered by the ESU concept," as suggested by Waples (1995: 25). Therefore, a biologically beneficial concept, such as the ESU, cannot be developed in a vacuum relative to discussions of species or the science of evolutionary biology. As such, the ESU is not immune to evaluation based on the same criteria used to evaluate concepts of species.

The terminology used in the conceptualization of the ESU is notably similar to the characterizations used for the reviewed species concepts. The reliance upon criteria such as "substantially reproductively isolated from other conspecific population units" and "important component in the evolutionary legacy of the species" incorporates attributes traditionally viewed as qualities of species (our deletions). Through the deletion of certain words, the definition by Waples (1991) is confused easily with definitions of species. The ESU combines the reproductive isolation or mate-recognition system of the nondimensional BSC and the RSC and invokes the evolutionary lineage perspective and the dimensionality of the PSC and ESC. The distinct lineage component of the ESC is carried implicitly by the necessity of ESUs being an "important component in the evolutionary legacy of a species." These components are nothing more than the "identities" of cohesive groups of organisms through time and over space that possess their own independent evolutionary fate and historical tendencies. If one of these "things" is discovered it qualifies as an evolutionary species. Likewise by identifying distinct populations, the ESU concept necessarily has unveiled the necessary biological attributes of a species. Although the ESU has been proposed as a concept targeted at revealing distinct populations within species (Waples 1991, 1995), the distinction between distinct populations and species as natural, evolutionary entities is not made clear. Thus, our epistemological and ontological understanding of diversity of natural biological entities cannot be developed fully without discussion and comparison of all of these concepts together.

Positive aspects.—We identify at least four positive attributes of this concept towards achieving its anticipated goals. First, similar to the ESC, the ESU advocates the evolutionary legacy of populations and species and thus can be considered a concept emphasizing lineages. In this sense, ESUs possess all of the qualities of individuals, as do species. In discussions of the ESU concept, Waples (1991, 1995) emphasizes that all attributes of organisms are appropriate for determining the evolutionary legacy of species, including behavior, life history,

phenotypes, ecology, and habitat. Second, the ESU is a multidimensional concept, applicable in sympatry or allopatry. Waples (1995:25) emphasizes the multidimensionality of the concept in that "consideration of allopatric populations as possible ESUs need only focus on the strength and duration of isolation that has actually occurred, not whether the allopatric units are hypothetically capable of reproducing successfully." Third, it is a nonrelational concept. Finally, like the BSC, the inclusion of the reproductive isolation criterion into the concept allows users to perceive more easily the "value" of the purity or integrity of the ESU.

Negative aspects.—Some qualities of this concept necessarily hinder its effectiveness in being biologically relevant for populations or species. First is the strong emphasis on genetics. Whereas Waples (1995) defends the use of genetic data, there is some reason for concern as to why the ESU concept relies heavily on genetic parameters. There are two ways of interpreting this emphasis. One, the ESU concept developed out of the perception that if we are looking at variability within species (e.g., populations) then the domain of population genetics and biology is appropriate. Alternatively, the concept could be emphasizing the importance of heritable attributes. Both of these could be considered positive aspects of the concept if it is understood that historical constraint (or evolutionary legacy) is not necessarily invoked in population genetic models. If, however, genetic data in the sense of protein electrophoresis or DNA analyses (raw data) are used as criteria for determining reproductive isolation and important components in the evolutionary legacy, then this should be viewed as a negative attribute of the concept. Waples (1995:22) discussed some of the common misconceptions regarding genetic information and emphasized that raw genetic data are important "primarily by providing information about reproductive isolation." Other attributes of organisms or populations are viewed as useful in identifying "adaptive genetic differences that generally must be inferred from nongenetic information." It is also noted that "if genetic data are not available, however, evidence to support an ESU must be found elsewhere, which inevitably places a greater burden of proof on other characters. Because data for other characters are often open to multiple interpretations, lack of genetic data may add complexity and contribute uncertainty to ESU determinations."

As discovered in our discussions on phylogenetic systematics and character evolution, divergence through descent can occur at multiple dimensions for a lineage. All heritable attributes being modified presumably are programmed genetically, and they can all be "open to multiple interpretations." There is nothing unique or less ambiguous about interpreting raw genetic data, especially when done outside a phylogenetic context. Because all types of attributes are being modified in the descent of populations or species in their environment, all should figure equally in interpretations of reproductive isolation and adaptive genetic differences. The problem is that the means usually employed to survey raw genetic variability and to infer other process information, such as run time, physiology, ecology, behavior, or coloration, are grossly trivial relative to the actual numbers of genes controlling either reproductive isolation or adaptive genetic differences. If we must rely so heavily upon raw genetic data to infer process, how can we be sure that we are examining variability encoding mate-recognition, run time, developmental physiology, behavior, etc.?

Second, the ESU recognizes only reproductive isolates, or populations that are substantially isolated. Whereas this may be viewed as an attractive aspect of a concept, it carries with it all the same negative connotations as reviewed under the BSC regarding restricted types of selection, divergence, and adaptation. As such, reliance upon this type of isolation in the concept will ultimately bias abilities to interpret the origin and maintenance of natural patterns and processes. This alone can be very detrimental to the achievement of objectives specified for biodiversity and the ESU.

Finally, with the reliance upon reproductive isolation, the ESU is necessarily incapable of being employed for either asexual or hybrid species. Although these types of species may represent the minority of vertebrate species, they nonetheless do exist, especially within fishes. If the concept of an ESU is to be more universally applicable to conserving evolutionarily important entities, either as distinct populations or as species, then it must be able to accommodate this type of diversity.

Recommendation.—The ESU made a valuable contribution towards the future preservation of biological diversity. This is especially true when considering that the biologists and agencies employing this concept hold as their operating principles the assumptions underlying polytypic biological species under the BSC. Without the guidelines spelled out in the ESU, there would be no sound justification for either recognizing or preserving many of these evolutionary entities under the ESA. Thus if one is operating within the framework of the BSC, not withstanding its notable shortcomings, then the ESU should be employed by federal and state agencies for identifying and preserving important biological entities.

However, there are two important concerns about this concept. First, is there really a need for this "biolegal" ESU concept when, by the nature of its conceptualization, the entities termed ESUs actually qualify as species? Many of them meet even the most stringent operational requirements of even the BSC. If these entities qualify as species, then they should be treated as such. As discussed at the beginning of this section, it is important to many disciplines, as well as to biological diversity itself, that equivalent biological entities be treated as such in our classifications and biotic inventories. When entities subsumed under one binomial are actually behaving as distinct evolutionary entities, we perform no service to them nor to the biologial community by treating them as a single species.

Second, if it is inconceivable that recognized ESUs can be considered equivalent to species under the ESC, then the concept must be modified. The ESU concept will be more successful in accomplishing its perceived objectives if it incorporates, in its future development and applications, the body of theory and empirical data behind the ESC and the field of systematic biology. As currently formulated, this concept, although well intended, will exclude known biodiversity and will unduly bias our perception of processes responsible for the observed diversity patterns. Incorrect assumptions about biological diversity we are trying to protect, brought about through misconceived formulations about the diversity, will only obstruct efforts to understand and preserve it.

Synthesis on Species Concepts

Today four different concepts of species are in practice, as well as the ESU for identifying distinct populations within a species. All of these concepts were evaluated on the same criteria for their abilities to accomplish the task of identifying naturally occurring biological diversity. It should be clear that the species concept employed really depends upon how interested we are in discovering natural patterns of biodiversity and using these patterns to reveal natural processes. Each of the concepts are derived from different inherent assumptions about descent with modification. Some concepts inherently handicap our abilities. Others are more accommodating of by-products of descent and permit a universal acceptance of all known entities that behave as species. These concepts provide us with greater assurance that our estimates of diversity will be accurate.

Unquestionably, the ESC is the most beneficial concept of species. The terms used in this concept adequately characterize our thoughts and perceptions of the things in nature viewed as species, without any nonessential overtones invoking operationalism or any processes other than descent. With any of the other concepts, including the ESU, portions of existing diversity will be intentionally excluded from inventories because they do not meet the criteria established for justifiable reality as either species or distinct populations. Equally devastating to our conservation efforts is the formal recognition of species or ESUs that are artificial constructs of nonnatural entities (e.g., heuristic example above involving Pacific Northwest salmonids). These taxa may never respond positively to classical conservation and management efforts because individually each taxon does not represent a natural biological product with a unique evolutionary history. Operationalism is an issue, of course, but it should not be an oppressive shroud distorting our view of natural systems, how they originated, or how they operate. Operational concerns are practical concerns, not theoretical concerns. If our theoretical and conceptual realms of biological science become subservient to, or are corrupted by, operational desires and mandates, then discovery and synthesis of the natural world will be doomed. Invoking a frequently used phrase by Dobzhansky, "Nothing in biology makes sense without evolution." Likewise, as Skolimowski (1974) has correctly articulated, "The difficulties of present biology are more conceptual than empirical. . . ." Given this perspective, the following observation by Paterson (1993:200) is particularly revealing about the way our current system largely operates with respect to the recognition and justification of biological diversity and understanding its evolution.

> Central to my theme is the fact that the [species] problem will not be solved by techniques alone, no matter how up-to-date they may be. The essential need is to be able to frame critical questions, informed by evolutionary insights. Only then do techniques become useful and important. It is often forgotten that techniques are tools that can be used by workers with skill and imagination, and what results from their use depends entirely on the conceptual grasp of the user.

Acknowledgments

We wish to thank Joel Cracraft, Darrell Frost, David Hull, and E. O. Wiley for enlightening discussions about species, and the National Science Foundation and Department of the Interior for support of this research.

References

Avise, J. C. 1994. Molecular markers, natural history and evolution. Chapman and Hall, New York.

Behnke, R. J. 1992. Native trout of western North America. American Fisheries Society Monograph 6.

Brooks, D. R. 1985. Historical ecology: a new approach to studying the evolution of ecological associations. Annals of the Missouri Botanical Garden 72:660–680.

Brooks, D. R., and D. A. McLennan. 1991. Phylogeny, ecology, and behavior: a research program in comparative biology. University of Chicago Press, Chicago.

Brooks, D. R., and D. A. McLennan. 1993. Historical ecology: examining phylogenetic components of community evolution. Pages 267–280 in R. E. Ricklifs and D. Schluter, editors. Species diversity in ecological communities. University of Chicago Press, Chicago.

Brooks, D. R., R. L. Mayden, and D. A. McLennan. 1992. Phylogeny and biodiversity: conserving our evolutionary legacy. Trends in Ecology & Evolution 7:55–59.

Burr, B. M., and R. L. Mayden. 1992. Phylogenetics and North American freshwater fishes. Pages 18–75 in Mayden (1992a).

Carson, H. L. 1957. The species as a field for gene recombination. Pages 23–38 in E. Mayr, editor. The species problem. American Association for the Advancement of Science Publication 50, Washington, DC.

Cashner, R. C., J. S. Rogers, and J. M. Grady. 1988. *Fundulus bifax*, a new species of the subgenus *Xenisma* from the Tallapoosa and Coosa river systems of Alabama and Georgia. Copeia 1988:673–683.

Chesser, R. T., and R. M. Zink. 1994. Modes of speciation in birds: a test of Lynch's (1989) method. Evolution 48:490–497.

Coyne, J. A., H. A. Orr, and D. J. Futuyma. 1988. Do we need a new species concept? Systematic Zoology 37: 190–200.

Cracraft, J. 1983. Species concepts and speciation analysis. Pages 159–187 in R. F. Johnson, editor. Current ornithology, volume 1. Plenum Press, New York.

Darwin, C. 1859. On the origin of species by means of natural selection, or the preservation of favoured races in the struggle for life. John Murray, London, UK.

de Queiroz, K., and M. J. Donoghue. 1988. Phylogenetic systematics and the species problem. Cladistics 4:317–338.

de Queiroz, K., and M. J. Donoghue. 1990. Phylogenetic systematics and species revisited. Cladistics 6:83–90.

DeMarais, B. D., T. E. Dowling, M. E. Douglas, W. L. Minckley, and P. C. Marsh. 1992. Origin of *Gila seminuda* (Teleostei: Cyprinidae) through introgressive hybridization: implications for evolution and conservation. Proceedings of the National Academy of Sciences 89:2747–2751.

Dobzhansky, T. 1935. A critique of the species concept in biology. Philosophy of Science 2:344–355.

Dobzhansky, T. 1950. Mendelian populations and their evolution. American Naturalist 74:312–321.

Dobzhansky, T. 1970. Genetics of the evolutionary process. Columbia University Press, New York.

Dowling, T. E., and B. D. DeMarais. 1993. Evolutionary significance of introgressive hybridization in cyprinid fishes. Nature 362:444–446.

Dowling, T. E., G. R. Smith, and W. M. Brown. 1989. Reproductive isolation and introgression between *Notropis cornutus* and *Notropis chrysocephalus* (Family Cyprinidae): comparison of morphology, allozymes, and mitochondrial DNA. Evolution 43:620–634.

Echelle, A. A., and A. F. Echelle. 1992. Mode and pattern of speciation in the evolution of inland pupfishes in the *Cyprinodon variegatus* complex (Teleostei: Cyprinodontidae): an ancestor-descendant hypothesis. Pages 691–709 in Mayden (1992a).

Echelle, A. A., A. F. Echelle, and C. D. Crozier. 1982. Evolution of an all female fish, *Menidia clarkhubbsi* (Atherinidae). Evolution 37:772–784.

Eldredge, N., and J. Cracraft. 1980. Phylogenetic patterns and the evolutionary process. Columbia University Press, New York.

Etnier, D. A., and J. D. Williams. 1989. *Etheostoma (Nothonotus) wapiti* (Osteichthyes: Percidae), a new darter from the southern bend of the Tennessee River system in Alabama and Tennessee. Proceedings of the Biological Society of Washington 102:987–1000.

Etnier, D. A., and W. C. Starnes. 1993. The fishes of Tennessee. University of Tennessee Press, Knoxville.

Frost, D., and D. M. Hillis. 1990. Species concepts and practice: herpetological applications. Herpetologica 46:87–104.

Funk, V. A., and D. R. Brooks. 1990. Phylogenetic systematics as the basis of comparative biology. Smithsonian Contributions to Botany 73:1–45.

Futuyma, D. J. 1986. Evolutionary biology, 2nd edition. Sinauer, Sunderland, Massachusetts.

Gould, S. J., and R. C. Lewontin. 1979. The spandrels of San Marco and the Panglossian paradigm: a critique of the adaptationist programme. Proceedings of the Royal Society of London Series B 205:581–598.

Grady, J. M., and W. H. LeGrande. 1992. Phylogenetic relationships, modes of speciation, and historical biogeography of the madtom catfishes, genus *Noturus* Rafinesque (Siluriformes: Ictaluridae). Pages 747–777 in Mayden (1992a).

Hennig, W. 1966. Phylogenetic systematics. University of Illinois Press, Champaign.

Hillis, D. M., J. J. Bull, M. E. White, M. R. Badgett, and I. J. Molineux. 1992. Experimental phylogenetics: generation of a known phylogeny. Science 255:589–592.

Humphries, C. J., R. I. Vane-Wright, and P. H. Williams. 1991. Biodiversity reserves: setting new priorities for the conservation of wildlife. Parks 2:34–38.

Lauder, G. V. 1986. Homology, analogy, and the evolution of behavior. Pages 9–40 in M. H. Nitecki and J. A. Kitchell, editors. Evolution of animal behavior. Oxford University Press, Oxford, UK.

Lauder, G. V. 1990. Functional morphology and systematics: studying functional patterns in an historical context. Annual Review of Ecology and Systematics 21:317–340.

Lynch, J. D. 1989. The gauge of speciation: on the frequencies and modes of speciation. Pages 527–553 in D. Otte and J. A. Endler, editors. Speciation and its consequences. Sinauer, Sunderland, Massachusetts.

Margulis, L. 1970. The origin of eukaryotic cells. Yale University Press, New Haven, Connecticut.

Masters, J., D. M. Lambert, and H. E. H. Paterson. 1984. Scientific prejudice, reproductive isolation, and apartheid. Perspectives in Biology and Medicine 28:107–116.

Mayden, R. L. 1985. Biogeography of Ouachita highland fishes. Southwestern Naturalist 30:195–211.

Mayden, R. L. 1986. Speciose and depauperate phylads and tests of punctuated and gradual evolution: fact or artifact? Systematic Zoology 35:591–602.

Mayden, R. L. 1987a. Historical ecology and North American highland fishes: a research program in community ecology. Pages 210–222 in W. J. Matthews and D. C. Heins, editors. Community and evolutionary ecology of North American stream fishes. University of Oklahoma Press, Norman.

Mayden, R. L. 1987b. Pleistocene glaciation and historical biogeography of North American central-highland fishes. Pages 141–152 in W. C. Johnson, editor. Quaternary environments of Kansas. Kansas Geological Survey Guidebook 5, Lawrence.

Mayden, R. L. 1988a. Systematics of the *Notropis zonatus* species group, with description of a new species from the interior highlands of North America. Copeia 1988:153–173.

Mayden, R. L. 1988b. Vicariance biogeography, parsimony, and evolution in North American freshwater fishes. Systematic Zoology 37:329–355.

Mayden, R. L., editor. 1992a. Systematics, historical ecology, and North American freshwater fishes. Stanford University Press, Stanford, California.

Mayden, R. L. 1992b. An emerging revolution in comparative biology and the evolution of North American freshwater fishes. Pages 866–890 in Mayden (1992a).

Mayden, R. L., and E. O. Wiley. 1992. The fundamentals of phylogenetic systematics. Pages 114–185 in Mayden (1992a).

Mayr, E. 1942. Systematics and the origin of species. Columbia University Press, New York.

Mayr, E. 1957. Species concepts and definitions. In E. Mayr, editor. The species problem. American Association for the Advancement of Science Publication 50, Washington, DC.

Mayr, E. 1963. Animal species and evolution. Harvard University Press, Cambridge, Massachusetts.

Mayr, E. 1970. Populations, species and evolution. Belknap Press, Cambridge, Massachusetts.

Mayr, E. 1982a. The growth of biological thought; diversity, evolution, and inheritance. Belknap Press, Cambridge, Massachusetts.

Mayr, E. 1982b. Processes of speciation in animals. Pages 1–19 in C. Barigozzi, editor. Mechanisms of speciation. A. R. Liss, New York.

Mayr, E. 1982c. Speciation and macroevolution. Evolution 36:1119–1132.

Mayr, E. 1988a. Toward a new philosophy of biology. Harvard University Press, Cambridge, Massachusetts.

Mayr, E. 1988b. The why and how of species. Biology & Philosophy 3:431–441.

Mayr, E. In press. The biological species concept. In

Q. D. Wheeler and R. Meier, editors. Species concepts and phylogenetic theory: a debate. Columbia University Press, New York.

Mayr, E. and P. D. Ashlock. 1991. Principles of systematic zoology. McGraw-Hill, New York.

McKitrick, M. C., and R. M. Zink. 1988. Species concepts in ornithology. Condor 90:1–14.

Meffe, G. K., and F. F. Snelson, Jr. 1989. Ecology and evolution of livebearing fishes (Poeciliidae). Prentice Hall, Englewood Cliffs, New Jersey.

Nelson, G., and N. Platnick. 1981. Systematics and biogeography: cladistics and vicariance. Columbia University Press, New York.

Nixon, K. C., and Q. D. Wheeler. 1990. An amplification of the phylogenetic species concept. Cladistics 6:211–223.

Office of Technology Assessment. 1987. Technologies to maintain biological diversity. OTA (Office of Technology Assessment)-F-330, Washington, DC.

Page, L. M. 1983. Handbook of darters. Tropical Fish Hobbiest Publications, Neptune City, New Jersey.

Page, L. M. 1985. Evolution of reproductive behaviors in percid fishes. Bulletin Illinois Natural History Survey, 33(3):275–295.

Paterson, H. E. H. 1993. Evolution and the recognition concept of species. In S. F. McEvey, editor. Johns Hopkins University Press, Baltimore, Maryland.

Robins, C. R., and six coauthors. 1991. Common and scientific names of fishes from the United States and Canada, 5th edition. American Fisheries Society Special Publication 20.

Robison, H. W., and T. M. Buchanan. 1988. Fishes of Arkansas. University of Arkansas Press, Fayetteville.

Rogers, J. S., and R. C. Cashner. 1987. Genetic variation, divergence, and relationships in the subgenus Xenisma of the genus Fundulus. Pages 251–264 in W. J. Matthews and D. C. Heins, editors. Community and evolutionary ecology of North American stream fishes. University of Oklahoma Press, Norman.

Rosen, D. E. 1978. Vicariant patterns and historical explanation in biogeography. Systematic Zoology 27:159–188.

Rosen, D. E. 1979. Fishes from the uplands and intermontane basins of Guatemala: revisionary studies and comparative biogeography. Bulletin of the American Museum of Natural History 162(5):267–376.

Simpson, G. G. 1951. The species concept. Evolution 5:285–298.

Simpson, G. G. 1961. Principles of animal taxonomy. Columbia University Press, New York.

Skolimowski, H. 1974. Problems of rationality in biology. Pages 205–224 in F. J. Ayala and T. Dobzhansky, editors. Studies in the philosophy of biology. McMillan, London.

Sloan, P. R. 1987. From logical universals to historical individuals: Buffond's idea of biological species. Pages 101–140 in J. Roger and J. L. Fischer, editors. Histoire du concept d'estèce les sciences de la vie. Singer-Polignac Foundation, Paris.

Stearley, R. F. 1992. Historical ecology of salmoninae, with special reference to Onchorhynchus. Pages 622–658 in Mayden (1992a).

Vane-Wright, R. I., C. J. Humphries, and P. H. Williams.

1991. What to protect?—systematics and the agony of choice. Biological Conservation 55:235–254.

Wallace, A. R. 1855. On the law which has regulated the introduction of new species. Annals and Magazine of Natural History 16(2):184–196.

Waples, R. S. 1991. Pacific salmon, Oncorhynchus spp., and the definition of "species" under the Endangered Species Act. U.S. National Marine Fisheries Service Marine Fisheries Review 53:11–22.

Waples, R. S. 1995. Evolutionarily significant units and the conservation of biological diversity under the Endangered Species Act. American Fisheries Society Symposium 17:8–27.

Warren, M. L., Jr. 1992. Variation of the spotted sunfish, Lepomis punctatus complex (Centrarchidae): meristics, morphometrics, pigmentation and species limits. Bulletin of the Alabama Museum Natural History 12:1–47, Tuscaloosa.

Wheeler, Q. D., and K. C. Nixon. 1990. Another way of looking at the species problem: a reply to de Queirox and Donoghue. Cladistics 6:77–81.

Wheeler, Q. D., and N. Platnick. In press. The argument for phylogenetic species. Pages in Q. D. Wheeler and R. Meier, editors. Species concepts and phylogenetic theory: a debate. Columbia University Press, New York.

Wiley, E. O. 1978. The evolutionary species concept reconsidered. Systematic Zoology 27:17–26.

Wiley, E. O. 1981. Phylogenetics: the theory and practice of phylogenetic systematics. Wiley, New York.

Wiley, E. O., and R. L. Mayden. 1985. Species and speciation in phylogenetic systematics, with examples from the North American fish fauna. Annals of the Missouri Botanical Garden 72:596–635.

Wiley, E. O., and R. L. Mayden. In press. The evolutionary species concept. Pages in Q. D. Wheeler and R. Meier, editors. Species concepts and phylogenetic theory: a debate. Columbia University Press, New York.

Wiley, E. O., D. Siegel-Causey, D. R. Brooks, and V. A. Funk. 1991. The compleat cladist. University of Kansas Publications of the Museum of Natural History 19:1–158.

Williams, G. C. 1966. Adaptation and natural selection. Princeton University Press, Princeton, New Jersey.

Williams, P. H., C. J. Humphries, and R. I. Vane-Wright. 1991. Measuring biodiversity: taxonomic relatedness for conservation priorities. Australian Systematic Botany 4:665–679.

Wilson, D. S. 1983. The group selection controversy: history and current status. Annual Review of Ecology and Systematics 14:159–188.

Wilson, E. O. 1992. The diversity of life. Norton, New York.

Wood, R. M. 1993. Phylogenetic systematics of the darter subgenus Nothonotus (Teleostei: Percidae). Doctoral dissertation. University of Alabama, Tuscaloosa.

Wood, R. M., and R. L. Mayden. 1992. Systematics, evolution and biogeography of Notropis chlorocephalus and N. lutipinnis. Copeia 1992:68–81.

Wood, R. M., and R. L. Mayden. 1993. Systematics of the Etheostoma jordani species group (Teleostei: Percidae), with descriptions of three new species. Bulletin of the Alabama Museum Natural History 16:31–44, Tuscaloosa.

American Fisheries Society Symposium 17:114–132, 1995

A Role for Molecular Systematics in Defining
Evolutionarily Significant Units in Fishes

LOUIS BERNATCHEZ[1]

Institut National de la Recherche Scientifique-Eau, 2800, rue Einstein
Sainte-Foy, Québec G1V 4C7, Canada

Abstract.—By summarizing the results of recent studies that have been conducted on coregonine, salmonine, and osmerid fishes, I illustrate how molecular systematics may improve recognition of evolutionarily significant units (ESUs). Three general observations are of importance: (1) comparison of population classifications based on molecular phylogenetic relationships and traditional taxonomy have indicated cases of polyphyletic and paraphyletic taxa, (2) unanticipated genetic discontinuities observed in all species studied have allowed the identification of distinct population groups that are presently ignored in taxonomy, and (3) the evolutionary significance of these units is supported in many instances by the observation of substantial reproductive isolation when these population groups are found in sympatry. These results emphasize that the goals of present conservation policies, notably preservation of genetic integrity and evolutionary processes, can be more effectively met if genetic relationships among populations are studied. An integrative approach accommodating different methods for assessing priority of protection for ESUs is introduced.

The goal of conservation biology is to preserve genetic integrity and evolutionary processes. The primary purpose of the U.S. Endangered Species Act (ESA; 16 U.S.C. §§ 1531 to 1544) is to prevent anthropogenic extinction of vulnerable groups of organisms in that conservation context. To qualify for ESA protection, groups must first appear on the official list of endangered species. Most efforts for listing groups have required that species under consideration are facing imminent threats to ecological and demographic status. Considerably less attention has been devoted to assessing the pertinence of listed species. However, there is a growing concern that given the limited resources devoted to conservation of biodiversity, priority recognition for protection should be commensurate with the evolutionary distinctiveness of the species (Rohlf 1991; Rojas 1992; Avise 1994).

Currently, the evolutionary distinctiveness of a species under consideration for ESA listing is largely determined by its hierarchical taxonomic recognition. For instance, bills House Resolution 1490 and Senate Resolution 1521 authorize the secretary of the Department of the Interior to determine and conserve endangered and threatened species in the following order of priority: (1) single-species genera, (2) species, (3) subspecies, and (4) distinct population segments. The acceptance of such a principle without further consideration may

detract conservation efforts from conservation's primary goal.

The most basic problem is that ESA listing, which is used to summarize biotic diversity, is dependent on traditional taxonomy, which is predominantly based on the analysis of phenotypic variation. Although such analysis often provides sufficient and unambiguous information to delineate well-defined groups of fishes (reviewed in Mayden 1992), its usefulness may be hampered by the limited numbers of characters available to assess relationships among taxa that are closely related or exhibit complex population structures and patterns of phenotypic variation. For example, detailed cladistic analyses of morphological traits commonly do not reveal more than one or two characters to delineate closely related taxa (Cavender and Coburn 1992; Simons 1992; Smith 1992). Sometimes, no informative characters are found, clustering several species in unresolved polytomy (e.g., Smith and Todd 1992). More importantly, rapidly evolving parallel populations may cluster together as polyphyletic groups when parallelisms are more probable than informative change (Felsenstein and Sober 1986). Recent studies suggest that replicate evolution of similar phenotypes may be common in north temperate fishes (Vuorinen et al. 1981; Foote et al. 1989; Hindar et al. 1986; McPhail 1993; Taylor and Bentzen 1993a). Consequently, there is growing evidence that traditional taxa may not always reflect natural evolutionary relationships (discussed in Avise 1994).

The second major problem of ESA listing con-

_____

[1]Present address: Département de biologie, GIROQ, Pavillon Vachon, Université Laval, Sainte-Foy, Québec G1K 7P4, Canada.

cerns identifying evolutionarily significant units (ESUs). This concept has recently been proposed in an effort to better define subunits of species for conservation purposes. According to this principle, two criteria must be satisfied by one or more populations to be considered an ESU. First, there must be substantial reproductive isolation between the population(s) and others. Second, the population unit must represent an important component of the species' evolutionary legacy. Few conservation biologists would argue the legitimacy of these criteria. However, efforts towards establishing clear guidelines by developing an integrative approach for determination of these criteria are just emerging (Avise and Ball 1990; Dizon et al. 1992; Riddell 1993).

A first step towards this general goal of establishing guidelines is to assess the usefulness of different types of data (e.g., phenotypic or genetic) and types of analyses (e.g., phenetic or phyletic) in defining an ESU. The potential usefulness of molecular systematics for this purpose has been clearly illustrated over the recent years (Avise 1994, and references therein). Particularly, phylogeography, the phylogenetic study of DNA sequence variation concomitant with the analysis of its geographic distribution (Avise et al. 1987) has proven useful for identifying evolutionarily distinct population units within a given species. Yet, despite the general claim that a combination of multiple characteristics adds rigor to the analysis of population divergence (e.g., Avise and Ball 1990; Behnke 1992; Dizon et al. 1992), few attempts have been made to use these informative evolutionary criteria. For example, in a book that probably represents the most important contemporary contribution to the study of systematics of North American fishes (Mayden 1992), no chapter was devoted to the use of DNA analysis.

The objective of this paper is twofold. First, by presenting the results of recent studies that have been conducted on mitochondrial DNA (mtDNA) variation of coregonine, salmonine, and osmerid fishes, I want to illustrate how molecular systematics may improve recognition of ESUs. Examples of mtDNA analysis were chosen because this approach largely dominates the literature related to DNA studies so far, whereas nuclear gene studies of fishes are still in their infancy. However, the case made here for integrating molecular tools to develop rigorous ESU criteria also applies to these emerging markers. Second, I want to propose a model that accommodates the integration of molecular systematic approaches with other types of data to develop analytical and survey protocols for ESU

analysis and decision-making policy, taking into account potential advantages and weaknesses of the different tools at hand.

Methods

This study summarizes recent research conducted on population complexes of whitefishes *Coregonus* sp., the brown trout *Salmo trutta*, and the rainbow smelt *Osmerus mordax* by means of the analysis of mtDNA differences. In each example, background information on phenotypic variation, geographic distribution, and traditional taxonomy is presented. The most relevant patterns of mtDNA variation are summarized and discussed in the context of ESU definition. Detailed treatments are presented in the original references cited for each example. When available, mtDNA results are compared with those obtained by other genetic studies involving allozymes. Congruence between genetic and phenotypic data is discussed, and population genetic divergence is compared with current taxonomy. Support for the evolutionary significance of population groups as revealed by mtDNA variation is provided by comparing genetic variation in contact zones.

Whitefish Population Complex

Background

The whitefish *Coregonus* (Salmonidae) has a continuous circumpolar distribution in the northern hemisphere, exhibiting extreme phenotypic variation in morphological and life history characteristics. In this paper, whitefish will refer to the Palearctic and Nearctic populations that have been divided into two major nomenclatural species complexes based on continental distribution: the European whitefish *C. lavaretus* L. (Reshetnikov 1988) and the North American lake whitefish *C. clupeaformis* Mitchill (Scott and Crossman 1973). Depending on authorship, these two groups may be either considered as a single polytypic species (e.g., Bodaly et al. 1991) or split into numerous taxa (e.g., Lelek 1987; Page and Burr 1991) as detailed below.

There is no clear pattern in the distribution of phenotypic variation in whitefish, except that Eurasian populations are generally phenotypically more variable than are North American ones (Himberg 1970; Lindsey et al. 1970). Numerous cases of phenotypic variation are found among distinct populations occurring sympatrically. Despite their great phenotypic diversity, the primary problem in classifying whitefish populations is that there is not a single morphological trait which provides diagnostic criteria for resolving relationships among them (Behnke 1972; Smith and Todd 1992). For example, even the widely recognized distinction between the European (*C. lavaretus*) complex and the North American lake whitefish does not rely on a valid diagnostic criteria. Indeed, this distinction has been largely based on continental origins of the populations (Reshetnikov 1975). In the absence of diagnostic differences, taxonomic recognition of whitefish groups has been based on modal counts of meristic characters that are highly susceptible to convergence, either from environmentally determined expression or selective pressure (Svärdson 1979; Lindsey 1981). Consequently, and despite extensive literature reflecting major efforts to understand their relationships (reviewed in McPhail and Lindsey 1970; Scott and Crossman 1973; Svärdson 1979; Smith and Todd 1992), the taxonomy of whitefish is still inconsistent, reflecting the difficulties encountered in the analysis of morphological variation to infer evolutionary relationships among whitefish populations.

Mitochondrial DNA Variation among Populations

Over the past several years, analyses of mtDNA variation among coregonine fishes have addressed questions of their patterns of radiation, phylogeography, stock structure, and population diversity (Bernatchez and Dodson 1990a, 1990b, 1991, 1995; Bernatchez et al. 1988, 1989, 1991a, 1991b; Vuorinen et al. 1993). Altogether, 1,075 whitefish specimens from 75 populations that represent the range of distribution and numerous phenotypic forms were analyzed. An average of 12 adult specimens per population were used. Purified mtDNA was digested with 13 restriction enzymes that generated an average of 90 restriction sites per fish. Restriction fragment patterns were coded for presence or absence of sites in binary matrices, which were used in character-based analyses to generate phylogenetic trees based on Wagner parsimony criteria. The MIX program from the PHYLIP 3.4 computer package was used[2]. Symplesiomorphic and autapomorphic characters were omitted from the site matrix for tree construction. A consensus tree was obtained by running 100 bootstrap replicates with the BOOT program from PHYLIP. Phylogenetic trees were rooted to a close relative of whitefish, the

vendace *C. albula* (Bernatchez et al. 1991a; Bodaly et al. 1991). We estimated nucleotide sequence divergence (P) among mtDNA genotypes and superimposed branch-length estimates on the phylogenetic tree.

One hundred and ten whitefish mtDNA genotypes were identified (Figure 1a). These genotypes clustered into five major clades that were each distinguished by two or three apomorphies represented by unique restriction sites (Bernatchez and Dodson 1994). Clades were supported at relatively high bootstrap levels (range 75–97%), and sequence divergence estimates among them varied between 0.4 and 1.2%. In comparison with levels of variation found in other fishes, these values correspond to those at an intraspecific population level (Billington and Hebert 1991).

All groupings corresponded to a geographic pattern of distribution (Figure 1b). Except for localized contact zones, populations from northeastern North America were fixed for group II of the phylogeny based on mtDNA, and all other populations across North America, excluding Beringia (Alaska and Yukon), were fixed for group I. Group III was found in only the Yukon River basin (Alaska and Yukon), where it overlapped in distribution with group IV. Group IV dominated northern Eurasia, constituting 88% of whitefish sampled. Group V contained 92% of all whitefish sampled from central alpine lake populations in Europe. Based on their genetic distinctiveness and their distributions, which corroborate patterns of dispersal from presumed refugia (McPhail and Lindsey 1970), it is most likely that these mtDNA clades identify population races that evolved in isolation during or before the Pleistocene and that secondary contacts among populations have been since limited (Bernatchez and Dodson 1994).

Congruence in Genetic Groups and Taxonomic Distinction

Geographic patterns of whitefish population genetic assemblages provided a basis to challenge the validity of traditional whitefish taxa based on the principle that taxa should identify groups of populations with distinct evolutionary origins. The tree constructed from frequency distribution and sequence divergence matrices suggested that several taxa inadequately reflect the evolutionary distinctiveness of whitefish populations (Figure 1c). In many cases, populations clustered independently of phenotypic similarity, suggesting polyphyletic ori-

---

[2]Felsenstein, J. 1992. PHYLIP (Phylogeny inference package) Version 3.4. Department of Genetics, SK-50, University of Washington, Seattle, 98195, USA.

gins. For example, mtDNA population analysis demonstrated that populations of lake whitefish from Beringia were more closely related to Eurasian populations than to other North American ones. This pattern of mtDNA variation in lake whitefish is congruent with nuclear genetic variation estimated from allozymes (Bodaly et al. 1991, 1992.). In the rest of North America, lake whitefish formed two evolutionarily distinct population groups distinguished by alternate fixation of groups I and II (except in the contact zone), that were previously undetected.

Within the Beringian–Eurasian cluster, populations from central and northern Europe formed two distinct groups, which largely corresponded to the geographic distribution of clades IV and V. No clear association was observed between these results and present taxonomic designations. Eurasian whitefish populations may be considered a unique, but polymorphic, taxon (Reshetnikov 1988). Alternatively, many authors refer to whitefish as *Coregonus* sp. (e.g., Chaumeton et al. 1989; Pedroli et al. 1991). Lumping all Eurasian variants under a single taxon fails to recognize the existence of two evolutionarily distinct assemblages in Eurasian populations. At the other extreme, other authors designate whitefish morphotypes as separate species according to modal counts of gill rakers (e.g., Svärdson 1957, 1979; Muus and Dahlstrom 1981; Lelek 1987), and many synonymous names are often used to designate the same phenotype (summarized in Lelek 1987). Taxonomic recognition has apparently resulted in designation of some polyphyletic taxa (Figure 1c). *C. lavaretus* found in northern and central Europe appear genetically closer to numerous taxa from the same region than to *lavaretus* populations from other regions. As in studies of allozyme variation (Vuorinen et al. 1986; Heinonen 1988), the analysis of mtDNA did not provide any basis for recognizing groups with different modal gill-raker counts as distinct taxa and revealed that groups based on gill raker counts were not genetically divergent. Altogether, the geographic partitioning of two mtDNA clades between northern and central Europe and the lack of significant genetic divergence among phenotypically distinct populations suggest that recent replicate phenotypic radiation within two distinct groups may partly explain the morphological diversity in whitefish. These observations add to the growing concern that some phenotypically distinct whitefish populations may not warrant taxonomic distinction (Heinonen 1988; Heese 1992).

Evolutionary Significance of Mitochondrial DNA Assemblages

Whitefish assemblages distinguished by distinct mtDNA clades may not simply represent groups of populations with different mtDNA labels but may also deserve recognition as ESUs. This is indicated by the study of a contact zone of whitefish populations having distinct mtDNA groupings. In northern Maine and southeastern Québec, phylogenetic groupings I and II overlap in distribution (Bernatchez and Dodson 1990a; Figure 1d). In several lakes of that region, two genetically separable whitefish populations are found to coexist sympatrically but are reproductively isolated and differ in ecological niche occupations, life history characteristics, and allelic variation at enzymatic loci (Fenderson 1964; Kirkpatrick and Selander 1979). In Cliff Lake (Maine) sympatric populations were fixed for either mtDNA group I or II. These results indicated that the allopatric evolution of these two groups of whitefish during the last glaciation events has led to the development of reproductive barriers.

Rainbow Smelt

Background

The rainbow smelt *Osmerus mordax* Mitchill is an osmerid species native to watersheds of the northwestern Atlantic Ocean, from New Jersey to Labrador (Scott and Crossman 1973). The rainbow smelt exhibits extensive life history diversity and is composed of anadromous and lacustrine, landlocked populations. Lacustrine populations are commonly found as normal- and dwarf-size phenotypes. Populations classified as dwarf or normal differ in several life history, meristic, and morphometric characteristics (Taylor and Bentzen 1993a). Furthermore, these dwarf and normal forms may be found sympatrically in several lakes, where they are apparently reproductively isolated. Consequently, normal- and dwarf-size rainbow smelt have been recognized by some as two distinct species, *O. mordax* and pygmy smelt *O. spectrum* (Cope 1870; Lanteigne and McAllister 1983; McAllister 1985; Mayden et al. 1992), whereas phenotypic dissimilarity between the two groups has been judged insufficient by others to justify taxonomic distinction (e.g., Robins et al. 1991).

Mitochondrial DNA Variation among Populations

Analyses of mtDNA variation among native rainbow smelt populations provided one way to docu-

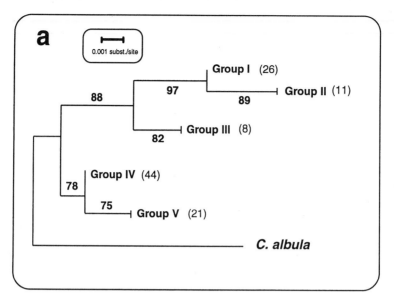

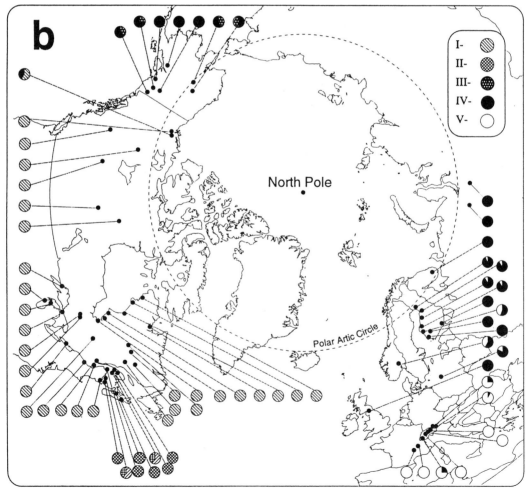

FIGURE 1.—See caption on facing page.

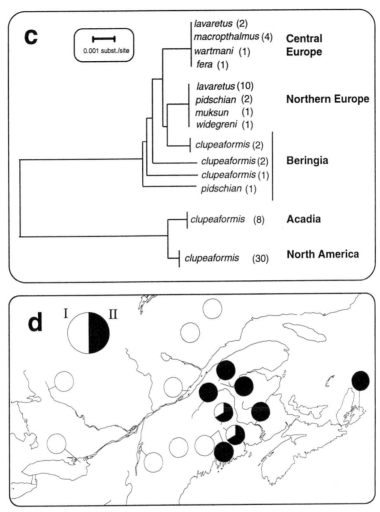

FIGURE 1.—Mitochondrial DNA (mtDNA) variation in whitefish. (**a**) Rooted majority-rule consensus tree clustering 110 genotypes into five monophyletic groups. Bootstrap estimates (%) are given along branches. Numbers in parentheses refer to number of genotypes. (**b**) geographic distribution of five mtDNA phylogenetic groupings. (**c**) Unweighted pair-group method with arithmetic averages phenogram clustering whitefish populations. Phenogram based on the distance matrix of the maximum likelihood estimation of the net average number of nucleotide substitutions per site between populations. Nomenclature follows Robins et al. (1991) and Lelek (1987). Numbers in parentheses refer to number of locations. (**d**) Contact zones between whitefish phylogenetic groups I and II in northeastern North America (after Bernatchez and Dodson 1990a, 1994).

ment their evolutionary relationships. The mtDNA of 1,165 individuals was screened by restriction fragment length polymorphism (RFLP) analysis, and 218 mtDNA genotypes were identified among 42 populations representing anadromous and lacustrine normal- and dwarf-size phenotypes from geographically separate sampling sites (Baby et al. 1991; Taylor and Bentzen 1993a, 1993b; Bernatchez 1995; Bernatchez et al. 1995a; L. Bernatchez, Universite Laval, unpublished). A Wagner parsimony analysis of restriction site variation (as described for whitefish) grouped the mtDNA genotypes into two major clades, A and B (Figure 2a), which are distinguished by six apomorphies. Distance-based analysis also resolved the same two groups, and sequence divergence was estimated at 0.8% between the two clades, which is relatively high among northern fishes (Billington and Hebert 1991). These two mtDNA groupings showed a strong geographic pattern of distribution (Figure 2b). Most eastern pop-

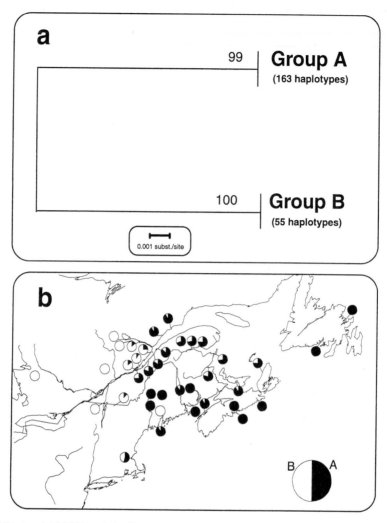

FIGURE 2.—Mitochondrial DNA variation in normal- and dwarf-size rainbow smelt. (a) Majority-rule consensus tree clustering 218 genotypes into two monophyletic groups. Details are presented in original references listed in introduction. Numbers above branches represent bootstrap estimates (%). (b) Geographic distribution of the two rainbow smelt phylogenetic groupings. (c) Population tree clustering 19 normal- and dwarf-size rainbow smelt populations fixed for either one of the phylogenetic groups A or B. Numbers in parentheses refer to number of populations. (d) Contact zone between the two rainbow smelt phylogenetic groups in the upper estuary of the St. Lawrence River.

ulations were either fixed or largely dominated (75% or more) by group A genotypes and western populations by group B. This phylogeographic pattern can best be explained by an allopatric origin of these monophyletic groupings (Avise 1989), presumably in different glacial refugia, followed by differential recolonization.

Congruence in Mitochondrial DNA Patterns and Phenotypic Differentiation

Both normal- and dwarf-size populations were fixed for either mtDNA group A or B, depending on their geographic location (Figure 2c). These results suggest a polyphyletic origin of phenotypic differentiation in rainbow smelt populations. Alternatively, it is possible that normal- and dwarf-size populations do represent monophyletic entities and that mtDNA variation may not correctly depict their evolutionary relationships. This scenario would imply that both dwarf and normal groups have dispersed extensively over the entire range of the species, that complete mtDNA introgression occurred between both groups within each region, or that homoplasy occurred at the six diagnostic

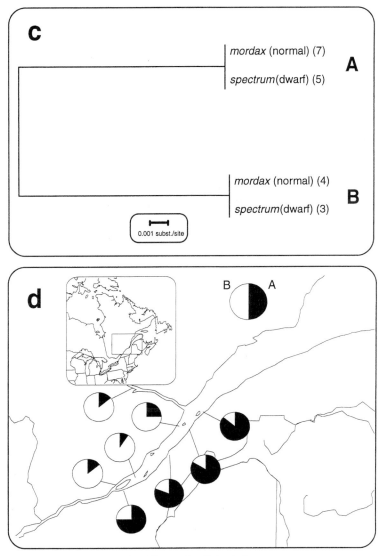

FIGURE 2.—Extended.

mutational sites distinguishing mtDNA groups A and B. At present, there are no nuclear gene data to confirm or dispute this hypothesis. Nevertheless, the scenario whereby mtDNA groups A and B represent two ancestral rainbow smelt assemblages that dispersed westward and eastward from separate glacial refugia, and within which dwarf and normal ecotypes evolved repeatedly, seems more parsimonious at present. Additional support of multiple origins of these phenotypes was indicated by detailed analyses within each grouping (Baby et al. 1991; Taylor and Bentzen 1993a). Taylor and Bentzen (1993a) suggested that replicate evolution of phenotypic patterns among rainbow smelt populations most likely result from strong selective pressures related to trophic ecology. Altogether, these results suggest that these phenotypes appear to be better indicators of adaptive processes than of phylogenetic tracers.

Evolutionary Significance of Mitochondrial DNA Assemblages

Although the taxonomic recognition of phenotypically distinct rainbow smelt populations apparently may not best reflect their evolutionary history, lumping the two genetic population groups in a single taxon may overlook divergence between them. As with whitefish, evolutionary significance of these groups is indicated by mtDNA variation in a contact zone (Bernatchez et al. 1995a). Statistical analysis of frequency distributions of mtDNA genotypes indicated that anadromous adult rainbow smelt sampled on the north and south shores of the upper estuary of the St. Lawrence River constituted two genetically distinct stocks (Figure 2d). Thus, no significant differences ($P > 0.05$) in the frequency distribution of genotypes was observed among samples ($N = 30$ to 36 per sample) from the north and south shore, respectively. In contrast, a marked genetic divergence ($P < 0.001$) was observed between south-shore (dominated by the A group) and north-shore (dominated by the B group) stocks (Figure 2d). Because most landlocked western populations were fixed for the B group and most landlocked eastern populations for the A group, two formerly allopatrically separated rainbow smelt assemblages may have recolonized the upper St. Lawrence estuary. Despite absence of a contemporary physical barrier and lack of apparent phenotypic variation between rainbow smelt that spawn on either shore of the upper St. Lawrence estuary, mtDNA gene flow appears restricted between the two groups. Because it is generally accepted that mtDNA crosses reproductive boundaries more easily than do nuclear genes (Takahata and Slatkin 1984; Dowling and Hoeh 1991; Dizon et al. 1992), this restriction suggests a reproductive constraint between these rainbow smelt assemblages. Mitochondrial DNA variation may thus fulfill the criteria for recognizing these groups as evolutionarily significant.

Brown Trout Population Complex

Background

The brown trout *Salmo trutta* is native to Eurasia and North Africa, where it has numerous forms and also exhibits complex patterns of phenotypic diversity and life history variation within geographical areas (reviewed in Blanc et al. 1971; Lelek 1987). There are few useful phenotypic criteria for classifying these divergent groups of brown trout populations (Balon 1968; Behnke 1972), although it has been common practice to recognize numerous brown trout taxa based on life history, morphometric, modal meristic count, and coloration pattern variations (e.g., Berg 1948). In addition to limitations of diagnostically distinguishing different taxa,

the potential for replicate expression of some characters further limits their usefulness for the assessment of brown trout population origins (Hindar et al. 1991). This is reflected by taxonomic disagreements persisting in the contemporary literature, in which the same populations may be treated as species, subspecies, or morpha, depending on authorship (Berg 1948; Blanc et al. 1971; Behnke 1972; Lelek 1987).

Since the late 1970s, the evolutionary history of brown trout has been addressed through genetic studies, particularly electrophoretic analysis of allozyme variation (reviewed by Ferguson 1989; Guyomard 1989). Allozyme studies have demonstrated that populations classified as subspecies according to ecological patterns of variation, that is, the anadromous (*S. t. trutta*), lacustrine (*S. t. lacustris*), and fluviatile (*S. t. fario*) modes of life, do not represent monophyletic groups (Guyomard et al. 1984; Hindar et al. 1991). Allozyme studies have also discerned divergence among populations into several geographical forms. In an analysis of French populations, Krieg and Guyomard (1985) demonstrated an important dichotomic differentiation between Atlantic and Mediterranean drainage populations and the genetic uniqueness of some Corsican populations. Osinov (1984) and Bernatchez and Osinov (1995) also found high levels of divergence between brown trout populations from the Baltic Sea, White Sea, and basins of the Black and Caspian seas basins. Hamilton et al. (1989) proposed that recolonization by two distinct races was responsible for the genetic diversity of brown trout in northwestern Europe.

Mitochondrial DNA Variation

In recent years the polymerase chain reaction (PCR) and direct sequencing were used to analyze mtDNA variation and infer phylogenetic relationships among different phenotypic forms of brown trout distributed throughout western Europe (Bernatchez et al. 1992; Hall 1992; Giuffra et al. 1994). Three hundred and seventy-six specimens representing 42 populations from the Atlantic, Adriatic, and Mediterranean basins were analyzed. Twelve mtDNA genotypes were found from sequencing 1,250 base pairs (bp) of mitochondrial genes: two segments of the control region (310 and 330 bp), cytochrome *b* (315 bp), and ATPase subunit VI (295 bp) (Figure 3a). Character-based analysis was conducted using the PHYLIP 3.5c computer package[3]. Sequence data were used to generate phylogenetic trees according to a maximum parsimony criterion by means of the DNAPARS program. Majority-rule consensus trees were constructed using the CONSENSE program, and confidence statements on branches were estimated by running DNAPARS on 1,000 bootstrap replicates obtained by the SEQBOOT program. Trees were rooted using Atlantic salmon *Salmo salar* as an outgroup. The 12 genotypes clustered into five distinct clades that were diagnostically differentiated by three to seven apomorphies. The distinctiveness of these groupings was also demonstrated by their net pairwise sequence divergence estimates, which varied from 0.64 to 1.52% (Giuffra et al. 1994).

This survey of mtDNA variation was recently enhanced by adding 147 additional individuals representing 36 populations from more eastern and southern parts of the native range (Bernatchez and Osinov 1995; L. Bernatchez, unpublished data). A PCR-RFLP analysis of a 5,000 bp segment that encompassed the entire control region, cytochrome *b* gene and the ND5/6 (subunits 5 and 6 of the NADH dehydrogenase gene complex) region was performed, following methods described by Bernatchez et al. (1995b). This analysis identified two to three apomorphies diagnostic for each of the five phylogenetic groupings, which unambiguously classified all brown trout into one of the groupings. This procedure was also used to characterize mtDNA variants ($N = 20$) found in an extensive study of northwestern European brown trout populations, in which more than 1,000 individuals from 35 populations were analyzed (A. Ferguson et al., University of Belfast, unpublished).

These data sets describe the geographic distribution of major mtDNA phylogenetic groupings in brown trout over most of its once native range; the data encompassing different geographic and ecological forms and reveal a pattern of geographic distribution among the five groups (Figure 3b). All brown trout from the Atlantic basin, the Moroccan populations, and a few individuals sampled in headwaters of the Danube drainage belonged to group IV. Group III was restricted to the eastern part of the range, in the basins of the Black, Caspian, and Aral seas. Group V was restricted to the northern Adriatic basin, particularly the Pô River drainages. Group II was found in only the Mediterranean basin, from Spain to Greece. Group I was also

exclusively found in drainages of the Mediterranean basin but had a broader distribution, from Spain to eastern Turkey. Although their distributions overlap, the three mtDNA groupings confined to the Mediterranean basin (I, II, and V) were usually found in distinct populations (Bernatchez et al. 1992; Giuffra et al. 1994; L. Bernatchez, unpublished). Due to their phylogenetic distinctiveness, and their congruent pattern of geographic distribution with that of allozymes (Osinov 1984; Krieg and Guyomard 1985; Hamilton et al. 1989; Giuffra 1993; Bernatchez and Osinov 1995), these monophyletic mtDNA groupings likely represent brown trout population assemblages of ancient allopatric origin. Subsequent natural intergradation among these was limited over the last several hundred thousand years (Bernatchez et al. 1992).

Congruence in Mitchondrial DNA and Phenotypic Patterns of Differentiation

In many instances, distributions of mitochondrial clusters lacked correspondence with recognized taxa (Figure 3c). A correspondence between phenotypic and phylogenetic divergence was observed in one taxon, *S. t. marmoratus*, for which monophyly was supported both by nuclear and mitochondrial data (Giuffra 1993; Giuffra et al. 1994). In contrast, a polyphyletic origin of the morphologically similar fluviatile populations, *S. t. fario*, was indicated by mtDNA (Figure 3c) and allozyme (Krieg and Guyomard 1985) variation. Polyphyly was also observed for *S. t. macrostigma*. This latter subspecies was traditionally composed of fluviatile populations from Morocco and southern Mediterranean drainages and is characterized by intense pigmentation and lower modal vertebral counts. Mitochondrial DNA analysis suggested that this taxon was composed of populations belonging to two distinct lineages, groups I and IV. An ongoing analysis of microsatellite DNA variation also corroborates this view (P. Presa, Institut National de la Recherche, personal communication). At present, there is insufficient data to assess whether other taxa belonging to mtDNA groups represent monophyletic lineages. Results to date do suggest that several taxonomically recognized groups of populations are genetically similar within major populations groupings not recognized in current taxonomy. Genetic similarity among some of these taxa, namely *S. t. labrax*, *S. t. caspius*, *S. t. oxianus*, and *S. t. ischchan* has also been indicated in nuclear genes studies (Osinov 1984, 1989, 1990a, 1990b).

---

[3]Felsenstein, J. 1993. PHYLIP (Phylogeny inference package) Version 3.5c. Department of Genetics, SK-50, University of Washington, Seattle, 98195, USA.

Evolutionary Significance of Mitochondrial
DNA Phylogenetic Groupings

The evolutionary distinctiveness of brown trout assemblages based on mtDNA phylogeny has been demonstrated in studies in which different groupings are found in parapatry. For instance, Giuffra et al. (1994) observed that *S. t. fario* and *S. t. marmoratus* from the same river and sampled less than 10 km apart were alternatively fixed for mtDNA groupings II and V (Figure 3d). Restricted gene flow between these populations was also indicated by allozyme studies of nuclear genes, which showed fixation for alternate alleles at six loci (Giuffra 1993). In drainages surrounding the Mediterranean basin in southern France, D. Beaudoud (University of Montpellier, unpublished data) found restricted gene flow for both mtDNA and nuclear DNA between native brown trout associated with mtDNA group II and introduced brown trout from the Atlantic basin, related to mtDNA group IV. In summary, the genetic analysis of brown trout populations across brown trout's native range revealed that four of its five most salient evolutionary lineages are not recognized by traditional taxonomy. The evolutionary distinctiveness of some population units is demonstrated by their apparent reproductive isolation when found sympatrically.

Towards an Applicable Recognition of Evolutionarily Significant Units in Fishes

Examples presented above highlight three points relevant to defining ESUs: (1) the use of molecular systematics identified major groups of populations exhibiting distinct geographic distribution and not taxonomically recognized; (2) the evolutionary distinctiveness of some population groups was suggested by evidence of reproductive isolation in sympatry; and (3) these results suggested that replicate evolution of similar phenotype may be a common phenomenon in these fishes, and, consequently, traditional taxonomies based on their analysis may not always reflect evolutionary relationships among populations. Because these observations are derived solely from the phylogeny of mtDNA, a strict confirmation of their veracity must await the analysis of nuclear gene phylogenies. Nevertheless, conclusions based on patterns of mtDNA variation were often supported with those obtained from allozymes when available and conformed to patterns of postglacial dispersal suspected from biogeographic studies. These examples do not represent exceptions to the rule, but add to the growing evidence that, in terms of defining what is an ESU, we

are still at the basic step of recognizing the most important natural assemblages of populations within species (e.g., Laerm et al. 1982; Avise and Nelson 1989; Daugherty et al. 1990; Bowen et al. 1991).

The analysis of DNA sequence variation is important for ESU definition because, unlike other genetic approaches such as allozyme and satellite DNA analysis, it allows clear phylogenetic interpretation of variation among alleles. Analysis of DNA sequence variation generates information defining cohesion among populations, whereas other methods often concentrate on discrimination without clearly assessing relatedness. When the rate of mutation is relatively fast, such as in mtDNA, easily detectable apomorphies may develop at the population level more frequently than they do in other markers. Because the vast majority of mutations detected at intraspecific levels do not translate into changes of protein expression (Meyer et al. 1990; Carr and Marshall 1991; Bernatchez and Danzmann 1993; Giuffra et al. 1994), which suggests neutrality, DNA sequence variation may be less susceptible to selective pressures and environmental influence than are phenotypic traits, which may blur historical information, as was indicated for whitefish, rainbow smelt, and brown trout.

Paradoxically, the apparent neutrality that provides power to DNA sequence variation in defining cohesion may limit its ability to discriminate among recently diverged populations, in which time of separation has been too short to allow detectable genetic differences to accumulate by mutations and drift (or stochastic lineage extinction). Characters strongly affected by selection may evolve over shorter times and thus be much more discriminant in such situations. Thus, in several species flocks, such as cichlids in African lakes and salmonids in recently deglaciated lakes, species and populations differ substantially in heritable adaptive characters (e.g., life history traits, feeding apparatus, and reproductive behavior) in absence of significant detectable genetic variation (Magnusson and Ferguson 1987; Meyer et al. 1990). Furthermore, neutral DNA variation cannot be interpreted in terms of local adaptation. Although the analysis of mtDNA variation may have been useful in assessing the evolutionary relationships of whitefish or rainbow smelt populations, it revealed nothing in terms of their adaptive uniqueness.

These counterexamples reemphasize that any model to be developed for optimal protection of biodiversity needs to be capable of integrating different types of information. The few attempts to do so have been based on the general principle that all types of information are equally important and that concordance among information is a major criterion for recognizing ESUs (Avise and Ball 1990; Dizon et al. 1992). Although conceptually appealing, the concordance principle may nevertheless be of limited use for conservation purposes. Indeed, because different types of markers are detected by different methods, reflect different types of information, and, most importantly, may be affected differentially by evolutionary forces, it is probably unrealistic to expect concordance in many cases. For instance, in the classic debate about the phylogenetic relationships between apes and humans, congruence among different studies has been insufficient to resolve the issue despite a tremendous amount of data (discussed in Hoelzer and Melnick 1994). A potential drawback of seeking concordance before recognizing an ESU is that decisions regarding the conservation of endangered groups of organisms may continue to be based on taxonomic distinctiveness that may not reflect evolutionary diversity.

The key for developing a more efficient and useful method for recognizing ESUs may lie in the acceptance that different approaches are not equally useful but, instead, are more or less better at doing different things. This is the general idea that guides the model outlined below which aims to integrate the use of different data, from DNA to life history variation, based on empirical evidence of their advantages and limitations (Figure 4).

Rationale for an Integrated Definition of the Evolutionarily Significant Unit

This procedure is based on the assumption that any differentiable population is an ESU potentially worth recognition. However, priority for conservation should be given to those populations exhibiting unique local adaptation, which assumes that their extinction would cause an irreversible loss of an adapted genome. Due to the apparent probability of replicate evolution of many phenotypes within species, adaptive uniqueness should be demonstrated within evolutionarily distinct population assemblages. Furthermore, the probability of recognizing uniqueness of adapted phenotypes will increase as the natural assemblage in which they are considered is reduced in size. For instance, there could be populations exhibiting similar, uncommon life history traits as a result of replicate evolution

within the whole species, yet they may constitute a unique variant worth recognition in a smaller evolutionarily distinct group. For these reasons, the model proposes that adaptive uniqueness of populations should be demonstrated for the smallest significant and cohesive groups of populations that can be detected. It may be argued that given the enormous genetic polymorphism within many taxa, such a criteria would create an unmanageable number of species subunits (e.g., Avise and Ball 1990). Nevertheless, empirical evidence to date shows that the intraspecific number of cohesive population assemblages which can be considered statistically significant is generally low. For instance, no more than five significant population assemblages can be recognized in both whitefish and brown trout over their range of distribution. Admittedly, there is the possibility that the increasing use of hypervariable loci (e.g., minisatellites and microsatellites) for population analysis will multiply the number of significant population assemblages detected. However, the few data available to date suggest that these powerful methods are more useful in discriminating populations than in assessing cohesion among them (Stevens et al. 1993; Taylor et al. 1994). Meanwhile, we are not much closer to defining intraspecific variation in an evolutionary sense than Linnaeus was at higher taxonomic levels two centuries ago. Consequently, we have to undertake actions with the best empirical evidence at hand before the increasing pace of genetic erosion makes the task simpler.

The criteria used to recognize natural assemblages need to be applicable to any type of data. For this reason, relative measures of distance (e.g., 1 or 2% sequence divergence estimate) are not suitable because (1) they represent a continuum within which the limits of categories become very arbitrary and (2) these measures are not easily comparable among different species or types of information. Instead, use of absolute criteria for defining hierarchical categories is proposed. This method progresses from defining the smallest natural assemblages of populations to demonstrating the adaptive uniqueness of populations within these (Figure 4). It implies that neutral markers are reliable in defining cohesion among populations, that is, assessing their evolutionary relationships, whereas neutral markers' utility declines as selected phenotypic traits become increasingly important in discriminating uniquely adapted populations within each natural assemblage.

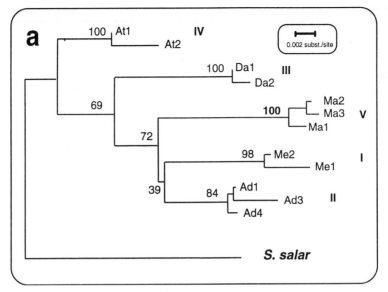

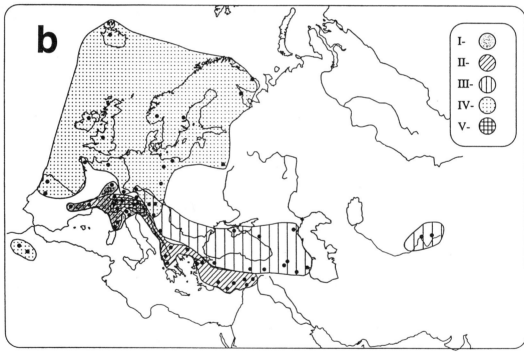

FIGURE 3.—Mitochondrial DNA (mtDNA) variation in subspecies of brown trout. (**a**) Rooted consensus tree among 12 genotypes resolved in sequence analysis of cytochrome b, ATPase subunit VI, and control region segments ($N = 1,250$ bp; after Giuffra et al. 1994). The 12 genotypes clustered into five distinct clades. Numbers above branches represent bootstrap estimates (%). (**b**) Geographic distribution of the five mtDNA phylogenetic groupings. (**c**) Population tree clustering 88 brown trout populations fixed for either one of the five phylogenetic groupings. Numbers in parentheses refer to number of populations. Nomenclature is after Berg (1948) and Lelek (1987). (**d**) Contact zone between *Salmo trutta marmoratus* (group V) and *S. t. fario* (group II) in Stura di Demonte, northern Italy.

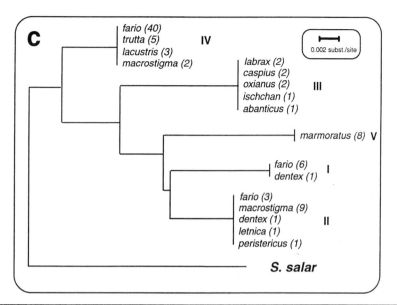

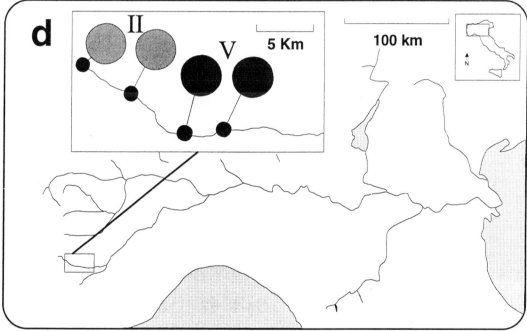

FIGURE 3.—Extended.

Four Steps to Assign Priority for Evolutionarily Significant Unit Protection

The first step is to define population cohesion by identifying the smallest monophyletic population assemblages among which all individuals possess at least one uniquely derived character and share a more recent evolutionary history with each other than with others. As exemplified in this paper, DNA sequence data may be the most appropriate tool to define these assemblages. If no additional data are available, monophyletic groups with one genetic marker should be considered as evidence for evolutionary distinctiveness in the context of conservation, keeping in mind that the results of different loci are more desirable.

Admittedly, the above criteria of diagnosability

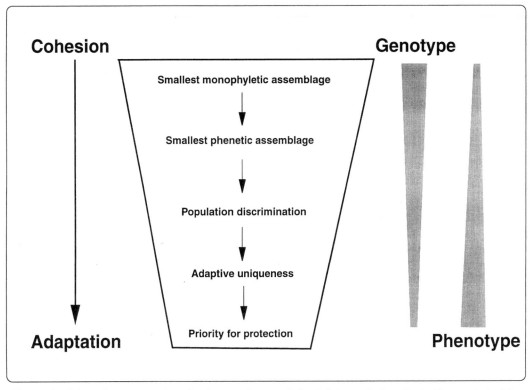

FIGURE 4.—Schematic representation of an integrated approach for defining evolutionarily significant units in fishes. The method progresses from defining cohesion among populations to demonstrating their adaptive uniqueness. Molecular genetic markers are more important in defining cohesion whereas the role of phenotypic data increases towards the recognition of adaptive uniqueness.

may be too restrictive for the identification of significant natural population assemblages that have diverged too recently for the development of apomorphies. Yet, it is apparent from the literature that many cohesive population groupings may be recognized on the basis of obvious frequency differences of characters. For this reason, in the second step significant groupings defined phenetically within monophyletic assemblages should be recognized as the smallest cohesive population unit within which adaptive uniqueness can be determined. This criteria would also allow the recognition of evolutionary distinctiveness of population assemblages evolved through reticulate processes, such as introgressive hybridization, which generally lack diagnostic definition but show distinct patterns of allele frequencies (Dowling and DeMarais 1994).

The third step is discriminating populations within the smallest detectable cohesive population unit. This may be found from molecular genetic data, but evidence based on heritable phenotypic variation may prove to be more useful in cases of recent divergence. Ecological observations and geographic distributions may be crucial at this step. For example, landlocked fish from separate drainages do not exchange genes and belong to different populations, regardless of whether or not they may be distinguished by the tools at hand.

The final step is to assess the adaptive uniqueness of discriminated populations to assign priority for protection and conservation. Adaptive uniqueness will be demonstrated from phenotypic traits with adaptive value (e.g., life history parameters, behavior, and physiological requirements). Populations showing unique or less frequently observed phenotypic variation should be assigned priority for protection, given equal threats to survival.

The benefits of this approach may be manifold. First, integrating different types of data used for defining the populations with highest priority for protection is accommodated without the constraints of having to demonstrate concordance among them. Second, populations would more likely be classified into evolutionarily cohesive units in this way than by traditional taxonomy. Third, the probability of according protection to given populations would be increased, even in the absence of direct evidence of genetic discrimination, due to the reduced size of the evolutionary lineage they represent. Thus, unusual ecotypic populations, such as large-sized cutthroat trout *Oncorhynchus clarki*, planktivorous whitefishes, spring spawning runs of chinook salmon *O. tshawytscha*, or benthic forms of Arctic char *Salvelinus alpinus*, could gain further recognition for protection despite the persistent lack of detectable genetic differentiation from other sympatric or allopatric populations.

Acknowledgments

I am indebted to A. Ferguson, University of Belfast, for kindly providing me with unpublished data on brown trout. I also thank S. Martin for laboratory assistance and figure preparation. I gratefully acknowledge J. J. Dodson, H. Glęmet, R. J. Behnke, C. Stepien, B. J. Turner, G. R. Smith, and J. L. Nielsen for their critical reviews and comments of earlier versions of the manuscript.

References

Avise, J. C. 1989. Gene trees and organismal histories: a phylogenetic approach to population biology. Evolution 43:1192–1208.

Avise, J. C. 1994. Molecular markers, natural history, and evolution. Chapman and Hall, New York.

Avise, J. C., and R. M. Ball. 1990. Principles of genealogical concordance in species concepts and biological taxonomy. Oxford Surveys in Evolutionary Biology 7:45–67.

Avise, J. C., and W. S. Nelson. 1989. Molecular genetic relationships of the extinct dusky seaside sparrow. Science 243:646–648.

Avise J. C., and seven coauthors. 1987. Intraspecific phylogeography: the mitochondrial DNA bridge between population genetics and systematics. Annual Review of Ecology and Systematics 18:489–522.

Baby, M. C., L. Bernatchez, and J. J. Dodson. 1991. Genetic structure and relationships among anadromous and landlocked populations of rainbow smelt, *Osmerus mordax*, Mitchill, as revealed by mtDNA restriction analysis. Journal of Fish Biology 39(Supplement A):61–68.

Balon, E. K. 1968. Notes to the origin and evolution of trouts and salmons with special reference to the Danubian trouts. Acta societatis zoologicae Bohemoslovacae XXXII:1–21.

Behnke, R. J. 1972. The systematics of salmonid fishes of recently glaciated lakes. Journal of Fisheries Research Board of Canada 29:639–671.

Behnke, R. J. 1992. Native trout of western North America. American Fisheries Society Monograph 6.

Berg, L. S. 1948. Freshwater fishes of the U.S.S.R. and adjacent countries. Zoological Institute Akademy Nauk Moscow USSR 1(27), volume 1. (In Russian) English translation, 1962: Office of Technical Services, Department of Commerce, Washington, DC.

Bernatchez, L. 1995. Détermination du degré de différenciation génétique entre échantillons d'Éperlan arc-en-ciel (*Osmerus mordax*) provenant de la pêcherie hivernale de Miguasha et de sites éloignés de la Baie-des-Chaleurs. Ministère de l'Environnement et de la Faune, Québec, Canada.

Bernatchez, L., F. Colombani, and J. J. Dodson. 1991a. Phylogenetic relationships among the subfamily Coregoninae as revealed by mitochondrial DNA restriction analysis. Journal of Fish Biology 39(Supplement A):283–290.

Bernatchez, L., and R. G. Danzmann. 1993. Congruence in control-region sequence and restriction-site variation in mitochondrial DNA of brook charr (*Salvelinus fontinalis* Mitchill). Molecular Biology and Evolution 10:1002–1014.

Bernatchez, L., and J. J. Dodson. 1990a. Allopatric origin of sympatric populations of lake whitefish (*Coregonus clupeaformis*) revealed by mitochondrial DNA restriction analysis. Evolution 44:1263–1271.

Bernatchez, L., and J. J. Dodson. 1990b. Mitochondrial DNA variation among anadromous populations of cisco (*Coregonus artedii*) as revealed by restriction analysis. Canadian Journal of Fisheries and Aquatic Sciences 47:533–543.

Bernatchez, L., and J. J. Dodson. 1991. Phylogeographic structure in mitochondrial DNA of the lake whitefish (*Coregonus clupeaformis*) and its relation to Pleistocene glaciations. Evolution 45:1016–1035.

Bernatchez, L., and J. J. Dodson. 1994. Phylogenetic relationships among palearctic and nearctic whitefish (*Coregonus* sp.) populations as revealed by mitochondrial DNA variation. Canadian Journal of Fisheries and Aquatic Sciences 51(Supplement 1):240–251.

Bernatchez, L., J. J. Dodson, and S. Boivin. 1989. Population bottlenecks: influence on mitochondrial DNA diversity and its effect in coregonine stock discrimination. Journal of Fish Biology 35(Suppl. A):233–244.

Bernatchez, L., T. A. Edge, J. J. Dodson, and S. U. Qadri. 1991b. Mitochondrial DNA and isozyme electrophoretic analyses of the endangered Acadian whitefish, *Coregonus huntsmani* Scott, 1987. Canadian Journal of Zoology 69:311–316.

Bernatchez, L., H. Glémet, C. C. Wilson, and R. G. Danzmann. 1995b. Fixation of introgressed mitochondrial genome of Arctic charr (*Salvelinus alpinus* L.) in an allopatric population of brook charr (*Salvelinus fontinalis* Mitchill). Canadian Journal of Fisheries and Aquatic Sciences 52:179–185.

Bernatchez, L., R. Guyomard, and F. Bonhomme. 1992. DNA sequence variation of the mitochondrial control region among geographically and morphologically remote European brown trout *Salmo trutta* populations. Molecular Ecology 1:161–173.

Bernatchez, L., and A. Osinov. 1995. Genetic diversity of trout (genus *Salmo*) from its most eastern native range based on mitochondrial DNA and nuclear gene variation. Molecular Ecology 4:285–297.

Bernatchez, L., L. Savard, J. J. Dodson, and D. Pallotta.

1988. Mitochondrial DNA sequence heterogeneity among James-Hudson Bay anadromous coregonines. Finnish Fisheries Research 9:17–26.

Bernatchez, L., and six coauthors. 1995a. Conséquences de la structure génétique de l'Éperlan arc-en-ciel (*Osmerus mordax*) pour la réhabilitation de l'espèce dans l'estuaire du Saint-Laurent. Saint-Laurent Vision 2000. Communications Branch, Fisheries and Oceans Canada, Quebec.

Billington, N., and P. D. N. Hebert. 1991. Mitochondrial DNA diversity in fishes and its implications for introductions. Canadian Journal of Fisheries and Aquatic Sciences 48:80–94.

Blanc, J. M., P. Benarescu, J. L. Gaudet, and J. C. Hureau, editors. 1971. European inland water fish, multilingual catalogue. Fishing News Books, Farnham.

Bodaly, R. A., J. W. Clayton, C. C. Lindsey, and J. Vuorinen. 1992. Evolution of lake whitefish (*Coregonus clupeaformis*) in North America during the pleistocene: genetic differentiation between sympatric populations. Canadian Journal of Fisheries and Aquatic Sciences 49:769–779.

Bodaly, R. A., J. Vuorinen, R. D. Ward, M. Luczynski, and J. D. Reist. 1991. Genetic comparisons of new and old world coregonid fishes. Journal of Fish Biology 38:37–51.

Bowen, B. W., A. B. Meylan, and J. C. Avise. 1991. Evolutionary distinctiveness of the endangered Kemp's ridley sea turtle. Nature 352:709–711.

Carr S. M., and H. D. Marshall. 1991. Detection of intraspecific DNA sequence variation in the mitochondrial cytochrome *b* gene of Atlantic cod (*Gadus morhua*) by the polymerase chain reaction. Canadian Journal of Fisheries and Aquatic Sciences 48:48–52.

Cavender, T. M., and M. M. Coburn. 1992. Phylogenetic relationships of north American cyprinidae. Pages 293–327 in Mayden (1992).

Chaumeton, H., P. Louisy, and T. Maitre-Allain. 1989. Les Poissons d'Europe. Guide Vert. Solar, Paris.

Cope, E. D. 1870. A partial synopsis of the fishes of the fresh waters of North Carolina. Proceedings of the American Philosophical Society 11:448–495.

Daugherty, C. H., A. Cree, J. M. Hay, and M. B. Thompson. 1990. Neglected taxonomy and continuing extinctions of tuatara (*Sphenodon*). Nature 347:177–179.

Dizon, A. E., C. Lockeyer, W. F. Perrin, D. P. Demaster, and J. Sisson. 1992. Rethinking the stock concept: a phylogeographic approach. Conservation Biology 6:24–36.

Dowling, T. E., and B. D. DeMarais. 1994. Evolutionary significance of introgressive hybridization in cyprinid fishes. Science 362:444–446.

Dowling, T. E., and W. R. Hoeh. 1991. The extent of introgression outside the contact zone between *Notropis cornutus* and *Notropis chrysocephalus* (Teleostei: Cyprinidae) Evolution 45:944–956.

Felsenstein, J., and E. Sober. 1986. Parsimony and likelihood: an exchange. Systematic Zoolology 35:617–626.

Fenderson, O. C. 1964. Evidence of subpopulations of lake whitefish, *Coregonus clupeaformis*, involving a dwarf form. Transactions of the American Fisheries Society 93:77–94.

Ferguson, A. 1989. Genetic differences among brown trout, *Salmo trutta*, stocks and their importance for the conservation and management of the species. Freshwater Biology 21:35–46.

Foote, C. J., C. C. Wood, and R. E. Withler. 1989. Biochemical genetic comparison of sockeye salmon and kokanee, the anadromous and nonanadromous forms of *Oncorhynchus nerka*. Canadian Journal of Fisheries and Aquatic Sciences 46:149–158.

Giuffra, E. 1993. Genetic identification and phylogeny of brown trout, *Salmo trutta* L., populations of the Pô River basin. Doctoral dissertation. University of Turin, Italy.

Giuffra, E., L. Bernatchez, and R. Guyomard. 1994. Mitochondrial control region and protein coding genes sequence variation among phenotypic forms of brown trout *Salmo trutta* from northern Italy. Molecular Ecology 3:161–172.

Guyomard R. 1989. Diversité génétique de la Truite commune. Bulletin Français de la Pêche et de la Pisciculture 314:118–135.

Guyomard R., G. Grevisse, F. X. Oury, and P. Davaine. 1984. Evolution de la variabilité génétique inter et intrapopulations de populations issues de mêmes pools géniques. Canadian Journal of Fisheries and Aquatic Sciences 41:1024–1029.

Hall, H. J. 1992. The genetics of brown trout (*Salmo trutta* L.) populations in Wales. Doctoral dissertation. University of Wales, UK.

Hamilton, K. E., A. Ferguson, J. B. Taggart, T. Tomasson, A. Walker, and E. Fahy. 1989. Post-glacial colonization of brown trout, *Salmo trutta* L.: *Ldh-5* as a phylogeographic marker locus. Journal of Fish Biology 35:651–664.

Heese, T. 1992. Systematics of Polish populations of European whitefish (L.), based on skull osteology. Polskie Archivum Hydrobiologii 39:491–500

Heinonen, M. 1988. Taxonomy and genetic variation of whitefish (*Coregonus* sp.) in Lake Saimaa. Finnish Fisheries Research 9:39–47.

Himberg, K. J. M. 1970. A systematic and zoogeographic study of some north European coregonids. Pages 219–250 in C. C. Lindsey and C. S. Woods, editors. Biology of coregonid fishes. University of Manitoba Press, Winnipeg.

Hindar, K., B. Jonsson, N. Ryman, and G. Stahl. 1991. Genetic relationships among landlocked, resident, and anadromous brown trout, *Salmo trutta* L. Heredity 66:83–91.

Hindar, K., N. Ryman, and G. Stahl. 1986. Genetic differentiation among local populations and morphotypes of Arctic charr, *Salvelinus alpinus*. Biological Journal of the Linnean Society 27:269–285.

Hoelzer, G. A., and D. J. Melnick. 1994. Patterns of speciation and limits to phylogenetic resolution. Trends in Ecology & Evolution 9:104–107.

Kirkpatrick, M., and R. Selander. 1979. Genetics of speciation in lake whitefishes in the Allegash basin. Evolution 33:478–485.

Krieg, F., and R. Guyomard. 1985. Population genetics

of French brown trout (*Salmo trutta* L.): large geographical differentiation of wild populations and high similarity of domesticated stocks. Genetique, Selection, Evolution 17:225–242.

Laerm, J., J. C. Avise, J. C. Patton, and R. A. Lansman. 1982. Genetic determination of the status of endangered species of pocket gopher in Georgia. Journal of Wildlife Management 46:513–518.

Lanteigne, J., and D. E. McAllister. 1983. The pygmy smelt, *Osmerus spectrum* Cope, 1870. A forgotten sibling species of eastern North American fish. Syllogeus 45:1–32.

Lelek, A. 1987. The freshwater fishes of Europe. Volume 9: threatened fishes of Europe. European Committee for the Conservation of Nature and Natural Resources. Aula-Verlag, Wiesbaden, Germany.

Lindsey, C. C. 1981. Stocks are chameleons: plasticity in gill-rakers of coregonid fishes. Canadian Journal of Fisheries and Aquatic Sciences 38:1497–1506.

Lindsey, C. C., J. W. Clayton, and W. G. Franzin. 1970. Zoogeographic problems and protein variation in the *Coregonus clupeaformis* species complex. Pages 127–146 *in* C. C. Lindsey and C. S. Woods, editors. Biology of coregonid fishes. University of Manitoba Press, Winnipeg.

Magnusson K. P., and M. M. Ferguson. 1987. Genetic analysis of four sympatric morphs of Arctic charr, *Salvelinus alpinus*, from Thingvallavatn, Iceland. Environmental Biology of Fishes 20:67–73.

Mayden, R. L., editor. 1992. Systematics, historical ecology, and north American fishes. Stanford University Press, Stanford, California.

Mayden, R. L., B. M. Burr, L. M. Page, and R. R. Miller. 1992. The native freshwater fishes of north America. Pages 827–863 *in* Mayden (1992).

McAllister, D. A. 1985. A list of threatened and endangered fishes in Canada. Syllogeus 51.

McPhail, J. D. 1993. Ecology and evolution of sympatric sticklebacks (*Gasterosteus*): origin of the species pairs. Canadian Journal of Zoology 71:515–523.

McPhail, J. D., and C. C. Lindsey. 1970. Freshwater fishes of northwestern Canada and Alaska. Fisheries Research Board of Canada Bulletin 173.

Meyer A., T. D. Kocher, P. Basasibwaki, and A. C. Wilson. 1990. Monophyletic origin of Lake Victoria cichlid fishes suggested by mitochondrial DNA sequences. Nature 347:550–553.

Muus, B. J., and P. Dahlstrom. 1981. Guide des poissons d'eau douce et pêche. Delachaux and Niestlé, Neuchâtel, Switzerland.

Osinov, A. G. 1984. Zoogeographical origins of brown trout, *Salmo trutta* (Salmonidae): data from biochemical genetic markers. Journal of Ichthyology 24:10–23.

Osinov, A. G. 1989. Brown trout (*Salmo trutta* L., Salmonidae) in basins of the Black and Caspian seas: a population genetic analysis. Genetika 24:2172–2186.

Osinov, A. G. 1990a. The level of genetic variation and differentiation of the brown trout (*Salmo trutta* L.) in Tadjikistan. Moscow University Biological Sciences Bulletin 45:37–41.

Osinov, A. G. 1990b. Low level of genetic variability and differentiation in ecological forms of Sevan trout *Salmo ischchan* Kessler. Genetika 25:1827–1835.

Page, L. M., and B. M. Burr. 1991. A field guide to freshwater fishes. Peterson field guide. Houghton Mifflin Company, Boston.

Pedroli, J. C., B. Zaugg, and A. Kirchhofer. 1991. Atlas de distribution des poissons et cyclostomes de Suisse. Centre Suisse de cartographie de la faune. Neuchâtel, Switzerland.

Reshetnikov, J. S. 1975. Relations between coregonid fishes of the USSR and North America. Report of the 13th Pacific Scientific Congress. Nauka, Moscow.

Reshetnikov, J. S. 1988. Coregonid fishes in recent conditions. Finnish Fisheries Research 9:11–16.

Riddell, B. E. 1993. Spatial organization of Pacific salmon: what to conserve? Pages 23–42 *in* J. G. Cloud and G. H. Thorgaard, editors. Genetic conservation of salmonid fishes. Plenum Press, New York.

Robins, C. R., and six coauthors. 1991. Common and scientific names of fishes from the United States and Canada. American Fisheries Society Special Publication 20.

Rohlf, D. J. 1991. Six biological reasons why the Endangered Species Act doesn't work—and what to do about it. Conservation Biology 5:273–282.

Rojas, M. 1992. The species problem and conservation: what are we protecting? Conservation Biology 6:170–178.

Scott, W. B., and E. J. Crossman. 1973. Freshwater fishes of Canada. Journal of Fisheries Research Board of Canada Bulletin 184.

Simons, A. M. 1992. Phylogenetic relationships of the *Boleosomona* species group (Percidae: *Etheostoma*). Pages 268–292 *in* Mayden (1992).

Smith, G. R. 1992. Phylogeny and biogeography of the Catostomidae, freshwater fishes of North America and Asia. Pages 778–826 *in* Mayden (1992).

Smith, G. R., and T. N. Todd. 1992. Morphological cladistic study of coregonine fishes. Polskie Archivum Hydrobiologii 39:479–490.

Stevens, T. A., R. E. Whitler, S. H. Goh, and T. D. Beacham. 1993. A new multilocus probe for DNA fingerprinting in chinook salmon (*Oncorhynchus tshawytscha*), and comparisons with a single-locus probe. Canadian Journal of Fisheries and Aquatic Sciences 50:1559–1567.

Svärdson, G. 1957. The coregonid problem VI. Institute of Freshwater Research of Drottningholm 38:267–356.

Svärdson, G. 1979. Speciation of Scandinavian *Coregonus*. Institute of Freshwater Research of Drottningholm 57:1–95.

Takahata, N., and M. Slatkin. 1984. Mitochondrial gene flow. Proceedings of the National Academy of Sciences of the United States of America 81:1764–1767.

Taylor, E. B., T. D. Beacham, and M. Kaeriyama. 1994. Population structure and identification of north Pacific Ocean chum salmon (*Oncorhynchus keta*) revealed by an analysis of minisatellite DNA variation.

Canadian Journal of Fisheries and Aquatic Sciences 51:1430–1442.

Taylor, E. B., and P. Bentzen. 1993a. Evidence for multiple origins and sympatric divergence of trophic ecotypes of smelt (*Osmerus*) in Northeastern North America. Evolution 47:813–832.

Taylor, E. B., and P. Bentzen. 1993b. Molecular genetic evidence for reproductive isolation between sympatric populations of smelt *Osmerus* in Lake Utopia, south-western New Brunswick, Canada. Molecular Ecology 2:345–357.

Vuorinen, J., M. K. J. Himberg, and P. Lankinen. 1981. Genetic differentiation in *Coregonus albula* (L.) (Salmonidae) populations in Finland. Hereditas 94:113–121.

Vuorinen, J., A. Champigneulle, K. Dabrowski, R. Eckmann, and R. Rosch. 1986. Electrophoretic variation in central European coregonid populations. Archiv für Hydrobiolie Beiheft Ergebnisse der Limnologie 22:291–298

Vuorinen, J. A., R. A. Bodaly, J. D. Reist, L. Bernatchez, and J. J. Dodson. 1993. Genetic and morphological differentiation between dwarf and normal size forms of lake whitefish (Coregonus clupeaformis) in Como Lake, Ontario. Canadian Journal of Fisheries and Aquatic Sciences 50:210–216.

American Fisheries Society Symposium 17:133–144, 1995

Evolutionary and Ecological Considerations in the Reestablishment of Great Lakes Coregonid Fishes

Ruth B. Phillips and Timothy J. Ehlinger

Department of Biological Sciences, University of Wisconsin
Post Office Box 413, Milwaukee, Wisconsin 53201, USA

Abstract.—The criteria for evolutionarily significant units (ESUs) formulated by the National Marine Fisheries Service emphasize the importance of past evolutionary processes in the persistence and lineage diversification of species. However, this historical framework may pose a philosophical problem when trying to apply the ESU concept to current issues in conservation biology. Phenotypic or genetic variation that was important historically may not be the same variation that will allow for continued success, persistence of, and adaptive changes in modern populations. Defining evolutionarily significant variation within a conservation context requires the collection and analysis of data on three levels. First, we must determine how intraspecific phenotypic variation contributes to success in growth, survival, and reproduction in a species' current ecological situation. Second, we must ask if this variation was important historically in lineage diversification and if the current variation reflects adaptations produced by the action of natural selection. Third, we can look at the patterns of phenotypic and genetic change within and among current lineages relative to changing ecological conditions and attempt to judge the importance of this phenotypic and genetic variation for the long-term persistence and adaptability of populations. This multilevel approach requires a variety of research techniques. We illustrate several approaches that we are using to study evolutionarily significant variation for reestablishment of Great Lakes coregonid fishes.

According to the National Marine Fisheries Service a population must both "have substantial reproductive isolation from conspecific populations" and "represent an important component in the evolutionary legacy of the species" in order to be considered an evolutionarily significant unit (ESU; Waples 1991). We therefore expect ESUs to possess distinct gene pools and exhibit adaptations to local environmental conditions. These criteria emphasize the importance of past evolutionary processes in the origination, persistence, and lineage diversification of species. However, this historical framework may pose a philosophical problem when trying to apply the ESU concept to current issues in conservation biology. Phenotypic or genetic variation that was important historically may not be the same variation that will allow for continued success, persistence of, and adaptive changes in modern populations.

Problems arise when trying to apply the ESU principles to reestablishment of disturbed systems. Background data on ecophenotypic and genetic variation is typically lacking for both the intraspecific and interspecific levels. Current variation may be adaptive, nonadaptive, or simply an artifact of disturbance (e.g., bottlenecks, drift, or hybridization). We often do not know if intrinsic reproductive isolating mechanisms have developed as barriers to gene flow among the original subspecies or stocks. It is critical to ask what types or amounts of variation should be promoted during recovery or reintroduction of populations. What morphological, behavioral, and genetic variation is significant for successful reestablishment (i.e., growth, survival, and reproduction)? In the case of polymorphic species, what variation is significant for evolutionary reradiation (i.e., restoration of niche and life history diversity)?

Historical Patterns of Coregonus in the Great Lakes

The endemic ciscoes *Coregonus* sp. (subgenus *Leucichthys*; also known commercially as chubs) of the Laurentian Great Lakes have been cited as an example of intralacustrine adaptive radiation into diverse trophic niches (Smith and Todd 1984), with affiliated morphological specializations for feeding in either pelagic or benthic habitats (Table 1). Parallel with this trophic divergence is evidence of segregation in spawning times and depths (Koelz 1929), which may have provided the reproductive isolation necessary for natural selection to produce genetically distinct species from unique stocks. The chameleon-like character of morphological variation within this group has made its taxonomy problematic. For example, Greene (1935) listed 11 subspecies of lake herring for Wisconsin, many restricted to single lakes. Today, it is believed more realistic to consider these 11 "subspecies" of coregonids as groups of related races or genetically isolated stocks (Todd 1981; Todd and Smith 1992).

TABLE 1. Distribution of species in the genus *Coregonus* among the Great Lakes. Abbreviations are present (P); threatened or rare (TR); and extinct (E). Sources are Koelz 1929; Todd and Smith 1992; and T. Todd, National Biological Service, Ann Arbor, Michigan, personal communication.

Species	Lake					
	Nipigon	Superior	Michigan	Huron	Erie	Ontario
Ubiquitous forms						
C. clupeaformis Lake whitefish	P	P	P	P	P	P
Shallow-water forms						
C. artedi Lake herring	P	P	TR	P	TR	TR
Mid-depth forms						
C. hoyi Bloater	P[a]	P	P	P		E
C. zenithicus Shortjaw cisco	P	TR	E	E		
C. alpenae Longjaw cisco		TR	E	E	E	
C. reighardi Shortnose cisco	P		E	TR		E
Deep-water forms						
C. nigripinnis Blackfin cisco	P		E	E		
C. kiyi Kiyi	P	P	E	E		E
C. johannae Deepwater cisco			E	E		

[a]Species similar to *C. hoyi* was called *C. nipigon* by Koelz (1929).

Whether or not the original species flock in Lake Michigan was composed of species or subspecies, it is clear that substantial reproductive isolation existed and that differential selection was high. The evidence for reduced gene flow is based on differences in time and place of spawning and the multiple chromosome rearrangements described by Booke (1968) for different species from Lake Superior. The evidence for strong differential selection is based on adaptive divergence in gill raker morphology for feeding (Smith and Todd 1984). However, the presence of intraspecific plasticity in the same characters (Hile 1937; Todd et al. 1981) indicates that the potential may exist for reestablishment of the coregonid complex from within a single species (Lindsey 1981).

The Great Lakes coregonid populations have undergone tremendous changes over the past century (Kitchell and Crowder 1986). Whereas they once supported a thriving commercial fishery in Lake Michigan, the combined pressures of intensive fishing, the invasion of the sea lamprey *Petromyzon marinus*, and resource competition with and predation by the invading alewife *Alosa pseudoharengus* and rainbow smelt *Osmerus mordax* have all but eliminated the once diverse assemblage of cisco species. The Lake Michigan populations have shifted to a point where only the smallest species,

the bloater, exists in significant numbers (Smith 1964). The aggressive stocking of Pacific salmon *Oncorhynchus* sp. in Lake Michigan since 1966 has resulted in a steady decline in alewife populations—associated with a 20-fold increase in bloater biomass between 1978 and 1988 (Kitchell 1991).

It is clear that the bloater in Lake Michigan today is different from the bloater of the 1930s. Crowder (1986) argued that competition with alewife resulted in an ecomorphological shift from pelagic to benthic feeding in bloater. However, the view that the present cisco population consists entirely of bloater is suspect (Todd and Stedman 1989). In the 1960s, Stanford Smith (1964) noted that many of the ciscoes were difficult to identify and had a combination of characteristics of both lake herring and bloater. A look at changes in bloater gill raker numbers from samples collected across the century is instructive. Gill raker numbers are used frequently to categorize coregonid species and are of manifest importance for intraspecific feeding ecology and interspecific niche segregation. In the early 1900s bloater exhibited a modal gill raker number near 41 and a range from approximately 38 to 46 (Figure 1A). Koelz (1929) did note differences in this distribution among some geographical races but all in all found bloater to fall within this range. Crowder (1986) showed a statistically significant

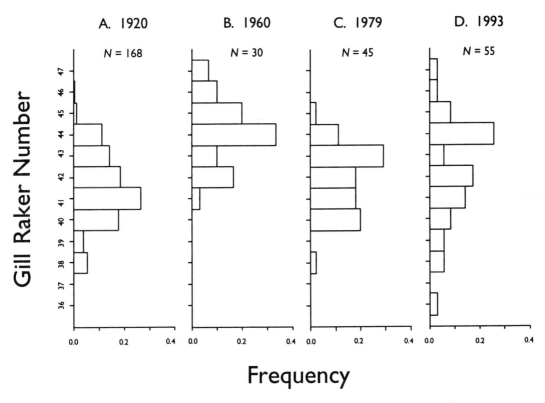

FIGURE 1.—Frequency distributions of gill raker number for bloater collected during the past 75 years. Data from 1920 taken from Koelz (1929) for fish collected near Alpena, Michigan. Data for 1960 and 1979 taken from Crowder (1986) for fish collected near Grand Haven, Michigan. Data for 1993 from T. Ehlinger (University of Wisconsin, Milwaukee, unpublished) for fish collected near Milwaukee, Wisconsin.

change in gill raker numbers between samples collected in 1960 and 1979 (Figure 1B and 1C). These data show a shift in both the modal gill raker number as well as a decrease in the range compared with historical populations (see also Todd and Stedman 1989). During the 1960s bloater had more gill rakers (resembling lake herring), and then in the 1970s shifted to fewer gill rakers. Crowder (1986) argued that this reflected an ecomorphological shift from pelagic to benthic feeding as a result of competition with alewife (Lindsey 1981). Gill raker counts that we have taken from bloaters sampled from commercial fishing boats in Milwaukee during 1991 and 1993 show a range rivaling the historical distribution (Figure 1D).

Lineage History, Evolutionarily Significant Units, and Adaptive Ecomorphology

The coregonids of the Laurentian Great Lakes provide a useful system for considering the complex matrix of ecological and evolutionary factors that

may have produced, may maintain, and may potentially influence the reestablishment of the species flock. The observed patterns of species extinctions, ecomorphological shifts, and reappearance of morphological variation cannot be explained simply by a single mechanism. Among the possible scenarios the most parsimonious include some combination of introgressive hybridization and environmentally induced phenotypic plasticity acting upon more-or-less genetically isolated species, races, or ESUs. The first extreme scenario (Figure 2, Lineage Selection) assumes historical eco-morphological diversity was the direct manifestation of distinct genetic lineages or species. The perturbed conditions of the mid-1900s (including both abiotic factors, such as habitat loss and pollution, as well as biotic factors, such as species invasions and overexploitation) may have changed the environment such that species may have needed to adapt to avoid extinction. If a lineage survived and if sufficient genetic variation was retained within the population, then we might ex-

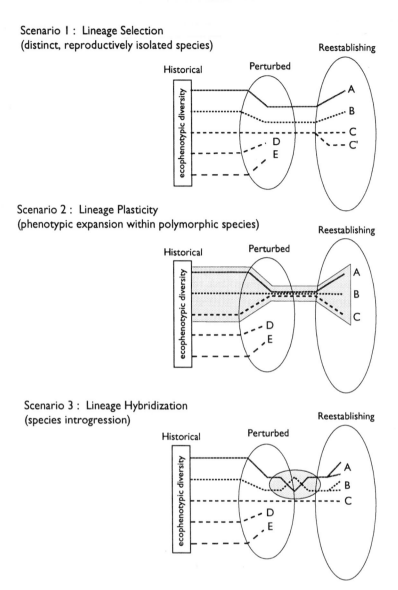

FIGURE 2.—Diagram of possible relationships between genetic lineages and ecophenotypic diversity due to changes in the environment during disturbance and reestablishment. Historical refers to patterns prior to the biotic and abiotic perturbations of the past century (described in text).

pect the species to return and fill its previous ecological niche when reestablishing. This, of course, assumes that the resource base has not changed and that neighboring ecophenotypes present prior to the disturbance are extant. If not, then the surviving species could expand to utilize the available unexploited resource within their morphological and behavioral capabilities (e.g., Species C′ in Figure 2).

The second extreme scenario (Figure 2, Lineage Plasticity) assumes that ecophenotypic diversity is the expression of environmental heterogeneity acting upon a plastic, nondifferentiated genotype. As such, differences among morphological lineages could be caused by variation in factors such as temperature or food abundance rather than underlying genetic variation. Ecological segregation could be viewed as an epiphenomena, the result of environmentally induced morphological variants choosing to forage on prey types or sizes that they capture most efficiently (see Mummert and Drenner 1986;

Ehlinger 1989). Under this scenario we might expect reestablishing lineages to reradiate without the need for genetic isolation among spawning populations (spatial or temporal). Furthermore, each spawning population could potentially, given the right environmental conditions, produce the breadth of ecophenotypic variation present historically.

Hile (1937) provides a concise review of numerous coregonid introductions, reestablishments and transplants that demonstrate the environmental effects on morphological development in individual lineages. The most common pattern is for lineages to develop greater numbers of gill rakers, longer gill rakers, shorter paired fins, and more fusiform bodies in deep lakes. Such traits correlate with a shift to a zooplanktonic feeding niche. Lineages transplanted to shallow lakes develop in the reverse direction with a concordant shift to a zoobenthic feeding niche. This common trend in ecomorphological adaptation to diverse lakes is typical of many holarctic species (Schluter and McPhail 1993) and may reflect an important mechanism for ecological and evolutionary radiation in species-poor lakes (Smith and Todd 1984). This extreme plasticity in morphology is often used to argue against the existence of genetic differentiation among coregonid lineages (Lindsey 1981). However, Hile (1937) points out that not all of the transplant studies resulted in extreme plastic responses. It is likely that examples of plastic responses received greater attention because they contradicted the view that the coregonids existed as distinct species.

A third scenario invokes hybridization between incompletely isolated lineages (Figure 2, Lineage Hybridization) to explain the high degree of phenotypic melding. In this case, weak reproductive isolating mechanisms (e.g., temporal or spatial segregation in spawning) break down under stress, and closely related species may interbreed to produce fertile offspring. Genetic assimilation is most likely to be a serious problem for small relict populations coming in contact with more numerous successful populations (as for cutthroat trout *Oncorhynchus clarki*; Allendorf and Leary 1988). Depending on the degree of genetic differentiation, hybrids may suffer a loss of fitness due to genetic and chromosomal incompatibilities. Developmental abnormalities may also be seen. For example, our 1993 collections of bloater revealed significant bilateral asymmetry in counts of gill-raker numbers. The female-biased sex ratio often observed in ciscoes (Bowen et al. 1991) could be attributed to Haldane's Rule, which states that in interspecific matings the heterogametic sex (males in salmonidae) will be less viable. Because no genetic analyses were conducted prior to the extirpation of many species in the 1960s, morphometric comparisons of museum samples to extant populations have been used to infer hybridization (Todd et al. 1981; Todd and Stedman 1989). This approach is less than ideal because coregonid morphology can be influenced by environmental factors independent of genotype (Hile 1937; Todd et al. 1981). Snyder et al. (1992) suggest, based on mitochondrial DNA (mtDNA) restriction fragment length polymorphism (RFLP) data, that bloater and lake herring share a recent common ancestor, but Snyder et al. argue against introgression as the mechanism for their genetic similarity.

The three scenarios are simplified extremes of a continuum of mechanisms accounting for the complex changes that have occurred and are continuing in the Great Lake coregonids. These scenarios can serve a valuable heuristic function by predicting different relationships among present day genetic, morphological, and spatial variation. If some genetically distinct lineages survived in small remnant populations, then the extreme morphological phenotypes reestablishing today should show genetic differences from other phenotypes. It would then be critical to identify populations related to historical species' lineages and actively manage their reestablishment. If, on the other hand, the variation appearing today is the consequence of a phenotypically plastic species reradiating into "vacant" ecological niches, then we would detect little or no genetic differentiation among morphological types. In this case, active management of spawning stocks, although important for protecting yearly recruitment, becomes less critical in promoting ecological reestablishment. Last, if hybridization among previously extant species resulted in genetic mixing, then the relationships between genetic and ecomorphological variation will be more complex and nearly impossible to predict. However, it is critical that the extent of hybridization be assessed because its effects on recruitment and reestablishment are potentially far reaching.

Genetic Approaches to the Analysis of the *Coregonus* Complex

Discriminating among the alternative scenarios described in the previous section will require two steps: (1) identification of clear genetic markers for the historically identified species and (2) collection of morphological and genetic data from extant populations separated spatially or temporally during spawning. These data should enable us to answer whether the cisco complex in Lake Michigan today consists of (1) several genetically unique species with environmentally plastic, convergent morphologies, (2) a single species, dominated by the previously recognized bloater, or (3) a genetic admixture of more-or-less introgressed hybrids that incorporate a pool of genotypes from previously extant lineages.

Although genetic analysis of extant and museum specimens would allow us to discriminate between the first two scenarios, species-specific genetic or cytogenetic markers are needed to determine with certainty whether extensive hybridization has taken place.

Our first task is to attempt to identify genetic markers for the historical species by examination of both extant populations and extinct populations (museum specimens). Among extant populations, that in Lake Nipigon, Ontario, may represent a nondisturbed species flock because it is physically isolated from the other lakes and not subjected to the same fishing pressures. Lake Nipigon fish might be a possible source for interspecific genetic markers, but they may not be part of the same ESU lineage that was in Lake Michigan prior to disturbance. Fortunately, museum specimens collected from both Lakes Michigan and Nipigon earlier in the century can be examined for morphological and genetic markers. The two other disturbed Great Lakes in proximity to Lake Michigan, Lakes Superior and Huron, may also retain some interspecific variation. In addition, eggs and fry from Lake Michigan were stocked in numerous northern Wisconsin lakes in the late 1900s (Greene 1935), prior to the disturbance, and may constitute remnant populations of now extinct lineages. These populations can also be compared morphologically and genetically with museum specimens from Lake Michigan. Finally, in smaller inland lakes in Wisconsin and Canada there are natural populations of some nonendemic species (lake herring and lake whitefish) that may provide additional markers for comparison.

Estimating Reproductive Isolation among Historical and Extant Populations

Salmonid fishes (trout, char, and salmon) are well known for their precise homing ability and tendency to split into genetically distinct stocks. Variation in location and timing of spawning and well-developed reproductive behaviors have resulted in rapid diversification of local populations. That these fishes are

ancestral tetraploids (reviewed in Allendorf and Thorgaard 1984; Hartley 1987) may enhance plasticity in development of morphology and alternative life histories. Their karyotypes also appear to evolve quickly, resulting in postzygotic isolation barriers. The coregonid fishes share most of these same characteristics, but they have less-developed reproductive behaviors, that is, they engage in broadcast spawning and do not build nests or provide parental care. It might be hypothesized that in the absence of strong sexual selection coregonids would not develop strong prezygotic reproductive isolation barriers as rapidly as do other salmonids.

Evidence for reduced gene flow in the coregonid species flocks found in the Great Lakes comes from two sources. First, considerable variation in location and time of spawning was found among the species in each lake (Koelz 1929). Time of spawning has been shown to be highly heritable in salmonids (Gharrett and Smoker 1993; P. Ihssen, Ontario Ministry of Natural Resources, personal communication). Second, the presence of multiple chromosomal rearrangements in several of the endemic species from Lake Superior suggested that they were reproductively isolated (Booke 1968). Unfortunately, Booke did not sample any fish from Lake Michigan, and his work was based on camera lucida drawings of early blastulas, which give very crude karyotypes. Recently, we have initiated a project to karyotype individuals from extant populations that differ in time and location of spawning and have been unable to confirm Booke's results (Phillips et al., in press a).

All *Coregonus* species in North America and many in Europe have a diploid chromosome number of 80, but arm numbers ranging from 92–108 (Viktorovsky et al. 1983; Frolov 1990, 1992; Rab and Jankun 1992). In these fishes, the total number of chromosomes stays constant, but the number of biarmed metacentric chromosomes compared with the number of uniarmed acrocentric chromosomes varies in different species. In other organisms the interconversion of metacentric chromosomes and acrocentric chromosomes has been shown to be the result of either pericentric inversions or a combination of unequal reciprocal translocations followed by inversions or centromere shifts. Usually, heterozygotes with such rearrangements produce genetically unbalanced gametes and have a reduction in fertility (White 1973). Although this may not be the case for single rearrangements (reviewed in Sites and Reed 1994), multiple inversions often result in reduced fertility in second-generation hybrids. In the case of heterozygotes with pericentric inversions, it is the gametes that result from recombination within the inverted region which are lost. A major effect of an inversion therefore is to maintain a certain combination of alleles as linked loci that may be advantageous for survival.

In order to determine if differences in the karyotypes of the Great Lakes coregonids are based on a series of chromosomal inversions, we have prepared revised karyotypes of bloater from Lake Michigan, lake herring and lake whitefish from Lake Huron, and lake whitefish, blackfin cisco, bloater, and shortjaw cisco from Lake Nipigon (R. Phillips, University of Wisconsin, Milwaukee, unpublished). We did not find any evidence for the major chromosomal inversions described by Booke (1968) in these species. All of the cisco species had the same diploid chromosome number ($2N = 80$) and the same number of metacentric and acrocentric chromosomes, although minor differences in the number of very short chromosome arms and the number of nucleolar organizing regions (NORs) were found (Phillips et al., in press a). The karyotype of lake whitefish differed from the karyotypes of the ciscoes in having one additional pair of metacentric chromosomes. However, C-banding (heterochromatin) revealed that the short arm of this chromosome is entirely heterochromatic. Thus the difference in chromosome arm number (NF number) between the karyotype of lake whitefish and the other ciscoes is the result of a heterochromatin addition, not an inversion. Multiple chromosomes with C-bands and NORs were found in several of the cisco species, so it is possible that further work will reveal species-specific differences in these chromosome banding patterns. However, the minor variations in karyotypes that we found would not be sufficient to cause postzygotic reproductive isolation.

The similarity between karyotypes of lake herring from Lake Huron and bloater from Lake Michigan, taken together with data on gill raker numbers, suggests that some of the fish presently identified as bloater in Lake Michigan could be the result of introgression between these two species. However, because all of the ciscoes from Lake Nipigon had essentially the same karyotype as that of bloater from Lake Michigan, it seems likely that the karyotypes of lake herring and bloater from Lake Michigan would be similar. Unless we can show that the lake herring from Lake Superior or the rare individuals still present in Lake Michigan have a different karyotype from that of bloater in Lake Michigan, karyotypic analysis on fish from Lake Michigan will not allow us to determine with certainty whether there are any hybrids between bloater and lake herring nor whether those hybrids are F_1 or introgressed hybrids. However, we believe it will be possible to find differences in nuclear DNA between the cisco species and that minor karyotypic differences in chromosome banding patterns, which are insufficient to serve as reproductive barriers but could be used as genetic markers, may be present. With combined morphological, cytogenetic, and DNA analysis we should be able to determine if the observed morphological variation is intraspecific or the result of interspecific hybridization and introgression.

Ideally one would like to compare karyotypes of extinct species with those of extant species. This may be possible in the future by means of telomeric and centromeric probes hybridized to sections of gonads. Other new cytogenetic techniques that could be used to differentiate coregonid chromosomes include in situ hybridization for localization of genes and microdissection of chromosome arms and cloning of the fragments by means of the polymerase chain reaction (PCR) to obtain paint probes (Bohlander et al. 1992). Recently we have produced a paint probe for the short arm of the Y chromosome in lake trout *Salvelinus namaycush* by means of this method of microdissection-based cloning (Reed et al. 1995). In the future it should be possible to examine genes found at specific chromosomal locations by means of this technique. Additionally, the coregonid species have a large amount of repetitive DNA that is visualized cytologically as C-bands on chromosomes. A combined cytogenetic and molecular analysis of this noncoding DNA may provide species-specific or population-specific markers for examining the relationships between current and extinct populations.

Estimating Genetic Diversity within Populations and Tracing Lineage Diversification

It should be possible to select DNA markers that can be amplified from small amounts of tissue so that DNA from current populations can be compared with that from museum collections of populations previously found in the lakes. Using PCR, we have successfully amplified DNA from specimens fixed briefly in formalin and stored in alcohol (L. Sajdak, University of Wisconsin, Milwaukee, unpublished). With genetic markers we hope to determine if present populations in the disturbed lakes contain a mixture of hybrids or if they represent relatively pure species. Museum specimens can be used as a baseline for comparison of current populations. By examining markers with a high degree of intraspecific variability, we should be able to determine if the current cisco population represents expansion of a single genetic type following a severe bottleneck. In the case of the relatively undisturbed Lake Nipigon, we would expect the current population to have a similar degree of genetic diversity to that of the museum specimens collected earlier in the century. By examining specific genes from both mitochondrial and nuclear DNA, we should be able to investigate the possibility that unidirectional hybridization occurred between some of the original species.

Analysis of RFLPs in mtDNA has shown that although diagnostic differences are present between the RFLP patterns of mtDNA from lake whitefish and lake herring, lake herring and bloater are very closely related (Bernatchez et al. 1991). A study of RFLPs of the mtDNA from 39 individuals of both lake herring and bloater from Lake Superior found differences in haplotype frequencies between the two species (Synder et al. 1992). No unique haplotypes were found, although 15% of the lake herring had a 100 base pair deletion in the mtDNA. A study of mtDNA variation among anadromous populations of lake herring from James and Hudson bays (Bernatchez and Dodson 1990b) found significant differences in frequencies of haplotypes between the two bays. A study of lake herring from three inland lakes in Minnesota (Shields et al. 1990) revealed large amounts of intrapopulation diversity and a unique profile of haplotypes from each lake. All of the published studies have involved restriction digestion of the entire mtDNA molecule, but this procedure requires high molecular weight DNA that cannot be obtained from museum specimens. To determine the genetic relationships between extant and extinct populations, we need to identify a mitochondrial gene with an appropriate amount of variation and that can be amplified and sequenced.

A major problem in using sequence data for species or subspecies comparisons is finding an appropriate gene or gene region that contains sufficient phylogenetically informative sites but is not saturated with change (Hillis and Huelsenbeck 1992). This is especially important for mtDNA data because mtDNA evolves rapidly in animals (reviewed in Moritz et al. 1987). In order to determine which mitochondrial gene regions are appropriate for species-level and subspecies-level questions in salmonid fishes, we have examined the percent sequence divergence and the transition:transversion ratio in four different mtDNA genes in a variety of salmonids (Domanico and Phillips, in press). Transitions are up to 20–30 times more common than are trans-

versions in closely related mtDNA sequences, but the ratio will decrease to 2:1 or 1:1 for saturated sequences because transversions erase transitions as the former accumulate in these unreliable regions (Holmquist 1983). For example, for salmonid fishes in the genus *Oncorhynchus*, the mitochondrial D loop is evolving too rapidly to be useful for interspecific phylogenetic analysis (Domanico and Phillips, in press). The rate of evolution was fastest in the D loop, followed by the *ATPase6* and *ND3* genes, and slowest in the *cytochrome b* gene. In preliminary work (Sadjak, University of Wisconsin, Milwaukee, unpublished) no variation in the *ATPase6* region was found among ciscoes, suggesting that all of these "species" are very closely related.

Although species-specific allozyme patterns have been identified in some coregonids (Bodaly et al. 1991), allozymes were not useful for distinguishing between different species of Lake Superior ciscoes (Todd 1981). In a first attempt to examine nuclear DNA markers in coregonids, the first intron of the growth hormone gene (B. Shields, Eastern Michigan University, personal communication) and the internal transcribed spacer regions (ITS1) of the ribosomal DNA (rDNA) have been sequenced in bloater, lake herring, shortjaw cisco, blackfin cisco, kiyi, and lake whitefish (Sajdak, personal communication). Differences in the sequence of the ITS1 were found between lake whitefish and the shortjaw and blackfin ciscoes, but no variation between the latter. This result is consistent with previous work in which we found that the ITS1 and ITS2 of the rDNA were useful for phylogenetic analysis of the major species in the genus *Salvelinus* (Phillips and Pleyte 1991) but did not contain enough information for discrimination of some of the subspecies relationships (Pleyte et al. 1992). Intraspecific variation exists in the 5' external transcribed spacer (5'ETS) which is evolving somewhat faster than the ITS1 and ITS2 in *Salvelinus* species (Phillips et al. 1992; Phillips et al., in press b). We are currently examining variation in the 5'ETS in Great Lakes ciscoes. Although analysis of sequence variation in this region was useful in differentiating subspecies of Arctic char *Salvelinus alpinus*, this region may not be evolving fast enough to resolve relationships among populations that have diverged within the last 10,000 years. No variation was found among any of the coregonid species in the sequence of first growth hormone intron A (B. Shields, personal communication). This intron is somewhat more conserved than is first growth hormone intron C (Devlin 1993), which exhibits considerable intraspecific variation in some *Oncorhynchus* species (S.

Forbes, University of Montana, personal communication). These results suggest that it may be necessary to look at hypervariable regions of the nuclear genome in order to determine the relationships among Great Lakes cisco populations.

Two candidate regions in the nuclear genome that are hypervariable and can be amplified in small segments are microsatellites and the hypervariable region of the major histocompatibility genes (class I and class II loci). Recently, microsatellite loci were used successfully to determine relationships among human populations (Bowcock et al. 1994) and populations of brown trout *Salmo trutta* (Estroup et al. 1993). Major histocompatibility class II B genes have recently been sequenced in a preliminary phylogenetic study of the species flock of cichlid fishes in Lake Malawi (Klein et al. 1993). We plan to examine a hypervariable region from both the mitochondrial and nuclear genome in extant and historical populations by means of PCR in order to reconstruct the history of the coregonid populations in Lake Michigan.

Linking Genetic Variation to Ecomorphology

The identification of lineage markers will not by itself indicate whether genetic diversification was or will once again be a significant component in ecological establishment or adaptive radiation in the coregonids. Phenotype is the functional bridge between genetic divergence among species and their ecological niches (Grant 1986; Smith and Todd 1984). On a strictly ecological level, by demonstrating a correlation between morphological variation (i.e., gill raker number, spacing and size; mouth, fin, and body shapes) and performance measures (e.g., habitat selection, diet, and growth rate) one might expect that the fit between phenotype and ecological factors will result in reestablishment of phenotypic variation. If the *Coregonus* complex does expand back into the trophic niches it once occupied, tracking lineages to determine the causal basis for the morphological variation related to feeding ecology and spawning time will be important.

Trophic radiation of morphologically plastic *Coregonus* could still occur in Lake Michigan even if the present ciscoes are a single, genetically undifferentiated species. Trophic variation can be produced and maintained by developmental heterochrony independent of genetic divergence (Meyer 1987, 1990). The transplant studies summarized by Hile (1937) suggest that plastic responses within lineages may play an important role in adaptation to pelagic or benthic feeding. Our analysis of data collected by Koelz (1931) compares gill raker morphology of 44 populations of lake herring with the maximum depth of each lake (Figure 3) and further strengthens the suggestion by Hile that intraspecific plasticity is not random. There is a growing literature linking morphological variation with adaptive trophic polymorphism (Ehlinger and Wilson 1988; Schluter and McPhail 1993). Although some studies have demonstrated genetic differences among trophic morphs (Bernatchez and Dodson 1990a; Taylor and Bentzen 1993), there is a definite paucity of data linking trophic polymorphism with adaptation within lineages (sensu Lauder et al. 1993). Are the trophic consequences of morphological differences the selective "cause" of radiation (i.e., lineage selection) or is the "need" to radiate the selective cause for developmentally plastic genotypes?

Morphology provides the backdrop for relating species that were distinguished by use of modern genetic techniques with the historical species that were distinguished on the basis of morphological techniques. In addition to museum specimens, extensive and detailed data sets on the Great Lakes ciscoes (Koelz 1929) and inland populations (Koelz 1931; Hile 1937) are available to compare with today's populations. Combined morphometric and genetic analyses will allow for testing the similarity among extant populations and historical lineages. Given that plasticity is known to occur, we expect that morphometric analyses will be more sensitive initially than will genetic analysis in detecting divergence among spatially or temporally separated spawning stocks (Todd 1981). The key question will be whether different lineages become more-or-less successful in different spatial or temporal niches. This is the main distinction between scenarios in Figure 2.

The diverse trophic morphologies, reproductive life histories, and pelagic life styles of the coregonid fishes make them a good system for studying the ecology of reestablishment. However, it is clear that the present lack of data on the lineage history of species makes it difficult to apply the ESU concept to this system. If the ESU concept is to be applied to the Great Lakes system, then increased efforts should be made to identify historical coregonid lineages that may be reappearing or expanding in the system. Such information should enter into decisions regarding management of the fishery. For example, active steps such as closed commercial fishing zones or seasons or active propagation programs could enhance coregonid chances for reestablishment. If coregonids do not reestablish to historic levels of variation, attempts should be made to identify popu-

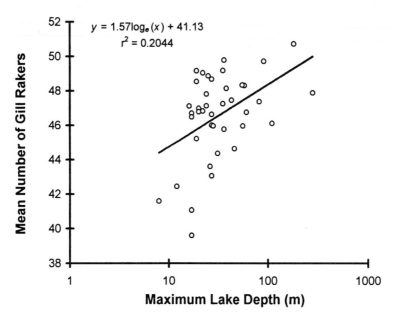

FIGURE 3.—Relationship between the mean number of gill rakers for populations of lake herring and the maximum depth of the lake from which the population was collected. Gill raker data are means from Koelz (1931) for lakes with sample sizes greater than 12 fish.

lations that are similar ecologically or genetically to those of the past and might increase chances for successful reintroduction.

The biotic community of Lake Michigan is in a critical stage of its history. Populations of introduced nonnative fishes are in decline, but new invaders are appearing. Water and habitat quality of confluent river systems is improving, but dams prevent the migration of potadromous fishes in many rivers. Stocked lake trout are growing well, but there is little or no evidence that they have reproduced naturally. By understanding the association between ecomorphology and genetic lineages we can gain insights into the factors that promoted and maintained diversity in the Great Lakes. In this context, the ESU concept may prove to be a helpful tool in guiding restoration efforts.

References

Allendorf, F. W., and G. H. Thorgaard. 1984. Tetraploidy and the evolution of salmonid fishes. Pages 1–53 in B. J. Turner, editor. Evolutionary genetics of fishes. Plenum Press, New York.

Allendorf, F. W., and R. F. Leary. 1988. Conservation and distribution of genetic variation in a polytypic species, the cutthroat trout. Conservation Biology 2:170–184.

Bernatchez, L., and J. J. Dodson. 1990a. Allopatric origin of sympatric populations of lake whitefish (*Coregonus*

clupeaformis) as revealed by mitochondrial-DNA restriction analysis. Evolution 44(5):1263–1271.

Bernatchez, L., and J. J. Dodson. 1990b. Mitochondrial DNA variation among anadromous populations of cisco (*Coregonus artedi*) as revealed by restriction analysis. Canadian Journal of Fisheries and Aquatic Sciences 47:533–543.

Bernatchez, L., F. Colobani, and J. H. Dodson. 1991. Phylogenetic relationships among the subfamily Coregoninae as revealed by mitochondrial DNA restriction analysis. Journal of Fish Biology 39(Supplement A):283–290.

Bodaly, R. A., J. Vuorinen, R. D. Ward, M. Luczynski, and J. D. Reist. 1991. Genetic comparisons of new and old world coregonid fishes. Journal of Fish Biology 38:37–51.

Bohlander, S. K., R. Espinosa III, M. M. Le Beau, J. D. Rowley, and M. O. Diaz. 1992. A method for the rapid-sequence-independent amplification of microdissected chromosomal material. Geomics 13:1322–1324.

Booke, H. E. 1968. Cytotaxonomic studies of coregonine fishes of the Great Lakes, USA: DNA and karyotype analysis. Journal of the Fisheries Research Board of Canada 25:1667–1687.

Bowcock, A. M., A. Ruiz-Linares, J. Tomfohrde, E. Minch, J. R. Kidd, and L. L. Cavilli-Sforza. 1994. High resolution of human evolutionary trees with polymorphic microsatellites. Nature 368:455–457.

Bowen, S. H., D. J. D'angelo, S. H. Arnold, M. J. Keniry, and R. J. Albrecht. 1991. Density-dependent maturation, growth, and female dominance in Lake Superior lake herring (*Coregonus artedi*). Canadian Journal of Fisheries and Aquatic Sciences 48(4):569–576.

Crowder, L. B. 1986. Ecological and morphological shifts in Lake Michigan fishes: glimpses of the ghost of competition past. Environmental Biology of Fishes 16:147–157.

Devlin, R. H. 1993. Sequence of sockeye salmon type 1 and 2 growth hormone genes and the relationship of rainbow trout with Atlantic and Pacific salmon. Canadian Journal of Fisheries and Aquatic Sciences 50:1738–1748.

Domanico, M. J., and R. B. Phillips. In press. Phylogenetic analysis of Pacific salmon (genus *Oncorhynchus*) using mitochondrial DNA sequences. Molecular Phylogenetics and Evolution.

Ehlinger, T. J. 1989. Foraging mode switches in the golden shiner (*Notemigonus crysoleucas*). Canadian Journal of Fisheries and Aquatic Sciences 46:1250–1254.

Ehlinger, T. J., and D. S. Wilson. 1988. Complex foraging polymorphism in bluegill sunfish. Proceedings of the National Academy of Sciences of the United States of America 85:1878–1882.

Estroup, A., P. Presa, F. Krieg, D. Vaiman, and R. Guyomard. 1993. (CT)n and (GT)n microsatellites: a new class of genetic markers for *Salmo trutta* (brown trout). Heredity 71:488–496.

Frolov, S. V. 1990. Differentiation of sex chromosomes in salmonidae: III. Multiple sex chromosomes in *Coregonus sardinella*. Tsitologiya 32(6):659–663.

Frolov, S. V. 1992. Some aspects of karyotype evolution in Coregoninae. Polskie Archivum Hydrobiologii 39(3–4):509–516.

Gharrett, A. J., and W. W. Smoker. 1993. Genetic components in life history traits contribute to population structure. Pages 197–202 *in* J. G. Cloud and G. H. Thorgaard, editors. Genetic conservation of salmonid fishes. Plenum Press, New York.

Grant, P. R. 1986. Ecology and evolution of Darwin's finches. Princeton University Press, Princeton, New Jersey.

Greene, C. W. 1935. The distribution of Wisconsin fishes. State of Wisconsin Conservation Commission, Madison.

Hartley, S. E. 1987. The chromosomes of salmonid fishes. Biological Review of the Cambridge Philosophical Society 62:197–214.

Hile, R. 1937. Morphometry of the cisco, *Leucichthys artedi* (Le Sueur), in the lakes of the northeastern highlands, Wisconsin. Internationale Revue der gesamten Hybrobiologie and Hydrographie 36:57–130, Leipzig, Germany.

Hillis, D. M., and J. P Huelsenbeck. 1992. Signal, noise, and reliability in molecular phylogenetic analyses. Journal of Heredity 83:189–195.

Holmquist, R. 1983. Transitions and transversions in evolutionary descent: an approach to understanding. Journal of Molecular Evolution 19:134–144.

Kitchell, J. F. 1991. Salmonid carrying capacity: estimates and experiences in the Great Lakes of North America. Pages *in* R. S. Svrjcek, editor. Marine ranching: proceedings of the seventh U.S.–Japan meeting on aquaculture. NOAA (National Oceanic and Atmospheric Administration) Technical Report NMFS (National Marine Fisheries Service) 102.

Kitchell, J. F., and L. B. Crowder. 1986. Predator-prey interactions in Lake Michigan: model predictions and recent dynamics. Environmental Biology of Fishes 16:205–211.

Klein, D., H. Ono, C. O'hUigin, V. Vincek, T. Goldschmidts, and J. Klein. 1993. Extensive MHC variability in cichlid fishes of Lake Malawi. Nature 364:330–334.

Koelz, W. 1929. Coregonid fishes of the Great Lakes. U.S. Bureau of Fisheries Bulletin 53:298–643.

Koelz, W. 1931. The Coregonid fishes of northeastern America. Papers of the Michigan Academy of Science, Arts and Letters 13:303–432.

Lauder, G. V., A. M. Leroi, and M. R. Rose. 1993. Adaptations and history. Trends in Ecology & Evolution 8:294–297.

Lindsey, C. C. 1981. Stocks are chameleons: plasticity in gill rakers of Coregonid fishes. Canadian Journal of Fisheries and Aquatic Sciences 38:1497–1506.

Meyer, A. 1987. Phenotypic plasticity and heterochrony in *Cichlasoma managuense* (Pices, Cichlidae), and their implications for speciation in cichlid fishes. Evolution 41(6):1357–1369.

Meyer, A. 1990. Ecological and evolutionary consequences of the trophic polymoromphism in *Cichlasoma citrinellum* (Pisces: Cichlidae). Biological Journal of the Linnean Society 39:279–299.

Moritz, C., T. E. Dowling, and W. M. Brown. 1987. Evolution of animal mitochondrial DNA: relevance for population biology and systematics. Annual Review of Ecology and Systematics 18:269–292.

Mummert, J. R., and R. W. Drenner. 1986. Effect of fish size on the filtering efficiency and selective particle ingestion of a filter-feeding Clupeid. Transactions of the American Fisheries Society 115:522–528.

Phillips, R. B., and K. A. Pleyte. 1991. Nuclear DNA and salmonid phylogenetics. Journal of Fish Biology 39(Supplement A):259–275.

Phillips, R. B., K. A. Pleyte, and M. R. Brown. 1992. Salmonid phylogeny inferred from ribosomal DNA restriction maps. Canadian Journal of Fisheries and Aquatic Sciences 49:2345–2353.

Phillips, R. B., K. M. Reed, and P. Rab. In press a. Revised karyotypes and chromosome banding of coregonid fishes from the Laurentian Great Lakes. Canadian Journal of Zoology.

Phillips, R. B., S. L. Sajdak, and M. J. Domanico. In press b. Evolutionary relationships among charrs inferred from ribosomal DNA sequences. Nordic Journal for Freshwater Research.

Pleyte, K. A., S. D. Duncan, and R. B. Phillips. 1992. Evolutionary relationships of the salmonid genus *Salvelinus* inferred from DNA sequences of the first internal transcribed spacer (ITS1) of ribosomal DNA. Molecular Phylogenetics and Evolution 1:223–230.

Rab, P., and M. Jankun. 1992. Chromosome studies of coregonine fishes: a review. Polskie Archiwum Hydrobioilogii 39:523–532.

Reed, K. M., S. K. Bohlander, and R. B. Phillips. 1995. Microdissection of the Y chromosome and FISH analysis of the sex chromosomes of lake trout, *Salvelinus namaycush*. Chromosome Research 3:221–226.

Schluter, D., and J. D. McPhail. 1993. Character displacement and replicate adaptive radiation. Trends in Ecology & Evolution 6:197–200.

Shields, B. A., K. S. Guise, and J. C. Underhill. 1990. Chromosomal and mitochondrial DNA characterization of a population of dwarf cisco (*Coregonus artedi*) in Minnesota. Canadian Journal of Fisheries and Aquatic Sciences 47:1562–1569.

Sites, J. W. Jr., and K. M. Reed. 1994. Chromosomal evolution, speciation, and systematics: some relevant issues. Herpatologica 50:237–249.

Smith, S. H. 1964. Status of the deepwater cisco population of Lake Michigan. Transactions of the American Fisheries Society 93:115–163.

Smith, G. R., and T. N. Todd. 1984. Evolution of species flocks of fishes in north temperate lakes. Pages 45–68 *in* A. Echelle and I. Kornfield, editors. Evolution of fish species flocks. University of Maine Press, Orono.

Snyder, T. P., R. D. Larsen, and S. H. Bowen. 1992. Mitochondrial DNA diversity among Lake Superior and inland lake ciscoes (*C. artedi* and *C. hoyi*). Canadian Journal of Fisheries and Aquatic Sciences 49:1902–1907.

Taylor, E. B., and P. Bentzen. 1993. Evidence for multiple origins and sympatric divergence of trophic ecotypes of smelt (*Osmerus*) in northeastern North America. Evolution 47:813–832.

Todd, T. N. 1981. Allelic variability in species and stocks of Lake Superior ciscoes (Coregoninae). Canadian Journal of Fisheries and Aquatic Sciences 38:1808–1813.

Todd, T. N., G. R. Smith, and L. E. Cable. 1981. Environmental and genetic contributions to morphological differentiation in ciscoes (Coregoninae) of the Great Lakes. Canadian Journal of Fisheries and Aquatic Sciences 38(1):59–61.

Todd, T. N., and R. M. Stedman. 1989. Hybridization of ciscoes (*Coregonus* spp.) in Lake Huron. Canadian Journal of Zoology 67:1679–1685.

Todd, T. N., and G. R. Smith. 1992. A review of differentiation in Great Lakes ciscoes. Polskie Archiwum Hydrobiologii 39:261–267.

Viktorovsky, R. M., L. N. Ermolenko, A. N. Makoedov, S. V. Frolov, and A. Shevchishin. 1983. Karyotype divergence in the genus Coregonus. Tsitologiya 25:1309–1315.

Waples, R. S. 1991. Pacific salmon, *Oncorhynchus* spp., and the definition of "species" under the Endangered Species Act. U.S. National Marine Fisheries Service Marine Fisheries Review 53:11–12.

White, M. J. D. 1973. Animal cytology and evolution. Cambridge University Press, London.

American Fisheries Society Symposium 17:145–148, 1995

Faunas of Isolated Regions as Principal Units in the Conservation of Freshwater Fishes

M. MINA

N. K. Koltzov Institute of Developmental Biology, Russian Academy of Sciences
Leninsky Prospect 33, 117171 Moscow, Russia

A. GOLUBTSOV

A. N. Severtzov Institute of Evolutionary Animal Morphology and Ecology
Russian Academy of Sciences

Abstract.—The preservation of biological diversity has traditionally been equated with the preservation of individual species. Conservation of a species, however, is rarely defined in any precise manner, especially when the species in question is represented by several spatially distinct populations. In such a case, if all but one of these populations were to go extinct, the species as a nominal taxon would still exist. One purported solution to this problem is to identify some populations as evolutionarily significant units. We still must begin with the demarcation of species, and the conservation status of a population would still depend on its taxonomic affiliation. Thus it seems reasonable to explore another approach to conservation of biological (genetic) diversity by using principal units of conservation that are not based on taxa. We suggest that conservation measures can be better planned and more effectively realized if faunas, rather than taxa, are considered as units of conservation. In our view such a unit would be composed of the fauna of an isolated region, within which any localized damage to fish populations could be repaired if the fauna of some other part of the region has been preserved. Drainages of large rivers or lakes and their tributaries would be primary examples of such isolated regions.

Conservation efforts have traditionally focused on preserving individual species. Consequently, the conservation of biological diversity is often reduced to the preservation of all of the items on a taxonomic list, and the loss of biological diversity is equated with the extinction of species. This has been the approach used in compiling Red Data Books and Conservation Lists published by the International Union for Conservation of Nature. It works only as long as the aim is to document the loss of species due to drastic effects on ecosystems caused by human activities or to draw attention to the likely negative results of such activities. However, if the actual goal is maintenance of the natural genetic diversity of organisms, the preservation of species as nominal taxa is obviously insufficient.

Our goal is to evaluate the expediency of using designated taxonomic groups as units of conservation. We propose a different approach, i.e., using entire faunas of isolated regions as the principal units of conservation.

Taxa as Units of Conservation

The taxa used most often as units of conservation are those of the lower taxonomic levels, i.e., the species or, occasionally, the subspecies. Conservation of a species, however, is rarely defined in any precise manner, especially when the species in question is represented by several spatially distinct populations. In such a case, if all but one of these populations were to go extinct, the species as a nominal taxon would still exist. The genetic makeup of the species, however, would be irretrievably altered. To avoid such outcomes it is necessary to shift attention from species to populations and to consider the population as the principal unit of conservation. One way of dealing with the problem of multiple populations would be to confer the taxonomic rank of species (or subspecies) upon every population that is considered worthy of conservation. This could, however, lead to biased taxonomic decisions motivated by conservational reasoning. The bias inherent in using taxa as units of conservation is not a problem that can be solved by simply substituting a different species concept or taxonomic level.

Although the biological species concept is the species concept most widely recognized by conservation biologists, it is often difficult to apply in practice (Scudder 1974). We do not see how the substitution of any alternative species concept could help, however, because no matter what concept is used, the fate of a population will depend on the decision of a taxonomist, and such decisions are always subjective. On many occasions populations have been described as separate species, subsequently lumped together into one species, and then split again. Such changes in classification are inevitable because they reflect the development of taxonomy. These changes can, however, be disastrous for the populations involved if the populations' conservation status depends on their taxonomic position.

The large barbs *Barbus* sp. of Lake Tana in Ethiopia are a good example of the difficulties inherent in the taxon-based approach. Ruppell (1836) described six species of this group in Lake Tana, whereas Bini (1940) recognized 10 species and 23 subspecies of large barbs in this lake. Subsequently, Nagelkerke et al. (1994) distinguished 13 forms (morphotypes), only some of which corresponded to previously described taxa. It is obvious that there is an extraordinary morphological diversity of Lake Tana barbs. However, Banister (1973) lumped all of these species and subspecies together with many other forms of East African large barbs into one species, *Barbus intermedius* Ruppell (1836). As a consequence, there are currently no recognized nominal species endemic to Lake Tana. Regardless of whether Banister's taxonomic decision is right or wrong, it should not undermine the fact that the Lake Tana barbs form a unique genetic system and are in need of protection in view of intensive development of the fishery in the lake. It seems evident, therefore, that the effect of arbitrary taxonomic decisions on conservation policy must be minimized.

One purported solution to this problem is to identify some populations as evolutionarily significant units. This, however, does not avoid the difficulties discussed above as long as the units of conservation are distinguished by their relations with allopatric, conspecific populations. We still must begin with the demarcation of species, and the conservation status of a population would then depend on its taxonomic affiliation. In the case of Lake Tana barbs, this would mean that recognition of one or all of the forms (morphotypes) as significant units may depend on taxonomic decision concerning their status as separate species or as component populations within *B. intermedius*.

Finally, all taxa that include allopatric populations are very inconvenient objects for conservation action. Any such action should take specific circumstances into account, and different actions are often necessary to conserve different populations composing a single taxon. Thus it seems reasonable to explore another approach to conservation of biological (genetic) diversity by using principal units of conservation that are not based on taxa.

Faunas of Isolated Regions as Units of Conservation

We suggest that conservation measures can be better planned and more effectively realized if faunas, rather than taxa, are considered as units of conservation. In our view such a unit would be composed of the fauna of an isolated region, within which any localized damage to fish populations could be repaired if the fauna of some other part of the region has been preserved (Mina 1992). Drainages of large rivers or lakes and their tributaries would be primary examples of such isolated regions.

We base this approach on a hierarchical view of Mendelian populations (Dobzhansky 1950, 1955). Within a geographically isolated region we find populations of a high rank reproductively isolated from each other, and each of them is in turn subdivided into populations of lower ranks. The high-rank populations can be considered to represent different biological species. It should be emphasized that in dealing with sympatric populations, the biological species concept is operational, though not always easy to apply. Any population, whatever be its rank, is a unit of evolution (sensu Dobzhansky 1955), because an evolutionary process occurring in one of its parts can influence the whole. Each population is also self-repairable, that is, able to compensate for damage inflicted on one part at the expense of its other parts.

It may appear that a high-rank population of a certain species, rather than an entire fauna, should be considered as the principal unit of conservation. This is not true, however, because the main goal of conservation should be, in our view, the maintenance of genetic diversity of the local fauna as a whole. Conservation measures should therefore be designed to protect all the high-rank populations (species) composing such a fauna, not only selected ones designated as significant in some aspect. It is especially important that conservation measures designed for one population not harm other populations in the same region. Ideally such measures should be equally beneficial for all populations, e.g., measures targeted at preventing the deleterious effects of human activities and at repairing damages that could not be prevented.

The larger the region to be protected from deleterious influences, the more difficult it is to organize and implement protection. The main advantage of local faunas as manageable units of conservation is that often the protection of a part (or parts) of an isolated region provides for the conservation of the whole fauna of the region. Such specially protected parts (reserves) act as refugia that allow fish to survive periods when living conditions in other parts of the region are unfavorable.

A fauna that is composed of populations which are self-repairable is also self-repairable. Our task is to promote this process of self-repair, protecting potential sources of recruitment to populations in affected localities.

General conservation measures may not be sufficient to protect all the species in a local fauna. In such a case auxiliary measures should be taken to aid individual species within the fauna. These measures may be the construction of fish passages at dams, hatcheries, or special fishery regulations. Examples of applications of such measures that have been highly effective are the preservation of stocks of beluga *Huso huso* and inconnu *Stenodus leucichthys* in the Volga River (Letichevsky 1976; Khodorevskaya et al. 1989). Once more it should be stressed that all of these measures should be planned so as to take into account their effect on all species constituting a given local fauna.

As for the relative significance of different local faunas, we do not think that one fauna is more significant than another if it presumes a priority in conservation. Some faunas can be more interesting as objects of ecological or evolutionary studies or they may be more economically important. However, these factors should not determine priority in conservation. To be sure, a system of priorities is necessary because of the limited resources of conservation agencies. In our view, however, the main factors that should determine priorities are the vulnerability of a fauna and the probable effectiveness of potential conservational measures. An example of the importance of taking both of these factors into account is the situation in the Aral Sea. However deplorable the destruction of the fish fauna of the Aral Sea (Plotnikov et al. 1991), its conservation was not possible unless major changes were made in the distribution of water resources of the Amu Darya and Syr Darya, the main tributaries to the sea.

Following this logic, conservation of endemic lacustrine fish faunas known to be exceptionally vulnerable (Kornfield and Echelle 1984) should be given high priority not because these faunas are endemic, but because they are so vulnerable. Those conservation efforts that have a high probability of success should then receive a high priority. For example, conservation measures can be very effective in Lake Tana right now. There are currently no major sources of water pollution and no introduced species. The only danger to the large barbs, the most diverse and interesting group of native fishes, is potential overfishing. Proper regulation of the fishery could provide protection for these barbs, and it seems expedient to introduce measures to limit the fishery as quickly as possible.

Problems and Perspectives

Special investigations are necessary to determine the optimal allocation of reserves in an isolated unit region. Both allocation and size of these reserves depends upon the geographical and ecological structure of a given region as well as on the nature of the deleterious influences threatening its fish fauna. For instance, in a lake where the main danger to fish is from overfishing, preservation of spawning grounds can be quite effective. If the main danger is from pollution, however, the only possibility is to protect the entire lake, because the whole lake, rather than simply a part of it, is often affected.

Unlike lake systems, harmful influences in river basins are often "vectorized," and are manifested downstream from the affected localities. The problem of allocating reserves is very complicated in rivers. In a large river, some species may be restricted in their distribution, occurring only in the upper or lower reaches. In such cases, the maintenance of both the species composition of the river fauna and the genetic diversity of individual species would require the creation of many reserves.

Introduced species that establish populations in colonized regions create a special problem. By protecting local faunas in these regions, we also protect nonnative species, and often there is little possibility to eliminate them. There is no general solution for this problem, and it must be considered for each particular case.

When a fauna undergoes restoration, the initial genetic diversity could probably be modified because both the relative abundance of species and the diversity of each of the species would probably not be the same as under natural conditions. Here we approach one of the central questions of nature conservation: should we define the principal goal of conservation as the maintenance of natural genetic diversity? On the one hand, it is obvious that this goal can hardly be achieved because it is linked to the preservation of the ecological context. On the other hand, it is not obvious that the genetic diversity that was characteristic of populations under natural environmental conditions is optimal after the changes brought about by human activities.

We believe that conservation (or restoration) of natural genetic diversity should be considered as a guideline simply because it is the least risky strategy. At the present level of knowledge it is much more risky to regulate genetic diversity. Because we are dealing with adaptive entities we should provide them with opportunities for adaptation rather than select and support only those species that appear to be the most promising in the new environment.

Acknowledgments

Our studies were supported by the Russian Foundation of Fundamental Sciences, Project 94-04-12701. We are grateful to C. A. Annett and R. Pierotti, whose valuable notes allowed us to improve the text considerably, as well as to W. W. Dimmick for helpful discussion of the earlier drafts of the manuscript. We thank R. Behnke and R. Reavis for critical comments on the manuscript. Kheryn Klubnikin, U.S. Forest Service, Forest Environment Research, made our participation in the conference possible.

References

Banister, K. E. 1973. A revision of the large *Barbus* (Pisces, Cyprinidae) of East and Central Africa. Bulletin of the British Museum of Natural History (Zoology) 26:1–148.

Bini, G. 1940. I pesci del Lago Tana. Missione di Studio al Lago Tana richerge limnologiche. B. Chimica e Biologia, volume 3. Reale Accademia d'Italia 18:401–418, Rome.

Dobzhansky, T. 1950. Mendelian populations and their evolution. American Naturalist 84:401–418.

Dobzhansky, T. 1955. A review of some fundamental concepts of population genetics. Cold Spring Harbor Symposium on Quantitative Biology 20:1–15.

Khodorevskaya, R. P., A. V. Pavlov, and G. F. Dovgopol. 1989. Role of sturgeon breeding at hatcheries in formation of commercial stocks of acipenserids in the Caspian basin. Pages 327–328 *in* Osetrovoje khozjajstvo vodoemov SSSR (in Russian).

Kornfield, I. L., and A. A. Echelle. 1984. Who's tending the flock? Pages 251–254 *in* A. Echelle and I. Kornfield, editors. Evolution of fish species flocks. University of Maine Press, Orono.

Letichevsky, M. A. 1976. Ways to increase efficiency of hatchery rearing and reproduction of *Stenodus leucichthys leucichthys* (Guld.) in the delta and the lower part of the Volga River. Pages 77–81 *in* O. A. Skarlato, editor. Ekologia i systematika lososevidnykh ryb. Zoological Institute of the USSR Academy of Sciences, Leningrad (in Russian).

Mina, M. V. 1992. Problems of protection of fish faunas in the USSR. Netherlands Journal of Zoology 42:200–213.

Nagelkerke, L. A. J., F. A. Sibbing, J. G. M. van den Boogaart, E. H. R. R. Lamens, and J. W. M. Osse. 1994. The barbs (*Barbus* spp.) of Lake Tana: a forgotten species flock? Environmental Biology of Fishes 39:1–22.

Plotnikov, I. S., N. V. Aladin, and A. A. Philippov. 1991. The past and present of the Aral Sea fauna. Zoologicheskii Zhurnal 70:5–15 (in Russian).

Ruppell, E. 1836. Neuer Nachtrag von Beschreibungen und Abbildungen neuer Fische im Nil entdeckt. Museum Senckenbergianum, Abhandlungen aus dem Gebiete der beschreibenden Naturgeschichte: 1–28, Band II, Heft 1, Frankfurt am Main.

Scudder, G. G. 1974. Species concepts and speciation. Canadian Journal of Zoology 52:1121–1134.

American Fisheries Society Symposium 17:149–165, 1995

Genetic Population Structure and History of Chinook Salmon of the Upper Columbia River

FRED M. UTTER

School of Fisheries, University of Washington
Seattle, Washington 98195, USA

DON W. CHAPMAN

Don Chapman Consultants, 3653 Rickenbacker, Suite 200
Boise, Idaho 83705, USA

ANNE R. MARSHALL

Washington Department of Fish and Wildlife
600 Capitol Way North, Olympia, Washington 98501, USA

Abstract.—Chinook salmon *Oncorhynchus tshawytscha* that return to the upper Columbia River (upstream from the confluence of the Yakima River) are considered from the perspectives of allelic variation at 32 polymorphic loci, historical activities within this region, and ancestral affinities to downstream populations. Collections of summer–fall-run fish are distinguished from spring-run fish by an eightfold greater genetic distance between groups than exists within either group. Each group was related to but remained distinct from adjacent downstream groups within different major ancestral units, previously identified throughout the Columbia River. Summer–fall-run fish are most closely related to fall-run fish of the mid-Columbia and Snake rivers, and spring-run fish to the spring–summer-run fish of the Snake River. In both groups, the present geographic distributions and genetic population structures within the upper Columbia River reflect translocations, confinements, and cultural activities between 1939 and 1943 under the Grand Coulee Fish Maintenance Project, and subsequent introductions and fish culture. The considerable genetic homogeneity within the summer–fall-run group appears to have been maintained through past and present interbreedings and strayings over a single continuous run. Some degree of genetic distinction persists between cultured and wild spring-run fish; the cultured fish are genetically indistinguishable from their ancestral source of the downstream Carson Hatchery, derived during the 1950s from fish returning to the upper Columbia and Snake rivers. The entire summer–fall-run group and the wild component of the spring-run group qualify for consideration as different evolutionarily significant units. Suggestions to conserve the genetic variation within these groups focus on measures that restrict excessive gene flow and permit maintenance and development of local adaptations.

The Columbia River is the largest river entering the Pacific Ocean from North America, draining 670,810 km^2 of the northwestern United States and southwestern Canada (Figure 1). Historically, this drainage supported the world's greatest runs of chinook salmon *Oncorhynchus tshawytscha*. The present distribution of returning fish in spring, summer, and fall modes contrasts with a continuum of returns and a summer mode recorded in the nineteenth century (Thompson 1951). This altered distribution and an overall numerical decline has been attributed to the combined effects of overharvest and habitat degradation (Mullan 1987; Nehlsen et al. 1991). The currently depleted number of summer-run fish has stimulated petitions for their protection under the U.S. Endangered Species Act (ESA; 16 U.S.C. §§ 1531 to 1544; Rohlf 1993). An adequate understanding of the ancestral relationships among geographically and temporally isolated chinook salmon populations, particularly within the drainages of the upper Columbia River, is a necessary component of response to these petitions (Waples 1991).

We examine relationships among chinook salmon populations of the upper Columbia River. Biochemical genetic data from 16 summer-run, fall-run, and spring-run collections identify two distinct groupings consistent with those indicated from previous studies in Figure 2. We relate these observations to historical fishery management in the region and discuss the relationships of these groups to other populations, their relevance as evolutionarily significant units (ESUs), and appropriate management strategies.

Background

Biochemical Genetic Studies

Biochemical genetic studies involving chinook salmon populations of the Columbia River have

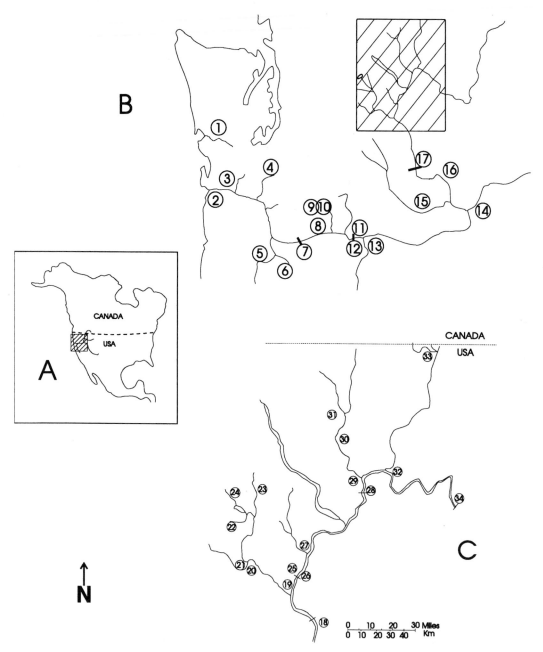

FIGURE 1.—(**A**) The Columbia River drainage relative to North America. (**B**) Enlargement of cross-hatched area of (**A**). The drainage is arbitrarily subdivided into the lower Columbia River below the Dalles Dam, the mid-Columbia River between the Dalles Dam and the confluence of the Yakima River, the Snake River, and the upper Columbia River upstream from the Yakima River. Numbers of specific locations (excluding cross-hatched area) indicate (1) Chehalis River, (2) Columbia River mouth, (3) Elokomin River, (4) Cowlitz River, (5) Willamette River, (6) Eagle Creek, (7) Bonneville Dam, (8) Spring Creek, (9) Carson Hatchery, (10) Wind River, (11) Little White Salmon River, (12) The Dalles Dam, (13) Deschutes River, (14) mouth of the Snake River, (15) Yakima River, (16) Hanford Reach, (17) Priest Rapids Hatchery. (**C**) Enlargement of cross-hatched area of (**B**). Numbers of locations indicate (18) Rock Island Dam, (19) Wenatchee River, (20) Leavenworth Hatchery, (21) Icicle Creek, (22) Nason Creek, (23) Chiwawa River, (24) White River, (25) Rocky Reach Dam, (26) Eastbank Hatchery, (27) Entiat River, (28) Wells Dam, (29) Methow River, (30) Carlton Pond, (31) Winthrop Hatchery, (32) Okanogan River, (33) Similkameen River, and (34) Grand Coulee Dam.

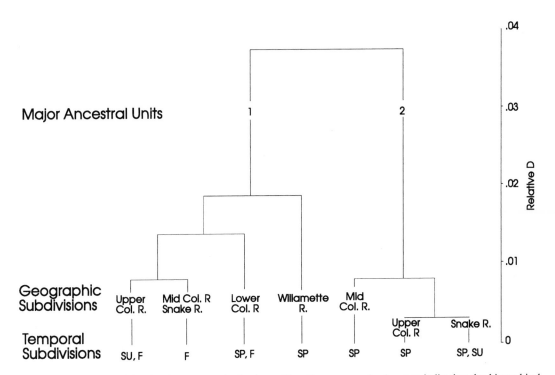

FIGURE 2.—An unweighted pair-group method using arithmetic averages dendrogram indicating the hierarchical subdivision of chinook salmon populations of the Columbia River. Derived from Matthews and Waples (1991: Figure 5) and based on pairwise measurements of genetic distance (*D*; Nei 1972) at 21 polymorphic loci. Multiple temporal subdivisions (SP = spring-run, SU = summer-run, and F = fall-run fish) included within some geographic subdivisions indicate the absence of distinguishing allele frequency patterns based on run timing. (Columbia = Col., River = R.)

revealed a complex population structure (Utter et al. 1989, 1992; Bartley et al. 1992; Waples et al. 1993) both between and within two major ancestral units (Figures 1 and 2). Subgroups of populations within unit 1 extend into major tributaries from the lower Columbia River through the Snake River and upper Columbia River, whereas the natural distribution of unit 2 subgroups is restricted to areas upstream from the lower Columbia River.

Relationships among temporally distinct runs vary with location. Dates for distinguishing spring and summer runs and summer and fall runs at Priest Rapids Dam are 23 June and 1 September (Public Utility Districts upstream from Hanford, Washington) or 17 June and 17 August (U.S. Army Corps of Engineers). In unit 1, spring- and fall-run fish of the lower Columbia River and summer- and fall-run fish of the upper Columbia River do not appear genetically differentiated (Utter et al. 1989). Except where otherwise noted, the convention of Mullan (1987) is followed here in reference to the latter subgroup as summer–fall-run fish. Within unit 2, genetic studies, coupled with juvenile life-history

and habitat data, of spring- and summer-run fish of the Snake River have concluded that these combined runs constitute a common evolutionary grouping (Matthews and Waples 1991).

Historical Modifications

The Grand Coulee Fish Maintenance Project.— Fishery management associated with the blockage to upstream migration by Grand Coulee Dam must be reviewed when considering the present genetic structure of chinook salmon populations of the upper Columbia River. Grand Coulee Dam permanently blocked access of anadromous salmonids to over 1,609 km of upstream spawning and rearing habitat in 1939. In compensation for this loss, the Grand Coulee Fish Maintenance Project (GCFMP) intercepted upstream migratory salmonids at Rock Island Dam, near Wenatchee (Figure 1), from 1939 through 1943 for relocation in tributaries downstream from Grand Coulee Dam. The details outlined in Table 1 indicate a 5-year period when almost all adult spring-run and summer–fall-run

TABLE 1.—A chronology of events affecting chinook salmon between Rock Island and Grand Coulee dams under the Grand Coulee Fish Maintenance Project (GCFMP) (from Fish and Hanavan 1948; Mullan 1987). Year given represents brood year.

1938—Normal spawning occurs upstream from Rock Island Dam, including areas upstream from Grand Coulee site. Juvenile summer–fall fish go to sea in 1939, and some possibly in 1940[a]. Spring-run progeny of the 1938 brood enter sea in 1940.

1939—All upstream migrants are trapped at Rock Island Dam, 1939–1943. Releases include mixed spring-run to racked area in Nason Creek (Wenatchee River tributary just downstream from Lake Wenatchee) and summer–fall fish to racked areas in the Entiat River and in the Wenatchee River between Lake Wenatchee and Tumwater. The same pattern of adult treatment continues through 1943, except for ending Entiat releases after 1940 brood.

1940—Juveniles of mixed-stock summer–fall chinook salmon and of spring-run chinook salmon are introduced to Methow River, Entiat River, and Icicle Creek (Wenatchee River tributary near Leavenworth, Washington). Fish culture begins in late August at Leavenworth Hatchery with spring-run and summer–fall-run parents.[b,c]

1941—Some mixed-origin juveniles of summer–fall chinook salmon are released to Entiat River.

1942—Mixed-stock juveniles of summer–fall chinook salmon are released to Methow River, Icicle Creek, and Entiat River. Release of spring-run juveniles of mixed origin is made to Methow River. Some mixed adults, presumably 3-year-old jacks from 1939 brood spawning, are released past Rock Island Dam.

1943—Releases of mixed-stock juveniles of summer–fall chinook salmon are made to Icicle Creek and Entiat River. Progeny of mixed spring-run fish are released in Icicle Creek and in Wenatchee, Methow, and Entiat rivers. About 15% of a limited release of jacks upstream from Rock Island Dam return to the Methow River.

1944—No trapping at Rock Island Dam, hence adults have access to tributaries and main Columbia River in the upper Columbia region. Returns include progeny of both mixed-stock juvenile releases and of natural spawners. Juveniles from mixed Rock Island summer–fall fish are released in Entiat River. Progeny of mixed spring-run fish are released in Methow River. "Mixed" here, and for subsequent brood years, means adults that returned to GCFMP hatcheries but that originated from adults trapped at Rock Island Dam.

1945—Returns include a few progeny of natural spawners in racked areas and hatchery-produced fish. Hatchery-reared spring-run fish are released to Methow River. Hatchery summer–fall chinook salmon are delivered to Entiat River.

1946—Returns are as for 1945 brood. Juveniles of mixed summer–fall chinook salmon are released in Icicle Creek, Methow River, and Entiat River. Mixed juvenile spring-run fish are released to Methow River.

1947—Returns are as for 1945 and 1946 broods. Mixed spring-run chinook salmon from Leavenworth Hatchery egg take are released in Icicle Creek, and those from Winthrop egg take are released in Methow River. Summer–fall juveniles from Entiat volunteers to trap are delivered to Leavenworth, with liberation point unknown but likely to be Icicle Creek. Remaining Entiat progeny probably are delivered to Entiat River.

[a] Role of stream-annulus summer–fall chinook salmon in pre-dam era is unknown. Present age distributions include some stream-annulus adults, but this may result from effects of hydropower system on seaward migration.
[b] Records do not separate progeny numbers in spring-run and summer–fall-run groups. We assume that maturation timing would prevent extensive mixing of the two groups at spawning time.
[c] Fish and Hanavan (1948) did not distinguish the fall-run component, calling all late-run chinook salmon "summer chinook." Chapman et al. (Don Chapman Consultants, Boise, Idaho, 1994) termed all late-run chinook salmon "summer–fall chinook."

fish, regardless of original destination, were respectively either confined to restricted areas for natural reproduction or used for cultural activities. These interceptions, translocations, and within-group admixtures permanently transfigured the populations of anadromous salmonids above Rock Island Dam and provided a foundation for the present population structures.

Cultural activities since the Grand Coulee Fish Maintenance Project.—Additional cultural activities that persist through the present have complicated modifications of population structures imposed through the GCFMP. The most obvious manipulations involve introductions of fish from regions downstream from the upper Columbia River. A review of such introductions (Table 2) indicates a continual influx from diverse geographic and ancestral origins.

The most persistent and extensive of these introductions involve spring-run fish from the Carson Hatchery on the Wind River (see Figure 1). This population was derived from spring-run fish that were destined for the Snake and the upper Columbia rivers and were intercepted at Bonneville Dam starting in 1955 (Ricker 1972). The spring-run fish of the adjacent Little White Salmon River Hatchery were largely derived from Carson fish (Howell et al. 1985).

Spring-run production of both the Leavenworth and Entiat hatcheries depended on eggs from downstream broodfish of mixed origin through the 1970s and into the 1981 brood year. These exoge-

TABLE 2.—Releases of chinook salmon in the upper Columbia River drainage from sources downstream from Rock Island Dam (compiled from Peven 1992); the latest spring-run releases were from the 1982 brood year. See Figure 1 for locations. Abbreviations are creek (Ck.), hatchery (H.), river (R.), subyearling (SY), and yearling (Y).

| Origin of released fish | Run of released fish | Release | | |
		Location	Numbers	Fish size
1960–1970				
Spring Ck. H.	Summer	Entiat R.	990,800	SY
	Fall	Icicle Ck.	2,922,000	SY, Y
	Spring	Icicle Ck.	251,000	Y
Eagle Ck. (Willamette R.)	Fall	Columbia R.	659,000	SY
	Spring	Icicle Ck.	86,000	Y
1971–1980				
Carson H.	Spring	Icicle Ck.	6,978,000	Y
	Spring	Entiat R.	1,677,000	Y
	Spring	Columbia R.	1,183,000	Y
Little White Salmon H.	Spring	Icicle Ck.	1,127,000	Y
	Spring	Entiat R.	1,161,000	Y
Cowlitz H.	Spring	Icicle Ck.	989,000	Y
	Spring	Entiat R.	436,000	Y
Simpson H. (Chehalis R.)	Fall	Columbia R.	715,000	SY, Y
1981–1990				
Carson H.	Spring	Icicle Ck.	155,000	Y
	Spring	Entiat R.	436,000	Y
	Spring	Columbia R.	762,000	Y
Little White Salmon H.	Spring	Entiat R.	622,000	Y
Elokomin H.	Fall	Columbia R.	296,000	Y
Bonneville H.	Fall	Columbia R.	226,000	Y
Snake R. × Priest Rapids H.	Fall	Columbia R.	1,136,000	SY, Y
Priest Rapids H.	Fall	Columbia R.	657,000	Y

nous infusions included intervals of five consecutive years in both hatcheries when all releases were of predominantly Carson Hatchery origins. This dependence on external sources ultimately gave way to full hatchery production from fish returning to the respective hatcheries in 1982 (Mullan 1987; Peven 1992).

The summer and fall hatchery programs of the upper Columbia River have been supported primarily through indigenous populations of this region, centering on the Wells Dam Hatchery for areas above Rocky Reach Dam and the Eastbank Hatchery for the Wenatchee River. The background of broodstocks has been reviewed by D. Chapman et al. (Don Chapman Consultants, Boise, Idaho, 1994). Returns to these hatcheries for a given year-class, in addition to fish released from the hatchery, included naturally produced upriver fish and substantial numbers of strays from downriver (Rocky Reach, Wenatchee, and Priest Rapids) hatchery releases of summer-run and fall-run fish. Since 1991, Priest Rapids Hatchery has supplied fall-run fish to the Rocky Reach Hatchery, and the Wells Hatchery has been the source of summer-run fish released in the Methow and Similkameen rivers. The more limited Bonneville Hatchery fall-run

chinook salmon releases (Figure 1; Table 2) represent fish of mixed upstream ancestry analogous to the spring-run fish of the Carson Hatchery. This stock was created during the 1980s for enhanced downstream production of "upriver bright" fall chinook salmon destined for the upper Columbia River (Smouse et al. 1990).

Materials and Methods

Sixteen collections, made between 1989 and 1992 under conditions detailed in Marshall and Young (1994), represented populations of adult or juvenile fish from 10 localities (Figure 1; Table 3). Attempts were made to sample equal numbers of male and female adults throughout the spawning period and over the contiguous spawning range of a given population. Juvenile progeny from separate brood years were sampled from several hatchery programs. Tissues dissected from adult fish in the field and placed directly on dry ice included approximately 1 cm^3 each of heart, liver, cheek muscle, and eye tissue; intact smolt-size juvenile fish were placed on dry ice when collected. All samples were transferred to a −80°C freezer for subsequent storage prior to electrophoresis.

TABLE 3.—Sampling locations for spring-run and summer–fall-run chinook salmon of the upper Columbia River. Collection year for juvenile samples reared at Eastbank Hatchery (†) or rearing ponds at Carlton (Methow River) or Similkameen (††) is year of parental spawning by adults collected at indicated locations. Designations for time of return are based on passage at Priest Rapids Dam: before 25 June, spring run (SP); between 25 June and 13 August, summer run (SU); and after 13 August, fall run (F). Maturity is either adult (A) or juvenile (J).

Collection number and name	Time of return	Collection years	Maturity	Sample size
1 Wenatchee River	SU	1989–1992	A	409
2 Wells Hatchery	SU	1991–1992	A	202
3 Wells Trap	SU	1991–1992	A	180
4 Wenatchee River†	SU	1992	J	86
5 Wells (Carlton)††	SU	1992	J	90
6 Wells (Similkameen)††	SU	1992	J	75
7 Similkameen River	SU	1991–1992	A	81
8 Hanford Reach	F	1990	A	99
9 Priest Rapids	F	1990–1991	A	200
10 Winthrop Hatchery	SP	1992	A	100
11 Leavenworth Hatchery	SP	1991	A	100
12 White River	SP	1989, 1991, 1992	A	113
13 Nason Creek	SP	1989–1992	A	71
14 Chiwawa River	SP	1989, 1991, 1992	A	133
15 Chiwawa River†	SP	1992	J	86
16 Chiwawa River†	SP	1991	J	100

Methods of tissue extraction, electrophoresis, and histochemical staining followed Aebersold et al. (1987). The loci and alleles screened (Table 4) and the laboratory protocol used are described in detail in Marshall and Young (1994). Phenotypes from all gels were independently double scored, and many were screened in two or more tissues and on two different buffers to ensure accuracy of the data and to resolve all known alleles. Genetic nomenclature followed the American Fisheries Society guidelines established by Shaklee et al. (1990). Allelic data for isolocus pairs (sAAT-1,2*, sMDH-A1,2*, and sMDH-B1,2*) were used in comparative analyses under the assumption that all of the variation occurred at one locus.

Genetic data were analyzed with the BIOSYS-1 computer program[1] for pairwise genetic distances (D; Nei 1972), and dendrograms were constructed through the unweighted pair-group method (Sneath and Sokal 1973). A G-test (log-likelihood ration test; Zar 1974) examined heterogeneity in pairwise comparisons of allele frequencies among samples; the G statistic approximates the chi-square distribution based on the same degrees of freedom and critical values. Samples were combined into pooled collections over two or more years for some locations when sample sizes in individual years were less than 50 and also when G-tests between or among years for larger samples from the same location were nonsignificant. Chi-square tests for departures of genotypes from the expected binomial distribution (Hardy–Weinberg equilibrium) were made on all disomic loci when frequencies of the common allele were less than 0.95.

Results and Discussion

Genetic Analyses

Genetic variability was detected within 17 classes of enzymatic proteins at 32 presumed single loci or isolocus pairs among the 16 collections of upper Columbia River chinook salmon examined in this study (Table 4). The allelic frequencies of the polymorphic loci (Appendix) provided the basis for genetic comparisons among these collections.

Thirteen out of 227 tests for deviations from Hardy–Weinberg equilibrium were significant at the 0.05 level. No pattern was discernable among the loci for which significant deviations were observed. Because this frequency of deviations would be expected by chance (sampling error) among 227 tests, conditions resulting in Hardy–Weinberg proportions (e.g., random mating and absence of strong selection) were presumed to underlie these collections.

Between-group distinctions.—The amounts and distributions of genetic variation varied considerably among loci (Appendix). The ranges of allele frequencies among the 17 most variable loci (Ta-

[1]Swofford, D. L., and R. B. Selander. 1989. BIOSYS-1: a computer program for the analysis of allelic variation in population genetics and biochemical systematics. Release 1.7, Illinois Natural History Survey, Champaign.

TABLE 4.—List of enzyme names and international numbers (IUBNC 1984) of variable enzymes, as well as loci designations, tissue distributions, and relative mobilities of variant allelic forms of the polymorphic loci. Tissue types are eye (E), heart (H), liver (L), and cheek muscle (M).

Enzyme name	Enzyme number	Locus	Relative mobilities of variants	Tissue distribution
Aspartate aminotransferase	2.6.1.1	sAAT-1,2*	85, 105	M, H
		sAAT-3*	90	E
		sAAT-4*	130, 63	L
		mAAT-1*	−77, −104	M, H
Adenosine deaminase	3.5.4.4	ADA-1*	83	M, E, H
		ADA-2*	105	M, E, H
Aconitate hydratase	4.2.1.3	sAH*	86, 112, 108	L
		mAH-4*	119	M, H
Dipeptidase	3.4.–.–	PEPA*	90, 81	M, E, H, L
Glucose-6-phosphate isomerase	5.3.1.9	GPI-B2*	60	M
		GPI-A*	105	M, E, H
Glutathione reductase	1.6.4.2	GR*	85	M, E, H, L
Hydroxyacylglutathione hydrolase	3.1.2.6	HAGH*	143, 131	M, H, L
Isocitrate dehydrogenase	1.1.1.42	mIDHP-2*	154, 50	M, E
		sIDHP-1*	74, 142, 94, 126	M, H, E, L
		sIDHP-2*	127, 83	H, E, L
L-Lactate dehydrogenase	1.1.1.27	LDH-B2*	112, 71	E, L
		LDH-C*	90, 84	E
Malate dehydrogenase	1.1.1.37	sMDH-A1,2*	27, 160	M, H, E
		sMDH-B1,2*	121, 70, 83, 126	M, H, L
		mMDH-2*	200	M, H
Malic enzyme (NADP⁺)	1.1.1.40	sMEP-1*	92	M, H
Mannose-6-phosphate isomerase	5.3.1.8	MPI*	109, 95	M, H, E
Proline dipeptidase	3.4.13.9	PEPD-2*	107	M
Leucyl-tyrosine dipeptidase	3.4.–.–	PEP-LT*	110	M, L
Phosphogluconate dehydrogenase	1.1.1.44	PGDH*	90	M, E
Phosphoglycerate kinase	2.7.2.3	PGK-2*	90, 74	M, E
Phosphoglucomutase	5.4.2.2	PGM-2*	136	M
Superoxide dismutase	1.15.1.1	sSOD-1*	−260, 580, −175	L
		mSOD*	142	H
Tripeptide aminopeptidase	3.4.–.–	PEPB-1*	130, −350	M, E
Triose-phosphate isomerase	5.3.1.1	TPI-2.2*	104	M, E

ble 5) reflect the actual divergence among collections. Typically, the greatest differences occurred between the summer–fall-run and spring-run groups, and no overlap of allele frequency occurred between groups for those nine loci with ranges exceeding 0.20. Particularly notable in this regard were sMEP-1* and PGK-2*, with respective overall ranges of 0.805 and 0.618.

The clear distinction between these two groups was apparent in other analyses. Significance levels of G-tests exceeded 0.001 in all comparisons, and between-group genetic distances ranged between 0.030 and 0.049 (Table 6). The dendrogram of pairwise genetic distances (Figure 3) separates the summer–fall-run and spring-run fish at an average between-group D value of 0.04, an eightfold greater distance than the largest pairwise D value observed among within-group separations.

Average relative heterozygosity values (based on only polymorphic loci; Appendix) did not overlap among collections representing different groups; differences between summer–fall-run (mean 0.101) and spring-run (mean 0.086) fish were highly signif-icant ($P < 0.001$ based on Mann–Whitney U-test). These differences comport with previous comparisons of heterozygosities among chinook salmon populations of the Columbia River (Utter et al. 1989; Winans 1989; Waples et al. 1991a). Heterozygosity values were consistently lower for either spring-run fish of the upper Columbia River or spring–summer-run fish of the Snake River when compared with other groups.

These major groupings fell within appropriate subdivisions of the different major ancestral groups inferred from previous studies in Figure 2. The summer–fall-run group of this study merged with the same subdivision indicated in unit 1, having affinities to fall-run fish of the mid-Columbia and Snake rivers, but were distinguished by differing allelic frequencies at several loci including sAH*, sSOD-1*, and PEPB-1* (Utter et al. 1989). Similarly, the spring-run group of this study coincides with the same subdivision of unit 2 in Figure 2, being distinguished from the closely related spring–summer-run subgrouping of the Snake River by

TABLE 5.—Ranges of allele frequencies for 17 most variable loci for which the common allele occurs at a frequency of less than 0.95 in one or more summer–fall-run or spring-run collections.

Locus	Summer–fall run		Spring run	
	Minimum	Maximum	Minimum	Maximum
ADA-1*	0.983	1.000	0.929	0.975
sAH*	0.741	0.820	0.991	1.000
mAH-4*	0.820	0.919	0.905	1.000
PEPA*	0.960	0.990	0.947	1.000
GPI-B2*	0.929	0.970	0.882	1.000
HAGH*	0.989	1.000	0.858	1.000
sIDHP-1*	0.980	1.000	0.665	0.810
sIDHP-2*	0.782	0.869	0.991	1.000
sMDH-B1,2*	0.938	0.995	0.938	0.990
mMDH-2*	0.973	1.000	0.695	0.900
sMEP-1*	0.756	0.839	0.034	0.095
MPI*	0.652	0.747	0.792	0.975
PEP-LT*	0.731	0.867	0.899	0.994
PGK-2*	0.560	0.698	0.080	0.180
sSOD-1*	0.482	0.550	0.574	0.872
PEPB-1*	0.691	0.747	0.779	0.847
TPI-2.2*	0.985	1.000	0.841	0.975

allelic frequency differences at sIDHP-1* and sSOD-1* (Waples et al. 1993).

Within-group comparisons.—The heterogeneity among the spring-run collections was considerably greater than that observed among summer–fall-run fish. Only one G-test for spring-run fish, that involving adults of the Chiwawa River (collection number 14) and Nason Creek (13), was nonsignificant (Table 6). The most divergent comparisons involved White River adults (12) and Chiwawa River juveniles (15 and 16), where the significance of all G-test values exceeded 0.001, and D values as high as 0.005 occurred; each of these three collections stand out as outliers in Figure 3. Two additional sub-

groups at lower levels of genetic divergence than that apparent in Figure 2 were adult fish returning to the Winthrop and Leavenworth hatcheries and to Nason Creek and the Chiwawa River.

A greater genetic uniformity was evident among the summer–fall-run collections. All G-test values (Table 6) exceeding 0.001 involved comparisons of juvenile fish that were released in the Similkameen River (6) and were progeny of adults returning to Wells Dam. Except for one, G-test values involving only adult samples were nonsignificant or significant at the 0.05 level; most (four out of five) of these significant tests included the Hanford Reach collection (8). No pairwise comparisons of genetic distance exceeded 0.001; 13 out of the 18 D values at the 0.001 level involved collections of hatchery-reared juvenile fish. Two such juvenile collections (5 and 6) diverged as a subcluster from the seven other summer–fall-run collections (Figure 3).

Biological implications of within-group divergences.—The most obvious feature of the genetic variation within the summer–fall-run and spring-run groups was that most of the divergence occurs from collections of juvenile progeny of hatchery-produced adults. This tendency to deviate from adult collections can be most easily explained as a consequence of limited numbers of parents or biased sampling of juvenile fish. For the Chiwawa River juvenile collections (15 and 16), genetic drift resulting from the small number of spawners used in 1989 (16 males and 37 females) and 1990 (7 males and 12 females) to generate these samples is assumed to be the major contributor to their outlying status (R. Bugert, Washington Department of Fish and Wildlife, personal communication). The small number of

TABLE 6.—Matrix of pairwise comparisons among collections (collections of juveniles [J]). Below diagonal are significance levels of G-tests over 36 loci. Significance level of total G-test value indicated by 0 for P > 0.05; * for 0.01 < P < 0.05; ! for 0.001 < P < 0.01; and # for P < 0.001. Above diagonal are values of genetic distance (D) × 1,000.

	Collection	1	2	3	4	5	6	7	8	9	10	11	12	13	14	15	16
1	Wenatchee River		0	0	0	1	1	0	1	1	38	35	45	40	39	38	47
2	Wells Hatchery	0		0	0	1	1	0	0	0	39	35	46	40	39	38	47
3	Wells Trap	0	0		0	0	1	1	0	0	38	34	45	39	38	37	46
4	Wenatchee J	0	0	!		0	1	1	0	0	36	33	43	37	37	36	44
5	Wells Carlton J	0	0	0	!		1	1	0	1	37	33	43	38	38	37	45
6	Wells Similkameen J	#	!	!	!	*		1	0	1	35	32	42	37	36	35	43
7	Similkameen	0	0	*	0	*	#		0	1	40	37	47	42	41	40	49
8	Hanford Reach	!	*	*	*	*	#	!		1	34	30	40	35	35	33	42
9	Priest Rapids Hatchery	*	0	0	*	0	!	0	0		37	34	44	39	38	37	46
10	Winthrop Hatchery	#	#	#	#	#	#	#	#	#		1	5	1	1	5	2
11	Leavenworth Hatchery	#	#	#	#	#	#	#	#	#	#		3	1	1	3	2
12	White River	#	#	#	#	#	#	#	#	#	#	#		3	2	5	3
13	Nason Creek	#	#	#	#	#	#	#	#	#	!	*	#		1	4	1
14	Chiwawa River	#	#	#	#	#	#	#	#	#	#	#	#	0		3	1
15	Chiwawa River 1992 J	#	#	#	#	#	#	#	#	#	#	#	#	#	#		5
16	Chiwawa River 1991 J	#	#	#	#	#	#	#	#	#	#	#	#	#	#	#	

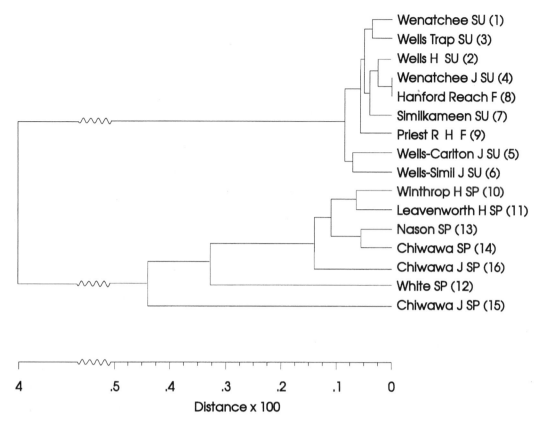

FIGURE 3.—An unweighted pair-group method using arithmetic averages dendrogram of genetic relationships among chinook salmon populations of the upper Columbia River based on pairwise genetic distance values (*D*; Nei 1972) for 36 polymorphic loci. Abbreviations given are fall (F), spring (SP), summer (SU), juveniles (J), hatchery (H), and river (R), and numbers in parentheses represent collection numbers (see Table 3 for sampling details).

males in particular would result in a reduced effective population size and a resultant opportunity for excessive genetic drift (e.g., Allendorf and Ryman 1987; Gall 1987).

At Wells Dam, sufficiently large numbers of adults were available for spawners to reduce the likelihood of genetic drift as a factor in the overall higher levels of significance of *G*-tests between Wells–Similkameen juveniles (6) and other summer–fall-run collections. These fish and those reared at Carlton ponds (collection 5) represented subsamples of the same year-class of Wells Hatchery juveniles. Thus, other factors, including differential subsampling of juveniles moved to the respective sites or differential mortalities at the sites, appear to be more likely explanations (Marshall and Young 1994). The presence of unique alleles at *sAH** in four individuals and at *sIDHP** and *sSOD** in single individuals supports the former possibility.

The most notable distinction involving adult col-

lections was the divergence within the spring-run group of the White River (12). Allele frequencies lie beyond the range of other adult spring-run samples at 11 of the 36 loci, although no unique alleles were found (Appendix). These differences indicate that this population should be considered distinct from the other sampled spring-run populations.

Some less obvious differences occurred among the adult samples. The clustering of the Winthrop (10) and Leavenworth (11) hatchery collections, and of the adjacent Nason Creek (13) and Chiwawa River (14) wild samples (Figure 3) supports the possibility of some persisting genetic isolation between these hatchery and wild spring-run populations.

Significant *G*-tests occurred between Hanford Reach (8) and all other summer–fall-run collections except Priest Rapids Hatchery (9). These differences were small and insufficient to affect measurements of genetic distance at the reported levels.

However, at loci such as *AAT-4**, *sIDPH-2**, *sMDH-A1,2**, *sMDH-B1,2**, and *TPI-2.2** (Appendix), slight outlying of allele frequencies for Hanford Reach from other collections within this group suggests the possibility of some persisting degree of isolation of these wild fall-run fish from upstream populations.

Synthesis from Genetic and Historical Information

The major points from the separate considerations of genetic and historical information presented separately in the preceding sections include

1. spring-run populations of the upper Columbia River are genetically distinct from summer–fall-run fish;
2. each group has genetic affinities with different major ancestral groups within the Columbia River;
3. the basis for all current distributions above Rock Island Dam lies in relocations over five consecutive years under the GCFMP;
4. releases of cultured fish under the GCFMP included crosses between summer- and fall-run fish and, possibly, between late-spawning spring-run chinook salmon and early-spawning summer-run chinook salmon;
5. extensive releases subsequent to the GCFMP have included origins from gene pools outside the upper Columbia River;
6. cultured summer–fall-run fish include parentage of broad temporal diversity within this group; and
7. there are no allele frequency differences from Priest Rapids Dam upstream that suggest genetic isolation of summer-run and fall-run populations.

Further conclusions can be derived from joint considerations of this information.

Interactions of distinct ancestral groups.—The historical information has identified numerous opportunities for breakdown of natural genetic structure through interbreedings among fish of distinct ancestral origins or establishment of exogenous gene pools. Effects of this nature were not apparent from the most diverged groups (Figure 2). There were no detectable residual effects from any possible interbeedings between late-maturing spring-run and early-maturing summer-run fish due to cultural activities under the GCFMP. No intermediate groups were evident to suggest persistence of a hybridized spring-run × summer–fall-run ancestry. Inspection of the allelic data of the somewhat distinct White

River spring-run population indicates a divergence from other upper Columbia River spring-run groups and not introgression from the summer–fall-run group.

Introductions from hatcheries on the Chehalis, Elokomin, and Cowlitz rivers and Eagle and Spring creeks (Figure 1) represent lineages distinct from either of the upper Columbia resident groups (Figure 2; Utter et al. 1989). These releases of purely exogenous fish (Table 2) appear to have left no detectable descendants.

Interactions with populations of mixed upriver ancestry.—The lower river populations derived from mixed upriver ancestry tell a different story. The presently self-sustaining spring-run populations of upper Columbia River hatcheries resulted from continued infusions of Carson and Carson-derived fish (Table 2) until returning adults were sufficient for an autonomous broodstock. The spring-run fish now returning to these hatcheries were genetically very similar to one another (Figure 3) and to Carson fish (Waples et al. 1991b). However, data of this study did not reflect a major contribution of fish of Carson ancestry, based on the distinction between these hatcheries and wild adult spring-run collections from Nason Creek, the Chiwawa River, and particularly the White River (Figure 3).

The less extensive introductions of fall-run hatchery fish of mixed ancestry (Table 2) presently preclude estimating possible genetic influences due to alleles common to upper Columbia River summer–fall-run and mixed ancestral groups. Sample sizes substantially larger than those reported here would be needed to detect shifts of allelic frequencies because of the close relationship to native fish of contributing exogenous gene pools and because of the high proportion of upper Columbia River ancestry (90%) estimated in the Bonneville fall-run population (Smouse et al. 1990).

However, the unique occurrence of the *sAH*112* allele in multiple individuals of one collection of juvenile offspring from Wells Hatchery parents (collection 6) could be a reflection of exogenous genes from introductions or strayings. This allele occurs at frequencies up to 0.03 in related fall-run populations of the mid-Columbia River (Smouse et al. 1990) and at slightly lower frequencies in Snake River and Yakima basin fall-run chinook salmon populations (A. Marshall, Washington Department of Fisheries and Wildlife, unpublished data). Its presence in these juvenile fish released in the Similkameen River raises the possibility of an exogenous component within this group and warrants close monitoring for persistence of this allele upon the return of these fish to their point of rearing and release.

Evolutionarily Significant Units

The focus of these proceedings on defining unique units in population conservation warrants a discussion of the status of the populations under consideration in this paper as evolutionarily significant units (ESUs). The ESA as amended in 1978 (16 U.S.C. § 1532[16]) mandates protection of "distinct population segments" of vertebrates as well as of recognized species and subspecies. The concept of the ESU provides a logical and biologically sound framework for defining such intraspecific segments (Waples 1991). To be considered as an ESU, populations (1) must be substantially reproductively isolated from other conspecific population units and (2) must represent an important component in the evolutionary legacy of the species. Four basic questions should be considered in defining an ESU.

1. Is the population genetically distinct from other conspecific populations?
2. Does the population occupy unique habitat?
3. Does the population show evidence of unique adaptation to its environment?
4. If the population became extinct, would this event represent a significant loss to the ecological or genetic diversity of the species?

Waples (1991) further suggested that ESUs should correspond to more comprehensive units unless there is clear evidence that evolutionarily important differences exist between smaller population segments. Being based on distinctions from other intraspecific groups, the ESU primary value is to provide a sound biological basis for proscribing admixtures beyond their boundaries, and definition of an ESU by no means implies a single panmictic unit. These questions and criteria guide considerations of chinook salmon populations of the upper Columbia River as possible ESUs.

Spring-run populations.—The clear genetic isolation of spring-run and summer–fall-run fish of the upper Columbia River (Figures 2, 3) qualifies them for separate consideration as ESUs. The mixed and partially exogenous ancestry within the hatchery component presumably precludes this segment of spring-run fish from ESU status (Hard et al. 1992). However, the differences between hatchery and adult wild collections of this study indicate that the latter group, and similarly distinct wild fish of this region, qualify for consideration as components of a common ESU. The geographic isolation and above-

noted genetic distinction from Snake River spring–summer-run fish (presently designated an ESU; Matthews and Waples 1991) would restrict the ESU to the upper Columbia River. The manipulations under the GCFMP limit any evolutionary divergence to the past 50 years and thus probably preclude subdivision in spite of the apparent distinction of the White River population.

The ESU status of spring-run populations of the Yakima River is unclear because of their exclusion from both the present study and investigations focused on spring–summer-run fish of the Snake River (e.g., Matthews and Waples 1991). Clusterings based on accumulated genetic information and providing the basis for Figure 2 (Waples et al. 1991b) suggest affinities of different populations within this drainage to either mid-Columbia River or upper Columbia River groups. Clarification of this issue awaits collection of more detailed information from within the Yakima River basin.

Summer–fall-run populations.—Different circumstances surround the ESU status of the summer–fall group of the upper Columbia River. Based on both genetic and historical data, the ancestry of existing populations appears to be predominantly—perhaps entirely—within the upper Columbia River drainage. The questions of interest converge on the possibility of genetic divergence within this drainage. The available information all points toward sufficient past and present admixture among temporal segments that would work against maintaining or establishing either temporal or geographic diversity.

The common elements within this group resulted in all summer–fall-run fish upstream from McNary Dam (50 km downstream from the confluence of the Columbia and Snake rivers, Figure 1), exclusive of the Snake River, recently being considered a single ESU (Waknitz et al. 1995). The inclusion of Yakima River populations within this ESU identifies a common need within this drainage for more detailed studies of fall-run fish as well as for spring-run fish noted above. Cumulatively, the summer–fall-run populations sampled in this study represent an important segment of the species' evolutionary legacy. They are reproductively isolated from, though related to, fall-run populations of the Snake River and the mid-Columbia River. In addition, upper Columbia River summer–fall-run and Snake River fall-run populations are distinguished by differing oceanic distributions, juvenile and adult sizes, and environmental features (summarized in Waples et al. 1991b).

Conclusions and Recommendations

The data and discussion to this point provide new information about distribution and relationships among chinook salmon populations of the upper Columbia River that is of potential value for the management of these fish. These perspectives pertain to different levels of genetic variation.

At greater levels of genetic distinction, the apparent failures of introductions from, or interbreedings with, more diverged subgroups contrast with successful introductions from more closely related (e.g., Carson) fish. These failures cause us to question the wisdom of further introductions or interbreedings at this level in the upper Columbia River; past efforts were apparently unsuccessful, and further introductions would continue to threaten the breakdown of existing adapted groups that may be displaced or interbred through initial numerical superiority of the exogenous fish, coupled with the likelihood of subsequent failure of hybrids or exogenous fish (Hindar et al. 1991; Waples 1995).

Introductions and interbreedings involving more closely related subgroups have been more successful and therefore pose a potentially greater threat to the stability of indigenous populations of the upper Columbia River. Two groups of closely related populations within which some genetic distinction exists are (1) the spring–summer-run fish of the Snake River, the mixed spring-run hatchery fish (e.g., Carson) and the spring-run fish of the upper Columbia River; and (2) the fall-run fish of the Snake River and mid-Columbia River, the hatchery populations (Bonneville) derived from mixed upriver fall-run fish, and the summer–fall-run fish of the upper Columbia River (Figure 2). Introductions and crosses among populations within either of these groups appear to be more amenable to producing reproductively viable progeny than introductions and crosses between these and other more distantly related groups. This increased potential viability makes it easier for transplants or strays to become established or to merge, potentially eroding adaptive distinctions between groups that may have arisen in both freshwater and marine habitats (e.g., Waples et al. 1991b). Thus, careful monitoring is necessary for detecting intrusions of closely related exogenous fish and for taking appropriate remedial actions.

The most problematic level of genetic variation is that for which differential adaptations have occurred among breeding groups which remain indistinguishable by biochemical or molecular genetic markers. Persistence or evolution of adaptive distinctions in the absence of conspicuous genetic differentiation is well documented (e.g., Gharrett and Smoker 1993). Such differences are possible because of the more rapid evolutionary time scale for genetic divergence of strongly adaptive characters (e.g., run timing) in contrast with more neutral characters such as the multiple polymorphic protein-coding loci used in this study (see Utter et al. 1993). Thus, sufficient flow or retarded drift of marker genes within these indistinguishable groupings may mask adaptive differences.

The persisting genetic affinity of the Leavenworth and Carson hatchery populations provides a possible example of this type of divergence. An initial dependence on Carson (and Carson-derived Little White Salmon Hatchery) eggs gradually subsided over three generations (Table 2). Numbers of returning releases from Leavenworth gradually increased from 1970 to the point where returning fish constituted the entire broodstock after 1983. Admittedly, these indirect data alone are not strong evidence for differential adaptation, and definitive data (e.g., based on reciprocal egg lots reared and released at both hatcheries) are needed. Better alternative explanations are presently lacking.

Indeed, most data supporting restricted movements of exogenous populations are indirect (e.g., Hindar et al. 1991; Campton 1995). With a few notable exceptions (e.g., Reisenbichler and McIntyre 1977; Chilcote et al. 1986; Campton et al. 1991), appropriate experiments have not been implemented to address theoretical arguments favoring restricted movements of exogenous populations. As such experimental data accumulate, local populations remain the best starting point for any enhancement activities (Hindar et al. 1991; Waples et al. 1991b).

Divergence of a single-source seeding of chinook salmon in New Zealand into diverse habitats and life history patterns (Quinn and Unwin 1993) attests to the evolutionary flexibility of chinook salmon. Similar population divergence has apparently occurred within the upper Columbia River since the disrupting and homogenizing effects of the GCFMP a half century ago. The White River spring-run fish appear to have diverged genetically and perhaps adaptively. Adaptive differences of hatchery stocks have been suggested in the absence of detectable genetic divergence from the source populations. These divergences have developed through continued breedings of adult fish returning to the locations and habitats of their parents.

However, continual infusions of individuals over wide geographic and temporal ranges, even within a genetically homogeneous group determined by essentially neutral markers, work against establishing both wild and hatchery adaptations—and thus promote inefficiency. For chinook salmon populations of the upper Columbia River then, the most effective strategy appears to be to reduce culturally induced straying and to permit existing populations to develop and adapt within local temporal, ecological, and geographic ranges.

The biological basis for this strategy has been apparent for a long time (e.g., Ricker 1972). The strategy makes sense from the perspectives of both conservation and production and should be implemented.

Acknowledgments

Sample collection, electrophoretic analyses, and portions of data analyses were carried out by Washington Department of Fisheries and Wildlife Laboratory through an agreement with the Chelan County Public Utility District (PUD). Efforts of F. M. Utter and D. W. Chapman included support from Chelan County PUD. Robert Bugert, Charles Peven, and two reviewers provided useful comments.

References

Aebersold, P. B., G. A. Winans, D. J. Teel, G. B. Milner, and F. M. Utter. 1987. Manual for starch gel electrophoresis: a method for the detection of genetic variation. NOAA (National Oceanic and Atmospheric Administration) Technical Report NMFS (National Marine Fisheries Service) 61, Seattle.

Allendorf, F. W., and N. Ryman. 1987. Genetic management of hatchery stocks. Pages 141–159 in N. Ryman and F. Utter, editors. Population genetics and fishery management. University of Washington Press, Seattle.

Bartley, D., B. Bentley, J. Brodziak, R. Gomulkiewicz, M. Mangel, and G. A. E. Gall. 1992. Geographic variation in population genetic structure of chinook salmon from California and Oregon. U.S. National Marine Fisheries Service Fishery Bulletin 90:77–100 (authorship amended per errata, Fishery Bulletin 90[3]:iii).

Campton, D. E. 1995. Genetic effects of hatchery fish on wild populations of Pacific salmon and steelhead: what do we really know? American Fisheries Society Symposium 15:337–353.

Campton, D. E., and six coauthors. 1991. Reproductive success of hatchery and wild steelhead. Transactions of the American Fisheries Society 120:816–827.

Chilcote, M. W., S. A. Leider, and J. J. Loch. 1986. Differential reproductive success of hatchery and wild summer-run steelhead under natural conditions. Transactions of the American Fisheries Society 115:726–735.

Fish, F. F., and M. G. Hanavan. 1948. A report on the Grand Coulee Fish Maintenance Project 1939–1947. U.S. Fish and Wildlife Service Special Scientific Report 55.

Gall, G. A. E. 1987. Inbreeding. Pages 47–87 in N. Ryman and F. Utter, editors. Population genetics and fishery management. University of Washington Press, Seattle.

Gharrett, A. J., and W. W. Smoker. 1993. A perspective on the adaptive importance of genetic infrastructure in salmon populations to ocean ranching in Alaska. Fisheries Research 18:45–58.

Hard, J. J, R. P. Jones, Jr., M. R. Delarm, and R. S. Waples. 1992. Pacific salmon and artificial propagation under the Endangered Species Act. NOAA (National Oceanic and Atmospheric Administration) Technical Memorandum NMFS (National Marine Fisheries Service) NWFSC-2, Northwest Fisheries Science Center, Seattle.

Hindar, K., N. Ryman, and F. Utter. 1991. Genetic effects of aquaculture on natural fish populations. Canadian Journal of Fisheries and Aquatic Sciences 48:945–957.

Howell, P., K. Jones, D. Scarnecchia, L. Lavoy, W. Kendra, and D. Ortmann. 1985. Stock assessment of Columbia River anadromous salmonids. I. Chinook, coho, chum and sockeye salmon stock summaries. Report (Contract No. DE-AI79-84BP12737) to Bonneville Power Administration, Portland, Oregon.

IUBNC (International Union of Biochemists, Nomenclature Committee). 1984. Enzyme nomenclature. Academic Press, San Diego, California.

Marshall, A. R., and S. Young. 1994. Genetic analysis of upper Columbia spring and summer chinook salmon for the Rock Island Hatchery Evaluation Program. Report to Chelan County Public Utility District, Wenatchee, Washington.

Matthews, G. M., and R. S. Waples. 1991. Status review for Snake River spring and summer chinook salmon. NOAA (National Oceanic and Atmospheric Administration) Technical Memorandum NMFS (National Marine Fisheries Service) F/NWC-200, Northwest Fisheries Science Center, Seattle.

Mullan, J. W. 1987. Status and propagation of chinook salmon in the mid-Columbia River through 1985. U.S. Fish and Wildlife Service Biological Report 87(3):111.

Nehlsen, W., J. E. Williams, and J. A. Lichatowich. 1991. Pacific salmon at the crossroads: stocks at risk from California, Oregon, Idaho, and Washington. Fisheries 16:4–21.

Nei, M. 1972. Genetic distance between populations. American Naturalist 106:283–292.

Peven, C. M. 1992. Population status of selected stocks of salmonids from the mid-Columbia River basin. Report, Chelan County Public Utility District, Wenatchee, Washington.

Quinn, T. P., and M. J. Unwin. 1993. Variation in life history patterns among New Zealand chinook salmon Oncorhynchus tshawytscha populations. Canadian Journal of Fisheries and Aquatic Sciences 50:1414–1421.

Reisenbichler, R. R., and J. D. McIntyre. 1977. Genetic differences in growth and survival of juvenile hatchery and wild steelhead trout. *Salmo gairdneri*. Journal of the Fisheries Research Board of Canada 34:123–128.

Ricker, W. E. 1972. Hereditary and environmental factors affecting certain salmonid populations. Pages 19–160 *in* R. Simon and P. Larkin, editors. The stock concept in Pacific salmon. H. R. MacMillan Lectures in Fisheries, University of British Columbia, Vancouver.

Rohlf, D. J. 1993. Petition for a rule to list mid-Columbia summer chinook salmon under the Endangered Species Act and to designate critical habitat. Northwest Environmental Defense Center, Portland, Oregon.

Shaklee, J. B., F. W. Allendorf, D. C. Morizot, and G. S. Whitt. 1990. Gene nomenclature for protein-coding loci in fish. Transactions of the American Fisheries Society 119:2–15.

Smouse, P. E., R. S. Waples, and J. A. Tworek. 1990. A genetic mixture analysis for use with incomplete source population data. Canadian Journal of Fisheries and Aquatic Sciences 47:620–634.

Sneath, P. H. A., and R. R. Sokal. 1973. Numerical taxonomy. W. H. Freeman, San Francisco, California.

Thompson, W. F. 1951. An outline for salmon research in Alaska. University of Washington School of Fisheries, Fisheries Research Institute Circular 18, Seattle.

Utter, F. M., G. B. Milner, G. Stahl, and D. J. Teel. 1989. Genetic population structure of chinook salmon in the Pacific northwest. U.S. National Marine Fisheries Service Fishery Bulletin 87:239–264.

Utter, F. M., J. E. Seeb, and L. W. Seeb. 1993. Complementary uses of ecological and biochemical genetic data in identifying and conserving salmon populations. Fisheries Research 18:59–76.

Utter, F. M., R. S. Waples, and D. J. Teel. 1992. Genetic isolation of previously indistinguishable chinook salmon populations of the Snake and Klamath rivers: limitations of negative data. U.S. National Marine Fisheries Service Fishery Bulletin 90:770–777.

Waknitz, F. W., G. M. Matthews, T. Wainright, and G. A. Winans. 1995. Status review for mid-Columbia River summer chinook salmon. NOAA (National Oceanic and Atmospheric Administration) Technical Memorandum NMFS (National Marine Fisheries Service) NWFSC-22, Northwest Fisheries Science Center, Seattle.

Waples, R. S. 1991. Pacific salmon, *Oncorhynchus* spp., and the definition of "species" under the Endangered Species Act. U.S. National Marine Fisheries Service Marine Fisheries Review 53:11–22.

Waples, R. S. 1995. Genetic effects of stock transfers of fish. Pages 51–69 *in* D. P. Philipp, J. M. Epifanio, J. E. Marsden, and J. E. Claussen, editors. Protection of aquatic biodiversity. Proceedings World Fisheries Congress, Theme 3. Oxford and IBH Publishing Co., New Delhi.

Waples, R. S., D. J. Teel, and P. B. Aebersold. 1991a. A genetic monitoring and evaluation program for supplemented populations of salmon and steelhead in the Snake River basin. Annual Report of Research to Bonneville Power Administration. Northwest Fisheries Science Center, Seattle.

Waples, R. S., R. P. Jones, Jr., B. R. Beckman, and G. A. Swan. 1991b. Status review for Snake River fall chinook salmon. NOAA (National Oceanic and Atmospheric Administration) Technical Memorandum NMFS (National Marine Fisheries Service) F/NWC-201, Northwest Fisheries Science Center, Seattle.

Waples, R. S., and six coauthors. 1993. A genetic monitoring and evaluation program for supplemented populations of salmon and steelhead in the Snake River basin. Annual report (Contract No. DE-A179–89BP00911) 1992 to Bonneville Power Administration, Portland, Oregon.

Winans, G. A. 1989. Genetic variability in chinook salmon stocks from the Columbia River basin. North American Journal of Fisheries Management 9:47–52.

Zar, J. H. 1974. Biostatistical analysis. Prentice-Hall, Inc. Englewood Cliffs, New Jersey.

Appendix: Allelic Frequencies

Allelic frequencies for 35 loci in 16 collections of chinook salmon from the upper Columbia River. Average heterozygosities (HET) are given for each collection. See Table 3 and Figure 1 for names and locations of numbered collections and Table 4 for relative allele mobilities.

Locus, allele, and sample size	Collection															
	1	2	3	4	5	6	7	8	9	10	11	12	13	14	15	16
sAAT-1,2*																
N	(409)	(404)	(180)	(86)	(180)	(75)	(81)	(99)	(200)	(100)	(100)	(113)	(71)	(133)	(86)	(100)
*100	0.999	0.998	1.000	0.994	0.997	1.000	1.000	0.995	1.000	1.000	1.000	1.000	0.993	0.994	0.988	0.975
*85	0.001	0.001	0.000	0.006	0.000	0.000	0.000	0.005	0.000	0.000	0.000	0.000	0.007	0.006	0.012	0.025
*105	0.001	0.001	0.000	0.000	0.003	0.000	0.000	0.000	0.000	0.000	0.000	0.000	0.000	0.000	0.000	0.000
sAAT-3*																
N	(405)	(190)	(170)	(86)	(90)	(75)	(76)	(89)	(199)	(98)	(97)	(112)	(68)	(130)	(86)	(100)
*100	0.999	1.000	1.000	1.000	1.000	1.000	1.000	1.000	0.997	1.000	1.000	1.000	1.000	1.000	1.000	1.000
*90	0.001	0.000	0.000	0.000	0.000	0.000	0.000	0.000	0.003	0.000	0.000	0.000	0.000	0.000	0.000	0.000
sAAT-4*																
N	(377)	(180)	(139)	(84)	(89)	(74)	(80)	(97)	(164)	(77)	(79)	(105)	(61)	(107)	(76)	(83)
*100	0.999	1.000	0.996	1.000	1.000	1.000	1.000	0.979	0.988	0.955	0.975	0.986	0.959	0.944	1.000	0.994
*130	0.000	0.000	0.000	0.000	0.000	0.000	0.000	0.021	0.000	0.000	0.000	0.000	0.000	0.000	0.000	0.000
*63	0.001	0.000	0.004	0.000	0.000	0.000	0.000	0.000	0.012	0.045	0.025	0.014	0.041	0.056	0.000	0.006
mAAT-1*																
N	(409)	(202)	(180)	(85)	(90)	(75)	(81)	(99)	(200)	(100)	(100)	(113)	(71)	(132)	(86)	(100)
*-100	0.980	0.990	0.992	0.994	0.989	0.987	0.969	0.995	0.988	0.990	0.985	0.996	1.000	0.985	1.000	0.980
*-77	0.001	0.002	0.000	0.000	0.000	0.013	0.000	0.000	0.000	0.000	0.000	0.000	0.000	0.000	0.000	0.000
*-104	0.018	0.007	0.008	0.006	0.011	0.000	0.031	0.005	0.013	0.010	0.015	0.004	0.000	0.015	0.000	0.020
ADA-1*																
N	(409)	(201)	(180)	(86)	(90)	(75)	(81)	(99)	(200)	(100)	(100)	(113)	(70)	(130)	(86)	(98)
*100	0.991	0.990	1.000	0.983	1.000	1.000	0.994	0.990	0.995	0.975	0.950	0.929	0.929	0.946	0.949	0.939
*83	0.009	0.010	0.000	0.017	0.000	0.000	0.006	0.010	0.005	0.025	0.050	0.071	0.071	0.054	0.051	0.061
ADA-2*																
N	(409)	(202)	(180)	(86)	(90)	(75)	(81)	(99)	(200)	(100)	(100)	(113)	(71)	(133)	(86)	(100)
*100	1.000	1.000	1.000	1.000	1.000	1.000	1.000	1.000	1.000	1.000	1.000	1.000	1.000	1.000	1.000	0.995
*105	0.000	0.000	0.000	0.000	0.000	0.000	0.000	0.000	0.000	0.000	0.000	0.000	0.000	0.000	0.000	0.005
sAH*																
N	(407)	(202)	(168)	(86)	(90)	(75)	(81)	(99)	(200)	(100)	(100)	(112)	(68)	(129)	(86)	(100)
*100	0.769	0.780	0.786	0.744	0.889	0.820	0.741	0.808	0.805	0.000	1.000	0.991	1.000	0.988	1.000	1.000
*86	.231	0.220	0.214	0.256	0.111	0.153	0.259	0.192	0.195	0.000	0.000	0.004	0.000	0.012	0.000	0.000
*112	0.000	0.000	0.000	0.000	0.000	0.027	0.000	0.000	0.000	0.000	0.000	0.000	0.000	0.000	0.000	0.000
*108	0.000	0.000	0.000	0.000	0.000	0.000	0.000	0.000	0.000	0.000	0.000	0.004	0.000	0.000	0.000	0.000
mAH-4*																
N	(409)	(201)	(169)	(86)	(90)	(75)	(81)	(99)	(200)	(100)	(100)	(113)	(69)	(129)	(84)	(100)
*100	0.911	0.873	0.908	0.895	0.894	0.820	0.846	0.919	0.895	0.985	0.975	1.000	1.000	1.000	0.905	1.000
*119	0.089	0.127	0.092	0.105	0.106	0.180	0.154	0.081	0.105	0.015	0.025	0.000	0.000	0.000	0.095	0.000
PEPA*																
N	(409)	(202)	(180)	(86)	(90)	(75)	(81)	(98)	(200)	(100)	(100)	(113)	(70)	(133)	(86)	(100)
*100	0.972	0.983	0.981	0.983	0.989	0.960	0.975	0.990	0.967	1.000	0.990	0.947	0.993	0.985	1.000	0.955
*90	0.027	0.017	0.019	0.017	0.011	0.040	0.019	0.010	0.030	0.000	0.005	0.035	0.007	0.004	0.000	0.000
*81	0.001	0.000	0.000	0.000	0.000	0.000	0.006	0.000	0.003	0.000	0.005	0.018	0.000	0.011	0.000	0.045
GPI-B2*																
N	(408)	(202)	(180)	(83)	(87)	(74)	(81)	(99)	(200)	(100)	(100)	(112)	(69)	(132)	(85)	(95)
*100	0.945	0.948	0.928	0.970	0.948	0.959	0.932	0.929	0.952	1.000	0.995	0.969	0.993	0.966	0.882	0.926
*60	0.055	0.052	0.072	0.030	0.052	0.041	0.068	0.071	0.047	0.000	0.005	0.031	0.007	0.034	0.118	0.074
GPI-A*																
N	(409)	(202)	(180)	(86)	(90)	(75)	(81)	(98)	(200)	(100)	(100)	(113)	(66)	(132)	(86)	(100)
*100	0.998	0.990	0.997	1.000	1.000	0.987	0.988	0.995	0.993	1.000	1.000	1.000	1.000	1.000	1.000	1.000
*105	0.002	0.010	0.003	0.000	0.000	0.013	0.012	0.005	0.007	0.000	0.000	0.000	0.000	0.000	0.000	0.000
GR																
N	(409)	(202)	(180)	(86)	(90)	(75)	(81)	(99)	(200)	(100)	(100)	(113)	(71)	(133)	(86)	(100)
*100	0.976	0.970	0.953	0.953	0.983	0.973	0.975	0.975	0.977	0.995	1.000	0.996	0.993	0.989	1.000	0.975
*85	0.024	0.030	0.025	0.007	0.017	0.027	0.025	0.025	0.023	0.005	0.000	0.004	0.007	0.011	0.000	0.025
HAGH*																
N	(409)	(202)	(180)	(86)	(90)	(75)	(81)	(99)	(200)	(100)	(100)	(113)	(71)	(133)	(86)	(100)
*100	0.998	0.998	0.994	1.000	0.989	1.000	1.000	1.000	0.993	0.900	0.910	0.858	0.887	0.914	1.000	0.870
*143	0.002	0.002	0.006	0.000	0.011	0.000	0.000	0.000	0.007	0.100	0.090	0.142	0.113	0.075	0.000	0.130
*131	0.000	0.000	0.000	0.000	0.000	0.000	0.000	0.000	0.000	0.000	0.000	0.000	0.000	0.011	0.000	0.000

Appendix.—Continued.

Locus, allele, and sample size	Collection															
	1	2	3	4	5	6	7	8	9	10	11	12	13	14	15	16
*mIDHP-2**																
N	(409)	(202)	(179)	(86)	(90)	(75)	(81)	(99)	(200)	(100)	(100)	(113)	(70)	(132)	(86)	(100)
*100	0.998	1.000	1.000	1.000	1.000	1.000	1.000	1.000	0.998	1.000	1.000	1.000	1.000	1.000	1.000	1.000
*154	0.000	0.000	0.000	0.000	0.000	0.000	0.000	0.000	0.002	0.000	0.000	0.000	0.000	0.000	0.000	0.000
*50	0.002	0.000	0.000	0.000	0.000	0.000	0.000	0.000	0.000	0.000	0.000	0.000	0.000	0.000	0.000	0.000
*sIDHP-1**																
N	(409)	(202)	(180)	(85)	(86)	(75)	(81)	(99)	(200)	(100)	(100)	(113)	(71)	(133)	(86)	(100)
*100	0.994	0.998	0.994	1.000	0.983	0.980	0.994	1.000	0.993	0.810	0.785	0.580	0.761	0.665	0.722	0.680
*74	0.006	0.002	0.006	0.000	0.011	0.007	0.000	0.000	0.005	0.170	0.210	0.345	0.225	0.327	0.278	0.285
*142	0.000	0.000	0.000	0.000	0.006	0.007	0.000	0.000	0.000	0.000	0.000	0.000	0.000	0.000	0.000	0.000
*94	0.000	0.000	0.000	0.000	0.000	0.000	0.006	0.000	0.002	0.020	0.005	0.075	0.014	0.008	0.000	0.035
*126	0.000	0.000	0.000	0.000	0.000	0.007	0.000	0.000	0.000	0.000	0.000	0.000	0.000	0.000	0.000	0.000
*sIDHP-2**																
N	(409)	(202)	(180)	(86)	(90)	(75)	(81)	(99)	(200)	(100)	(100)	(113)	(71)	(133)	(86)	(100)
*100	0.782	0.829	0.817	0.807	0.822	0.833	0.821	0.869	0.805	1.000	0.995	0.991	0.993	0.992	1.000	1.000
*127	0.218	0.171	0.183	0.193	0.172	0.167	0.179	0.126	0.192	0.000	0.005	0.009	0.007	0.008	0.000	0.000
*83	0.000	0.000	0.000	0.000	0.006	0.000	0.000	0.000	0.002	0.000	0.000	0.000	0.000	0.000	0.000	0.000
*LDH-B2**																
N	(409)	(202)	(180)	(86)	(90)	(75)	(81)	(99)	(200)	(100)	(100)	(113)	(71)	(133)	(86)	(100)
*100	1.000	1.000	1.000	1.000	1.000	1.000	1.000	1.000	0.998	0.980	1.000	0.996	1.000	1.000	1.000	1.000
*112	0.000	0.000	0.000	0.000	0.000	0.000	0.000	0.000	0.000	0.020	0.000	0.004	0.000	0.000	0.000	0.000
*71	0.000	0.000	0.000	0.000	0.000	0.000	0.000	0.000	0.002	0.000	0.000	0.000	0.000	0.000	0.000	0.000
*LDH-C**																
N	(405)	(196)	(176)	(86)	(90)	(75)	(80)	(96)	(200)	(98)	(100)	(112)	(68)	(130)	(86)	(100)
*100	0.974	0.982	0.972	0.994	0.961	0.987	0.975	0.984	0.983	1.000	1.000	0.996	1.000	0.996	1.000	1.000
*90	0.026	0.018	0.028	0.006	0.039	0.013	0.025	0.016	0.018	0.000	0.000	0.000	0.000	0.000	0.000	0.000
*84	0.000	0.000	0.000	0.000	0.000	0.000	0.000	0.000	0.000	0.000	0.000	0.004	0.000	0.000	0.000	0.000
*sMDH-A1,2**																
N	(409)	(202)	(180)	(86)	(180)	(75)	(81)	(99)	(200)	(100)	(100)	(113)	(71)	(133)	(86)	(100)
*100	1.000	1.000	1.000	1.000	1.000	1.000	1.000	0.995	1.000	1.000	1.000	1.000	1.000	0.997	1.000	0.995
*27	0.000	0.000	0.000	0.000	0.000	0.000	0.000	0.005	0.000	0.000	0.000	0.000	0.000	0.000	0.000	0.000
*160	0.000	0.000	0.000	0.000	0.000	0.000	0.000	0.000	0.000	0.000	0.000	0.000	0.000	0.003	0.000	0.005
*sMDH-B1,2**																
N	(409)	(202)	(180)	(86)	(180)	(75)	(81)	(99)	(200)	(100)	(100)	(113)	(71)	(133)	(86)	(100)
*100	0.951	0.960	0.961	0.959	0.956	0.960	0.934	0.980	0.966	0.990	0.990	0.938	0.979	0.971	0.971	0.945
*121	0.020	0.019	0.014	0.015	0.017	0.007	0.019	0.015	0.018	0.000	0.000	0.044	0.000	0.003	0.000	0.000
*70	0.026	0.019	0.019	0.026	0.028	0.033	0.037	0.005	0.015	0.000	0.000	0.000	0.000	0.000	0.000	0.000
*83	0.000	0.000	0.000	0.000	0.000	0.000	0.000	0.000	0.000	0.000	0.000	0.000	0.000	0.003	0.000	0.000
*126	0.002	0.002	0.006	0.000	0.000	0.000	0.000	0.000	0.001	0.010	0.100	0.013	0.014	0.024	0.029	0.050
*mMDH-2**																
N	(408)	(202)	(180)	(86)	(90)	(75)	(81)	(98)	(200)	(100)	(100)	(113)	(71)	(132)	(85)	(99)
*100	0.993	0.990	0.981	1.000	0.983	0.993	0.994	0.990	0.973	0.695	0.750	0.885	0.803	0.803	0.900	0.717
*200	0.007	0.010	0.019	0.000	0.017	0.007	0.006	0.010	0.027	0.305	0.250	0.115	0.197	0.197	0.100	0.283
*sMEP-1**																
N	(409)	(202)	(180)	(86)	(90)	(75)	(81)	(99)	(200)	(100)	(100)	(113)	(71)	(132)	(86)	(98)
*100	0.811	0.809	0.828	0.756	0.839	0.807	0.796	0.758	0.777	0.095	0.095	0.027	0.035	0.072	0.034	0.041
*92	0.189	0.191	0.172	0.244	0.161	0.193	0.204	0.242	0.222	0.905	0.905	0.973	0.965	0.928	0.966	0.959
*MPI**																
N	(409)	(201)	(180)	(85)	(90)	(75)	(81)	(99)	(200)	(100)	(100)	(113)	(71)	(132)	(86)	(99)
*100	0.655	0.672	0.683	0.688	0.678	0.747	0.660	0.652	0.685	0.940	0.860	0.792	0.915	0.913	0.807	0.975
*109	0.344	0.323	0.317	0.312	0.317	0.247	0.340	0.348	0.308	0.060	0.140	0.208	0.085	0.087	0.193	0.025
*95	0.001	0.005	0.000	0.000	0.006	0.007	0.000	0.000	0.007	0.000	0.000	0.000	0.000	0.000	0.000	0.000
*PEPD**																
N	(409)	(201)	(180)	(84)	(89)	(75)	(81)	(97)	(200)	(100)	(100)	(113)	(71)	(132)	(86)	(100)
*100	0.996	1.000	0.994	1.000	0.989	1.000	0.975	0.985	0.988	0.990	1.000	0.969	0.993	0.996	1.000	1.000
*107	0.004	0.000	0.006	0.000	0.011	0.000	0.025	0.015	0.013	0.010	0.000	0.031	0.007	0.004	0.000	0.000
*PEP-LT**																
N	(406)	(201)	(178)	(86)	(90)	(75)	(80)	(98)	(200)	(99)	(100)	(113)	(69)	(133)	(86)	(91)
*100	0.771	0.786	0.829	0.795	0.828	0.867	0.731	0.801	0.783	0.914	0.950	0.960	0.899	0.936	0.994	0.962
*110	0.229	0.214	0.171	0.205	0.172	0.133	0.269	0.199	0.218	0.086	0.050	0.040	0.101	0.064	0.006	0.038
*PGDH**																
N	(409)	(202)	(180)	(86)	(90)	(75)	(81)	(99)	(200)	(100)	(100)	(113)	(71)	(130)	(86)	(100)
*100	1.000	1.000	1.000	1.000	1.000	1.000	1.000	1.000	0.998	1.000	1.000	1.000	1.000	1.000	1.000	1.000
*90	0.000	0.000	0.000	0.000	0.000	0.000	0.000	0.000	0.002	0.000	0.000	0.000	0.000	0.000	0.000	0.000

Appendix.—Continued.

Locus, allele, and sample size	Collection															
	1	2	3	4	5	6	7	8	9	10	11	12	13	14	15	16
PGK-2																
N	(409)	(202)	(180)	(86)	(90)	(75)	(81)	(99)	(199)	(99)	(100)	(113)	(71)	(133)	(86)	(100)
*100	0.560	0.587	0.581	0.602	0.589	0.600	0.605	0.591	0.608	0.106	0.180	0.119	0.127	0.117	0.080	0.080
*90	0.439	0.411	0.414	0.398	0.411	0.400	0.395	0.404	0.392	0.894	0.820	0.881	0.873	0.883	0.920	0.920
*74	0.001	0.002	0.006	0.000	0.000	0.000	0.000	0.005	0.000	0.000	0.000	0.000	0.000	0.000	0.000	0.000
PGM-2																
N	(409)	(202)	(180)	(86)	(90)	(75)	(81)	(97)	(200)	(100)	(100)	(113)	(71)	(133)	(86)	(100)
*100	1.000	0.998	1.000	1.000	1.000	1.000	0.994	1.000	1.000	1.000	1.000	1.000	1.000	1.000	1.000	1.000
*136	0.000	0.002	0.000	0.000	0.000	0.000	0.006	0.000	0.000	0.000	0.000	0.000	0.000	0.000	0.000	0.000
sSOD-1																
N	(409)	(202)	(180)	(86)	(90)	(75)	(81)	(99)	(200)	(100)	(100)	(113)	(71)	(133)	(86)	(100)
*100	0.482	0.485	0.511	0.506	0.550	0.540	0.494	0.535	0.507	0.755	0.755	0.872	0.782	0.737	0.574	0.820
*-260	0.517	0.512	0.489	0.494	0.450	0.453	0.506	0.465	0.493	0.245	0.245	0.128	0.218	0.263	0.426	0.180
*580	0.001	0.002	0.000	0.000	0.000	0.000	0.000	0.000	0.000	0.000	0.000	0.000	0.000	0.000	0.000	0.000
*-175	0.000	0.000	0.000	0.000	0.000	0.007	0.000	0.000	0.000	0.000	0.000	0.000	0.000	0.000	0.000	0.000
mSOD																
N	(409)	(202)	(179)	(86)	(90)	(75)	(81)	(98)	(200)	(100)	(100)	(112)	(71)	(133)	(86)	(100)
*100	1.000	1.000	1.000	1.000	1.000	1.000	1.000	1.000	1.000	0.985	0.990	1.000	0.993	0.996	1.000	1.000
*142	0.000	0.000	0.000	0.000	0.000	0.000	0.000	0.000	0.000	0.015	0.010	0.000	0.007	0.004	0.000	0.000
PEPB																
N	(409)	(202)	(180)	(86)	(90)	(75)	(81)	(99)	(199)	(100)	(100)	(113)	(71)	(133)	(86)	(100)
*100	0.729	0.696	0.717	0.739	0.706	0.727	0.691	0.747	0.741	0.840	0.840	0.779	0.803	0.812	0.847	0.845
*130	0.260	0.300	0.269	0.261	0.289	0.247	0.309	0.237	0.241	0.110	0.085	0.080	0.085	0.090	0.051	0.045
*-350	0.011	0.005	0.014	0.000	0.006	0.027	0.000	0.015	0.018	0.050	0.075	0.142	0.113	0.098	0.097	0.110
TPI-2.2																
N	(409)	(202)	(180)	(86)	(90)	(75)	(81)	(99)	(200)	(100)	(100)	(113)	(71)	(133)	(86)	(100)
*100	0.998	1.000	0.997	1.000	1.000	1.000	0.994	0.985	0.993	0.915	0.905	0.841	0.965	0.955	0.960	0.975
*104	0.002	0.000	0.003	0.000	0.000	0.000	0.006	0.015	0.007	0.085	0.095	0.159	0.035	0.045	0.040	0.025
HET	0.116	0.114	0.111	0.112	0.1907	0.109	0.124	0.108	0.112	0.081	0.085	0.093	0.079	0.086	0.076	0.081

PART THREE

BEHAVIOR AND LIFE HISTORY

American Fisheries Society Symposium 17:169–175, 1995

SESSION OVERVIEW
The Relevance of Behavior and Natural History to Evolutionarily Significant Units

GEORGE W. BARLOW

Department of Integrative Biology and Museum of Vertebrate Zoology
University of California, Berkeley, California 94720, USA

The final sentence of Darwin's *On the Origin of Species* closes with "From so simple a beginning endless forms most beautiful and most wonderful have been, *and are being, evolved*" (italics added). Nowhere are we better reminded of the ongoing nature of evolution than in the study of the behavior and natural history of animals. Such research illuminates the dynamic, hierarchical nature of breeding units of animals, starting with the reproducing individuals, whether a pair or a small group, and ranging upward through ever larger assemblages of individuals sharing genetic material. The question recurs: where does one draw the line? What is the evolutionarily significant unit (ESU)?

Part of the difficulty in coming to grips with ESUs is inherent in how we think. We need categories to communicate our perception of the world around us, even if the categories that we invent reflect stages in a continuum, such as baby, child, adolescent and so on. Our limitation is significant in the process of producing regulations because any resulting legislation has enormous consequences for conservation. Legislators, however, seldom understand the nuances of organic evolution, and, indeed, some biologists don't either. Legislators are compelled to rely on knowledgeable biologists to advise them, and biologists find giving such advice difficult without resorting to discrete categories such as the ESU.

Looking back, we see that the Endangered Species Act (ESA; 16 U.S.C. §§ 1531 to 1544) quickly confronted regulators with the problem of what constitutes the appropriate cluster of individuals for conservation. The species, as the proper unit, has enormous utility. But even the ESA recognizes that the concept has to embrace some clusters at a finer level of resolution than that of species, else economically important populations, such as some runs of Pacific salmon *Oncorhynchus* spp., would be lost. Thus, the notion of a hierarchical arrangement of significant units was imbedded in the legislation from the start. Delimiting ESUs is, consequently, a portentous undertaking, and some have prescribed concise definitions (e.g., Moritz 1994). Although

such definitions are biologically reasonable, other considerations may arise that require a wider or a narrower framework (e.g., the management unit of Moritz 1994). In that context, no consistent boundary exists between local demes at the one extreme and the species at the other.

Expanding on this theme, Healey and Prince (1995, this volume) apply the hierarchical scheme proposed by Hughes and Noss (1992). Beginning at the level of large-scale geographic variation (the landscape), they work down to ever finer levels of comparison and illustrate how genotypic and phenotypic differences create a different picture depending on the particular level of analysis. The appropriate ESU for a Pacific salmon, for instance, may depend on whether the analysis is done at the broad landscape level or at that of the local habitat. Healey and Prince conclude, however, that the most informative level for studies of genetic differentiation lies at that of the population.

At higher levels, the agreement between behavioral and morphological markers is often good and has been useful in recognizing ESUs. But as the degree of resolution has become finer and finer, the morphological approach has encountered difficulties in the case of cryptic species (see also Stauffer et al. 1995, this volume, for supporting and countering examples). Molecular genetics has then often been embraced as the method of choice to tease out genetically distinct clusters of individuals.

Unfortunately, molecular genetics does not always resolve the problem. As brought out in the session on genetics in this symposium, application of different methods in molecular genetics may lead to different conclusions. Although the meshing of morphology and molecular genetics has been productive in that the two approaches usually complement one another, still more is needed to understand the dynamics of population structure and especially to give meaning to the reality of ESUs.

The behavioral aspects of natural history can help in recognizing ESUs by alerting fisheries biologists to potential differences between populations. Historically, the timing of spawning runs may have been the single most important behavioral difference pointing to separate populations of Pacific salmon (see Wood 1995, this volume). Differences in color pattern have also been useful, though some might prefer to think of coloration as morphology. The behavior that is relevant here, however, is the response of conspecifics and heterospecifics to those differences in coloration (see Stauffer et al. 1995).

Behavioral differences do not always indicate genetic differentiation (nor do differences in coloration; Stauffer et al. 1995). For example, Baylis (1995, this volume) documents how large breeding males of smallmouth bass *Micropterus dolomieu* alternate generation with small breeding males by means of differences in time of hatching and rate of growth during the first year of life.

Nonetheless, if two morphs differ in the timing or nature of salient activities, such as feeding, reproducing, or use of habitat, chances are good that the morphs are discreet populations. Arctic char *Salvelinus alpinus* in one Icelandic lake, for example, have differentiated into four coexisting morphs (Skúlason et al. 1989; Noakes et al. 1995, this volume). They can be separated by morphology, habitat, and feeding behavior as well as life history styles and reproductive behavior. Molecular genetic techniques (allozyme, mitochondrial DNA, and nuclear DNA polymorphism), however, revealed only small differences. Laboratory rearing experiments confirmed that the differences are indeed genetic (Skúlason et al. 1989).

A cautionary note: Arctic char vary enormously within and between lakes. Whether such differences constitute genetically based ESUs depends on where in the process of evolution the populations lie. For example, Hindar and Jonsson (1993) have found a polymorphic population of Arctic char in a Norwegian lake with parallels to the Icelandic Arctic char discussed by Noakes et al. (1995). The Norwegian Arctic char are characterized as dwarf and normal morphs that differ in color as well as size. But when fish were raised under constant conditions in the lab, the differences disappeared. Some difference in growth rates were detected, suggesting slight genetic separation.

Behavioral studies of all stages of a life history, not just the adult stage, lead to an understanding of how the putative ESUs are adapted to their environment. Relevant here is the classic study by Hoar (1976) on the trenchant differences in behavioral mechanisms that move juvenile Pacific salmon from freshwater habitats to the ocean. Information of

this type is enormously useful in planning how best to protect ESUs.

The aspect of behavior that has received the most attention for sorting out sympatric ESUs is that of reproductive isolation. This reproductive isolation can be striking, as when subtly different forms breed at the same time yet retain their integrity. The forms are typically recognized by small differences in morphology or color. Largely unexplored, and perhaps therefore undetected, are isolating mechanisms based on odor or acoustic signals.

Behavior can also be valuable in generating hypotheses about ESUs, as Stauffer et al. (1995) point out. Thus, if lekking, mouthbrooding cichlid fishes build different bowers in which to spawn, the differences in bowers suggest the cichlids are different species, even if they look alike. Closer morphological and genetic analyses confirm the differences. The discovery of behavioral differences invites closer examination of morphology and genetics.

Understanding the behavior and demographics of a population can suggest how the rate of evolution, and hence the production of ESUs, can be affected. Both the size of the population and the strength of selection are important factors driving change (Wade and Goodnight 1991), and the behavior of the fish in question bears on these two variables. For instance, if the mating system is highly polygynous, the effective population size is reduced to something close to the number of males reproducing. I'll return to this point. At the risk of belaboring the obvious, predation is a powerful behavioral selective agent. Different populations of threespine stickleback *Gasterosteus aculeatus* respond differently to predators, depending on whether the populations have historically experienced piscivorous pike *Esox lucius* (Huntingford and Wright 1992).

Further, movements of fish determine the rate of gene flow between populations, which can dampen the development of ESUs if the invading fish can breed successfully and in sufficient numbers with the local population (Wade and Goodnight 1991). Likewise, highly patrilocal behavior promotes the isolation of gene pools and hence the development of ESUs. Thus Wood (1995) reports that different populations of anadromous sockeye salmon *O. nerka*, whose spawning sites are as much as 180 km apart, are relatively genetically homogeneous compared with genetically distinct populations in the same region that live in isolated lakes.

Baylis (1995) has studied the natural history of a population of smallmouth bass sequestered in a small seepage lake in Wisconsin. His research program contrasts with the other reports of freshwater fishes in this volume because the others center mainly on migratory salmonids or on cichlids living in lakes so large as to be compared in some respects with small oceans. Yet most lakes are small, and some are partially or totally isolated.

Baylis did not analyze the smallmouth bass genetically, but his findings bear importantly on the issue of what constitutes the breeding population, or deme, of a freshwater species, which has genetic and evolutionary ramifications (Wade and Goodnight 1991). The first remarkable finding, compared with other sunfishes (e.g., Jennings and Philipp 1992), is that each male smallmouth bass apparently reproduces but once in its lifetime, and most males mate with only one female. If the males regularly spawned with multiple females, the effective size of the population would be reduced. The second finding is that only about 33% of the male smallmouth bass in Baylis' study lake reproduce in a given year, though about 45% of the females do. The effective population size is thus much smaller than the naive biologist would usually assume.

The absolute number of breeding female smallmouth bass may fall so low as to increase the possibility of genetic drift (Wright 1932). As Baylis (1995) points out, based on Wright's theory, a population that experiences little or no genetic exchange with other populations could produce a unique ensemble of genes—through drift when the size of the breeding population falls and through local adaptation as the population grows (see also Wade and Goodnight 1991).

The nature of the mating system, together with the proportion of fish of each sex that reproduces, bears on the variation in reproductive success, hence fitness, of its individuals. That variation in reproductive success can have significant consequences for what Arnold and Wade (1984) call the opportunity for selection (see Baylis 1995, for further details). The opportunity for selection can be helpful in evaluating candidate populations for ESUs. Thus Baylis (1995) writes that the mating system can be even more important than drift in its consequences for the rate of differentiation among small populations.

Behavior can therefore suggest that different species may be more or less prone to speciate, or at least to produce genetically distinctive populations. That salmon and trout adapt to the local conditions in the tributaries where they spawn is now well known (Healey and Prince 1995; Wood 1995). Less appreciated is the tendency of Arctic char to do the same. Noakes et al. (1995) report a mini-radiation of Arctic char in a single lake in Iceland. These morphs were detected on the basis of behavior as well as morphology (Skúlason et al. 1989). They dwell in different microhabitats, have different diets, and mate assortatively. Noakes et al. (1995) suggest that the situation may be general for Arctic char in other lakes.

Recent work (Bell and Foster 1994; McPhail 1994) indicates a similar phenomenon in the threespine stickleback; it appears to be radiating in parallel in a number of lakes in British Columbia and Alaska. In discussing that radiation, Bell and Foster (1994) presented a model called a "raceme," taken from botany. Here the central stem represents the basal species, passing through time. The derived species, or ESUs, are the flowers radiating off that main stem. As time progresses, the derived species go extinct, and new but strikingly similar ones spring forth further up the central stem. The kokanee (lacustrine sockeye salmon) in the Pacific Northwest show a similar pattern of local adaptation and parallel evolution during postglacial times (Wood 1995).

Morphological differences do not always indicate profound genetic differentiation. Pink salmon *O. gorbuscha*, as Noakes et al. (1995) mention, have been thought to occur in two genetically distinct male morphs: one form is large and has a conspicuous hump on its forehead; the other is small and has no hump. However, although male pink salmon vary greatly, they do so continuously. Thus they are not polymorphic in the strict sense of the term (Ford 1945). Noakes et al. (1995) and Wood (1995) make the further point, one well worth keeping in mind, that variation per se ought to be conserved.

One of the difficulties with ESUs derives from the dynamics of adaptation. Healey and Prince (1995) report profound differences between populations of chinook salmon *O. tshawytscha*. They describe heritable components to spawning date, egg size and rate of development. Healey and Prince go on to mention that the chinook salmon recently planted in the Great Lakes have shown a remarkable capacity to adapt quickly to their chosen spawning sites. That capacity creates the perplexing problem of deciding whether populations having these recently acquired adaptations constitute ESUs and therefore deserve the full protection of the law. That capacity to adapt quickly also leaves the reader with the impression that chinook salmon are exceedingly plastic and hence adaptable to new habitats.

As is so often the case, however, species differ. Wood (1995) reports that the anadromous sockeye salmon do not transplant well; they are tightly adapted to their natal streams and have little capacity to cope with new circumstances. In general, Pacific salmon do not transplant well, despite the few remarkable successes and their apparent phenotypic plasticity (Healey and Prince 1995). But again we have the problem of risking an overgeneralization. The small, lake-dwelling kokanee is generally regarded as the same species as the larger anadromous sockeye salmon. Wood (1995) informs us that kokanee, in contrast to sockeye salmon, are relatively easy to transplant and establish in other lakes.

Pacific salmon provide us with another precautionary note about behavior. The usual assumption is that reproductively isolated populations maintain the separation by means of differences in reproductive behavior. However, Healey and Prince (1995) remind us that male mating tactics are much the same in different populations of coho salmon *O. kisutch*. Though variation on the basic pattern exists, the different species of salmon do have much the same basic mating system, which may be telling us that the tactics themselves are not important in keeping the species or forms separate (but see comments on cichlids, below). Rather, the separation may be maintained by other factors or combinations of them, such as timing, small differences in habitat preferences, or chromatic, olfactory, or acoustic signals that key female choice.

One of the risks of a symposium such as this one is that our perception of the problem will be dominated by the information coming from salmonids and other freshwater fishes of North America. However, fishes from other parts of the world can provide valuable lessons. Our attempts to regulate the trade in endangered species beyond our borders is but one illustration of the concern of our society for environmental issues that transcend our provincial interests.

One area of concern is East Africa, in whose rift lakes cichlid fishes have undergone the most spectacular evolutionary radiations known for any kind of vertebrate animal. McKaye and Stauffer and their collaborators (see Stauffer et al. 1995) estimate that Lake Malaŵi alone may harbor around 1,000 endemic species of cichlids.

The aquarium trade has placed heavy demands on those colorful cichlids, most of them being exported to Europe and North America. More importantly, many of those fishes are vital to the people who live in that part of Africa. Stauffer et al. (1995) mention that the people near Lake Malaŵi derive about 70% of their dietary proteins from fish. A similar situation prevailed in the recent past at Lake Victoria. But, due to the ill-advised introduction

about 30 years ago of the Nile perch *Lates niloticus*, a large and highly effective predator, the cichlid fishery there has been nearly destroyed, and many species have been driven to near extinction (Kaufman 1992).

However, I wished to have the cichlids represented here for a reason that is more relevant to this symposium. I wanted to contrast the freshwater fishes in the cool temperate parts of North America, where by comparison relatively few species are found, with those of the extraordinarily species-rich lakes in the tropics. The differences are apparent and could carry a message to ichthyologists from temperate-zone countries who are not familiar with freshwater tropical communities. The tropical systems present the same sorts of problems when we contemplate ESUs, though the solutions are bound to be to some degree different.

Systematists struggle to make meaning out of the diversity of fishes they encounter in Lakes Malaŵi and Tanganyika in Africa. A given species is often polymorphic, and regional variation is rampant. In some examples, the differentiation is enough to persuade a systematist that the types merit species status, in other cases not. Because the species have evolved so recently, genetic differences are difficult to detect (Stauffer et al. 1995).

Studying the behavior of those fishes in the field has been instructive in sorting out the problems in systematics. In some instances behavior indicates that sympatric morphs are good species (Holzberg 1978; Schröder 1980; Stauffer et al. 1995).

Stauffer et al. (1995) provide an example of clear interspecific differences in the courtship "dance" of some similar cichlids in Lake Malaŵi. The males pile up sand to build small volcanos, which the authors call bowers, after the bower-building birds. The males of different species all swim in circular patterns in the mouth of the bower when displaying to approaching females. But the males of some species move in a figure-eight pattern whereas those of another species follow an S-shaped pattern. The bowers themselves differ in shape and size. Minimally, behavioral studies produce useful hypotheses for further analysis, especially when the species do not co-occur.

I had hoped to have at least one speaker address the relevance of ESUs to marine shore fishes, but that did not work out. The behavior of marine fishes in the pelagic realm and those that live over open bottom has not been much studied and deserves more attention. In situ research on the behavior of reef fishes, on the other hand, has increased rapidly over the last few decades, facilitated by the use of scuba diving equipment.

Coastal regions have the vast majority of species of marine fishes. Most of the research on them has been concentrated in the tropics on coral reefs, at least in part for reasons that are easy to grasp: coral reefs have a spectacular array of fish species living in warm, clear, shallow water. Some research on ecology and behavior is being done, however, in the less inviting temperate coastal waters of North America and Europe (e.g., Potts 1984; Goulet and Green 1988; Hobson and Chess 1988). The lack of discussion of shore fishes in the context of ESUs is understandable because the oceanic fauna is still casually regarded in nonscientific circles as limitless, even though most of us know that view is false. Moreover, from what is known generally about shore fishes, conservation of them would not seem to be an issue in North America.

In contrast to freshwater fishes, but in common with marine invertebrates (Strathmann 1974), the eggs and larvae of marine coastal fishes are scattered into the ocean as propagules (Waples and Rosenblatt 1987; Leis 1991). In most cases, the offspring have little chance of returning to precisely the same reef from which they originated (Victor 1991), though some interesting exceptions are emerging (see below). In general, a large portion of the propagules never get back to the home reef, though some adaptations might favor a portion of them doing so (Barlow 1981; Lobel and Robinson 1988; Leis 1993). What a strong contrast with Pacific salmon returning to a specific tributary of their natal stream.

Consequently, when the fauna of a reef of limited size is destroyed, the expectation is that it will recover quickly. The plants and animals are recruited from the immense pool of propagules that have originated on other reefs (Pulliam 1988). But the situation can be confounded by other considerations, such as population density of the source population in relation to strength of reproduction, spacing of refugia, patterns of coldwater upwelling and ability to disperse (Waples and Rosenblatt 1987; Polocheck 1990; Quinn et al. 1993).

Evidence is also accumulating that local differentiation may exist to some degree in coastal reef fishes, which would indicate some degree of genetic isolation. Molecular genetic analysis of 10 fish species off the coast of southern California and Baja California revealed a complex pattern of similarity and differentiation (Waples and Rosenblatt 1987; see also Stepien and Rosenblatt 1991). Those findings were interpreted in terms of differences in life

history, ability to disperse, remoteness of the populations, and the possibility of local adaptation.

By implication, gene flow between populations, or lack of it, can be influenced by behavior. In British Columbia, J. B. Marliave (Vancouver Aquarium, unpublished) has observed in larval fishes close to shore behavior that indicates resistance to dispersal. Some cottoid larvae manage to stay right next to the reef, descending and ascending depending on tides and currents, in order to maintain station. Similar behavior was found in other species, raising the possibility of local genetic differentiation.

One of the most apparent families of coastal reef fishes in the northern Pacific is the Embiotocidae, the surfperches. S. J. Holbrook and R. J. Schmitt (University of California, Santa Barbara, unpublished) are finding evidence that recruitment in some surfperches depends on local production of offspring. That suggests, again, restricted dispersal and hence the possibility of local differentiation (see Haldorson 1980). The surfperches are viviparous and give birth to large young that are miniature adults. As a consequence, the young are not well adapted to life as planktonic propagules but instead could be expected to remain near the protective reef.

The loss of just one keystone predator can have profound effects for vast reef-system communities. The marine otter, for instance, feeds on sea urchins. In the absence of otters, sea urchins prosper and multiply. Sea urchins become a keystone predator by destroying the kelp bed. With the loss of kelp much of the community of invertebrates and fishes disappears (Estes et al. 1989). Similarly, on coral reefs environmental disturbances, such as excess runoff from rivers full of silt, can lead to outbreaks of the crown-of-thorns starfish (Birkeland 1989; Birkeland and Lucas 1990). The starfish literally eat their way through the coral reef, causing massive destruction of many of the reef corals. Coralivorous fishes drop in numbers drastically, as do many organisms that depend on the live coral. On the other hand, dead coral provides a substrate for marine algae and sometimes promotes the abundance of herbivores such as surgeonfishes *Acanthurus* sp. (Birkeland 1989; personal observation at Moorea, French Polynesia).

Humans enter the above equations. They can either foster marine otters or remove them; likewise, the new fishery for sea urchins along the Pacific coast has called for intelligent management. Human agricultural practices and cutting passes in reefs can promote populations of crown-of-thorn starfish by increasing survival of their larvae (Birke-land 1989; Birkeland, University of Guam, personal communication). In each case, entire communities are threatened, recalling the argument that habitats may be considered ESUs.

Persistent threats to shore fishes also come from the pollution emitted by industrial plants and sewer outfalls as well as the activities of commercial harbors. Some countries in the tropics (e.g., Sri Lanka) mine the coral reefs for limestone, a building material. G. P. Jones (James Cook University, North Queensland, unpublished) commented that we are experiencing worldwide declines in coral reefs and beds of sea grass as well as kelp beds, salt marshes, and mangroves, and these declines "are now viewed with growing alarm."

I hope more attention will be directed to shore fishes in the future. We cannot afford to wait so long that conservation efforts become desperate, as we are now witnessing in the salmonids of southwestern British Columbia and the Columbia River basin (see Wood 1995).

References

Arnold, S. J., and M. J. Wade. 1984. On the measurement of natural and sexual selection: theory. Evolution 38:709–719.

Barlow, G. W. 1981. Patterns of parental investment, dispersal and size among coral-reef fishes. Environmental Biology of Fishes 6:65–85.

Baylis, J. R. 1995. The population-level consequences of individual reproductive competition: observations from a closed population. American Fisheries Society Symposium 17:217–226.

Bell, M. A., and S. A. Foster, editors. 1994. Evolutionary biology of the threespine stickleback. Oxford University Press, Oxford, UK.

Birkeland, C. 1989. The Faustian traits of the crown-of-thorns starfish. American Scientist 77:154–163.

Birkeland, C., and J. S. Lucas. 1990. *Acanthaster planci*: major management problems of coral reefs. CRC Press, Boca Raton, Florida.

Estes, J. A., D. O. Duggins, and G. B. Rathbun. 1989. The ecology of extinction in kelp forest communities. Conservation Biology 3:252–264.

Ford, E. B. 1945. Polymorphism. Biological Reviews of the Cambridge Philosophical Society 20:73–88.

Goulet, D., and J. M. Green. 1988. Reproductive success of male lumpfish (*Cylopterus lumpus* L. Pisces: Cyclopteridae): evidence against female mate choice. Canadian Journal of Zoology 66:2513–2519.

Haldorson, L. 1980. Genetic isolation of Channel Islands fish populations: evidence from two embiotocid species. Pages 433–442 *in* D. M. Powers, editor. The Channel Islands: proceedings of a multidisciplinary symposium. Santa Barbara Museum of Natural History, Santa Barbara, California.

Healey, M. C., and A. Prince. 1995. Scales of variation in life history tactics in Pacific salmon and the conservation of phenotype and genotype. American Fisheries Society Symposium 17:176–184.

Hindar, K., and B. Jonsson. 1993. Ecological polymorphism in Arctic char. Biological Journal of the Linnean Society 48:63–74.

Hoar, W. S. 1976. Smolt transformation: evolution, behavior and physiology. Journal of the Fisheries Research Board of Canada 33:1233–1252.

Hobson, E. S., and J. R. Chess. 1988. Trophic relations of the blue rockfish, *Sebastes mystinus*, in a coastal upwelling system off northern California. U.S. National Marine Fisheries Service Fishery Bulletin 86:715–743.

Holzberg, S. 1978. A field and laboratory study of the behaviour and ecology of *Pseudotropheus zebra* (Boulenger), an endemic cichlid of Lake Malawi (Pisces: Cichlidae). Zeitschrift für Zoologische Systematic und Evolutions Forschung 16:171–187.

Hughes, T. P., and R. F. Noss. 1992. Biological diversity and biological integrity: current concerns for lakes and streams. Fisheries 17:11–19.

Huntingford, F. A., and P. J. Wright. 1992. Inherited population differences in avoidance conditioning in three-spined sticklebacks, *Gasterosteus aculeatus*. Behaviour 122:264–273.

Jennings, M. J., and D. P. Philipp. 1992. Female choice and male competition in longear sunfish. Behavioral Ecology 3:84–94.

Kaufman, L. 1992. Catastrophic change in species-rich freshwater ecosystems. The lessons of Lake Victoria. Bioscience 42:846–858.

Leis, J. M. 1991. The pelagic stage of reef fishes: the larval biology of coral reef fishes. Pages 183–230 *in* P. F. Sale, editor. The ecology of fishes on coral reefs. Academic Press, New York.

Leis, J. M. 1993. Larval fish assemblages near Indo-Pacific coral reefs. Bulletin of Marine Science 53:362–392.

Lobel, P. S., and A. R. Robinson. 1988. Larval fishes and zooplankton in a cyclonic eddy in Hawaiian waters. Journal of Plankton Research 10:1209–1223.

McPhail, J. D. 1994. Speciation and the evolution of reproductive isolation in the sticklebacks (*Gasterosteus*) of south-western British Columbia. Pages 404–437 *in* M. A. Bell and S. A. Foster, editors. The evolutionary biology of the threespine stickleback. Oxford University Press, New York.

Moritz, C. 1994. Defining 'evolutionarily significant units' for conservation. Trends in Ecology & Evolution 9:373–375.

Noakes, D. L. G., M. M. Ferguson, B. Ashford, and W. Stott. 1995. Size and shape variation in Laurentian Great Lakes pink salmon. American Fisheries Society Symposium 17:185–194.

Polocheck, T. 1990. Year around closed areas as a management tool. Natural Resources Modeling 4:327–354, Tempe, Arizona.

Potts, G. W. 1984. Parental behaviour in temperate marine teleosts with special reference to the development of nest structures. Pages 223–244 *in* G. W. Potts and R. J. Wootton, editors. Fish reproduction. Strategies and tactics. Academic Press, New York.

Pulliam, H. R. 1988. Sources, sinks, and population regulation. American Naturalist 132:652–661.

Quinn, J. F., S. R. Wing, and L. W. Botsford. 1993. Harvest refugia in marine invertebrate fisheries: model and applications to the red sea urchin *Strongylocentrotus franciscanus*. American Zoologist 33:537–550.

Schröder, J. H. 1980. Morphological and behavioural differences between the BB/OB and B/W colour morphs of *Pseudotropheus zebra* Boulenger (Pisces; Cichlidae). Zeitschrift für Zoologische Systematic und Evolutions Forschung 18:69–76.

Skúlason, S., D. L. G. Noakes, and S. S. Snorrason. 1989. Ontogeny of trophic morphology in four sympatric morphs of Arctic char *Salvelinus alpinus* in Thingvallavatn, Iceland. Biological Journal of the Linnean Society 38:281–301.

Stauffer, J. R., Jr., N. J. Bowers, K. R. McKaye, and T. D. Kocher. 1995. Evolutionarily significant units among cichlid fishes: the role of behavioral studies. American Fisheries Society Symposium 17:227–244.

Strathmann, R. 1974. The spread of sibling larvae of sedentary marine invertebrates. American Naturalist 108:29–44.

Stepien, C. A., and R. H. Rosenblatt. 1991. Patterns of gene flow and genetic divergence in the northeastern Pacific Clinidae (Teleostei: Blennioidei). Copeia 1991:873–896.

Victor, B. C. 1991. Settlement strategies and biogeography of reef fishes. Pages 231–260 *in* P. F. Sale, editor. The ecology of fishes on coral reefs. Academic Press, New York.

Wade, M. J., and C. J. Goodnight. 1991. Wright's shifting balance theory: an experimental study. Science 253: 1015–1018.

Waples, R. S., and R. H. Rosenblatt. 1987. Patterns of larval drift in southern California marine shore fishes inferred from allozyme data. U.S. National Marine Fisheries Service Fishery Bulletin 85:1–11.

Wood, C. C. 1995. Life history variation and population structure in sockeye salmon. American Fisheries Society Symposium 17:195–216.

Wright, S. 1932. The roles of mutation, inbreeding, crossbreeding, and selection in evolution. Proceedings of the 6th International Congress of Genetics 1:356–366.

American Fisheries Society Symposium 17:176–184, 1995

Scales of Variation in Life History Tactics of Pacific Salmon and the Conservation of Phenotype and Genotype

M. C. HEALEY AND ANGELA PRINCE

Westwater Research Center and Fisheries Center, University of British Columbia
Vancouver, British Columbia V6T 1Z2, Canada

Abstract.—In this paper we explore patterns of genotypic and phenotypic variation in Pacific salmon *Oncorhynchus* spp. on population, ecosystem, and landscape scales. Many examples demonstrate that genotypic and phenotypic variation occur at all these scales. However, we argue that documented genotypic variation maps mainly on the population scale, whereas phenotypic variation also maps strongly on the ecosystem and landscape scales. This means that loss of local populations may have a much greater effect on phenotypic diversity than on overall genotypic diversity. We also note that Pacific salmon have very labile phenotypes, and probably genotypes, so that when the species are transplanted or raised artificially, new phenotypes (and perhaps genotypes) emerge quickly. These features of Pacific salmon make conservation of the population in its habitat a necessity if particular phenotypes are to be conserved. These features also make possible the recapturing of phenotypic variety, if not specific phenotypes, as long as appropriate habitat opportunities can be provided to Pacific salmon. The frequent failure of attempts to transplant anadromous runs, however, should caution managers against too great a reliance on reestablishing extinct runs.

The genotypic discreetness of individual populations or stocks of Pacific salmon *Oncorhynchus* spp. and the importance of such discreetness has been a subject of debate for more than two decades (Ricker 1972; Larkin 1981; Scudder 1989; Taylor 1991). This debate has taken on a new urgency with the documentation of the rates of loss of Pacific salmon populations in the Pacific Northwest (Nehlsen et al. 1991). The principal concern in earlier writings was about the loss of unique genotypes and genetic diversity. More recently, however, the focus of attention has shifted to the maintenance of biodiversity. Hughes and Noss (1992) defined biodiversity as biological diversity on five scales: the genotype, the species, the species assemblage, the ecosystem, and the landscape scales. Inclusion of the larger scales is important for a number of reasons. It elevates the idea of diversity from an esoteric concern with genetic code, something invisible and ineffable, to a concern for attributes (species and landscapes) that people can see and that directly touch their lives. However, inclusion of larger scales also complicates the problem of conservation because nothing seems to be excluded and acceptable levels of alteration are undefined. This dilemma is well illustrated in Pacific salmon. Every spawning population is potentially a unique genotype and is undeniably a unique phenotype. There are few, if any, locations in the Pacific Northwest where the species are not affected by human activities. How are we to accomplish effective conservation of biodiversity without eliminating much of our economic infrastructure? What can we do to ensure that the rich phenotypic and

genotypic variety of these species is protected while still carrying on reasonable levels of human economic activity?

Questions such as these challenge us to define the critical units for conservation. For Pacific salmon, a key element of that definition is the way in which phenotypic and genotypic diversity map on the scales of biodiversity as defined by Hughes and Noss (1992). This paper represents our attempt at such a mapping. Our message is that although both phenotypic and genotypic diversity exist at all scales, the majority of the genotypic diversity that has been measured is contained within stocks, whereas phenotypic diversity is high at ecosystem and landscape scales. Too strong a focus on the preservation of genetic code for presumed unique populations diverts attention from the fact that the uniqueness of those populations is a consequence of their unique habitats. The appropriate conservation unit, therefore, is the population with its habitat. Maintaining a rich diversity of Pacific salmon genotypes and phenotypes depends on maintaining habitat diversity and on maintaining the opportunity for the species to take advantage of that habitat diversity.

Phenotypic and Genotypic Diversity on Different Scales

The Landscape Scale

On a large geographic or landscape scale, one of the best documented examples of phenotypic diversity is that of the stream- and ocean-type forms of chinook salmon *O. tshawytscha*. Originally described by Gilbert in 1913 on the basis of scale patterns, these two forms are now known to have dramatically different life histories and distinct geographic distributions (Healey 1983, 1991). Stream-type chinook salmon are characterized by having a long freshwater phase as a juvenile. These chinook salmon spend a year, sometimes two, in freshwater before migrating to sea. They undertake extensive oceanic migrations. During their marine phase, stream-type chinook salmon are found primarily in the open waters of the North Pacific Ocean. When mature, they return to their natal streams in spring and early summer. In southern rivers, stream-type chinook salmon may enter their spawning river up to 4 months prior to spawning. Ocean-type chinook salmon, in contrast, are characterized by having a short freshwater phase as a juvenile. They typically migrate to sea within 3 months of emerging from the spawning gravels and undertake less extensive oceanic migrations. Ocean-type chinook salmon are found primarily in the waters over the continental shelf of North America during their marine phase. They have late summer and fall spawning migrations, and they usually enter their spawning rivers only a few weeks before spawning. Stream-type chinook salmon make up virtually 100% of chinook salmon populations throughout Alaska, northern British Columbia, and Asia. The Nass, Skeena, and Yakoun rivers on the northern British Columbia coast have chinook salmon populations that are approximately evenly split between stream and ocean types. South of the Skeena River to the southernmost rivers inhabited by chinook salmon, ocean-type chinook salmon predominate. South of the Skeena River, stream-type chinook salmon occur primarily in the headwaters of the larger rivers, such as the Fraser, Columbia, and Sacramento, but are generally less than 25% of the spawning escapement in any river (Healey 1991).

Early experiments by Rich and Holmes (1928) suggested that the stream- and ocean-type phenotypes were inherited, and studies of enzyme polymorphisms have confirmed genetic differences between stream- and ocean-type populations (Kristiansson and McIntyre 1976). More recently, Clarke et al. (1994) found that differences in photoperiod response between stream- and ocean-type juveniles could be attributed to a single gene with two alleles.

Utter et al. (1989), on the other hand, found little genotypic differentiation related to adult run timing among a widely distributed collection of chinook salmon populations. They did not specifically compare stream- and ocean-type life histories, but, be-

cause the life histories are linked to adult run timing, Utter et al.'s failure to find a relationship suggests a small or inconsistent genetic difference between stream and ocean types. Utter et al. (1989) did find significant differences in allele frequencies among regional groupings of chinook salmon and among populations, levels of differentiation also reported by other authors (Gharrett et al. 1987; Bartley and Gall 1990).

A second example of phenotypic dichotomy on a large geographic scale occurs between even- and odd-year lines of pink salmon *O. gorbuscha*. Because virtually all pink salmon mature at 2 years of age, there is almost complete genetic separation between even- and odd-year lines, and the lines have diverged genetically (Aspinwall 1974). From central Alaska south to Washington, pink salmon maturing in odd years are larger in size than are those maturing in even years, but this dichotomy is not apparent in Asia (Heard 1991). The lines also differ morphologically, odd-year lines having larger heads and thicker caudal peduncles (Beacham 1985). The genetic divergence is, therefore, accompanied by morphological divergence.

The Ecosystem Scale

On a smaller geographic scale (the ecosystem scale), considerable phenotypic variation has been observed among local populations of Pacific salmon. Indeed, it was this kind of variation that first attracted the attention of biologists interested in local adaptation (Ricker 1972). One dramatic example is the phenotypic variation shown by three populations of chum salmon *O. keta* spawning in two small streams flowing into a small inlet on Vancouver Island (Tallman and Healey 1991). The two streams flow into the head of Ladysmith Harbor less than 2 km apart. One stream (Bush Creek) has two runs of chum salmon, an early run that spawns in late October and early November in the lower reaches of the creek and a late run that spawns in late November and December farther upstream. The other stream (Walker Creek) has one run of chum salmon that spawns in December and early January (Figure 1). The populations are, thus, characterized by differences in location and time of spawning. During 1981 and 1982 the populations also differed in average age at maturity, the late-spawning chum salmon (Bush and Walker creeks) being younger in 1982 than were the early-spawning chum salmon; in average body size, the late-spawning Bush Creek chum salmon being larger than were the early-spawning Bush Creek

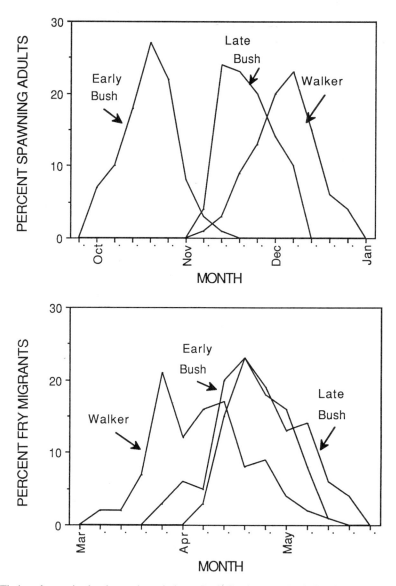

FIGURE 1.—Timing of spawning by chum salmon in lower Bush Creek, upper Bush Creek, and Walker Creek (upper panel) and timing of downstream migration of fry from the same creeks (lower panel). Adapted from Tallman and Healey 1991.

chum salmon; and in egg size, the late-spawning Bush Creek chum salmon having smaller eggs and a lower slope for the regression of egg weight on female size compared with the other two populations.

Despite the distinctly different spawning times of the adults and the differences in egg size, fry from each run emerged relatively synchronously in the spring (Figure 1). In fact, Walker Creek fry, which were spawned latest, emerged somewhat earlier than did Bush Creek fry. Fry of the late-spawning populations emerged after fewer accumulated degree-days than did the early spawning population. When eggs from each stock were reared under controlled temperature conditions in the laboratory, development times were found to be stock specific, the later-spawning fish having a more rapid embryo development rate. The synchrony in fry emergence was, therefore, due to inherent differences in embryo development rates among populations (Tallman 1986).

Differences in development rate were maintained, presumably by selection, despite apparent rates of straying as high as 46% among some populations (Tallman and Healey 1994). The late-spawning population in Bush Creek was particularly vulnerable to straying, receiving temporal strays from the early Bush Creek run and geographic strays from adjacent Walker Creek. Nevertheless, this population was able to maintain its integrity and resist introgression. It is also noteworthy that the apparent rate of gene exchange among populations, as revealed by electrophoretic analysis, was less than 5%, much less than that predicted from the straying of individuals, as revealed by tagging studies. This suggests that most fish that strayed to a nonnatal spawning population were unsuccessful at reproducing.

The Population Scale

At the finest geographic scale is phenotypic variation within populations. Age and size variation among mature salmon is the most obvious kind of within-population variation, and this variation is well documented (Healey 1986, 1987). Recently, however, variation in reproductive tactics has attracted attention (Schroder 1981; Gross 1985; Holtby and Healey 1986, 1990). In particular, for male coho salmon *O. kisutch*, Gross (1985) argued that precocial "jack" males coexisted with large "hooknose" males in a mixed evolutionarily stable strategy. Recent work that we have conducted on the breeding behavior of male coho salmon suggests a more complicated mixture of tactics and a more complicated system to maintain those tactics than that proposed by Gross (1985). We found that large, dominant males defended a segment of stream channel against other large males even when no females were present. These large males had a restricted range of movement in the spawning tributaries, although they did move among receptive females within their territory (Figure 2). Dominant males may not have competed for females but, instead, had access to females that constructed nests within their defended area. Individual jack males also tended to remain within a restricted area of the spawning tributary (Figure 2). At times the jack males used cover near the nest to avoid aggression from the dominant male, but often we found jack males stationed within the spawning nest together with the dominant male and female. These jack males did not attract aggression from the dominant male and would even court the female. Smaller 3-year-old males seldom challenged the dominant

male but appeared to adopt a satellite position downstream from the spawning nest by choice. These males were highly mobile, ranging widely among spawning areas and visiting many widely dispersed spawning groups during a single day (Figure 2). These observations will be reported in detail elsewhere. However, they suggest that there are at least three distinct male reproductive tactics (dominant, jack, and satellite) and that all three may exist in a mixed evolutionarily stable strategy (Maynard Smith 1982). Holtby and Healey (1986) suggested a similar explanation for variation in female size within coho salmon populations. Furthermore, Holtby and Healey (1990) postulated that size variation between males and females was a consequence of different demands placed on the sexes by the breeding environment coupled with sex-specific differences in the trade-off between foraging and predator avoidance. In combination, these results suggest a rich mosaic of adult phenotypes within individual coho salmon populations maintained, at least partially, by alternating disruptive and stabilizing selection.

The Emergence of New Phenotypes and Genotypes

Much of the adaptive variation that we presently observe among Pacific salmon populations is of relatively recent origin. Pacific salmon populations in Washington, British Columbia, and Alaska must have invaded since the most recent glaciation, that is within the last 10,000 years, presumably from refugia in Oregon, California, and Beringia (Gharrett et al. 1987). The degree of phenotypic and genotypic divergence among these populations represents a truly remarkable blossoming of variation in the short time since the retreat of the glaciers. However, there is evidence that, under the right circumstances, phenotypic and genotypic divergence can occur on time scales of decades rather than centuries.

Pacific salmon and other salmonids have been subject to hatchery culture for many decades. The effect of hatchery culture on the morphology of salmon is evident (Taylor 1986; Fleming and Gross 1989), and the likelihood that genetic selection is occurring in hatcheries has long been a concern. Indeed, changes in genetic composition of salmonids raised in hatcheries are documented (Reisenbichler and McIntyre 1977; Ryman and Stahl 1981; Swain et al. 1991). Hindar et al. (1991) have cautioned strongly against unrestricted release of cultured fish into natural populations because of the

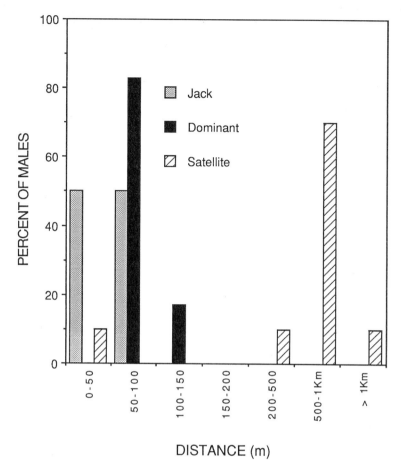

FIGURE 2.—Frequency with which male coho salmon displaying different breeding tactics moved various distances in the spawning grounds each day. Dominant males are large 3-year-old males that consort directly with females. Satellites are smaller 3-year-old males that wait downstream from the breeding pair and enter the nest during spawning. Jacks are small 2-year-old males that hide near or within the nest to sneak spawning opportunities.

potential for adverse genetic alteration to the wild population.

New phenotypes have also emerged in very short periods of time among Pacific salmon transplanted to new environments. In the Great Lakes, an unplanned introduction of approximately 20,000 pink salmon resulted in the establishment of naturally spawning populations. From the initial introduction of the odd-year line in the 1950s, an even-year line developed in the 1970s. The even-year line resulted from the production of 3-year-old spawners by the odd-year line (Kwain and Chappel 1978; Wagner and Stauffer 1980). Because 3-year-old spawners are extremely rare among Pacific coast pink salmon (Heard 1991), their occurrence in the Great Lakes was a dramatic departure from the normal phenotype for the species.

Chinook salmon have also been introduced to the Great Lakes and had developed naturally spawning populations by the mid 1970s (Carl 1982). Although the introduced chinook salmon were all from fall-spawning populations, in at least one instance chinook salmon were found spawning in the spring (Kwain and Thomas 1984). Again, this suggests the rapid emergence of divergent phenotypes from fish that, in their natural habitat, display a more restricted phenotype.

The most recent documentation of phenotypic divergence from a limited source population is the divergence evident in New Zealand chinook salmon (Quinn and Unwin 1993). Chinook salmon were introduced to New Zealand early in this century, and naturally spawning populations have become established in several rivers of the south island. The self-sustaining populations are all believed to have originated from a single stock, possibly from Battle Creek in California. Analysis of age and size at seaward migration, age and size at maturity, and time of river entry revealed significant divergence in these characters among New Zealand chinook salmon. Modal age at maturity was 3 years, and 2-year-old females were common; the original stock had a modal age of maturity of 4 years and 2-year-old females are extremely rare in North America. Age at maturity and size at age also varied significantly among the populations in New Zealand. All the populations consisted of mixtures of fish that migrated to sea as underyearlings or spent one year in freshwater before migrating to sea. The proportion of yearling smolt migrants also varied among populations, ranging from about 29 to 76%. Quinn and Unwin (1993) speculate that these and other differences between New Zealand chinook salmon and the donor population, as well as differences among populations within New Zealand, may reflect rapid genetic adaptation. The genotypic contribution to these phenotypes is still under investigation, although results of mitochondrial DNA analysis do not suggest much divergence among New Zealand populations (T. Quinn, University of Washington, personal communication). Nevertheless, the New Zealand chinook salmon illustrate the rapid divergence of phenotype in less than a century.

The evidence for rapid divergence among populations of Pacific salmon introduced to new environments suggests that in some circumstances the species can adjust rapidly to new environments. Nevertheless, the success of deliberate introductions of Pacific salmon to new habitats has, in general, been poor. Many attempts have been made over the last century to introduce Pacific salmon to apparently suitable but unused habitats within or outside their range. And considerable effort has been expended attempting to reintroduce Pacific salmon to previously occupied habitats from which they had been extirpated (see chapters in Groot and Margolis 1991 for species-specific summaries of these attempts). The difficulty of establishing or reestablishing Pacific salmon in suitable habitats seems to fly in the face of the species' apparent phenotypic plasticity and opportunism. It also seems to fly in the face of the relatively rapid colonization of suitable habitats following the last glaciation and following successful introduction to the Great Lakes and New Zealand. Whatever it is that blocks successful introduction, the history of failure in this regard should warn that recovery of lost

TABLE 1.—Percentage of variation in genotype and phenotype within populations, among populations within a region, and among regions.

Salmon species and character	Variation (%)		
	Within population	Among population	Among region
Chinook			
Genotype[a]	87.7	4.6	7.7
Genotype[b]	94.1	3.3	2.6
Pink[c]			
Fry length	30.3	41.4	28.2
Adult weight	13.3	1.5	85.2
Chum[c]			
Fecundity	7.9	5.1	87.0
Age	14.0	0.5	85.5

[a]From Utter et al. 1989.
[b]From Gharrett et al. 1987.
[c]From Groot and Margolis 1991.

phenotypes may not follow directly from conservation of the genotype.

Mapping Genetic and Phenotypic Variation on Geographic Scales

Although the evidence is strong that phenotypic variation on ecosystem and landscape scales reflects an underlying genetic variation (Tallman and Healey 1994; Utter et al. 1989), these scales do not appear to be where the majority of the observed genetic information resides. Evidence from allozymic variation suggests that the majority of genetic variation maps onto the population, not the ecosystem or landscape, scale. Utter et al. (1989), for example, found that 87.7% of the electrophoretically detectable variation in chinook salmon was within populations (Table 1). Among-population and among-region differences accounted for only a small amount of variation. Gharrett et al. (1987) found more than 90% of genetic variation in chinook salmon was within populations (Table 1). Wood (1995, this volume) reports a similar distribution of genetic variation within and among populations of sockeye salmon O. nerka. Furthermore, the electrophoretically detectable variation among populations and regions is virtually all due to variation in allelic frequencies, not to the presence of unique alleles.

If electrophoretically detectable variation reflects the genome as a whole, then each population of Pacific salmon contains most of the genetic information for the species. This is not to say that the genetic variation among populations is unimportant and can be ignored. Clearly it is important, and a number of significant phenotypic differences among populations have been shown to have a quantitative genetic basis. Nor can it be argued categorically that the electrophoretically detectable variation is a good index of total genetic variation. Much electrophoretically detectable variation is presumed to be neutral to selection. Furthermore, the amount of the genome sampled is a tiny fraction of the total. The problems of drawing inferences about genome structure from the measurements available to us were clearly laid out by Lewontin (1974).

Phenotypic variation has quite a different distribution across scales. Although many phenotypes are variations on a theme, some are so different as to capture our attention immediately: the winter-run chinook salmon in the Sacramento River, for example, the summer and autumn chum salmon of Asia, or the stream- and ocean-type chinook salmon. These are ecosystem and landscape scales of variation.

We partitioned the variation in fry size and adult body size for pink salmon and fecundity and age at maturity for chum salmon into components for within-population, among-populations-within-a-region, and among-region variation to provide a comparison with the partitioning of genetic variation reported by Utter et al. (1989) and Gharrett et al. (1987). We obtained the data from tables in Groot and Margolis (1991) and analyzed them in a nested analysis of variance (ANOVA; Sokal and Rohlf 1981). Variation within populations ranged from about 8% of total variation for chum salmon fecundity to 30% for pink salmon fry size, and variation attributed to regional differences ranged from 28% for pink salmon fry size to 87% for chum salmon fecundity (Table 1). Although within-population variation is somewhat underestimated in these analyses, because the data reflected interannual rather than interindividual variation, we conclude that much phenotypic variation exists at the ecosystem (among-population) and landscape (among-region) scales.

Discussion

If the distributions of genotypic and phenotypic variation are as different as the data presented above suggest, then the loss of populations, such as that documented by Nehlsen et al. (1991), has quite different implications for phenotypic and genotypic diversity. The loss of a few populations within a landscape or an ecosystem could have only a small effect on overall genetic diversity but could have a large effect on phenotypic diversity (Table 1). Because phenotypic diversity is a consequence of the

genotype interacting with its particular environment, preserving unique habitats and ensuring their accessibility to Pacific salmon are crucial to conservation. Yet, this has not been the focus of recent conservation efforts.

When particular phenotypes are threatened with imminent extinction, extreme measures taken to preserve them appear to emphasize genetic salvage (such as the recent efforts to conserve Red Lake sockeye salmon and Sacramento River winter-run chinook salmon). Although some form of artificial propagation program may seem to be the only option when stocks are as depleted as these, we are skeptical of the long-term outcome of such salvage operations. Given the importance of habitat in molding Pacific salmon phenotypes (including their genetic underpinning), the risk that these measures will fail to conserve the desired traits is high. In the case of the Sacramento winter-run chinook salmon, the phenotype has already had to adjust to a virtually complete loss of its traditional spawning habitat. Thus, it is not clear exactly what is being salvaged. Nor is it clear what the salvaged populations will be like if the opportunity to reintroduce them arrives.

Regardless of whether genetic salvage succeeds in preserving a viable captive population of Pacific salmon, the reintroduction of these fishes into the natural environment is likely to be problematic. As noted above, the success of such reintroductions has been low, although previous attempts at reintroduction did not involve the presumed original genetic strain. Successful reintroduction will, nevertheless, depend on ensuring that there is suitable habitat and habitat access for the reintroduced fishes. Our willingness to give up major alternative uses of river systems that have lead to the demise of Pacific salmon populations, such as hydroelectric power generation and irrigation, is still very much in doubt as is our willingness to consign a significant portion of the floodplain to natural river channel activity and to fish and wildlife habitat. To the extent that it distracts us from the importance of habitat conservation in maintaining Pacific salmon diversity, genetic salvage undermines our ability to have successful conservation programs.

These observations emphasize that to be successful conservation efforts for Pacific salmon must focus as much on habitat and habitat opportunity as on genotype. This is because conservation is, at present, stimulated by the anticipated loss of particular phenotypes, not the loss of a species as a whole and because the preservation of phenotype is as dependent on Pacific salmon habitat as on their genetics. The critical conservation unit, therefore, is the population within its habitat. Once the habitat that made a particular genotype–phenotype combination possible is gone, preservation or re-creation of the phenotype is probably impossible.

On the other hand, Pacific salmon species are capable of rapidly developing new phenotypes when the right opportunity presents itself. This means that new genotype–phenotype combinations will emerge naturally provided habitat opportunity for the Pacific salmon can be maintained. Thus, although it may not be possible to prevent the loss of phenotypes due to human activity, it may be possible to enhance phenotypic diversity by expanding habitat opportunity elsewhere. The colonizing ability of Pacific salmon coupled with the apparently high genetic diversity within populations is the one positive aspect of an otherwise disturbing history of population extinction.

This is not to say that loss of habitats can be condoned on the basis that new habitat opportunities can be created elsewhere. Nor is it to say that loss of populations can be condoned on the basis that much genetic diversity resides within populations. Rather, it is an argument for maintaining and expanding habitat opportunity for salmon as well as preserving genotypes. These habitat opportunities must be provided on the landscape as well as the local scale if the range of diversity is to be maintained. The argument is also an expression of our hope that, if we can achieve a more balanced and sensitive approach to our use of river systems, the species themselves have the potential to recapture the variety of their phenotype if not the specific phenotypes that we have extirpated.

Acknowledgments

Research on which this paper was based was supported in part by a Natural Sciences and Engineering Research Council grant to M. Healey. M. Henderson and R. Tallman read and commented on a draft of the manuscript.

References

Aspinwall, N. 1974. Genetic analysis of North American populations of pink salmon, *Oncorhynchus gorbuscha*, possible evidence for the neutral mutation-random drift hypothesis. Evolution 28:295–305.

Bartley, D. M., and G. A. E. Gall. 1990. Genetic structure and gene flow in chinook salmon populations of California. Transactions of the American Fisheries Society 119:55–71.

Beacham, T. 1985. Meristic and morphometric variation in pink salmon (*Oncorhynchus gorbuscha*) in southern British Columbia and Puget Sound. Canadian Journal of Zoology 63:366–372.

Carl, L. M. 1982. Natural reproduction of coho salmon and chinook salmon in some Michigan streams. North American Journal of Fisheries Management 2:375–380.

Clarke, W. C., R. E. Withler, and J. E. Shelbourn. 1994. Inheritance of smolting phenotypes in backcrosses of hybrid stream-type × ocean-type chinook salmon (*Oncorhynchus tshawytscha*). Estuaries 17:13–25.

Fleming, I. A., and M. R. Gross. 1989. Evolution of adult female life history and morphology in a Pacific salmon (coho: *Oncorhynchus kisutch*). Evolution 43: 141–157.

Gharrett, A. J., S. M. Shirley, and G. R. Tromble. 1987. Genetic relationships among populations of Alaskan chinook salmon (*Oncorhynchus tshawytscha*). Canadian Journal of Fisheries and Aquatic Sciences 44: 765–774.

Gilbert, C. H. 1913. Age at maturity of the Pacific coast salmon of the genus *Oncorhynchus*. Bulletin of the U.S. Bureau of Fisheries 32:1–22.

Groot, C., and L. Margolis, editors. 1991. Pacific salmon life histories. University of British Columbia Press, Vancouver.

Gross, M. 1985. Disruptive selection for alternative life histories in salmon. Nature 313:47–48.

Healey, M. C. 1983. Coastwide distribution and ocean migration patterns of stream- and ocean-type chinook salmon, *Oncorhynchus tshawytscha*. Canadian Field-Naturalist 97:427–433.

Healey, M. C. 1986. Optimum size and age at maturity in Pacific salmon and effects of size-selective fisheries. Canadian Special Publication in Fisheries and Aquatic Sciences 89:39–52.

Healey, M. C. 1987. The adaptive significance of age and size at maturity in female sockeye salmon (*Oncorhynchus nerka*). Canadian Special Publication in Fisheries and Aquatic Sciences 96:110–117.

Healey, M. C. 1991. Life history of chinook salmon (*Oncorhynchus tshawytscha*). Pages 311–394 in C. Groot and L. Margolis, editors. Pacific salmon life histories. University of British Columbia Press, Vancouver.

Heard, W. R. 1991. Life history of pink salmon (*Oncorhynchus gorbuscha*). Pages 119–230 in C. Groot and L. Margolis, editor. Pacific salmon life histories. University of British Columbia Press, Vancouver.

Hindar, K., N. Ryman, and F. Utter. 1991. Genetic effects of cultured fish on natural fish populations. Canadian Journal of Fisheries and Aquatic Sciences 48:945–957.

Holtby, L. B., and M. C. Healey. 1986. Selection for adult size in female coho salmon. Canadian Journal of Fisheries and Aquatic Sciences 43:1946–1959.

Holtby, L. B., and M. C. Healey. 1990. Sex specific foraging strategies and risk taking in coho salmon. Ecology 71:678–690.

Hughes, R. M., and R. F. Noss. 1992. Biological diversity and biological integrity: current concerns for lakes and streams. Fisheries 17:11–19.

Kristiansson, A. C., and J. D. McIntyre. 1976. Genetic variation in chinook salmon (*Oncorhynchus tshaw-*

ytscha) from the Columbia River and three Oregon coastal rivers. Transactions of the American Fisheries Society 105:620–623.

Kwain, W., and J. A. Chappel. 1978. First evidence for even-year spawning pink salmon, *Oncorhynchus gorbuscha*, in Lake Superior. Journal of the Fisheries Research Board of Canada 35:1373–1376.

Kwain, W., and E. Thomas. 1984. First evidence of spring spawning by chinook salmon in Lake Superior. North American Journal of Fisheries Management 4:227–228.

Larkin, P. A. 1981. A perspective on population genetics and salmon management. Canadian Journal of Fisheries and Aquatic Sciences 38:1469–1475.

Lewontin, R. C. 1974. The genetic basin of evolutionary change. Columbia University Press, New York.

Maynard Smith, J. 1982. Evolution and the theory of games. Cambridge University Press, Cambridge, UK.

Nehlsen, W., J. E. Williams, and J. A. Lichatowich. 1991. Pacific salmon at the crossroads: stocks at risk from California, Oregon, Idaho, and Washington. Fisheries 16:4–21.

Quinn, T. P., and M. J. Unwin. 1993. Variation in life history patterns among New Zealand chinook salmon (*Oncorhynchus tshawytscha*) populations. Canadian Journal of Fisheries and Aquatic Sciences 50:1414–1421.

Reisenbichler, R. R., and J. D. McIntyre. 1977. Genetic differences in growth and survival of juvenile hatchery and wild steelhead trout, *Salmo gairdneri*. Journal of the Fisheries Research Board of Canada 34:123–128.

Rich, W. H., and H. B. Holmes. 1928. Experiments in marking young chinook salmon on the Columbia River, 1916 to 1927. Bulletin of the U.S. Bureau of Fisheries 44:215–264.

Ricker, W. E. 1972. Hereditary and environmental factors affecting certain salmonid populations. Pages 19–160 *in* R. Simon and P. Larkin, editors. The stock concept in Pacific salmon. H. R. MacMillan Lectures in Fisheries, University of British Columbia, Vancouver.

Ryman, N., and G. Stahl. 1981. Genetic perspectives of the identification and conservation of Scandinavian stocks of fish. Canadian Journal of Fisheries and Aquatic Sciences 38:1562–1575.

Schroder, S. L. 1981. The role of sexual selection in determining overall mating patterns and mate choice in chum salmon. Doctoral dissertation. University of Washington, Seattle.

Scudder, G. G. E. 1989. The adaptive significance of marginal populations: a general perspective. Canadian Special Publication in Fisheries and Aquatic Sciences 105:180–185.

Sokal, R. R., and F. J. Rohlf. 1981. Biometry, 2nd edition. Freeman, New York.

Swain, D. P., B. E. Riddell, and C. B. Murray. 1991. Morphological differences between hatchery and wild populations of coho salmon (*Oncorhynchus kisutch*): environmental versus genetic origin. Canadian Journal of Fisheries and Aquatic Sciences 48:1783–1791.

Tallman, R. F. 1986. Genetic differentiation among seasonally distinct spawning populations of chum salmon, *Oncorhynchus keta*. Aquaculture 57:211–217.

Tallman, R. F., and M. C. Healey. 1991. Phenotypic differentiation in seasonal ecotypes of chum salmon, *Oncorhynchus keta*. Canadian Journal of Fisheries and Aquatic Sciences 48:661–671.

Tallman, R. F., and M. C. Healey. 1994. Homing, straying and gene flow among seasonally separated populations of chum salmon (*Oncorhynchus keta*). Canadian Journal of Fisheries and Aquatic Sciences 51:577–588.

Taylor, E. B. 1986. Differences in morphology between wild and hatchery populations of juvenile coho salmon. Progressive Fish-Culturist 48:171–176.

Taylor, E. B. 1991. A review of local adaptation in Salmonidae with particular reference to Pacific and Atlantic salmon. Aquaculture 98:185–207

Utter, F., G. Milner, G. Stahl, and D. Teel. 1989. Genetic population structure of chinook salmon (*Oncorhynchus tshawytscha*) in the Pacific northwest. U.S. National Marine Fisheries Service Fishery Bulletin 87:239–264.

Wagner, W. C., and T. M. Stauffer. 1980. Three-year-old pink salmon in Lake Superior tributaries. Transactions of the American Fisheries Society 109:458–460.

Wood, C. C. 1995. Life history variation and population structure in sockeye salmon. American Fisheries Society Symposium 17:195–216.

American Fisheries Society Symposium 17:185–194, 1995

Size and Shape Variation in Laurentian Great Lakes Pink Salmon

DAVID L. G. NOAKES, MOIRA M. FERGUSON, BLAIR ASHFORD,
AND WENDYLEE STOTT[1]

Institute of Ichthyology, Department of Zoology
University of Guelph, Guelph, Ontario N1G 2W1, Canada

Abstract.—Male pink salmon *Oncorhynchus gorbuscha* seek fertilizations by means of two behavioral methods: guarding females and sneaking. Two distinct categories of males have been hypothesized corresponding to these behavioral methods: larger males with well-developed dorsal humps (alpha) and smaller males without dorsal humps (gamma). We found that hump arc length, hump area, and the eye to hypural length in sexually mature males from one population showed no biologically important clustering and that the frequencies of these measurements were not significantly different from normality. Although males mated to the same female produced young of different sizes and conditions, these differences were not related to behavioral category (i.e., progeny of alpha males were not larger than those of gamma males). We conclude that within-population morphological variation among male pink salmon is continuous and isometric. Intraspecific variation in salmonids must be examined on a case-by-case basis and include studies of ontogenetic development as well as genetic analyses.

Intraspecific variation in salmonid fishes, including behavioral, morphological, and genetic measures, has attracted a lot of attention because of its possible relevance to evolutionarily significant units (ESUs; e.g., Noakes et al. 1989). There is convincing evidence in some cases of a genetic basis for such variation, lending support to hypotheses of the evolution and maintenance of this variation (e.g., Skulason et al. 1989a, 1989b). However, other cases of intraspecific variation in salmonids have been less well studied in terms of the mechanisms involved. A principal example concerns the genetic basis of differences between sexually mature males (Gross 1985). It has been hypothesized, based upon empirical evidence from several species, that males belong to either one of two alternative life histories. Larger, later-maturing males compete aggressively for access to spawning females. Smaller, earlier-maturing males gain access to spawning females by sneaking close at the moment gametes are released. These differences between males are suggested to be an evolutionarily stable strategy, evolved through frequency-dependent disruptive selection (Gross 1985). Genetically based alternative life histories, and corresponding phenotypic specializations, would be obvious candidates for consideration as ESUs.

We addressed this question with pink salmon *Oncorhynchus gorbuscha* as an extreme case. Pink salmon are notable among salmonids for a virtually invariant 2-year life cycle and extreme dimorphism in sexually mature adults (Heard 1991). The common English name for the species, hump-backed salmon, describes the appearance of sexually mature males (Beacham and Murray 1985). The invariant 2-year life cycle eliminates age differences at maturity as a confounding variable in this species. In other *Oncorhynchus* species some males mature at an earlier age and a smaller size. Pink salmon is semelparous, further eliminating complications with repeated spawning by some males. Studies of pink salmon both in their native range in the Pacific basin (Keenleyside and Dupuis 1988) and where they have established sustaining populations after accidental release in the Laurentian Great Lakes (Noltie 1988) indicate that males seek fertilizations using two tactics. The first is guarding: larger males with well-developed dorsal humps cluster around digging females and fight among themselves for proximity to the female. The second tactic is sneaking: smaller males with little or no humps position themselves as satellites and do not join in the cluster prior to the actual moment of gamete release. Both types of males rush in during oviposition presumably to release their gametes (Keenleyside and Dupuis 1988; Noltie 1990).

Davidson (1935) found that only the hump size and snout length were significantly different between prespawning and spawning males. Snout length was longer in sexually mature than in sexually immature males but was uniform within size-classes of mature males. In contrast, hump size was

---

[1]Present address: Department of Biology, McMaster University, Hamilton, Ontario L8S 4L8, Canada.

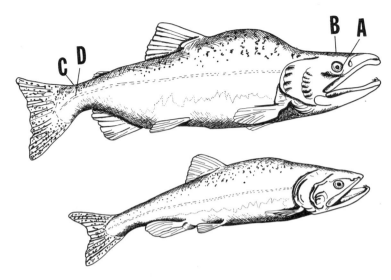

FIGURE 1.—An alpha (upper fish) and gamma (lower fish) adult male pink salmon, drawn from a photograph to illustrate differences in size and shape. Three morphological characters measured in photographs of adult male pink salmon were eye–hypural length (A–D), from anterior of orbit to middle of hypural plate; hump arc length (B–C), from above orbit of eye along the dorsal surface to the point above the hypural plate; and hump area (A–B–C–D), the area bounded by the eye–hypural length below and the hump arc above and the two line segments (A–B) and (C–D) at either end.

significantly greater in mature males and was not uniform among males of different sizes. This non-uniform morphology and the behavioral observations are consistent with the hypothesis (Gross 1985) that there are two genetically distinct classes of male pink salmon which differ both in morphology and behavior: small sneakers with a morphology like females (gamma) and large hump-backed guarders (alpha; Figure 1). This hypothesis predicts bimodalities in length and hump-size distributions among adult males, with significant clusters of small, little-hump males and large, big-hump males. A second prediction is that the size and shape of young pink salmon depends on the size and shape of the male parent.

As a null hypothesis, we propose that the size and shape distributions of adult male pink salmon are continuous, with hump size an isometric function of length. The behavioral observations would still be valid under this hypothesis and in accordance with earlier conclusions (Gross 1985), but the tactics of the males would not be fixed and would be decided for any particular spawning event of an individual by the size distribution of the competing males at that event. In addition, the null hypothesis predicts that size and shape of young would not depend totally on the size and shape of the father.

All previous work to differentiate between male pink salmon has used an a priori classification, with males divided into two classes (large males and small males). Measurements were then made, and the results were used to discriminate between the two classes (Noltie 1988). This approach assumes, rather than tests, the hypothesis of distinct categories of males. We do not assume any such a priori classification but instead test for bimodality in the frequency distributions and for the number of clusters in the body-size versus hump-size plots on the basis of our null hypothesis.

The immediate practical question is how to measure hump size (e.g., Bookstein et al. 1985; Lovich and Gibbons 1992). To take body depth as a measure of hump size, the visceral cavity must be assumed constant among fish; that assumption depends on several factors including whether the fish has spawned or not. As well, the deepest point of the fish is not always a homologous point between fish, that is, it is not always at the anterior dorsal fin insertion but could be midway between the snout and dorsal fin.

Douglas (1993) attempted to solve the problem of measuring the hump size of the humpback chub *Gila cypha* by superimposing a truss (Strauss and Bookstein 1982) on an image of the fish. To compensate for the lack of landmarks for the endpoints of the truss on the hump, Douglas constructed landmarks by radiating lines out from the dorsal insertion of the pectoral fin. This approach has problems of its own, including whether the constructed landmarks are homologous among fish or whether the landmarks have any meaningful biological interpretation. Most importantly, is the variation in fish shape reflected in the various measurements?

We believe the best way to measure hump size is to analyze the area of the hump, or the hump perimeter. This measurement not only quantifies the hump size but constructs a hump measurement that can be homologous between fish (with a proper description of the area to be measured that is independent of the fish's morphology).

Methods

Influence of Male Parent on Body Size of Young

Sexually mature pink salmon were collected between 14 and 29 September 1989 from Fishers and Young creeks, two adjacent tributaries to Long Point Bay in Lake Erie (separated by about 5 km). Previous information from fish assessment efforts (Jim Collins, Ontario Ministry of Natural Resources, personal communication) indicated that a significant population of "odd-year" pink salmon (fish that regularly complete their 2-year life cycles in odd-numbered years) was established concurrently in each of these streams as part of the invasion of pink salmon from the upper Great Lakes. Ova and semen were collected in dry containers and transported on wet ice to the University of Guelph. Males were categorized as alpha (large body size, well-developed dorsal hump) or gamma (small body size, no dorsal hump; Figure 1). The ova from each female were divided into lots, and each lot was combined with the semen of a single male (designated alpha or gamma) to produce a mating set of maternal half-sib families. Three females were mated to four different males (2 alpha, 2 gamma), and six females were used twice each with different males (1 alpha, 1 gamma). The half-sib families from each mating set were reared in a common tray of a vertical incubating rack in darkness at 6–8°C. After hatching, and at the onset of exogenous feeding, each family of alevins was moved into troughs with flowing well water that varied between 6 and 12°C depending on the season. The photoperiod simulated natural conditions. All families of alevins were fed more commercial trout food (Martin's Feed Mills, Elmira, Ontario; BioDiet, Olympia, Washington) than they were able to consume at each feeding.

The fish were reared until 11 May 1990. Then all fish from each family were removed from the troughs. A subsample of 20 fish (if available) from each family was anaesthetized in MS-222 (tricaine methanesulfonate; 0.25 mg/4 L). The fork lengths (mm) and wet weights (g) of each fish were determined. All measured fish were allowed to recover and returned to the trough. The density of fish in each trough was increased to 30 with the addition of excess (and unmeasured) fish. All fish were reared for an additional 2.5 months (until 31 July 1990) at which time they were killed by an overdose of MS-222, weighed, and measured as described previously.

The data set was divided into two components for analysis because of the unbalanced numbers of males mated to each female. Mating sets for which females were mated with four males (females 15, 9, and 2) were analyzed separately from those females mated to a single gamma and single alpha male (females 24, 27, 29, 26, 36, and 40). We analyzed the progeny data in two ways to address two questions. First, nested analyses of variance (ANOVAs; sources of variation: female, male nested within female) tested whether different males mated to common females produced different size (length or weight) or shape (condition) progeny. Condition was calculated as the residual from the relationship between $\log_{10}$ body length and $\log_{10}$ body weight in both May and July 1990. This nested analysis tested for a genetic basis for size and shape differences among males but ignored alpha or gamma classifications. Second, two-way ANOVAs (sources of variation: female, male type [alpha or gamma]) with interaction were used to determine if the progeny of alpha and gamma males differed in fork length, wet body weight, or condition in either May or July. The interaction mean square was used as the denominator in the F-test of male type effects when significant interaction effects were detected.

Shape and Size in Adult Males

Sexually mature pink salmon collected from Fishers and Young creeks in September of 1989 were photographed beside a ruler. Black and white copies of these photographs were scanned into digital format and analyzed on a Macintosh LCIII computer using the public domain Image program[2] part number PB93–504868. Measurements were made

---

[2]Rasband, W. Image version 1.54. U.S. National Institutes of Health, NTIS, 5285 Port Royal Rd., Springfield, VA 22161, USA.

with a precision of 0.25 cm. In order to avoid an a priori bias with the two extreme male groups, none of the males used as parents for experimental crosses (13 alpha and 13 gamma males) was measured. A total of 87 males, selected from across the complete size range of males in our collection, were measured.

Because only hump size has been reported to be nonuniform between size-classes (Davidson 1935), we used only that character as a potentially discriminating feature (our analysis of snout length also showed no significant difference related to body size of males; unpublished observations). To account for allometric size and shape differences, the length of the fish was also measured. The length measured was from the anterior orbit of the eye to the hypural plate, because the standard length would include the elongated snout (Keenleyside and Dupuis 1987).

We calculated two variables that describe hump size: hump arc length and hump area. The hump arc length is the length from above the anterior orbit of the eye along the dorsal surface of the hump to the point above the hypural plate. The hump area was measured as the area bounded by the hump arc above, the line from the anterior orbit of the eye to the middle of the hypural plate below, and the vertical lines defined at the anterior orbit and the hypural plate at either end (Figure 1). The hump arc lengths and hump areas were plotted against the eye–hypural length, and the linear regressions were calculated. The frequency distributions of the hump arc lengths, hump areas, and eye–hypural lengths were tested for normality using a Kolmogorov–Smirnov normality test, and the number of clusters was assessed by the Wong and Schaack (1982) kth-nearest-neighbor method, as outlined by Statistical Analysis Systems (SAS; SAS Institute 1985). All statistical analysis was performed on SAS; an alpha value of 0.05 was considered significant.

Results

Inheritance of Body Size

We could not determine the sex of the pink salmon progeny at the time of sampling, so data are reported for all progeny as a single group for each set of parents (Table 1). The nested ANOVAs detected significant variation in size and condition among the progeny of different males to the same female. In May, length ($P = 0.001, P < 0.001$ for the two component data sets—female with four males versus two males—respectively), weight ($P = 0.083, P < 0.001$), and condition ($P < 0.001, P < 0.001$)

differed significantly in most comparisons. Similar results were detected for fish measured in July (length: $P = 0.002, P = 0.001$; weight: $P = 0.03, P < 0.001$; condition: $P < 0.001, P > 0.05$). Despite the suggestion of paternal effects on size and shape, the factorial ANOVAs indicated that progeny of gamma and alpha males did not differ in fork length, wet body weight, or condition on either sampling date in either component data set ($P > 0.05$). In fact, gamma progeny were slightly larger on average than were alpha progeny.

Analyses of Size and Shape in Adult Males

Eye–hypural lengths of males ranged from 27 to 42 cm for the 87 males measured (Figure 2). This length-frequency distribution, and the frequency distributions for hump area (Figure 3) and hump arc length (Figures 4) did not differ significantly from normal distributions (Kolmogorov–Smirnov normality test, $P > 0.05$). Figures 5 and 6 are the plots of the two calculated variables (hump arc length and hump area) versus eye–hypural length. In both cases the simple linear regression is highly significant. Thus hump arc length and hump area are isometric (Gould 1977). No clustering is observed, and the Wong and Schaack (1982) kth-nearest-neighbor test corroborates this observation on the hump arc length versus eye–hypural length plot (Figure 7). If clustering were present, the Wong and Schaack test should plateau at a modal cluster number for a wide range of k values (on the order of the number of members in the smallest cluster). Instead it quickly reaches a modal cluster number of 1. Together with the Wong and Schaack test for the number of clusters, this implies the absence of biologically important clusters.

Discussion

We found no evidence for any heritable effect of the category of the male parent (alpha or gamma) on size or condition of his progeny during their first year. Paternal genetic effects were detected, but these were not related to category of male (i.e., alpha male progeny were not always larger). The detection of paternal genetic effects independent of male category agrees with the findings of Beacham and Murray (1988a, 1988b), who detected such effects in the second year of life when the young fish were maturing. Beacham and Murray used fish from native, anadromous (sea-run) populations that had a larger size difference among males than did males in our study. They raised their fish in salt water. Our fish were from a landlocked, introduced

TABLE 1.–Mean fork lengths (mm), wet body weights (g), and condition (weight:length), with standard errors in parentheses for each measure, of the progeny of female pink salmon mated to either alpha (A) or gamma (G) males.

Female number and male type	May 1990				July 1990			
	N	Length	Weight	Condition	N	Length	Weight	Condition
15[a]								
A	40	64.7 (1.4)	2.22 (0.14)	0.011 (0.008)	62	131.3 (1.8)	24.35 (1.06)	0.014 (0.006)
G	40	67.3 (1.1)	2.37 (0.13)	−0.016 (0.006)	57	135.1 (2.0)	25.98 (1.13)	0.006 (0.007)
9[a]								
A	42	68.7 (1.4)	2.69 (0.20)	0.003 (0.006)	62	127.9 (2.2)	21.81 (1.12)	−0.004 (0.007)
G	29	67.3 (1.5)	2.57 (0.21)	0.016 (0.010)	34	133.8 (2.0)	24.94 (1.35)	0.003 (0.006)
2[a]								
A	40	68.3 (1.1)	2.44 (0.13)	−0.022 (0.007)	53	137.9 (1.3)	26.14 (0.92)	−0.016 (0.005)
G	41	69.9 (1.4)	2.76 (0.16)	−0.004 (0.005)	58	133.7 (1.8)	23.15 (0.91)	−0.029 (0.006)
24								
A	20	54.4 (1.2)	1.10 (0.09)	−0.049 (0.009)	32	118.8 (2.3)	17.04 (0.98)	−0.005 (0.006)
G	21	56.0 (1.8)	1.27 (0.13)	−0.035 (0.009)	30	125.1 (2.4)	19.60 (1.20)	−0.013 (0.007)
27								
A	20	56.1 (2.3)	1.37 (0.23)	−0.037 (0.010)	21	127.1 (3.1)	20.62 (1.81)	−0.013 (0.005)
G	22	51.7 (2.1)	1.09 (0.14)	0.000 (0.013)	32	120.3 (2.2)	16.63 (0.90)	−0.023 (0.004)
29								
A	21	59.1 (1.4)	1.49 (0.10)	−0.028 (0.006)	31	127.3 (2.4)	21.16 (1.27)	−0.007 (0.006)
G	20	58.5 (2.2)	1.63 (0.15)	0.011 (0.012)	32	132.1 (2.1)	23.36 (1.26)	−0.007 (0.005)
26								
A	20	61.7 (1.2)	1.79 (0.12)	−0.008 (0.009)	25	123.6 (2.3)	19.99 (1.09)	0.018 (0.007)
G	20	75.7 (1.9)	3.76 (0.27)	0.025 (0.010)	31	135.9 (2.6)	28.26 (1.74)	0.030 (0.009)
36								
A	20	68.6 (1.8)	2.78 (0.29)	0.024 (0.008)	30	115.8 (1.6)	15.66 (0.68)	0.011 (0.006)
G	20	65.8 (1.6)	2.32 (0.19)	0.006 (0.009)	30	114.7 (2.5)	15.11 (0.90)	−0.006 (0.005)
40								
A	20	66.1 (2.5)	2.50 (0.34)	0.008 (0.008)	29	119.7 (2.7)	17.69 (1.28)	0.001 (0.006)
G	20	63.9 (2.2)	2.23 (0.23)	0.018 (0.006)	30	117.4 (2.2)	16.63 (0.99)	0.008 (0.003)
Mean								
A	243	64.1	2.14	−0.009	345	127.1	21.35	−0.000
G	233	64.8	2.28	0.001	334	128.8	22.06	−0.005

[a]Mated to two different alpha and two different gamma males.

population with overall smaller-size fish and a smaller range of male sizes. Our fish were raised in freshwater for a shorter period of time (see also Berg 1979).

The significance of these procedural differences between our and Beacham and Murray's studies remains to be determined because these differences might affect conclusions as to paternal effects on progeny size. Effects of male type on progeny size might have been detected after a longer rearing period. Adult size of pink salmon is known to be highly variable, depending on geographic location, latitude, spawning year, oceanic feeding conditions, and, possibly, population size (Ricker 1972; Heard 1991). However, our conclusion that dorsal hump size of males is an isometric function of body size lessens the concerns related simply to differences in body size among populations.

Our results of no biologically important clustering of hump size and body size in adult male pink salmon are unequivocal and consistent with our interpretation of the results of our study of early growth and development. We feel these results to-

gether allow us to reject the hypothesis of distinct morphological categories of males and accept the predictions of the null hypothesis, i.e., continuous shape and size distributions.

Can the continuous variance hypothesis be consistent with the behavioral observations of Keenleyside and Dupuis (1987) and Noltie (1990)? If only large males with large humps and small males with small humps were successful in reproduction, the result should be disruptive selection, acting to remove the intermediate males. Any heritable basis for this size and shape distinction among males would be strengthened (Beacham and Murray 1985).

One possible explanation for the behavioral observations is the handicap principle, in which hump size is an honest signal of the male's condition (Grafen 1991). Because the males do cluster around the females, an honest signal would reduce aggression within the cluster dominance hierarchy, with smaller-hump males responding to the larger-hump males as being in superior condition. This hypothesis would also explain why the snout length tends to be uniform within size-classes—because males

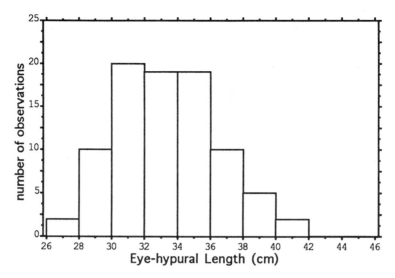

FIGURE 2.—Frequency distribution of eye–hypural lengths for 87 adult male pink salmon.

would still need to fight regardless of their ranks in the cluster. Presumably condition is correlated to size and thus the strong correlation between length and hump size.

Another possible explanation is the hump is a hydrodynamic feature that allows the males to maintain their position behind the females with less effort (Soin 1954). Larger males tend to be found more in areas with higher water velocity, where increased hydrodynamic efficiency could be significant (Heard 1991). However, we should note that swimming capability is a function of body size, and

so for that simple reason alone we would expect to find larger fish in faster water currents (Brett and Groves 1979).

Chebanov (1980) reported on a confined spawning population of pink salmon with an equal sex ratio. He found that movement is inversely correlated with body size, but movement ending in leadership is positively correlated with body size. Hence, smaller males move more on the spawning grounds, but the larger males, when they do move, have a greater chance of becoming a leader in a spawning cluster. Although all males are capable of becoming

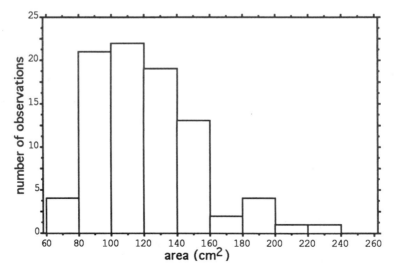

FIGURE 3.—Frequency distribution of hump area for 87 adult male pink salmon.

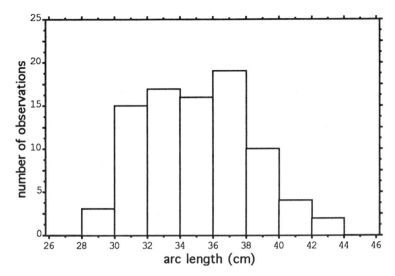

FIGURE 4.—Frequency distribution of hump arc length for 87 adult male pink salmon.

either the first male or a male lower down in the spawning cluster hierarchy, most spawning males are of a similar size as the spawning female. Only the largest males are leaders in clusters around both small and large females. Thus, the large males maximize their fitness by staying with one female for the whole spawning procedure, and they need a large hump to maintain their position. The smallest males have no need of a hump, since they are moving the most of all sizes, and have the least chance of becoming a leader (and so presumably may sneak).

Yet another explanation is suggested by recent detailed behavioral observations of spawning behavior of unconfined coho salmon *Oncorhynchus kisutch* in their native range (Healey and Prince 1995, this volume). Large, dominant males de-

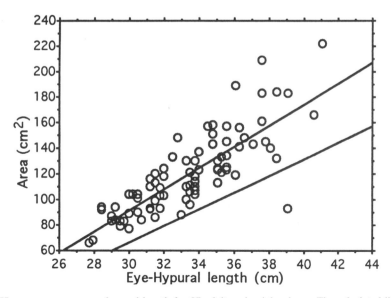

FIGURE 5.—Hump area versus eye–hypural length for 87 adult male pink salmon. The calculated linear regression (upper line) is highly significant ($P = 0.001$, $r^2 = 0.663$, $N = 87$). The regression line for a comparable number of female fish (lower line, no data points plotted) is included for comparison with undifferentiated fish.

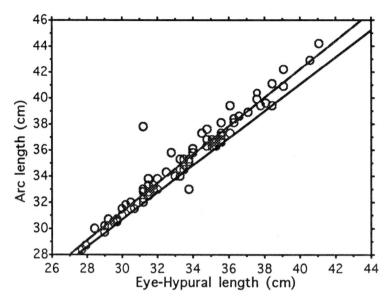

FIGURE 6.—Hump arc length versus eye–hypural length for 87 adult male pink salmon. The calculated linear regression (upper line) is highly significant ($P = 0.001$, $r^2 = 0.946$, $N = 87$). The calculated female regression (lower line, no data points plotted) is included for comparison with undifferentiated fish.

fended against other large male stretches of the stream as territories, even when no females were present. They attempted to mate with any female spawning within that section of stream. Small males were seldom directly attacked by large males. Small males distributed themselves throughout the river or stream and, with little direct response from large males, participated in spawning events with large males and females. Intermediate males, as satellites, moved among different sections of the stream

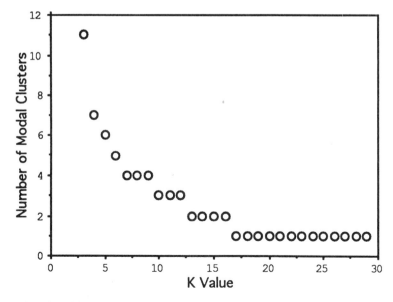

FIGURE 7.—Number of modal clusters versus kth nearest neighbors in hump arc length versus eye–hypural length for sample of 87 adult male pink salmon (details in text).

and attempted to participate in spawning events with large males and females. Intermediate males were sometimes threatened or attacked by large males, but they moved frequently to areas occupied by other large males and appeared to be successful in spawning. Healey and Prince (1995) concluded this indicated there were at least three distinct male reproductive tactics.

Thus the conclusion for morphological variation among pink salmon appears to be one of phenotypic variation, largely determined by environmental factors. Noltie (1990) suggested that alternative breeding tactics in male pink salmon appear to be linked more to proximate factors (density and body condition) than to divergent life history strategies (Gross 1985). It is known, for example, that there is a heritable component to size in pink salmon, independent of any classification of males (Beacham and Murray 1988a), and there are genetically based differences in body size, morphology and egg size among populations of pink salmon (Beacham et al. 1988). There are differences between males identified as the largest and smallest in a given population, as we have previously reported for somatic allocation by large and small male pink salmon (Ferguson et al. 1990).

We do not wish to imply anything about the genetic basis for other examples of sympatric morphs in salmonids—and hence any general conclusion as to ESUs in salmonids. We deliberately chose pink salmon for this study because of a number of unique characteristics, and so the conclusions might not apply to other species. For example, we have reported elsewhere on our studies of the dramatic phenotypic diversity developed within populations of Icelandic Arctic char *Salvelinus alpinus* that invaded freshwater systems following the most recent glacial retreat (Magnusson and Ferguson 1987; Noakes et al. 1989; Skulason et al. 1989a, 1989b; Ferguson et al. 1990). Four sympatric morphs coexist in the landlocked Thingvallavatn, differing substantially in habitat, food, morphology, reproduction, and life history styles.

We have carried out studies on those Arctic char comparable to the studies reported here for pink salmon. Laboratory rearing experiments indicated a significant genetic basis for much of the Arctic char's phenotypic differentiation. Morphological differences appeared soon after hatching and are correlated with significant differences in foraging behavior (Skulason et al. 1993). Differences in coloration, age of first maturation, and growth rate among the morphs are consistent in progeny reared under controlled laboratory conditions (Skulason

1990). Genetic distances calculated from allozyme (Magnusson and Ferguson 1987), mitochondrial DNA (Danzmann et al. 1991), and nuclear DNA polymorphisms (Volpe 1994) among the morphs are relatively small, so reproductive isolation is relatively recent or incomplete (Sigurjonsdottir and Gunnarsson 1989). Our conclusion from our Arctic char studies is that indeed the four sympatric morphs are ESUs, quite possibly incipient species. As we have stated above, we believe the intrapopulational variation among our male pink salmon is an isometric phenomenon not ESUs.

Our overall conclusion from our studies of pink salmon and Arctic char, as perhaps two extreme cases, is that variation in salmonids must be studied on a case-by-case basis. We strongly recommend that any such studies include investigation of both ontogeny and genetics to understand the mechanisms involved.

Acknowledgments

For assistance with collection of fish, care and feeding of fish, and photography of fish we thank Lenore Drahuschak, Jan Den Dulk, Robert Frank, Andrew Ludwig, and Douglas Noltie. Financial support was provided by the Ontario Ministry of Natural Resources (OMNR) through their Renewable Resources Research Program. Fish were collected under a scientific collecting permit from the OMNR. Live fish were held and raised under conditions approved by the Animal Care Committee of the University of Guelph. The advice, encouragement, and assistance of Jim Collins and other staff members of the Simcoe office of the OMNR was very much appreciated. For comments, discussion, and suggestions on earlier versions of the manuscript we thank Roy Danzmann, Mike Healey, Rob McLaughlin and Skuli Skulason. We thank Jennifer Nielsen and George Barlow for the invitation to participate in this symposium.

References

Beacham, T. D., and C. B. Murray. 1985. Variation in length and body depth of pink salmon (*Oncorhynchus gorbuscha*) and chum salmon (*O. keta*) in southern British Columbia. Canadian Journal of Fisheries and Aquatic Sciences 42:312–319.

Beacham, T. D., and C. B. Murray. 1988a. A genetic analysis of body size in pink salmon (*Oncorhynchus gorbuscha*). Genome 30:31–35.

Beacham, T. D., and C. B. Murray. 1988b. Genetic analysis of growth and maturity in pink salmon (*Oncorhynchus gorbuscha*). Genome 30:529–535.

Beacham, T. D., R. E. Withler, C. B. Murray, and L. W. Barker. 1988. Variation in body size, morphology, egg size, and biochemical genetics of pink salmon in British Columbia. Transactions of the American Fisheries Society 117:109–126.

Berg, R. E. 1979. External morphology of the pink salmon, *Oncorhynchus gorbuscha*, introduced into Lake Superior. Journal of the Fisheries Research Board of Canada 36:1283–1287.

Bookstein, F., B. Chernoff, R. Elder, J. Humphries, G. Smith, and R. Strauss. 1985. Morphometrics in evolutionary biology. Academy of Natural Sciences, Special Publication 15, Philadelphia.

Brett, J. R., and T. D. Groves. 1979. Physiological energetics. Pages 279–352 *in* W. S. Hoar, D. J. Randall, and J. R. Brett, editors. Fish physiology, volume 9. Academic Press, New York.

Chebanov, N. A. 1980. Spawning behavior of the pink salmon, *Oncorhynchus gorbuscha*. Journal of Ichthyology 20:64–73.

Danzmann, R. G., M. M. Ferguson, S. Skulason, S. S. Snorrason, and D. L. G. Noakes. 1991. Mitochondrial DNA diversity among four sympatric morphs of Arctic charr, *Salvelinus alpinus* L. from Thingvallavatn, Iceland. Journal of Fish Biology 39:649–659.

Davidson, F. A. 1935. The development of the secondary sexual characters in the pink salmon (*Oncorhynchus gorbuscha*). Journal of Morphology 57:169–183.

Douglas, M. E. 1993. Analysis of sexual dimorphism in an endangered cyprinid fish (*Gila cypha* Miller) using video image technology. Copeia 1993:334–343.

Ferguson, M. M., D. L. G. Noakes, S. Skulason, and S. S. Snorrason. 1990. Life-history styles and somatic allocation in iteroparous arctic charr and semelparous pink salmon. Environmental Biology of Fishes 28:267–272.

Gould, S. J. 1977. Ontogeny and phylogeny. Harvard University Press, Cambridge, Massachusetts.

Grafen, A. 1991. Modelling in behavioural ecology. Pages 5–31 *in* J. R. Krebs and N. B. Davies, editors. Behavioural ecology, an evolutionary approach, 3rd edition. Blackwell, Oxford, UK.

Gross, M. R. 1985. Disruptive selection for alternative life histories in salmon. Nature 313:47–48.

Healey, M. C., and A. Prince. 1995. Scales of variation in life history tactics in Pacific salmon and the conservation of phenotype and genotype. American Fisheries Society Symposium 17:176–184.

Heard, W. R. 1991. Life history of pink salmon (*Oncorhynchus gorbuscha*). Pages 121–230 *in* C. Groot and L. Margolis, editors. Pacific salmon life histories. University of British Columbia Press, Vancouver.

Keenleyside, M. H. A., and H. M. C. Dupuis. 1988. Courtship and spawning competition in pink salmon (*Oncorhynchus gorbuscha*). Canadian Journal of Zoology 66:262–265.

Lovich, J. E., and J. W. Gibbons. 1992. A review of techniques for quantifying sexual size dimorphism. Growth Development and Aging 56:269–281.

Magnusson, K. P., and M. M. Ferguson. 1987. Genetic analysis of four morphs of arctic charr (*Salvelinus alpinus* L.) from Thingvallavatn, Iceland. Environmental Biology of Fishes 20:67–73.

Noakes, D. L. G., S. Skulason, and S. S. Snorrason. 1989. Alternative life-history styles in salmonine fishes with emphasis on arctic charr, *Salvelinus alpinus*. Pages 329–346 *in* M. N. Bruton, editor. Alternative life-history styles of animals. Kluwer Academic, Dordrecht, Netherlands.

Noltie, D. B. 1988. The breeding ecology of pink salmon (*Oncorhynchus gorbuscha* Walbaum) from the Carp River, eastern Lake Superior. Doctoral dissertation. University of Western Ontario, London, Ontario.

Noltie, D. B. 1990. Intrapopulation variation in the breeding of male pink salmon (*Oncorhynchus gorbuscha*) from a Lake Superior tributary. Canadian Journal of Fisheries and Aquatic Sciences 47:174–179.

Ricker, W. E. 1972. Heredity and environmental factors affecting certain salmonid populations. Pages 27–160 *in* R. C. Simon, editor. The stock concept in Pacific salmon. H. R. MacMillan Lectures in Fisheries, University of British Columbia, Vancouver.

SAS Institute. 1985. SAS/STAT user's guide, version 6 edition. SAS Institute, Cary, North Carolina.

Sigurjonsdottir, H., and K. Gunnarsson. 1989. Alternative mating tactics of arctic charr, *Salvelinus alpinus*, in Thingvallavatn, Iceland. Environmental Biology of Fishes 26:159–176.

Skulason, S. 1990. Variation in morphology, life history and behaviour among sympatric morphs of arctic charr: an experimental approach. Doctoral dissertation. University of Guelph, Guelph, Ontario.

Skulason, S., D. L. G. Noakes, and S. S. Snorrason. 1989a. Ontogeny of trophic morphology in four sympatric morphs of arctic charr *Salvelinus alpinus* in Thingvallavatn, Iceland. Biological Journal of the Linnean Society 38:281–301.

Skulason, S., S. S. Snorrason, D. L. G. Noakes, M. M. Ferguson, and H. J. Malmquist. 1989b. Segregation in spawning and early life history among polymorphic arctic charr, *Salvelinus alpinus*, in Thingvallavatn, Iceland. Journal of Fish Biology 35:225–232.

Skulason, S., S. S. Snorrason, D. Ota, and D. L. G. Noakes. 1993. Genetically based differences in foraging behaviour among sympatric morphs of arctic charr (Pisces: Salmonidae). Animal Behaviour 45:1179–1192.

Soin, S. G. 1954. Pattern of development in summer chum, masu and pink salmon. Tr. Soveshch. Ikhtiol. Kom. Akad. Nauk SSSR 4:144–155 (In Russian.) English translation, 1961: Israel Program for Scientific Translations, Jerusalem.

Strauss, R. M., and F. L. Bookstein. 1982. The truss: body form reconstruction in morphometrics. Systematic Zoology 31:113–135.

Volpe, J. P. 1994. A molecular genetic examination of the polymorphic arctic charr, *Salvelinus alpinus*, of Thingvallavatn, Iceland. Master's thesis. University of Guelph, Guelph, Ontario.

Wong, M. A., and C. Schaack. 1982. Using the *k*th nearest neighbour clustering procedure to determine the number of subpopulations. Proceedings of the Statistical Computing Section, American Statistical Association [1982]:40–48.

American Fisheries Society Symposium 17:195–216, 1995

Life History Variation and Population Structure in Sockeye Salmon

CHRIS C. WOOD

Department of Fisheries and Oceans, Pacific Biological Station
Nanaimo, British Columbia V9T 5K6, Canada

Abstract.—Biochemical and molecular genetic surveys reveal that genetic diversity in sockeye salmon *Oncorhynchus nerka* is extensively subdivided among major geographic regions and among different lake systems. Differences among samples from northwestern, coastal Canadian, and southeastern parts of the species' range are attributed to prolonged isolation of three colonizing races that survived the last Pleistocene glaciation in different refuges. Hierarchical analyses of gene diversity within river systems indicate that the nursery lake is an appropriate and convenient unit for defining distinct sockeye salmon populations although additional subpopulation structure has been recognized within many lakes. Corresponding differences in the occurrence of life history and behavioral phenotypes that are known to be heritable and adaptive strongly suggest that reproductively isolated populations are typically adapted to local conditions and are in some sense evolutionarily significant. Furthermore, there is direct evidence that anadromous behavior is adapted to local conditions in that attempts to establish self-sustaining anadromous runs have failed with few exceptions, yet the nonanadromous form (kokanee) has been transplanted successfully to many new locations both within and beyond the species' natural range. These observations imply that genetic diversity within the species and the productivity of sockeye salmon as a resource for human exploitation are best protected by conserving as many locally adapted populations as possible. However, the history of recolonization following cycles of glaciation suggests populations adapted to local conditions in specific lakes are often evolutionary dead ends and that recolonization following glaciation events may depend on sea-type or river-type sockeye salmon that are common in glacially influenced habitats today.

Conservation biologists generally agree that rational management of wild fish populations should prevent the loss of genetic diversity. This goal implies several objectives: the preservation of genetic variation, the maintenance of subpopulation structure, and the avoidance of artificial selection and hybridization (Nelson and Soule 1987). Thus, the greatest challenge in preserving the genetic diversity of salmonid fishes is to identify and protect nontaxa (Behnke 1993). Nehlsen et al. (1991) identified 214 salmonid stocks of concern on the Pacific coast of North America, 159 of which faced a moderate to high risk of extinction. In southwestern British Columbia, one-third of spawning runs known since the early 1950s have now been lost or have decreased to such low numbers that spawners are not consistently monitored (Riddell 1993). These trends in number and magnitude of spawning runs imply a loss of genetic diversity, through the loss of both locally adapted subpopulations and genetic variation due to low effective population sizes.

The National Marine Fisheries Service has developed the policy that a population or group of populations warrant special protection under the U.S. Endangered Species Act (16 U.S.C. §§ 1531 to 1544) if they represent an evolutionarily significant unit (ESU; Utter 1981; Waples 1991). To be an ESU, a population must be reproductively isolated and contribute substantially to the ecological and genetic diversity of the species in the sense that it is both a distinct product of past evolutionary events and a unique reservoir for adaptation and evolutionary change in the future (Waples 1995, this volume). Such decisions about evolutionary significance require information about population structure (gene flow) and the extent of unique local adaptation (e.g., Utter et al. 1993).

My objective here is to summarize recent information on population structure and to reexamine evidence for local adaptation in sockeye salmon *Oncorhynchus nerka* in the context of the ESU policy. Although it is now widely accepted that salmonids occur as locally adapted populations, much of the evidence supporting this view is circumstantial or based on plausible assumptions about natural selection and reproductive isolation (e.g., Ricker 1972; papers in Billingsley 1981; Taylor 1991). I focus on studies that have investigated both the extent of gene flow among different spawning runs and corresponding phenotypic differences in heritable and adaptive traits affecting life history and spawning behavior. These results help to substantiate previous inferences based on more circumstantial evidence about the potential for finely subdivided population structure in sockeye salmon and other salmonids with comparable homing abilities and diversity of life history types.

Sockeye salmon are semelparous and occur throughout the Pacific Ocean between 41 and 61° north latitude. Unlike other Pacific salmon *Oncorhynchus* spp., sockeye salmon typically spawn in tributaries that provide access to lake habitat for juvenile rearing. Even so, the species exhibits remarkable variation in life history and adaptation to a wide variety of spawning or juvenile rearing habitats (Burgner 1991). Because of the commercial importance of sockeye salmon, its biology has been researched intensively for many years. In particular, recent surveys of genetic variation throughout the species' range have provided new insights about colonization patterns following the last glaciation, the extent of reproductive isolation among spawning locations, and the feasibility of developing stock identification techniques to manage individual populations in the face of current mixed-stock harvesting policies.

Data Sources and Methods

I have relied primarily on biochemical genetic data recently summarized by Varnavskaya et al. (1994a, 1994b, which includes previously published data from Utter et al. 1984), Winans et al. (in press), and Wood et al. (1994) and recent molecular genetic data found in Beacham et al. (1995), Bickham et al. (1995), and Taylor et al. (in press). In several cases I recomputed statistics to facilitate comparisons. Data on the occurrence of transplanted, self-perpetuating kokanee (lacustrine sockeye salmon) populations were obtained both from publications and personal communications from investigators in various agencies (see Appendix).

Principal components for Figure 1 were calculated separately for nuclear gene frequencies (both protein-coding and variable-number tandem repeat [VNTR] minisatellite loci) and for mitochondrial DNA (mtDNA) haplotype frequencies by means of the CLUSTER program of SYSTAT version 5.03 (SYSTAT, Evanston, Illinois). Frequencies of alleles at the five polymorphic protein-coding loci (*ALAT**, which codes for alanine aminotransferase, enzyme number 2.6.1.2 [IUBNC 1984]; *GPI-B1,2**, glucose-6-phosphate isomerase, 5.3.1.9; *LDH-B2**, L-lactate dehydrogenase, 1.1.1.27; *PGM-1**, phosphoglucomutase, 5.4.2.2; and *PGM-2**) available for all samples (Varnavskaya et al. 1994b; Wood et al. 1994), of mtDNA cytochrome-*b* haplotypes (Bick-

ham et al. 1995), and of alleles revealed by single-locus VNTR probes (Beacham et al. 1995) were arcsine-transformed to stabilize variances prior to analyzing principal components. The average number of bands per bin revealed by the multilocus VNTR probe OtPBS1 (Beacham et al. 1995) were not transformed. Because VNTR data were unavailable for Pierre Creek, all data for Pierre Creek and Fulton River (separate sites within Babine Lake in the Skeena River) were pooled to permit comparison with other samples.

Phylogenies were reconstructed with unrooted neighbor-joining trees (Saitou and Nei 1987) based on Cavalli-Sforza and Edwards' (1967) chord distances by means of PHYLIP version 3.5p[1]. A neighbor-joining tree was considered most appropriate for reconstructing a sockeye salmon phylogeny because of its superior performance for cases in which lineages evolve at different rates and divergence is not a continuous process (Kim et al. 1993). Pairwise chord distances were calculated from allozyme allele frequencies at six polymorphic loci (those already mentioned plus *MDH-B1,2**, malate dehydrogenase, 1.1.1.37) available for sockeye salmon samples from 96 different lakes and major spawning sites not associated with lakes throughout the species' range. Sites are identified by number and letter in Figure 2; the letter denotes the reference source and the number the sample number within the reference. Pairwise chord distances between samples from the Stikine River were computed from allozyme allele frequencies at the 30 nonselected loci reported by Wood et al. (1994).

The software BIOSYS-1 (Swofford and Selander 1981) was used to estimate F_{ST} (Wright 1951) and to perform hierarchical analyses of gene diversity (Chakraborty 1980) based on allozyme allele frequencies at polymorphic loci for which the frequency of the common allele was less than 95% in at least one sample under consideration. The geographical hierarchy analyzed included 4 political regions (Russia, Alaska, Canada, and Washington), 29 river systems, including comparable areas of coastline (coastal Washington, Vancouver Island, Queen Charlotte Islands, south coastal British Columbia, north coastal British Columbia, southeast Alaska, and the Alaska Peninsula), and 65 lakes, including one spawning site (6B) not associated with a lake. Because our version of the BIOSYS program could not handle more than 65 taxonomic

---

[1]Felsenstein, J. 1993. PHYLIP (Phylogeny inference package) Version 3.5p. Department of Genetics, SK-50 University of Washington, Seattle, 98195, USA.

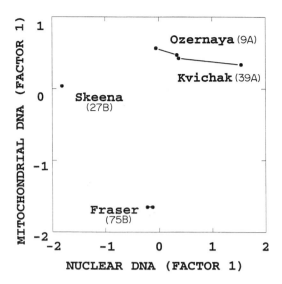

FIGURE 1.—Principal components of genetic variation revealed by biochemical and molecular genetic analysis of sockeye salmon from northwestern (Ozernaya and Kvichak rivers), coastal Canadian (Skeena River), and southeastern (Fraser River) parts of the species' range. All rivers except Skeena are represented by individual spawning sites to illustrate relative divergence. Numbers identify corresponding samples in Figure 2 (9A, Kuril Lake; 39A, Iliamna Lake; 27B, Babine Lake; 75B, Shuswap Lake).

units for this analysis, only samples marked with an asterisk in Figure 2 were included. To provide representative coverage across regions and postulated glacial refuges (see below), all samples from Russia, southeast Alaska, the Queen Charlotte Islands, Vancouver Island, the Fraser River, and Washington State were included; the three smallest samples (43B, 47B, and 55B) from coastal British Columbia and all except the two largest samples from each river system in northern Canada and western Alaska were excluded. Samples from different spawning sites within a single lake system were pooled to represent the lake, designated by the lowest numbered sample in the series (e.g., 71B represents the pooled data for sites 71, 72, 73, and 74 in Wood et al. 1994). Although Okanagan River sockeye salmon spawn primarily in Canada, the nursery lake straddles the Canada–U.S. border, and this sample (77B) was included in the Washington region together with the other Columbia River sample.

Gene flow was estimated as the average number of migrants (Nm) exchanged between sites per generation based on the equilibrium relationship $F_{ST} = 1/(1 + 4 \, Nm \cdot a)$ where F_{ST} was estimated from allele frequencies at polymorphic loci. The variable a is given by $a = [n/(n - 1)]^2$, derived from Wright's (1951) infinite-neutral-alleles island model of gene flow modified to account for the number of samples, n (see Slatkin and Barton 1989).

Geographical Patterns in Genetic Diversity

Surveys of genetic variation at protein-coding loci, mtDNA, and nuclear VNTR loci indicate that genetic diversity of sockeye salmon is highly subdivided among populations and regions throughout the species' range. Relatively large genetic differences exist among the largest sockeye salmon stocks in northwestern, coastal Canadian, and southeastern parts of the species' range. This is illustrated in Figure 1 where the first principal component for mtDNA haplotype frequencies at cytochrome-b is plotted against the first principal component for the combined nuclear DNA data (allele frequencies or number of bands for protein-coding loci and VNTR loci). Samples from individual spawning sites within Kuril Lake (Ozernaya River, Russia), Iliamna Lake (Kvichak River, Alaska) and Shuswap Lake (Fraser River, British Columbia) are represented separately to illustrate the relative divergence between spawning sites and among lakes and regions. Unfortunately, it was necessary to pool samples from Babine Lake (Skeena River, British Columbia) for this analysis (see Methods).

The unrooted neighbor-joining tree (Figure 2) based on the more extensive surveys of variation at six polymorphic protein-coding loci illustrates a similar pattern of subdivision of genetic diversity, with samples from the same region generally connected together in one or more clusters. However, the regional structuring is imperfect, and genetic differences among samples from neighboring lakes are often relatively large (long horizontal branches). Hierarchical gene-diversity analysis of the data represented in Figure 2 indicates that 17.1% of the total genetic variation is associated with geographical subdivision—1.9% with differences among political regions (scheme 1 in Table 1), 8.2% with differences among river systems (within regions), and 7.0% with differences among lakes (within river systems). The remaining 82.9% is associated with variation among individuals within lakes. Comparatively little variation (<1%) is associated with samples from different spawning sites within the same lake system (but this variation can be highly significant) or from the same sites in different years (Varnavskaya et al. 1994b; Wood et al. 1994).

The regional clustering in Figures 1 and 2 is

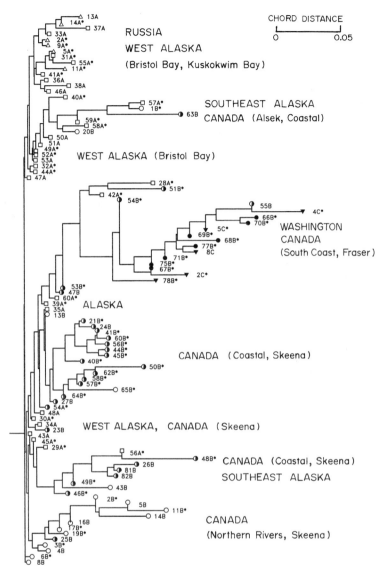

FIGURE 2.—Unrooted phylogenetic tree of 96 sockeye salmon populations based on the neighbor-joining method (Saitou and Nei 1987) and Cavalli-Sforza and Edwards' (1967) chord distances computed from allozyme allele frequencies at six polymorphic loci. Length of horizontal branches indicates extent of differentiation from a postulated common ancestor (note scale). Numbers correspond to sample numbers in references denoted by letters (A, Varnavskaya et al. 1994a; B, Wood et al. 1994; and C, Winans et al., in press). Symbol shape denotes political region (△ Russia, □ Alaska, ○ Canada, ▽ Washington) and symbol fill denotes postulated colonizing race (open, Beringian; half-open, coastal Canadian; solid, Cascadian). Asterisks indicate samples included in the hierarchical gene-diversity analysis (Table 1).

almost certainly associated with colonization events following the last glaciation. At the peak of the Wisconsin glaciation 15,000 years ago, ice sheets covered all of mainland British Columbia, much of Washington State and Alaska, and parts of Kam-

chatka; the coastline was very different because of reduced sea level (Figure 3). Evidence from geological studies and the distribution of freshwater fish assemblages (McPhail and Lindsey 1970, 1986; Lindsey and McPhail 1986) and genetic data (Var-

TABLE 1.—Hierarchical gene-diversity analysis of sockeye salmon samples from 65 lakes (or sea- and river-type sockeye salmon spawning sites) throughout the species' range. Lakes included are marked by asterisks in Figure 2 (see "Data Sources and Methods" for details). Relative gene diversity among rivers or areas within regions (or races) can be found by subtracting the sum of values in columns A, B, and C (or D for races) from 1. Average G_{ST} is across six loci.

Locus and statistic	Absolute gene diversity		Relative gene diversity (G_{ST})			
	Total	Within lakes	(A) Within lakes	(B) Among lakes within rivers or areas	(C) Among political regions (scheme 1)	(D) Among colonizing races (scheme 2)
ALAT*	0.6120	0.5453	0.893	0.059	−0.007	0.019
PGM-1*	0.4223	0.2964	0.671	0.098	0.049	0.173
PGM-2*	0.3155	0.2848	0.900	0.056	0.039	0.003
LDH-B2*	0.1418	0.1201	0.840	0.062	0.006	0.058
GPI-B1,2*	0.0194	0.0176	0.907	0.095	−0.001	0.008
MDH-B1,2*	0.0058	0.0057	0.988	0.011	0.002	0.001
Average G_{ST}			0.829	0.070	0.019	0.062

navskaya et al. 1994a; Wood et al. 1994; Taylor et al., in press) strongly suggests that modern sockeye salmon populations are derived primarily from a northern race which survived glaciation in the Bering Sea area (Beringia) and a southern race which survived south of the cordilleran ice sheet in the Columbia River (Cascadia). There is also evidence to suggest the existence of other small refuges in the coastal islands of British Columbia (Clague et al. 1980; Pojar 1980; Warner et al. 1982; Wood et al. 1994) and in ice-free, landlocked lakes in Kamchatka (Braitseva et al. 1968; Varnavskaya et al. 1994a) and perhaps on Kodiak Island (Karlstrom 1969). When political regions are replaced by three postulated colonizing races (Beringian, Cascadian, and coastal British Columbian) as the first hierarchy in the gene-diversity analysis (scheme 2 in Table 1), race accounts for 6.2% of total genetic variation within the species, over one-third of that attributed to geographic subdivision. Thus modern populations founded by different races are genetically distinct owing to prolonged reproductive isolation during the last glaciation.

Surprisingly, even more genetic variation (7.0%) is associated with differences among sockeye salmon using different nursery lakes within the same river system presumably colonized by the same race. Gene frequency differences among samples from different lakes are almost always statistically significant. If these genes are neutral to selection so that frequency differences have arisen primarily by genetic drift or founder effects (as is usually assumed, but see Altukhov and Salmenkova 1991; Verspoor et al. 1991; Varnavskaya et al. 1994a), it follows that sockeye salmon in different lakes are typically reproductively isolated populations. Averaged over all regions, the number of

effective migrants (Nm) exchanged among lakes within the same river system or local area is estimated to be only 3.2 individuals per generation based on the equilibrium relationship between F_{ST} and Nm. Such restricted gene flow among lakes underscores the fidelity with which sockeye salmon home to their natal lake system. These results confirm previous inferences based on tagging studies (Foerster 1936, 1968) and estimates of gene flow and straying among lakes (Quinn et al. 1987) that the nursery lake is the primary geographic unit of population structure in sockeye salmon.

Indirect Evidence for Local Adaptation

Reproductive isolation will lead to genetic divergence of populations even in the absence of natural selection because of random genetic drift. For this reason surveys of selectively neutral genetic characters are useful for estimating gene flow among populations, but they provide no direct evidence of divergence in evolutionarily significant traits (Clayton 1981). However, local adaptation is implied indirectly in cases in which populations exhibiting different adaptive and heritable phenotypes are known to be reproductively isolated. Many researchers have suggested that any phenotypic variation exhibited by salmonids represents local adaptation if the phenotypes have a genetic basis and influence fitness and a plausible mechanism for the natural selection of the phenotype has been identified (Taylor 1991). For example, a genetic difference in susceptibility to infection (and subsequent mortality) by the haemoflagellate *Cryptobia salmositica* (Bower and Margolis 1984) strongly suggests local adaptation within the two reproductively isolated sockeye salmon populations studied (one

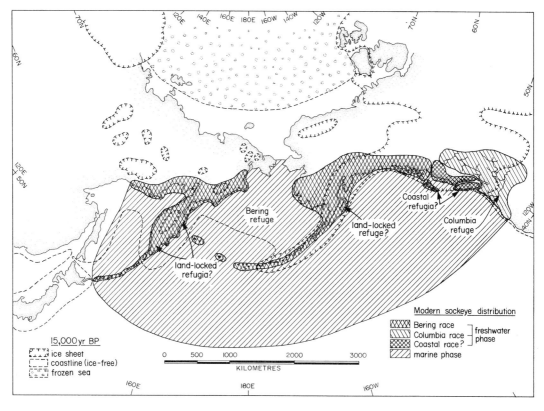

FIGURE 3.—Modern distribution of sockeye salmon (adapted from Burgner 1991) superimposed on postulated location of ice sheets, coastlines, and refuges at the height of the Wisconsin glaciation (15,000 years before present) (adapted from Pielou 1994).

from the Skeena River and one from the Fraser River). In the context of defining ESUs, I examine indirect evidence for local adaptation by restricting attention to studies that address the extent of reproductive isolation and differentiation of sockeye salmon populations with respect to heritable and adaptive traits affecting life history and spawning behavior.

In the typical life history (reviewed by Burgner 1991), sockeye salmon spawn in an inlet tributary to a lake. Upon emergence the fry migrate downstream to the lake and adopt a planktivorous existence involving diel vertical migrations in the limnetic zone. The juveniles migrate downstream to the ocean as smolts in the spring of their second or third year (age 1 or 2). After spending three or four summers at sea, sockeye salmon return to their natal stream to spawn and die. However, sockeye salmon exhibit remarkable variation in life history associated with differences in spawning and freshwater rearing behavior.

Sea-Type and River-Type Forms

Considerable variation occurs in the type of habitat used by underyearlings and in the duration of freshwater residence (Table 2). For example, in the Stikine River (Figure 4A), only Tahltan Lake provides a "typical" clear, productive lacustrine environment for juvenile sockeye salmon; the only other lakes accessible to anadromous salmon, Chutine and Christina, are both glacially turbid. About half of the adult sockeye salmon returning to the Stikine River spawn in tributaries without access to lakes, and their progeny rear in slackwater side channels of the lower Stikine and Iskut rivers for several months (sea-type sockeye salmon) or 1 to 2 years (river-type sockeye salmon) before migrating to sea (Craig 1985; Wood et al. 1987a). Sea-type or river-type life histories occur only rarely in southern rivers (Gilbert 1918; Birtwell et al. 1987) but are common and occasionally predominant in many northern river systems including the Nass (Ricker

TABLE 2.—Occurrence of different juvenile life histories of sockeye salmon.

Rearing habitat	Age (year) at seaward migration		Nonanadromous (kokanee or residual)
	0 (sea type)	1 or 2	
Lake	rare	typical (lake type)	common
River	common	common (river type)	not observed

1966; Rutherford et al. 1994), Taku (Murphy et al. 1991; Eiler et al. 1992), East Alsek (Pahlke and Riffe 1988), Copper (Sharr et al. 1988), Nushagak and Mulchatna (Russell et al. 1989), Kamchatka (Bugaev 1984), Paratunka (Krogius 1958), Bolshaya (Semko 1954), and other Alaskan rivers listed by McPherson and McGregor (1986).

Seawater adaptability in underyearling sockeye salmon is typically very limited although it is influenced by photoperiod (Clarke et al. 1978) and is positively correlated with fry size (Heifetz et al. 1989; Foote et al. 1992). Rice et al. (1994) reported that underyearling sea-type sockeye salmon from the East Alsek River exhibited superior seawater adaptability (at the same size) and faster growth than did river-type and lake-type fry from other river systems when reared under controlled conditions. In similar seawater challenge experiments to compare the seawater adaptability of Stikine River sockeye salmon from lake-type (Tahltan Lake) and sea- and river-type (Scud River) populations, the pure and hybrid progeny of Scud River females were significantly more likely to survive exposure to 33 parts per thousand seawater as underyearlings than were the pure and hybrid progeny of Tahltan Lake females (C. C. Wood and C. J. Foote, Department of Fisheries and Oceans, unpublished data). In this case, the superior seawater adaptability of sea-type sockeye salmon resulted from their larger mean size at age—egg size was significantly larger in Scud River females (mean, 54.1-mm fresh diameter) than in Tahltan Lake females (mean, 49.4 mm; $P < 0.001$) and this conferred a size advantage to both the pure and hybrid progeny of Scud River females. Poor survival of fish to maturity prevented us from investigating the heritability of egg size in sockeye salmon as originally intended, but egg size is known to be heritable in rainbow trout O. mykiss (Bromage et al. 1990). Sea-type sockeye salmon in the Harrison Rapids (Fraser River) also have larger eggs than other neighboring lake-type populations (Beacham and Murray 1993), yet large egg size does

not appear to be characteristic of the East Alsek River sea-type population (Rice et al. 1994). Collectively, these findings indicate that sea-type sockeye salmon exhibit heritable physiological or morphological adaptations for seaward migration as underyearlings but that these adaptations differ among populations.

Surveys of genetic variation at 30 protein-coding loci indicate that Tahltan Lake and Scud River are genetically distinct, reproductively isolated populations ($P < 0.001$, G-test; $F_{ST} = 0.079$ averaged over the six most polymorphic loci, implying $Nm = 0.73$; data from Wood et al. 1994). Within the Stikine River, statistically significant differences in allele frequencies exist among all lake-type populations ($P < 0.001$, G-test, six polymorphic loci) but not between two of the sea–river sockeye salmon samples (Scud River and Chutine River; $P > 0.70$). Small but significant differences between the Iskut River site and the other sea–river sites were detected in part because sample sizes from the Iskut River were several times larger. Overall, much less differentiation was evident among sea–river-type sockeye salmon (note horizontal branch lengths in Figure 4B) even though distances between some sea–river-type spawning sites exceeded 180 km, farther than between lakes (Wood et al. 1987b). This pattern is corroborated by mtDNA haplotypes: the same mtDNA haplotype (in cytochrome b and ND1 genes) was dominant in both sea–river-type spawning sites sampled (Chutine River and Scud River) but found in only one (Chutine Lake) of the two lakes sampled and the river sites shared another haplotype not found in Chutine or Tahltan lakes (Wood, Department of Fisheries and Oceans, J. W. Bickham, Texas A&M University, and J. C. Patton, LGL Ecological Genetics, Inc., unpublished data). The occurrence of heritable adaptations for alternate juvenile life histories in these reproductively isolated populations strongly suggests genetic distinctiveness of the sort implied in the definition of an ESU.

Nonanadromous Forms

At the other extreme, some lake-rearing sockeye salmon are nonanadromous. The adaptive value of nonanadromy appears to depend primarily on two factors affecting the relative fitness of anadromous and nonanadromous forms: lake productivity, which typically limits the food supply (and hence reproductive potential) of the nonanadromous form, and the difficulty (fitness cost) of anadromous migration (Figure 5). Anadromy should be advantageous in

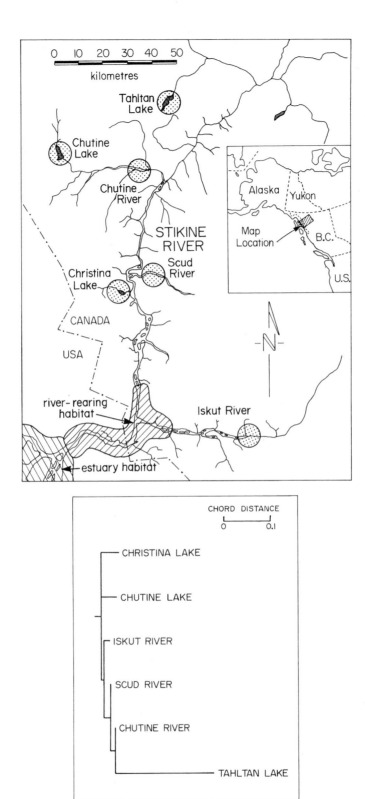

FIGURE 4.—Top: Map of the Stikine River showing the location of principal sockeye salmon spawning sites (circles) and rearing areas for sea-type and river-type forms (hatched). Bottom: Unrooted phylogenetic tree of sockeye salmon in the Stikine River based on the neighbor-joining method (Saitou and Nei 1987) and Cavalli-Sforza and Edwards' (1967) chord distances computed from allele frequencies at 30 protein-coding loci (Wood et al. 1994).

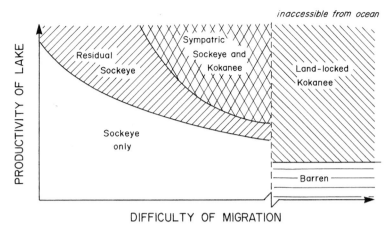

FIGURE 5.—Schematic illustration of the influence of lake productivity and difficulty of anadromous migration on the occurrence of anadromous and nonanadromous life histories in sockeye salmon.

situations in which food is much less abundant in freshwater than in the ocean and migration is neither hazardous nor energetically costly. For example, nonanadromous sockeye salmon rarely (if ever) occur in easily accessible, ultraoligotrophic coastal lakes. However, nonanadromous progeny of anadromous females sometimes occur in more productive lakes with easy access from the ocean, such as Cultus Lake (Fraser River; Ricker 1938), Cheewhat Lake (Vancouver Island; Wood, personal observation), Sopochnaya Lake (Kuril Islands; Ivankov 1984), and Lake Dal'nee (Kamchatka; Krokhin 1967); these are called "residuals" and are usually male (Ricker 1938). The self-perpetuating, nonanadromous form known as kokanee occurs commonly in landlocked, balanced sex-ratio populations in lakes throughout the species' range where access from the ocean has become difficult or impossible, typically following the isostatic rebound of land masses (Nelson 1968). Anadromous sockeye salmon and kokanee often occur together in the same lakes where access from the ocean is possible and the lake is sufficiently productive as in many interior lakes of the Columbia, Fraser, and Skeena rivers (Nelson 1968; Foote et al. 1989).

Sockeye salmon and kokanee exhibit differences in a number of heritable traits that appear adaptive for their respective anadromous and nonanadromous life histories. Kokanee typically have more gill rakers than do sockeye salmon (Nelson 1968), and this trait is known to be inherited (Leary et al. 1985) and associated with ecological differentiation in salmonids (Fenderson 1964; Bodaly 1979; Lindsey 1981). Inherited differences in the rate of embry-

onic development in sympatric sockeye salmon and kokanee from Shuswap Lake are adaptive in that the different rates compensate for differences in mean egg size between the forms. In hybrids, egg size is mismatched with development rate so that yolk reserves are depleted significantly earlier (kokanee eggs) or later (sockeye salmon eggs) than in the pure forms (Wood and Foote 1990). Underyearling sockeye salmon exhibit superior swimming ability (Taylor and Foote 1991), grow faster on average, and exhibit less variation in growth rate than do kokanee reared under identical conditions (Wood and Foote 1990, in press). Large size and superior swimming ability would be advantageous in sockeye salmon smolts undertaking seaward migration (Henderson and Cass 1991). Both sockeye salmon and kokanee exhibit heritable circannual cycles in seawater adaptability (Foote et al. 1992, 1994), but the cycle of seawater adaptability in kokanee from Shuswap Lake is not synchronous with that in sockeye salmon from the same lake, either in the wild or when reared under identical conditions (Foote et al. 1992). Under natural conditions kokanee typically mature at the same age but at a smaller size than do sockeye salmon (McCart 1970; Wood and Foote, in press); when the progeny of age-4 parents are reared under identical conditions, mean age of maturity is lowest in kokanee, intermediate in hybrids, and greatest in sockeye salmon (Shuswap Lake, Foote et al. 1992; Takla Lake, Wood and Foote, in press; Babine Lake, Wood and Foote, Department of Fisheries and Oceans, unpublished data). Early maturation appears to be triggered in fish that exceed an inherited size threshold at a critical stage of development (Hankin et al. 1993). Kokanee exhibit a lower minimum size threshold than do sockeye salmon, and consequently a higher proportion mature early (Wood and Foote, in press). This is likely adaptive because kokanee cannot attain a large size feeding on the zooplankton available in Shuswap and Takla lakes. A similar threshold response in anadromous sockeye salmon would be maladaptive (at least in females) in that early maturing fish would pay all the costs of anadromy without fully exploiting the opportunity to increase their size and fecundity while at sea. In summary, sockeye salmon and kokanee exhibit a suite of heritable differences in morphology, early development, seawater adaptability, growth, and maturation that appear to be divergent adaptations arising from different selective regimes associated with anadromous versus nonanadromous life histories.

Not surprisingly, sockeye salmon and kokanee inhabiting different lakes appear to be genetically distinct, reproductively isolated populations (see data presented by Foote et al. 1989; Taylor et al., in press). It is remarkable, however, that the two forms can be reproductively isolated where they occur together, spawning in the same tributaries at about the same time (Foote et al. 1989; Wood and Foote, in press). The average number of effective migrants (*Nm*) exchanged between populations of sockeye salmon and kokanee spawning in the same tributaries to Takla Lake has been estimated at less than two individuals per generation compared with about 35–56 individuals per generation among sockeye spawning in different tributaries to Takla Lake (Wood and Foote, in press). Clearly, sockeye salmon and kokanee can exist as ecologically distinct, reproductively isolated populations and warrant consideration as separate ESUs. In fact, sockeye salmon and kokanee in Takla Lake ought to be considered as separate biological species were it not for the fact that similar divergence appears to have occurred between sympatric sockeye salmon–kokanee pairs in numerous different lake systems, suggesting parallel evolution and sympatric speciation (Nelson 1968; Foote et al. 1989; Taylor et al., in press; Wood and Foote, in press).

Variation in Spawning Behavior

Anadromous sockeye salmon within individual lake systems also exhibit considerable variation in spawning behavior (Figure 6; Burgner 1991). Sockeye salmon typically use spawning habitat in inlet or

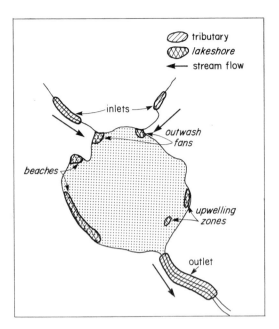

FIGURE 6.—Schematic illustration of variations in spawning habitat used by sockeye salmon.

outlet tributaries to the nursery lake, but they may also spawn within the lake itself, usually in the outwash fans of tributaries or in upwelling springs where inflowing water percolates through the redds (e.g., in the Wood River system, Rogers 1987; Tustumena Lake, Burger et al. 1995). In some lakes sockeye salmon spawn along beaches where the gravel or rocky substrate is free of fine sediments and where lake water oxygenates the eggs by diffusion or wave-driven circulation (e.g., in Iliamna Lake, Kerns and Donaldson 1968).

Variations in spawning behavior often require corresponding adaptations in fry behavior or embryonic development. For example, the rheotactic response of sockeye salmon fry reared under identical conditions was negative in progeny of sockeye salmon from inlet tributaries but positive in progeny of sockeye salmon from the outlet tributary of Karluk Lake, Alaska (Raleigh 1967). Similarly, a heritable rheotactic response in newly emerged fry has been demonstrated for sockeye salmon in Chilko and Fraser lakes (Brannon 1967, 1972). Presumably rheotaxis in sockeye salmon fry is an adaptation that enables sockeye salmon from outlet tributaries to migrate upstream and those from inlet tributaries to move downstream to reach the nursery lake. Because lakes cool more slowly than do small streams in the autumn, eggs deposited in lake water (or in the outlet tributary) will experience a much warmer thermal regime than will eggs deposited in small inlet tributaries. Thus, corresponding adaptations in egg size, development rate, and, more commonly, in spawning time, are required to synchronize fry emergence with the onset of forage production in the spring (Goodlad et al. 1974; Godin 1982; Brannon 1987). Presumably this is why inlet tributary spawners typically spawn earlier than do lakeshore or outlet tributary spawners (Burgner 1991). Selective regimes in the tributary and lakeshore incubation habitats are likely to differ in other respects too. Redds of sockeye salmon spawning in lakeshore habitat without upwelling may be adversely affected by reductions in water level and scouring by ice over winter if they are situated too shallow (Blair et al. 1993) and by low dissolved oxygen concentrations if they are too deep (Wood, unpublished data for Babine Lake). Lake-spawning males differ morphologically from river-spawning males, especially in their larger hump size; this hump-size variation has been attributed to tradeoffs between sexual and natural selection in the two spawning habitats (Blair et al. 1993; Quinn and Foote 1994).

Where genetic differentiation of sockeye salmon subpopulations within lakes occurs, it is typically associated with variations in spawning behavior (Wilmot and Burger 1985; Varnavskaya et al. 1994b) or with other obvious specialization in life history (e.g., sockeye salmon versus kokanee; Foote et al. 1989; Winans et al., in press). Estimates of gene flow (Nm) in Table 3 indicate that on average less than 10 migrants are exchanged per generation between lake and tributary subpopulations or between inlet and outlet tributary subpopulations in most lakes and in most years studied (except Dvu-Yurta Lake, where no differentiation was detected at the two protein-coding loci examined). Genetic differentiation is also evident between subpopulations with early and late run timing in the Russian River (1.1–5.1 effective migrants), to a lesser extent in Kuril Lake (6.9–31.2 effective migrants), but not in Babine Lake (>31.2 migrants). Overall, these estimates of gene flow imply very low effective straying rates among subpopulations (<1% in most cases), given that census populations in these spawning sites are typically measured in thousands of individuals. Moreover, these patterns of genetic differentiation confirm that locally adapted, partially isolated subpopulations of anadromous sockeye salmon can coexist within a surprisingly small spatial scale.

Direct Evidence for Local Adaptation

The most direct evidence for local adaptation in reproductively isolated sockeye salmon populations is provided by examining the record of attempts to transplant sockeye salmon from one location to another (Ricker 1972; MacLean and Evans 1981; Quinn 1985). Transplant activities have succeeded in establishing a number of self-perpetuating runs of nonanadromous sockeye salmon within and beyond the species' natural range (Figure 7). Even so, some of these self-perpetuating runs, such as those in Ontario, persisted for less than 10 generations (E. Crossman, Royal Ontario Museum, personal communication).

Despite numerous attempts and considerable effort to transplant anadromous sockeye salmon (see Figure 8), self-perpetuating anadromous runs have been established at only three sites (Figure 7), all within the species' natural range. Two of the sites became accessible to anadromous sockeye salmon after artificial modification of the habitat (Lake Washington, Royal and Seymour 1940; Frazer Lake, Blackett 1979); the other site (upper Adams River) had previously supported a large anadromous sockeye salmon population that disappeared following a rock slide and rock dumping in the Fraser River Canyon in 1913–1914 (Williams 1987). In each case, the successful transplants involved donor populations only 15, 20, and 90 km distant (by air) from the transplant site and in many ways resembled a natural colonization of adjacent habitat. In several other cases, transplanted sockeye salmon persisted only as a nonanadromous run (New Zealand, Scott 1984; Hokkaido, Kaeriyama et al. 1992; Walkus Lake, British Columbia, Aro 1979). The failure to transplant anadromous sockeye salmon despite considerable success with nonanadromous kokanee provides direct evidence that the anadromous migratory behavior of sockeye salmon is a local adaptation within the donor population. A number of studies on sockeye salmon (Ricker 1972; Quinn 1985) and other Pacific salmon (Bams 1976; Reisenbichler 1988) suggest that both the distance and direction of freshwater migration are critical factors that must be considered in choosing an appropriate donor population.

Evolutionary Significance of Sockeye Salmon Populations

Precise homing behavior shapes the population structure of sockeye salmon by promoting reproductive isolation, which in turn facilitates adaptation to local conditions (Wright 1931). Some re-

TABLE 3.—Evidence for genetic differentiation of subpopulations of sockeye salmon inhabiting the same nursery lake.

Lake	Data source[a]	Year	Number of sites	Number of loci	Mean F_{ST}	Estimated gene flow (Nm)	References for other phenotypic comparisons
Lake versus tributary spawning							
Kuril	1, 2, 5	1989	2	6[b]	0.026	2.3	
	5	1989	23	5	0.007	8.9	
	5	1990	18	6	0.022	2.8	
Nachiki	5	1982	9	2	0.042	1.4	
Dvu-Yurta	5	1982	3	2	0		
Iliamna	5	1991	6	5	0.007	8.9	Blair et al. 1993
	1, 2, 5	1991	2	8[b]	0.021	2.9	
Tatsamenie	8	1993	2	6	0.015	4.1	Pacific Salmon Commission 1994
Meziadin	5	1986	3	4	0.007	8.9	Rutherford et al. 1994
Quesnel	4	1992	2	2[b]	0.013	4.7	
Inlet versus outlet tributary spawning							
Kuril	5	1989	11	5	0.004	15.6	
	5	1990	11	5	0.010	6.2	
Karluk	6	1982	6	2	0.002	31.2	Raleigh 1967
	6	1983	6	2	0.012	5.1	Gard et al. 1987
	6	1984	6	2	0.012	5.1	
	6	1985	6	2	0.011	5.6	
Tustumena	3	1991	8	1[b]	0.210	0.2	Burger et al. 1995
Babine	5	1987	9	4	0.009	6.9	McCart 1967
Harrison[c]	7	1992	2	7	0.030	2.0	Cave 1978
Early versus late inlet tributary spawning							
Kuril	5	1989	10	5	0.002	31.2	
	5	1990	10	5	0.009	6.9	
Lower Russian	6	1978	2	2	0.023	2.7	
	6	1979	2	2	0.054	1.1	
	6	1980	2	1	0.050	1.2	
	6	1981	2	3	0.012	5.1	
Babine	5	1985	4	3	0		
	5	1987	7	4	0.001	62.4	
	5	1988	5	5	0.002	31.2	

[a]Data sources are (1) Beacham et al. 1995; (2) Bickham et al. 1995; (3) Burger et al. 1995 and unpublished data; (4) Taylor et al., in press; (5) Varnavskaya et al. 1994b; (6) Wilmot and Burger 1985; (7) Wood et al. 1994; (8) P. Milligan and D. Rutherford, Department of Fisheries and Oceans, unpublished data.
[b]Includes molecular genetic data.
[c]Compares Weaver Creek versus Birkenhead River spawning sites.

searchers (e.g., Hanamura 1966; Quinn 1985) have suggested that homing behavior is more precise in sockeye salmon than in other Pacific salmon. This inference is supported by geographical patterns of population subdivision revealed by biochemical genetic surveys. Whereas sockeye salmon exhibit a mosaic pattern of differentiation because populations in different lakes within a river system are reproductively isolated (e.g., Utter et al. 1984; Varnavskaya et al. 1994a; Wood et al. 1994), other Pacific salmon exhibit more pronounced clines or regional structuring in allele frequencies (e.g., Beacham et al. 1985a, 1985b; Gharrett et al. 1987; Wehrhahn and Powell 1987; Utter et al. 1989; Wilmot et al. 1994). Ryman (1983) reported that gene diversity was more highly subdivided among populations in brown trout *Salmo trutta* than in sockeye salmon but emphasized that the sockeye salmon

data (from Grant et al. 1980) used in his analysis were collected from a comparatively restricted area.

Presumably fidelity to natal streams is especially adaptive in sockeye salmon because it ensures that juvenile sockeye salmon, which unlike other Pacific salmon are typically adapted for limnetic life, have access to a suitable rearing area. Natural selection for precise homing may be relaxed in sea–river-type sockeye salmon in rivers like the Stikine and Taku in which fry from spawning areas throughout the watershed migrate downstream to slackwater rearing areas in the lower river and estuary (Wood et al. 1987a; Murphy et al. 1991). Thus adult sea–river-type sockeye salmon may stray from natal spawning sites at a rate more characteristic of other Pacific salmon. This conjecture is consistent with the apparent lack of reproductive isolation between widely separated spawning sites in both the Stikine

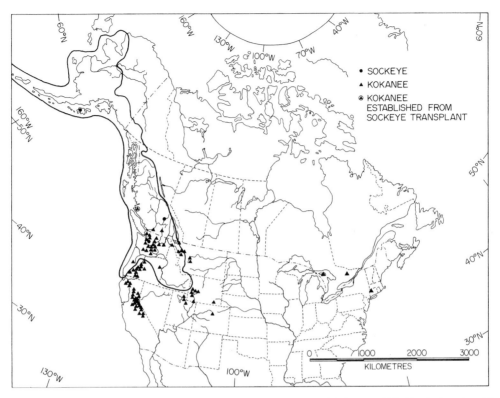

FIGURE 7.—Distribution of successful transplants of sockeye salmon and kokanee within North America. All are recorded as having been self-perpetuating at one time although some populations have not persisted to the present day (see Appendix for sources). Enclosed area indicates native range of sockeye salmon in North America.

River (Wood et al. 1987b) and the main stem of the Taku River (Guthrie et al., in press).

Sockeye salmon, like other temperate freshwater fishes, have evolved in regions where ice sheets continually advance and retreat with interglacial periods that typically last 10,000–40,000 years (Pielou 1994). Glaciation events are primarily a consequence of cycles in the tilt and precession of the Earth's spin axis and the shape of the Earth's orbit; these orbital cycles affect the intensity of insolation and timing of seasons at particular latitudes independently of average global temperature (Imbrie and Imbrie 1979; Broecker and Denton 1990). Virtually all modern sockeye salmon populations in Canada and most of Alaska were established within the last 10,000 years since the retreat of the cordilleran ice sheet in the current interglacial period. Similarly, much of the existing freshwater habitat used by sockeye salmon will likely be lost as ice sheets readvance within the next 10,000–30,000 years (assuming we are more than halfway through the current interglacial period, Pielou 1994). It is

therefore tempting to speculate that a myriad of reproductively isolated sockeye salmon populations adapted to conditions in specific lakes, or to different habitats within lakes, will be lost within a relatively short time (in evolutionary terms) regardless of human activities in the next few centuries. Thus long-term evolution of sockeye salmon is perhaps best represented by the "bottle-brush" model: homing behavior facilitates divergence and adaptation to a wide range of new habitats during interglacial periods (lineages represented by the bristles), most of which are destroyed by the ensuing ice age. Lineages that persist through ice ages (the shaft of the bottle brush) are the most evolutionarily significant in the long term. From this perspective, specialized adaptations in lake-type sockeye salmon (and kokanee) may be evolutionary dead ends in areas of the species' range vulnerable to glaciation. Long-term persistence in these regions may depend upon colonizing ability, not homing (Larkin 1981; Quinn 1985); an ability to survive without lake habitat; and heritable polymorphism in adaptive traits that will

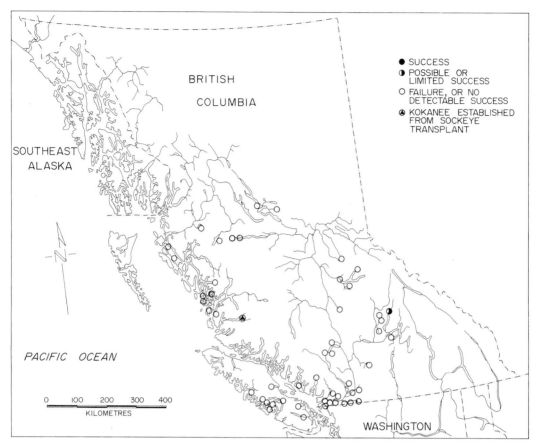

FIGURE 8.—Distribution and outcome of attempts to transplant anadromous sockeye salmon in British Columbia (data from Aro 1979; Withler 1982).

facilitate rapid adaptation to new opportunities which arise as ice sheets recede. Most extant populations of sea–river-type sockeye salmon live in close proximity to glaciers or in glacially influenced drainages, suggesting that sea–river-type sockeye salmon are the principal colonists of new habitat following the retreat of glaciers (see Milner and Bailey 1989). Such a colonizing role is consistent with higher straying rates inferred among sea–river-type spawning sites within the Stikine and Taku rivers and with evidence that allele frequencies in these sea–river-type sockeye salmon do not differ significantly between the Stikine and Taku rivers ($P > 0.9$, $F_{ST} < 0.001$, based on data for 10 polymorphic loci reported in Wood et al. 1994 and Guthrie et al., in press). It is also worth noting from the phylogenetic tree that the Stikine River sea–river type populations (6B and 8B located at the bottom of Figure 2) show little divergence based on summed horizontal branch lengths from the postu-

lated ancestral form or from many west Alaskan populations.

A surprising feature of genetic diversity in sockeye salmon is that very few extant populations, even very small ones, exhibit fixation of alleles at protein-coding and VNTR loci that are highly polymorphic in other populations; moreover, no two populations are fixed for alternative alleles (Varnavskaya et al. 1994a; Wood et al. 1994; Taylor et al., in press; Winans et al., in press). A similar pattern is evident in the distribution of maternally inherited mtDNA haplotypes (Wood, Bickham, and Patton, unpublished data for cytochrome-b and ND1 sequences in more than 50 populations from the Columbia River to the Kuril Islands). The occurrence of multiple shared allozyme and VNTR alleles and mtDNA haplotypes within different populations indicates that these populations were not founded by single pairs of spawning fish. Rather the pattern suggests that there have been multiple colonization events, emphasizing the importance of colonizing forms, and that the genetic diversity of the species was largely retained despite the last glaciation. Perhaps lake-type sockeye salmon adapted to new habitats quickly (relative to the rate of change in climate and local topography) during the last glaciation such that neutral alleles from individual populations were preserved by descent despite compression of the species' range and extinction of the original populations.

The preceding discussion illustrates that the definition of an ESU for conservation implicitly involves assumptions about timescale. From a long-term perspective (>10,000 years), it seems unwarranted to define all locally adapted, reproductively isolated sockeye salmon populations as ESUs. In regions vulnerable to glaciation, it may be more appropriate to restrict the application of ESUs to sea–river-type populations and other large, genetically diverse populations comprising complex subpopulations adapted to a variety of local conditions. On the other hand, the history of transplant attempts suggests very strongly that most sockeye salmon populations (and their unique adaptations) are irreplaceable within a human lifetime and therefore significant to other human values. The critical point for discussion and further research concerns the potential rate of adaptation to new conditions. Rates of divergence in phenotypic characteristics among transplanted populations established from a common donor population may provide opportunities for direct measurement (e.g., Quinn and Unwin 1993). A more systematic comparison, across populations and species, of the extent of divergence in selectively neutral traits versus corresponding divergence in adaptive traits may also offer valuable insight into past rates of adaptation (Clayton 1981).

In the present, we must attempt to conserve genetic diversity to protect the productive potential of the sockeye salmon resource for human exploitation and for aesthetic and ecological reasons. The hierarchical structure of sockeye salmon populations implies that the best way to conserve genetic diversity within the species is to preserve sockeye salmon in as many different lake systems as possible (MacLean and Evans 1981; Altukhov and Salmenkova 1991; Riddell 1993). Local adaptation of sockeye salmon populations to conditions within individual lakes enhances the productivity of populations and hence potential sustainable yield to fisheries (Lannan et al. 1989). Thus, wherever possible we should also endeavour to conserve subpopulation structure within individual lakes by protecting runs using different spawning habitats or exhibiting other characteristics that may have adaptive value. Sockeye salmon populations remain relatively healthy throughout most of their range (Gharrett et al. 1993). Modest efforts to conserve genetic diversity in productive populations before they are threatened are preferable to heroic measures to preserve them once they are endangered.

Acknowledgments

I am grateful to T. D. Beacham, C. V. Burger, J. W. Bickham, C. J. Foote, A. J. Gharrett, C. M. Guthrie, III, J. H. Helle, P. Milligan, J. C. Patton, D. T. Rutherford, E. B. Taylor, R. L. Wilmot, and G. A. Winans for contributing unpublished data or data from recently submitted manuscripts that have helped to make this article current and comprehensive. I also thank all those listed in the Appendix as well as K. Kostow, G. Anderson, J. Banks, D. Barry, P. Bourque, W. Bowers, C. Burley, M. Everson, T. Greene, P. Hulbert, S. Jackson, T. Jackson, S. Perry, B. Rieman, T. Steinwand, C. Wall, and T. Wiggins for providing helpful information about the status and distribution of transplanted kokanee populations in North America. I also wish to thank D. T. Rutherford, K. Chapman, and L. Fitzpatrick for technical assistance; C. V. Burger, T. P. Quinn, W. E. Ricker, R. E. Withler, and an anonymous reviewer for thoughtful comments on the manuscript; and J. W. Bickham, E. Bermingham, C. J. Foote, E. B. Taylor, F. M. Utter, N. V. Varnavskaya, R. S. Waples, and G. A. Winans for helpful discussions.

References

Altukhov, Y. P., and E. A. Salmenkova. 1991. The genetic structure of salmon populations. Aquaculture 98:11–40.

Aro, K. V. 1979. Transfers of eggs and young of Pacific salmon within British Columbia. Canada Fisheries and Marine Service Technical Report 861:145.

Bams, R. A. 1976. Survival and propensity for homing as affected by presence or absence of locally adapted paternal genes in two transplanted populations of pink salmon (Oncorhynchus gorbuscha). Journal of the Fisheries Research Board of Canada 33:2716–2725.

Beacham, T. D., and C. B. Murray. 1993. Fecundity and egg size variation in North American Pacific salmon (Oncorhynchus). Journal of Fish Biology 42:485–508.

Beacham, T. D., R. E. Withler, and A. P. Gould. 1985a. Biochemical genetic stock identification of chum salmon (Oncorhynchus keta) in southern British Columbia. Canadian Journal of Fisheries and Aquatic Sciences 42:437–448.

Beacham, T. D., R. E. Withler, and A. P. Gould. 1985b.

Biochemical genetic stock identification of pink salmon (*Oncorhynchus gorbuscha*) in southern British Columbia and Puget Sound. Canadian Journal of Fisheries and Aquatic Sciences 42:1474–1483.

Beacham, T. D., R. E. Withler, and C. C. Wood. 1995. Stock identification of sockeye salmon (*Oncorhynchus nerka*) using minisatellite DNA variation. North American Journal of Fisheries Management 15:249–265.

Behnke, R. J. 1993. Status of biodiversity of taxa and nontaxa of salmonid fishes: contemporary problems of classification and conservation. Pages 43–48 *in* J. G. Cloud and G. H. Thorgaard, editors. Genetic conservation of salmonid fishes. Plenum, New York.

Bickham, J. W., C. C. Wood, and J. C. Patton. 1995. Biogeographic implications of cytochrome *b* sequences and allozymes in sockeye (*Oncorhynchus nerka*). Journal of Heredity 86(2):140–144.

Billingsley, L. W., editor. 1981. Proceedings of the stock concept international symposium. Canadian Journal of Fisheries and Aquatic Sciences 38:1457–1921.

Birtwell, I. K., M. D. Nassichuk, and H. Beune. 1987. Underyearling sockeye salmon (*Oncorhynchus nerka*) in the estuary of the Fraser River. Pages 25–35 *in* Smith et al. (1987).

Blackett, R. F. 1979. Establishment of sockeye (*Oncorhynchus nerka*) and chinook (*O. tshawytscha*) salmon runs at Frazer Lake, Kodiak Island, Alaska. Journal of the Fisheries Research Board of Canada 36:1265–1277.

Blair, G. R., D. E. Rogers, and T. P. Quinn. 1993. Variation in life history characteristics and morphology of sockeye salmon in the Kvichak River system, Bristol Bay, Alaska. Transactions of the American Fisheries Society 122:550–559.

Bodaly, R. A. 1979. Morphological and ecological divergence within the lake whitefish (*Coregonus clupeaformis*) species complex in Yukon Territory. Journal of the Fisheries Research Board of Canada 36:1214–1222.

Bower, S. M., and L. Margolis. 1984. Detection of infection and susceptibility of different Pacific salmon stocks (*Oncorhynchus* spp.) to the haemoflagellate *Cryptobia salmositica*. Journal of Parasitology 70:273–278.

Braitseva, O. A., I. V. Melekestsev, I. C. Evteyeva, and E. G. Lupikina. 1968. Stratigraphy of Pleistocene geology and glaciation of Kamchatka. Nauka, Moscow.

Brannon, E. L. 1967. Genetic control of migrating behavior of newly emerged sockeye salmon fry. International Pacific Salmon Fisheries Commission Progress Report 16.

Brannon, E. L. 1972. Mechanisms controlling migration of sockeye salmon fry. International Pacific Salmon Fisheries Commission Bulletin 21:86.

Brannon, E. L. 1987. Mechanisms stabilizing salmonid fry emergence timing. Pages 120–124 *in* Smith et al. (1987).

Broecker, W. S., and G. H. Denton. 1990. What drives glacial cycles? Scientific American 262(1):48–56.

Bromage, N., P. Hardiman, J. Jones, J. Springate, and V.

Bye. 1990. Fecundity, egg size, and total egg volume differences in 12 stocks of rainbow trout, *Oncorhynchus mykiss* Richardson. Aquaculture and Fisheries Management 21:269–294.

Bugaev, V. F. 1984. Method for identification of sockeye salmon *Oncorhynchus nerka* (Salmonidae), of different spawning populations in the Kamchatka River basin. Journal of Ichthyology 24:47–53.

Burger, C. V., J. E. Finn, and L. Holland-Bartels. 1995. Pattern of shoreline spawning by sockeye salmon in a glacially turbid lake: evidence for subpopulation differentiation. Transactions of the American Fisheries Society 124:1–15.

Burgner, R. L. 1991. Life history of sockeye salmon (*Oncorhynchus nerka*). Pages 1–117 *in* C. Groot and L. Margolis, editors. Pacific salmon life histories. University of British Columbia Press, Vancouver.

Buss, K. 1957. The controversial kokanee—a salmon for the lakes of northeastern United States. Pennsylvania Fish Commission Special Purpose Report, Harrisburg.

Cavalli-Sforza, L. L., and A. W. F. Edwards. 1967. Phylogenetic analysis: models and estimation procedures. Evolution 21:550–570.

Cave, J. D. 1978. The contribution of environment and heredity to differences in freshwater growth between Birkenhead River and Weaver Creek sockeye salmon (*Oncorhynchus nerka*). Master's thesis. University of British Columbia, Vancouver.

Chakraborty, R. 1980. Gene diversity analysis in nested subdivided populations. Genetics 96:721–726.

Clague, J. J., J. E. Armstrong, and W. H. Mathews. 1980. Advance of the late Wisconsin cordilleran ice sheet in southern British Columbia since 22,000 yr BP. Quaternary Research 13:322–326.

Clarke, W. C., J. E. Shelbourn, and J. R. Brett. 1978. Growth and adaptation to sea water in 'underyearling' sockeye (*Oncorhynchus nerka*) and coho (*O. kisutch*) salmon subjected to regimes of constant or changing temperature and day length. Canadian Journal of Zoology 56:2413–2421.

Clayton, J. W. 1981. The stock concept and the uncoupling of organismal and molecular evolution. Canadian Journal of Fisheries and Aquatic Sciences 38:1515–1522.

Craig, P. C. 1985. Identification of sockeye salmon (*Oncorhynchus nerka*) stocks in the Stikine River based on egg size measurements. Canadian Journal of Fisheries and Aquatic Sciences 42:1696–1701.

Curtis, B., and J. C. Fraser. 1948. Kokanee in California. California Fish and Game 34:111–114.

Eiler, J. H., B. D. Nelson, and R. F. Bradshaw. 1992. Riverine spawning by sockeye salmon in the Taku River, Alaska and British Columbia. Transactions of the American Fisheries Society 121:701–708.

Fedorenko, A. Y., and B. G. Shepherd. 1986. Review of salmon transplant procedures and suggested transplant guidelines. Canadian Technical Report of Fisheries and Aquatic Sciences 1479.

Fenderson, O. C. 1964. Evidence of subpopulations of lake whitefish, *Coregonus clupeaformis*, involving a dwarf form. Transactions of the American Fisheries Society 93:77–94.

Foerster, R. E. 1936. The return from the sea of sockeye salmon (*Oncorhynchus nerka*) with special reference to percentage survival, sex, proportions and progress of migration. Journal of the Biological Society of Canada 3:26–42.

Foerster, R. E. 1968. The sockeye salmon. Bulletin of the Fisheries Research Board of Canada 34.

Foote, C. J., I. Mayer, C. C. Wood, W. C. Clarke, and J. Blackburn. 1994. On the developmental pathway to nonanadromy in sockeye salmon, *Oncorhynchus nerka*. Canadian Journal of Zoology 72:397–405.

Foote, C. J., C. C. Wood, W. C. Clarke, and J. Blackburn. 1992. Circannual cycle of seawater adaptability in *Oncorhynchus nerka*: genetic differences between sympatric sockeye salmon and kokanee. Canadian Journal of Fisheries and Aquatic Sciences 49:99–109.

Foote, C. J., C. C. Wood, and R. E. Withler. 1989. Biochemical genetic comparison of sockeye salmon and kokanee, the anadromous and nonanadromous forms of *Oncorhynchus nerka*. Canadian Journal of Fisheries and Aquatic Sciences 46:149–158.

Fraser, J. C., and A. F. Pollitt. 1951. The introduction of kokanee red salmon (*Oncorhynchus nerka kennerlyi*) into Lake Tahoe, California, and Nevada. California Fish and Game 37:125–127.

Fulton, L. A., and R. E. Pearson. 1981. Transplantation and homing experiments on salmon, *Oncorhynchus* spp., and steelhead trout, *Salmo gairdneri*, in the Columbia River system: fish of the 1939–1944 broods. NOAA (National Oceanic and Atmospheric Administration) Technical Memorandum NMFS (National Marine Fisheries Service) F/NWC 12, Northwest Fisheries Science Center, Seattle.

Gard, R., B. Drucker, and R. Fagen. 1987. Differentiation of subpopulations of sockeye salmon (*Oncorhynchus nerka*), Karluk River system, Alaska. Pages 408–418 *in* Smith et al. (1987).

Gharrett, A. J., B. E. Riddell, J. E. Seeb, and J. H. Helle. 1993. Status of genetic resources of Pacific Rim salmon. Pages 295–301 *in* J. G. Cloud and G. H. Thorgaard, editors. Genetic conservation of salmonid fishes. Plenum, New York.

Gharrett, A. J., S. M. Shirley, and G. R. Tromble. 1987. Genetic relationships among populations of Alaskan chinook salmon (*Oncorhynchus tshawytscha*). Canadian Journal of Fisheries and Aquatic Sciences 44: 765–774.

Gilbert, C. H. 1918. Contributions to the life history of sockeye salmon. Annual Report of the British Columbia Fisheries Department 5:26–52, Vancouver.

Godin, J. G. J. 1982. Migrations of salmonid fishes during early life phases: daily and annual timing. Pages 22–50 *in* E. L. Brannon and E. O. Salo, editors. Salmon and trout migratory behavior symposium. University of Washington, College of Fisheries, Seattle.

Goodlad, J. C., T. W. Gjernes, and E. L. Brannon. 1974. Factors affecting sockeye salmon (*Oncorhynchus nerka*) growth in four lakes of the Fraser River sys-

tem. Journal of the Fisheries Research Board of Canada 31:871–892.

Grant, W. S., G. B. Milner, P. Krasnowski, and F. M. Utter. 1980. Use of biochemical genetic variants for identification of sockeye salmon (*Oncorhynchus nerka*) stocks in Cook Inlet, Alaska. Canadian Journal of Fisheries and Aquatic Sciences 37:1236–1247.

Guthrie, C. M., III, J. H. Helle, and A. J. Gharrett. In press. Genetic relationships among populations of sockeye salmon (*Oncorhynchus nerka*) in southeast Alaska and northern British Columbia. Canadian Journal of Fisheries and Aquatic Sciences.

Hanamura, N. 1966. Salmon of the north Pacific Ocean. Part III. A review of the life history of the north Pacific salmon 1. Sockeye salmon in the far east. International North Pacific Fisheries Commission Bulletin 18:1–27.

Hankin, D. G., J. W. Nicholas, and T. W. Downey. 1993. Evidence for inheritance of age at maturity in chinook salmon (*Oncorhynchus tshawytscha*). Canadian Journal of Fisheries and Aquatic Sciences 50:347–358.

Heifetz, J., S. W. Johnson, K. V. Koski, and M. L. Murphy. 1989. Migration timing, size, and salinity tolerance of sea-type sockeye salmon (*Oncorhynchus nerka*) in an Alaska estuary. Canadian Journal of Fisheries and Aquatic Sciences 46:633–637.

Henderson, M. A., and A. J. Cass. 1991. Effect of smolt size on smolt-to-adult survival for Chilko Lake sockeye salmon (*Oncorhynchus nerka*). Canadian Journal of Fisheries and Aquatic Sciences 48:988–994.

Imbrie, J., and K. P. Imbrie. 1979. Ice ages: solving the mystery. Enslow, Short Hills, New Jersey.

IUBNC (International Union of Biochemistry, Nomenclature Committee). 1984. Enzyme nomenclature. Academic Press, San Diego, California.

Ivankov, V. N. 1984. Anadromous and non-anadromous forms of sockeye salmon (*Oncorhynchus nerka* Walbaum) on Iturup Island (Kuril Islands). Pages 65–73 *in* V. N. Ivankov, editor. Biological investigations of fish in the Far East. University Press, Vladivostok, USSR.

Kaeriyama, M., S. Urawa, and T. Susuki. 1992. Anadromous sockeye salmon (*Oncorhynchus nerka*) derived from nonanadromous kokanee: life history in Lake Toro. Scientific Reports of the Hokkaido Salmon Hatchery 46:157–174.

Karlstrom, T. N. V., and G. E. Ball, editors. 1969. The Kodiak Island refugium: its geology, flora, and history. Ryerson Press, Toronto.

Kerns, O. E., Jr., and J. R. Donaldson. 1968. Behavior and distribution of spawning sockeye salmon on island beaches in Iliamna Lake, Alaska, 1965. Journal of the Fisheries Research Board of Canada 25:485–494.

Kim, J., F. J. Rohlf, and R. R. Sokal. 1993. The accuracy of phylogenetic estimation using the neighbor-joining method. Evolution 47:471–486.

Kimsey, J. B. 1951. Notes on kokanee spawning in Donner Lake, California, 1949. California Fish and Game 37:273–279.

Krogius, F. V. 1958. On scale pattern of Kamchatka

sockeye of different local populations. Fisheries Research Board of Canada Translation Series 181.

Krokhin, E. M. 1967. A contribution to the study of dwarf sockeye, *Oncorhynchus nerka*, Walb., in Lake Dal'nee (Kamchatka). Voprosy Iktiologii 7(3) (Journal of Ichthyology 44:433–455).

Lannan, J. E., G. A. E. Gall, J. E. Thorpe, C. E. Nash, and B. E. Ballachey. 1989. Genetic resource management of fish. Genome 31:798–804.

Larkin, P. A. 1981. A perspective on population genetics and salmon management. Canadian Journal of Fisheries and Aquatic Sciences 38:1469–1475.

Leary, R. F., F. W. Allendorf, and K. L. Knudsen. 1985. Inheritance of meristic variation and the evolution of developmental stability in rainbow trout. Evolution 39:308–314.

Lindsey, C. C. 1981. Stocks are chameleons: plasticity in gill rakers of coregonid fishes. Canadian Journal of Fisheries and Aquatic Sciences 38:1497–1506.

Lindsey, C. C., and J. D. McPhail. 1986. Zoogeography of fishes of the Yukon and Mackenzie basins. Pages 639–674 *in* C. H. Hocutt and E. O. Wiley, editors. Zoogeography of North American freshwater fishes. Wiley, New York.

MacLean, J. A., and D. O. Evans. 1981. The stock concept, discreetness of fish stocks, and fisheries management. Canadian Journal of Fisheries and Aquatic Sciences 38:1889–1898.

McCart, P. 1967. Behaviour and ecology of sockeye salmon fry in the Babine River. Journal of the Fisheries Research Board of Canada 24:375–428.

McCart, P. 1970. A polymorphic population of *Oncorhynchus nerka* at Babine Lake, B.C. involving anadromous (sockeye) and non-anadromous (kokanee) forms. Doctoral dissertation. University of British Columbia, Vancouver.

McPhail, J. D., and C. C. Lindsey. 1970. Freshwater fishes of northwestern Canada and Alaska. Bulletin of the Fisheries Research Board of Canada 173:1–381.

McPhail, J. D., and C. C. Lindsey. 1986. Zoogeography of freshwater fishes of Cascadia (the Columbia system and rivers north to the Stikine). Pages 615–637 *in* C. H. Hocutt and E. O. Wiley, editors. Zoogeography of North American freshwater fishes. Wiley, New York.

McPherson, S. A., and A. McGregor. 1986. Abundance, age, sex, and size of sockeye salmon (*Oncorhynchus nerka* Walbaum) catches and escapements in southern Alaska in 1985. Alaska Department of Fish and Game Technical Data Report 188:1–222, Juneau.

Milner, A. M., and R. G. Bailey. 1989. Salmonid colonization of new streams in Glacier Bay National Park, Alaska. Aquaculture and Fisheries Management 20:179–192.

Mullan, J. W. 1986. Determinants of sockeye salmon abundance in the Columbia River, 1880's–1982: a review and synthesis. U.S. Fish and Wildlife Service Biological Report 86:1–136.

Murphy, M. L., J. M. Lorenz, and K. V. Koski. 1991. Population estimates of juvenile salmon downstream migrants in the Taku River, Alaska. NOAA (National Oceanic and Atmospheric Administration) Technical Memorandum NMFS (National Marine Fisheries Service) F/NWC 203, Northwest Fisheries Science Center, Seattle.

Nehlsen, W., J. E. Williams, and J. A. Lichatowich. 1991. Pacific salmon at the crossroads: stocks at risk from California, Oregon, Idaho, and Washington. Fisheries 16(2):4–21.

Nelson, J. S. 1968. Distribution and nomenclature of North American kokanee, (*Oncorhynchus nerka*). Journal of the Fisheries Research Board of Canada 25:409–414.

Nelson, K., and M. Soule. 1987. Genetical conservation of exploited fishes. Pages 345–368 *in* N. Ryman and F. M. Utter, editors. Population genetics and fishery management. University of Washington Press, Seattle.

Pacific Salmon Commission. 1994. Transboundary River sockeye salmon enhancement activities: final report for fall 1990 to spring 1992. Transboundary Technical Committee Report TCTR 94-1:1–79, Vancouver.

Pahlke, K. A., and R. R. Riffe. 1988. Compilation of catch, escapement, age, sex, and size data for salmon returns to the Yakutat area in 1986. Alaska Department of Fish and Game Technical Data Report 224:151, Juneau.

Parsons, J. W. 1973. History of salmon in the Great Lakes, 1850–1970. United States Bureau of Sport Fisheries and Wildlife Technical Papers 68.

Pielou, E. C. 1994. After the ice age: the return of life to glaciated North America. University of Chicago Press, Chicago.

Pojar, J. 1980. Brooks Peninsula: possible Pleistocene glacial refugium on northwestern Vancouver Island. Botanical Society of America Miscellaneous Series Publication 158:89.

Quinn, T. P. 1985. Homing and the evolution of sockeye salmon (*Oncorhynchus nerka*). Pages 353–366 *in* M. A. Rankin, editor. Migration: mechanisms and adaptive significance. Contributions in Marine Science, Supplement to Vol. 27, University of Texas, Port Aransas.

Quinn, T. P., and C. J. Foote. 1994. The effects of body size and sexual dimorphism on the reproductive behaviour of sockeye salmon (*Oncorhynchus nerka*). Animal Behaviour 48:751–761.

Quinn, T. P., and M. J. Unwin. 1993. Life history patterns of New Zealand chinook salmon (*Oncorhynchus tshawytscha*) populations. Canadian Journal of Fisheries and Aquatic Sciences 50:1414–1421.

Quinn, T. P., C. C. Wood, L. Margolis, B. E. Riddell, and K. D. Hyatt. 1987. Homing in wild sockeye salmon (*Oncorhynchus nerka*) populations as inferred from differences in parasite prevalence and allozyme allele frequencies. Canadian Journal of Fisheries and Aquatic Sciences 44:1963–1971.

Raleigh, R. F. 1967. Genetic control in the lakeward migrations of sockeye salmon (*Oncorhynchus nerka*) fry. Journal of the Fisheries Research Board of Canada 24:2613–2622.

Reisenbichler, R. R. 1988. Relation between distance transferred from natal stream and relative return rate for hatchery coho salmon. North American Journal of Fisheries Management 8:172–174.

Rice, S. D., R. E. Thomas, and A. Moles. 1994. Physiological and growth differences in three stocks of underyearling sockeye salmon (Oncorhynchus nerka) on early entry into seawater. Canadian Journal of Fisheries and Aquatic Sciences 51:974–980.

Ricker, W. E. 1938. "Residual" and kokanee salmon in Cultus Lake. Journal of the Fisheries Research Board of Canada 4:192–218.

Ricker, W. E. 1966. Sockeye salmon in British Columbia. Salmon of the north Pacific Ocean-Part III: a review of the life history of North American salmon. International North Pacific Fisheries Commission Bulletin 18:59–70.

Ricker, W. E. 1972. Hereditary and environmental factors affecting certain salmonid populations. Pages 27–160 in R. C. Simon and P. A. Larkin, editors. The stock concept in Pacific salmon. H. R. MacMillan Lectures in Fisheries, University of British Columbia, Vancouver.

Riddell, B. E. 1993. Spatial organization of Pacific salmon: what to conserve? Pages 23–41 in J. G. Cloud and G. H. Thorgaard, editors. Genetic conservation of salmonid fishes. Plenum, New York.

Rogers, D. E. 1987. The regulation of age at maturity in Wood River sockeye salmon (Oncorhynchus nerka). Pages 78–89 in Smith et al. (1987).

Royal, L. A., and A. Seymour. 1940. Building new salmon runs. Progressive Fish-Culturist 52:1–7.

Russell, R. B., D. L. Bill, and W. A. Bucher. 1989. Salmon spawning ground surveys in the Bristol Bay area, 1988. Alaska Department of Fish and Game Regional Information Report 2K88-14, Anchorage.

Rutherford, D. T., C. C. Wood, A. L. Jantz, and D. R. Southgate. 1994. Biological characteristics of Nass River sockeye salmon (Oncorhynchus nerka) and their utility for stock composition analysis of test fisheries. Canadian Technical Report of Fisheries and Aquatic Sciences 1988:65.

Ryman, N. 1983. Patterns of distribution of biochemical genetic variation in salmonids: differences between species. Aquaculture 33:1–21.

Saitou, N., and M. Nei. 1987. The neighbor-joining method: a new method for reconstructing phylogenetic trees. Molecular Biology and Evolution 4:406–425.

Scott, D. 1984. Origin of the New Zealand sockeye salmon, Oncorhynchus nerka (Walbaum). Journal of the Royal Society of New Zealand 14:245–249.

Scott, W. B., and E. J. Crossman. 1973. Freshwater fishes of Canada. Bulletin of the Fisheries Research Board of Canada 184:966.

Seeley, C. M., and G. W. McCammon. 1966. Kokanee. Pages 274–294 in A. Calhoun, editor. Inland fisheries management. California Department of Fish and Game, Sacramento.

Semko, R. S. 1954. The stocks of West Kamchatka salmon and their commercial utilization. Izvestiia TINRO 41:3–109 (Translated 1960, Fisheries Research Board of Canada Translation Series 288.)

Sharr, S., C. Peckham, and G. Carpenter. 1988. Catch and escapement statistics for Copper River, Bering River, and Prince William Sound salmon, 1986. Alaska Department of Fish and Game Technical Fishery Report 88-17, Anchorage.

Slatkin, M., and N. H. Barton. 1989. A comparison of three indirect methods for estimating average levels of gene flow. Evolution 43:1349–1368.

Smith, H. D., L. Margolis, and C. C. Wood, editors. 1987. Sockeye salmon (Oncorhynchus nerka) population biology and future management. Canadian Special Publication of Fisheries and Aquatic Sciences 96.

Swofford, D. L., and R. B. Selander. 1981. BIOSYS-1: a FORTRAN program for the comprehensive analysis of electrophoretic data in population genetics and systematics. Journal of Heredity 72:281–283.

Taylor, E. B. 1991. A review of local adaptation in Salmonidae, with particular reference to Pacific and Atlantic salmon. Aquaculture 98:185–207.

Taylor, E. B., and C. J. Foote. 1991. Critical swimming velocities of juvenile sockeye salmon and kokanee, the anadromous and non-anadromous forms of Oncorhynchus nerka (Walbaum). Journal of Fish Biology 38:1–13.

Taylor, E. B., C. J. Foote, and C. C. Wood. In press. Molecular genetic evidence for parallel life history evolution within a Pacific salmon (sockeye salmon and kokanee, Oncorhynchus nerka). Evolution.

Utter, F. M. 1981. Biological criteria for definition of species and distinct intraspecific populations of anadromous salmonids under the U.S. Endangered Species Act of 1973. Canadian Journal of Fisheries and Aquatic Sciences 38:1626–1635.

Utter, F., P. Aebersold, J. Helle, and G. Winans. 1984. Genetic characterization of populations in the southeastern range of sockeye salmon. Pages 17–32 in J. M. Walton and D. B. Houston, editors. Proceedings of the Olympic Wild Fish Conference, Peninsula College, Port Angeles, Washington.

Utter, F., G. Milner, G. Stahl, and D. Teel. 1989. Genetic population structure of chinook salmon, Oncorhynchus tshawytscha, in the Pacific Northwest. U.S. National Marine Fisheries Service Fishery Bulletin 87: 239–264.

Utter, F. M., J. E. Seeb, and L. W. Seeb. 1993. Complementary uses of ecological and biochemical genetic data in identifying and conserving salmon populations. Fisheries Research 18:59–76.

Varnavskaya, N. V., C. C. Wood, and R. J. Everett. 1994a. Genetic variation in sockeye salmon (Oncorhynchus nerka) populations of Asia and North America. Canadian Journal of Fisheries and Aquatic Sciences 51 (Supplement 1):132–146.

Varnavskaya, N. V., and six coauthors. 1994b. Genetic differentiation of subpopulations of sockeye salmon (Oncorhynchus nerka) within lakes of Alaska, British Columbia, and Kamchatka, Russia. Canadian Journal of Fisheries and Aquatic Sciences 51(Supplement 1): 147–157.

Verspoor, E., N. H. C. Fraser, and A. F. Youngson. 1991. Protein polymorphism in Atlantic salmon within a Scottish river: evidence for selection and estimates of gene flow between tributaries. Aquaculture 98:217–230.

Waples, R. S. 1991. Definition of "species" under the Endangered Species Act: application to Pacific salmon. NOAA (National Oceanic and Atmospheric Administration) Technical Memorandum NMFS (National Marine Fisheries Service) F/NWC 194, Northwest Fisheries Science Center, Seattle.

Waples, R. S. 1995. Evolutionarily significant units and the conservation of biological diversity under the Endangered Species Act. American Fisheries Society Symposium 17:8–27.

Warner, B. C., R. W. Mathews, and J. J. Clague. 1982. Ice-free conditions on the Queen Charlotte Islands, British Columbia, at the height of late Wisconsin glaciation. Science 218:675–677.

Wehrhahn, C. F., and R. Powell. 1987. Electrophoretic variation, regional differences, and gene flow in the coho salmon (*Oncorhynchus kisutch*) of southern British Columbia. Canadian Journal of Fisheries and Aquatic Sciences 44:822–831.

Williams, I. V. 1987. Attempts to re-establish sockeye salmon (*Oncorhynchus nerka*) populations in the upper Adams River, British Columbia, 1949–84. Pages 235–242 *in* Smith et al. (1987).

Wilmot, R. L., and C. V. Burger. 1985. Genetic differences among populations of Alaskan sockeye salmon. Transactions of the American Fisheries Society 114:236–243.

Wilmot, R. L., R. J. Everett, W. J. Spearman, R. Baccus, N. V. Varnavskaya, and S. V. Putivkin. 1994. Genetic stock structure of Western Alaska chum salmon and a comparison with Russian Far East stocks. Canadian Journal of Fisheries and Aquatic Sciences 51 (Supplement 1):84–94.

Winans, G. A., P. B. Aebersold, and R. S. Waples. In press. Allozyme variability in selected populations of *Oncorhynchus nerka* with special consideration of Redfish Lake, Idaho. Transactions of the American Fisheries Society.

Withler, F. C. 1982. Transplanting Pacific salmon. Canadian Technical Report of Fisheries and Aquatic Sciences 1079:27.

Wood, C. C., and C. J. Foote. 1990. Genetic differences in the early development and growth of sympatric sockeye salmon and kokanee (*Oncorhynchus nerka*), and their hybrids. Canadian Journal of Fisheries and Aquatic Sciences 47:2250–2260.

Wood, C. C., and C. J. Foote. In press. Evidence for sympatric genetic divergence of anadromous and non-anadromous morphs of sockeye salmon (*Oncorhynchus nerka*). Evolution.

Wood, C. C., B. E. Riddell, and D. T. Rutherford. 1987a. Alternative juvenile life histories of sockeye salmon (*Oncorhynchus nerka*) and their contribution to production in the Stikine River, northern British Columbia. Pages 12–24 *in* Smith et al. (1987).

Wood, C. C., B. E. Riddell, D. T. Rutherford, and K. L. Rutherford. 1987b. Variation in biological characters among sockeye salmon populations of the Stikine River with potential application for stock identification in mixed-stock fisheries. Canadian Technical Report of Fisheries and Aquatic Sciences 1535.

Wood, C. C., B. E. Riddell, D. T. Rutherford, and R. E. Wither. 1994. Biochemical genetic survey of sockeye salmon (*Oncorhynchus nerka*) in Canada. Canadian Journal of Fisheries and Aquatic Sciences 51(Supplement 1):114–131.

Wright, S. 1931. Evolution in Mendelian populations. Genetics 16:97–159.

Wright, S. 1951. The genetical structure of populations. Annals of Eugenics 15:323–354, Cambridge, UK.

Appendix: Sockeye Salmon Transplants

Sources of information about sockeye salmon and kokanee transplants within North America.

State or province and lake[a]	Reference or authority[b]
Published sources	
Alaska	
Frazer	Blackett (1979)
British Columbia	
Walkus	Aro (1979)
Adams	Williams (1987)
Vidette	Fedorenko and Shepherd (1986)
Jones	Fedorenko and Shepherd (1986)
Ontario and Great Lakes	
Huron	Parsons (1973)
Ontario	Parsons (1973)
Boulter	Scott and Crossman (1973)
Washington	
Washington	Royal and Seymour (1940)
Icicle Ck	Fulton and Pearson (1981), Mullan (1986)
California	
Shasta	Seeley and McCammon (1966)
Donner	Curtis and Fraser (1948), Kimsey (1951)
Salt Springs R	Curtis and Fraser (1948), Buss (1957)
Strawberry	Curtis and Fraser (1948)
Tahoe	Fraser and Pollitt (1951)
Echo	Curtis and Fraser (1948)
Montana	
Koocanusa	Fedorenko and Shepherd (1986)
Unpublished sources (1994)	
Ontario	
Huron	E. Crossman, Royal Ontario Museum, Toronto, Ontario
Ontario	
Boulter	
Washington	
Banks	J. Foster, WDFW, Ephrata
Palmer	
F. D. Roosevelt	
Bumping	J. Cummins, WDFW, Yakima
Chelan	
Cle Elum	
Kachess	
Keechelus	
Lost	
Rimrock	
Wenatchee	
American	C. Philips, WDFW, Mill Creek
Mountain	
Samish	
Stevens	
Washington	
Merwin	E. Anderson, WDFW, Vancouver
Yale	
Oregon	
Elk	T. Fies, ODFW, Bend
Crescent	
Crane Prairie R	
Wickiup	
Waldo	M. Wade, ODFW, Springfield
Billy Chinook	S. Thiesfeld, ODFW, Prinevilee
Odell	J. Fortune, ODFW, Klamath Falls
Miller	
Fourmile	
Lake of the Woods	
Olive	T. Unterwegner and M. Gray, ODFW, John Day
Hemlock R	D. Loomis, ODFW, Roseburg
Lemolo R	

Appendix.—Continued.

State or province and lake[a]	Reference or authority[b]
Unpublished sources (1994)	
Triangle	G. Westfall, ODFW, Florence
California	
Trinity R	R. Barnhart, California Cooperative Fishery Research Unit, Humboldt State University, Arcata
Lewiston R	D. Weidlein, CDFG, Redding
Shasta	
Whiskeytown R	
Bucks	F. Reynolds, CDFG, Sacramento
Donner	
Bullards Bar R	
Little Grass Valley R	
Hell Hole R	
Stampede R	
Boca R	
Union Valley R	
Don Pedro R	
Pardee R	
Tahoe	
Fallen Leaf	
Folsom R	
Echo	
Bass	
Shaver	
Huntington	
Upper Twin	
Lower Twin	
Idaho	
Pend Oreille	D. Pittman, Idaho Department of Fish and Game, Boise
Nevada	
Tahoe	J. Curran, Nevada Department of Wildlife, Reno
Montana	
Holter R	T. Dotson, Montana Department of Fish, Wildlife and Parks, Helena
Swan	
Hauser R	
Ronan	
Flathead	
Little Bitterroot	
Wyoming	
Fremont	R. Wiley, Wyoming Game and Fish Department, Laramie
New Fork	
Boulder	
Middle Piney	
Flaming Gorge R	
Fontenelle R	
Hattie	
Utah	
Causey R	T. Miles and D. Archer, Utah State Department of Natural Resources, Salt Lake City
Porcupine R	
Strawberry R	
Flaming Gorge R	
Colorado	
Dillon R	P. Martinez, State of Colorado Division of Wildlife, Fort Collins
Connecticut	
East Twin	W. Hyatt, Connecticut Department of Environmental Protection, Fisheries Division, Hartford

[a]Site names refer to lakes unless indicated by R (river) or Ck (creek).
[b]Abbreviations are Washington Department of Fish and Wildlife (WDFW), Oregon Department of Fish and Wildlife (ODFW), and California Department of Fish and Game (CDFG).

American Fisheries Society Symposium 17:217–226, 1995
© Copyright by the American Fisheries Society 1995

The Population-Level Consequences of Individual Reproductive Competition: Observations from a Closed Population

Jeffrey R. Baylis

Department of Zoology, University of Wisconsin 250 North Mills Street,
Madison, Wisconsin 53706, USA

Abstract.—I review 12 years of research on reproduction of smallmouth bass *Micropterus dolomieu* in a closed population. This population is isolated and small, likely characteristics of evolutionarily significant units (ESUs). In such a population, the mating system has profound effects on effective population size and increases the probability of genetic drift. Life history characteristics are sensitive to events early in development. The mating system can affect these early events, leading to nongenetic parent–offspring correlations that may mimic local adaptations. Both of these effects should be considered when reviewing populations for ESU status.

In 1932, Sewall Wright proposed a sweeping model for the genetic basis of evolutionary change, the shifting balance model (Wright 1932). He posited that most animal populations are small and isolated and experience restricted gene flow among demes. In such a population structure, genetic drift would hold equal sway with natural selection as a mechanism for genetic change. Each semi-isolated population would then evolve its own unique mix of genes through drift and local adaptation; this process would greatly accelerate and amplify the effects of natural selection by increasing the diversity and variation of phenotypic types, which are the fuel of evolution (Wright 1932, 1980).

Fishes breeding in freshwater lakes or rivers are a clear example of just the sort of small and semi-isolated populations envisioned by Wright. The vast majority of freshwater bodies are small and isolated and, as such, contain small and isolated populations of fishes (Wetzel 1983). Even in streams or rivers, some degree of isolation of fish populations is evident. Whereas the world's great oceans are all interconnected to some degree by chemically and physically similar water masses, physical and chemical barriers are more profound in freshwater systems already partially isolated by terrestrial barriers. For major river systems, the saline oceans into which they drain constitute an impassible obstacle to many freshwater fishes and hence are a barrier to migration between systems. For anadromous and potamodromous migratory forms, the prevalence of philopatry in fishes preserves a degree of genetic isolation even in the absence of physical barriers. Within drainage basins, the upper reaches of small feeder streams are ecologically very different habitats from the lower reaches of major rivers, and physical barriers such as waterfalls or rapids may act as one-way barriers that genetically isolate up-

stream forms. Factors such as water temperature, dissolved oxygen, and dissolved solids may prevent or inhibit exchange of fishes among headwater populations, even though a continuous aquatic path exists within the drainage. Such a view of population structure and evolution may go far towards explaining why freshwater bodies, with only 0.01% of the world's water volume, contain 41% of the earth's fish species (Horn 1972). Virtually all freshwater fish populations are potential candidates to exhibit substantial reproductive isolation from conspecific populations in the absence of human manipulation, and, in an evolutionary framework, the very existence of each deme is an important component in the evolutionary legacy of the species. These are the two key criteria that must be satisfied for a population to be listed as an evolutionarily significant unit (ESU) by the National Marine Fisheries Service (see Waples 1995, this volume).

In this paper I will present some of my laboratory's findings from 12 years of studying reproduction in a small and isolated population of smallmouth bass *Micropterus dolomieu*. I make no claims about the genetic "purity" or isolation of the population under study; rather, I wish to examine the consequences of reproduction in small, demographically closed populations—typical of most freshwater fish populations—in the hopes that we might better understand the properties of such semiclosed genetic systems and improve our management of them. I submit it is necessary to study reproduction in a healthy, isolated population to observe its normal function; endangered or declining populations may exhibit many reproductive pathologies.

The mating system in this species severely reduces the effective population size: the breeding population is often only a fraction of the adult population. Reproductive competition that affects the timing of individual reproduction within a breeding season can produce parent–offspring correlations in life history characteristics that may be either positive or negative in sign. Such life history characteristics can appear to be local adaptive strategies, when in fact they may be mere phenotypic consequences of reproductive competition combined with environmental effects. Although the local variation in life history characteristics may not itself be an adaptation, it has profound effects on population dynamics and demography and may be expected to greatly contribute to secondary local genetic adaptation and differentiation from other populations.

Smallmouth Bass Reproductive Competition in Nebish Lake

The Study Site and Methods

Nebish Lake is a 45-ha, clear seepage lake in the Northern Highland State Forest of north-central Wisconsin. The lake is managed and intensively studied by the Wisconsin Department of Natural Resources (WDNR). In 1967, the lake was poisoned with rotenone and restocked; in addition to smallmouth bass, Nebish Lake now supports yellow perch *Perca flavescens* and bluntnose minnows *Pimephales notatus*.

The adult smallmouth bass in the population were sampled annually from 1967 until 1993 by fykenetting and electroshocking (Baylis et al. 1993). All fish captured were weighed, measured, and aged by scale annuli. Captured individuals 20.3 cm total length and larger were tagged with a uniquely numbered Floy FD-67C anchor tag, and about 45% of the population was tagged each year (see Raffetto et al. 1990). Aging of smallmouth bass from temperate lakes by means of scale annuli is particularly reliable (Carlander 1977), and the WDNR verified the technique for Nebish Lake at the beginning of the population study by resampling tagged individuals in later years. Fishing is permitted only on a special permit basis, and a mandatory creel census is continuously in effect. These data were used to estimate the number of individuals in each year-class (Peterson and Schnabel estimates; Ricker 1975), size at age, growth, mortality, and exploitation rates for the population.

During the spawning season, we conducted an intensive census of the total number of nests in the lake and their status over the course of the season. A subsample of nests was sampled to obtain egg counts and estimates of the number of free-swimming fry produced. The guarding male was captured

to obtain size, weight, and age from scale samples and to read the tag number. Untagged males were tagged at this time as well. This subsample of breeding males ranged from 10 to 40% of all breeding males, depending on the year of the study. See Raffetto et al. (1990) and Wiegmann et al. (1992) for detailed methodology.

In Nebish Lake male smallmouth bass mature at an age of 3 years and 20 cm or greater total length. Females mature 1 year later, at an age of 4 years and 22 cm or greater total length (Raffetto et al. 1990). Smallmouth bass in Nebish Lake become active as temperatures rise above 11°C in the spring, and nest building and spawning commence at about 15°C. Spawning continues as long as the temperature remains between about 15°C and 20–22°C, a period ranging from about 1 to 6 weeks, depending upon the year and thermal regime (J. R. Baylis, University of Wisconsin, unpublished data). Ridgway et al. (1991) have found a similar relationship between temperature and smallmouth bass reproduction in Ontario. The larvae become free swimming and start exogenous feeding approximately 10–14 d after spawning, at about 9 mm total length, at which time they are still being guarded by the parental male. Male nest defense, territoriality, and parental care make it relatively easy to obtain detailed information on male reproductive success in this species and relatively difficult to obtain the same information on females. However, because the population is effectively monogamous, the male reproductive data will accurately reflect the population effect (Wiegmann et al. 1992). Because the lake is a closed system, knowledge of which males breed can reveal, by subtraction, which males are failing to breed in a given year. This is a key advantage to studying reproduction in closed freshwater lakes; under normal circumstances the only way for a fish to get into the lake is to be hatched and the only way for it to leave the lake is to die.

The Mating System and Effective Population Size

In the following discussion of mating systems and effective population size, I will employ Crow's *I*, the index of the opportunity for selection, as a useful metric (Crow 1958; Wiegmann 1990). This index is defined as

$$I = \frac{\text{variance}}{\text{mean}^2};$$

where the values of the variance and mean are measures of the individual fitness with respect to some trait. Here I use the number of matings obtained by individual males as both the trait and the measure of fitness (Wade and Arnold 1980; Wiegmann 1990). The variable I assumes all the variation observed is genetic in origin; thus it is an index of the maximum effect natural selection could have on the trait in one generation. The square root of this index value is the upper limit in standard deviations that the mean fitness can move due to selection in a single generation (Arnold and Wade 1984). However, it is a misnomer to call I the opportunity for selection. It is actually the opportunity for genetic change in fitness from one generation to the next; I could also be called the index of the opportunity for genetic drift. It is a useful metric in this context because it will reflect the potential for genetic change (i.e., loss of genetic diversity) from one generation to the next. Because some males leave more offspring than others, I_m (m = males; I_m = index of the opportunity for selection on males due to the mating system) measures the potential for genetic change due to the effects of the mating system. It would have its minimum value, 0, when all males in the population obtained exactly the same number of matings. Its maximum value would be obtained when only one male in the population mated with all the females. A low I_m value would imply that mating system competition would have little effect on genetic change from one generation to the next; a high I_m value would indicate that the mating system could strongly affect the genetic composition of the next generation.

The most striking aspect of the smallmouth bass mating system is the relatively small proportion of adults that breed in a given year (Table 1). Only about one third of the adult male population spawned in a typical year, and no more than 50%

TABLE 2.—Indices of opportunity for selection (I) measured for male smallmouth bass in Nebish Lake. Index based on number of matings from Wiegmann 1990; index based on number of offspring from Raffetto et al. 1990.

	I	
Year	Based on matings	Based on offspring
1982	8.58	14.06
1983	2.00	5.24
1984	2.37	5.02
1985	2.28	5.34

even attempted to nest. Females showed a similar pattern, but a higher percentage (about 45%) managed to breed in a given year (females mature a year later than do males; Raffetto et al. 1990, Wiegmann et al. 1992). These data for males yield the I_m values shown in the first column of Table 2. Although the population is monogamous, the index values shown here rival or exceed such male values for a wide variety of polygynous mating systems (Table 3). Indeed, for some years, the index values are among the highest ever measured in natural populations. This is a direct result of the large "zero class" consisting of males who fail to mate in a given year and hence fail to have their genes represented in the resulting year-class.

Even a random mating system has a significant potential for facilitating genetic change. The panmixis assumption of population genetics, often called random mating, is not truly random mating. A random mating system for most sexually reproducing dioecious organisms would have a Poisson

TABLE 1.—The proportion of adult male smallmouth bass that breed in a given year in Nebish Lake, Wisconsin. Data from Raffetto et al. 1990; Wiegmann 1990; Baylis et al. 1993.

	Year			
Population estimates	1982	1983	1984	1985
Number of adult males[a]	525	755	503	629
Number of nests[b]	88	413	191	291
Nests with eggs[b]	56	252	157	193
Number of males mated[c]				
Monogamously	47	84	36	44
Bigamously	1	0	2	0
Percentage of adult males breeding[d]	11%	33%	31%	31%

[a]Annual population estimate based on tagging and aging fish.
[b]Based on spawning season census.
[c]Based on subsample of breeding adult males.
[d]Number of nests with eggs divided by number of adult males.

TABLE 3.—Values of opportunity for selection indices (I) based on number of matings or number of offspring for males in polygynously mating populations (from Downhower et al. 1987).

	I	
Species	Based on matings	Based on offspring
Sage grouse *Centrocercus urophasianus*		
I	3.81	
II	8.74	
Prairie chicken *Tympanuchus cupido*	6.72	
Ruff *Philomachus pugnax*		
I	4.66	
II	1.57	
Village indigobird *Vidua chalybeata*		
I	3.77	
II	7.08	
Bullfrog *Rana catesbeiana*		2.34
Red-winged blackbird *Agelaius phoeniceus*		
I		4.78
II		6.22

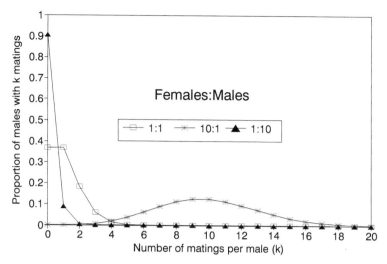

FIGURE 1.—The expected distribution of matings per male in a population in which each female mates only once per season, assuming random mating (Poisson distribution) and three different sex ratios.

distribution of number of matings per individual male and would be a polygynous mating system because males can mate more frequently than can females (also see Nunney 1993).

The proportion (Pr) of individual males in the population expected to obtain x matings under random mating is given by the equation

$$Pr(x = k) = \frac{m^k e^{-m}}{k!};$$

where $k = 0, 1, 2, 3, \ldots$ and $m > 0$. The number of matings obtained by individual males is given by k; m is the mean number of matings expected, the single Poisson parameter; and e is the natural base of logarithmic functions. In the case of females mating only once each, m is determined solely by the sex ratio and is obtained by dividing the number of adult females in the population by the number of adult males. Figure 1 is a plot of the Poisson distribution for three different values of m based on sex ratio. In the case of an equal sex ratio, the mean number of matings a male can expect is 1, yet under random mating about one-third of the males obtain no mates, one-third obtain one mate, and about one-third obtain more than one mate. For this case I_m is equal to 1; panmixis would have I_m equal to 0. Note that we need not assume females mate only once; the 10 females to 1 male curve (Figure 1) would also describe the distribution expected if the sex ratio were equal and each female mated an average of 10 times. This case has an I_m equal to 0.1,

much closer to panmixis. The case of 1 female to 10 males (Figure 1) has an I_m equal to 10, a very high potential for change. Thus, random mating is quite far from panmixis in real populations. The mating system can make small populations even more susceptible to genetic change than one might expect from population size alone; in a small population this is likely to mean the loss of the very genetic diversity one is attempting to preserve. However, the potential size of this effect is easily estimated from the mean and variance in mating success; hence I_m could be a useful tool in evaluation of the genetic stability of candidate ESU populations.

The examples in Figure 1 also show how deceptive it can be to sample just the breeding population. All three mating systems shown are random, yet if one observed only the breeders, the 1:1 sex ratio would appear to be a polygynous mating system, with one-third of the males obtaining more than one mate; the 1:10 sex ratio would appear to be monogamous, because over 90% of the males who mate, mate with only one female; and, of course, the 10:1 sex ratio would appear to be a strongly polygynous system.

Life History Characteristics and Reproductive Competition

In smallmouth bass, whether or not an individual breeds at a given age is strongly size and age dependent (Figure 2). For age-3 males, the largest size-classes are significantly more likely to breed

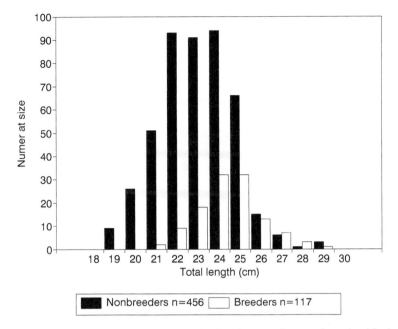

FIGURE 2.—The size distribution of age-3 males captured as breeders guarding nest (open bars) in the 1989 breeding season and during the general population census (closed bars) in 1989 in Nebish Lake. Fish captured during the census who later showed up as breeders ($N = 80$) do not appear in the nonbreeder histogram. The two distributions are significantly different by a Kolmogorov–Smirnov two sample test, $P < 0.05$.

during their third year. Smaller males within a cohort first breed at age 4 or later. Although the data shown here are for the 1986 cohort, the pattern holds for every cohort we have followed (1982–1987; 1988 and later cohorts still have living members). Ridgway et al. (1991) found a similar relationship in a separate study on a population of smallmouth bass in Ontario. Back-calculated growth data from scales obtained from breeding males in our population are also consistent with this. Males breeding for the first time at age 4 were smaller at age 3 than were males from the same cohort who bred at age 3 (Raffetto et al. 1990; Baylis et al. 1993). Indeed, the size difference between breeders at age 3 and fish who breed at age 4 or more within a cohort are consistent over their entire lives (Lee's phenomenon; Busacker et al. 1990). Sizes back-calculated to age 1 show a significantly different distribution between these two groups in size at age 1 (Kolmogorov–Smirnov two-sample test, maximum difference = 0.58, $P < 0.001$; Figure 3). There is also a positive relationship between back-calculated size at age 1 and size at breeding for the cohort within each breeding season ($r^2 = 0.24$, 0.26, and 0.67 for ages 3, 4, and 5 respectively; Figure 4). Thus, size at age 1 is a

predictor of both the age at which a male will first breed and its size at first breeding. However, absolute size of the males in each cohort is larger with age (Figure 4). Although the data shown are for the 1983 cohort, these trends are consistent for age-3 and age-4 fish of all cohorts we have examined (1982–1987).

Within a breeding season, larger males spawn earlier in the season than do smaller males (Figure 5; see also Ridgway et al. 1991; Baylis et al. 1993). Because size is strongly correlated with age, this results in older fish breeding earlier in a season than younger fish (Wiegmann et al. 1992; Baylis et al. 1993). At any given time, four to six overlapping cohorts of males are breeding within any given season (Baylis, University of Wisconsin, unpublished data). Thus, the large fish within the 1983 cohort that breed in 1986 (Figures 3 and 4) are small males when compared with all the males breeding within the 1986 breeding season. Moreover, the evidence of Lee's phenomenon described above is easily explained by the high mortality rate (95%) observed among breeding males (Baylis et al. 1993; see also Ridgway and Shuter 1994). The largest fish within a cohort are the most likely to breed at any given age, and the cost of breeding is an extremely high

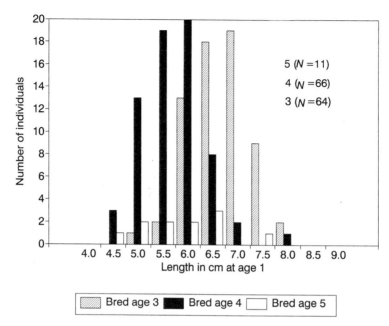

FIGURE 3.—Back-calculated size estimates (scale data) at age 1 for individual breeding males from the 1983 year-class ($N = 141$) in Nebish Lake. Males who bred at age 3 were significantly larger at age 1 than were males who bred for the first time at age 4 or age 5.

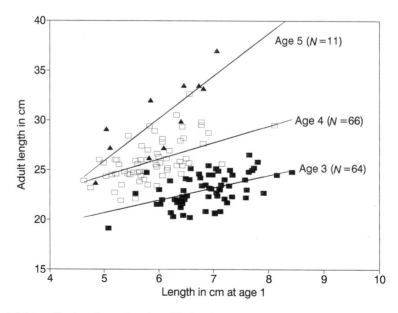

FIGURE 4.—Adult breeding length as a function of back-calculated length at age 1 for males who first bred at age 3 ($y = x(1.26) + 14.35$; $r^2 = 0.241$), age 4 ($y = x(1.66) + 16.04$; $r^2 = 0.262$), or age 5 ($y = x(4.27) + 4.48$; $r^2 = 0.67$). Data are from the same individual fish as in Figure 3.

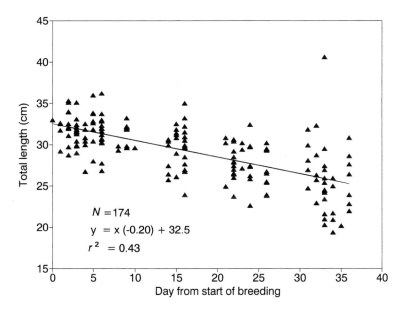

FIGURE 5.—The relationship between size of the males breeding during the 1993 breeding season in Nebish Lake and the number of days from the first nesting male (20 May 1993). The males are from five overlapping cohorts.

mortality rate. Other centrarchids also exhibit this relationship between growth, reproduction, and mortality (Gross 1982; Danylchuk and Fox 1994).

Baylis et al. (1993) suggest that smallmouth bass in Nebish Lake have an alternating life history that is the result of an inverse relationship between the life history characteristics of offspring and their parents. Newly independent larval smallmouth bass grow at about 1 mm/d for the first 30 d of life (Carlander 1977). Eggs spawned early in the season can produce larvae that are over 20 mm long, but eggs spawned late in the season produce larvae 9 mm long at the end of breeding. If this size and spawning date relationship contributes to size variation at age 1 (see above), then large fish within a cohort would tend to come from early breeders. Because early breeders within a season are older fish who were smaller at age-1 within their cohort, they in turn would tend to be the products of younger, later spawners within a season. This results in an alternating life history (see Figure 5 of Baylis et al. 1993; see also Tsukamoto et al. 1987). Fish spawned late within a season tend to produce small fish within a cohort, who in turn breed later in life at a large size. Because these fish are large, they breed early within a season and produce offspring who are large within a cohort. These larger offspring breed at a young age but at a small absolute adult size.

Thus they spawn late in a season and as a result produce small young within a cohort.

Such a life history pattern depends on microvariation in age affecting growth rate and size at reproduction over the whole life of the fish. I have presented evidence that such microvariation may have just such a macroscopic effect in smallmouth bass; evidence in the literature suggests this effect in other fishes (Tsukamoto et al. 1987; Goodgame and Miranda 1993) and even in humans (Dudink 1994). In the case of the smallmouth bass and possibly other centrarchids, the size-related competition within a breeding season and overlapping generations produce a negative parent–offspring correlation in life history characteristics. However, a positive correlation could just as easily occur. If smallmouth bass were an annual species, for example, then large fish breeding early in a season and small fish breeding late would leave offspring with the same characteristics as the adults that produced them. This could easily produce two distinct forms whose characteristics would appear to be heritable, although the characteristics would not be genetically transmitted.

Discussion

This 12-year study of smallmouth bass reproduction in Nebish Lake has two important implications for the designation of ESUs as a management tool. The first is that the mating system of the population should be examined when considering a population for ESU status. As we have shown in Nebish Lake, the breeding population may be only a small fraction of the adult fishes, and the mating system can profoundly reduce effective population size. This in turn can influence the susceptibility of the population to genetic change from chance forces that may cause the loss of the very genetic diversity one is trying to preserve by ESU designation. However, quantitative tools such as Crow's index of the opportunity for selection can be used to estimate such effects of the mating system. In general, the mating system almost always acts to reduce effective population size and may impose significant intergenerational changes in gene frequency, even in the case of random mating. However, a best-case scenario is easily calculated by estimating the mean number of matings for males (m, the Poisson parameter). By multiplying the mean number of times a female mates by the number of females in the population and dividing by the number of males, m can be estimated and I_m can be calculated assuming Poisson mating (the variance is equal to the mean). If this I_m value is unacceptably high, one may not want to depend upon natural reproduction, and intervention may be required to preserve genetic diversity in the population.

Because most adults fail to breed, stocking would not help increase reproduction in a population like the one in Nebish Lake; it would only increase competition. Improvement of spawning habitat would be necessary to increase the number of breeding adults.

The second implication is the observation that there may be strong nongenetic effects of parental life history traits on offspring traits, and these effects can be positive or negative in sign. Because of rapid early development in fishes, minor differences in age within a year-class or minor differences in early physical environment may produce major life history variation in the subsequent adult population. Differences between water bodies in physical variables such as rate of temperature rise or fall may produce life history differences between populations that are not due to genetic variation. Caution should be used when assigning ESU status based on life history characteristics alone. Timing of reproduction and the number of growing days within a season may have particularly strong effects, and when these vary from basin to basin they may falsely appear to be adaptive life history strategies. For example, the prevalence of sneakers among some

hatchery salmonids may simply be due to early maturation as a consequence of rapid growth (Gross 1991). The relationship between rapid growth, reproduction at an early age, and small adult body size may be quite general in animals (Charnov 1993).

In spite of this caveat, the population-level implication of life history effects based on early developmental history is that many native fish populations are probably uniquely differentiated populations. In an alternating life history, there is a negative correlation between the life history characteristics of parents and offspring. This would tend to be inherently stabilizing in a population with a sufficiently long breeding season, especially one with overlapping generations. The adult population would then tend to be a smeared average of several cohorts that would buffer the demography of the population as a whole from year-to-year variation in environmental effects on development. However, one can picture a scenario of a smallmouth bass population in a lake that consistently has a rapid and early temperature rise in the spring to temperatures above 20°C and a long season of water temperatures above 11°C. Smallmouth bass in this lake would breed early, and their offspring would grow rapidly and tend to breed early in life (age 3 or younger); only part of the alternating cycle would be present, as only early breeders can spawn. These fish would tend to show rapid growth, early maturation, but relatively small adult body size. Another lake with a much later temperature rise and a shorter growing season above 11°C would tend to produce smallmouth bass that are slow growing, breed later in life (age 4 or older), and have a large adult body size. Thermal conditions that vary from lake to lake could cause the resulting populations to skew towards one or the other of the two life histories on a consistent basis. Based on thermal regime alone, we might expect basin-to-basin variation in life history traits and resulting population demography. Even though the life history traits may not be due to genetic variation themselves, they are strongly correlated with components of fitness and create very different demographic and social environments, and hence generate sources of variation through which natural selection can act on the gene pool. Even if populations started with a common gene pool, environmentally induced variation in life history traits from population to population should produce rapid change in coadapted complexes of genes, leading to differentiation of isolated populations. As noted above, in small populations the mating system can be expected to exacerbate this effect.

In many species the environmental correlation between adult and offspring life history characteristics could be positive. For example, if a fish matured at a specific age or was an annual species, early spawned fish would be larger, grow more during the year, and spawn early in the next season; late spawned fish would be smaller, grow less, and breed late in the next season. In this case, like would produce like. Any selection for early and late spawning and against spawning in the middle of the season could split the population into two distinct forms based on life history attributes consequent to spawning date. Depending upon the degree of reproductive isolation produced by the asynchronous spawning, the population could fission into sibling species, but the differences in life history would precede the speciation. Such a mechanism could be a partial explanation for the apparent frequency of cases of sympatric speciation in freshwater lakes; it is interesting that the firmest evidence for sympatric speciation is coming from the smallest lakes (Schliewen et al. 1994).

Wright's (1932) shifting balance theory seems fraught with implication for freshwater fishes. Semi-isolation and small effective population size have been the rule for many of these fish populations over much of their evolutionary history. Whereas the historical degree of isolation among fish populations has varied greatly depending on habitat and drainage, human activities and management practices have served to disrupt much of the genetic isolation between fishes in different water bodies, and we treat fish species as if they are megapopulations with little meaningful variation among component demes. Much damage has already been done to native populations. Damming rivers has disrupted spawning migration patterns in anadromous and potamodromous forms, producing one-way barriers to gene flow where none existed before. In other instances, "bait bucket" introductions and purposeful management practices such as stocking of fishes between water bodies and hatchery rearing are suspect enterprises that may have already compromised the genetic uniqueness of many populations.

Acknowledgments

I thank the University of Wisconsin Department of Zoology, the WDNR, and Trout Lake Biological Station for their support of this work. Much of the credit for this work goes to my collaborators on this project, Mike Hoff, Nancy Raffetto, and Dan Wiegmann. Very special thanks go to all those unpaid undergraduate volunteers who have worked as the "Nebish Lake Cavaliers" to collect the data. Bill Feeny prepared the figures. The comments of Mart Gross, June Mire, and Jennifer Nielsen greatly improved the manuscript. This research was supported in part by the Graduate School of the University of Wisconsin and the Electric Power Research Institute.

References

Arnold, S. J., and M. J. Wade. 1984. On the measurement of natural and sexual selection: theory. Evolution 38:709–719.

Baylis, J. R., D. D. Wiegmann, and M. H. Hoff. 1993. Alternating life histories of smallmouth bass. Transactions of the American Fisheries Society 122:500–510.

Busacker, G. P., I. R. Adelman, and E. M. Goolish. 1990. Growth. Pages 363–387 in C. B. Schreck and P. B. Moyle, editors. Methods for fish biology. American Fisheries Society, Bethesda, Maryland.

Carlander, K. D. 1977. Handbook of freshwater fishery biology, volume 2. Iowa State University Press, Ames.

Charnov, E. L. 1993. Life history invariants. Oxford University Press, Oxford, UK.

Crow, J. F. 1958. Some possibilities for measuring selection intensities in man. Human Biology 30:1–13.

Danylchuk, A. J., and M. G. Fox. 1994. Age and size-dependent variation in the seasonal timing and probability of reproduction among mature female pumpkinseed, *Lepomis gibbosus*. Environmental Biology of Fishes 39:119–127.

Downhower, J. F., L. S. Blumer, and L. Brown. 1987. Opportunity for selection: an appropriate measure for the potential for selection? Evolution 41:1395–1400.

Dudink, A. 1994. Birth date and sporting success. Nature 368:592.

Goodgame, L. S., and L. E. Miranda. 1993. Early growth and survival of age-0 largemouth bass in relation to parental size and swim-up time. Transactions of the American Fisheries Society 122:131–138.

Gross, M. R. 1982. Sneakers, satellites and parentals: polymorphic mating strategies in North American sunfishes. Zeitschrift fuer Tierpsychologie 60:1–26.

Gross, M. R. 1991. Salmon breeding behavior and life history evolution in changing environments. Ecology 72(4):1180–1186.

Horn, M. M. 1972. The amount of space available for marine and freshwater fishes. U.S. National Marine Fisheries Service Fishery Bulletin 70:1295–1297.

Nunney, L. 1993. The influence of mating system and overlapping generations on effective population size. Evolution 47:1329–1341.

Raffetto, N. S., J. R. Baylis, and S. L. Serns. 1990. Complete estimates of reproductive success in a closed population of smallmouth bass (*Micropterus dolomieui*). Ecology 71:1523–1535.

Ricker, W. E. 1975. Computation and interpretation of biological statistics of fish populations. Fisheries Research Board of Canada Bulletin 191.

Ridgway, M. S., and B. J. Shuter. 1994. The effects of supplemental food on reproduction in parental smallmouth bass. Environmental Biology of Fishes 39:201–207.

Ridgway, M. S., B. J. Shuter, and E. E. Post. 1991. The relative influence of body size and territorial behaviour on nesting asynchrony in male smallmouth bass, *Micropterus dolmieui* (Pisces: Centrarchidae). Journal of Animal Ecology 60:665–681.

Schliewen, U. K., D. Tautz, and S. Paabo. 1994. Sympatric speciation suggested by monophyly of crater lake cichlids. Nature 368:629–632.

Tsukamoto, K., R. Ishida, K. Naka, and T. Kajihara. 1987. Switching of size and migratory pattern in successive generations of landlocked ayu. American Fisheries Society Symposium 1:492–506.

Wade, M. J., and S. J. Arnold. 1980. The intensity of sexual selection in relation to male sexual behaviour, female choice, and sperm precedence. Animal Behaviour 28(2):446–461.

Waples, R. S. 1995. Evolutionarily significant units and the conservation of biological diversity under the Endangered Species Act. American Fisheries Society Symposium 17:8–27.

Wetzel, R. G. 1983. Limnology, 2nd edition. Saunder & Co., Philadelphia.

Wiegmann, D. D. 1990. On assessing the potential for evolutionary change due to male–male competition and female choice in territorial species. Journal of Theoretical Biology 144:203–208.

Wiegmann, D. D., J. R. Baylis, and M. H. Hoff. 1992. Sexual selection and fitness variation in a population of smallmouth bass, *Micropterus dolomieui* (Pisces: Centrarchidae). Evolution 46:1740–1753.

Wright, S. 1932. The roles of mutation, inbreeding, crossbreeding, and selection in evolution. Proceedings of the 6th International Congress of Genetics 1:356–366.

Wright, S. 1980. Genic and organismic selection. Evolution 34(5):825–843.

American Fisheries Society Symposium 17:227–244, 1995

Evolutionarily Significant Units among Cichlid Fishes: The Role of Behavioral Studies

JAY R. STAUFFER, JR. AND N. J. BOWERS[1]

School of Forest Resources, The Pennsylvania State University
University Park, Pennsylvania 16802, USA

KENNETH R. MCKAYE

Appalachian Environmental Laboratory, University of Maryland
Frostburg, Maryland 21532, USA

THOMAS D. KOCHER

Department of Zoology, University of New Hampshire
Durham, New Hampshire 01823, USA

Abstract.—Cichlid fishes represent an outstanding case of explosive evolution and offer extraordinary opportunities to investigate the evolutionary processes that have led to such diversity. Throughout the world, however, these fishes are threatened by overfishing, introduction of exotics, habitat destruction, and pollution of the environment. Determination of the specific status of local taxonomic units is critical for the development of programs both to conserve and to utilize these fishes for food, tourism, disease control, and scientific investigations. Rapid speciation within these fishes, however, has resulted in a paucity of characters for discriminating among species. Our experiences in Africa and Central America demonstrate that in situ behavioral studies, integrated with morphological and genetic analysis of taxonomic units, are vital to determining the specific status and relationships among evolutionarily significant units (ESUs). The critical element in determining whether a taxon is an ESU is knowledge of its reproductive biology; therefore, it is imperative that we develop a multidisciplinary emphasis in biodiversity studies.

The phrase evolutionarily significant unit (ESU) implies that (1) a heritable difference exists among populations; (2) an important statistical difference exists in a group of characters among units; and (3) a classification system is being used. From a pure conservation point of view, any such ESU must be protected. We are not suggesting that the term ESU replace our concept of a species or other formally recognized taxonomic category but that it be used to recognize unique entities that need protection. For example, Waples (1991) suggested that a population should be considered an ESU if it is reproductively isolated from other conspecific populations and if it represents an important component in the evolutionary trajectory of the species. Evolutionarily significant units may also be defined geographically, in that they may be a particular community or ecosystem that harbors a highly diverse fauna or flora or is a site of high endemism. Portions of a widespread population that has a disjunct distribution may be designated as an ESU. For example, the longnose sucker *Catostomus catostomus* (Forster) is panmictic; however, there exists a

small disjunct population in the Monongahela River system in West Virginia, Maryland, and Pennsylvania (Stauffer et al. 1995). If this disjunct population were designated as an ESU, then perhaps a vehicle would be in place to protect this unique population of a widely dispersed species.

Minimally, an ESU may be a population that exhibits a distinctive behavior. The importance of behavior in distinguishing among fish taxa was pioneered by Trewavas (1983), who used behavioral characters when delimiting three genera of tilapiine fishes. In many cases, behavioral studies are instrumental in recognizing novel entities, assigning populations to taxa (Brooks and McLennan 1991), and estimating phylogenetic relationships among taxa (Wenzel 1992; deQueiroz and Wimberger 1993).

Nowhere is the designation and protection of ESUs needed more than in tropical ecosystems. It is estimated that as many as half of the extant species inhabit the approximately 6% of the earth covered with tropical rain forests (Myers 1988). With respect to fishes, there are 66 families endemic to tropical freshwaters, whereas only 18 are endemic to temperate freshwaters (Berra 1981); moreover, greater than 70% of the described species of fishes inhabit the tropics (Moyle and Cech 1988). One of the most speciose families of freshwater fishes is the

[1]Present address: Environmental Sciences and Resources, Portland State University, Portland, Oregon 97207, USA

Cichlidae, thus many of the examples that follow will be from this family.

Species Concepts

In part, the concept of the ESU involves grouping individuals or populations into distinct taxa, which, in turn, depends on the definition of species or some lower hierarchical taxon. Subsequent to the evolutionary synthesis (Mayr 1982a; Eldridge 1985) there has been much debate concerning species concepts (e.g., Simpson 1961; Wiley 1981; Donoghue 1985; Paterson 1985; Templeton 1989; Mayr 1992; van Devender et al. 1992). This debate can be attributed to a certain degree to some biologists treating species as epiphenomena, whereas others regard species as participants in the evolutionary process (Mayr and Ashlock 1991). We would agree with Mayr (1992) that a nondimensional (nonhistorical) concept of the species is the one with which most biologists are concerned and which is probably the most applicable to conservation and protection programs. We argue, however, that it is difficult to develop an unambiguous species definition given the mixture of conspecific populations, incipient species, and good species that predominate in allopatric populations of freshwater fishes, such as the cichlids. Hence, the ESU provides an effective concept upon which to base conservation practices when dealing with rapidly evolving groups, such as the cichlids.

Speciation

The concept of speciation involves the origin of a unique gene pool. The processes responsible for the ecological separation and reproductive isolation of populations have long been debated. Intralacustrine allopatric speciation has been widely purported to account for the rapid and extensive speciation by cichlid fishes in the African Great Lakes (Fryer 1959; Fryer and Iles 1972; Mayr 1982b). The first stage in allopatric speciation is geographical segregation of a single population into two or more subpopulations. Speciation culminates with the development of reproductive isolating mechanisms that prevent interbreeding even if the geographical barriers are removed and the populations experience secondary contact (Mayr 1942). Both pre- and postmating isolating mechanisms influence reproductive isolation among heterospecific populations. Postmating isolating mechanisms include gametic mortality, zygotic mortality, hybrid inviability, and hybrid sterility; premating isolating mechanisms include incompatible reproductive anatomy, ecological separation, ethological isolation, and allochronic mating. The development of many premating barriers are the direct consequence of changes in behavioral characters.

Several investigators have suggested that speciation of cichlids may have occurred sympatrically as well as allopatrically (Fryer and Iles 1972; McKaye et al. 1990). In sympatric speciation models, reproductive isolating mechanisms originate within the dispersal area of the offspring produced by a single deme (Hartl and Clark 1989) and premating isolation develops before populations inhabit distinct niches (Bush 1975). Controversy over the mechanisms of sympatric speciation center around the question of how reproductive isolation can arise prior to a barrier to gene flow (Mayr 1982b). Kosswig (1963) suggests that populations can be isolated ecologically without overt geographical barriers, due to differences in habitat preference in a varied environment. Factors that may contribute to ecological isolation of populations include competitive isolation (McKaye 1980), seasonal isolation (Lowe-McConnell 1959), mate selection isolation (Trewavas et al. 1972; Barlow and Munsey 1976), and runaway sexual selection (Dominey 1984; McKaye 1991; McKaye et al. 1993). In addition, intrapopulational variation in the expression of a given genotype due to environmental conditions permits the maximum use of a heterogenous habitat (Liem and Kaufman 1984; Via and Lande 1985). Within the cichlids, alternative adaptations (polymorphisms) may also have contributed to the extensive adaptive radiation and sympatric coexistence of closely related forms (West-Eberhard 1983).

Cichlid fishes throughout the tropics and specifically in the Great Lakes of Africa are generally recognized as one of the most dramatic examples of extensive trophic radiation and explosive speciation. Discrimination among species of Cichlidae can be difficult because differences among species may be very small and intraspecific variation may be relatively large (Fryer and Iles 1972; Ribbink et al. 1983). The acquisition of reproductive isolation without significant morphological change makes it difficult to distinguish African haplochromine cichlids (Lewis 1982). Attempts to use starch gel electrophoresis have been inconclusive for delimiting species (Kornfield 1974, 1978). McKaye et al. (1982) electrophoretically examined three color morphs of *Petrotilapia tridentiger* Trewavas (a cichlid endemic to Lake Malaŵi) that could not be distinguished morphometrically. They found no fixed alleles at any of the 25 loci studied, although allele frequencies were heterogeneous among taxa, suggesting that the color morphs represented isolated gene pools or incipient species. Marsh (1983) subsequently described these morphs as distinct species.

Mitochondrial DNA (mtDNA) has been widely recognized as an important tool for resolving relationships among closely related species. Mitochondrial DNA has also been used to delimit higher taxonomic categories. Meyer et al. (1990) used mtDNA sequence divergence to demonstrate the monophyly of the Lake Victoria cichlid species flock, and Meyer et al.'s data suggest the possible monophyly of the Lake Malaŵi flock. Monophyly of the Lake Malaŵi flock has been implied by morphological studies (Stiassny 1981) and supported by additional mtDNA analyses (Kocher et al. 1993). Moran et al. (1994) conducted studies of phylogenetic relationships among African cichlids by means of restriction fragment length polymorphism (RFLP) analysis of mtDNA. Recent work based on DNA sequencing indicates that mtDNA may be adequate for discriminating among Lake Malaŵi cichlids in some lineages (Bowers et al. 1994). Moran and Kornfield (1993) caution, however, that the rapid speciation of Malaŵian cichlids may have prevented sorting of mitochondrial lineages, allowing distantly related species of Lake Malaŵi cichlids to share mtDNA polymorphisms derived from a common ancestor. These results suggest that mtDNA data alone cannot delimit certain Lake Malaŵi taxa.

Detailed behavioral studies, however, have consistently proven useful in distinguishing among species. Many morphologically and genetically similar species can be separated based on breeding coloration and behavioral characteristics (Ribbink et al. 1983; Witte 1984; McKaye and Stauffer 1986; Stauffer 1988; Stauffer and McKaye 1988; Stauffer and Bolts 1989; Stauffer et al. 1993). Holzberg (1978) and Schröder (1980) first used behavioral observations to conclude that the blue-black color form of *Pseudotropheus zebra* (Boulenger) was reproductively isolated from the blue color morph *Pseudotropheus callainos* Stauffer and Hert (*Pseudotropheus* abbreviated as *P.* hereinafter). That many cichlid species, when artificially crossed under laboratory conditions, can produce viable hybrid offspring forces the taxonomist to rely solely on the study of premating isolating mechanisms when delimiting species. Thus, behavior plays a significant role in defining sympatric species and is essential in inferring whether or not allopatric species would potentially exhibit reproductive isolation.

Sexual Selection

Both natural and sexual selection have contributed to speciation within Cichlidae. The frequently conflicting forces of natural and sexual selection were first noticed by Darwin (1871). Natural selection arises from differential viability and fertility, whereas sexual selection results from differential mate acquisition. In effect, a particular male trait can be a handicap in terms of survival but result in more fertilizations (Trivers 1972; Nur and Hasson 1984). Sexual selection pressures can shift mean male character values far from their equilibria attained under natural selection alone (Kirkpatrick and Ryan 1991). Although sexual dimorphism can arise from other causes (Lande 1980; Hedrick and Temeles 1989), it is often a useful indicator of the magnitude of sexual selection acting on a character. Commonly observed dimorphisms in body size, plumage, coloration, or weaponry can often be ascribed to this force.

In a recent review, Kirkpatrick and Ryan (1991) classified models of female-choice selection according to whether selection on preferences was direct or indirect. They concluded that in many species preferences evolve in response to direct selection on female fitness. For example, female convict cichlids, *Cichlosoma nigrofasciatum* (Günther) consistently prefer larger males when given a choice between two mates (Noonan 1983; Keenleyside et al. 1985). This preference may be interpreted as direct selection for reproductive success because larger males provide better defense and resources for the young. Female preferences for males with larger nuptial gifts (Thornhill and Alcock 1983) or for those carrying a lower load of a communicable disease (Borgia and Collis 1990) have a direct positive effect on female fitness.

Several models can be classified as invoking indirect sexual selection on male and female preferences. In the "good genes" models, female preference is derived from the improved fitness of a female's progeny because of genes acquired from the male. One such model postulates that females prefer males carrying genes that make those males resistant to parasites (Hamilton and Zuk 1982).

Conversely, "nonadaptive" models have been postulated in which female preference is not related to the forces of natural selection acting on the population. Hert (1989) demonstrated that the egg spots of male *Astatotilapia elegans* Trewavas could stimulate spawning and that female *P. aurora* Burgess spawned more frequently with males possessing higher numbers of egg spots (Hert 1991). Fisher

(1930) was the first to propose a "runaway" process, which has since been extensively modeled (O'Donald 1980; Lande 1981; Kirkpatrick 1982) and discussed (Arnold 1983; Kirkpatrick 1987). One feature in the nonadaptive models is that the runaway process can be initiated by arbitrary female preferences, and several recent studies have shown that female preference for particular male characters can evolve long before the characters themselves. Basolo (1990, 1991) demonstrated a preexisting preference for caudal swords in swordless species of the poeciliid *Xiphophorus*. Meyer et al. (1994), however, provided genetic evidence suggesting that the ancestor of this genus possessed a sword. Preferences may frequently arise from sensory biases (Ryan and Keddy-Hector 1992) and may be an inherent property of sensory systems (Enquist and Arak 1993). Kirkpatrick and Ryan (1991) interpret this to mean that direct selection was responsible for the evolution of female preferences. Ryan and Rand (1993) have stressed the importance of recognizing that sexual selection and species recognition are elements of a single process: the matching of male signal traits to female preference function.

The existence of speciose flocks of animals restricted to isolated habitats may best be explained by sexual selection in many cases. The large number of *Drosophila* species endemic to Hawaii led Ringo (1977) to elaborate on the hypothesis of Spieth (1974) that sexual selection can accelerate the divergence of populations. Carson (1978) suggested that sexual selection could create coevolutionary races between particular male characters and female preferences, leading to the evolution of increasingly complex courtship behaviors. Possible interaction of founder effects and sexual selection during speciation was suggested by Kaneshiro (1989) as an explanation for the *Drosophila* species flock. Dominey (1984) generalized these hypotheses to account for rapid speciation in African cichlids and recognized that the cichlids share many characteristics with the Hawaiian *Drosophila*, including sexual dimorphism, lek-based breeding systems invoking a high degree of female choice, and isolated local populations.

We propose that the variations observed in male coloration, bower size (breeding platform), and courtship behavior among closely related cichlid species are the result of intraspecific sexual selection (McKaye 1991). In many instances, morphologically similar populations may in fact be subspecies, sibling species, or incipient species at various stages of speciation (Mayr 1963). Divergence in female preference for male secondary sexual traits

may lead to assortative mating of populations prior to a sympatric speciation event or during secondary contact following allopatric speciation; thus, one or several sexually selected traits may become differentiated with each speciation event. Strong sexual selection may cause differentiation of breeding behaviors even in the face of considerable gene flow and among diverging populations in secondary sympatry. Natural selection may act to differentiate morphological and behavioral traits further. Therefore, it is our contention that the use of both morphological and behavioral data to delimit closely related species, such as the Lake Malaŵi cichlids, is essential. Below we discuss the use of color, bower shape, courtship behavior, and feeding behaviors to discriminate among cichlid species. We consider color form and bower shape to be manifestations of behavioral characteristics via female choice. In many cases, behavioral studies may first identify novelties that indicate which specific forms might be valid species.

Case Histories Demonstrating the Value of Behavioral Studies

Role of Color in Delimiting Species

The incredible variety of color patterns within the haplochromine cichlids of the African Rift lakes is well known (see Figures 1a–f; Fryer and Iles 1972; Greenwood 1981; Ribbink et al. 1983; McKaye and Stauffer 1986), and we consider it to be essential in female mate selection. The existence of unique color patterns is recognized to be suitable for delimiting species (Barlow 1974; Barel et al. 1977; Greenwood 1981; Hoogerhoud and Witte 1981; McKaye et al. 1982, 1984), and in many cases new species have been recognized solely on the basis of male color pattern (McElroy et al. 1991). Although color is certainly a morphological character, we regard it as a manifestation of female preference, which is a behavioral trait.

The following rock-dwelling (mbuna) taxa were first hypothesized to be valid species based on male breeding color and later substantiated based on morphometrics and meristic data: *P. aurora* (Burgess 1976), *P. barlowi* (McKaye and Stauffer 1986), *P. flavus* (see Figure 1a), *P. ater*, *P. cyanus* (Stauffer 1988), *P. xanstomachus* (Stauffer and Boltz 1989), and *P. callainos* (Stauffer and Hert 1992), among others.

Holzberg (1978) and Schröder (1980) demonstrated that color patterns of females may also be useful in delimiting species, such as within the *P. zebra* species complex. Male *P. callainos* are pale

FIGURE 1.—Representative examples of the diverse color patterns exhibited by Lake Malaŵi cichlids: (**a**) *Pseudotropheus flavus* from Chinyankwazi Island; (**b**) orange blotch (OB) morph of *Labeotropheus trewavasae* from Thumbi West Island; (**c**) *Melanochromis heterochromis* from Chinyankwazi Island; (**d**) blue-black (BB) color form of *P. zebra* from Thumbi West Island; (**e**) *Cynotilapia afra* from Thumbi West Island; and (**f**) *Chilotilapia* c.f. *rhodesii* from Kanjedza Island.

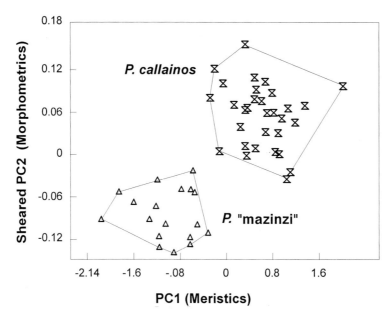

FIGURE 2.—A plot of the second principal component (PC2; morphometrics) and the first principal component (PC1; meristics) based on data from *P. callainos* and *P.* c.f. *zebra* "mazinzi."

blue (Stauffer and Hert 1992) and closely resemble an undescribed *P.* c.f. *zebra* from Mazinzi Reef, Lake Malaŵi. Many female *P. callainos* are white, whereas white females of *P.* c.f. *zebra* from Mazinzi Reef have never been collected. This observed difference in female color pattern prompted us to complete a more detailed morphological study of these two forms. Based on sheared principal component analysis of morphometric data and principal component analysis of meristic data (see Stauffer 1993 for an explanation of the methods employed), the two taxa were shown to be heterospecific (Figure 2).

The importance of color pattern is not limited to the haplochromine fishes in the Great Lakes of Africa. Our work over the past 3 years throughout the Great Lakes basin in Nicaragua has impressed upon us, as it has earlier workers (Meek 1907; Barlow and Munsey 1976), the great variation among cichlids in coloration and body form in isolated water bodies (see Figures 3a–c). For example, in the midas cichlid *Cichlasoma citrinellum* group, several species have been described. With respect to this commercially important group of cichlids, Meek (1907:122) stated, "Of all the species of fishes in these lakes, this one is by far the most variable. I made many repeated efforts to divide this material ... in from two to a half-dozen or more species, but in all cases I was unable to find any tangible con-

stant characters to define them. To regard them as more than one species meant only to limit the number of material at hand, and so I have lumped them all in one." Three species of this group are presently recognized by Barlow and Munsey (1976), although Villa (1982) only recognized two. Our behavioral work, however, confirms that the three species recognized by Barlow and Munsey (1976) are, in fact, valid. Furthermore, our direct underwater observations that these forms assortatively mate by color and that their habitat preferences and nest forms differ suggest that at least three additional undescribed species are also present. Preliminary morphological analyses of two of these forms (Figure 4) confirm that they are distinct from the type specimens housed in the Natural History Museum (London).

Similarity in color patterns, however, may be misleading. For example, many authors (e.g., Fryer and Iles 1972; Ribbink et al. 1983) regard the two populations of the Lake Malaŵi blue-black (BB; Figure 1d) color form of *P. zebra* at Nkhata Bay and Thumbi West Island to be conspecific. McKaye et al. (1984) found differences in allele frequencies between northern and southern populations of BB *P.* c.f. *zebra* although there were no fixed allelic differences. Examination of the morphological data (Figure 5) suggests that these two populations are actually heterospecific. Another example includes

a.

b.

c.

FIGURE 3.—Representatives of the species complex of the midas cichlid *Cichlasoma citrinellum* Günther. (**a**) *Cichlasoma* c.f. *citrinellum* "Xiloa" from Laguna de Xiloa, Nicaragua; (**b**) *Cichlasoma* c.f. *citrinellum* "amarillo" from Laguna de Xiloa, Nicaragua; and (**c**) *Cichlasoma* c.f. *citrinellum* "chancho" from Laguna de Apoyo, Nicaragua.

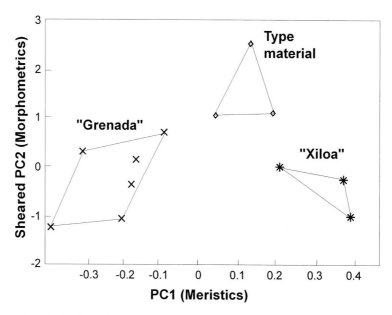

FIGURE 4.—A plot of the sheared second principal component (PC2; morphometrics) and the first principal component (PC1; meristics) based on data from three members of the *Cichlasoma citrinellum* species group: *Cichlasoma citrinellum* type material from the British Museum (Natural History), *Cichlasoma* c.f. *citrinellum* "grenada" from Lake Nicaragua, and *Cichlasoma* c.f. *citrinellum* "Xiloa" from Laguna de Xiloa.

the orange blotch (OB) morphs of many Lake Malaŵi cichlids (e.g. *P. zebra*, *Labeotropheus trewavasae* Fryer [see Figure 1b], and *P. tropheops* Regan). Color differences would initially suggest that these forms are heterospecific with the similarly shaped, normally colored individuals; however, closer examination shows that all OB morphs are female, suggesting that these color forms are not valid species.

Color differences in allopatric populations may also be misleading. For example, two populations of a *Melanochromis* species occur at Chinyamwezi and Chinyankwazi Islands in Lake Malaŵi. Because male coloration differed between the two populations, Ribbink et al. (1983) regarded these taxa to be heterospecific. Examination of the morphometrics and meristics of 13 populations of this form from other locations within Lake Malaŵi revealed slight clinal variation in shape pattern (Bowers and Stauffer 1993), suggesting that these populations are conspecific. This conclusion was supported by allozyme analysis, which showed very low variation at 3 polymorphic loci out of 24 loci that were assayed. Because the morphological evidence indicated no differences in shape among the populations, and variation in male coloration tended to be greater within than among populations, Bowers and

Stauffer (1993) described this form as a single species, *Melanochromis heterochromis* Bowers and Stauffer (Figure 1c).

Color pattern may also provide insight into the phylogeny of certain groups, although care should be taken when interpreting the results. For example, the prevalence of the BB color morph in most of the rock-dwelling cichlid genera (see Figure 1d–e) and some sand-dwelling forms (Figure 1f) throughout Lake Malaŵi suggests that this color pattern is primitive, whether one uses the commonality principle or outgroup comparisons (Smith and Koehn 1971; Watrous and Wheeler 1981). Conversely, the presence of the red dorsal fin within *P.* c.f. *zebra* "red dorsal," *P.* c.f. *zebra* "cobalt mbenji," and *Labeotropheus trewavasae* implies that this character state is a product of convergent or parallel evolution.

In their recent monograph of non-mbuna haplochromines endemic to Lake Malaŵi, Eccles and Trewavas (1989) suggested that similarity of color patterns among species may reflect phyletic relationships. For example, the "polystigma" pattern, which consists of three longitudinally arranged features of either stripes or a series of spots or blotches, is restricted to the genus *Nimbochromis* (Figure 6d). Conversely, the following melanin pat-

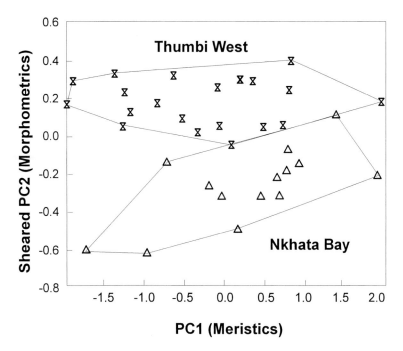

FIGURE 5.—A plot of the sheared second principal component (PC2; morphometrics) and the first principal component (PC1; meristics) based on data from two populations of the blue-black color form of zebra mbuna *P. zebra* from Nkhata Bay (northern population) and Thumbi West Island (southern population), Lake Malaŵi, Malaŵi.

terns are found in more than one genus: "kirkii" pattern (Figure 6a), which emphasizes the horizontal elements of the very common and hence perhaps pleisiomorphic color pattern, is represented by *Nyassachromis breviceps* (Regan), *Lethrinops lethrinus* Günther, and *Protomelas kirkii* (Günther); transverse bars (Figure 6b), is represented by *Placidochromis johnstoni* (Günther), *Lethrinops gossei* Burgess and Axelrod, and *Alticorpus peterdaviesi* (Burgess and Axelrod); "dimidiatus" pattern (Figure 6c), which is a simple, straight, midlateral band, is represented by *Dimidiachromis dimidiatus* (Günther) and *Taeniochromis holotaenia* (Regan); oblique band (Figure 6e), which consists of an oblique band or series of spots from nape to middle of the caudal base, is represented by *Docimodus evelynae* Eccles and Lewis, *Mylochromis anaphyrmus* (Burgess and Axelrod), and *Taeniolethrinops praeorbitalis* (Regan); three-spot patterns (Figure 6f), which consists of a series of spots that appear along the position of the midlateral component of the horizontal element of the plesiomorphic pattern, is represented by *Otopharynx ovatus* (Trewavas) and *Copadichromis quadrimaculatus* (Regan), which have the spots below the upper lateral line, and *Cyrtocara moori* Boulenger and *Ctenopharynx pictus* (Trewavas), which have the spots above or on the

upper lateral line; and "rostratus" pattern (Figure 6g), which consists of three series of large spots approximately in the position of the stripes or rows that constitute the kirkii pattern, is represented by *Fossochromis rostratus* (Regan) and *Eclectochromis festivus* (Trewavas). Consequently, Eccles and Trewavas (1989) considered the rostratus color pattern a result of parallelism and thus uninformative.

Role of Bower Shape in Delimiting Species

Research on the breeding behavior of several Lake Malaŵi sand-dwelling fishes has demonstrated that the process by which females choose mates is complex. McKaye et al. (1990) found a preference for males with larger bowers in female *Copadichromis conophorus* Stauffer, LoVullo, and McKaye. Males of this species form huge leks that may have more than 50,000 males at the height of the breeding season (McKaye 1983, 1984). In comparisons between paired bowers, males on larger bowers received a two- to threefold increase in female attention (bower entry and circling behavior) over males on smaller bowers. In a smaller lek occupied by 20 to 50 *Otopharynx argyrosoma* (Regan) males, the males occupying bowers closest to the center of

FIGURE 6.—Examples of the color patterns recognized by Eccles and Trewavas (1989) as being phylogenetically informative for Lake Malaŵi cichlids: (**a**) kirkii pattern, (**b**) transverse bars, (**c**) dimidiatus pattern, (**d**) polystigma pattern, (**e**) oblique band, (**f**) three-spot pattern, and (**g**) rostratus pattern.

the lek received approximately three times as many matings as did the males around the periphery (McKaye 1991). In order to separate the effect of bower size and bower location, we substituted artificial bowers in the lek of *Lethrinops* c.f. *parvidens* (Trewavas). Several tagged males located on the periphery of the lek had not been observed to fertilize any eggs during a 3-week period during which approximately 1,800 eggs were laid in other areas of the lek. The same tagged males were observed fertilizing between 15 and 30 eggs per day when large (approximately 22 cm in height) bowers were placed on top of the tagged males existing ones. In another arena, female *Lethrinops auritus* (Regan) preferred to mate with males whose bowers contained more peripheral bumps. In general, these data suggest that within several species, a specific character, bower size, can influence the mate preference of females and that males will evolve behaviors that increase the size, shape, or position of their bower in order to attract more females.

In Lake Malaŵi, 10 major bower forms, which vary in size from small depressions in the sand to elaborate castles, have been identified (McKaye 1991). Within each class of bower shape, significant quantitative variation in bower dimensions occurs. Among bowers within a lek, height varies depending on the age of the bower and the activities of the male. Some dimensions of the bower remain constant, despite variation in height, strongly suggesting a genetic basis to bower form. The diameter of the breeding platform of the bowers of *Copadichromis conophorus* appears to be species specific (Stauffer et al. 1993). We demonstrated that three closely related species in the *Copadichromis eucinostomus* group had differently shaped bowers, and we used these data to aid in the differentiation of these species. Similarly, McKaye et al. (1993) studied five leks of *Tramitichromis* near Nankumba Peninsula in Lake Malaŵi and demonstrated significant differences in bower shape among these leks. These data are discussed in more detail in the section

"Congruence of Behavioral, Genetic, and Morphological Data."

Preliminary evidence indicates that members of different genera not only have different bower forms but also exhibit different courtship dances. Male *C. conophorus*, *C. cyclicos* Stauffer, LoVullo, and McKaye, and *C. thinos* Stauffer, LoVullo, and McKaye all exhibit circular courtship patterns. Sympatric *Tramitichromis* species exhibit a figure-eight courtship pattern (Stauffer et al. 1993), and the taxa closely allied to *Tramitichromis praeorbitalis* have an S-shaped courtship dance. Thus, differences in courtship dances may be taxonomically informative.

Role of Feeding Behavior in Delimiting Species

The tremendous trophic diversity exhibited by the fishes of Lake Malaŵi suggests that these haplochromines must be extremely efficient at partitioning food resources. Documented unusual feeding strategies of Lake Malaŵi fishes include death feigning (McKaye 1981), paedophagy (McKaye and Mackenzie 1982; McKaye and Kocher 1983; Stauffer and McKaye 1986), lepidophagy (Ribbink 1984), foraging associations (Fryer and Iles 1972; Kocher and McKaye 1983), and cleaning (Ribbink and Lewis 1982; Stauffer 1991). Additionally, recent observations have shown that closely related mbuna species orient to the substrate at similar angles when feeding on the lithophilous algae. Underwater video footage has shown that members of the *P. zebra* complex approached the substrate at an angle ranging between 63–72°, members of the elongate mbuna *P. elongatus* Fryer complex at 27–30°, *P. tropheops* at 37–46°, *Melanochromis* spp. at 65–70°, *Labeotropheus* spp. at 27–30°, and *Petrotilapia* spp. at 50–55°. Preliminary analyses of jaw morphology indicate that feeding angles may be constrained by jaw construction. Thus, feeding angles are indicative of groups of closely related species, and, in fact, a diver can identify the genus or species complex of many of these rock-dwelling species simply by noting the angle at which they feed.

Some of the above behaviors appear to be species specific, such as the death feigning observed in *Nimbochromis livingstonii*. Other behaviors appear to be a result of parallel or convergent evolution. For example, lepidophagy is noted in the genera *Docimodus*, *Corematodus*, *Melanochromis*, and *Genyochromis*; ramming behavior in two species of *Caprichromis*, *Diplotaxodon greenwoodi* Stauffer and McKaye, and an undescribed *Protomelas* species; foraging associations in *Cyrtocara moori* and *Protomelas annectens* (Regan); and cleaning in two mbuna genera, *Melanochromis crabo* Ribbink and Lewis and *P. pursus* Stauffer.

Congruence of Behavioral, Genetic, and Morphological Data

The extremely high diversity of cichlid fishes throughout the tropics, although an excellent example of explosive vertebrate speciation, causes a multitude of problems for the practicing taxonomist. Morphology has played a dominant role in the alpha taxonomy of fishes. In fact, the use of behavioral observations in delimiting species has been criticized because of the influence of the environment on learned behavior. Phenotypic plasticity, which is defined as environmental modification of the phenotype (Bradshaw 1965), also confounds the interpretation of morphological data. Phenotypic plasticity has been documented in a wide range of organisms including fishes (Barlow 1961; Behnke 1972; Chernoff 1982; Meyer 1987), amphibians (Calhoon and Jameson 1970; Newman 1988), birds (James 1983), insects (Atchley 1971), and other diverse animal (Gould and Johnston 1972) and plant (Bradshaw 1965; Schlicting and Levin 1986) groups. Historically, there was a tendency to assume that the morphology of many cichlid species is very rigid and that there is little intraspecific variation (van Oijen 1982). This is in direct contrast with the morphological plasticity observed in other fishes (Barlow 1961; Crossman 1966; Behnke 1970; Baltz and Moyle 1981; Chernoff 1982; Matthews 1987). Only two studies have directly tested the effects of diet on the phenotypic plasticity of New World cichlids (Meyer 1987; Wimberger 1988). Meyer's (1987) study examined the effect of diet on head morphology of the jaguar guapote *Cichlasoma managuense* (Günther). He fed siblings two different diets for 8.5 months and then quantified head shape. Those individuals that were fed flake food and oligochaetes had a short, blunt snout, reflecting the biting mode required to capture their prey (obturorostral), whereas those individuals that were fed nauplii of brine shrimp *Artemia* sp. had a long, pointed snout, reflecting the suction mode (acuturostral). Meyer (1987) hypothesized that similar trends in head shape occur in Old World cichlids but presented no data to substantiate this idea. He does mention that because Old World cichlids are predominantly mouthbrooders, the young would be better developed at the onset of feeding, thus diet-induced change in morphology might not be as pronounced.

In a similar study, Wimberger (1988) examined

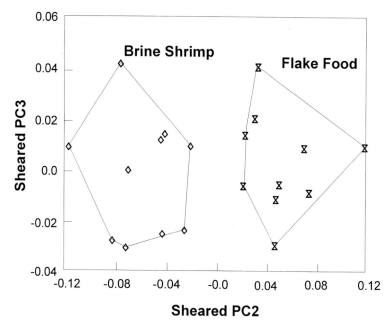

FIGURE 7.—A plot of the second and third sheared principal components based on eight head measures collected from a split brood of *P.* c.f. *zebra* "red-top" fed two different diets.

the effects of diet on the phenotype of two New World cichlids: a mouthbrooder, the redhump eartheater *Geophagus steindachneri* (Eigenmann and Hildebrand), and a substrate spawner, the pearl eartheater *Geophagus braziliensis* (Quoy and Gaimand). The experimental design was similar to that of Meyer's (1987) and both species exhibited the expected trend. Based on this study it would appear that mouthbrooding may not greatly alter the phenotypic plasticity induced by diet in substrate-spawning cichlids.

Similar studies have not been conducted on Old World cichlids, but there have been some important observations. Witte (1984) reported that wild-caught (Lake Victoria) and domesticated *Haplochromis squamipinnis* Trewavas had differently shaped premaxillaries. The difference was attributed to the fact that those individuals kept in aquaria dug in the sand with their mouths, thus increasing the power of their bite over that of the wild-caught ones, which did not exhibit this digging behavior. In addition, Witte (1984) noted that the change in premaxillary shape was not limited to young fish, indicating that it was not strictly controlled by some ontogenetic factor. A second important observation was reported by Greenwood (1965) for *Astatoreochromis alluadi* Pellegrin. Individuals feeding on thick-shell snails had stronger

pharyngeal bones and larger molariform teeth than did those individuals that ate snails with thinner shells. In preliminary experiments conducted in our laboratory, we used F_1 siblings derived from wild-caught *P.* c.f. *zebra* "red top" and randomly divided them into two dietary treatments: (1) brine shrimp nauplii and *Daphnia magna* and (2) commercial flake food and tubifex worms. After 18 weeks the fish were sacrificed and morphometric measurements were recorded. A sheared principal components analysis, in which cheek depth, head depth, and snout length accounted for most of the variability, resulted in complete separation of the two groups (Figure 7).

Clearly, approaches integrating morphological, genetic, ecological, and ethological data are required for species-level description of these fishes. In a study of five putative populations of *Tramitichromis* species in the vicinity of Nankumba Peninsula in Lake Malaŵi, McKaye et al. (1993) examined protein electromorphs of 24 enzyme loci and compared these data with bower shape of each of the five populations. No fixed differences were found for any of the alleles. Frequency differences indicated that the two populations found at Cape Maclear were distinct from the populations from Kanjedza Island, Mpandi Island, and Nkudzi Point. The population inhabiting Nkudzi Bay, which is located between Kanjedza Island and Mpandi Island was intermediate between these islands and the populations at Cape Maclear. Shape analysis of bower forms produced two major groupings, which showed that the bowers from populations located at Cape Maclear were distinct from those found at the other three localities. A critical examination of the lower pharyngeal bone and the gill rakers located on the ceratobranchial showed that populations from Nkudzi Bay, Mpandi Island, and Kanjedza Island were *Lethrinops* c.f. *parvidens*, whereas those located at Cape Maclear were *Tramitichromis* c.f. *lituris*. Hence, the results suggested by the morphological, genetic, and behavioral data were congruent.

Another example of congruence among genetic, morphological, and behavioral data is found in the three species *Copadichromis conophoros*, *C. cyclicos*, and *C. thinos*, which were recently described by Stauffer et al. (1993). For over a decade, extensive research on the ecology and behavior of these sand-dwelling fishes indicated that at least three populations, which fit the original description of *Copadichromis eucinostomus*, constructed bowers with three different population-specific shapes. The clusters formed by plotting the principal component analysis scores (see Humphries et al. 1981 and Bookstein et al. 1985 for a discussion of shape analysis) of the morphometric and meristic data for *Copadichromis conophorus* and *C. cyclicos* did not overlap. *Copadichromis thinos*, although intermediate, was significantly different ($P < 0.05$) from the other two species. Shape analysis also confirmed that bower shapes for the three species were significantly different ($P < 0.05$), although data from the bowers of *Copadichromis conophorus* were intermediate. Subsequent to the description of these taxa, mtDNA haplotype frequencies in *Copadichromis conophorus*, *C. cyclicos*, *C. thinos*, and an undescribed species of *Copadichromis* from Thumbi West Island were examined (Table 1). Haplotype frequencies were significantly different ($P < 0.05$) among populations. Males on the small lek at Thumbi East Island are nearly fixed for a single mtDNA haplotype. These data confirm the genetic uniqueness of *Copadichromis conophorus*, *C. thinos*, and *C. cyclicos*, which had been inferred from morphological evidence.

Conclusions

In sympatric situations behavior can provide direct evidence for reproductive isolation or cohesion. In both allopatric and sympatric circumstances, de-

TABLE 1.—Distribution of mitochondrial DNA haplotypes among four populations of *Copadichromis* once suspected to be conspecific. All populations are from southern Lake Malaŵi.

Population	Mitochondrial DNA haplotypes								
	A	B	C	D	E	F	G	H	I
Thumbi West Island	16	3	0	0	0	0	0	0	0
Cape Maclear	1	11	0	4	2	1	0	0	0
Kanchedza Island	0	0	14	1	0	0	2	0	0
Mazinzi Reef	2	3	0	2	1	0	0	1	3

tectable behavioral differences may have initiated the speciation process through assortative mating, which may, in turn, lead to runaway sexual selection. Thus, behavioral data are extremely valuable and, at least with some groups such as cichlids, are essential and can (1) initially identify distinct taxa or identify novelties which prompt further investigation; (2) confirm or support genetic and morphological data needed to delimit taxa; and (3) provide needed information to speculate on phylogenies.

It is our contention that if ESUs are recognized at the population level, the population designated should possess some heritable atypical trait, such as an unusual behavior pattern. Perhaps an ESU can be designated on a temporary basis because of an unresolved taxonomic status. We are not proposing that the ESU replace existing taxonomic categories but that these units be given standard nomenclatural status when possible, so that they are formally recognized by the scientific community. Such distinction provides the necessary framework to initiate and foster debate on the significance and reality of such discrimination. We realize that species definitions and concepts are difficult and sometimes burdensome, but we urge investigators not to regard these varied concepts as mutually exclusive. We also conclude that behavioral data are essential to delimit species.

We further propose that the ESU be defined in geographical terms, so that areas of high diversity or endemism can be designated as ESUs. Such a unit may consist of crater lakes in Nicaragua or particular islands or shorelines in Lake Malaŵi. For example, in the southeast arm of Lake Malaŵi more than one-third and one-half of the species native to the Maleri Islands and to Chinyankwazi and Chinyamwezi Islands, respectively, are endemic (Figure 8). Such a geographical approach to conservation must permit the continued use of the lake by Malaŵians, who derive about 70% of the animal protein consumed from fish, and must also preserve those areas that harbor high concentrations of ge-

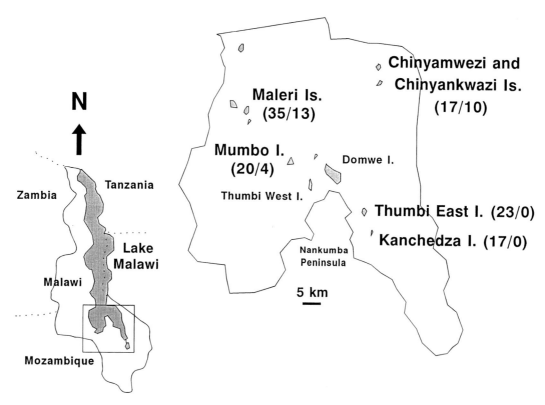

FIGURE 8.—A map of the southern region of Lake Malaŵi showing the total number of species of mbuna at a given location and the number of those species that are endemic to the location.

netic diversity. As stated by Orville Freeman (former U.S. Secretary of Agriculture), "We make a potentially dangerous mistake when we assume that we must choose between serving humanity or serving the environment. It must be a priority to bring these goals into harmony. They need not and they must not be mutually exclusive."

Acknowledgments

This work was funded in part by the U.S. Agency for International Development Program in Science and Technology Cooperation, Office of Science Advisor (Grant Number 10.069, Com-5600-G-00-0017-00; Grant Number 11.204, DHR-5600-G-1043-00); the National Science Foundation (BNS86-06836; IBN-9225060; BSR-9007015); and Fulbright Research Awards (Council for International Exchange of Scholars) to Jay Stauffer and Kenneth McKaye. The authors appreciate the critical review of the manuscript by Deborah McLennan and Allan de Queiroz.

References

Arnold, S. J. 1983. Sexual selection: the interface of theory and empiricism. Pages 67–108 in P. Bateson, editor. Mate choice. Cambridge University Press, Cambridge, Massachusetts.

Atchley, W. R. 1971. A comparative study of the causes and significance of morphological variation in adults and pupae of Culicoides: a factor analysis and multiple regression study. Evolution 25:563–583.

Baltz, D. M., and P. B. Moyle. 1981. Morphometric analysis of tule perch (*Hysterocarpus traski*) populations in three isolated drainages. Copeia 1981:305–311.

Barel, C. D. N., M. J. O. Van Oijen, F. Witte, and E. L. M. Witte-Maas. 1977. An introduction to the taxonomy and morphology of the haplochromine Cichlidae from Lake Victoria. Part A: text. Netherlands Journal of Zoology 27:333–389.

Barlow, G. W. 1961. Causes and significance of morphological variation in fishes. Systematic Zoology 10:105–117.

Barlow, G. W. 1974. Contrasts in social behavior between Central American cichlid fishes and coral reef surgeon fishes. American Zoologist 14:9–34.

Barlow, G. W., and J. W. Munsey. 1976. The red devil-midas arrow cichlid species complex in Nicaragua. Pages 359–370 in T. B. Thorsen, editor. Investigations of the ichthyofauna of Nicaraguan lakes. University of Nebraska Press, Lincoln.

Basolo, A. L. 1990. Female preference predates the evolution of the sword in swordtail fish. Science 250:808–810.

Basolo, A. L. 1991. Male swords and female preferences. Science 253:1426–1427.

Behnke, R. J. 1970. The application of cytogenetic and biochemical systematics to phylogenetic problems in the family Salmonidae. Transactions of the American Fisheries Society 99:237–248.

Behnke, R. J. 1972. Systematics of salmonid fishes of recently glaciated lakes. Journal of the Fisheries Research Board of Canada 29:639–671.

Berra, T. M. 1981. An atlas of distribution of the freshwater fish families of the world. University of Nebraska Press, Lincoln.

Bookstein, F., B. Chernoff, R. Elder, J. Humphries, G. Smith, and R. Strauss. 1985. Morphometrics in evolutionary biology. Academy of Natural Sciences, Special Publication 15, Philadelphia.

Borgia, G., and K. Collis. 1990. Parasites and bright male plumage in the satin bowerbird (Ptilonorhynchus violaceus). American Zoologist 30:203–219.

Bowers, N. J., T. D. Kocher, and J. R. Stauffer. 1994. Intra- and inter-specific mitochondrial DNA sequence variation within two species of rock-dwelling cichlids (Teleostei: Cichlidae) from Lake Malaŵi, Africa. Molecular Phylogenetics and Evolution 3(1):75–82.

Bowers, N. J., and J. R. Stauffer, Jr. 1993. New species of rock-dwelling cichlid (Pisces: Cichlidae) from Lake Malaŵi, Africa, with comments on Melanochromis vermivorus Trewavas. Copeia 1993:715–722.

Bradshaw, A. D. 1965. Evolutionary significance of phenotypic plasticity in plants. Advances in Genetics 13:115–155.

Brooks, D. R., and D. A. McLennan, editors. 1991. Phylogeny, ecology, and behavior. University of Chicago Press, Chicago.

Burgess, W. E. 1976. Studies on the family Cichlidae: 5. Pseudotropheus aurora, a new species of cichlid fish from Lake Malaŵi. Tropical Fish Hobbyist 24:52–56, Jersey City, New Jersey.

Bush, G. L. 1975. Modes of animal speciation. Annual Review of Ecology and Systematics 6:339–364.

Calhoon, R. E., and D. L. Jameson. 1970. Canonical correlation between variation in weather and variation in size in the Pacific tree frog, Hyla regilla, in southern California. Copeia 1970:124–134.

Carson, H. L. 1978. Speciation and sexual selection in Hawaiian Drosophila. Pages 93–107 in P. F. Brussard, editor. Ecological genetics: the interface. Springer-Verlag, New York.

Chernoff, B. 1982. Character variation among populations and the analysis of biogeography. American Zoologist 22:425–439.

Crossman, E. J. 1966. A taxonomic study of Esox americanus and its subspecies in eastern North America. Copeia 1966:1–20.

Darwin, C. 1871. The descent of man and selection in relation to sex. John Murray, London.

de Queiroz, A., and P. H. Wimberger. 1993. The usefulness of behavior for phylogeny estimation: levels of homoplasy in behavioral and morphological characters. Evolution 47:46–60.

Dominey, W. J. 1984. Effects of sexual selection and life history on speciation: species flocks in African cichlids and Hawaiian Drosophila. Pages 231–249 in A. A. Echelle and I. Kornfield, editors. Evolution of fish species flocks. University of Maine Press, Orono.

Donoghue, M. J. 1985. A critique of the biological species concept and recommendations for a phylogenetic alternative. The Bryologist 88:172–181, Lancaster, Pennsylvania.

Eccles, D. H., and Trewavas, E. 1989. Malaŵian cichlid fishes: the classification of some haplochromine genera. Lake Fish Movies, Herten, West Germany.

Eldridge, N. 1985. Unfinished synthesis: biological hierarchies and modern evolutionary thought. Oxford University Press, New York.

Enquist, M., and A. Arak. 1993. Selection of exaggerated male traits by female aesthetic senses. Nature 361:446–448.

Fisher, R. A. 1930. The genetical theory of natural selection. Oxford University Press, Dover, New York.

Fryer, G. 1959. Some aspects of evolution in Lake Nyasa. Evolution 13:440–451.

Fryer, G., and T. Iles. 1972. The cichlid fishes of the Great Lakes of Africa. Oliver and Boyd, London.

Gould, S. J., and R. F. Johnston. 1972. Geographic variation. Annual Review of Ecology and Systematics 3:457–498.

Greenwood, P. H. 1965. Environmental effects on the pharyngeal mill of a cichlid fish, Astatoreochromis alluadi, and their taxonomic implications. Proceedings of the Linnean Society of London 176:1–10.

Greenwood, P. H. 1981. The haplochromine fishes of East African lakes. Cornell University Press, Ithaca, New York.

Hamilton, W. D., and M. Zuk. 1982. Heritable true fitness and bright birds: a role for parasites? Science 218:384–387.

Hartl, D. L., and A. G. Clark, editors. 1989. Principles of population genetics. Sinauer, Sunderland, Massachusetts.

Hedrick, A. V., and E. J. Temeles. 1989. The evolution of sexual dimorphism in animals: hypotheses and tests. Trends in Ecology & Evolution 4:136–138.

Hert, E. 1989. The function of egg-spots in an African mouth-brooding cichlid fish. Animal Behaviour 37:726–732.

Hert, E. 1991. Female choice based on egg-spots in Pseudotropheus aurora Burgess 1976, a rock-dwelling cichlid of Lake Malaŵi, Africa. Journal of Fish Biology 38:951–953.

Holzberg, S. 1978. A field and laboratory study of the behaviour and ecology of Pseudotropheus zebra (Boulenger), an endemic cichlid of Lake Malaŵi (Pisces: Cichlidae). Zeitschrift fuer Zoologische Systematik und Evolutions Forschung 16:171–187.

Hoogerhoud, R. J. C., and F. Witte. 1981. Revision of

species from the *"Haplochromis" empodisma* group. Revision of the haplochromine species (Teleostei, Cichlidae) from Lake Victoria, part II. Netherlands Journal of Zoology 31:232–274.

Humphries, J., F. Bookstein, B. Chernoff, G. Smith, R. Elder, and S. Poss. 1981. Multivariate discrimination by shape in relation to size. Systematic Zoology 30: 291–308.

James, F. C. 1983. Environmental component of morphological variation in birds. Science 221:184–186.

Kaneshiro, K. Y. 1989. The dynamics of sexual selection and founder effects in species formation. Pages 279–296 *in* L. V. Giddings, K. Y. Kaneshiro, and W. W. Anderson, editors. Genetics, speciation, and the founder principle. Oxford University Press, Oxford, UK.

Keenleyside, M. H. A., R. W. Rangley, and B. U. Kuppers. 1985. Female mate choice and male parental defense behaviour in the cichlid fish *Cichlasoma nigrofasciatum*. Canadian Journal of Zoology 63:2489–2493.

Kirkpatrick, M. 1982. Sexual selection and the evolution of female choice. Evolution 36:1–12.

Kirkpatrick, M. 1987. Sexual selection by female choice in polygynous animals. Annual Review of Ecology and Systematics 18:43–70.

Kirkpatrick, M., and M. J. Ryan. 1991. The evolution of mating preferences and the paradox of the lek. Nature 350:33–38.

Kocher, T. D., J. Conroy, K. R. McKaye, and J. R. Stauffer, Jr. 1993. Similar morphologies of cichlid fish in lakes Tanganyika and Malaŵi are due to convergence. Molecular Phylogenetics and Evolution 2:158–165, Orlando, Florida.

Kocher, T. D., and K. R. McKaye. 1983. Territorial defense of heterospecific cichlids by *Cyrtocara moori* in Lake Malaŵi, Africa. Copeia 1983:544–547.

Kornfield, I. 1974. Evolution genetics of endemic African cichlids. Doctoral dissertation. State University of New York, Stony Brook.

Kornfield, I. 1978. Evidence for rapid speciation in African cichlid fishes. Experientia 34:335–336.

Kosswig, C. 1963. Ways of speciation in fishes. Copeia 1963:238–244.

Lande, R. 1980. Sexual dimorphism, sexual selection, and adaptation in polygenic characters. Evolution 34: 292–305.

Lande, R. 1981. Models of speciation by sexual selection on polygenic traits. Proceedings of the National Academy of Sciences of the United States of America 78:3721–3725.

Lewis, D. S. C. 1982. Problems of species definition in Lake Malaŵi cichlid fishes (Pisces: Cichlidae). Ichthyological Bulletin of the J. L. B. Smith Institute of Ichthyology 23:1–5.

Liem, K. F., and L. S. Kaufman. 1984. Intraspecific macroevolution: functional biology of the polymorphic cichlid species *Cichlasoma minckleyi*. Pages 203–216 *in* A. Echelle and I. Kornfield, editors. Evolution of fish species flocks. University of Maine Press, Orono.

Lowe-McConnell, R. H. 1959. Breeding behavior patterns and ecological differences between *Tilapia* species and their significance for evolution within the genus *Tilapia* (Pisces: Cichlidae). Proceedings of the Zoological Society of London 132:1–30.

Matthews, W. J. 1987. Geographic variation in *Cyprinella lutrensis* (Pisces: Cyprinidae) in the United States, with notes on *Cyprinella lepida*. Copeia 1987:616–637.

Mayr, E. 1942. Systematics and the origin of species. Columbia University Press, New York.

Mayr, E. 1963. Animal species and evolution. Harvard University Press, Cambridge, Massachusetts.

Mayr, E. 1982a. The growth of biological thought: diversity, evolution, inheritance. Harvard University Press, Cambridge, Massachusetts.

Mayr, E. 1982b. Speciation and macroevolution. Evolution 36:1119–1132.

Mayr, E. 1992. A local flora and the biological species concept. American Journal of Botany 79:222–238.

Mayr, E., and P. D. Ashlock. 1991. Principals of systematic zoology. McGraw Hill, New York.

McElroy, D. M., I. Kornfield, and J. Everett. 1991. Coloration in African cichlids: diversity and constraints in Lake Malaŵi endemics. Netherlands Journal of Zoology 41:250–268.

McKaye, K. R. 1980. Seasonality in habitat selection by the gold color morph of *Cichlasoma citrinellum* in Lake Jiloa, Nicaragua. Environmental Biology of Fishes 5:75–78.

McKaye, K. R. 1981. Death feigning: a unique hunting behaviour by the predatory cichlid, *Haplochromis livingstonni* of Lake Malaŵi. Environmental Biology of Fishes 6:361–365.

McKaye, K. R. 1983. Ecology and breeding behavior of a cichlid fish, *Cyrtocara eucinostomus*, or a large lek in Lake Malaŵi, Africa. Environmental Biology of Fishes 8:81–96.

McKaye, K. R. 1984. Behavioural aspects of cichlid reproductive strategies: patterns of territoriality and brood defense in Central American substratum spawners versus African mouth brooders. Pages 245–273 *in* R. J. Wooton and G. W. Potts, editors. Fish reproduction: strategies and tactics. Academic Press, New York.

McKaye, K. R. 1991. Sexual selection and the evolution of the cichlid fishes of Lake Malaŵi, Africa. Pages 241–257 *in* M. H. A. Keenleyside, editor. Cichlid fishes: behavior, ecology and evolution. Chapman and Hall, London.

McKaye, K. R., J. H. Howard, J. R. Stauffer, Jr., R. P. Morgan II, and F. Shonhiwa. 1993. Sexual selection and genetic relationships of a sibling species complex of bower building cichlids in Lake, Malaŵi, Africa. Japanese Journal of Ichthyology 40:15–21.

McKaye, K. R., and T. D. Kocher. 1983. Head ramming behavior by three paedophagous cichlids in Lake Malaŵi, Africa. Animal Behaviour 31:206–210.

McKaye, K. R., T. Kocher, P. Reinthal, R. Harrison, and I. Kornfield. 1982. A sympatric sibling species complex of *Petrotilapia* Trewavas from Lake Malaŵi analyzed by electrophoresis (Pisces: Cichlidae). Zoological Journal of the Linnean Society 76:91–96.

McKaye, K. R., T. Kocher, P. Reinthal, R. Harrison, and I. Kornfield. 1984. Genetic variation among color morphs of a Lake Malaŵi cichlid fish. Evolution 31: 215–219.

McKaye, K. R., S. M. Louda, and J. R. Stauffer, Jr. 1990. Bower size and male reproductive success in a cichlid fish lek. American Naturalist 135:597–613.

McKaye, K. R., and C. Mackenzie. 1982. *Cyrtocara liemi*, a previously undescribed paedophagous cichlid fish (Teleostei: Cichlidae) from Lake Malaŵi, Africa. Proceedings of the Biological Society of Washington 95:398–402.

McKaye, K. R., and J. R. Stauffer, Jr. 1986. Description of a gold cichlid, *Pseudotropheus barlowi* (Teleostei: Cichlidae), from Lake Malaŵi, Africa. Copeia 1986: 870–875.

Meek, S. E. 1907. Synopsis of the fishes of the Great Lakes of Nicaragua. Field Columbian Museum Publication 121, Zoology Series 7:97–132, Chicago.

Meyer, A. 1987. Phenotypic plasticity and heterochrony in *Cichlasoma managuense* (Pisces: Cichlidae) and their implications for speciation in cichlid fishes. Evolution 41:1357–1369.

Meyer, A., T. D. Kocher, P. Basasibwaki, and A. Wilson. 1990. Monophyletic origin of Lake Victoria cichlid fishes suggested by mitochondrial DNA sequences. Nature 347:550–553.

Meyer, A., J. M. Morrissey, and M. Schartl. 1994. Recurrent origin of a sexually selected trait in *Xiphophorus* fishes inferred from a molecular phylogeny. Nature 368:539–542.

Moran, P., and I. Kornfield. 1993. Retention of an ancestral polymorphism in the mbuna species flock (Pisces: Cichlidae) of Lake Malaŵi. Molecular Biology and Evolution 10:1015–1029.

Moran, P., I. Kornfield, and P. Reinthal. 1994. Molecular systematics and radiation of the haplochromine cichlids (Teleostei: Cichlidae) of Lake Malaŵi. Copeia 1994:274–288.

Moyle, P. B., and J. J. Cech, Jr. 1988. Fishes: an introduction to ichthyology. Prentice Hall, Englewood Cliffs, New Jersey.

Myers, N. 1988. Tropical-forest species: going, going, going. . . . Scientific American 259:132.

Newman, R. A. 1988. Adaptive plasticity in development of *Scaphiopus couchi* tadpoles in desert ponds. Evolution 42:774–783.

Noonan, K. C. 1983. Female mate choice in the cichlid fish *Cichlasoma nigrofasciatum*. Animal Behaviour 31:1005–1010.

Nur, N., and O. Hasson. 1984. Phenotypic plasticity and the handicap principle. Journal of Theoretical Biology 110:275–297.

O'Donald, P. 1980. Genetic models of sexual selection. Cambridge University Press, Cambridge, Massachusetts.

Paterson, H. E. H. 1985. The recognition concept of species. Transvall Museum Monograph 4:21–29.

Ribbink, A. J. 1984. The feeding behaviour of a cleaner and scale, skin and fin eater from Lake Malaŵi (*Dociomodus evelynae*; Pisces, Cichlidae). Netherlands Journal of Zoology 34:182–196.

Ribbink, A. J., and D. S. C. Lewis. 1982. *Melanochromis crabro*, sp. nov.: a cichlid fish from Lake Malaŵi

which feeds on ectoparasites and catfish eggs. Netherlands Journal of Zoology 32:72–87.

Ribbink, A. J., B. A. Marsh, A. C. Marsh, A. C. Ribbink, and B. J. Sharp. 1983. A preliminary survey of the cichlid fishes of rocky habitats of Lake Malaŵi. South African Journal of Zoology 18:149–310.

Ringo, J. M. 1977. Why 300 species of Hawaiian *Drosophila*? the sexual selection hypothesis. Evolution 31:694–696.

Ryan, M. J., and A. Keddy-Hector. 1992. Directional patterns of female mate choice and the role of sensory biases. American Naturalist 139:S4–S35,

Ryan, M. J., and A. S. Rand. 1993. Species recognition and sexual selection as a unitary problem in animal communication. Evolution 47:647–657.

Schlicting, C. D., and D. A. Levin. 1986. Effects of inbreeding on phenotypic plasticity in cultivated *Phlox*. Theoretical and Applied Genetics 72:114–119.

Schröder, J. H. 1980. Morphological and behavioural differences between the BB/OB and B/W colour morphs of *Pseudotropheus zebra* Boulenger (Pisces: Cichlidae). Zeitschrift fuer Zoologische Systematik und Evolutions Forschung 18:69–76.

Simpson, G. G. 1961. Principles of animal taxonomy. Columbia University Press, New York.

Smith, G. R., and A. K. Koehn. 1971. Phenetic and cladistic studies of biochemical and morphological characteristics of *Catostomus*. Systematic Zoology 20:282–297.

Spieth, H. T. 1974. Mating behavior and evolution of the Hawaiian *Drosophila*. Pages 94–101 in M. J. D. White, editor. Genetic mechanisms of speciation in insects. Australia and New Zealand Book Co., Boston.

Stiassny, M. L. J. 1981. Phylogenetic versus convergent relationship between piscivorous cichlid fishes from Lakes Malaŵi and Tanganyika. Bulletin of the British Museum (Natural History) Zoology 40:67–101.

Stauffer, J. R., Jr. 1988. Descriptions of three rock-dwelling cichlids (Teleostei: Cichlidae) from Lake Malaŵi, Africa. Copeia 1988:663–668.

Stauffer, J. R., Jr. 1991. Description of a facultative cleanerfish (Teleostei: Cichlidae) from Lake Malaŵi, Africa. Copeia 1991:141–147.

Stauffer, J. R., Jr. 1993. A new species of *Protomelas* (Teleostei: Cichlidae) from Lake Malaŵi, Africa. Ichthyological Exploration of Freshwaters 4:343–350, München, Germany.

Stauffer, J. R., Jr., and J. M. Boltz. 1989. Description of a new species of Cichlidae, from Lake Malaŵi, Africa. Proceedings of the Biological Society of Washington 102:8–13

Stauffer, J. R., Jr., J. M. Boltz, and L. R. White. 1995. The fishes of West Virginia. Proceedings of the Academy of Natural Sciences of Philadelphia 146:1–389.

Stauffer, J. R., Jr., and E. Hert. 1992. *Pseudotropheus callainos*, a new species of mbuna (Cichlidae), with analyses of changes associated with two intralacustrine transplantations in Lake Malaŵi, Africa. Icthyological Explorations of Freshwaters 3:253–264.

Stauffer, J. R., Jr., T. J. LoVullo, and K. R. McKaye. 1993. Three new sand-dwelling cichlids from Lake Malaŵi, Africa, with a discussion of the status of the genus

Copadichromis (Teleostei: Cichlidae). Copeia 1993: 1017–1027.

Stauffer, J. R., Jr., and K. R. McKaye. 1986. Description of a paedophagous deep-water cichlid (Teleostei: Cichlidae) from Lake Malaŵi, Africa. Proceedings of the Biological Society of Washington 99:29–33.

Stauffer, J. R., Jr., and K. R. McKaye. 1988. Description of a genus and three deep-water species of fishes (Teleostei: Cichlidae) from Lake Malaŵi, Africa. Copeia 1988:441–449.

Templeton, A. R. 1989. The meaning of species and speciation: a genetic perspective. Pages 3–27 *in* D. Otte and J. A. Endler, editors. Speciation and its consequences. Sinauer, Sunderland, Massachusetts.

Thornhill, R., and J. Alcock. 1983. The evolution of insect mating systems. Harvard University Press, Cambridge, Massachusetts.

Trewavas, E., J. Green, and S. Corbert. 1972. Ecological studies of crater lakes in West Cameroon, fishes of Barombi. Journal of Zoology 167:41–95.

Trewavas, E. 1983. Tilapiine fishes of the genera *Sarotherodon*, *Oreochromis*, and *Danakilia*. British Museum (Natural History) Publication No. 878, London.

Trivers, R. L. 1972. Parental investment and sexual selection. Pages 136–179 *in* B. Campbell, editor. Sexual selection and the descent of man. Aldine, Chicago.

van Devender, T. R., C. H. Lowe, H. K. McCrystal, and H. E. Lawler. 1992. Viewpoint: reconsider suggested systematic arrangements for some North American amphibians and reptiles. Herpetological Review 23: 10–14.

van Oijen, M. J. P. 1982. Ecological differentiation among the piscivorous haplochromine cichlids of Lake Victoria (East Africa). Netherlands Journal of Zoology 32:336–363.

Via, S., and R. Lande. 1985. Genotype-environment interaction and the evolution of phenotypic plasticity. Evolution 39:505–522.

Villa, J. 1982. Peces Nicaraguenses de aqua dulce. Banco de America, Serie Geografia y Naturalez No. 3, Managua, Nicaragua.

Waples, R. S. 1991. Definition of "species" under the Endangered Species Act: application to Pacific salmon. NOAA (National Oceanic and Atmospheric Administration) Technical Memorandum NMFS (National Marine Fisheries Service) F/NWC-194, Northwest Fisheries Science Center, Seattle.

Watrous, L. E., and Q. D. Wheeler. 1981. The out-group comparison method of character analysis. Systematic Zoology 30:1–11.

Wenzel, J. W. 1992. Behavioral homology and phylogeny. Annual Review of Ecology and Systematics 23: 361–381.

West-Eberhard, M. J. 1983. Sexual selection, social competition, and speciation. Quarterly Review of Biology 58:155–183.

Wiley, E. O. 1981. Phylogenetics: the theory and practice of phylogenetic systematics. Wiley, New York.

Wimberger, P. H. 1991. Plasticity of jaw and skull morphology in the neotropical cichlids *Geophagus brasiliensis* and *G. steindachneri*. Evolution 45:1545–1563.

Witte, F. 1984. Ecological differentiation in Lake Victoria haplochromines: comparison of cichlid species flocks in African lakes. Pages 155–168 *in* A. Echelle and I. Kornfield, editors. Evolution of fish species flocks. University of Maine Press, Orono.

PART FOUR

GENETICS

American Fisheries Society Symposium 17:247–248, 1995

Genetics: Defining the Units of Conservation

FRED W. ALLENDORF

Division of Biological Sciences, University of Montana
Missoula, Montana 59812, USA

This section contains five papers that deal with genetics and defining units of conservation. The first detailed studies of the genetics of natural populations of fishes were by Schmidt in a series of papers describing the genetic basis of morphological variation in populations of the eelpout *Zoarces viviparus* (e.g., Schmidt 1917). W. F. Thompson (1959, 1965) was apparently the first to argue for preserving local populations of fishes for conservation on the basis of genetic differences among populations. Thompson was ahead of his time; he anticipated and discussed insightfully many current issues we face today in conservation of Pacific salmon, *Oncorhynchus* spp., including metapopulation structure and the misuse of hatcheries to address symptoms rather than causes of decline (Thompson 1959, 1965). He urged protection for local populations as units of conservation (Thompson 1965).

> Each stream or lake has its own extremely complex characteristics, and if salmon live in one of them we find that these salmon are adapted in an equally complex way to that environment.

Thompson was hired as the director of the Department of Fisheries at the University of Washington in 1930 and began teaching a course entitled "Conservation" (Fisheries 159) the next year (Stickney 1989:25). A memo of Thompson's (when he was hired) to the president of the University of Washington states that the "reorganized Department of Fisheries will lay equal stress upon the basic sciences and upon specialized studies to provide training in fisheries biology" (Stickney 1989:24). Perhaps we would not be facing the current crises in management of fish stocks throughout the world if basic genetics had played a greater role in the training of fisheries biologists over the last 50 years (see Hallerman 1994 for an analysis of genetics in fisheries education).

With one exception (Jeffrey Hard), all of the papers in this section deal with the use of molecular genetic methods of defining genetic variation. Two additional papers describing molecular genetic methods were given at the symposium by Linda Park and Paul Bentzen. The dominance of molecular papers in this section, and the general literature, reflects the remarkable progress over the last 25 years in describing molecular genetic variation among natural populations. Fred Utter has recently reviewed the use of molecular genetic methods in fishery management (Utter 1991). On the negative side, however, the preponderance of molecular papers in this section also reflects lack of progress in understanding the adaptive significance of genetic variation in populations of fishes and other species (Lewontin 1991).

Genetics is more than the description of molecular variation. An understanding of the genetics of fish populations requires understanding the genetic basis of variation in phenotypes that are of evolutionary and conservation importance (e.g., behavior, morphology, physiology, and life history). Geneticists have focused on molecular variation because of the simplicity of interpreting its genetic basis. Nevertheless, it is the genetic variation that underlies adaptive phenotypic variation which is of central importance in conservation. We are trying to conserve organisms, not molecules.

Many papers in other sections of this symposium are concerned with the genetics of complex phenotypic traits that are important for conservation (e.g., Wood 1995, this volume; Healey and Prince 1995, this volume; Smith et al. 1995, this volume). Several of these authors come to conclusions that sometimes conflict with the conclusions of the papers that deal largely with molecular genetic variation. The disagreement can usually be traced to the difficulties in detecting the effects of natural selection.

Study of proteins and DNA provides a very sensitive method to detect the effects of genetic drift, isolation, and patterns of gene flow through genetic markers that are assumed to act as largely selectively neutral markers of the genome. This approach is very powerful for reconstructing the phylogenetic relationships of species because genetic divergence between species is largely determined by the time since two species shared a common ancestor (Avise 1994).

However, molecular markers may not accurately reflect the patterns of genetic divergence between populations within species because of natural selection. That is, the patterns of divergence within species are also molded by natural selection so that patterns of neutral markers may not reflect patterns of genetic divergence for traits related to adaptation. This problem is discussed in much more detail in the contribution of Hard (1995, this volume). Carvalho (1993) has recently reviewed the evidence for the effects of natural selection and adaptation on patterns of genetic divergence in fish species.

For example, imagine two subpopulations that are connected by regular gene flow but are subjected to very different regimes of natural selection. Limited gene flow will be sufficient to keep the two subpopulations genetically similar for markers that are not affected by natural selection. However, selection may maintain substantial genetic divergence between the subpopulations at the genes responsible for the traits under selection (Allendorf 1983). Therefore, an examination of molecular genetic variation will not detect the substantial genetic divergence that exists for adaptive differences between the subpopulations.

Some recent studies also have suggested that the patterns of genetic divergence among populations at different types of molecular markers also differ because of natural selection. In some cases, greater genetic divergence has been detected at nuclear DNA markers than at allozyme loci. This has been suggested to result from natural selection maintaining similar genotypic frequencies at allozyme loci whereas the frequencies of nuclear DNA markers diverge through isolation and genetic drift (Karl and Avise 1992; Pogson et al. 1995).

This is an exciting time for the study of the genetics of fish populations. Spectacular advances in our understanding of genetic variation at the level of DNA sequences throughout the entire genome will certainly come about in the next few years. However, this is not enough. Such studies must be combined with study of the genetic basis of phenotypic variation associated with adaptation within and between local populations if we are to get at the important question of the evolutionary significance of genetic variation. This integrated understanding of genetic variation, phenotypic variation, and adaptation is urgently needed as we face the crisis of trying to conserve fish populations.

Acknowledgments

I thank Jennifer Nielsen for inviting me to participate in this symposium; K. Knudsen, N. Kanda, R. Leary, and D. Tallmon for their comments on this note; and Jim Seeb for a copy of Stickney's monograph. F. W. Allendorf was supported by a grant from the National Science Foundation (DEB-9300135) while preparing this paper.

References

Allendorf, F. W. 1983. Isolation, gene flow, and genetic differentiation among populations. Pages 51–65 in C. Schonewald-Cox, S. Chambers, B. MacBryde, and L. Thomas, editors. Genetics and conservation. Benjamin/Cummings, Menlo Park, California.

Avise, J. C. 1994. Molecular markers, natural history, and evolution. Chapman and Hall, New York.

Carvalho, G. R. 1993. Evolutionary aspects of fish distribution: genetic variability and adaptation. Journal of Fish Biology 43(Supplement A):53–73.

Hallerman, E. M. 1994. Assessment of needs for education opportunities in fisheries genetics. Fisheries 19(3):6–12.

Karl, S. A., and J. C. Avise. 1992. Balancing selection at allozyme loci in oysters: implications from nuclear RFLPs. Science 256:100–102.

Lewontin, R. C. 1991. Twenty-five years ago in genetics. Electrophoresis in the development of evolutionary genetics: milestone or millstone? Genetics 128:657–662.

Pogson, G. H., K. A. Mesa, and R. G. Boutilier. 1995. Genetic population structure and gene flow in the Atlantic cod Gadus morhua: a comparison of allozyme and nuclear RFLP loci. Genetics 139:375–385.

Schmidt, J. 1917. Racial investigations. I. Zoarces viviparus L. and local races of the same. Comptes rendus Travaux du Laboratoire Carlsberg 14:1–14, Copenhagen.

Stickney, R. R. 1989. Flagship: a history of fisheries at the University of Washington. Kendall/Hunt, Dubuque, Iowa.

Thompson, W. F. 1959. An approach to population dynamics of the Pacific red salmon. Transactions of the American Fisheries Society 88:206–209.

Thompson, W. F. 1965. Fishing treaties and salmon of the north Pacific. Science 150:1786–1789.

Utter, F. M. 1991. Biochemical genetics and fishery management: an historical perspective. Journal of Fish Biology 39(Supplement A):1–20.

American Fisheries Society Symposium 17:249–262, 1995

Using Allele Frequency and Phylogeny to Define Units for Conservation and Management

CRAIG MORITZ, SHANE LAVERY,[1] AND ROB SLADE

Department of Zoology and Centre for Conservation Biology
The University of Queensland, Queensland 4072, Australia

Abstract.—The recognition of appropriate population units is an important step in managing threatened taxa but has been plagued by uncertainty about criteria and conservation goals. This uncertainty applies in particular to the concept of evolutionarily significant units (ESUs) which enables populations to be listed under the U.S. Endangered Species Act (ESA). We suggest that some of the conflicts which have arisen over ESUs in practice can be resolved by recognizing two types of conservation unit, each type being important for practical conservation of natural populations. Evolutionarily significant units can be defined to consist of historically isolated sets of populations for which a stringent and qualitative criterion is reciprocal monophyly for mitochondrial DNA (mtDNA) combined with significant divergence in frequencies of nuclear alleles. Such ESUs complement described species and are identified in order to contribute to the setting of conservation priorities. In contrast, management units (MUs) are demographically independent sets of populations identified to aid short-term management of the larger entities and are delimited by differences in frequencies of mtDNA or nuclear alleles, irrespective of allele phylogeny. The strengths and limitations of these concepts are illustrated by application to several marine organisms: the southern elephant seal *Mirounga leonina*, the green turtle *Chelonia mydas*, and the coconut crab *Birgus latro*. It is suggested that both ESUs and the MUs that constitute ESUs or described species should be eligible for listing under the ESA.

More than most groups of wildlife managers, fisheries biologists have long recognized the importance of identifying separate population units for management, and the use of genetic information for this purpose has a substantial history (see Ryman and Utter 1987). The motivations for recognizing stocks and using genetic information vary. One common goal is to maximize production by modeling maximum sustainable yield, taking into account the existence of different stocks, or by monitoring and maintaining genetic diversity within hatchery stocks. A more conservation-oriented goal is to maintain natural systems by minimizing human effects on ecological and evolutionary processes. The concepts and criteria developed in this paper are more concerned with the latter.

The aim of this conference is to consider the theoretical basis for identifying conservation units and the definitions and criteria that might operate in practice. The concepts of stocks and evolutionarily significant units (ESUs) are widely used, but there is little consensus on their nature or on how to identify them (Gauldie 1991; Dizon et al. 1992). This ambiguity is a serious impediment because we cannot design effective conservation programs without clearly defined entities and goals (Meffe and

Vrijenhoek 1988). It also complicates efforts to have specific entities recognized under conservation legislation (e.g., Belsky and Kerr 1993). No doubt much of the difficulty arises because we are attempting, for the purpose of management, to impose divisions on an evolutionary continuum.

Most authors have adopted an eclectic approach to identifying population units for conservation, suggesting that information on molecular genetic variation should be combined with data on life history, morphology, and behavior, particularly as these nonmolecular characters may be more relevant to current adaptation than are neutral molecular variants. (Figure 1; e.g., Waples 1991, 1995, this volume; Dizon et al. 1992; Vogler and DeSalle 1994). Such a holistic approach would indeed be preferable, if the characters have been shown to be both heritable and adaptive and if their use to recognize conservation units can be related to clear and attainable conservation goals. This is rarely the case. Accordingly, in this paper we focus on ways in which molecular information might be used as the primary step in defining different types of conservation units. Subsequently, nonmolecular characters can be mapped onto the genetically defined conservation units to test for congruence (e.g., Vogler and DeSalle 1994).

A reevaluation of genetic approaches is timely because of the rapid development of new tech-

---

[1]Queensland Agriculture Biotechnology Centre, The University of Queensland

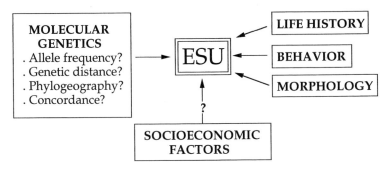

FIGURE 1.—The sources of information that may influence the recognition of a population as an evolutionarily significant unit (ESU) under current interpretations of the U.S. Endangered Species Act.

niques for assaying variation. The most notable is the use of gene amplification methods to assay variation in DNA sequences directly (e.g., Slade et al. 1993; Palumbi and Baker 1994) or to assay single hypervariable loci, such as microsatellites (Queller et al. 1993). Although traditional techniques such as allozyme electrophoresis remain important and cost-effective (Utter 1991), there is no doubt that the new technologies will increase both the range and power of genetic assays (see Avise 1994 for examples). Especially important is the new ability to discern the phylogenetic relationships or molecular differences among alleles, as well as their frequency and distribution (Avise 1989).

Genetic criteria used for identifying conservation units have ranged from requiring significant divergence in allele frequencies (e.g., Waples 1991; Leary et al. 1993) to requiring congruence of phylogeographic structure across genes (Avise and Ball 1990). A major and recurrent problem with using quantitative measures of genetic difference to define "significant" units is that there is no theoretically sound answer to the question, How much difference is enough? A common approach is to assess the variation between populations in relation to that between species or to that expected through isolation by distance within the taxon (e.g., Bowen et al. 1991). This approach has some attraction but can be circular and assumes a consistent relationship between the amount of genetic divergence and attainment of reproductive isolation (see, for critiques, Avise and Aquadro 1982; Vogler and DeSalle 1994). We suggest that lack of consensus on appropriate genetic criteria is partly due to inadequate understanding of the reasons for identifying conservation units in the first place.

Concepts

Genetic information can be used for conservation purposes in two distinct ways: (1) gene conservation, in which genetic diversity is identified and managed for its own sake, and (2) molecular ecology, in which patterns of genetic variation are used as a guide and complement to ecological studies (Moritz 1994a). The former has a strong evolutionary component; the aim is to maintain the amount and structure of genetic diversity in order to preserve evolutionary potential (Frankel and Soulé 1981; Hedrick and Miller 1992). Maintaining this genetic diversity is of major concern in fisheries management where anthropogenic modifications of genetic population structure are rife (Allendorf and Leary 1988; Ryman 1991). The latter set of applications of genetic information is more relevant to short-term management of populations and seeks to resolve questions about managed populations that are difficult to address by means of ecological methods—examples include analysis of mating systems (e.g., Amos et al. 1993), identification of the source of individuals in migratory species (e.g., Pella and Milner 1987), and the analysis of population structure (see below). The distinction between these two areas of application is fundamental because it affects the way in which data are collected and interpreted (Moritz 1994a). Further, the excessive focus on gene conservation has contributed to the suggestion that genetics is only marginally relevant to practical wildlife conservation because of the perception that demographic fluctuations are a more immediate threat to the survival of small populations (e.g., Lande 1988; compare Mills and Smouse 1994).

The dichotomy between gene conservation and molecular ecology is pertinent to identifying popu-

ALLELE FREQUENCIES

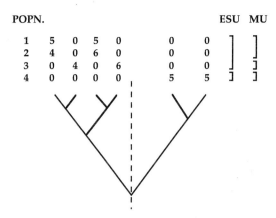

POPN.							ESU	MU
1	5	0	5	0		0	0	
2	4	0	6	0		0	0	
3	0	4	0	6		0	0	
4	0	0	0	0		5	5	

FIGURE 2.—Hypothetical distribution of six alleles among four populations (popn) and allele frequencies and phylogeny in relation to the definitions of evolutionarily significant units (ESU) and management units (MU) suggested in the text. Populations 1 and 2 have similar allele frequencies and are part of the same MU. Both populations 3 and 4 have fixed differences relative to populations 1 and 2, but only population 4 shows phylogenetic separation and would be classified as a separate ESU. Reprinted from Moritz (1994a).

lation units because it suggests two complementary approaches to identifying units of significance to conservation (Moritz 1994b). First, in relation to gene conservation, we suggest that the objective should be to identify sets of populations that have had a long period of independent evolution and to define these as ESUs for the purpose of conservation planning and the setting of priorities. Based on the results of simulation studies (Neigel and Avise 1986) and recent empirical studies, Moritz (1994b) suggested that patterns of mitochondrial DNA (mtDNA) allele phylogeny combined with information on nuclear allele frequencies be used to define a qualitative, relational criterion for an ESU: different ESUs should be reciprocally monophyletic for mtDNA alleles and show significant divergence of allele frequencies at nuclear loci (Figure 2). Note that the branch length (Figure 2) is irrelevant; it is the phylogenetic pattern that is important. By contrast, some (e.g., Crozier 1992) have suggested that conservation value be ascribed in relation to the amount of sequence divergence separating clades on a molecular phylogeny.

The proposal to use phylogeographic pattern to define ESUs is not new. Dizon et al. (1992) suggested that conservation significance be scaled ac-

cording to the degree of phylogenetic sorting of alleles between populations; Vogler and DeSalle (1994) suggested that ESUs be defined by derived character states; and Avise and Ball (1990) advocated the need for congruence of phylogeographic structure across loci. In our view, the definition we propose is restrictive enough to be meaningful as a complement to targeting recognized species for conservation, but is not so restrictive as to be unreasonable. In their simulations Neigel and Avise (1986) found that it usually took at least $4N_e$ (where N_e is the effective population size) generations from the time of separation for the processes of lineage extinction and divergence to result in reciprocal monophyly of mtDNA alleles between two populations. Sorting of nuclear alleles takes much longer because of larger effective population size and lower mutation rate (Takahata 1989) and may be rare within species (although see Bernardi et al. 1993). Accordingly, we do not require that nuclear gene phylogenies are concordant, but we do require significant divergence in allele frequencies at nuclear loci to avoid falsely identifying populations with female, but not male, philopatry as separate ESUs.

Second, in relation to the molecular ecology dimension of conservation genetics, we propose that the relevant conservation unit is the set of populations that exchange substantial numbers of individuals between each other but which are functionally separate from other such sets. The purpose of identifying functionally independent sets of populations is to establish the appropriate geographic scale for monitoring and managing populations (Baverstock et al. 1994). This is similar to one use of the term stock, i.e., "a relatively homogeneous and self-contained population whose losses by emigration and accessions by immigration, if any, are negligible in relation to the rates of growth and mortality" (Harden-Jones 1992). Given the multiple uses of the term stock (reviewed by Gauldie 1991), we suggest the term management unit (MU) to contrast with ESU (Moritz 1994b). Populations connected by levels of gene flow that are trivial in comparison with birth rates within populations are expected to diverge in allele frequency (Slatkin 1987) but do not necessarily show phylogeographic structure. Therefore an MU consists of one or more homogeneous populations that show statistically significant divergence of allele frequencies at nuclear or mitochondrial loci from other such units (Figure 2). In contrast to the criterion for an ESU, the emphasis for MUs is on differences in allele frequency, irrespective of allele phylogeny.

Neither of the definitions above incorporate information on characters that might be expected to demonstrate local adaptation, a factor often considered important for recognizing conservation units (e.g., Waples 1991; Dizon et al. 1992; Vogler and DeSalle 1994). This is an issue that needs more debate. The definition of an ESU suggested above puts the emphasis on historical population structure rather than on current adaptation on the grounds that we can assess "significance" of particular populations or genotypes in terms of the past but not the future (Moritz 1994b). In the context of gene conservation, we suggest that the goal is to maintain the potential to adapt, rather than the full array of current adaptations. On one hand, the natural process of evolution proceeds by the replacement of one set of (former) adaptations with a new set, and to preserve all currently adapted forms would retard this process. On the other hand, the rate of anthropogenic habitat modification far exceeds that of natural landscape evolution, so that we need to protect individual, differently adapted populations to enable the natural evolutionary process to operate. We see no immediate solution to this dilemma.

Whatever their evolutionary significance, the presence of differently adapted forms is important for short-term management of populations that constitute ESUs. This is because specific adaptations, by definition, are expected to affect survival and reproduction and thus population persistence. Presumably such forms should not be mixed or translocated to new habitats (Meffe and Vrijenhoek 1988), although our ability to predict the outcomes for fitness is limited (e.g., Allendorf and Leary 1988; Ferguson et al. 1988). One must also recall that variation in phenotype alone does not necessarily indicate differential adaptation (Gould and Lewontin 1979), nor is it necessarily genetic. The phenotypic plasticity of fishes can make it difficult to distinguish between environmental and genetic causes of differences in behavior, morphology, or life history (Allendorf et al. 1987; Meffe and Vrijenhoek 1988). Given that the variation is genetic and adaptive, it is likely that, under the definitions proposed, differently adapted populations would be recognized as separate MUs based on patterns of neutral variation. The exception is if selection is spatially heterogeneous and sufficiently strong to cause divergence in frequencies of selected alleles despite high gene flow (Endler 1977).

Different types of information are needed to identify the two types of conservation unit as defined above. Evolutionarily significant units can be delineated by phylogenetic analysis of mtDNA restriction sites or sequences obtained from relatively few individuals (but see "Caveats and Comments") combined with analysis of allozymes or nuclear restriction fragment length polymorphisms (RFLPs). By contrast, the identification of MUs will typically require analysis of large numbers of individuals to assess differences in allele frequency and can be done using allozymes, micro- or minisatellite loci, or nuclear or mitochondrial RFLPs, perhaps targeted by preliminary sequencing (Slade et al. 1993). Under some circumstances large-scale sequencing may increase the power to detect population subdivision (Hudson et al. 1992), but it is neither necessary nor cost-effective in many cases. This is illustrated in the following examples.

Case studies

The Southern Elephant Seal

The southern elephant seal *Mirounga leonina* has major breeding populations located southeast of Australia (Macquarie Island, MQ), in the south Indian Ocean (Heard Island, HD), and off the southwest coast of South America (South Georgia, SG); there are smaller colonies near the last two locations. Although subject to harvesting, the species never reached such small numbers as its famous congener, the northern elephant seal *M. angustirostris* (Hoelzel et al. 1993). The populations recovered to large numbers by the 1950s but since then declines of up to 50% have been reported in the Indian Ocean and Macquarie Island populations (Hindell and Burton 1987).

Variation in mtDNA control region and nuclear gene sequences was analyzed by R. W. Slade (University of Queensland, unpublished), and the results are summarized in Figure 3. Phylogenetic analysis of the mtDNA sequences from the three major populations and from a smaller mainland Argentinian population (Peninsula Valdez, PV; reported by Hoelzel et al. 1993) revealed a phylogenetic disjunction between the five MQ samples versus the others, although the branch defining the monophyly of SG, HD, and PV was weakly supported (Figure 3A). Screening diagnostic restriction sites of an additional 96 samples confirmed the distinctiveness of mtDNA alleles between MQ and other locations—however, inclusion of additional SG alleles detected by Hoelzel et al. (1993) resulted in a basal polytomy that joins several clades of the SG and HD samples and the MQ clade. In contrast, the HD and SG sequences were polyphyletic and were also paraphyletic with respect to the PV samples, which

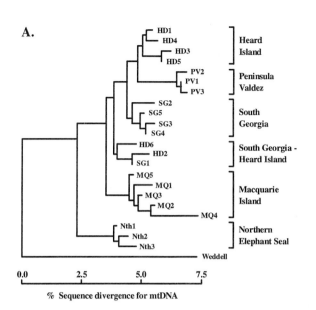

A.

% Sequence divergence for mtDNA

B.

Populations	Tests for heterogeneity			
	mtDNA	Nuclear loci		
		DQA	Ald-A	β-globin
MQ versus HD	***	ns	ns	*
MQ versus SG	***	ns	ns	***
HD versus SG	*** a	ns	ns	ns
PV versus others	***			

a Not significant based on analysis of molecular variance (AMOVA) or G_{ST}.

C.

Evolutionarily significant units	Management units
Northern elephant seal	California
Southern elephant seal	
Macquarie Island	Macquarie Island
Rest (HD + SG + PVa)	Heard Island
	South Georgia
	Argentina

a The status of the PV population is provisional.

FIGURE 3.—Summary of information on allele frequency and phylogeny among breeding populations of southern elephant seals (R. W. Slade, unpublished data). (A) Phylogeny of southern elephant seal mitochondrial DNA (mtDNA) control region sequences created by neighbor-joining method. This phylogeny is identical to the shortest parsimony tree, although a tree one step longer puts the MQ clade between the two major SG and HD clades. The Peninsula Valdez breeding populations are represented by PV. The elephant seals are compared with the Weddell seal *Leptonychotes weddelli*. (B) Results of randomized χ^2 tests for heterogeneity of allele frequencies for mtDNAs assayed by means of restriction fragment length polymorphisms and introns from three nuclear genes. * = $P < 0.05$, *** = $P < 0.001$; ns is not significant. (C) Suggested classification of the populations as evolutionarily significant units or management units on the basis of the information on allele frequency and phylogeny.

were themselves monophyletic (Figure 3A). Analysis of the allele frequency data generated by the RFLP analysis of the control region segment revealed highly significant differences (tested using observed versus randomized χ^2 values) between all populations (Figure 3B), except for the comparison between HD and SG when information on sequence divergence among alleles was included.

For nuclear loci, the alleles of the southern elephant seal are paraphyletic with respect to its congener, the northern elephant seal (Slade et al. 1994), and no phylogeographic structure was observed among the different populations of the southern elephant seal. However, the MQ population showed significantly different allele frequencies at one nuclear locus—a β-globin microsatellite (Figure 3B; R. W. Slade, unpublished)—confirming a previous report of divergence in allozyme allele frequencies between the MQ and HD populations (Gales et al. 1989).

In combination, these data permit an evaluation of conservation units (Figure 3C). Despite the poor

resolution at the base of the mtDNA phylogeny (see below), we suggest that the MQ population be considered a separate ESU from the remainder. The status of the PV population is uncertain; it was nested well within the SG–HD group, and, in the absence of data on nuclear genes, it is included in the ESU that consists of SG, HD, and PV populations. On the basis of the mtDNA allele frequency data, SG, HD, and PV each represent separate MUs within the larger ESU. Sampling of other small colonies near HD and SG may reveal more MUs, but it is unlikely that additional ESUs will be detected. This application serves to illustrate the appropriateness of sequence data for defining ESUs and RFLP assays for detecting MUs. It also makes it clear that to require congruence of mtDNA and nuclear gene phylogenies for an ESU is too restrictive.

The Green Turtle

The green turtle *Chelonia mydas*, like other marine turtles, undergoes extensive migrations during its development and, as an adult, between its nesting and feeding grounds (Carr 1967; Limpus et al. 1992). The species is subject to a large harvest as well as other anthropogenic effects, and many populations have declined or have been extirpated. Because of the conservation problems and the unusual life history, green turtle rookeries have been extensively surveyed for variation in mtDNA (Bowen et al. 1992; Norman et al. 1994), allozymes (Bonhomme et al. 1987), RFLPs of anonymous nuclear loci (Karl et al. 1992), and, most recently, microsatellite loci (FitzSimmons et al. 1995).

On a global scale, reciprocal monophyly of mtDNA alleles is observed between the Atlantic and Pacific–Indian oceans but, as a rule, not between regions within either ocean (Bowen et al. 1992). This, together with the nuclear gene data (Karl et al. 1992) suggests that the Atlantic and Pacific–Indian ocean populations represent separate ESUs and that the eastern Pacific black turtle *C. m. aggasizi* is part of the latter (Bowen et al. 1992; Moritz 1994b).

The data for the green turtle are particularly interesting in relation to the definition of MUs. The mtDNA polymorphisms are very highly structured among regions, more so than nuclear gene polymorphisms, leading Karl et al. (1992) to suggest that males may be less philopatric than are females. However, because the behavior of females is paramount for population management, the mtDNA heterogeneity has special significance, and MUs recognized on the basis of divergent mtDNA allele frequencies typically consist of one or a few nearby rookeries (Bowen et al. 1992; Norman et al. 1994). This interpretation is consistent with extensive recapture data (e.g., Limpus et al. 1992) and with the failure of several Carribean rookeries to be recolonized after being overharvested, despite the presence of other substantial rookeries in the region (Parsons 1962).

The ability to distinguish MUs in the green turtle is clearly technique dependent (Figure 4). For just the populations in Australian waters, intensive RFLP analysis of whole mtDNA identified three MUs, of which two were parts of geographically broader units (see Figure 4 for definitions of acronyms). More detailed analysis, based on a combination of sequencing and RFLP or denaturing-gradient gel electrophoresis analysis of a 380-base pair control region segment, greatly increased sensitivity and separated the GULF population from the two West Coast samples (NWC + LAC), those from the other east Indian Ocean rookeries (IND + MAL),

as well as PNG from SGBR (Norman et al. 1994). Analysis of the distribution of alleles at four microsatellite loci has revealed each of the WA (= NWC + LAC) and GULF populations to be unique but did not reveal any difference between the NGBR and SGBR stocks (N. FitzSimmons and coworkers, University of Queensland, unpublished data).

The Coconut Crab

The coconut crab *Birgus latro* is predominantly terrestrial and was formerly distributed throughout tropical islands of the Indo-Pacific region. Because of its large size (up to 4 kg) and tasty flesh, it has been overharvested to the point where populations have declined across the range and disappeared from many islands (see papers in Brown and Fielder 1992). The species has a short (2–4 week) pelagic larval stage and otherwise is completely terrestrial, prompting concern that movement between islands may be insufficient to promote recolonization.

Geographic variation has been examined for allozymes (Lavery et al. 1995) and mtDNA (S. Lavery, unpublished data), the latter based on RFLP analysis. The pattern of genetic variation is remarkably congruent across the two data sets (Figure 5). The Christmas Island population in the Indian Ocean has mtDNA alleles that are reciprocally monophyletic relative to Pacific Ocean populations and shows much greater allozyme divergence than that predicted by its geographic separation from the Pacific islands. Accordingly, we recognize two ESUs—one in the Indian Ocean and the other in the Pacific.

Within the Pacific, both mtDNA and allozymes showed isolation-by-distance effects (Figure 5). For allozymes, allele frequencies were relatively homogeneous among nearby populations along the island chain from Papua New Guinea to Vanuatu but were significantly different between these and the more isolated, peripheral populations in the Pacific. The mtDNA data exhibited the same pattern, with haplotype statistics (e.g., χ^2 tests for heterogeneity of allele frequencies) showing greater sensitivity in discriminating population heterogeneity than did measures that included sequence divergence.

Lavery et al. (1995) estimated that genetic neighborhoods (i.e., the geographic area "from which the parents of central individuals may be treated as if drawn from random," Wright 1969:295) in the coconut crab span an average distance of the order of 2,000 km. Given that this is larger than the distance between most adjacent island groups in the Pacific,

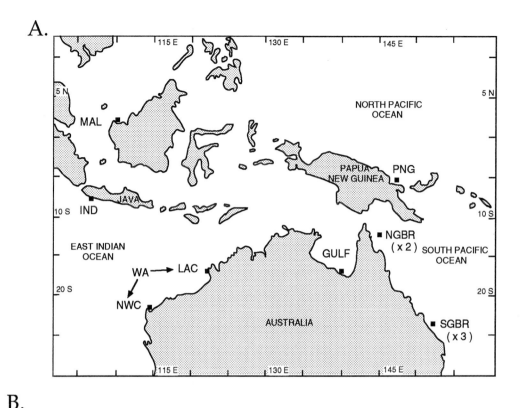

FIGURE 4.—Definition of management units of the green turtle in the waters of Australia and adjacent nations (data from Norman et al. 1994; N. Fitzsimmons and coworkers, University of Queensland, unpublished data). (**A**) Map showing the location of regional breeding populations sampled for genetic analysis. Regions are Papua New Guinea (PNG), southern Great Barrier Reef (SGBR), northern Great Barrier Reef (NGBR), Gulf of Carpentaria (GULF), Western Australia (WA), Indonesia (IND), and Malaysia (MAL). The WA region includes the North West Cape (NWC) and Lacepede Island (LAC) rookeries. (**B**) Grouping (indicated by continuous asterisks) of sampled rookeries on the basis of differences in the frequency of alleles detected by (1) intensive restriction fragment length polymorphism (RFLP) analysis of whole mitochondrial DNA (mtDNA), (2) a combination of sequencing and RFLP analysis of a section of mtDNA control region, and (3) amplification of four microsatellite loci.

it is more appropriate to consider overlapping management areas rather than discrete MUs (Figure 5). This is still of use for conservation management because it establishes the appropriate geographical scale for monitoring population trends and provides an indication of the maximum distance over which migration can be expected to restore overharvested populations in the short term.

Allozymes

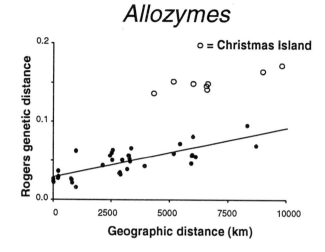

mtDNA

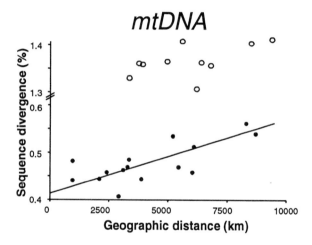

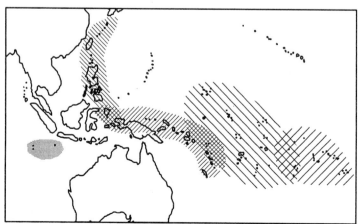

FIGURE 5.—Patterns of genetic differentiation in the coconut crab revealed by allozyme electrophoresis (Lavery et al. 1995), and restriction fragment length polymorphism analysis of mitochondrial DNA (mtDNA) (S. Lavery, University of Queensland, unpublished data), and a map illustrating the nature of management areas inferred from the patterns of isolation by distance. The open circles in the plots are the comparisons involving the Christmas Island population from the Indian Ocean, suggested to represent a separate evolutionarily significant unit.

Caveats and Comments

The criteria proposed for ESUs and MUs are intended to be theoretically well founded and explicit, and they appear to be useful in practice. However, as with most evolutionary properties of populations, ambiguities and anomalies will occur, and the criteria have to be applied with common sense. Further, it must be acknowledged that any delineation of conservation units is a hypothesis to be tested through further analysis and, if necessary, modified. This is entirely consistent with the practice of adaptive management.

Evolutionarily Significant Units

A difficulty that may arise with defining ESUs on the basis of reciprocal monophyly of mtDNA alleles is the effect of hybridization. Natural hybridization, a common feature in aquatic systems, can occur in two ways. A taxon may arise by hybridization but subsequently be isolated. If the period of isolation is sufficient, then taxa of hybrid origin would be recognized as ESUs and protected accordingly. A possible example is the cyprinid *Gila seminuda*, which appears to have arisen by hybridization but now has distinct and apparently monophyletic mtDNA (DeMarais et al. 1992). Alternatively, between otherwise isolated lineages there may be occasional episodes of hybridization interrupting an otherwise strongly structured allele phylogeny (e.g., Dowling and DeMarais 1993). Rare hybridization events between genetically distinct lineages can have a major effect on the genetic variance and evolutionary dynamics of populations (Grant and Grant 1992; Dowling and DeMarais 1993), in which case it seems reasonable not to classify such interacting forms as historically isolated populations or ESUs. Nonetheless, it is important to retain the interacting entities (most recognized as species) to maintain the overall evolutionary dynamics of the system (Dowling et al. 1992).

Anthropogenic hybridization due to translocations or changes in drainage patterns has a major influence on the genetic structure of many freshwater taxa (e.g., Allendorf and Leary 1988) and, tragically, destroys the natural distribution of variation and, along with it, the information on population history needed to assess status. Populations with known or suspected histories of nonnatural introgression should be excluded in ESU assessment; if no natural populations exist, then alternative approaches are needed.

The requirement for reciprocal monophyly of mtDNA among ESUs may seem overly restrictive given that many phenotypically divergent or reproductively isolated forms have been shown to have paraphyletic or polyphyletic mtDNA. However, it needs to be kept in mind that the definition of ESUs complements rather than replaces protection of species recognized under broader biological criteria. For example, several of the Lake Malaŵi cichlids have paraphyletic or polyphyletic mtDNAs (Moran and Kornfield 1993; Bowers et al. 1994) and are also polyphyletic for major-histocompatibility-complex alleles (Ono et al. 1993; Klein et al. 1993). However, most are recognized as separate species, and, in any case, conservation efforts are directed towards the community rather than individual species (Reinthal 1993). The Death Valley pupfishes *Cyprinodon* spp. are another interesting example. They occupy relict water sources thought to have been isolated since the late Pleistocene, and they show substantial differences in morphology, behavior, and life history, but several subspecies have mtDNAs that are indistinguishable in RFLP analyses (Echelle and Dowling 1992). Again, these are currently protected and managed as separate species or subspecies, regardless of their status as ESUs.

The identification of ESUs is prone to error through artifacts of sampling. Insufficient sampling of individuals or populations could provide a false impression of reciprocal monophyly, leading to oversplitting (i.e., a type I error where the null hypothesis is a single ESU). In this context, sampling of multiple populations within a hypothesized ESU is more important than obtaining large numbers of individuals per population, and pilot studies can be used to great advantage (Baverstock and Moritz 1990). However, it may not be necessary to obtain sequences from all individuals. A reasonable compromise, illustrated by the study on elephant seals (above), is to sequence a few individuals per population and, if populations appear to be monophyletic, then screen for restriction sites that diagnose the hypothesized monophyletic groups of alleles in larger samples (see also Worthington Wilmer et al. 1994).

We have argued for strict reciprocal monophyly of mtDNA alleles because it provides a qualitative and thus powerful operational criterion. However, it can also be argued that allowance should be made for cases with strong phylogeographic structure interrupted by rare "stray" alleles, such as might be detected in large-scale sequencing studies (e.g., Pacific versus Atlantic humpback whales *Megaptera novaeangliae*; Baker et al. 1993). Given the stochastic nature of lineage sorting it is difficult to argue that such populations are intrinsically of less evolutionary "significance" than those with full reciprocal monophyly; yet to make allowance for the former weakens the operational definition. With some reservations, Moritz (1994a) suggested that the significance of phylogeographic structuring could be assessed by comparing the phylogeographic distribution of alleles to randomized data, perhaps using a modification of the Maddison and Slatkin (1991) test for panmixia.

The second potential sampling error occurs if too few variable nucleotides are assayed to reveal the underlying phylogenetic structure, in which case an existing ESU may not be recognized (i.e., type II error). A related problem is the lack of resolution sometimes seen at the base of intraspecific phylogenies (e.g., Figure 3A). In this context it may be sufficient to recognize a monophyletic group as an ESU so long as it does not create statistically supported paraphyly among the remaining populations. Further consideration needs to be given to these grey areas. Despite the potential difficulties described above, mtDNA in several aquatic species (e.g., Bermingham and Avise 1986; Cronin et al. 1994) and many terrestrial taxa does show strong phylogeographic structure, indicating that the proposed criteria for an ESU can be usefully applied.

Management Units

Because it is based on quantitative rather than qualitative criteria, the recognition of MUs could suffer from the "how much is enough?" syndrome. This question is raised because of the sensitivity of tests for population subdivision to the methods used to assay (e.g., Figure 4B) and analyze (e.g., Hudson et al. 1992) variation. Sample size also has a powerful effect; as samples become larger, the probability of detecting small differences in allele frequency increases (Richardson et al. 1986). Ultimately, if all individuals from a random mating population are assayed, any two arbitrarily defined divisions of that population are likely to differ in allele frequency (in this case, the parameter rather than the estimate). The use of species-specific population models with demographically meaningful rates of exchange may be one way to determine thresholds of genetic divergence beyond which MUs are recognized. In general, the chance of drawing fundamentally incorrect conclusions about population structure is reduced by integrating genetic and ecological evidence.

Where sampled populations differ in allele frequency, the level of interchange is assumed to be low relative to intrinsic processes; however, genetic drift can also occur in small populations with high rates of immigration. Using simulations, Allendorf and Phelps (1981) found that significant differences in allele frequency were common even with the exchange of 10 individuals per generation among populations ranging in size from 25 to 100. Bowen et al. (1995) reported significant differences in mtDNA allele frequency between populations of loggerhead turtle *Caretta caretta* for which occasional exchanges of tagged nesting females have been recorded. However, these loggerhead turtle populations behave as separate entities (e.g., they have separate feeding grounds).

Oversplitting, that is incorrectly identifying interacting units as separate entities, could result if there is nonrandom sampling of family groups within larger interbreeding populations (e.g., Amos et al. 1991, 1993) or substantial temporal variations in allele frequency. The latter is exemplified by the significant divergence in allozyme allele frequency before and after a population crash in chubs (DeMarais et al. 1993). Such cases emphasize the importance of resampling populations or sites across years, although unperturbed populations of salmonids typically have stable allele frequencies (Altukhov and Salmenkova 1987).

The opposite problem is the failure to recognize demographically independent populations because of a lack of power in the genetic analyses. In a management context, type II errors may be more significant than are type I as the consequence of the former is to fail to recognize functionally independent units, which may lead to local extinctions. Where the populations actually differ by small shifts in allele frequency, the sample sizes needed to have reasonable statistical power (i.e., the probability of a type II error less than 0.5) may be enormous, especially if type I error is maintained at 5% (Richardson et al. 1986). This problem can be overcome by analyzing additional loci (e.g., Utter et al. 1992) or by moving to a genetic system in which differences in allele frequency are enhanced. For example, analyses of mtDNA often reveal population structuring where nuclear markers do not (e.g., Figure 3B; Bermingham et al. 1990) because of the lower effective population size of mtDNA, male-biased gene flow, or both.

The identification of MUs can be problematic in species with a large continuous or "stepping stone" distribution. Unless there are strong and distinct barriers to migration due to current or historical physical boundaries (e.g., between Indian Ocean and Pacific Ocean coconut crabs, Figure 5), there will often be no clear-cut physical limits to MUs. Yet in species in which there is an isolation-by-distance effect (e.g., among Pacific Ocean coconut crabs, Figure 5) it is not appropriate to consider the entire distribution as one MU. Migration between adjacent habitats is likely to be great enough to overcome (in ecological time) local population declines but too low to compensate for widespread population reductions. For management purposes, this critical geographic scale can be defined as the distance over which genetic homogeneity is maintained and beyond which significant genetic heterogeneity is detected in either nuclear or mtDNA alleles. It may be useful to view such a population structure as consisting of a series of overlapping management areas (e.g., Figure 5).

Conclusions

The purpose of this paper is to suggest ways of using genetic information to define conservation units that are theoretically sound, empirically feasible, and relevant to conservation needs. Building on previous suggestions that different depths of genetic structuring might have different implications for population management (e.g., Avise 1994), we suggest many of the ambiguities experienced previously can be overcome by recognizing two types of conservation unit: ESUs and MUs. It is fundamental to our approach that ESUs and MUs are each recognized as being significant for conservation but for different purposes: one to set priorities for conservation action and the other to manage properly the entities so recognized.

The Conservation Significance of Management Units

The presence of the term "significant" in ESU but not MU could leave the impression that, somehow, recognition of the former is more important for conservation. This is not so. It is important to identify the MUs, demographically independent populations, in order to maintain the larger units, whether species or ESUs.

1. The loss of an MU through overharvesting or habitat destruction is unlikely to be reversed by natural immigration in the short term (e.g., hundreds of years for the green turtle). This provides managers with an important predictive tool, allowing a decision about whether it is necessary to translocate individuals to reestablish a population.
2. The population response to manipulations or impacts should be monitored at the geographic scale of an MU and separately for different MUs in order to avoid masking declines in some MUs because of increases in others.
3. Some MUs also constitute source populations, periodically supporting others (sinks) within a metapopulation, and these two classes need to be recognized and considered separately for management.
4. By virtue of having distinctive genetic profiles, an MU can represent a substantial proportion of a species' genetic diversity, even though it does not qualify as an ESU.

A clear example is provided by the Sonoran topminnow *Poeciliopsis occidentalis*. A survey of allozyme variation indicated the presence of three divergent groups of populations that with one exception correspond to different drainages (Vrijenhoek et al. 1985). Within one of these groups (a potential ESU) were genetically distinct populations (presumably MUs) of varying heterozygosity. Quattro and Vrijenhoek (1989) showed that the most heterozygous population was also the most productive. The management agency subsequently accepted Quattro and Vrijenhoek's suggestion that this population be used as the source for local translocations rather than the less variable and lower fitness population previously used.

Biology and Realpolitik

Decisions about which populations to manage are made on the basis of biological and socioeconomic considerations, and this reality has lead some conservation biologists to despair. One reviewer of this paper suggested that ESU be replaced by PSU—politically sensitive unit; another suggested that the genetics is redundant because "MUs will be implemented anyway when and if politics, economics, or geography demands it, regardless of the genetics, and will not be implemented if none of these forces apply, despite what the genetics might suggest." This may be true, but the need for rigorous and theoretically defensible biological input is all the more important given the conflicting agendas involved.

A major stimulus for this conference is an impending reauthorization of the Endangered Species Act (ESA; 16 U.S.C. §§ 531 to 1544). At present the ESA provides for the protection of "any distinct population segment of a vertebrate fish or wildlife which interbreeds when mature." (This may seem vague, but the corresponding Australian legislation is worse: populations are listed at "the minister's discretion"!) Given the direction to invoke this clause sparingly and only when the data dictate, Waples (1991) restricted usage to ESUs, requiring evidence that the population(s) have "substantial reproductive isolation" and represent "an important component in the evolutionary legacy of the species." Waples is to be applauded for attempting to suggest workable criteria, but, inevitably, ambiguity arises in defining what qualifies as an "important component" of the evolutionary legacy of a species. We suggest that a solution to this dilemma is to expand the definition of protected populations to include stringently defined ESUs together with the MUs that are judged important to the survival of the larger entity, be it an ESU or a species. This is entirely justifiable and would extend the reach of the ESA to include populations that are an important component of the listed entity even if they do not themselves qualify as an ESU.

Acknowledgments

Our thanks to Jennifer Nielsen and Fred Allendorf for the invitation to contribute; to Paul Bentzen, Sandie Degnan, Andrew Dizon, Tom Dowling, Nancy FitzSimmons, Steve Palumbi, and Tom Smith for critical reviews of the manuscript; to Anita Heideman for illustrations; and to Nancy FitzSimmons, Lisa Pope, and Col Limpus for permission to report unpublished data. Support was provided by the Australian Antarctic Commission, the Queensland Department of Environment and Heritage, the Australian Nature Conservation Agency, the Australian Council for International Agricultural Research, and the Australian Research Council.

References

Allendorf, F. W., and R. F. Leary. 1988. Conservation and distribution of genetic variation in a polytypic species, the cutthroat trout. Conservation Biology 2:170–184.

Allendorf, F. W., and S. R. Phelps. 1981. Use of allele frequencies to describe population structure. Canadian Journal of Fisheries and Aquatic Sciences 38: 1507–1514.

Allendorf, F. W., N. Ryman, and F. M. Utter. 1987. Genetics and fishery management: past, present, and future. Pages 1–19 in Ryman and Utter (1987).

Altukhov, Y. P., and E. A. Salmenkova. 1987. Stock transfer relative to natural organization, management, and conservation of fish populations. Pages 333–343 in Ryman and Utter (1987).

Amos, B., J. Barrrett, and G. A. Dover. 1991. Breeding behaviour of pilot whales revealed by DNA fingerprinting. Heredity 67:49–55.

Amos, B., C. Schlotterer, and D. Tautz. 1993. Social structure of pilot whales revealed by analytical DNA profiling. Science 260:670–672.

Avise, J. C. 1994. Molecular markers, natural history and evolution. Chapman and Hall, New York.

Avise, J. C. 1989. Gene trees and organismal histories: a phylogenetic approach to population biology. Evolution 43:1192–1208.

Avise, J. C., and C. F. Aquadro. 1982. A comparative summary of genetic distances in the vertebrates. Evolutionary Biology 15:151–185.

Avise, J. C., and R. M. Ball. 1990. Principles of genealogical concordance in species concepts and biological taxonomy. Oxford Surveys in Evolutionary Biology 7:45–68.

Baker, C. S., and thirteen coauthors. 1993. Abundant mitochondrial DNA variation and world-wide population structure in humpback whales. Proceedings of the National Academy of Sciences of the United States of America 90:8239–8243.

Baverstock, P. R., L. Joseph, and S. Degnan. 1994. Units of management in biological conservation. Pages 287–293 in C. Moritz and J. Kikkawa, editors. Conservation biology in Australia and Oceania. Surrey Beatty and Sons, Chipping Norton, Australia.

Baverstock, P. R., and C. Moritz. 1990. Sampling design. Pages 13–24 in D. M. Hillis and C. Moritz, editors. Molecular systematics. Sinauer, Sunderland, Massachusetts.

Belsky, J., and A. Kerr. 1993. ONRCs steelhead strategy: a lesson in ESA politics. The Osprey 19:6–16.

Bermingham, E., and J. C. Avise. 1986. Molecular zoology of freshwater fishes in southeastern United States. Genetics 113:939–965.

Bermingham, E., S. H. Forbes, K. Friedland, and C. Pla. 1990. Discrimination between Atlantic salmon (*Salmo salar*) of North American and European origin using restriction analyses of mitochondrial DNA. Canadian Journal of Fisheries and Aquatic Sciences 48:884–893.

Bernardi, G., P. Sordino, and D. A. Powers. 1993. Concordant mitochondrial and nuclear DNA phylogenies for populations of the teleost fish *Fundulus heteroclitus*. Proceedings of the National Academy of Science of the United States of America 90:9271–9274.

Bonhomme, F., S. Salvidio, A. Lebeau, and G. Pasteur. 1987. Comparison genetique des tortues vertes (*Chelonia mydas*) des Oceans Atlantique, Indien et Pacifique. Genetica 74:89–94.

Bowen, B. W., A. B. Meylan, and J. C. Avise. 1991. Evolutionary distinctiveness of the endangered Kemp's Ridley sea turtle. Nature 352:709–711.

Bowen, B. W., A. B. Meylan, J. P. Ross, C. J. Limpus, G. H. Balazs, and J. C. Avise. 1992. Global population structure and natural history of the green turtle (*Chelonia mydas*) in terms of matriarchal phylogeny. Evolution 46:865–881.

Bowen, B. W., J. C. Avise, J. I. Richardson, A. B. Meylan, D. Margaritoulis, and S. R. Hopkins-Murphy. 1993. Population structure of loggerhead turtles (*Caretta caretta*) in the northwestern Atlantic Ocean and Mediterranean Sea. Conservation Biology 7:834–844.

Bowers, N., J. R. Stauffer, and T. D. Kocher. 1994. Intra-

and interspecific mitochondrial DNA sequence variation within two species of rock-dwelling cichlids (Teleostei: Cichlidae) from Lake Malaŵi, Africa. Molecular Phylogenetics and Evolution 3:75–82.

Brown, I. W., and D. R. Fielder. 1992. The coconut crab: aspects of *Birgus latro* biology and ecology in Vanuatu. ACIAR, Canberra, Australia.

Carr, A. 1967. So excellent a fish: a natural history of green turtles. Scribner, New York.

Cronin, M. A., S. Hills, E. K. Born, and J. C. Patton. 1994. Mitochondrial DNA variation in Atlantic and Pacific walruses. Canadian Journal of Zoology 72:1035–1043.

Crozier, R. 1992. Genetic diversity and the agony of choice. Biological Conservation 61:11–15.

DeMarais, B. D., T. E. Dowling, and W. L. Minckley. 1993. Post-perturbation genetic changes in populations of endangered Virgin River chubs. Conservation Biology 7:334–341.

DeMarais, B. D., T. E. Dowling, M. E. Douglas, W. L. Minckley, and P. C. Marsh. 1992. Origin of *Gila seminuda* (Teleostei: Cyprinidae) through introgressive hybridization: implications for evolution and conservation. Proceedings of the National Academy of Science of the United States of America 89:2747–2751.

Dizon, A. E., C. Lockyer, W. F. Perrin, D. P. Demaster, and J. Sisson. 1992. Rethinking the stock concept: a phylogeographic approach. Conservation Biology 6:24–36.

Dowling, T. E., and B. D. DeMarais. 1993. Evolutionary significance of introgressive hybridization in cyprinid fishes. Nature 362:444–446.

Dowling, T. E., B. D. DeMarais, W. L. Minckley, and M. E. Douglas. 1992. Use of genetic characters in conservation biology. Conservation Biology 6:7–8.

Echelle, A. A., and T. E. Dowling. 1992. Mitochondrial DNA variation and evolution of the Death Valley pupfishes (*Cyprinodon*, Cyprinodontidae). Evolution 46:193–2206.

Endler, J. A. 1977. Geographic variation, speciation and clines. Princeton University Press, Princeton, New Jersey.

Ferguson, M. M., R. G. Danzmann, and F. W. Allendorf. 1988. Developmental success of hybrids between two taxa of salmonid fishes with moderate structural gene divergence. Canadian Journal of Zoology 66:1389–1395.

FitzSimmons, N. N., C. Moritz, and S. S. Moore. 1995. Conservation and dynamics of microsatellite loci over 300 million years of marine turtle evolution. Molecular Biology and Evolution 12:432–440.

Frankel, O. H., and M. E. Soulé. 1981. Conservation and evolution. Cambridge University Press, Cambridge, UK.

Gales, N. J., M. Adams, and H. R. Burton. 1989. Genetic relatedness of two populations of the southern elephant seal (*Mirounga leonina*). Marine Mammal Science 5:57–67.

Gauldie, R. W. 1991. Taking stock of genetic concepts in fisheries management. Canadian Journal of Fisheries and Aquatic Sciences 48:722–731.

Gould, S. J., and R. C. Lewontin. 1979. The spandrels of San Marco and the Panglossian paradigm: a critique of the adaptationist programme. Proceedings of the Royal Society of London B205:581–598.

Grant, P. R., and B. R. Grant. 1992. Hybridization of bird species. Science 256:193–197.

Harden-Jones, R. 1992. An overview—the needs of management. Pages 3–5 *in* P. I. Dixon, editor. Population genetics and its application to fisheries management and aquaculture. Center for Marine Science, University of New South Wales, Sydney, Australia.

Hedrick, P. W., and P. S. Miller. 1992. Conservation genetics: techniques and fundamentals. Ecological Applications 2:30–46.

Hindell, M. A., and H. R. Burton. 1987. Past and present status of the southern elephant seal (*Mirounga leonina*) at Macquarie Island. Journal of Zoology, London 213:365–380.

Hoelzel, A. R., and seven coauthors. 1993. Elephant seal genetic variation and the use of simulation models to investigate historical population bottlenecks. Journal of Heredity 83:443–449.

Hudson, R. R., D. D. Boos, and N. L. Kaplan. 1992. A statistical test for detecting geographic subdivision. Molecular Biology and Evolution 9:138–151.

Karl, S. A., B. W. Bowen, and J. C. Avise. 1992. Global population genetic structure and male mediated gene flow in the green turtle (*Chelonia mydas*): RFLP analysis of anonymous nuclear loci. Genetics 131:163–173.

Klein, D., H. Ono, C. O'hUigun, V. Vincek, T. Goldschmidt, and J. Klein. 1993. Extensive MHC variability in cichlid fishes of Lake Malaŵi. Nature 364:330–334.

Lande, R. 1988. Genetics and demography in biological conservation. Science 241:1455–1460.

Lavery, S., C. Moritz, and D. R. Fielder. 1995. Changing patterns of population structure and gene flow at different spatial scales in the coconut crab (*Birgus latro*). Heredity 74:531–541.

Leary, R. F., F. W. Allendorf, and S. H. Forbes. 1993. Conservation genetics of the bull trout in the Columbia and Klamath river drainages. Conservation Biology 7:856–865.

Limpus, C. J., J. D. Miller, C. J. Parmenter, D. Reimer, N. McLachlan, and R. Webb. 1992. Migration of green (*Chelonia mydas*) and loggerhead (*Caretta caretta*) turtles to and from eastern Australian rookeries. Australian Wildlife Research 19:347–358.

Maddison, W. P., and M. Slatkin. 1991. Null models for the number of evolutionary steps in a character on a phylogenetic tree. Evolution 45:1184–1197.

Meffe, G. K., and R. C. Vrijenhoek. 1988. Conservation genetics in the management of desert fishes. Conservation Biology 2:157–169.

Mills, L. S., and P. E. Smouse. 1994. Demographic consequences of inbreeding in remnant populations. American Naturalist 144:412–431.

Moran, P., and I. Kornfield. 1993. Retention of an ancestral polymorphism in the mbuna species flock (Teleostei: Cichlidae) of Lake Malaŵi. Molecular Biology and Evolution 10:1015–1060.

Moritz, C. 1994a. Applications of mitochondrial DNA analysis on conservation: a critical review. Molecular Ecology (3):401–411.

Moritz, C. 1994b. Defining evolutionary significant units for conservation. Trends in Ecology & Evolution 9:373–375.

Neigel, J., and J. C. Avise. 1986. Phylogenetic relationships of mitochondrial DNA under various demographic models of speciation. Pages 515–534 in E. Nevo and S. Karlin, editors. Evolutionary processes and theory. Academic Press, New York.

Norman, J., C. Moritz, and C. J. Limpus. 1994. Mitochondrial DNA control region polymorphisms: genetic markers for ecological studies of marine turtles. Molecular Ecology 3:363–373.

Ono, H., C. O'Huigin, H. Tichy, and J. Klein. 1993. Major-histocompatibility-complex variation in two species of cichlid fishes from Lake Malaŵi. Molecular Biology and Evolution 10:1060–1073.

Palumbi, S. R., and C. S. Baker. 1994. Opposing views of humpback whale population structure using mitochondrial and nuclear sequences. Molecular Biology and Evolution 11:426–435.

Parsons, J. 1962. The green turtle and man. University of Florida Press, Gainesville.

Pella, J. J., and G. B. Milner. 1987. Use of genetic marks in stock composition analysis. Pages 247–276 in Ryman and Utter (1987).

Quattro, J. M., and R. C. Vrijenhoek. 1989. Fitness differences among remnant populations of the endangered Sonoran topminnow. Science 245:976–978.

Queller, D. C., J. E. Strassman, and C. R. Hughes. 1993. Microsatellites and kinship. Trends in Ecology & Evolution 8:285–288.

Reinthal, P. 1993. Evaluating biodiversity and conserving Lake Malaŵi's cichlid fish fauna. Conservation Biology 7:712–718.

Richardson, B. J., P. R. Baverstock, and M. Adams. 1986. Allozyme electrophoresis: a handbook for animal systematics and population studies. Academic Press, Sydney, Australia.

Ryman, N. 1991. Conservation genetics considerations in fishery management. Journal of Fish Biology 39 (Supplement A):211–224.

Ryman, N., and F. Utter, editors. 1987. Population genetics and fishery management. University Washington Press, Seattle.

Slade, R. W., C. Moritz, A. Heideman, and P. T. Hale. 1993. Rapid assessment of single-copy nuclear DNA variation in diverse species. Molecular Ecology 2:359–373.

Slade, R. W., C. Moritz, and A. Heideman. 1994. Multiple nuclear gene phylogenies: applications to pinnipeds and a comparison with a mtDNA gene phylogeny. Molecular Biology and Evolution 11:341–356.

Slatkin, M. 1987. Gene flow and the geographic structure of animal populations. Science 236:787–792.

Takahata, N. 1989. Gene geneology in three related populations: consistency probability between gene and population trees. Genetics 122:957–966.

Utter, F. M. 1991. Biochemical genetics and fishery management: an historical perspective. Journal of Fish Biology 39(Supplement A):1–20.

Utter, F. M., R. S. Waples, and D. J. Teel. 1992. Genetic isolation of previously indistinguishable chinook salmon populations of the Snake and Klamath rivers: limitations of negative data. U.S. National Marine Fisheries Service Fishery Bulletin 90:770–777.

Vogler, A. P., and R. DeSalle. 1994. Diagnosing units of conservation management. Conservation Biology 8:354–363.

Vrijenhoek, R. C., M. E. Douglas, and G. K. Meffe. 1985. Conservation genetics of endangered fish populations in Arizona. Science 229:400–402.

Waples, R. S. 1991. Pacific salmon, Oncorhynchus spp., and the definition of "species" under the Endangered Species Act. U.S. National Marine Fisheries Service Marine Fisheries Review 53:11–22.

Waples, R. S. 1995. Evolutionarily significant units and the conservation of biological diversity under the Endangered Species Act. American Fisheries Society Symposium 17:8–27.

Worthington Wilmer, J., C. Moritz, L. Hall, and J. Toop. 1994. Extreme population structuring in the threatened ghost bat, Macroderma gigas: evidence from mitochondrial DNA. Proceedings of the Royal Society of London B257:193–198.

Wright, S. 1969. Evolution and the genetics of populations, volume 2. The theory of allele frequencies. Chicago University Press, Chicago.

American Fisheries Society Symposium 17:263–287, 1995

Population Genetic Divergence and Geographic Patterns from DNA Sequences: Examples from Marine and Freshwater Fishes

CAROL A. STEPIEN

Department of Biology, Case Western Reserve University
Cleveland, Ohio 44106, USA

Abstract.—Determining levels of genetic divergence within and among populations and developing strategies for maintaining this genetic diversity are fundamental for successful fisheries and environmental management. The present study compares overall levels of population genetic divergence discerned by DNA sequencing and the geographic components of this divergence in four commercially important marine and freshwater fishes. These fishes are from three North American aquatic ecosystems having different potential barriers to gene flow. Species and populations examined include the pleuronectid Dover sole *Microstomus pacificus* and the scorpaenid shortspine thornyhead *Sebastolobus alascanus* from three biogeographic provinces of the Pacific continental slope, the serranid spotted sand bass *Paralabrax maculatofasciatus* from disjunct populations in warm temperate kelp forests of the outer Pacific coast and the northern Gulf of California, and the percid walleye *Stizostedium vitreum* from river and reef spawning sites in Lake Erie. The first objective is to determine whether DNA sequence data from the mitochondrial (mt) control region resolves greater amounts of overall genetic polymorphism than have been found in prior allozyme or restriction fragment length polymorphism studies of these species. The second objective is to test whether genetic divergence in these species is partitioned in patterns corresponding to geographic regions or potential barriers to gene flow. For all species, results show that mtDNA sequencing of the control region produces considerably greater numbers of useful genetic characters. Results indicate that regional genetic divergence is pronounced in the spotted sand bass populations, which are separated by an apparent temperature barrier to gene flow (the southern Gulf of California). Walleye genotypes differ among spawning locations in Lake Erie, supporting the hypothesis that walleyes home to natal spawning sites. Despite high mtDNA diversity, the genotypes of continental slope fishes are widely dispersed and have fewer geographic groupings, suggesting higher levels of gene flow. Greater gene flow is probably due to extended periods of larval dispersal and lack of geographic barriers, excepting the divisions of major current systems. The ecosystems examined in the present study and the fisheries they support are presently facing significant anthropogenic disturbances, which renders baseline genetic data critically important to evaluating evolutionarily significant units.

Allendorf et al. (1987) stated, "The classical and most frequently encountered problem in fisheries management is the identification of genetically meaningful management units, i.e., stocks." Fishery stocks (i.e., population units) represent unique breeding groups that often possess novel forms of genetic, physiological, and ecological variation, which maintain diversity within a species. Identification of stocks thus delineates unique populations within a species that may each possess a distinct set of biological characters and differ in life history parameters (including growth rates, fecundity, total biomass, spawning biomass, numerical abundance, age and size distribution, recruitment and mortality rates, and length–weight relationships). Due to such variation, each stock or population may respond to environmental factors in a unique way (Brown et al. 1987). Genetic diversity may enable a species to inhabit a variety of environments and presumably augments its ability to withstand perturbations, such

as fishing pressure, habitat degradation, and competition from invading species. Maintaining this genetic variability thus appears fundamental for the long-term ability of a species to adapt to various habitats and changing environments and thus constitutes a primary goal for modern fisheries management (Allendorf et al. 1987).

For management purposes, an evolutionarily significant unit (ESU) of a species has been defined as a population unit that (1) is reproductively isolated from conspecific population units, and (2) contributes substantially to the ecological or genetic diversity of the species (Waples 1991). Degree of genetic divergence among population units and the existence of geographic barriers may be important criteria in determining degree of genetic isolation. In the present study, several examples are given that illustrate various degrees of genetic divergence, possible stock structure, and geographic delineation among populations of fishes.

Genetic Techniques

Intraspecific genetic differences have most often been examined using indirect molecular approaches, either (1) through analysis of charge differences in enzyme variants (allozyme and isozyme electrophoresis), which has been the most common method for the past 20 years, or (2) by restriction fragment length polymorphism (RFLP) analysis, which compares shared patterns of DNA fragments and has been used over the last 10 years (reviewed by Avise 1994). In comparison with allozyme and RFLP analyses, DNA sequencing provides a direct measure of genetic divergence (DNA sequences) rather than an indirect estimate based on the products of DNA sequences and often yields a markedly greater number of variable "characters" among populations for genetic analysis.

Several population studies of fishes have apparently been limited by the lack of resolving power of allozyme and RFLP techniques to discern the levels of polymorphism that are necessary to address small-scale differences among populations. For example, Siebenaller (1978) found few regional genetic differences in populations of the shortspine thornyhead *Sebastolobus alascanus* and the longspine thornyhead *S. altivelis* along the Pacific coast continental slope due to low overall allozyme polymorphism. The present study compares Siebenaller's (1978) results with those obtained from mitochondrial DNA (mtDNA) sequence data from the same species. A study by Graves et al. (1990) examined RFLP variation in the Pacific coast kelp forest sea basses, the spotted sand bass *Paralabrax maculatofasicatus*, the barred sand bass *P. nebulifer*, and the kelp bass *P. clathratus*, and resolved interspecific characters but found little intraspecific variation. Resolution of mtDNA sequences and allozyme variation for discerning intraspecific variation and divergence between putatively isolated populations of the spotted sand bass are tested in this investigation. Similarly, studies of genetic divergence of walleye *Stizostedium vitreum* in the Great Lakes based on allozymes (Ward et al. 1989; Todd 1990) and RFLPs (Ward et al. 1989; Billington et al. 1992) yielded characters apparently limited to discerning large-scale divergences among lakes. Ability to detect finer-scale geographic differences from mtDNA sequences is tested in this study.

Mitochondrial DNA evolves at a rate 5 to 10 times that of most nuclear DNA regions and is thus particularly useful for analyzing genetic relationships among populations. Mitochondrial DNA is haploid and maternally inherited in most species

that have been studied and, consequently, genetic lineages are usually less complicated to trace than they are with nuclear DNA (Attardi 1985). The region sequenced in examples of this study is the "control" or "D-loop" region, which contains the origin site of heavy-strand replication and is the most variable portion of the mtDNA molecule in many fishes (Brown et al. 1979; Avise 1991, 1994; Lee et al. 1995). This region does not code for proteins, and its variable domains are presumably largely selectively neutral, accumulating mutations randomly over time (Attardi 1985; Avise 1994). Relative numbers of individuals possessing a given mutation in a population thus change from generation to generation through the random processes of gene flow and genetic drift. These random changes are regulated by degree of population subdivision, population size, migration rate, barriers to dispersal, and biogeographic heterogeneity.

The present study examines and compares degree of mtDNA control sequence differences and population structure in fishes from continental slopes, kelp forests, and the Great Lakes. My first objective is to test whether greater numbers of informative genetic characters can be detected in species in which prior allozyme or RFLP analyses have been unsuccessful in resolving appreciable intraspecific differences for addressing questions of geographic and stock structure. My second objective is to test for geographic genetic divergence by discerning whether related genotypes are grouped by geographic region, which would indicate barriers to dispersal and gene flow.

Genetic Divergence in Deep-Sea Continental Slope Fishes

Biogeographic patterns of genetic differences are compared among populations of two sympatric groundfishes of the northeastern Pacific continental slope (Figure 1), the Dover sole *Microstomus pacificus* and the shortspine thornyhead. Adults live on the seafloor of the deep slope (600–1,000 m) in a narrow coastal band (usually less than 50 km wide) spanning three biogeographic provinces (Briggs 1974; Figure 1).

The continental slope is a relatively uniform and stable environment characterized by temperatures of 3–6°C, low oxygen levels (concentrations 0.27–0.36 mL/L) that distinguish it as the oxygen-minimum zone, and soft substrates (Hunter et al. 1990). The fishes are believed to be highly K-selected, with extremely slow metabolic and growth rates (Yang et al. 1992). Large Dover soles have been aged at 60

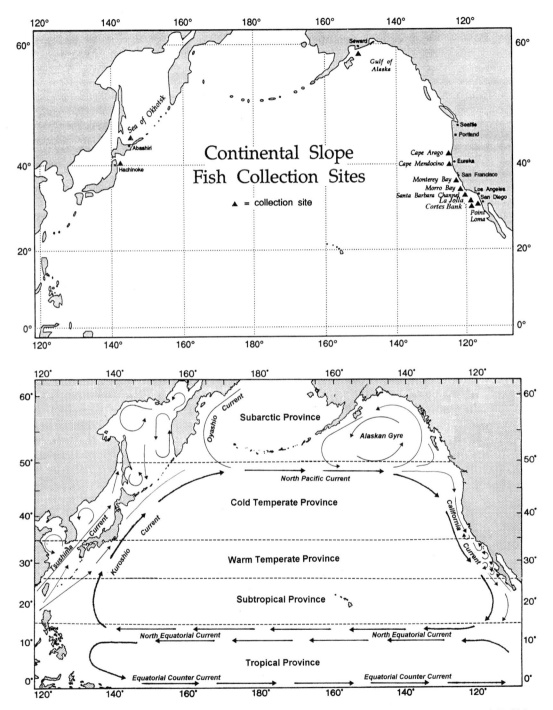

FIGURE 1.—Top: Map showing collection sites of continental slope fishes in the North Pacific. Not shown is Del Mar, off San Diego. Bottom: Biogeographic provinces and major currents in regions of collection sites of continental slope fishes. Provinces are adapted from Briggs (1974). Current patterns are adapted from Wyllie (1966) and Owen (1980).

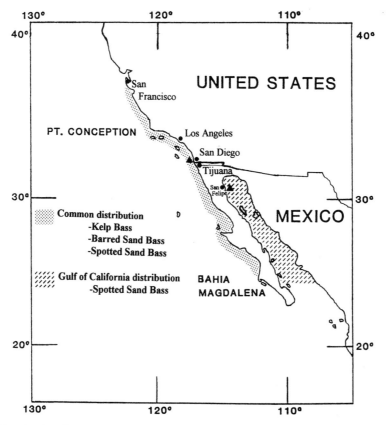

FIGURE 2.—Map showing distributions (stippling) and collection sites (triangles) of the kelp forest sea bass in the northeastern Pacific warm temperate province.

years and shortspine thornyheads at 120 years (Hunter et al. 1990; Loh-Lee 1991; J. Butler, National Marine Fisheries Service, personal communication). These slow growth rates render these populations essentially nonreplenishable on the timescale of fishery exploitation, and they are being rapidly depleted by commercial trawling (Pacific Fishery Management Council, unpublished data).

Latitudinal ranges of the thornyheads and Dover sole are so extensive that these fishes may provide a test of whether population genetic relationships are influenced by major oceanic circulation patterns (Figure 1). For example, tagging studies suggest that these adult groundfishes move little (J. R. Hunter, National Marine Fisheries Service, personal communication). Quirollo and Kalvass (1987) found 45% of tag returns for Dover soles within 1 km of the release point. Geographic population genetic patterns of these species may thus be largely influenced by dispersal during larval stages, which is primarily due to oceanic currents (Moser 1974;

Hunter et al. 1990). Major current patterns might create barriers to larval dispersal, producing nonuniform distribution of mtDNA haplotypes. These barriers may be indicated by geographic clustering and regional divergence of haplotypes. Alternatively, geographic patterns may be produced by selection. If the larvae are panmictically distributed, then genetic patterns may be produced by selective survival of juveniles or adults in given areas.

Genetic Divergence in Kelp Forest Spotted Sand Bass

The spotted sand bass is distributed in subtidal kelp forest areas along the warm temperate eastern Pacific coast, where it is sympatric with two congeners, the barred sand bass and the kelp bass (Figure 2). These species constitute important sport fisheries in California and Mexico. Commercial fishing is banned in California due to overexploitation (Oliphant 1979). The spotted sand bass is also found in the northern Gulf of California, Mexico (Figure 2), and may be reproductively isolated from the populations of the outer coast by the warmer subtropical waters of the southern Gulf. The southern Gulf is believed to form a temperature barrier to dispersal and gene flow between the two areas (Present 1987; Figure 2) that has persisted since the Gulf of California's warming subsequent to the end of the Pleistocene, approximately 10,000 years before present (Durham and Allison 1960). This hypothesis of a warmwater barrier to migration and gene flow is tested in the present investigation.

Separation of the northern Gulf from the outer coastal populations of spotted sand bass would result in accumulation of neutral mutations, assuming that little or no gene flow (and migration) has occurred between them. This hypothesis of genetic divergence is tested by comparing results from two molecular genetic approaches involving (what are believed to be) selectively neutral mutations (Brown et al. 1979; Allendorf et al. 1987): analyses of mtDNA control region sequences and electrophoretic mobilities of allozyme variants from 40 presumptive gene loci. These data are compared with differences in the related species, the barred sand bass and the kelp bass.

Genetic Divergence in Great Lakes Walleye

Population genetic differences in Great Lakes populations of the commercially and sportfished walleye are being examined by J. E. Faber and C. A. Stepien. Data from regions of Lake Erie are reported here. The three basins of Lake Erie (Figure 3B) differ in depth and are believed to house putatively different stocks of walleye (Colby and Nepszy 1981). Genetic divergence of these basin populations is tested in this study. We are testing (1) whether stock structure of walleye exists within lake regions (basins) and (2) whether there is evidence that walleyes home to specific spawning areas.

Populations of walleye in Lake Erie were severely depressed by 1960 due to exploitation and environmental stresses. This reduction in population sizes may be evidenced by low genetic diversity, which would indicate a bottleneck and is examined in the present study. Closure of sport and commercial fisheries from 1970 to 1976 and subsequent establishment of quotas on catches have significantly improved population numbers (Hatch et al. 1987).

Tagging studies of walleye have suggested homing of adults to putative natal areas for spawning (Ferguson and Derksen 1971; Bodaly 1980), signifying potential for stock structuring. However, past studies of stock structure based on morphological and physiological characters have not detected patterns of divergence in walleyes from the Great Lakes (Colby and Nepszy 1981; Riley and Carline 1982). More recently, based on allozyme and RFLP data, Billington and Hebert (1988), Ward et al. (1989), Todd (1990), and Billington et al. (1992) have found some genetic divergence among walleyes from widely spaced sites, but levels of genetic variability were too low to resolve stock structure. For example, Ward et al. (1989) found only one allozyme locus that showed evidence of geographic patterning of allele frequencies in populations from the Great Lakes and found broadscale genetic patterns by means of RFLP analysis of mtDNA. Todd (1990) found four allozyme frequency differences between walleyes in Lake Erie versus Lake St. Clair and no significant differences among putative stocks in the western basin of Lake Erie. Billington et al. (1992) also described regional haplotype differences in RFLPs of mtDNA among walleyes from the eastern, central, and northwest portion of the walleye's North American range and between the Mississippi and Gulf Coast drainage systems. These studies support the hypothesis that walleyes from widely separated sites are genetically heterogeneous due to isolation of spawning groups by geographic distance.

The question of whether there are smaller-scale differences among putative stocks and potentially isolated spawning populations is explored in the present investigation by means of mtDNA sequencing. Results from a prior study of RFLPs of mtDNA by Woodruff and Merker (reported in Merker 1994) are compared here with those from mtDNA control region sequences of the same walleye individuals, which were collected during spawning runs in three Lake Erie tributaries—the western basin Maumee and Sandusky rivers and the central basin Grand River (Figure 3). These DNA sequence data are also compared with sequence data from the Van Buren Bay spawning population in the eastern basin (Figure 3).

Methods

Sampling.—Genetic divergences in continental slope groundfishes are compared among five areas (Figure 1) whose geographic classifications have been adapted from Briggs (1974): the subarctic (Alaska), northern cold temperate (Cape Arago, Oregon, and Cape Mendocino, Northern California), southern cold temperate (Monterey and Morro bays, California), northern warm temperate

STEPIEN

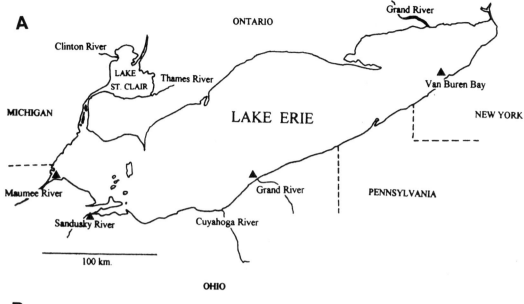

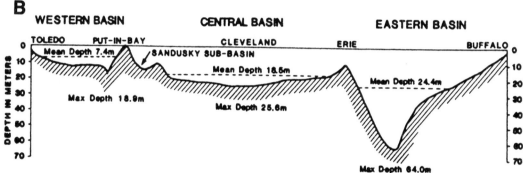

FIGURE 3.—(**A**) Map showing collection sites (triangles) of walleye in Lake Erie. (**B**) Divisions of the lake into three regions: western, central, and eastern basins (adapted from Bolsenga and Herdendorf 1993).

(Santa Barbara, California), and central warm temperate (Del Mar, Point Loma, and Cortes Bank off San Diego, California; see Figure 1). Samples include 102 Dover soles (N = 15–22 individuals per site) and 80 shortspine thornyheads (N = 9–15 individuals per site). Comparisons are made with overall levels of genetic variability in the other two species of the genus *Sebastolobus*: the sympatric longspine thornyhead (N = 50) and the broadbanded thornyhead *S. macrochir* (N = 8) from Japan. Divergence in Dover sole is compared with that of its hypothesized sister species (Sakamoto 1984), the slime flounder *M. achne* (N = 6) from Japan.

Twenty specimens of the spotted sand bass from the outer coastal region (San Diego, California) and 16 from the Gulf of California (San Felipe, Baja California, Mexico; see Figure 2) are compared for allozyme electrophoretic variability at 40 presumptive gene loci (see Stepien and Rosenblatt 1991 and Stepien et al. 1994 for details of allozyme methodology). Of these, 11 individuals from the Gulf and 11 from the coastal site, along with two kelp basses and one barred sand bass, have been sequenced to date. (This study is being completed and will be published in its entirety by Stepien and Rosenblatt).

Mitochondrial DNA from walleyes collected during 1993 spring spawning runs in river and reef sites from the three basins of Lake Erie (Figure 3) are analyzed and compared in this paper. Sites and samples include eggs from the Sandusky (N = 15) and Maumee ($N = 15$) rivers in the western basin and the Grand River ($N = 14$) in the central basin and fin clips from Van Buren Bay ($N = 6$) in the eastern basin (Figure 3). Sequence data for walleye are compared with RFLP data of mtDNA collected from the same individuals in the Sandusky, Maumee, and Grand rivers by Merker and Woodruff (Merker 1994).

Preservation of tissues and preparation of DNA.—Tissues, which included muscle, liver, fin clips, and eggs, were either fresh frozen in liquid nitrogen or on dry ice and stored at −80°C or, alternatively, minced into 95% ethanol. (In the latter procedure, the ethanol was changed twice over the following 24 h). Ethanol-preserved samples accounted for approximately 10% of the data of these studies.

Frozen tissues were ground in liquid nitrogen by use of a cylindrical stainless steel mortar and pestle. The DNA from frozen and ethanol-preserved samples was extracted in a guanidine buffer to circumvent degradation, purified using proteinase K, RNase, phenol, and chloroform, and then precipitated, following standard methods (Stepien et al. 1993). A small sample of the DNA was run on a minigel to verify relative amounts and quality.

Amplification and sequencing of DNA.—Procedure for amplifying the variable left domain of the mtDNA control region by means of the polymerase chain reaction followed Kocher et al. (1989) and Meyer et al. (1990). Primers for the heavy mitochondrial chain were end-labeled with biotin (Hultman et al. 1989) for later separation of the strands by means of Dynal streptavidin magnetic beads. Amplified DNA was bound to Dynabead M-180 streptavidin (Dynal Corp., Lake Success, New York) and denatured, producing high yields of purified single-stranded template DNA for sequencing (Hultman et al. 1989; Uhlen 1989).

Sanger dideoxy sequencing (Sanger et al. 1977) by means of Sequenase II (U.S. Biochemical Corp., Cleveland, Ohio) was performed using the purified single-stranded DNA as a template and the complementary primer. Samples from sequencing reactions were run on 6% acrylamide gels with a constant temperature of 50°C at approximately 2,500 V. Samples were usually run on three separate gels in order to resolve approximately 500 base pairs of sequence for these regions at various distances from the primer. Gels were transferred to blotting paper, dried for 2 h, and visualized by autoradiography after 72 h of exposure on Kodak X-OMAT film. Sequences from gels were read into a computer by

means of an IBI–Kodak digitizer and AssemblyLIGN software[1].

Data analysis.—The data sets consist of aligned sequences, and the variable characters include base substitutions (transitions, changes from purine to purine or pyrimidine to pyrimidine, and transversions, changes from purine to pyrimidine and vice versa) and insertions and deletions of sections of sequence. The individually variable genotypic patterns of bases are termed haplotypes. In the present study, sequence differences among individual haplotypes and patterns of geographic distribution are analyzed in two ways: (1) by clustering of overall percent sequence divergences among haplotype bases on pairwise genetic distances and the neighbor-joining algorithm (Saitou and Nei 1987) in MEGA[2], and (2) by calculating with PAUP[3] parsimonious relationships among haplotypes based on a character state method that minimizes overall number of changes. The sister species or a closely related species is used as the outgroup to root the tree in all cases except for the walleye, for which the sister species has not yet been sequenced.

Bootstrap tests are used to compute support for individual nodes of the neighbor-joining trees (MEGA). These tests randomly sample with replacement from the original nucleotide data set and construct a new neighbor-joining tree from the set of resampled data during each bootstrap replication. This process is repeated several hundred times. For each node (a discrete grouping of genotypes) on the original neighbor-joining tree, the proportion of bootstrap trees that contain the same grouping (component taxa) is calculated, yielding a percent of support for the relationships at that node.

Consensus analyses of the shortest trees from PAUP based on the 50% majority rule (Margush and McMorris 1981) criterion are employed to compare relationships among haplotypes that are supported by multiple equally parsimonious trees.

---

[1]IBI (International Biotechnologies, Inc.) Kodak. 1992. AssemblyLIGN sequence assembly software. New Haven, Connecticut.

[2]Kumar, S., K. Tamura, and M. Nei. 1993. MEGA molecular evolutionary genetics analysis. Version 1.01. Institute of Molecular Evolutionary Genetics, The Pennsylvania State University, University Park.

[3]Swofford, D. L. 1993. PAUP (phylogenetic analysis using parsimony). Version 3.1 for MacIntosh Computers. Illinois Natural History Survey, Champaign.

TABLE 1.—Relative frequencies of four nucleotide bases (G, guanine; A, adenine; T, thymine; and C, cytosine) in the mitochondrial DNA control region of eight fish species. The entire haplotype gene pool sampled for each species, upon which percentages are based, are given as N.

Species	N	Nucleotide (%)			
		G	A	T	C
Shortspine thornyhead	560	16.6	33.4	28.2	21.8
Longspine thornyhead	506	15.8	34.4	30.2	19.6
Broadbanded thornyhead	439	15.0	36.4	29.2	19.4
Dover sole	543	19.2	31.9	26.3	22.7
Slime flounder	458	13.1	35.4	27.9	23.6
Spotted sand bass	436	15.8	33.7	29.8	20.6
Barred sand bass	420	18.1	34.5	28.3	19.0
Walleye	822	13.3	33.1	33.6	19.7

Swofford's BIOSYS 2.0[4] is used to analyze allozyme data for the spotted sand bass in comparison with the mtDNA sequence results.

Results

Overall Genetic Divergence in the Mitochondrial DNA Control Region

Sequence data from the proline transfer RNA end (left domain) of the mtDNA control region heavy strand complement for the eight species studied are summarized in Appendix 1. Alignments among groups are for purposes of rough comparison only and do not necessarily correspond to evolutionary homology. Relative frequencies of the four nucleotide bases per taxon are summarized in Table 1. In all cases, guanine is the rarest base, followed by cytosine, thymine, and adenine. Overall number and relative frequencies of transitional and transversional substitutions are summarized in Table 2. Transitions range from 64 to 88% of the total number of nucleotide substitutions. There are some insertions and deletions between taxa separated at the species level, including *Paralabrax*, *Microstomus*, and *Sebastolobus* (see Appendix 1). Most individuals examined (except in *Paralabrax*) have distinct mtDNA sequence haplotypes.

Population Genetic Divergences in Continental Slope Fishes

Overall diversity of haplotypes is extensive among the continental slope groundfishes, *Microstomus* spp. and *Sebastolobus* spp. Only two pairs of exact shared sequence haplotypes occur among the Dover soles examined, and all are from sites in the northern cold temperate region (a shared pair in Oregon and another in northern California; see Figure 1). There are also two shared pairs of exact sequence haplotypes among the shortspine thornyhead (one pair in cold temperate northern California and one in the central warm temperate site off Del Mar, San Diego).

A greater degree of sequence divergence distinguishes species-level separations from those at the population level. For example, the Dover sole and the slime flounder which are hypothesized to be sister groups based on morphology (Sakamoto 1984) are distinguishable by 38 fixed base differences and by one insertion–deletion (Appendix 1). There are 66 additional sites with base substitutions in this group. All six slime flounder examined have unique sequence haplotypes.

Phylogenetic parsimony analysis and genetic similarity analysis of the mtDNA data indicate that the North American shortspine and longspine thornyheads are sister species and diverge from the Asian broadbanded thornyhead by one insertion–deletion and fixed differences at 13 nucleotide positions (Appendix 1). All eight broadbanded thornyheads sequenced have unique haplotypes, varying by 13 transitional and 3 transversional substitutions.

---

[4]Swofford, D. L., and R. B. Selander. 1989. BIOSYS-1: a computer program for the analysis of allelic variation in population genetics and biochemical systematics. Release 1.7, Illinois Natural History Survey, Champaign.

TABLE 2.—Relative frequencies of substitution types per species. Nucleotide bases are guanine (G), adenine (A), thymine (T), and cytosine (C).

Species	Total number of substitutions	Transitions (%)			Transversions (%)				
		G–A	T–C	Total	G–T	G–C	A–T	A–C	Total
Dover sole	76	46	25	71	8	7	9	5	29
Slime flounder	13	39	46	85	0	0	0	15	15
Shortspine thornyhead	75	32	45	77	5	0	3	15	23
Longspine thornyhead	36	42	39	81	3	3	14	0	20
Broadbanded thornyhead	16	35	47	82	6	6	6	0	18
Spotted sand bass	8	63	25	88	0	0	13	0	13
Walleye	17	35	29	64	6	18	0	12	36

There are 45 nucleotide sites that are variable only in shortspine thornyhead, 11 in longspine thornyhead, and 8 in broadbanded thornyhead. Twenty of the same positions are substituted in both shortspine and longspine thornyheads (but not in broadbanded thornyhead), and 18 of these have the same base substitution and 2 have a different base substitution (Appendix 1).

Neighbor-joining analyses of pairwise distances (MEGA) among Dover sole haplotypes show several regional groupings (Figure 4). For example, node 4A designates a cluster of nine cold temperate haplotypes, and all six in group B are from the northern site (Cape Arago, Oregon; the terminal clusters are significant based on bootstrap analysis). Cold temperate haplotypes also group in branch 5 at nodes G (5 cold temperate haplotypes, with 1 warm temperate haplotype at node H), B (4 haplotypes), and F (2); in branch 6 at nodes F (4 southern cold temperate haplotypes as sister group to 3 haplotypes from Alaska at node G), U (2), B (4 southern cold temperate haplotypes as sister group to 2 from Alaska at node C), L (3), M (2), and N (5); in branch 7 (3), and in branch 8 at nodes C (2), and H (2 cold temperate haplotypes with 1 being sister to 1 haplotype from Alaska). More Alaskan haplotypes are similar to haplotypes from the cold temperate region than to haplotypes from other locations, except Alaska; examples include nodes 3C, 5A, 5L, 6G, 6C, 6T, 8B, and 9E. There are several clusters of haplotypes from the central warm temperate including nodes 5E (3 haplotypes), 8A (9), 9B (2 plus 1 from the southern cold temperate), and 9C (2, plus 1 from the southern cold temperate). Groups such as 5D, 5H, 8J, and 9B show relationships among haplotypes from the warm temperate and southern cold temperate regions.

A PAUP analysis of the shortest trees from Dover sole haplotypes also shows more Alaskan haplotypes to be grouped as related to the northern cold temperate haplotypes than to haplotypes from other sites. Consensus trees based on PAUP support some clades of Alaskan haplotypes (among haplotypes m, n, and k by the 75% majority rule; this grouping is also supported by neighbor-joining bootstrap analysis: see node 6G on Figure 4). As with neighbor-joining, PAUP consensus analysis supports clades from the southern cold temperate (Monterey Bay site, 100%; see node 6H on Figure 4), central warm temperate (two clades at 100%; see nodes 8A and 9A on Figure 4), and northern cold temperate sites (Oregon haplotypes, 98% [node 4B, Figure 4]; northern California, one clade of 98% [node 6L] and two at 100% [nodes 4A, 6O]).

For the shortspine thornyhead, neighbor-joining trees (Figure 5) suggest genetic distinctiveness of the following groups: central warm temperate (node 5B; 4 haplotypes from Cortes Bank); cold temperate (nodes 2B, 3 haplotypes; 4A, 4; 6B, 2 northern cold temperate; 10A, 4; 10D, 3 southern cold temperate; 12B, 2 southern; and 13C, 2); Alaskan (node 4C, 3 haplotypes); Alaskan with cold temperate haplotype groups (nodes 4A, 3 Alaskan and 3 northern cold temperate haplotypes; 6A, 1 Alaskan and 2 northern; and 7A, 1 Alaskan with 2 southern haplotypes); and those exclusive to the California current system (nodes 2A, 9 haplotypes; 10A, 14; 12, 11 haplotypes; 13A, 3; and 13F, 4).

The clade of five shortspine thornyhead haplotypes from Cortes Bank (group 5C on the neighbor-joining tree Figure 5) is also strongly supported (100% of the shortest trees) by majority rule consensus of maximum parsimony PAUP analysis, as well as by neighbor-joining bootstrap analysis. Other clades supported by the PAUP consensus trees include the Alaskan–cold temperate group 4C (100%; also supported by neighbor-joining bootstrap analysis), the southern cold temperate group 10D (98%), and the northern cold temperate group at node 6B (100%).

Population Genetic Divergence of the Spotted Sand Bass

Mean heterozygosity and percent polymorphism from allozyme variants of the spotted sand bass are, respectively, 0.085 ($\pm$0.012) and 42.50% for San Diego populations, and 0.090 ($\pm$0.011) and 42.50% for the Gulf of California populations. There are 17 polymorphic allozyme loci. Six alleles are unique to the Gulf of California population and three to the outer coastal samples. Frequencies of three polymorphic allozyme loci are significantly different between the outer coast and Gulf of California populations based on contingency table G-tests (Sokal and Rohlf 1981). The mean F_{ST} value (Wright 1978) averaged across all polymorphic allozyme loci (which measures differences in heterozygosities among the two populations in comparison with overall heterozygosity) is 0.021, suggesting relatively low genetic divergence between the two regions. The genetic distance (Nei 1972) separating these populations is 0.004 $\pm$ 0.0002 and the modified Roger's distance (Wright 1978) is 0.061.

Mitochondrial DNA control region sequences revealed four unique haplotypes, two in the Gulf of California and two in the outer coast, among the 22 individuals sequenced to date (Appendix 1). Ten of

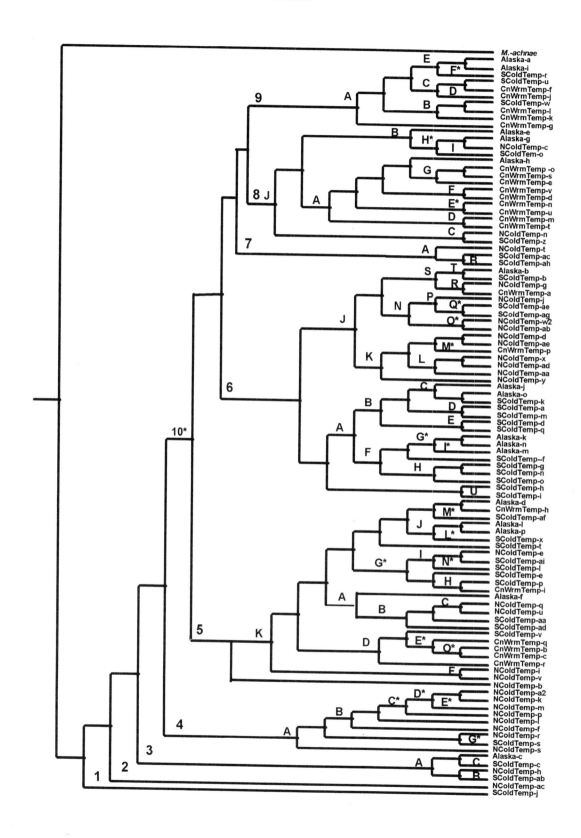

the 11 Gulf of California individuals are identical and one is unique. The latter differs at four nucleotides, of which three are unique substitutions (at bases 228, 323, and 425) and one (364) is shared with the outer coast individuals. Ten of the 11 outer coast individuals sequenced share identical haplotypes, and the other is distinguished by a single unique transition (at base 221; see Appendix 1).

Three fixed differences distinguish all of the Gulf of California individuals from those of the outer coast—the former have adenines at base positions 225, 284, and 386 and the latter have guanines). A MEGA analysis based on pairwise distances groups the two coastal haplotypes together (node C in Figure 6), which is supported by 98% of the data set in bootstrap analysis. The more common Gulf of California haplotype is sister to those of the coast, and the Gulf of California haplotypes are more closely placed to the outgroup species, the barred sand bass and the kelp bass (Figure 6). The barred sand bass differs from the spotted sand bass by 33 fixed base substitutions and 2 single base insertion–deletions.

An exhaustive search of all possible trees by means of PAUP yields three most parsimonious trees, all of which group the coast haplotypes together (length = 23 steps, consistency index excluding uninformative characters = 0.85) and one of which is identical to the tree shown in Figure 6. The same PAUP trees are obtained with and without differential weighting of transversions:transitions, and PAUP bootstrap support for the coastal haplotype clade is 80%.

Population Genetic Divergence of Walleye

Most walleye individuals sampled have unique mitochondrial control region sequence haplotypes (Appendix 1). A sequence pattern, near the beginning of the control region, consists of 11 bases (GCAAATATTTT) that are repeated 7 to 13 times in each individual (bases 071 to 082; Appendix 1). This perfect repeat is followed by a series of two "imperfect" repeats, which vary in base sequence both within and among individual haplotypes (bases 083 to 104 of Appendix 1). The lengths of the imperfect repeats range from 9 to 13 bases each and are repeated in individually varying motifs from 12 to 19 times (Appendix 1).

The tree from neighbor-joining analyses based on pairwise distances (MEGA) is shown in Figure 7. This tree suggests that there are primary groupings of genotypes which are predominant in each of the three basins and differ among spawning sites. Individuals having identical haplotypes were almost always found in the same spawning site (Figure 7; node B—two from the Sandusky River, node C—four from Van Buren Bay, nodes D and E—two each from the Maumee River, and node F—two from Van Buren Bay). Several clusters (B, E, I, J, and L), supported by bootstrapping, are almost entirely composed of western basin individuals (at node I there is one individual, from the Grand River spawning group, that has a distinct haplotype). Genetic distinctiveness of Sandusky River walleyes is indicated by the cluster at node B (supported by bootstrapping) of four haplotypes. Node M haplotypes (supported by bootstrapping) are found in only the Grand and Maumee rivers, and group L is composed almost entirely of Maumee River fish (four of five, supported by bootstrapping). Grand River individuals cluster at two primary nodes, H and M. Group H is almost entirely composed of Grand River walleyes (six of seven, supported by bootstrapping), with one walleye from the Sandusky River sharing an identical haplotype with one from the Grand River.

Character state analyses from PAUP also suggest that there are major groupings of walleye haplotypes and that, within these, many haplotype groupings are specific to the Sandusky and Maumee rivers, and Van Buren Bay spawning sites. Haplotypes of the Grand River (central basin) suggest population affinities to the western basin. The two groups of Van Buren Bay haplotypes are distinct in the PAUP consensus tree.

FIGURE 4.—Neighbor-joining tree (MEGA analysis) based on mitochondrial DNA control region sequence haplotypes of the Dover sole. Branches discussed are numbered and nodes are lettered. Haplotypes are numbered alphabetically in lower case, by regions. Haplotypes from regions containing more than one collection site are Cape Arago, Oregon—northern cold temperate (NColdTemp) a–p; Northern California—northern cold temperate q–ae; Monterey Bay, California—southern cold temperate (SColdTemp) a–q; Morro Bay, California—southern cold temperate r–ai; Del Mar, California—central warm temperate (CnWrmTemp) m–v; Point Loma, California—central warm temperate e–l; and Cortes Bank—central warm temperate a–d. Those haplotypes marked with "2" have two individuals per haplotype. Nodes supported by bootstrap reshufflings of the data set have asterisks. Clades supported by PAUP repeated heuristic searches are given in "Results".

Discussion

Sequencing the mtDNA control region of fishes from deep-sea, kelp forest, and Great Lakes habitats reveals greater numbers of intraspecifically variable characters than were obtained in previous allozyme and RFLP studies. Differences in the mtDNA control region are extensive in all species examined, and most individuals sequenced have unique haplotypes. Variation is, however, conservative in that most sites within a species have a single substitution, and there appear to be few reversals in the phylogenetic signal. As in data sets from other groups of fishes (Johansen et al. 1990; Digby et al. 1992; Sturmbauer and Meyer 1992; Lee et al. 1995), there are low numbers of guanine and cytosine bases in the heavy strand complements (Table 1). Transitions outnumber transversions by approximately 3:1 in most of the species examined (Table 2), a bias found in other vertebrate control region data sets (Brown et al. 1982; Lee et al. 1995). There is no apparent bias in type of transition or transversion, and the types of transitions and transversions are not similar in congeners (Table 2).

The detection of greater divergence among individual genotypes allows for testing of patterns of geographically related differences. Geographic divergence is a dynamic process governed by the origination of new genetic diversity through mutation, the rate of dispersal of individuals, barriers to migration, and selective forces.

Genetic Divergence in Populations of Continental Slope Fishes

This investigation demonstrates extensive genetic differences among individual fishes of the little-studied continental slope. For example, in contrast to a prior allozyme study of shortspine thornyhead that discerned few population genetic characters and no significant geographic divergence (Siebenaller 1978), mtDNA sequence data from adult shortspine thornyheads and Dover soles indicate high variability among individuals and some patterns of genetic differences among populations from different geographic regions, despite extended pelagic larval periods. The predominance of unique haplotypes in these groundfishes suggests there has been considerable time for genetic diversification and that high historical diversity persists despite present exploitation.

There is considerable evidence for high gene flow among most areas of the slope, presumably due to larval dispersal, because many similar groups contain haplotype sequences from different geographic areas (Figures 4 and 5). These data support the null hypothesis of panmixia, while some selected haplotype groups suggest geographic patterning. It can thus be argued that potential for maintaining high genetic diversity in these fisheries is good because in most areas potential for recolonization by juveniles having diverse genotypes is high. This recolonization potential is dependent upon enough adults remaining in the gene pool in broadscale areas.

The distinguishable geographic groupings of sequences suggest that there also may be some barriers to larval dispersal in some areas, which may limit gene flow. Alternatively, these patterns of differences may be regulated by selection (for example, due to temperature). Populations that cluster together in discrete haplotype groups included the shortspine thornyhead population from Cortes Bank, San Diego, which is geographically separated from the main coastal slope region, and the Dover sole population from Monterey Bay. Future study is necessary to determine the degree to which these unique populations are being exploited. It may be advisable to restrict catches in these unique areas to preserve diversity. In both species, Alaskan groups of haplotypes show limited discreetness and appear to be related to a number of groups of haplotypes in other regions. In other words, there appear to have been several founding incidents of haplotype exchange between the Alaskan gyre and California current system. In both species, haplotypes from the California current system show separation from the Alaskan gyre (Figure 1) because many groupings are unique to the Californian system (Figures 4 and 5).

Sequence data from adults demonstrate some patterns of genetic differences among populations from different geographic regions, despite extended pelagic larval periods. A relatively recent concept in fish larval recruitment research involves larval retention areas (member–vagrant hypothesis; Sinclair 1988); the concept explains differences in the abundance and population richness of marine species by relating the number of population genetic units to the number of discrete larval retention areas over the species' ranges. Marine larvae that survive to settle in appropriate habitats may be retained in major coastal current patterns, resulting in genetic divergence among regions.

Specific groups of related haplotypes, as shown in the present study, may suggest larval retention, such as in the California current system (Figure 1). Patterns of haplotypes suggest that there may be barriers to larval dispersal which limit gene flow between the major current systems. Alternatively,

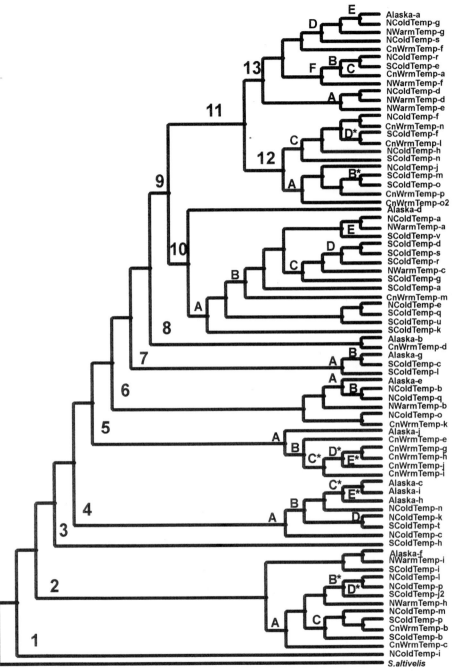

FIGURE 5.—Neighbor-joining tree (MEGA analysis) based on mitochondrial DNA control region sequence haplotypes of the shortspine thornyhead. Branches discussed are numbered and nodes are lettered. Haplotypes are numbered alphabetically in lower case, by regions. Haplotypes from regions containing more than one collection site are: Cape Arago, Oregon—northern cold temperate (NColdTemp) a–i; Northern California—northern cold temperate j–q; Monterey Bay, California—southern cold temperate (SColdTemp) a–l; Morro Bay, California—southern cold temperate m–v; Santa Barbara, California—northern warm temperate (NWarmTemp) a–i; Del Mar, California—central warm temperate (CnWrmTemp) k–p; Point Loma, California—central warm temperate a–e; and Cortes Bank—central warm temperate f–j. Those haplotypes marked with "2" have two individuals per haplotype. Nodes supported by bootstrap reshufflings of the data set have asterisks. Clades supported by PAUP repeated heuristic searches are given in "Results".

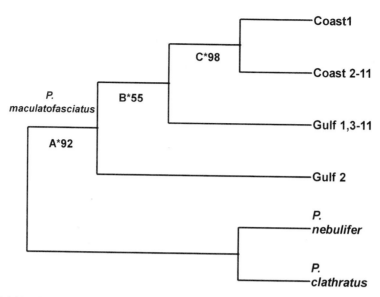

FIGURE 6.—Neighbor-joining tree of mitochondrial DNA control region sequence haplotypes of the spotted sand bass *Paralabrax maculatofasicatus* (MEGA analysis). Tree is rooted to the barred sand bass *P. nebulifer* and the kelp bass *P. clathratus*. Nodes discussed are lettered and bootstrap values follow asterisks. This tree is also one of three most parsimonious trees from an exhaustive search using PAUP, in which the length is 21 steps and the consistency index excluding uninformative characters is 0.85. All of the 3 most parsimonious trees in the PAUP analysis cluster coast 1 and 2–11 haplotypes as sister groups.

these patterns of differences may be regulated by selection (for example, due to temperature, which is variable between primary geographic regions and to which marine fish larvae are particularly sensitive; see Stepien and Rosenblatt 1991). These ground fisheries of the continental slope are being seriously affected by trawling and are expected to be decimated within the next 15 years, rendering data on genetic variability critically important to understanding this community.

Relationships of the faunas across the North Pacific have been of interest to zoogeographers for many years (Briggs 1974; Springer 1982). The thornyhead genus *Sebastolobus* contains three species, all in the North Pacific. The northeastern Pacific shortspine thornyhead and longspine thornyhead are largely sympatric and closely resemble each other in overall morphology. In the present investigation, sequences of all three thornyhead species demonstrate markedly greater divergence between the Asian broadbanded thornyhead and the North American species. In a phenetic analysis of the four species in the genus *Microstomus*, Sakamoto (1984) hypothesized a possible sister relationship between the North American Dover sole and the Asian slime flounder based on osteological data. The present study discerns several fixed differences

and significant frequency differences in base substitutions distinguishing slime flounder and Dover sole (Appendix 1).

It is probable that ancestral distributions of the genera *Microstomus* and *Sebastolobus* were continuous throughout the northern subarctic and cold temperate provinces and these distributions were disrupted. Present southerly extension through the warm temperate provinces by these fishes is presumably due to their occupation of cold, deep waters along the continental slope. Divergence between sister groups of *Microstomus* and *Sebastolobus* from the eastern and western Pacific suggests that the deep waters of the northern Aleutian Basin off the Siberian coast may form a significant barrier to east–west dispersal. Faunas on either side of the Bering Sea show marked separation, and approximately 60% of the taxa are distinct (Briggs 1974). In the eastern and western North Pacific faunas there are a number of geminate and twin species pairs whose distributions mirror those of *Microstomus* and *Sebastolobus*. These occur in the clam genus *Mya* and the fish families Cottidae, Liparidae, and Zoarcidae (many species in the latter two families are located along the continental shelf and slope; Briggs 1974; Springer 1982).

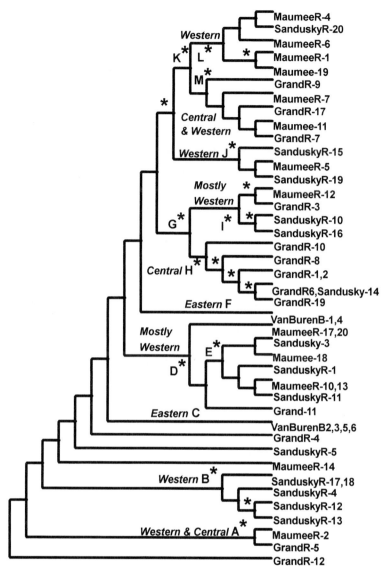

Western K L M
Central & Western
Western J
Mostly Western
G I
Central H
Eastern F
Mostly Western D E
Eastern C
Western B
Western & Central A

MaumeeR-4
SanduskyR-20
MaumeeR-6
MaumeeR-1
Maumee-19
GrandR-9
MaumeeR-7
GrandR-17
Maumee-11
GrandR-7
SanduskyR-15
MaumeeR-5
SanduskyR-19
MaumeeR-12
GrandR-3
SanduskyR-10
SanduskyR-16
GrandR-10
GrandR-8
GrandR-1,2
GrandR6,Sandusky-14
GrandR-19
VanBurenB-1,4
MaumeeR-17,20
Sandusky-3
Maumee-18
SanduskyR-1
MaumeeR-10,13
SanduskyR-11
Grand-11
VanBurenB2,3,5,6
GrandR-4
SanduskyR-5
MaumeeR-14
SanduskyR-17,18
SanduskyR-4
SanduskyR-12
SanduskyR-13
MaumeeR-2
GrandR-5
GrandR-12

FIGURE 7.—Neighbor-joining tree of mitochondrial DNA control region sequence haplotypes, including repeat regions and base substitutions for walleyes from Lake Erie (MEGA analysis). Nodes and branches discussed are lettered, and those supported by a majority of bootstrap replications are marked with asterisks. Clusters from the three lake basins (western, central, and eastern) are marked. The 50% majority rule consensus trees generated from repeated heuristic searches of the data (PAUP) support many of these haplotype groupings and discrete clades of the Van Buren Bay fishes and Grand River haplotypes.

Genetic Divergence in Populations of the Spotted Sand Bass

Mean heterozygosity and percent polymorphism from the allozyme portion of the study approximate previous estimates from allozymes for other marine fishes (Stepien and Rosenblatt 1991; Ward et al. 1994), including the kelp bass and the barred sand bass (Kirpichnikov 1981; Beckwitt 1983; Waples and Rosenblatt 1987; Waples 1987; Graves et al. 1990). The present study discerns 17 polymorphic allozyme loci in the spotted sand bass in comparison with four found by Graves et al. (1990), who sampled only two to eight individuals per locus from a San Diego population. The difference is most probably due to increased sample size in the present investigation.

Genetic divergence between the outer Pacific coast and Gulf of California populations of the spotted sand bass is indicated by the presence of unique allozyme alleles and unique mtDNA haplotypes in the two areas. Together, these data suggest that although there may be some slight gene flow, the warm waters of the southern Gulf may form a barrier that results in genetic divergence. Nei's (1972) D from the present allozyme study suggests genetic isolation of the two populations for approximately 8,000 years (based on the calibration of Grant 1987); this period is consistent with warming subsequent to the end of the Pleistocene. These results suggest that the two populations may constitute distinct ESUs, because their gene pools show genetic separation. This is being further tested by sequencing additional individuals and sites.

Approximately 15% of the common nearshore fish species in the northern Gulf of California share this disjunct population pattern with the outer Pacific coast (Walker 1960; Thomson and Gilligan 1983). Present (1987) analyzed allozyme divergence of disjunct Gulf of California and Pacific outer coast populations of the intertidal mussel blenny *Hypsoblennius jenkinsi* and found larger values of Nei's D (0.035–0.038) separating them than were found for the spotted sand bass in the present allozyme study. Present also found six statistically significant differences in allelic frequencies between the Gulf and coastal populations, which suggests limited gene flow and is similar to results of the present study.

In this investigation, mtDNA control region sequence data reveal markedly more intraspecifically informative characters than were found by Graves et al. (1990) in an RFLP study of mtDNA in the spotted sand bass, the kelp bass, and the barred sand bass. In our study, genetic divergence between the Gulf of California and the outer Pacific coast populations of the spotted sand bass is more pronounced in the mtDNA sequences than in allozyme data from the same individuals. Of the 13 individual spotted sand bass sequenced, four unique mtDNA haplotypes were found (Figure 6). In contrast to mtDNA sequence data from continental slope groundfishes (*Sebastolobus* spp. and *Microstomus* spp.) in the same geographical region and from the Lake Erie walleyes (in which most individuals have unique genotypes; Appendix 1), intraspecific genetic diversity is markedly lower in the spotted sand bass. These data suggest that population numbers and genetic diversity of spotted sand bass popula-

tions along both the outer Pacific coast and the northern Gulf of California may have undergone bottlenecks, presumably due to exploitation. It may be advisable for management agencies to protect genetic diversity of potential ESUs in this species, as well as in its congeners, pending further study.

Genetic Divergence in Populations of Lake Erie Walleye

Considerably more genetic characters are also discerned from sequencing the mtDNA control region of walleye than were found in prior allozyme (Ward et al. 1989; Todd 1990) and RFLP studies (Billington et al. 1992; Merker 1994). Almost all (44) of the 50 Lake Erie walleyes sequenced to date have unique mtDNA sequence genotypes, and many of these clearly group according to spawning location (Figure 7). Many walleyes in spawning populations are thus genetically closer to each other than to those in other spawning locations, supporting the natal homing hypothesis. High levels of genetic diversity in walleye populations of Lake Erie suggest that, despite exploitation, natural historic spawning populations have been maintained to the present day.

Sequence haplotype groups of walleyes in the Maumee and Sandusky rivers (Figure 7, nodes B and L), which are separated by approximately 55 km in the western basin of Lake Erie (Figure 3), suggest that these rivers have different breeding stocks, apparently maintained by homing during spawning. Van Vooren (1978) suggested that spawning populations of the two major tributaries in the western basin, the Sandusky and Maumee rivers, are discrete (Colby and Nepszy 1981), which is supported by the present investigation. Samples from the Maumee, Sandusky, and Grand rivers also differ in frequencies of two RFLP patterns of mtDNA (data reported in Merker 1994), and all three river spawning areas appear genetically different in the present study. The central basin Grand River spawning population clusters in two primary groups and appears most closely related to western basin groups (Figure 7, nodes H and M).

The central basin of Lake Erie was dominated until the mid-twentieth century by a now-extinct form, the blue pike *Stizostedion vitreum glaucum*, which was morphologically distinguished from the walleye by placement and size of the eye, adult size, and color pattern (Trautman 1981). The central basin was believed to have been repopulated by western and eastern walleye stocks following the overexploitation and demise of the blue pike (Colby and Nepszy 1981; Hatch et al. 1987). This hypothesis is tested in the present study by comparing the genetic composition and relationships of the Grand River central basin spawning population with the spawning populations of the eastern and western basins. Results show that there are unique haplotypes in the Grand River, and their genetic relationships mirrors their geographic location between the populations of the eastern and western basins. The distinct haplotype grouping of many Grand River spawning individuals sampled (at node H of Figure 7) suggests that these represent historic Grand River spawning run populations which persist to the present day.

The eastern basin population from Van Buren Bay, New York (branches C and F of Figure 7) constitutes two genetically discrete groups in both the neighbor-joining (MEGA) and parsimony (PAUP) analyses. Other studies have suggested that the walleye population in the New York waters of Lake Erie is a geographically discrete stock confined to the inshore waters of the eastern basin (Wolfert 1977; Wolfert and Van Meter 1978; Colby and Nepszy 1981), which is supported by the present results.

In conclusion, gene flow is apparently reduced among these spawning populations and the basins of Lake Erie (Figure 7), supporting the natal homing hypothesis. Genetic grouping of spawning-run sequence haplotypes supports fidelity of homing from generation to generation. These results suggest that it may be important to protect spawning habitats in order to maintain genetic diversity in the walleye. Stepien and J. Faber are presently sequencing additional samples from these and other locations in the Great Lakes. We are also investigating historic genetic diversity of walleye from remains of Indian midden material and dried scales of the extinct blue pike to compare with present-day genetic diversity.

Summary

Together, these studies demonstrate that mtDNA sequencing reveals a substantial amount of genetic variability, which can be useful for testing for geographic patterns and stock structure of fishes. This technique appears to be especially applicable in cases in which traditional allozyme and RFLP approaches have been unable to discern enough variability to address fine-scale problems. In the cases examined, genetic divergence is discerned among geographic regions separated by major current patterns (continental slope fishes), temperature barriers (the Gulf of California versus outer Pacific coast disjunct populations of sandbass), and homing patterns (walleye in tributaries and basins of the Great Lakes). In regards to discerning ESUs, it may be critical to protect genetic diversity across the range of a species in order to enable populations to withstand pressures from overfishing, environmental degradation, habitat loss, and introductions of exotic species. Areas housing unique mtDNA haplotype groups suggest genetic divergence of these populations and may be especially important as components of overall species diversity. Unique haplotypes may also indicate lineages having lower dispersal capabilities or may be the by-products of natural selection. Of the examples shown in the present study, the walleye homing areas and the disjunct populations of spotted sand bass suggest greatest potential for constituting discrete ESUs. However, it may certainly be argued that continental slope fishes possess some regional patterns of population structure that are necessary components for maintaining species genetic diversity.

Acknowledgments

The following students in my laboratory helped to collect sequence data: Allyson N. Hubers (Dover sole, slime flounder, and the shortspine and longspine thornyheads), Katarina Kvassay (broadbanded thornyhead), Jennifer Melnyck (slime flounder), Brian A. Bargmeyer (spotted sand bass), and Joseph E. Faber (walleye). A. N. Hubers assisted with data analysis, and J. E. Faber modified Figure 1B and drew Figures 2 and 3. The Dover sole and shortspine thornyhead sequence data were collected by me, 75% during a National Research Council Research Associateship at National Marine Fisheries Service Southwest Fisheries Science Center (SWFSC), La Jolla, California, which provided 75% of the support for this work. The remainder was funded by an Ohio Board of Regents grant to the Department of Biology, Case Western Reserve University (CWRU), and by my George P. Mayer assistant professorship funds. Approximately 25% of the sequencing work and all of the data analyses and writing were completed by me at Case Western Reserve University. The continental slope fish research at SWFSC was sponsored by John Roe Hunter, and technician Carol Kimbrell assisted in running sequence gels. The following students helped me in various stages of the project: Glenn C. Johns, Leonard Naftalin, Nicholas Valtz, and Helen Strick. The National Research Council funded a collection expedition to Japan for broadbanded thornyhead and slime flounder; in Japan I was graciously hosted and assisted by Kunio Amaoka, Kouichi Kawaguchi, Tokiharu Abe, and K. Okamura. Research on the sea basses was funded by a California Sea Grant Project R/CZ-94 to Richard H. Rosenblatt and Jeffrey Graham and by a Howard Hughes fellowship to the Department of Biology, CWRU. Research on Lake Erie walleye was funded by the Ohio State Sea Grant Project RF-7267 (1994) and the Lake Erie Protection Fund (1995–1998) to C. Stepien. Collections of walleye were arranged by Carl Baker and Roger Knight, Division of Fisheries and Wildlife, Ohio Department of Natural Resources, and Donald Einhouse, New York Department of Environmental Conservation. I gratefully acknowledge suggestions contributed by Richard H. Rosenblatt, John R. Hunter, Andrew Dizon, Patricia Rozell, H. Geoffrey Moser, John Butler, Axel Meyer, Robin Waples, David M. Hillis, Michael Dixon, Andrew Martin, Allyson Hubers, Joseph Faber, Ronald Woodruff, Robert Merker, Roy Stein, Jeffrey Reutter, Gary Isbell, Roger Knight, and Joseph Koonce. This manuscript benefited substantially from critical reviews by Douglas Marckle, E. O. Wiley, and Jennifer Nielsen.

References

Allendorf, F., N. Ryman, and F. Utter. 1987. Genetics and fishery management: past, present, and future. Pages 1–20 in N. Ryman and F. Utter, editors. Population genetics and fishery management. University of Washington Press, Seattle.

Attardi, G. 1985. Animal mitochondrial DNA: an extreme example of genetic economy. International Review of Cytology 93:93–145.

Avise, J. C. 1991. Ten unorthodox perspectives on evolution prompted by comparative population genetic findings on mitochondrial DNA. Annual Review of Genetics 25:45–69.

Avise, J. C. 1994. Molecular markers, natural history, and evolution. Chapman and Hall, New York.

Beckwitt, R. 1983. Genetic structure of Genyonemus lineatus (Sciaenidae) and Paralabrax clathratus (Serranidae) in southern California. Copeia 1983:691–696.

Billington, N., R. J. Barrette, and P. D. N. Ward. 1992. Management implications of mitochondrial DNA variation in walleye stocks. North American Journal of Fisheries Management 12:276–284.

Billington, N., and P. D. N. Hebert. 1988. Mitochondrial DNA variation in Great Lakes walleye (Stizostedion vitreum) populations. Canadian Journal of Fisheries and Aquatic Sciences 45:643–654.

Bodaly, R. A. 1980. Pre- and post-spawning movements of walleye, Stizostedion vitreum, in Southern Indian Lake, Manitoba. Canadian Technical Report of Fisheries and Aquatic Sciences 931:1–30.

Bolsenga, S. J., and C. E. Herdendorf. 1993. Lake Erie and Lake St. Clair handbook. Wayne State University Press, Detroit, Michigan.

Briggs, J. C. 1974. Marine zoogeography. McGraw-Hill, New York.

Brown, B. E., G. H. Darcy, and W. Overholtz. 1987. Stock assessment/stock identification: an interactive process. NOAA (National Oceanic and Atmospheric Administration) Technical Memorandum NMFS (National Marine Fisheries Service)-SEFC-199, Southeast Fisheries Center, Miami.

Brown, W. N., M. George, Jr., and A. C. Wilson. 1979. Rapid evolution of animal mitochondrial DNA. Proceedings of the National Academy of Sciences of the United States of America 76:1967–1971.

Brown, W. N., E. M. Prager, A. Wang, and A. C. Wilson. 1982. Mitochondrial DNA sequences of primates: tempo and mode of evolution. Journal of Molecular Evolution 18:225–239.

Colby, P. J., and S. J. Nepszy. 1981. Variation among stocks of walleye (Stizostedion vitreum vitreum): management implications. Canadian Journal of Fisheries and Aquatic Sciences 38:1814–1831.

Digby, T. J., M. W. Gray, and C. B. Lazier. 1992. Rainbow trout mitochondrial DNA: sequence and structural characteristics of the non-coding control region and flanking tRNA genes. Gene 113(1992): 197–204.

Durham, J. W., and E. C. Allison. 1960. The geologic history of Baja California and its marine faunas. Part I. Geologic history. Symposium: the biogeography of Baja California and adjacent seas. Systematic Zoology 9:47–91.

Ferguson, R. G., and A. J. Derksen. 1971. Migrations of adult and juvenile walleyes (Stizostedion vitreum vitreum) in southern Lake Huron, Lake St. Clair, Lake Erie, and connecting waters. Journal of the Fisheries Research Board of Canada 28:1133–1142.

Grant, W. S. 1987. Genetic divergence between congeneric Atlantic and Pacific Ocean fishes. Pages 225–246 in N. Ryman and F. Utter, editors. Population genetics and fishery management. University of Washington Press, Seattle.

Graves, J. E., M. J. Curtis, P. A. Oeth, and R. S. Waples. 1990. Biochemical genetics of southern California basses of the genus Paralabrax: specific identification of fresh and ethanol-preserved individual eggs and early larvae. U.S. National Marine Fisheries Service Fishery Bulletin 88:59–66.

Hatch, R. W., S. J. Nepszy, K. M. Muth, and C. T. Baker. 1987. Dynamics of the recovery of the western Lake Erie walleye (Stizostedion vitreum vitreum) stock. Canadian Journal of Fisheries and Aquatic Sciences 44:15–22.

Hultman, T., S. Stahl, E. Hornes, and M. Uhlen. 1989. Direct solid phase sequencing of genomic and plasmid DNA using magnetic beads as solid support. Nucleic Acids Research 17:4937–4946.

Hunter, J. R., J. L. Butler, C. Kimbrell, and E. A. Lynn. 1990. Bathymetric patterns in size, age, sexual maturity, water content, and caloric density of Dover sole, Microstomus pacificus. CalCOFI (California Cooperative Oceanic Fisheries Investigations) Report 31: 132–144.

IBI (International Biotechnologies, Inc.). Kodak. 1992. Assembly LIGN Sequence Assembly Software, New Haven, Connecticut.

Johansen, S., P. H. Guddal, and T. Johansen. 1990. Organization of the mitochondrial genome of Atlantic cod, *Gadus morhua*. Nucleic Acids Research 18(3): 411–419.

Kirpichnikov, V. S. 1981. Genetic basis of fish selection. Springer-Verlag, New York.

Kocher, T. D., and six coauthors. 1989. Dynamics of mitochondrial DNA evolution in animals: amplification and sequencing with conserved primers. Proceedings of the National Academy of Sciences of the United States of America 86:6196–6200.

Lee, W.-J., J. Conroy, W. H. Howell, and T. D. Kocher. 1995. Structure and evolution of teleost mitochondrial control regions. Journal of Molecular Evolution 41:54–66.

Loh-Lee, L., editor. 1991. Status of living marine resources off the Pacific coast of the United States as assessed in 1991. NOAA (National Oceanic and Atmospheric Administration)/NMFS (National Marine Fisheries Service) Technical Memorandum F/NWC-210, Northwest Fisheries Science Center, Seattle.

Margush, T., and F. R. McMorris. 1981. Consensus n-trees. Bulletin of Mathematical Biology 43:239–244.

Merker, R. 1994. Molecular evidence for discrete breeding stocks of walleye (*Stizostedion vitreum*) in the Maumee, Sandusky, and Grand rivers. Master's thesis. Bowling Green State University, Bowling Green, Ohio.

Meyer, A., T. D. Kocher, P. Basasibwaki, and A. C. Wilson. 1990. Monophyletic origin of Lake Victoria cichlid fishes suggested by mitochondrial DNA sequences. Nature 347:550–553.

Moser, H. G. 1974. Development and distribution of larvae and juveniles of *Sebastolobus* (Pisces; family Scorpaenidae). U.S. National Marine Fisheries Service Fishery Bulletin 7(3–4):865–884.

Nei, M. 1972. Genetic distance between populations. American Naturalist 106:283–292.

Oliphant, M. S. 1979. California marine fish landings for 1976. California Department of Fish and Game Fish Bulletin 170:1–56.

Owen, R. W. 1980. Eddies of the California current system: physical and ecological characteristics. Pages 237–263 *in* D. M. Powers, editor. The Channel Islands: proceedings of a multidisciplinary symposium. Santa Barbara Museum Natural History, Santa Barbara, California.

Present, T. M. C. 1987. Genetic differentiation of disjunct Gulf of California and Pacific outer coast populations of *Hypsoblennius jenkinsi*. Copeia 1987(4): 1010–1024.

Quirollo, L. F., and P. Kalvass. 1987. Results of Dover sole tagging in waters off northern California, 1969–1971. California Department of Fish and Game Marine Research Administrative Report No. 87-4, Sacramento.

Riley, L. M., and R. F. Carline. 1982. Evaluation of scale

shape for the identification of walleye stocks from western Lake Erie. Transactions of the American Fisheries Society 111:736–741.

Saitou, N., and M. Nei. 1987. The neighbor-joining method: a new method for reconstructing phylogenetic trees. Molecular Biology and Evolution 4:406–425.

Sakamoto, K. 1984. Interrelationships of the family Pleuronectidae (Pisces: Pleuronectiformes). Memoirs of the Faculty of Fisheries Hokkaido University 31(1–2):95–215.

Sanger, F., S. Nicklen, and A. R. Coulson. 1977. DNA sequencing with chain-terminating inhibitors. Proceedings of the National Academy of Sciences of the United States of America 74:5463–5467.

Siebenaller, J. F. 1978. Genetic variability in deep-sea fishes of the genus *Sebastolobus* (Scorpaenidae). Pages 95–122 *in* B. Battaglia and J. A. Beardmore, editors. Marine organisms: genetics, ecology, and evolution. Plenum Press, New York.

Sinclair, M. 1988. Marine populations: an essay on population regulation and speciation. University of Washington Press, Seattle.

Sokal, R. R., and F. J. Rohlf. 1981. Biometry: the principle and practice of statistics in biological research, 2nd edition. Freeman, San Francisco.

Springer, V. G. 1982. Pacific plate biogeography with special reference to shorefishes. Smithsonian Contributions to Zoology 367:1–182.

Stepien, C. A., M. T. Dixon, and D. M. Hillis. 1993. Evolutionary relationships of the fish families Clinidae, Labrisomidae, and Chaenopsidae: congruence between DNA sequence and allozyme data. Bulletin of Marine Science 52(1):873–921.

Stepien, C. A., and R. H. Rosenblatt. 1991. Patterns of gene flow and genetic divergence in the northeastern Pacific clinid fishes, based on allozyme and morphological data. Copeia 1991(4):873–896.

Stepien, C. A., J. Randall, and R. H. Rosenblatt. 1994. Genetic and morphological divergence of a circumtropical complex of goatfishes. Pacific Science 48(1): 44–56.

Sturmbauer, C., and A. Meyer. 1992. Genetic divergence, speciation and morphological stasis in a lineage of African cichlid fishes. Nature 58:578–581.

Thomson, D. A., and M. R. Gilligan. 1983. The rocky-shore fishes. Pages 98–129 *in* T. J. Case and J. L. Cody, editors. Island biogeography in the Sea of Cortez. University of California Press, Berkeley.

Todd, T. N. 1990. Genetic differentiation of walleye stocks in Lake St. Clair and western Lake Erie. U.S. Department of the Interior Fish and Wildlife Service Fish and Wildlife Technical Report 28:1–19.

Trautman, M. B. 1981. The fishes of Ohio. Revised edition. Ohio State University Press, Columbus.

Uhlen, M. 1989. Magnetic separation of DNA. Nature 340:733–734.

Van Vooren, A. R. 1978. Characteristics of walleye spawning stocks. Ohio Department of Natural Resources, Federal Aid in Fish Restoration Project F-35-R-4, Final Report, Columbus.

Walker, B. W. 1960. The distribution and affinities of the marine fish fauna of the Gulf of California. Systematic Zoology 9(3):123–133.

Waples, R. S. 1987. A multispecies approach to the analysis of gene flow in marine shore fishes. Evolution 41(2):385–400.

Waples, R. S. 1991. Pacific salmon, *Oncorhynchus* spp., and the definition of "species" under the Endangered Species Act. U.S. National Marine Fisheries Service Marine Fisheries Review 53(3):11–22.

Waples, R. S., and R. H. Rosenblatt. 1987. Patterns of larval drift in southern California marine shore fishes inferred from allozyme data. U.S. National Marine Fisheries Service Fishery Bulletin 88(1):1–11.

Ward, R. D., M. Woodwark, and D. O. F. Skibinski. 1994. A comparison of genetic diversity levels in marine, freshwater, and anadromous fishes. Journal of Fish Biology 44:213–232.

Ward, R. D., N. Billington, and P. D. N. Hebert. 1989. Comparison of allozyme and mitochondrial variation in populations of walleye, *Stizostedion vitreum*. Canadian Journal of Fisheries and Aquatic Sciences 46: 2074–2084.

Wolfert, D. R. 1977. Age and growth of the walleye in Lake Erie, 1963–1968. Transactions of the American Fisheries Society 106:569–577.

Wolfert, D. R., and H. D. Van Meter. 1978. Movements of walleyes tagged in eastern Lake Erie. New York Fish Game Journal 25:16–22.

Wright, S. 1978. Evolution and the genetics of populations. Volume 4. Variability within and among natural populations. University of Chicago Press, Chicago.

Wyllie, J. G. 1966. Geostrophic flow of the California current at the surface and at 200 meters. CalCOFI (California Cooperative Oceanic Fisheries Investigations) Atlas No. 4. State of California, Marine Research Committee, La Jolla.

Yang, T.-H., N. C. Lai, J. B. Graham, and G. N. Somero. 1992. Respiratory, blood, and heart enzymatic adaptations of *Sebastolobus alascanus* (Scorpaenidae; Teleostei) to the oxygen minimum zone: a comparative study. Biological Bulletin 183:490–499.

Appendix: Nucleotide Sequence Data

Summary of genetic differences in the complement of the heavy strand of the mtDNA control region (beginning at base 50; 1–49 is transfer RNA proline) for eight fish species. Abbreviations are guanine (G), adenine (A), thymine (T), cytosine (C), gap or deletion (-), and unresolved bases (N). The consensus sequences are a summary of the most common bases among haplotypes. Base substitution variations among haplotypes at given base positions are indicated below the consensus sequence. Alignments between species groups are for purposes of rough comparison only and do not necessarily indicate evolutionary homology. Lowercase letters and associated numbers 1–3 denote the three tandem repeat regions.

```
Base Position         0000000000111111111122222222223333333333444444
                      0000000000111111111122222222223333333333444444
                      1234567890123456789012345678901234567890123456

Dover sole            TCAAAGAAAGGAGATTTCAACTCCTACCCCTAACTCCAAAGCTAGGATT
                                                  C                C
Slime flounder        TCAAAGAAAGGAGATTTCAACTCCTACCCCTAACTCCAAAGCTAGGATT
                                              C  G
Shortspine thornyhead TCAAAGAAAGGAGATTTAACTCCGGCTCCCACCCCTAACTCCAAAGCCAGGATT
                                        G
Longspine thornyhead  TCAAAGAAAGGAGATTTAACTCCGGCTCCCACCCCTAACTCCAAAGCCAGGCTT
                                        G
Broadbanded
  thornyhead          TCAAAGAAAGGAGATTTAACTCCCTCCCACCCCTAACTCCAAAGCCAGGATT
Spotted sand bass     TCAAAGAAAGGAGATTCGAACTCCCTCCCACCCCTAACTCCAAAGCCAGGATT
Barred sand bass      TCAAAGAAAGGAGATTCGAACTCCCTCCCACCCCTAACTCCAAAGCTAGGATT
Walleye               CAACCGGAAGGAGATTTAACTCCCACCCCTAACTCCAAAGCTAG--TT
```

```
Base Position         000000000000000000000000000000000000000
                      555555555566666666667777777777888888888900000
                      123456789012345678901234567890123456789012345

Dover sole            CTAGCACTAAACTATTCTTTGCGGCACAACATATG-TCCATG-----
                         GGGT      A     A   C              T
                                                            G
Slime flounder        CTAGCACTAAACTATTCTTTGCGGCGTACTACATG-TACAAT-
                                                G
Shortspine thornyhead CTGAG-TTAAACTATTCTTTGTATAATATAA------TACATG-
Longspine thornyhead  CTGAG-TTAAACTATTCTTTGTATAATATAA------TACATG-
Broadbanded
  thornyhead          CTGCG-TTAAACTATTCTTTGTATAATATAA------TACATG-
Spotted sand bass     CTAAA-TTAAACTATTCTTTGAATG-TATTTTAAA-TACATA-
Barred sand bass      CTAAA-TTAAACTATTCTTTGAATG-TATTTTAAA-TACATA-
Walleye               CTAAA-TTAAACTATTCTTTgcaaa-tattt-gcaaatt-gcaaagca
                                           c    gc-g   cggt gt-
                                           g            -c   a
                              1           2              3
```

```
Base Position          11111111111111111111111111111111111111111111111111
                       00000000011111111112222222222333333333344444444445
                       12345678901234567890123456789012345678901234567890
Dover sole             ----AAAGATTTTTCATGTACATGTATGTAATAACACCATATATTTATAG
                              C
Slime flounder         ----AAAGAATTTT-ATGTACATGTATGTAATAACACCATATATTTATAG
                                              A               C
Shortspine thornyhead  ----TATGTATTATCACCATTA--ATTTATATTAA-CCATATC--AATAG
Longspine thornyhead   ----TATGTATTATCACCATTA--ATTTATATTAA-CCATATC--AATAG
Broadbanded            ----TATGTATAATCACCATTA--ATTTATATTAA-CCATATC--AATAG
  thornyhead                         T                      G
Spotted sand bass      ----TATGTATATACACC-ATA-CATTTATATTAAA-CATATC--AATAG
Barred sand bass       ----TATGTATATACACCTATA-CATTTATATTAAA-CATATC--AATAG
Walleye                ttt-TATGTATTTACACC-ATA-CATCTATATTAA--CCATATC-AATGG
                       c -t
Base Position          11111111111111111111111111111111111111111111111112
                       55555555556666666666777777777788888888889999999990
                       12345678901234567890123456789012345678901234567890
Dover sole             TAACCATTTT-ATATAATGCATTAGGACATTCATGTATAATAACCTAATC
                               G  TG                 CT
Slime flounder         TAACCATTTT-ATATAATGCATTAGGACATTCATGTATAATAACCTAATC
Shortspine thornyhead  CATTCAAGTACATA-CATGTTTTATCCACATATGTAGGTTTTAAA-----
                            G         T   C  CGCT    C C  G    C   G
Longspine thornyhead   CAATCAAGTACATA-CATGTTTTATCCACATATGTAGGTTTTAAG-----
                                                T                  A
Broadbanded            CAATCAAGTACATA-TATGTTTTATCCACATATGTAGGTTTTAAG-----
  thornyhead                          C            T               A
Spotted sand bass      CATTCAAGGACATAACATGTTTAATCAACAAAACTAGGATTACCC-----
Barred sand bass       CATTCAAGGACATATCATGTTTAATCACCAAATCTAGGATTACCC-----
Walleye                CATTCAAGTACATAT-ATGTTTTATCAACATATCTAGGATTAACA-----
```

Base Position (columns 2001–2450)

```
Base Position       2 2 2 2 2 2 2 2 2 2 2 2 2 2 2 2 2 2 2 2 2 2 2 2 2 2 2 2 2 2 2 2 2 2 2 2 2 2 2 2 2 2 2 2 2 2 2 2 2 2
                    0 0 0 0 0 0 0 0 0 1 1 1 1 1 1 1 1 1 1 2 2 2 2 2 2 2 2 2 2 3 3 3 3 3 3 3 3 3 3 4 4 4 4 4 4 4 4 4 4 5
                    1 2 3 4 5 6 7 8 9 0 1 2 3 4 5 6 7 8 9 0 1 2 3 4 5 6 7 8 9 0 1 2 3 4 5 6 7 8 9 0 1 2 3 4 5 6 7 8 9 0

Dover sole          T A G T A A T A T A G C A C T C A T T C A T C A A C A T T T T A - - - A T A A G A A A T T A C T
Slime flounder      T A G T A A T A C A G . . . . . . . . . . . . . . . . C . . . . . . . C . . . . G . A G G G . . .
Shortspine          - - - - - - - - - C A T T C A C T T A T C A A C A T A C T - - - T A - . . . . . . . . . . C T
  thornyhead                                                        G                    C
Longspine           - - - - - - - - - C A T T C A C T T A T C A A C A T A T T - - - . . . . . . . . . T A A G A T A T C C A T
  thornyhead                          C                          G C                        C           T
Broadbanded         - - - - - - - - - . . . . . . . . . . . . . . . . . . . . . . . - - - . . . . . . C A A G A T A T C C A T
  thornyhead                                                                  G                            T A
Spotted sand bass   - - - - - - - - - C A T T C A C T T G T C A A C A T A A A A C - - - T A A T A T A T A C A C
                                                                  G G                                     C G
Barred sand bass    - - - - - - - - - C A T T C A - - T A - C A A C A C A T G A A A C C - A A A G G T T T A C A
                                          G                      G
Walleye             - - - - - - - - - - C A T T C A T A T C A C C A G C A T G A T A C C - T A C C C T A A G G G T T - A C A
                                        C
```

Base Position (columns 2551–2790)

```
Base Position       2 2 2 2 2 2 2 2 2 2 2 2 2 2 2 2 2 2 2 2 2 2 2 2 2 2 2 2 2 2 2 2 2 2 2 2 2 2 2 2 2 2 2 2 2 2 2 2
                    5 5 5 5 5 5 5 5 6 6 6 6 6 6 6 6 6 6 7 7 7 7 7 7 7 7 7 7 8 8 8 8 8 8 8 8 8 8 9 9 9 9 9 9 9 9 9 9
                    3 4 5 6 7 8 9 0 1 2 3 4 5 6 7 8 9 0 1 2 3 4 5 6 7 8 9 0 1 2 3 4 5 6 7 8 9 0 1 2 3 4 5 6 7 8 9 0

Dover sole          - A A A C C T G C T T A A T T A C T A A C C T - T A C A T T A G T G A A G A T C C A G G A C T A
                                T               G G C C             C G               A A               G C
Slime flounder      - - A A A A C C T A T T T A A C C A T T A A A T C T - C A C A T A T G T G A A A A C C C A G G A C C A
                                        T T C                              A G C                   T
Shortspine          - - A A G C A T G A A C T T A T A A A C - - - - - - C A C A T A T A A A T T A C A T T A A A C - - - A
  thornyhead                              G                                      T T G G          G
Longspine           - - A A A G C A T G A A C T T A T A T A A C - - - . . . . . . C A C A A T A A A T T A C A - - - A
  thornyhead                          C                                                          C
Broadbanded         - - A G A G C A T G A A C T C A T A A A C - - - . . . A A C A A T A A G T T A T G T G A A A C - - - A
  thornyhead                A                                                        T A
Spotted sand bass   T T A A A - C A T A A A A T T G T T T T T - - - . . . C A A T A A T G T A T T T T G A G C G C A
                                                                                      G
Barred sand bass    T T A A A - C A T A A A A C T G C T T A T T - - - . . C A A T A A T A G A A T T T A G G T C G A
Walleye             T - A A A G C A T A - - - T A G A C C T T T A T C T A A C A A T A T T A A A T C A A G G A T - A
```

```
Base Position   3 3 3 3 3 3 3 3 3 3 3 3 3 3 3 3 3 3 3 3 3 3 3 3 3 3 3 3 3 3 3 3 3 3 3 3 3 3 3 3 3 3 3 3 3 3 3 3 3 3
                0 0 0 0 0 0 0 0 0 0 1 1 1 1 1 1 1 1 1 1 2 2 2 2 2 2 2 2 2 2 3 3 3 3 3 3 3 3 3 3 4 4 4 4 4 4 4 4 4 5
                1 2 3 4 5 6 7 8 9 0 1 2 3 4 5 6 7 8 9 0 1 2 3 4 5 6 7 8 9 0 1 2 3 4 5 6 7 8 9 0 1 2 3 4 5 6 7 8 9 0

Dover sole      C C C G A T T T - A A G A C C G A C C A C A A - - C A C T C - A T C A G T C G A G T T A C A C C
                                                                T                                     A

Slime flounder  G T C G A A A C T T - A A G A C C G A C C A C A A - C A C T C - A T C G G T C A A G T T A T A C C
                    C                                 G                   T
                                                                         G

Shortspine      G G C G A A A C T T - A A G A C C T A A C A C A A T - A A A T C - A T G A G T T A A G T T A T A C C
  thornyhead        A                                           A               C C T G
                    T                                                           C

Longspine       G G C G A A A C T T - A A G A C C T A A C A C A A T - A A A T C - A T A A G T T A A G T T A T A C C
  thornyhead                        T                                           A         G G

Broadbanded     G G C G A A A G T C - A A G G C C T A A C A C A A T - A A A T C - A T C A A G T C A A G T T A T A C C
  thornyhead                                A                         G                                     G T

Spotted sand    A G C G A A A T T T - A A G A A C T C A - A T C C T T A A G T C C T T A A G T A A G G T T A T A C G
  bass                                                C

Barred sand     G G C G A G A C T T - A A G A A C T C A - A T C C T T A C G T T C T T A A G T A A G G T T A T A C G
  bass

Walleye         G G C G A T T A T T T A A G A - C C G A - A C A C T T C T A C T C A T A A G T T A A G T T A T A C C
                                                                                      G


Base Position   3 3 3 3 3 3 3 3 3 3 3 3 3 3 3 3 3 3 3 3 3 3 3 3 3 3 3 3 3 3 3 3 3 3 3 3 3 3 3 3
                5 5 5 5 5 5 5 5 5 6 6 6 6 6 6 6 6 6 6 7 7 7 7 7 7 7 7 7 7 8 8 8 8 8 8 8 8 8 8 9 9 9
                1 2 3 4 5 6 7 8 9 0 1 2 3 4 5 6 7 8 9 0 1 2 3 4 5 6 7 8 9 0 1 2 3 4 5 6 7 8 9 0 1 2 3 4 5 6 7 8 9 0

Dover sole      A A G A C T C A A A A T C T C G C C A G C C G C A A A C C A C - - T G T G T A G - T A A G A G C -
                            A                           A                              G A

Slime flounder  A A G A C T C A A A A T C T C G C C A A T C G C A A A A T C C - - T A T G T A G - T A A G A G C -

Shortspine      T T T A C T C A A A C T C T C G T C A A T T - T A A A A T C T T A A T G T A G - T A A G A G C C G
  thornyhead            T                   C T                       C G

Longspine       T T T A C T C A A A C T C T C G C C A A T T - T A A A A T C T T A A T G T A G - T A A G A G C C G
  thornyhead            T                                                                             A T

Broadbanded     T T T A C T C A A A C T C T C G C C A A T T - T A A A A T C T T A A T G T A G - T A A G A G C C G
  thornyhead                                                                                             C

Spotted sand    T T T A C C T G A C A T C T C G C C A T A C C T A A T A G A T A T C A A C C C A A T A A G A - - -
  bass                  C                                                                          G

Barred sand     T T T A C T T G A C A T C T C G C C A T C T C A C A T C T A A T A G A C A T C A G C C C A A T A A G A - - -
  bass

Walleye         T T T A C C C A A C A T C T C G C C A T A C C T C A A A A T C T T A A T G T T A G A G A G - - - -
                                      T
```

Base Position (401–450)

Dover sole	- C T A C C A A C C G G T G A T T C C T G A A T G A T A A C C T C T T A T T G A G G G T G A G G G A
Slime flounder	- C T A C C A A C C A G T G A T C C C T T A A T G A T A A C C T C T T A T T G A G G G T G A G G G A
Shortspine thornyhead	A C C A A C A A - - G T C C A T T C - T T A A T G C C A A - C G G T T A T T G A A G G T G A G G G A
Longspine thornyhead	A C C A A C A A - - G T C C A T T C - T T A A T G C C A A - C G G T T A T T G A A G G T G A G G G A
Broadbanded thornyhead	A C C A A C A A - - G T C C A T T T - T T A A T G C C A A - C G G T T A T T G A A G G T G A G G G A
Spotted sand bass	A G T A A C A T C G G T T G A T A C C T C G A T A C C T A - C G G T T A T T G A G G - T G A G G G A
Barred sand bass	A G T A A C A T C G G T T G A T A T C T A G A T A C C T A - C G G T T A T T G A G G - T G A G G G A
Walleye	C C T A C C A T C A G T T G A T T T C T T A A T G C T A A - C G G T T A T T G A G G - T G A G G G A

Base Position (451–502)

Dover sole	C C A A A G A T C G T G G G G T T - - C A C C G G T G A A C C T A T T C C T A T T C C T G G C A T C T G G T T C C T
Slime flounder	C C A A A A A T C G T G C G G G T T - - C A C T C A G T G C A C T A T T C C T G G C A T C T G G T T C C T
Shortspine thornyhead	C A A T G A T T C G T G G G G G T T T - C A C A C A G T G A A C T A T T C C T G G C A T C T G G T T C C T
Longspine thornyhead	C A A T G A T T C G T G G G G G T T T - C A C A C A G T G A A C T A T T C C T G G C A T C T G G T T C C T
Broadbanded thornyhead	C A A T A A T T C G T G G G G G T T - C A C A C A G T G A A C T A T T C C T G G C A T C T G G T T C C T
Spotted sand bass	C A - G T A T T T G T - G G G G T A A C - - - G A G T G A A C T A T - - C T G - C A T C T G G T T C C T
Barred sand bass	C A - G T A T T C G T - G G G G T G A C - - - T A G T G C A C T A T - - C T G - C A T C T G G T T C C T
Walleye	C A A C T A T T - G T G G G G G T T T C T T A C C - - - - - A G T C N - - - - - C A T C T G G T T C C T

American Fisheries Society Symposium 17:288–294, 1995

Why Statistical Power is Necessary to Link Analyses of Molecular Variation to Decisions about Population Structure

Andrew E. Dizon, Barbara L. Taylor, and Gregory M. O'Corry-Crowe

National Marine Fisheries Service, Southwest Fisheries Science Center
Post Office Box 271, La Jolla, California 92038, USA

Abstract.—This paper demonstrates the importance of calculating statistical power when genetic data are used to estimate population structure. We provide an example of a posteriori estimates of statistical power of mitochondrial DNA sequence studies based on two species of pelagic dolphins. We address the management implications of making α (type I) and β (type II) errors in a risk-averse conservation situation. The management implications are given in the context of the Marine Mammal Protection Act, which seeks to balance the health and safety of marine mammal populations against minimizing costly restrictions on fishing industries known to kill marine mammals incidentally. For the results of scientific investigations to be properly applied to management decision making, both the probability of incorrectly rejecting the null hypothesis (α) and the probability of correctly rejecting the null hypothesis ($1-\beta$, or power) must be calculated. Calculations of statistical power are necessary to determine the costs of management decisions. In our examples, we show that if we use the typical type I error level ($\alpha = 0.05$), power is quite low. Given the data, if the subpopulations were actually genetically subdivided (the null hypothesis false), either large sample sizes would have been required to reject the null hypothesis at an α level of 0.05 or the α level would have to be relaxed. The latter alternative might be appropriate for a risk-averse management strategy that assigns greater costs to potentially losing an important subpopulation than to overrestricting commercial fisheries.

When an experimenter fails to find evidence of population subdivision in a typical intraspecific genetic study, the results are rarely published. The experimenter reevaluates his or her experimental design or hypotheses and tries again. Although career costs may accrue, in a typical academic context such results are not costly to the study organism. In a management context, however, such results are often taken as reason to manage a population as a single unit. This is an appropriate decision if the experiment in question failed to find evidence of population subdivision because in reality it was not present. Alternatively, inappropriate decisions may be made if the experimental design lacked power to discriminate a subdivision. For example, a scientist may decide not to gather further samples because he or she believes the population is panmictic. In addition, a manager may decide to allow fisheries to remove a number of animals based on an abundance that is incorrect for the area being fished because the abundance estimate includes more than one population. Under such circumstances, it is important to estimate the probability of incorrectly failing to reject the null hypothesis of population panmixia in repeated comparable studies. Yet power, the probability of this incorrect decision, is an often neglected topic in genetic analyses and fisheries research (Peterman 1990a). Although many speakers during this conference described the process of making management decisions based on

negative evidence ("if no evidence of . . . is found, then the population should be managed as a single unit"), not one of the approximately 33 speakers addressed the power of his or her study to resolve the competing hypotheses (see Peterman 1990b for a similar example in forestry).

Statistical Errors in Hypothesis Testing

In order to test hypotheses, a rule must be chosen to decide when to reject the null hypothesis (H_0). Typically scientists take precaution against deciding an alternate hypothesis is true when, in fact, the outcome was a result of chance. For this reason, hypothesis testing is usually driven by the desire to have a low probability of falsely rejecting H_0: α is usually set at 0.05. Taylor and Gerrodette (1993) argued that power, the probability of correctly rejecting H_0, was the more important criterion for hypothesis testing in a conservation context because the cost of incorrectly not rejecting H_0 can be high. There is a large body of literature on the use of power in ecological applications that is referenced in Peterman 1990a and Taylor and Gerrodette 1993. Table 1 defines the statistical terms. In a risk-averse conservation policy (such as followed under the Marine Mammal Protection Act 16 U.S.C. §§ 1361 to 1407) the cost of failing to reject a false H_0 (a type II error) is, perhaps, greater than the costs associated with the familiar type I error. The number of marine mammals that may be re-

TABLE 1.—The four possible outcomes of a statistical test of a null hypothesis (H_0) assuming reality is known (after Peterman 1990).

Validity of H_0	If the decision is to . . .	
	Not reject H_0	Reject H_0
H_0 in reality is true	Correct decision is made with probability $(1 - \alpha)$	Type I error is made with probability α
H_0 in reality is false	Type II error is made with probability β	Correct decision is made with probability $(1 - \beta)$

moved from a population is calculated based on abundance. Many fisheries occur in only a portion of a marine mammal species' range. If a manager decides that abundance should be based on the whole range, but, in fact, the population is subdivided, evolutionarily significant local subpopulations may be reduced to low levels or destroyed. A type I error, rejecting a true H_0, does not carry this risk of inadvertently extirpating a potential evolutionarily significant unit (ESU). The cost of such an error is that if population structure is arbitrarily subdivided too finely, local fisheries may be unnecessarily closed down so that the artificial subpopulation may be preserved. Estimating the costs associated with such management decisions is a difficult task beyond the scope of this paper. The estimation, however, cannot proceed without the calculation of both α and β, which should be standard procedure whenever statistical tests fail to reject the null hypothesis.

In the following examples, we calculate power for two marine mammal genetic studies by extending an analysis of variance methodology (AMOVA) developed by Excoffier et al. (1992). We then concentrate on the interpretation of the results. Any power analysis requires the comparison of two specific hypotheses. Because we are almost always limited to sampling populations, the hypotheses take the form of distributions that give the probability of observing various values given, for example, our sample size or experimental design (Winer 1971). Often the choice of null and alternate hypotheses are clear for experimental situations. Consider the case in which an experimenter wants to know whether an experimental feed can increase mean body size by 5% (the size necessary to offset the additional costs of the feed). The experimenter obtains a distribution of body sizes on standard feed and on the experimental feed. It is a simple matter, then, to calculate whether the increase in body size was significant, the power of the test, and the associated costs of type I and type II errors. We are not so fortunate in conservation issues. We do not know

what level of genetic differentiation between subpopulations is meaningful either in an evolutionary context or in a management context.

Consider Figure 1, which shows distributions for the test statistic Φ_{ST} (the statistic calculated in an AMOVA). The statistic Φ_{ST} is the ratio of the variance between populations to the total variance of pairwise genetic distances. The shape of the distribution of Φ_{ST} depends on the shapes of the distributions of pairwise genetic distances within and between populations. The shapes of distributions of pairwise genetic distances vary for different populations and species and depend on the evolutionary history of the organisms. We show normal distributions here for simplicity in illustrating the concept of estimating power. The distribution of Φ_{ST} values that could be obtained from a panmictic population (the null distribution) is indicated by the dashed line (Figure 1A, B). Population subdivision always results in higher expected Φ_{ST} values, so our test will be one-tailed. A 5% probability of falsely rejecting H_0 corresponds to a critical Φ_{ST} of 0.08 (5% of the distribution is greater than this value—the crosshatched area beneath the null hypothesis distribution in Figure 1A). The distributions shown with the solid lines represent possible distributions when some population structure is present (alternate distributions), i.e., genetic distances are on average greater between subpopulations than within subpopulations. Power is the proportion of the distribution for the alternate hypothesis that is at least as great as the critical Φ_{ST}: the lined area beneath the distribution for the alternate hypothesis in Figure 1A when α is 0.05. The differences between the subpopulations are greater in Figure 1A than in 1B. This is evidenced by the smaller difference between the means of the null and alternate distributions (often called the effect size) that can be seen in Figure 1B. Comparing these two figures, one can also see that the greater the difference between the null and alternate distributions (effect size), the greater the power. Power also increases as α increases, i.e., as we increase the probability of committing a type I error. What we cannot determine from these figures is whether the level of population subdivision depicted in Figure 1A or 1B represents significance in either a biological or management context. We return to this problem in the discussion.

Methods

To illustrate power calculations based on genetic data, we calculate power for dolphin populations. Data used to calculate pairwise distances are from previous studies (Dizon et al. 1994; García Rodríguez 1994). The genetic or evolutionary distance separating each pair of sequences (haplotypes) are estimated using the Tamura–Nei method as implemented in the computer package, MEGA[1]; the authors recommend this method as especially appropriate for mitochondrial DNA (mtDNA) sequences. The derived matrix of $N(N-1)/2$ pairwise genetic distances were examined for evidence of a geographic structure, i.e., concordance between genetic distance and geographic strata, by means of the AMOVA (Excoffier et al. 1992). The linear model employed assumes individuals are arranged into subpopulations and defined by nongenetic criteria. The program computes estimates of variance components and F_{ST} analogues designated, as mentioned previously, as Φ_{ST}. The method of Excoffier et al. (1992) employs a permutation technique to test the significance of the variance components and statistics, thus avoiding assumptions of normality. As outlined by Excoffier et al. (1992), we estimate the distribution of Φ_{ST} values by randomly reassigning subpopulation identity according to the frequency of each subpopulation because H_0 is assumed true (i.e., that the subpopulation stratification is artificial). A distribution is formed from 1,000 Φ_{ST}'s calculated from random subpopulation identity permutations. Because the test is one tailed, critical values are simply the percentiles of an ascending vector of the Φ_{ST}' s of the permutation trials that correspond to the chosen α levels. For example, if α is 0.05, the critical Φ_{ST} value would be the 950th Φ_{ST} value in the sorted vector of estimated Φ_{ST} values.

In order to estimate β, the probability that an analysis falsely fails to reject H_0, we extend the Excoffier et al. (1992) method so that we resample the subpopulations individually; i.e., we do not pool

the sample because H_0 is assumed false. Individuals are separated into subpopulations based on a geographic or morphologic basis. The population is resampled with replacement, but now the individual subpopulations are kept separate during the resampling. Type I error is estimated from the distribution of resampled, nonpooled, Φ_{ST}'s by determining the proportion of the resampled populations in which H_0 fails to be rejected, assuming population subdivision. Power is simply the complement of β.

We are concerned about possible bias introduced into the calculation of the alternate distribution due to resampling from the often very small subpopulation samples. The correct unbiased technique for estimating power for Φ_{ST} statistics is the subject of ongoing research, but the examples here should still serve as illustrations of the importance of power calculations to decision making.

Results

The genetic variation within and between strata of 60 spinner dolphins *Stenella longirostris* from the eastern tropical Pacific Ocean (García Rodríguez 1994) and within and between 65 northern right whale dolphins *Lissodelphis borealis* from the north temperate Pacific Ocean (Dizon et al. 1994) was analyzed previously in our laboratory by means of 400 base-pair sequences from the D-loop region of mtDNA. The spinner dolphin sample was stratified (among other schemes) based on fishery stock type, which was defined by the appearance of the majority of the adult dolphins within a school, whether of the "whitebelly" morphotype (20 animals) or "eastern" morphotype (40 animals). The northern right whale dolphins were stratified based on geographic location when sampled, either offshore (52 animals) or coastal (13). The abundances of these populations are very large (tens to hundreds of thousands), and samples were obtained over large areas, so it is unlikely that samples were taken from close relatives. No genetic structure was detected by applying the Excoffier et al. (1992) methods for either species; the H_0 of no population subdivision could not be rejected at the α level of 0.05. So a posteriori power analyses are appropriate for both cases.

Critical Φ_{ST} values for both species were determined at α levels of 0.05, 0.10, 0.15, and 0.20 based on Φ_{ST} distributions calculated from random permutations of subpopulation identity. Estimates of β for each critical value were made by determining the proportion of the alternate distribution of Φ_{ST} estimates (generated by resampling individuals within a population with replacement) that were

---

[1]Kumar, S., K. Tamura, and M. Nei. 1993. MEGA: molecular evolutionary genetics analysis, version 1.0, 1.0 ed. The Pennsylvania State University, University Park.

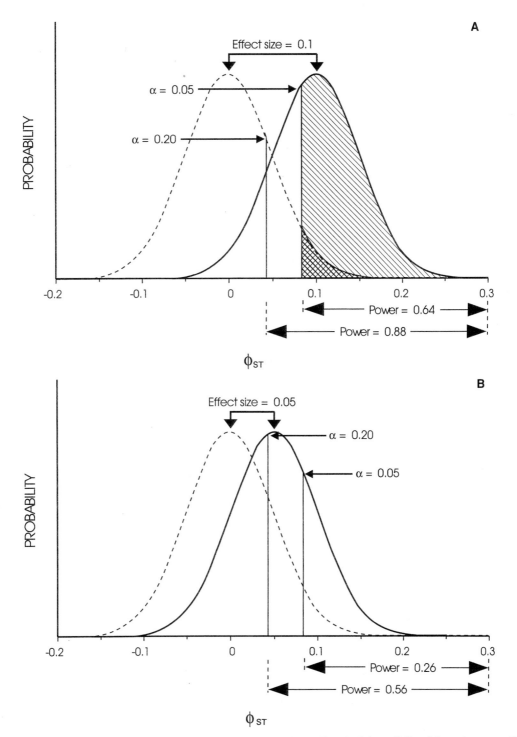

FIGURE 1.—(A) A hypothetical one-tailed power analysis in which the effect size is large (0.1) and Φ_{ST} values normally distributed. The distribution of Φ_{ST} values, calculated in analysis of variance methodology, that could be obtained from a panmictic population is indicated by the dashed line; the distribution that could be obtained from a subdivided population is indicated by the solid line. This alternate distribution is displaced to the right because genetic distances are on average greater between subpopulations than within subpopulations. Beta is the proportion of the alternate distribution that is less than the critical Φ_{ST} (H_0 fails to be rejected). Power equals $1-\beta$, that is, the proportion of the alternate distribution greater than the critical Φ_{ST}. (B) A hypothetical one-tailed power analysis in which the effect size is smaller (0.5) than that in (A). Power is subsequently reduced.

smaller than the critical Φ_{ST} (Table 2). Given an α of 0.05, power is quite low for detecting population subdivision if it existed, particularly for spinner dolphins. However, by relaxing α to 0.20, statistical power increases to 0.94 for the northern right whale dolphin, and to 0.91 for the spinner dolphin. This discrepancy is likely due to differences in both the effect size and the dispersion of the resampled Φ_{ST} values (Figure 2A, B). The effect size (the difference between the means of the pooled and the subdivided Φ_{ST} distributions) is greater in the northern right whale dolphin sample than in the spinner dolphin sample. Although this leads, as expected, to greater power to detect the subdivision (Table 2), the picture is complicated by the greater variance in Φ_{ST} values for the northern right whale dolphin. The effect of this greater variance is revealed by the relatively smaller gain in power as α is increased (Table 2): an increase in α of 0.15 results in increasing power by 0.39 in the case of the spinner dolphin but by only 0.17 in the case of the northern right whale dolphin. The greater variance in the distribution for the alternate hypothesis for the northern right whale dolphin is probably due to the greater imbalance in sample sizes between the two subsamples (52 and 13 for the northern right whale dolphins versus 40 and 20 for the spinner dolphins).

Discussion

Although Table 2 provides a great improvement over simply rejecting or not rejecting a null hypothesis of no population structure given an α of 0.05, managers would still be baffled as to how to make a decision given these results. As biologists, estimating the costs of various decisions is likely beyond our purview, but we should provide advice on the biological meaning of different levels of genetic differentiation, e.g., when does statistical evidence of population subdivision cease to have utility in making management decisions? Conservation managers cannot be expected to translate from Φ_{ST} values to appropriate choice of management units. This is a scientific problem that needs addressing in its own right.

What does the effect size mean both in a biological and a management context? In the spinner dolphin sample, the effect size is likely minuscule (perhaps smaller than 0.03) which indicates that only about 3% of the total within-population variation in pairwise genetic distance is due to stratification into subpopulations. The challenge remains to determine the relevance of small degrees of subdivision in both an evolutionary and a management

TABLE 2.—Critical Φ_{ST}, β, and power for various α levels. Critical values are determined by pooling the sample without regard to subpopulation membership and forming a distribution of estimated Φ_{ST} values from subpopulations resampled from the pool. Critical Φ_{ST} values are the percentiles of an ascending vector of the Φ_{ST} values of the resampling trials that correspond to the chosen α levels (a single-tailed test; see Figure 1). Beta is estimated from the distribution of resampled, but nonpooled, Φ_{ST} values by determining the proportion of those resampled populations in which H_0 fails to be rejected. Power is $1 - \beta$.

α	Critical Φ_{ST} value	β	Power
Northern right whale dolphin			
0.05	0.033	0.23	0.77
0.1	0.024	0.14	0.86
0.15	0.017	0.08	0.92
0.2	0.013	0.06	0.94
Spinner dolphin			
0.05	0.033	0.48	0.52
0.1	0.021	0.23	0.77
0.15	0.016	0.16	0.84
0.2	0.012	0.09	0.91

context. In order to interpret genetic data in these two contexts, we need to gain an understanding of the expected distribution of haplotypes or genotypes within and between subpopulations.

We suggest two approaches to achieve this goal. The first approach is to conduct simulation studies in which population parameters, such as movement rates and social structure, are known, and the probability of reconstructing the population structure based on various genetic data can be determined. One such approach is to use models to formulate expectations regarding genetic distributions under various migratory, life history, and population size assumptions. The simulated population can then be sampled. Although such a modeling effort is a formidable task, the relatively small population sizes involved in most risk-averse management situations is tractable for such models, and we anticipate progress in this area especially when coupled with accumulating future data on actual haplotypic and genotypic distributions. This modeling approach is appropriate when management goals can be defined in terms that can be parameterized in a population dynamics model. For example, consider the scenario if ESUs were defined to be no more than one disperser per generation between subpopulations. The effect size could then be determined in a modeling exercise with the appropriate population dynamics.

The second approach relies on using empirical

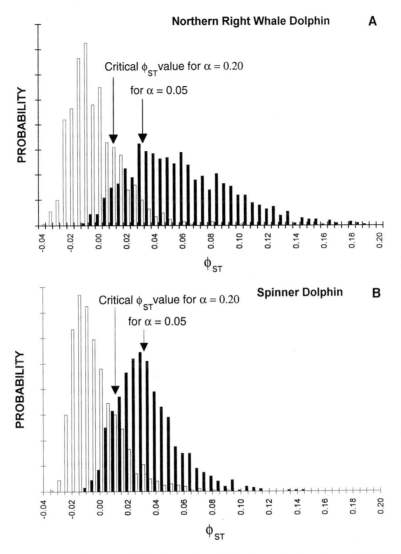

FIGURE 2.—(**A**) An a posteriori power analysis of a genetic analysis of a sample of northern right whale dolphins. Open bars represent the pooled distribution (H_0) and filled bars the subdivided distribution (alternate distribution). Critical Φ_{ST} values ranged between 0.033 and 0.013 for α of 0.05 to 0.20. Power ranged from 0.77 to 0.94, respectively (Table 2). (**B**) An a posteriori power analysis of a genetic analysis of a sample of spinner dolphins. Critical Φ_{ST} values ranged between 0.033 and 0.012 for α of 0.05 to 0.20. Power ranged from 0.52 to 0.91, respectively (Table 2).

data to define the appropriate effect size. One of the criteria a population must satisfy to be considered an ESU (Waples 1991) is that it must represent an important component in the evolutionary legacy of the species. Not only genetic distinctness but evidence of unusual or distinctive adaptations bear on whether this criterion is met. Therefore, for determination of ESU status, it is relevant to examine levels of difference for subpopulations that are considered separate based on nongenetic data. The

smallest, statistically significant ($\alpha = 0.05$) Φ_{ST} values we have observed between pairs of subpopulations (which are defined as subpopulations by nongenetic criteria) are short-beaked common dolphins *Delphinus delphis* in the Black Sea and the eastern, tropical Pacific Ocean ($\Phi_{ST} = 0.15$; Rosel 1992), and the exclusively Washington–central California clade of harbor porpoise *Phocoena phocoena*, which can be geographically separated into northern and southern subpopulations ($\Phi_{ST} = 0.18$; Rosel 1992 and additional unpublished data from our laboratory). The harbor porpoise pair are less obviously separate subpopulations because there are no geographic barriers, but recent contaminant analyses indicate little exchange between the two geographic regions (Calambokidis and Barlow 1991). In a paper that calculates a priori power, Dizon et al. (1994) considered the minimum value expected for Φ_{ST} for similar mtDNA analyses among similar taxa to be about 0.15. Thus, actual populations known to be subdivided from other data sources can be used to establish what level of genetic differences should constitute a "bona fide" subpopulation. Such a comparative approach could be used to "calibrate" future power studies. This empirical calibration technique to establish effect size should be used with caution. Ongoing research has shown that effect size is influenced by abundance; genetic drift in small populations can lead to seemingly greater population distinctness. Researchers should also bear in mind that this technique does not attempt to estimate the degree of mixing between populations. Thus, if a management objective requires detection of relatively high degrees of mixing, such as 1% per year, then using an effect size of Φ_{ST} equal to 0.15 (which implies a low degree of mixing) is likely to be inappropriate.

To summarize, (1) for genetic data to be applied properly to conservation decisions, biologists must routinely report power, and (2) a literature must be developed to ease interpretation of genetic results in terms of significance in both an evolutionary and a management context. In order to calculate power, biologists must estimate the effect size that meets the management objective. Fine-scale genetic tools can readily distinguish individuals; clearly, simple genetic subdivision is not a meaningful criterion on which to solely base the conservation unit. Biologists must, therefore, translate statistics that measure genetic differentiation into terms which allow managers to make decisions based on the probability that management objectives will or will not be met.

References

Calambokidis, J., and J. Barlow. 1991. Chlorinated hydrocarbon concentrations and their use for describing population discreteness in harbor porpoises from Washington, Oregon, and California. Pages 101–110 in J. E. Reynolds, III, and D. K. Odell, editors. Marine mammal strandings in the United States: proceedings of the second marine mammal stranding workshop. NOAA (National Oceanic and Atmospheric Administration) NMFS (National Marine Fisheries Service) Technical Report 98.

Dizon, A. E., C. A. LeDuc, and R. G. LeDuc. 1994. Intraspecific structure of the northern right whale dolphin (Lissodelphis borealis). California Cooperative Oceanic Fisheries Investigations Reports 35:61–67.

Excoffier, L., P. E. Smouse, and J. M. Quattro. 1992. Analysis of molecular variance inferred from metric distances among DNA haplotypes: application to human mitochondrial DNA restriction data. Genetics 131:479–491.

García Rodríguez, A. I. 1994. Population structure of the spinner dolphin (Stenella longirostris) from the eastern tropical Pacific in terms of matriarchal lineages. Master's thesis. Scripps Institution of Oceanography, San Diego, California.

Peterman, R. M. 1990a. Statistical power analysis can improve fisheries research and management. Canadian Journal of Fisheries and Aquatic Sciences 47:2–15.

Peterman, R. M. 1990b. The importance of reporting statistical power: the forest decline and acidic deposition example. Ecology 71:2024–2027.

Rosel, P. E. 1992. Genetic population structure and systematic relationships of some small cetaceans inferred from mitochondrial DNA sequence variation. Doctoral dissertation. University of California, San Diego.

Taylor, B. L., and T. Gerrodette. 1993. The uses of statistical power in conservation biology: the vaquita and northern spotted owl. Conservation Biology 7:489–500.

Waples, R. S. 1991. Definition of "species" under the Endangered Species Act: application to Pacific salmon. NOAA (National Oceanic and Atmospheric Administration) Technical Memorandum NMFS (National Marine Fisheries Service) F/NWC-194, Northwest Fisheries Science Center, Seattle.

Winer, B. J. 1971. Statistical principles in experimental design. McGraw-Hill, New York.

American Fisheries Society Symposium 17:295–303, 1995
© Copyright by the American Fisheries Society 1995

Mixed DNA Fingerprint Analysis Differentiates Sockeye Salmon Populations

Gary H. Thorgaard, Paul Spruell, and Shawn A. Cummings

Department of Zoology and Department of Genetics and Cell Biology
Washington State University, Pullman, Washington 99164, USA

Andrew S. Peek

Department of Genetics and Cell Biology, Washington State University

Ernest L. Brannon

Aquaculture Institute, University of Idaho, Moscow, Idaho 83843, USA

Abstract.—We used DNA fingerprint analysis to assess relationships of populations of sockeye salmon *Oncorhynchus nerka*. These studies were facilitated by using mixed DNA samples; DNA from a number of individuals per population was pooled and composite DNA fingerprint patterns of each population were generated and analyzed. The patterns of relationships determined by analysis with three variable numbers of tandem repeats and two short interspersed nuclear element (SINE) probes were compared; four of the probes generated trees of relationship that were statistically similar to each other based on symmetric-difference distance analysis, and one SINE probe yielded a divergent tree. Our results demonstrate the utility of mixed DNA fingerprint analysis for assessing population relationships.

Deoxyribonucleic acid (DNA) fingerprints are highly variable restriction fragment length polymorphism (RFLP) patterns that typically show individual specificity (Jeffreys et al. 1985). Polymorphisms in these patterns result from variations in the number of tandem repeat sequences in a region (Jeffreys et al. 1985; Nakamura et al. 1987). Fingerprint patterns of DNA have been useful in the analysis of relatedness of individuals (e.g., Kuhnlein et al. 1990; Geyer et al. 1993; Lang et al. 1993). The technique has also been applied to the analysis of population relationships (e.g., Kuhnlein et al. 1989; Gilbert et al. 1990; Wirgin et al. 1991; Ellegren et al. 1992; Baker et al. 1993; Alberte et al. 1994; Moersch and Leibenguth 1994).

In poultry, analysis of fingerprint patterns of mixed DNA samples has facilitated studies of population relationships (Dunnington et al. 1990) and identification of loci linked to single genes that have major phenotypic effects (Dunnington et al. 1992; Plotsky et al. 1993). Mixtures of DNA prepared from particular populations or groups can facilitate the interpretation of band-pattern differences. The technique recently has been used in a study of relationships of three populations of rainbow trout *Oncorhynchus mykiss* (Spruell et al. 1994).

Genetic markers are needed to differentiate populations of salmonid fishes to address a variety of evolutionary, conservation, and management questions (Ryman and Utter 1987; Parker et al. 1990). Protein electrophoresis, the principal technique used for such studies to date, has numerous advantages, including low cost, relative ease of analysis, and a large existing database for comparisons. However, additional markers may be needed to resolve population relationships in some species and situations in which groups are closely related or genetic variability is limited. Multilocus DNA fingerprints may help fill this need.

We used five multilocus DNA fingerprint probes to examine population relationships in 14 western North American populations of sockeye salmon *Oncorhynchus nerka*. Samples of DNA were pooled among individuals within the populations, and mixed DNA fingerprint patterns were analyzed.

Materials and Methods

Blood samples were obtained from 14 populations of sockeye salmon (Figure 1) and DNA was prepared from the blood samples as previously described (Cummings and Thorgaard 1994). Populations of both resident (kokanee) and anadromous individuals were sampled (Table 1). The best information about past introductions suggests that 6 of the 14 populations are predominantly of native origin, 2 are certainly of introduced origin, and many of the kokanee populations have had introductions from common sources (Table 1).

Fingerprint patterns of DNA from mixed DNA samples were prepared as previously described (Spruell et al. 1994). The number of individuals per population in the DNA mixes ranged from 5 to 30 and was 15 or greater for every population except

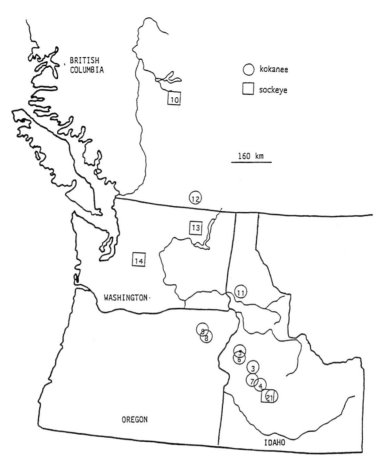

FIGURE 1.—Map showing locations in Washington, Oregon, Idaho, and British Columbia from which sockeye salmon samples were obtained. Kokanee is lacustrine sockeye salmon. Numbers refer to population identifications given in Table 1.

the Redfish Lake sockeye salmon, an endangered species for which few individuals were available (Waples et al. 1991). Briefly, equal quantities of DNA from individuals from each population were combined. Deoxyribonucleic acid fingerprinting involved digestion of DNA with *Hae* III restriction enzyme (New England Biolabs, Beverly, Massachusetts), gel electrophoresis in 1% agarose at 50 V for 24 h, Southern transfer to a nylon membrane (Magnagraph, Micron Separations, Inc., Westboro, Massachusetts), hybridization with an alkaline phosphatase-conjugated oligonucleotide probe (Lightsmith I or II, Promega, Madison, Wisconsin), and detection of band patterns on X-ray film by means of a chemiluminescent reaction. The following oligonucleotide probes were used: 33.6 (×2) (Edman et al. 1988); M13 (×1) (Vassart et al. 1987); CAC (×5) repeat (Schafer et al. 1988); Hpa I 5′,GGCAGGG TAGCCTAGTGGTT (Kido et al. 1991); and Fok I

5′,CGTGTGGCTCAGTTGGTAG (Kido et al. 1991). The nylon membranes were stripped of the probe and reprobed successively with each oligonucleotide probe as described by Spruell et al. (1994). The CAC probe was hybridized and washed in solutions at 45°C as previously described.

Band patterns were analyzed using a customized software program as described by Spruell et al. (1994). The X-ray films were placed on a light box, and visible bands between 3 and 30 kilobase (kb) in size were traced onto a transparent overlay by an individual with no knowledge of which lanes represented which populations. Traced bands were scanned using a Microtek 600ZS scanner. Bands were then grouped into size-classes by the software program (Spruell et al. 1994) and each band-class was scored as being present or absent in each population. Greater similarity of band pattern of populations in adjacent lanes is a potential concern with

TABLE 1.—Sampling information for individuals used in DNA mixes for DNA fingerprint analysis of sockeye salmon populations.

Sample location	Sample abbreviation and number	Migratory tendency[a]	Number of individuals in DNA mix (year sampled)	Probable ancestry[b]
Idaho				
Redfish Lake	RK, 1	K	19 (1990)	N[c]
	RS, 2	S	5 (1991)	N[c]
Warm Lake	WA, 3	K	16 (1992)	U[d]
Stanley Lake	ST, 4	K	20 (1992)	U[e]
Payette Lake				
Lower	PL, 5	K	29 (1992)	U[d]
Upper	PU, 6	K	30 (1992)	U[d]
Deadwood Reservoir	DE, 7	K	30 (1992)	I[d]
Dworshak Reservoir	DW, 11	K	18 (1990)	I[f]
Oregon				
Wallowa Lake				
Lower	WL, 8	K	30 (1992)	U[d]
Upper	WU, 9	K	28 (1992)	U[d]
British Columbia				
Horsefly River	HO, 10	S	15 (1991)	N
Peachland Creek	PE, 12	K	18 (1990)	N
Washington				
Okanogan River	OK, 13	S	18 (1991)	N
Wenatchee Lake	WE, 14	S	20 (1992)	N

[a]Individuals that spend their entire life in freshwater are classified as kokanee (K), and those that migrate to the ocean for a portion of their life cycle are classified as sockeye salmon (S).

[b]Groups are classified as native (N; probably representative of the original stock indigenous to the area), introduced (I; no sockeye salmon were originally found in the area), or of uncertain origin (U; sockeye salmon were native to the area but substantial introductions may have affected the original population).

[c]Redfish Lake, Idaho, has had a number of outside introductions of sockeye salmon, but protein electrophoresis results (Waples et al. 1991; G. Winans, National Marine Fisheries Service, personal communication) indicate that both the kokanee and anadromous sockeye salmon there are likely of native origin.

[d]State agency records for these lakes indicate substantial past introductions, including some that trace back to a common kokanee stock originally native to Lake Whatcom, Washington.

[e]The Idaho Department of Fish and Game introduced sockeye salmon from Babine Lake, British Columbia, into Stanley Lake in the early 1980s.

[f] This reservoir received substantial introductions of kokanee from Grandby Lake, Colorado, in the 1970s.

this approach if mobilities of the same band differ regionally in the gel. However, we are confident that band comparisons across the blot were appropriate in this study because marker bands that appeared present in all populations were scored as such in the output matrix and because populations in adjacent lanes sometimes proved to be very divergent in our subsequent analyses.

Band-sharing values and distance values were calculated among the populations for all five probes. Band sharing is calculated as twice the number of bands shared between populations divided by the sum of the number of bands in each ($2N_{ab}/([N_a + N_b])$; Wetton et al. 1987). Distance equals 1 minus the band-sharing value and thus ranges from a minimum of 0 to a maximum of 1. This statistic is the same as the "average percent difference" of Gilbert et al. (1990) expressed as a decimal.

Binary, unrooted dendrograms of relationship among the 14 populations were prepared based on

the distance values for each of the five probes by unweighted pair-group method using arithmetic averages (UPGMA) cluster analysis by use of version 3.4 of the PHYLIP program[1]. The branching patterns of the five dendrograms and of a dendrogram of relationships based on protein electrophoretic results for eight of the populations (Waples et al. 1991; Winans, National Marine Fisheries Service, personal communication) were compared using the symmetric-difference distance (SDD) metric of Penny et al. (1982). This method quantifies tree differences by comparing the branching pattern of trees and allows calculation of the probability that two trees, selected at random from all possible unrooted trees, would exhibit that degree of difference. The PHYLIP tree-file output of UPGMA

[1]Felsenstein, J. 1992. PHYLIP (Phylogeny inference package) Version 3.4. Department of Genetics, SK-50, University of Washington, Seattle, 98195, USA.

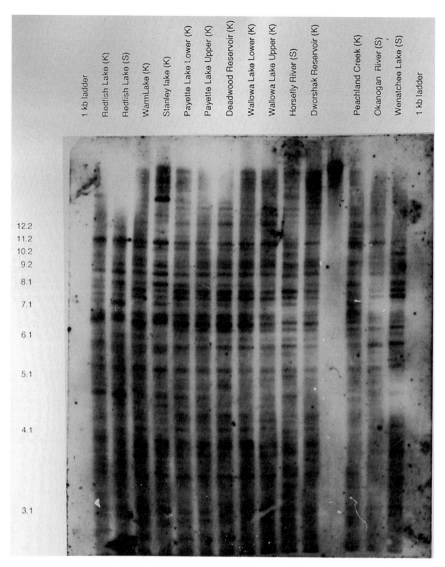

FIGURE 2.—Sockeye salmon population fingerprint patterns of mixed DNA samples hybridized with M13 probe. DNA fragment size in kilobase (kb) pairs is given at left. Populations identified in Figure 1 and Table 1.

clustering was converted into NEXUS format, and SDDs were calculated using version 3.0 of the PAUP program[2]. The SDD metric was analyzed for binary, unrooted trees of 14 taxa that yielded 11 significant tree nodes to compare among trees for the five DNA probes. Trees based on the protein electrophoretic analysis (Waples et al. 1991) of eight taxa yielded five significant tree nodes to com-

pare with trees based on the DNA probe data. The probability distribution of similarities between trees with 11 significant nodes is provided by Hendy et al. (1984) and that for trees with 5 significant nodes is provided by Penny et al. (1982).

Results

High-resolution DNA fingerprint patterns were obtained using all five probes tested (see Figures 2 and 3). The number of bands detected with each probe ranged from 11 to 46 per population (Table 2). Previous results have indicated that most

[2]Swofford, D. 1991. PAUP (Phylogenetic analysis using parsimony) Version 3.0. Illinois Natural History Survey, Champaign.

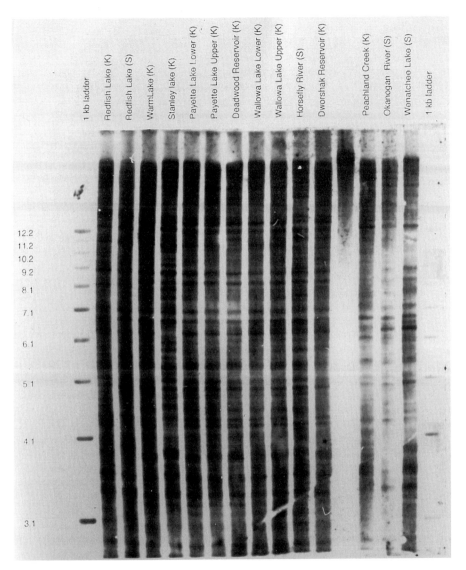

FIGURE 3.—Sockeye salmon population fingerprint patterns of mixed DNA samples hybridized with Fok I 5′ probe. The DNA fragment size in kb pairs is given at left. Populations identified in Figure 1 and Table 1.

bands are not detected in mixed samples unless half or more of the individuals share the band (Spruell et al. 1994). Thus, comparisons of bands among populations should be indicative of band frequency differences among individuals in the populations. The number of population-specific bands was highest (20) for the CAC probe and lowest (1) for the Hpa I 5′ probe (Table 2). Distance values varied considerably in pairwise comparisons of the populations, with the following ranges for each probe: 33.6, 0.16–0.75; M13, 0.21–0.83; CAC, 0.21–0.90; Hpa I 5′, 0.14–0.63; Fok I 5′, 0.21–0.84.

Trees of relationship among the populations were prepared for the five probes based on the distance values and the UPGMA method (Figure 4). Certain clusterings were apparent in many of the trees: the kokanee and sockeye salmon from Redfish Lake, Idaho, clustered in three of five trees; the kokanee from Peachland Creek, British Columbia, and the sockeye salmon from the nearby Okanogan River, Washington, clustered in four of five trees; and a number of the kokanee populations with histories of introductions from related sources (populations 5–9) clustered together in most of the trees.

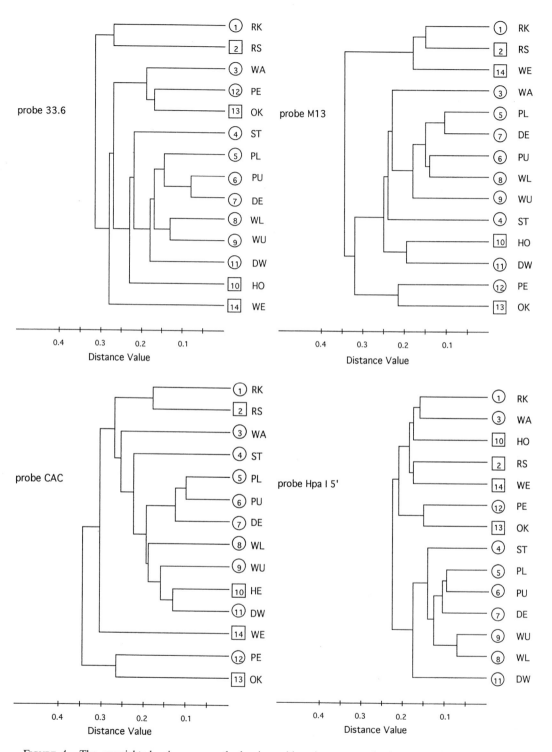

FIGURE 4.—The unweighted pair-group method using arithmetic averages dendrograms of 14 sockeye salmon populations based on analysis of distance values of mixed DNA fingerprint patterns for the probes indicated. Population designations are as in Table 1; population numbers in squares represent anadromous populations.

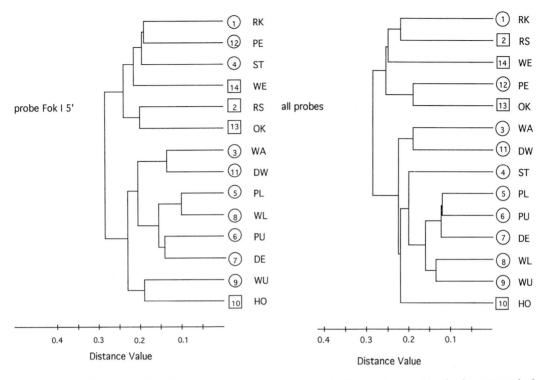

probe Fok I 5'

all probes

0.4   0.3   0.2   0.1

Distance Value

0.4   0.3   0.2   0.1

Distance Value

FIGURE 4.—Extended.

FIGURE 5.—Composite unweighted pair-group method using arithmetic averages cluster tree of 14 sockeye salmon populations based on analysis of distance values of band patterns detected using all five probes. Population designations are as in Table 1; population numbers in squares represent anadromous populations.

TABLE 2.—Number of DNA fingerprint bands and population-specific bands (in parentheses) identified in 14 sockeye salmon populations (see Table 1 for explanation of abbreviations). The mean number of bands detected per population (mean) and total number of population-specific bands (total) detected for each probe are also provided.

Population and summary statistic	DNA probe				
	33.6	M13	CAC	Hpa I 5'	Fok I 5'
RK, 1	20 (1)	24	35 (5)	38	29 (1)
RS, 2	24	16 (1)	25 (1)	37	26 (2)
WA, 3	26	24	26 (1)	33	26
ST, 4	38	28 (2)	39 (6)	46	33 (1)
PL, 5	39 (1)	28	33	38	25
PU, 6	38	27 (1)	28	34	26 (1)
DE, 7	38	28	32 (1)	38	30
WL, 8	34	23 (1)	27	31	28 (1)
WU, 9	31	19 (1)	27 (1)	32	27
HO, 10	35 (1)	23 (1)	33 (1)	40	34 (2)
DW, 11	33 (1)	26 (1)	29	31	29
PE, 12	27	16	23 (3)	32	26
OK, 13	28	12	11	34	29 (1)
WE, 14	38 (2)	17 (3)	15 (1)	43 (1)	31 (2)
Mean	32	22	27	36	29
Total	6	11	20	1	11

The SDD metric of Penny et al. (1982) was used to compare the trees of the 14 sockeye salmon populations obtained using the five probes (Table 3). Pairwise comparisons of the trees generated using the different probes demonstrated that all probes except Fok I 5' generated trees that were more similar than would be predicted due to chance. The tree prepared based on distance values for the Fok I 5' probe (Figure 4E), however, had a markedly divergent pattern from that for the other four probes. A composite tree combining data from the five probes was constructed and had the major features discussed above (Figure 5).

Comparison of trees for the eight populations common to our study and a protein electrophoretic study (Waples et al. 1991; Winans, personal communication) showed that the trees based on DNA fingerprint data were generally divergent from that based on the protein data. The tree for only one of the five probes (Hpa I 5') was marginally more similar to the protein tree (SDD = 6, $P = 0.045$)

TABLE 3.—Symetric-difference distances and random tree probabilities (in parentheses) of trees for 14 sockeye salmon populations based on analysis with five different DNA fingerprint probes. With 14 populations, the symetric-difference distances range from a minimum of 0 to a maximum of 22.

DNA probe	DNA probe			
	M13	CAC	Hpa I 5′	Fok I 5′
33.6	14 (9×10^{-5})	14 (9×10^{-5})	12 (7×10^{-6})	20 (0.155)
M13		14 (9×10^{-5})	18 (1.5×10^{-2})	20 (0.155)
CAC			16 (1.2×10^{-3})	22 (1)
Hpa I 5′				22 (1)

than would be predicted due to chance, whereas the other four were no more similar than would be predicted due to chance ($P > 0.05$).

Discussion

Notable differences were detected among DNA fingerprint patterns of the mixed DNA samples from the 14 sockeye salmon populations. The trees we generated based on the distance values reflect the zoogeography and stocking histories of the populations. Kokanee and anadromous sockeye salmon within a region (Redfish Lake and Okanogan region) clustered as closely related. This finding is consistent with previous studies of genetic relationship among such populations (Foote et al. 1989). A number of kokanee populations with histories of introductions from common sources also clustered together. The general concordance of the results based on four of the five probes as confirmed by the SDD metric increases our confidence in the approach.

The divergence of the Fok I 5′ tree from the trees generated using the other probes was interesting. The Fok I sequence we used is based on a retroposon first identified by Kido et al. (1991) in salmonids, which they reported to be specific to the genus *Salvelinus*. One possible explanation for the surprising divergence of the Fok I tree is that the Fok I retroposon sequences may be continuing to move within the genomes of at least some sockeye salmon populations.

Fingerprint analysis of mixed DNA samples appears to have considerable potential value for assessing population relationships. A recent study comparing the analysis of DNA mixtures and DNA from individuals of the same three rainbow trout populations found the mixtures to be more powerful in discriminating these populations (Spruell et al. 1994). The obvious limitation of the approach is the inconsistencies that will result because scoring bands as present or absent at the population level loses some information by simplifying a more complex situation. However, as more loci are scored this limitation should be less of a problem, and the merits of the approach may outweigh the limitation. The method detects substantial differences among populations and is more likely to detect divergence than is protein electrophoresis because of the large number of loci detected and their high levels of variability.

The use of mixed samples greatly facilitated this study. Many times more effort would have needed to be expended to analyze individual DNA samples instead of DNA mixtures. A study comparing large numbers of populations might not be feasible if based solely on individual analyses, and difficulties could also be encountered in making quantitative comparisons between independent blots. Mixing blood rather than extracted DNA (Dunnington et al. 1992), might further facilitate the use of mixed samples and make comparing several populations not appreciably more difficult than comparing the same number of individuals.

An additional benefit of DNA mixing approaches combined with DNA fingerprint analysis is that DNA mixing may facilitate the isolation of markers diagnostic for particular populations or groups of populations. A number of population-specific bands (Table 2) and group-specific bands were identified in this study. Targeted cloning of markers that show population or group associations from size-selected genomic libraries could be a general approach for isolating diagnostic probes. These simple, single-locus probes could then be used to document allele differences among individuals and populations to address evolutionary, conservation, or management problems.

Acknowledgments

This study was supported by funds from Bonneville Power Administration Project 90–93 and Washington Sea Grant Project NA 89 AA/D/SG022 from the National Oceanic and Atmospheric Administration, Project R/B-4. We thank Ann Setter and Keya Collins for their assistance in sample collection and handling, and the Oregon Department of Fish and Wildlife, National Marine Fisheries Service, Byril Kurtz of Fisheries and Oceans Canada, and Brian Janse of the British Columbia Ministry of the Environment for facilitating sample collections.

References

Alberte, R. S., G. K. Suba, G. Procaccini, R. C. Zimmerman, and S. R. Fain. 1994. Assessment of genetic diversity of seagrass populations using DNA fingerprinting: implications for population stability and management. Proceedings of the National Academy of Sciences of the United States of America 91:1049–1053.

Baker, C. S., and seven coauthors. 1993. Population characteristics of DNA fingerprints in humpback whales (*Megaptera novaeangliae*). Journal of Heredity 84:281–290.

Cummings, S. A., and G. H. Thorgaard. 1994. Extraction of DNA from fish blood and sperm. BioTechniques 17:426–430.

Dunnington, E. A., and nine coauthors. 1990. DNA fingerprints of chickens selected for high and low body weight for 31 generations. Animal Genetics 21:247–257.

Dunnington, E. A., A. Haberfeld, L. C. Stallard, P. B. Siegel, and J. Hillel. 1992. Deoxyribonucleic acid fingerprint bands linked to loci coding for quantitative traits in chickens. Poultry Science 71:1251–1258.

Edman, J. C., M. E. Evans-Holm, J. E. Marich, and J. L. Ruth. 1988. Rapid DNA fingerprinting using alkaline phosphatase-conjugated oligonucleotides. Nucleic Acids Research 16:6235.

Ellegren, H., L. Andersson, M. Johansson, and K. Sandberg. 1992. DNA fingerprinting in horses using a simple $(TG)_n$ probe and its application to population comparisons. Animal Genetics 23:1–9.

Foote, C. J., C. C. Wood, and R. E. Withler. 1989. Biochemical genetic comparison of sockeye salmon and kokanee, the anadromous and nonanadromous forms of *Oncorhynchus nerka*. Canadian Journal of Fisheries and Aquatic Sciences 46:149–158.

Geyer, C. J., O. A. Ryder, L. G. Chemnick, and E. A. Thompson. 1993. Analysis of relatedness in the California condors from DNA fingerprints. Molecular Biology and Evolution 10:571–589.

Gilbert, D. A., N. Lehman, S. J. O'Brien, and R. K. Wayne. 1990. Genetic fingerprinting reflects population differentiation in the California Channel Island fox. Nature 344:764–766.

Hendy, M. D., C. H. C. Little, and D. Penny. 1984. Comparing trees with pendant vertices labelled. Society for Industrial and Applied Mathematics Journal on Applied Mathematics 44:1054–1065, Philadelphia.

Jeffreys, A. J., V. Wilson, and S. L. Thein. 1985. Hypervariable "minisatellite" regions in human DNA. Nature 314:67–73.

Kido, Y., and six coauthors. 1991. Shaping and reshaping of salmonid genomes by amplification of tRNA-derived retroposons during evolution. Proceedings of the National Academy of Sciences of the United States of America 88:2326–2330.

Kuhnlein U., Y. Dawe, D. Zadworny, and J. S. Gavora. 1989. DNA fingerprinting: a tool for determining genetic distances between strains of poultry. Theoretical and Applied Genetics 77:669–672.

Kuhnlein, U., D. Zadworny, Y. Dawe, R. W. Fairfull, and J. S. Gavora. 1990. Assessment of inbreeding by DNA fingerprinting: development of a calibration curve using defined strains of chickens. Genetics 125:161–165.

Lang, J. W., R. K. Aggarwal, K. C. Majumdar, and L. Singh. 1993. Individualization and estimation of relatedness in crocodilians by DNA fingerprinting with a Bkm-derived probe. Molecular & General Genetics 238:49–58.

Moersch, G., and F. Leibenguth. 1994. DNA fingerprinting in roe deer using the digoxigenated probe (GTG). Animal Genetics 25:25–30.

Nakamura, Y., and ten coauthors. 1987. Variable number of tandem repeat markers for human gene mapping. Science 235:1616–1622.

Parker, N. C., A. E. Giorgi, R. C. Heidinger, D. B. Jester, Jr., E. D. Prince, and G. A. Winans. 1990. Fish-marking techniques. American Fisheries Society Symposium 7.

Penny, D., L. R. Foulds, and M. D. Hendy. 1982. Testing the theory of evolution by comparing phylogenetic trees constructed from five different protein sequences. Nature 297:197–200.

Plotsky, Y., A. Cahaner, A. Haberfeld, U. Lavi, S. J. Lamont, and J. Hillel. 1993. DNA fingerprint bands applied to linkage analysis with quantitative trait loci in chickens. Animal Genetics 24:105–110.

Ryman, N., and F. Utter, editors. 1987. Population genetics and fishery management. University of Washington Press, Seattle.

Schafer, R. H. Zischler, and J. T. Epplen. 1988. $(CAC)_5$, a very informative probe for DNA fingerprinting. Nucleic Acids Research 16:5196.

Spruell, P., S. A. Cummings, Y. Kim, and G. H. Thorgaard. 1994. Comparison of three anadromous rainbow trout (*Oncorhynchus mykiss*) populations using DNA fingerprinting and mixed DNA samples. Canadian Journal of Fisheries and Aquatic Sciences 51 (Supplement 1):252–257.

Vassart, G., M. Georges, R. Monsieur, H. Brocas, A. S. Lequarre, and D. Christophe. 1987. A sequence in M13 phage detects hypervariable minisatellites in human and animal DNA. Science 235:683–684.

Waples, R. S., O. W. Johnson, and R. P. Jones, Jr. 1991. Status review for Snake River sockeye salmon. NOAA (National Oceanic and Atmospheric Administration) Technical Memorandum NMFS (National Marine Fisheries Service) F/NWC-195, Northwest Fisheries Science Center, Seattle.

Wetton, J. H., R. E. Carter, D. T. Parken, and D. Walters. 1987. Demographic study of a wild house sparrow population by DNA fingerprinting. Nature 327:147–149.

Wirgin, I. I., C. Grunwald, and S. J. Garte. 1991. Use of DNA fingerprinting in the identification and management of a striped bass population in the southeastern United States. Transactions of the American Fisheries Society 120:273–282.

American Fisheries Society Symposium 17:304–326, 1995

A Quantitative Genetic Perspective on the Conservation of Intraspecific Diversity

JEFFREY J. HARD

National Marine Fisheries Service, Northwest Fisheries Science Center
Coastal Zone and Estuarine Studies Division, 2725 Montlake Boulevard East
Seattle, Washington 98112, USA

Abstract.—Identifying appropriate units for the conservation of intraspecific variation requires consideration of the genetic basis of this variation. Most of the information available on the genetic basis of intraspecific variation has come from variation in molecular (primarily single-locus) traits. However, the genome is also composed of polygenic traits (e.g., life history characters) that, unlike these molecular traits, exhibit quantitative phenotypes that are often highly sensitive to natural selection. Genetic and phenotypic variation in quantitative traits determines the evolutionary significance of adaptive differences among units of intraspecific variation, and the maintenance of genetic and ecological diversity within a species depends on recognition and preservation of these differences. Although variation in molecular characters can be useful in identifying reproductively isolated units, it is of limited use in assessing the evolutionary significance of these units. The correspondence between variation in molecular markers and variation in quantitative traits is poorly understood. This paper summarizes evidence for deficiencies in an approach to conservation genetics that relies entirely on the interpretation of molecular variation to identify units of intraspecific variation. The major problem with a strictly molecular approach is that conservation units may be identified without seriously addressing whether these units represent adaptive variation. Conservation genetics and quantitative genetics have developed largely independently of one another, a situation that I believe has led to misrepresentations of the power of molecular genetic interpretations and has fostered a false sense of security in efforts to conserve genetic resources. Conserved populations are arguably the least appropriate populations in which to neglect adaptive variation. The uncertain relationship between molecular genetic variability and quantitative genetic and phenotypic variability suggests that conservation which relies on molecular approaches alone may neither identify appropriate conservation units nor minimize genetic risks to them. Because it provides the link between genetic and phenotypic variability and adaptation, assessment of quantitative genetic variation in life history, physiological, and behavioral traits should be integrated where possible in conservation strategies.

Molecular biology and evolutionary biology are in constant danger of diverging totally, both in the problems with which they are concerned . . . and as scientific communities ignorant and disdainful of each other's methods and concepts. (From R. C. Lewontin 1991)

Molecular techniques do not survey most loci that are adaptively important. . . . We caution that molecular genetic data generally give a picture of only a small part of the genetic variation, and in fact may not always be a good indicator of adaptive genetic differences. (From P. W. Hedrick and P. S. Miller 1992)

Identification and explanation of the processes that influence patterns of genetic variability in natural populations are fundamental problems in evolution and conservation. Conservation genetics has grown rapidly in the last two decades, but its development has been largely independent of relevant progress in evolutionary genetics. The purpose of this paper is to examine the practice of conservation genetics, as it has been applied to identifying intraspecific units of genetic and ecological diversity, and to evaluate its prospects for preserving this diversity from the perspective of evolutionary quantitative genetics. In addition to relying heavily on genetic variability in presumably neutral molecular markers as an indicator of overall genomic variability within populations and species (Cheverud et al. 1994), conservation geneticists have depended on the distribution of these markers to identify intraspecific units for conservation. Less attention has been paid to the genetic basis of phenotypic polymorphisms. The majority of efforts to characterize patterns of genetic variability are deficient because they do not yet routinely include direct methods to assess adaptive differentiation.

The typical assertion that the maintenance of genetic variability within and among natural populations constitutes a species' foremost evolutionary defense against extinction has almost assumed the stature of an axiom in conservation genetics. Although the general validity of this assertion is seldom questioned, nature offers several examples of populations lacking detectable molecular genetic variation (e.g., Bonnell and Selander 1974; O'Brien et al. 1987; Waller et al. 1987; Lesica et al. 1988;

Soltis et al. 1992; Hoelzel et al. 1993). The evolutionary fates of populations and species like these are clearly of considerable interest to conservation geneticists. What, if anything, distinguishes these populations from those with large amounts of detectable genetic variation? For example, some of these species, like elephant seals *Mirounga* spp., are currently thriving. One possibility is that molecular variation is a flawed surrogate for genetic variation in adaptive potential. It is worth noting that nearly all the empirical papers on conservation genetics published in *Conservation Biology* since that journal's inception in 1987 have dealt with the evaluation of genetic variability through the analysis of protein-coding loci detected by allozyme electrophoresis or DNA fingerprinting. As illustrated by cases like those cited above, however, whether the consequences of such variability extend to the viability of natural populations in the short term is unclear. As Caughley (1994) has argued for the study of demography in conservation biology, additional or alternative methods in conservation genetics may shed light on the limitations of more commonly used methods and promote more powerful multidisciplinary approaches.

This paper addresses two questions that are relevant to the effective conservation of genetic diversity within and among conspecific populations. (1) How reliably do molecular measures of genetic variability reflect genetic variation for traits that affect fitness and adaptation? (2) How well do these molecular measures represent the distribution of adaptive variation among units of diversity? After describing the essential properties of an intraspecific conservation unit, I examine the evidence for concordance of molecular and quantitative genetic measures of genetic variability. I then discuss mechanisms of adaptation and argue that only one of these, genetic polymorphism, is addressed by molecular approaches. Finally, I argue that the integration of molecular and quantitative genetic approaches is essential to effective conservation of genetic resources, and I describe several empirical examples to support a case for incorporating quantitative genetic approaches into conservation biology.

The Structure of Intraspecific Diversity

Effective conservation of genetic diversity demands attention to biological units smaller than taxonomic species (Antonovics 1991). Genetically based phenotypic variants often characterize different conspecific populations and are often unique to particular environments. The maintenance of intraspecific variants requires that conservation be active on two fronts: among-population and within-population diversity. Maintaining genetic variation within populations is in principle a straightforward process, and its theoretical framework is well developed, even for quantitative traits (Lande and Barrowclough 1987). Knowledge of the genetic variation among conspecific populations is necessary to define the focus of conservation efforts. Within a species, a conservation unit may be a population or group of conspecific populations (in this paper, unit and population[s] are used synonymously). An appropriate unit has at least two desirable properties. First, it should be discrete, a quality that implies that intraspecific variation exists and is persistent, at least over short-term evolutionary time. This property acknowledges that intraspecific variation generally has some genetic basis, and it therefore reflects the reliability with which conservation units can be identified. Second, an appropriate unit should be distinctive, a quality that implies that a detectable unit is, to some extent, a novel component of the phenotypic variation expressed in the species as a whole. This property acknowledges the ecological and evolutionary significance of intraspecific variation and constitutes the contextual framework for assessing this variation.

The properties of discreteness and distinctiveness suggest two hypotheses for the maintenance of intraspecific variation among conservation units: (1) genetically based differentiation among units results from some degree of genetic cohesiveness within them, and (2) observed differences among units result from evolutionary responses to environmental differences experienced by them (i.e., local adaptation to different regimes of natural selection), as well as from restricted gene flow. These ideas are not new; the conceptual foundation for distinguishing between conspecific populations was established by Mayr (1963), Ehrlich and Raven (1969), and many others (see Endler 1977 and references therein).

More recently, Ryder's (1986) concept of the evolutionarily significant unit (ESU) was developed by Waples (1991; see also Waples 1995, this volume) as a means of interpreting the meaning of the term "distinct population segment," as used in the U.S. Endangered Species Act of 1973 (ESA; 16 U.S.C. §§ 1531 to 1544) to define intraspecific units that qualify for federal protection under the ESA. The properties of discreteness and distinctiveness used here to define a conservation unit are related to the ESU's two fundamental attributes. That is, a discrete and distinctive unit "is substantially reproductively isolated from other conspecific population units" and "represents an important component in the evolutionary legacy of the species" (Waples 1991). However, discreteness and distinctiveness amplify the ESU's attributes in a subtle but important way: as used here, the terms emphasize the central role of phenotypic variation in adaptive evolution. Evolution may proceed by changes in gene frequencies, but natural selection acts on phenotypes, which result from a complex interaction of heritable and environmental sources of variation (Barker and Thomas 1987).

Furthermore, the distinctiveness criterion described here provides a means of evaluating an ESU's evolutionary significance. Waples (1991) outlined several factors (including biogeographic, phenotypic, behavioral, and life history characteristics and habitat use) that can be used to assess significance, but he noted that ignorance of the genetic variation which underlies these factors makes it difficult to interpret this information reliably. Determining the genetic basis of phenotypic variation provides a direct means of assessing whether an ESU represents a novel component of intraspecific variation and, consequently, is significant in the evolution of the species as a whole. The interpretation of molecular genetic variation alone to assess an ESU's evolutionary significance employs indirect methods whose validity is questionable (see below).

Approaches to the Genetic Conservation of Intraspecific Diversity

The premise that a population's ability to cope with environmental variation and avoid extinction is proportional to its genetic variability is widely accepted (e.g., Soulé 1987; Lande 1988a; Lynch and Lande 1993). The assumption implicit in most surveys of molecular genetic variability is that the level of this variation indicates, at least roughly, the level of adaptive genetic variation. Given my concern that few relevant empirical data may be available to support this assumption, it is appropriate to ask whether this is a reasonable practice. If molecular genetic variability is only a weak correlate of adaptive potential, then current reliance on molecular techniques to minimize genetic risks to conserved populations should be reconsidered unless there are no feasible alternatives.

Many conservation geneticists appear to make an additional assumption in evaluating the distribution of genetic variation among conspecific populations. In interpreting patterns of genetic structure, geneticists using molecular techniques generally assume that these patterns or changes in them are largely a consequence of gene flow and genetic drift. Again, a potential problem with exclusively molecular approaches is that adaptive processes are addressed only indirectly by relying entirely on the interpretation of molecular data. A more direct approach would involve quantitative genetics, the study of traits that typically do not exhibit discrete phenotypes but instead a more or less continuous phenotypic range. By contrast with molecular genetic approaches, which concentrate for the most part on single-locus variation, quantitative genetics focuses on polygenic traits that are controlled by variation at multiple loci.

Quantitative genetics has often been viewed as being useful primarily for achieving specific phenotypic goals from the selective breeding of domesticated plants and animals (e.g., Lush 1945), but the use of quantitative genetic methods to understand the genetic processes underlying adaptive evolution has become widespread in the last 20 years (e.g., Lande 1976, 1980, 1982; Turelli 1984; Lynch and Hill 1986; Clark 1987; Charlesworth 1990). The hallmark of quantitative genetic analyses lies in their focus on the composite statistical properties of the genes underlying phenotypic variation. From the phenotypic resemblance of individuals of known average relationship, these analyses permit inferences about the inheritance and evolution of these characters, processes that are controlled by hidden variation in gene frequences and effects (Falconer 1989). In this way, observed patterns of means, variances, and covariances in a population, when combined with appropriate breeding designs and statistical techniques, can be used to estimate the genetic parameters that determine the population's response to selection. Because quantitative genetic analyses are designed to detect composite genetic and environmental effects on the phenotype rather than individual gene effects, these analyses are more synoptic and less mechanistic than molecular genetic analyses. Despite the inability to identify individual genes, quantitative genetic approaches are well suited for investigating patterns of adaptive evolution and comparing them in populations existing over environmental or geographic gradients (Lande 1988b). Quantitative genetic approaches differ from molecular genetic approaches in that they are prospective rather than retrospective (Ewens 1979): quantitative genetics deals primarily with predicting the potential evolutionary consequences of particular genetic states rather than determining the genetic states that have resulted from past evolution. Quantitative genetics therefore lends itself well to generating hypotheses about what the consequences of genetic change, at least in the short term, will be. These consequences are relevant to the conservation of genetic diversity.

Despite the remarkable advances in molecular genetics in the last decade, evolutionary quantitative genetics remains the tool of choice for analyzing variation in quantitative traits, many of which have important consequences for fitness (Barker and Thomas 1987; Lande 1988b). Topics currently addressed in evolutionary quantitative genetics include the inheritance and genetic basis of polygenic traits, the components of phenotypic variation, factors affecting genetic variation, the genetic basis of fitness variation, estimation of the intensity of natural selection and the response to selection, and mechanisms of population divergence and local adaptation (see Barton and Turelli 1989).

It is not yet clear whether the current techniques commonly used by empirical conservation geneticists are able to support many of the interpretations and recommendations that are being made to evaluate and manage genetic risk in conserved populations. One way this issue has been approached in the past is simply to examine the correlation between molecular genetic variation, based primarily on samples of single loci, and average fitness (e.g., Mitton and Grant 1984). This approach raises a serious concern. It assumes a direct correspondence between molecular and quantitative genetic variation and ignores variation in fitness. Because this approach inappropriately simplifies the presumed genetic basis for fitness variation, it can lead to erroneous conclusions about genetic risks to conserved populations.

Based on current knowledge of the genetics of adaptation (reviewed by Barker and Thomas 1987; Barton and Turelli 1989; Orr and Coyne 1992), the correspondence between molecular and quantitative genetic variation would appear to be tenuous. Lande and Barrowclough (1987) argued on theoretical grounds that this correspondence would be weak. Consider the following points that distinguish polygenic from single-locus variation.

1. Traits that contribute to fitness are typically polygenic. Estimates of the number of genes may be in the hundreds or more for some of these traits (Wright 1968; Lande 1981). However, the possibility that many adaptations involve a few genes of large effect cannot be dismissed on the basis of available evidence (Lande 1983; Orr and Coyne 1992).

2. A hallmark of phenotypic expression in polygenic traits that affect fitness is the dependence of the phenotype on nongenetic variation (Falconer 1989).

3. Even in the absence of environmental factors, gene expression for polygenic traits may take a variety of forms: additivity (independence of effects of loci and alleles within loci), dominance (interaction among alleles within loci), and epistasis (interaction among loci). In addition, these loci may exhibit linkage (nonindependent inheritance) (Wright 1978; Goodnight 1988; Gimelfarb 1989).

The comparison of molecular and quantitative genetic polymorphisms points to other fundamental differences in their expression as well. First, the nature of quantitative genetic variation is that a distribution of genic effects on the phenotype exists; rather than a single locus with a predictable effect, several to many loci with large-to-minute effects may exist. Most quantitative genetic models of evolution assume that the number of constituent loci is infinite (and, consequently, that the effects of these loci are equivalent and small; e.g., Fisher 1958; Bulmer 1971; Lande 1976; but see Turelli 1984). Regardless of the distribution of genic effects, however, the phenotypic distribution results from two main factors: simultaneous segregation at all constituent loci and differences in environmental sensitivity among different genotypes. The expression of quantitative traits, especially those that affect viability and reproductive success, is generally quite sensitive to the environment. This sensitivity often varies among different genotypes, sometimes in direction as well as magnitude (Falconer 1989).

One consequence of an environmental contribution to observed variation is that phenotypic variation in adaptive traits may exist without detectable genetic variation (e.g., Waller et al. 1987; Lesica et al. 1988; Soltis et al. 1992). In addition, genotypic sensitivity to environmental variation can complicate the relationship between genetic and phenotypic variation (Comstock and Moll 1963; Gupta and Lewontin 1982; Via 1993).

Another feature of polygenic variation in traits that affect fitness is that the constituent genes often affect multiple traits, a phenomenon referred to as pleiotropy (Williams 1966; Rose 1982). The pleiotropic effects of these genes can, in combination with different modes of gene expression, cause the genetic "architecture" of adaptation to be extraordinarily complex (Hegmann and Dingle 1982; Lande and Arnold 1983).

A main corollary of the difference between molecular and quantitative genetic polymorphisms is their response to selection. With some notable exceptions, many molecular polymorphisms appear to be effectively neutral with respect to selection (Dykhuizen and Hartl 1980; Turelli and Ginzburg 1983; Yamazaki et al. 1983; Eanes 1987; but see Zouros and Pogson 1994). Most polygenic traits, on the other hand, respond to selection (Wright 1968, 1969, 1977, 1978; Falconer 1989)—often quite rapidly—and some phenotypic changes in polygenic traits under selection can far exceed many changes observed in the fossil record (Kurtén 1959; Gingerich 1983). Consequently, molecular polymorphisms may be poor indicators of adaptive potential.

Molecular and quantitative genetic polymorphisms also appear to differ with respect to the generation of novel variation. The appearance, accumulation, and fixation or loss of mutations in the genome can have profound consequences for the evolution and fate of populations, but these issues are only now being appreciated (Lande 1983, 1994; Lynch et al., in press). For example, available evidence indicates that the effective mutation rate per generation differs substantially between single-locus and polygenic traits. Mutation rates appear to be 10^2 to 10^4 times higher for polygenic traits than they are for single-locus traits; estimates of the "mutational heritability" per generation for polygenic traits range from 10^{-3} to 10^{-2}, whereas corresponding estimates of the single-locus mutation rate are approximately 10^{-7} to 10^{-5} (Dobzhansky 1970; Lande 1976; Lynch 1988; Keightley and Hill 1992). The higher mutation rates for polygenic traits potentially contribute to increases in genetic variation that would not be detectable with molecular methods.

Based on these considerations, there is little reason to expect, a priori, close concordance of the evolutionary behavior of molecular and quantitative genetic polymorphisms. In the next section, I examine evidence for the correlation between single-locus molecular and quantitative genetic variation and their relationships to phenotypic variation.

Phenotypic Patterns of Molecular and Quantitative Genetic Variation

Within-Population Variation

A common procedure used to justify the use of molecular genetic techniques as tools in conservation genetics is to examine the correlation between molecular genetic variation and average phenotypes. One estimate of molecular genetic variation is average heterozygosity, often designated H (e.g., Hartl and Clark 1989). Other, more sensitive measures of changes in molecular variation exist, but H is the most widely used. Many investigators have found positive correlations between allozyme heterozygosity and morphological phenotypes affecting fitness (reviews by Mitton and Grant 1984; Mitton 1993). The mechanism usually cited for this relationship is heterozygote advantage or functional overdominance (Houle 1989). However, in many other cases such correlations are not found.

Schaal et al. (1991) reviewed several plant studies that compared allozyme, DNA, and quantitative genetic variation with the average values of morphological traits and found that the sign of the correlation fluctuated from study to study, even within the same species. Positive correlations that do appear to exist may result from other factors, including the level of inbreeding, variation in which tends to be positively correlated with genomic heterozygosity. Strauss (1986) found that experimental reduction of genetic variation through inbreeding in a plantation of the knobcone pine *Pinus attenuata* removed most of an observed correlation between allozyme heterozygosity and mean growth or reproduction.

Correlations have also been observed between heterozygosity at specific loci and mean fitness. One of the best representatives of this work is Ward Watt's work on sulphur butterflies *Colias* spp. For example, Carter and Watt (1988) found that heterozygosity at the polymorphic phosphoglucose isomerase, phosphoglucomutase, and glucose-6-phosphate dehydrogenase loci was strongly associated with mating activity in *Colias eurytheme* and *Colias philodice eriphyle* (Figure 1). All three enzymes share a substrate important in metabolism and flight activity. Carter and Watt (1988) could not attribute the differential mating activity observed between heterozygotes and homozygotes to assortative mating. It is important to recognize, however, that this and similar studies examined the ability of molecular heterozygotes to achieve higher average performance. This is different from asking whether molecular variation is correlated with phenotypic variation. In addition, although these enzymes have important metabolic functions with potentially marked fitness consequences, the opportunity for mating itself may be only weakly correlated with reproductive success and, hence, with fitness.

Fewer studies have examined the more germane relationship between molecular genetic variation and phenotypic variation. Zink et al. (1985) found a

Colias eurytheme

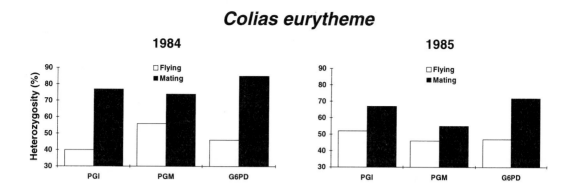

Colias philodice eriphyle

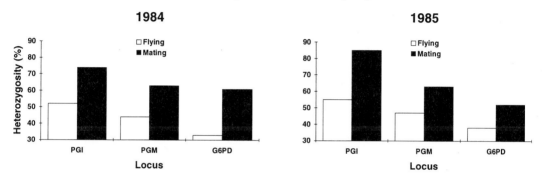

FIGURE 1.—Comparison of allozyme heterozygosity at three loci (phosphoglucose isomerase [PGI]; phosphoglucomutase [PGM]; and glucose-6-phosphate dehydrogenase [G6PD]) and flying versus mating frequency in two species of sulphur butterflies. Data are from two consecutive years (after Carter and Watt 1988).

weak negative relationship between protein heterozygosity and morphological variability in both the fox sparrow *Passerella iliaca* and the pocket gopher *Thomomys bottae*. By contrast, Beacham and Withler (1985) showed that allozyme heterozygosity and meristic variability (coefficient of variation in gill raker number) in populations of pink salmon *Oncorhynchus gorbuscha* were positively correlated. Strauss (1991) found that allozyme heterozygosity and variability in body shape in 32 populations of freshwater sculpins *Cottus* spp. were positively related. Enzyme heterozygosity and morphological variability in male rufous-collared sparrows *Zonotrichia capensis*, presumably from 25 different populations, appeared to be positively correlated, but this relationship was quite weak (Yezerinac et al. 1992). The lack of a consistent pattern among these studies indicates that reliable generalizations about the relationship between genetic and phenotypic variability cannot yet be made.

A more direct line of inquiry would be to investigate the relationship between levels of molecular genetic and quantitative genetic variation. Two representative parameters for these measures are average H and its quantitative genetic analog, heritability. When used in its narrow sense (Falconer 1989), heritability (designated h^2) is the proportion of phenotypic variance that is due to the variance in average effects of genes acting additively in their expression. This additive genetic variance is the chief source of resemblance between parents and their offspring (Falconer 1989). Combined with the intensity of selection, this variance limits a quantitative trait's rate of evolution (Robertson 1968). Inspection of the relationship between H and h^2 provides a means to test the assumption that molecular genetic variation is a reasonable proxy for quantitative genetic variation (i.e., the genetic basis for fitness variation).

Some concordance might be expected between H and h^2 in their response to reductions in population size but only under certain restrictive conditions (reviewed by Bryant et al. 1986). The large sampling variances of both H and h^2 further complicate prediction of their relationship because these estimates can depart widely from expectation in particular circumstances (Falconer 1989; Hartl and Clark 1989). For characters effectively neutral with respect to selection, the additive genetic variance and h^2 in quantitative characters are predicted to respond to population bottlenecks in qualitatively the same way (i.e., loss of low-frequency alleles and decrease in heterozygosity) that genetic variation at single loci declines (Lande 1980). However, these similarities between single loci and quantitative characters break down when genes underlying quantitative traits exhibit dominance or epistasis or when these traits are affected by selection, as is often the case with fitness-related traits (Barton and Turelli 1989). When dominance or epistasis exists, the consequences of bottlenecks for quantitative genetic variance are generally unpredictable, but, in the absence of selection, dominance and epistasis may contribute to increases in genetic variation after bottlenecks due to the conversion of dominance and epistatic variance to additive variance (Goodnight 1987, 1988). Consequently, if gene expression is not additive or if these genes are linked, the correspondence between H and h^2 is expected to be poor, even for neutral characters. A trait's h^2 is generally far more sensitive to selection than is H; selection generally acts to reduce h^2, but the realized effects of selection on h^2 depend on the relationship between phenotype and fitness (i.e., the form of the fitness function) and the intensity of selection (Lande 1988b). However, these effects on h^2 are compounded in most populations by variation in the magnitude and direction of selection, pleiotropic interactions between genes, and other factors, which can all act to increase quantitative genetic variation (Barton and Turelli 1989). For many quantitative characters, then, the combination of nonadditive underlying gene action and sensitivity of these characters to selection make the consequences of reductions in population size for this variation difficult to predict.

The heritability of a trait is thought to be closely tied to its contribution to fitness. Fisher (1958) was first to argue that, in the absence of mutation, genetic variance in fitness cannot be maintained in a closed population at evolutionary equilibrium. The generally low heritabilities of life history, physiological, and behavioral traits that have been estimated in a variety of taxa have been used to infer that these traits have lower additive genetic variance than do other quantitative traits (Allendorf et al.

1987; Mousseau and Roff 1987). However, low heritability of these traits may reflect high phenotypic variance more than low additive genetic variance (Price and Schluter 1991), suggesting that h^2 at best is a weak indicator of a population's ability to respond to selection. Furthermore, several studies on natural populations have demonstrated that quantitative traits under strong stabilizing or directional selection often have moderate and even high levels of additive variance (e.g., Hegmann and Dingle 1982; Istock 1983; Lynch 1984; Roff 1990; Hard et al. 1993). As a result, some workers have called into question the utility of Fisher's model of adaptive evolution. However, in a study of collared flycatchers *Ficedula albicollis* in the Baltic Sea, Gustafsson (1986) showed that a trait's h^2 declined strongly as the proportion of lifetime reproductive success explained by that trait increased (Figure 2), lending empirical support to Fisher's theory (see also Mousseau and Roff 1987). Apparent inconsistencies with the theory posed by the presence of high additive variance in a "fitness" trait may be explained by factors such as mutation–selection balance, antagonistic pleiotropy, or fluctuating selection (Lande 1976; Rose 1982; Travis 1989).

Few studies provide estimates of both H and h^2 that can be used to examine their correspondence (Table 1). These studies have looked at a variety of traits in widely divergent taxa, and the interpretation of the estimates of average H and h^2 is complicated by several factors. (1) Some studies have examined wild and others captive populations. Evidence is growing that domestication can alter the genetic basis of adaptation (Rose 1984; Service and Rose 1985; Holloway et al. 1990), potentially confounding comparisons among wild and captive groups. (2) In the majority of these cases, the estimates of H reported are based on very few ($N < 5$) polymorphic loci. Hence, these estimates are unreliable. (3) The h^2 estimates reported are for quantitative traits that have an unknown correlation with fitness (especially the studies on houseflies and on cotton-top tamarins). (4) Patterns of estimates compared across distantly related taxa and under different experimental conditions can have difficulties in interpretation. In any case, the correlation between H and h^2 from the data in Table 1 is weak, if present at all.

Briscoe et al. (1992) estimated H from nine allozyme loci and the h^2 of sternopleural bristle number in several different populations of the fruit fly *Drosophila melanogaster*. The observed correlation between these estimates exceeded 0.8 and was highly significant (Briscoe et al. 1992). However,

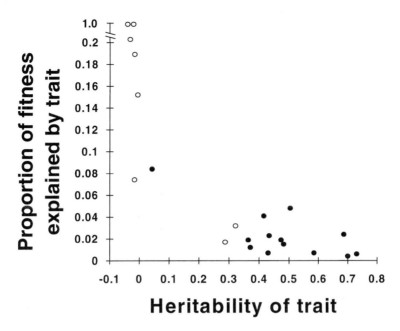

FIGURE 2.—Relationship between narrow-sense heritability of several life history (light symbols) and morphometric (dark symbols) traits and the proportion of lifetime reproductive success explained by these traits in collared flycatchers. Adapted from Gustafsson (1986).

sternopleural bristle number in *Drosophila* is probably effectively neutral with respect to selection (Kearsey and Barnes 1970; Falconer 1989), and the correlation between *H* and h^2 for a life history, physiological, or behavioral trait could be far lower.

The most comprehensive comparison available of heterozygosity and heritability comes from the work of Michael Lynch and Ken Spitze on several species of water fleas *Daphnia* (Lynch et al. 1989; Spitze 1993; Lynch and Spitze 1994). This work is signifi-

cant because it involves life history traits with direct consequences for fitness. It is interesting that Lynch et al. (1989) found inconsistent rankings of genetic variability between two sexual populations of *Daphnia pulex* when quantitative genetic variation in 20 life history traits was compared with variation at six polymorphic isozyme loci and nine variable mitochondrial DNA haplotypes.

However, it should be noted that these species of *Daphnia* are cyclically parthenogenetic, which has

TABLE 1.—Summary of data relating molecular and quantitative genetic indices of genetic variation within populations. Studies are on houseflies *Musca domestica*, pitcher-plant mosquitoes *Wyeomyia smithii*, rainbow trout *Oncorhynchus mykiss*, and cotton-top tamarins *Saguinus oedipus*. For each estimate of *H*, the subscript refers to the number of loci used in its calculation; for each estimate of h^2, the subscript denotes a quantitative trait: wing length (WL), development time (DT), critical photoperiod for dormancy (CP), growth rate (GR), fluctuating asymmetry (FA), and body weight (BW).

Species	Type	Heterozygosity (*H*)	Heritability (h^2) ± SE	References
Housefly	Wild	0.380_4	$0.58_{WL} \pm 0.10$	Bryant et al. (1986); McCommas and Bryant (1990)
Pitcher-plant mosquito	Wild	0.154_2	$0.47_{DT} \pm 0.07$ $0.44_{CP} \pm 0.07$	Istock and Weisburg (1987); Bradshaw and Holzapfel (1990)
		0.250_2	$0.33_{DT} \pm 0.03$	Istock and Weisburg (1987); Bradshaw and Holzapfel (1990)
Rainbow trout	Captive	0.065_{42}	$0.67_{GR} \pm 0.11$ $0.02_{FA} \pm 0.07$	Leary et al. (1992)
Cotton-top tamarin	Captive	0.014_{41}	$0.35_{BW} \pm 0.05$	Cheverud et al. (1994)

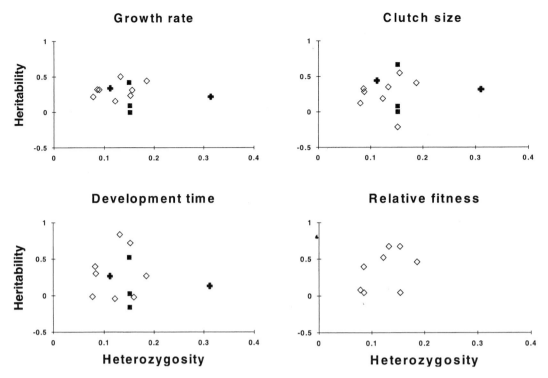

FIGURE 3.—Relationships between allozyme heterozygosity (H) and broad-sense heritabilities (H^2) for relative fitness and three of its components (growth rate, clutch size, and development time) in three species of water fleas: open diamonds, *Daphnia obtusa*; solid squares, *Daphnia pulicaria*; and solid crosses, *Daphnia pulex*. Data from Lynch et al. (1989), Spitze (1993), and Lynch and Spitze (1994).

mixed consequences for the purposes of comparing heterozygosities and heritabilities. The primary advantage of using an organism that reproduces clonally for this purpose is that heritabilities can be obtained quickly and easily from the phenotypic differences among clonal lines. The tradeoff is that the heritabilities obtained include variance due to maternal and cytoplasmic effects on the phenotype, as well as to nonadditive genetic sources of variance; these estimates are referred to as broad-sense heritabilities, usually designated H^2 (Falconer 1989).

The *Daphnia* spp. data are summarized in Figure 3. For relative fitness and for three of its components (growth rate, clutch size, and development time), no apparent correlation exists between average H (estimated from a half-dozen polymorphic loci) and H^2, either within or across three species. Taken together, the data in Table 1 and Figure 3 suggest that the concordance of molecular and quantitative genetic measures of genetic variation is weak.

Among-Population Variation

Many studies have drawn attention to correlations observed between genetic and phenotypic measures of distance among populations. The correlations that have been observed between genetic and phenotypic distance are often statistically significant but have questionable predictive power. Antonovics (1991) conducted a principal components analysis on an extensive set of data from Adams (1977) on 22 cultivars of the bean *Phaseolus vulgaris* that shows a correlation of 0.66 between genetic diversity (based on pedigrees) and phenotypic diversity (based on 18 chemical and agronomic traits). A study of 21 populations of pink salmon reported correlations between genetic distance (based on eight allozyme loci) and Mahalanobis distance (based on four morphometric and one meristic trait); these correlations ranged from about 0.3 to 0.8 (Beacham and Withler 1985). In these studies, a large amount of scatter is usually evident. Therefore, although these correlations are undoubtedly influenced by the genetic basis of phenotypic differentiation, it is dangerous to conclude from such a correlation that phenotypic differences among populations are an accurate reflection of their evolutionary divergence. Nevertheless, phenotypic differences among disparate populations grown in a common environment should provide some indication of the minimum evolutionary divergence among them.

What is unclear from these sorts of investigations is whether phenotypic measures of population structure correspond to quantitative genetic measures of structure. A more direct approach is to examine the genetic diversity among populations at the molecular and quantitative trait levels. Very few studies that have been published have taken this approach, using the gene diversity indices F_{ST} or G_{ST} for allozymes (Wright 1969; Crow 1986) and the analogous index Q_{ST} for quantitative traits (Wright 1951; Lande 1992; Spitze 1993). For quantitative traits, Q_{ST} estimates the genetic component of phenotypic differentiation among populations. For genetic divergence at six allozyme loci among 19 populations of *Drosophila buzzatii*, Prout and Barker (1993) estimated F_{ST} ($\pm$ 1 SE) at 0.03 $\pm$ 0.01; for differentiation in body size among these populations they estimated Q_{ST} at 0.15 $\pm$ 0.05. For genetic divergence at six isozyme loci among eight populations of *Daphnia obtusa*, Spitze (1993) and Lynch and Spitze (1994) estimated G_{ST} at 0.28 $\pm$ 0.11; for differentiation in five life history traits among these populations, they estimated Q_{ST} at 0.29 $\pm$ 0.06. Leibold and Tessier (1991) and Tessier et al. (1992) provided data to estimate somewhat higher levels of Q_{ST} for three morphometric characters among six populations of *Daphnia pulicaria* (0.46 $\pm$ 0.05) and seven populations of *Daphnia galeata* (0.61 $\pm$ 0.24), but they did not estimate molecular divergence.

Higher values of Q_{ST} relative to F_{ST} or G_{ST} might be expected if selection is acting to diversify populations, but no firm conclusions can be made about the relationship of these metrics based on the available data. This is particularly true because the studies that estimated both parameters used such different characters to estimate Q_{ST}. The life history traits examined in *Daphnia obtusa* have strong contributions to reproductive success, whereas body size probably has a complex effect on fitness in *Drosophila buzzatii* (see discussion in Falconer 1989, Chapter 20). Thus, empirical evidence for the relationship between molecular genetic and quantitative genetic measures of population diversity is mixed and far from definitive.

In the following section, I review briefly some basic mechanisms of adaptation and show that quantitative genetic approaches can address mechanisms that are beyond the scope of molecular techniques and their interpretation.

Mechanisms of Adaptation

Adaptation is a complex process that can occur through any of several mechanisms. A great deal of research has been conducted over the last 30 years to identify genetic polymorphisms that explain phenotypic variation (e.g., Lewontin 1974; Mitton and Grant 1984; Clegg and Epperson 1985; see also Lewontin 1991; Watt 1994). Polymorphisms represent the genotypic variation that is directly responsible for different phenotypes which result when two or more alleles exist at each constituent locus (Hartl and Clark 1989). Not surprisingly, efforts to characterize genetic polymorphisms were originally stimulated by the availability of novel biochemical techniques to detect them. These techniques are still useful for a variety of situations when direct detection of molecular variation is desired. However, these techniques are most appropriate to address problems associated with gene flow and random genetic drift. Many other factors can also generate and maintain genetic variation, including environmental variation, mutation–selection balance, overdominance, antagonistic pleiotropy, nonadditive gene action, frequency-dependent selection, linkage disequilibrium, and variation in allelic effects (Hartl and Clark 1989). Genetic variation maintained in these ways can be difficult to detect with biochemical methods alone.

An additional problem that is not well understood is how the relationship between genetic and phenotypic variation changes with environmental variability. Genetic polymorphism reflects an evolutionary ability to respond to environmental change. But phenotypic variation may also result from a genotype's ability to produce an appropriate phenotype when environmental variation occurs during development. This ability can be considered from an evolutionary point of view to be tantamount to "ignoring" or "compensating" for environmental change rather than "responding" directly to it (see Orzack 1985), and it contributes to a population's evolutionary homeostasis. Phenotypic variation can also be produced by a combination of these strategies.

At least two mechanisms of adaptation are not amenable to molecular genetic analysis. The first is phenotypic plasticity, which is generally defined to

TABLE 2.—Prominent features of three mechanisms of adaptation. See text for discussion and references.

Mechanism of adaptation	Primary advantages	Primary disadvantages	Examples
Genetically based polymorphism	Somewhat stable; avoids acclimation costs	Incurs costs of recombination	Several protein and DNA variants
Phenotypic plasticity	Avoids costs of recombination	Ineffective for large or unpredictable environmental variation; requires appropriate cue	Fecundity, development rate, size and age at maturity; other life history traits
Range variation	Avoids costs of recombination and environmental sensitivity; does not require cue	Lower mean fitness in "average" environment	Egg and seed dormancy, egg size

be the predictable phenotypic response of a genotype to a specific environmental change (Stearns 1989; Via 1993). Phenotypic plasticity can provide a means for a population to increase its phenotypic repertoire without the genetic costs incurred by polymorphism. Phenotypic plasticity is widespread in plants and animals and appears to be favored in variable environments (Bradshaw 1965; Via 1987). However, its value to fitness depends in part on how reliably organisms can anticipate environmental change (Bradshaw 1986).

Another adaptive mechanism effectively uncouples phenotypic and genetic variability. Range variation (Bonner 1965), a tactic in which the range of phenotypic variation is under genetic control but the phenotypes of individual progeny are not, may allow a population to spread its risk (den Boer 1968; Stearns 1976) by producing a wide array of offspring phenotypes, only a fraction of which might be able to cope with the environment they encounter (see also Kaplan and Cooper 1984; Bull 1987). Range variation may act as a fail-safe against extreme environmental uncertainty. Tactics such as range variation are not well known; those examples that are consistent with this hypothesis are associated with highly unpredictable or ephemeral environments (Wourms 1972; Arthur et al. 1973; Wipking and Neumann 1986; see also Crump 1984; Meffe 1987; Kaplan 1987). Table 2 compares some of the major advantages and disadvantages of polymorphism, plasticity, and range variation.

The most effective tactic that a population can employ may depend on the degree of environmen-

tal uncertainty (Orzack 1985; Bradshaw 1986). The tactics depicted in Table 3 suggest that unless the environment is highly unstable, high levels of genetic polymorphism may be sufficient to respond to it. However, under some patterns of environmental variation, genetic polymorphisms may be augmented or replaced by plasticity or range variation. This shifting of tactics might help to account for instances of low molecular genetic variation in wild populations that continue to exist in highly variable environments (e.g., Waller et al. 1987; Soltis et al. 1992). The consequences of environmental uncertainty for quantitative genetic variation are unclear but likely to be more complex, because these consequences depend on the form of the fitness function as well as on the effects of selection on the genetic variance.

Relevance of Quantitative Genetics to Conservation Biology

In this section, I discuss three examples that illustrate how quantitative genetic approaches can contribute to the conservation of genetic resources.

Population Bottlenecks, Quantitative Genetic Variation, and Inbreeding Depression

The reduction of within-population genetic variability through inbreeding and genetic drift is evidently considered by many geneticists to be the leading genetic risk in conserving wild and captive populations (Ralls and Ballou 1983; Lande 1988a; Hedrick 1992). Although a great deal of research

TABLE 3.—Adaptive responses to environmental unpredictability (after Bradshaw 1986).

Level of environmental uncertainty	Genetic variability required for persistence	Response to environmental change
Very low	Low	Physiological or behavioral adjustment
Low to moderate	Low to moderate	Phenotypic plasticity, genetic polymorphism, or both
Moderate to high	Low to high	Genetic polymorphism (coupled with phenotypic plasticity)
Very high	Low to high	Genetic polymorphism, range variation, or both

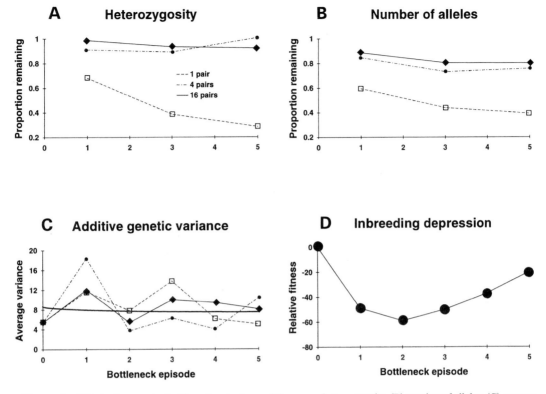

FIGURE 4.—Effects of successive bottleneck episodes on (**A**) average heterozygosity, (**B**) number of alleles, (**C**) average additive genetic variance for eight morphometric traits, and (**D**) average relative fitness in the housefly. Bottleneck sizes in **A–C** are 1, 4, and 16 mating pairs; the line in (**D**) results from a composite regression including all bottleneck sizes (see Bryant et al. 1990); and the heavy line in (**C**) shows the upper 95% confidence limit for the neutral expectation of additive genetic variance for the bottleneck size of 16. Data from Bryant et al. (1990), McCommas and Bryant (1990), and Bryant and Meffert (1992).

has been conducted on inbreeding depression (especially in *Drosophila*; Simmons and Crow 1977; Charlesworth and Charlesworth 1987; Miller and Hedrick 1993), it is still commonly assumed that inbreeding depression can be predicted from inbreeding coefficients, which are far more easily measured (Cavalli-Sforza and Bodmer 1971; Hedrick 1992). Unfortunately, the situation is complicated by the lack of information on the correlation between molecular and quantitative genetic variation. A series of studies on houseflies by Edwin Bryant and his colleagues (Bryant et al. 1986, 1990; McCommas and Bryant 1990; Bryant and Meffert 1992) showed that the reduction in allozyme heterozygosity observed after successive experimental bottlenecks was a poor correlate of changes in either quantitative genetic variation or mean fitness. The average heterozygosity and the number of alleles declined in accordance with the drift expecta-

tion when populations of 1, 4, and 16 pairs of breeding adults were forced through five successive bottleneck episodes (Figure 4A–B), but the average additive genetic variance for eight morphometric traits did not decline and, in at least the two larger bottleneck treatments, were at levels comparable to that in the control (Figure 4C). Mean reproductive success relative to the control declined after the first and second bottlenecks but began to recover by the third bottleneck, suggesting that increases in the frequencies of deleterious recessives in these populations through genetic drift were being offset by the purging of these alleles through mortality of homozygotes (Figure 4D). It is unclear how variance in reproductive success responded to these changes in genetic variance.

However, two points are significant with respect to these studies. First, the observed recovery in mean fitness was incomplete and undoubtedly is

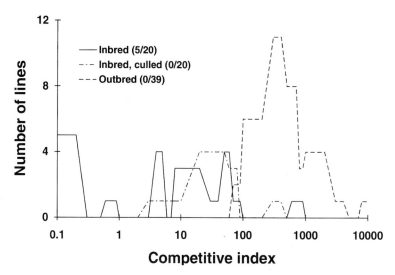

FIGURE 5.—Comparison of fitness, measured with a competitive index based on survivorship and fecundity, between unselected but inbred lines (solid curve), inbred lines culled to remove the least fit lines (dashed–dotted curve), and outbred controls (dashed curve) of the fruit fly *Drosophila melanogaster*. Inbreeding occurred via brother–sister mating for seven generations. Ratios in the legend are the ratios of extinct lines to the initial number. Note the log scale for the competitive index. Data from Frankham et al. (1993).

attributable, at least in part, to the extinction of some replicate experimental lines. Second, the additive genetic variance observed in the study was based on characters with an unclear effect on fitness. The risks of inbreeding depression are therefore considerable. Nevertheless, although the reduction in molecular genetic variation may indicate higher risk to a population, it is likely to be unreliable in forecasting the extent of inbreeding depression experienced by the population, much less its likelihood of imminent extinction.

Frankham et al. (1993), working with *Drosophila melanogaster*, found that brother–sister mating for seven generations reduced average fitness (measured with a competitive index that takes into account survivorship and fecundity) two orders of magnitude relative to an outbred control and also led to the extinction of one quarter of the inbred lines. None of the outbred lines went extinct. Artificial selection for high fitness in inbred lines prevented line extinctions and increased average fitness more than tenfold over the unselected inbred lines (Figure 5). The artificial selection was thus able to ameliorate the inbreeding depression but did not eliminate it entirely (see Frankham et al. 1988; Willis and Orr 1993).

My interpretation of these studies is that reductions in population size have somewhat unpredictable fitness consequences. Despite the opportunity for bottlenecks to reduce genetic variability, the consequences of bottlenecks for quantitative genetic variation, especially for traits that affect fitness, are far from clear. Goodnight (1987, 1988) showed theoretically how epistatic variance (the nonadditive component of genetic variance that results from interactions between different loci) could be converted to additive variance in a population subject to strong genetic drift. Bryant et al. (1986) and López-Fanjul and Villaverde (1989) found that experimental bottlenecks can produce increases in the additive variance of life history traits. These increases in variance were nevertheless accompanied by substantial reductions in viability. Willis and Orr (1993) argued that increases in additive variance resulting from bottlenecks are likely to be accompanied by inbreeding depression resulting from dominance. Consequently, sharp or recurrent reductions in population size should still be considered substantial risks, both in the short term and over longer evolutionary time.

Phenotypic Plasticity

Because of their sensitivity to environmental variation, many quantitative traits show evidence of phenotypic plasticity. Phenotypic plasticity can facilitate rapid adaptive differentiation, particularly if no single genotype is superior to all others in all

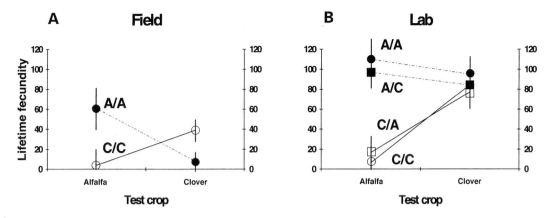

FIGURE 6.—Evidence for phenotypic plasticity in lifetime fecundity of the pea aphid grown on two crop plants in the field (**A**) and laboratory (**B**). Pea aphids initially collected from alfalfa and conditioned for three generations on alfalfa are represented by line A/A; pea aphids collected from and conditioned on clover by C/C; pea aphids collected from alfalfa and conditioned on clover by A/C; and pea aphids collected from clover and conditioned on alfalfa by C/A. Vertical bars are ± 2 SD. Adapted from Via (1991, 1994).

environments (Barton and Turelli 1989; Via 1994). Phenotypic plasticity in a trait is often expressed graphically as a "norm of reaction" (Stearns 1989). In a linear norm of reaction, nonzero slopes of lines relating phenotypes to environments and genotypes (i.e., variation in genotypic response to environment) indicate plasticity, and differences in the intercepts of these lines indicate genotypic variation. Differences between lines in their slopes indicate variation among genotypes in sensitivity to environmental variation (genotype-by-environment interaction, or G × E). However, norms of reaction may generally be nonlinear (Stearns 1989), so genotypic sensitivity to environment can be complex. The existence of G × E in genetically variable populations indicates the potential for genetically based, local adaptations to arise (Via 1994).

Via (1991) conducted a lab and field study on pea aphids *Acyrthosiphon pisum* developing on alfalfa and clover in an attempt to determine whether the pea aphids had evolved local adaptations to these crop hosts. In transplant experiments in the field, pea aphids taken from one host had higher lifetime fecundity on that host; reciprocal transplants showed lower fecundity of pea aphids moved to the alternate host (Figure 6A). This experiment indicates that G × E for fecundity existed in the pea aphids across the two hosts. Via (1991) then conditioned pea aphids taken from either alfalfa or clover on each host for three generations in the laboratory, producing four groups of pea aphids. She repeated the field experiment in the laboratory with these groups. The results, illustrated in Figure 6B, show

that conditioning had no qualitative effect on pea aphid fecundity on the two hosts, indicating that the plasticity observed in the field had evolved and was not simply a consequence of environmental variation.

It is noteworthy that the fecundities of pea aphids collected from alfalfa and conditioned and grown on both hosts, and of pea aphids collected and conditioned on both hosts and grown on clover, were higher in the lab than in the field. This pattern indicates the importance of considering the effects of both experience and environment on fitness. The consequences of these effects for adaptation to different habitats are poorly understood. An issue gaining more attention in conservation biology is the domestication of wild animals during captive propagation (Foose et al. 1986; Loebel et al. 1992; Allendorf 1993; Borlase et al. 1993). This issue is one that will be difficult to address satisfactorily with molecular methods. Via's studies suggest that local adaptations may be widespread, genetically based, and susceptible to environmental perturbation, features that should be considered carefully during conservation and recovery efforts.

Outbreeding Depression

Closely tied to local adaptation is the concept of outbreeding depression, that is, the reduction in fitness that results from mating between unrelated or distantly related individuals (Lynch 1991). The loss of genetic structure among populations can result from artificially elevated rates of gene flow between previously isolated populations. If the progeny of such crossbreeding survive to reproduce successfully, these populations will tend to become more similar. Extinction will become a greater threat if crossbreeding is frequent and the crossbred progeny show reduced fitness. Such outbreeding depression might result if the locally adapted parental genomes have diverged sufficiently to become genetically incompatible (Dobzhansky 1948; Hedrick et al. 1978) or if reproduction occurs in an environment unsuitable for their hybrids (Templeton 1986).

Like inbreeding depression, outbreeding depression is generally thought to be a product of nonadditive gene action. Inbreeding depression typically results from dominance within loci, and outbreeding depression usually reflects interactions among loci (although interactions among loci involving dominance may also produce inbreeding depression; Lynch 1991). Generally, outbreeding depression that results from the breakdown of coadapted gene complexes is expected to manifest itself in the F_2 or later generations through reduced trait means and increased trait variances with respect to fitness. When two interbreeding populations are so distantly related that their genomes have diverged considerably, the genic interactions that result can be so strong that outbreeding depression may be expressed in the F_1, before genomic breakdown can occur via recombination. However, it should be recognized that outbreeding depression in the F_1 may also be expressed in the absence of a genomic breakdown if the hybrid offspring (with a mean phenotype intermediate to that of their parents) are poorly adapted to the habitat they occupy.

Evidence exists for outbreeding depression in many species, but the quality of this evidence varies and it relies heavily on a few, primarily invertebrate, species (Endler 1977). Hard et al. (1992a and unpublished data) measured egg viability and larval diapause in populations of pitcher-plant mosquitoes from Florida and Ontario and their first- and second-generation hybrids to determine whether outbreeding depression would result from the hybridization and, if so, whether it appeared to involve the breakup of coadapted gene complexes. Percent hatch in eggs from the F_1 and F_2 hybrids was significantly lower than that of the parental mean for one cross but not the other (Figure 7A, B). However, larval diapause showed different results. A comparison of observed means and variances in the parental populations, their F_1 and F_2 hybrids, and their first-generation backcrosses with the expectations of additive and additive-dominance genetic

models (Hard et al. 1992a) suggests that reproduction between the Florida and Ontario populations would result in outbreeding depression with possibly disastrous fitness consequences. The deviation of observed means and variances from the additive expectation (Figure 7C, D) is large enough so that adaptation to local climatic conditions would be affected; the departure from expectation also indicates that independent assortment (and the consequent effect on genic interactions across loci) would almost certainly break apart any coadapted complexes that exist in each parental population (Figure 7C, D). A more comprehensive study involving larval diapause in populations along a latitudinal gradient led to similar conclusions (Hard et al. 1993). What these studies indicate is that outbreeding depression can be trait dependent, associated with nonadditive gene action, and can involve changes in trait variance as well as the mean.

Critical Research

Future research in conservation genetics should address several issues in the management of genetic resources. Theoretical exploration of these issues will always be important, but theory should complement, and cannot substitute for, empirical work involving relevant organisms. Below are five prominent areas that need investigation.

1. The distinction between neutral and selective components of genetic variability is important but poorly characterized. However, it is unclear whether describing the relationship between neutral genetic variability and variation in fitness deserves the degree of attention that has been called for (e.g., Meffe and Vrijenhoek 1988); the uncertainty in this relationship (Hedrick et al. 1986) suggests that direct investigation of genetic variation in fitness is probably a wiser investment. Additional evidence for a relationship between molecular genetic variation and correlates of mean fitness does not preempt the need for quantitative genetic evaluation of fitness variation and its relationship to molecular variation.

2. The relationship between a population's genetic variability, demographic potential, and capacity for adaptation to environmental change remains ill defined despite extensive study (Selander 1983; Lande 1988b; Kareiva et al. 1993; Lynch and Lande 1993). Phenotypic and developmental plasticity in fitness obscures this relationship, indicating that further empirical research is needed on traits sensitive to environmental per-

Egg viability

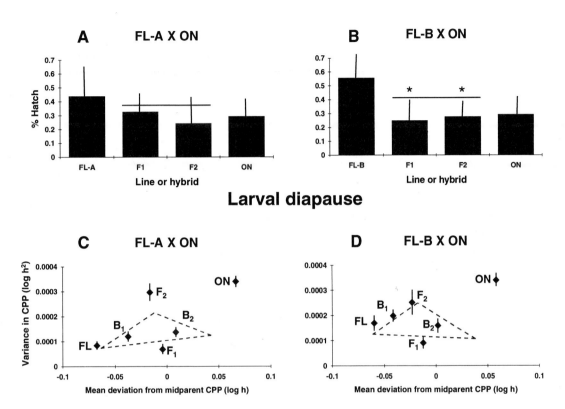

FIGURE 7.—Evidence for outbreeding depression in first-generation (F_1) and second-generation (F_2) crosses and backcrosses (B_1 and B_2) of three conspecific populations of the pitcher-plant mosquito. Comparison of egg viability among crosses relative to the midparent (horizontal line) for a cross between two different Florida populations (FL-A and FL-B) each crossed to the same Ontario (ON) population (panels **A** and **B**, J. J. Hard, W. E. Bradshaw, and C. M. Holzapfel, unpublished data). Vertical bars show ± 2 SE. Asterisks indicate significant differences from the midparent at the 5% level of significance. Panels **C** and **D** (Hard et al. 1992a) show, for all six parental and first- and second-generation lines in each of the same crosses, the observed values (± 2 SE) of the mean and variance of $\log_{10}$ critical photoperiod (CPP; in hours) for larval diapause termination relative to the expected values if composite gene action were additive (denoted by the dashed triangles). See text for discussion.

turbation, particularly for populations in sensitive or fragmented habitat.

3. The genetic consequences of small population size and population bottlenecks for life history, physiological, and behavioral variation require further investigation, especially in light of research suggesting that genetic architecture may be altered, perhaps even "reorganized" following a founder event (Lints and Bourgois 1982; Bryant et al. 1986; Goodnight 1988; López-Fanjul and Villaverde 1989). Bottlenecks may not necessarily produce the dramatic declines in genetic variability that are generally thought to occur (e.g., Powell 1978; Schwaegerle and Schaal

1979; Carson and Wisotzkey 1989), but they may yield other genetic changes that can have profound effects on population structure (and, consequently, the focus of conservation efforts). Analysis and detection of bottlenecks requires a diversity of techniques, including assessment of genetic variation in both neutral and selective traits, to determine the genetic consequences of these perturbations for subsequent adaptation and persistence.

4. The correspondence between molecular genetic and quantitative genetic differentiation among populations has received almost no empirical attention. This relationship has strong implica-

tions for efforts to conserve or recover wild populations through artificial propagation. A lack of molecular divergence between a population requiring conservation and another to be used to supplement the conserved population does not provide any assurance that negative genetic interactions will not occur upon reproduction between these populations in the wild. Consequently, such supplementation should be considered as a conservation tool only after options requiring less intervention have been exhausted (e.g., Hard et al. 1992b). If a decision is made to implement supplementation for conservation, adequate monitoring must be implemented to detect genetic problems that can arise upon interbreeding (see Hard 1995).

5. Unlike the effects of inbreeding on genetic variability, which have received much attention, the effects of outbreeding on patterns of genetic diversity are less well understood (Charlesworth and Charlesworth 1987; Lynch 1991). Aspects of outbreeding that merit investigation include its consequences for local adaptation as well as its capacity to offset inbreeding depression. Outbreeding between wild populations and between wild and captive populations under quasi-natural conditions is particularly worthy of attention. Also of interest is whether populations currently experiencing outbreeding depression can evolve to overcome it.

The techniques necessary to address these issues are well developed and described in detail in several texts. Falconer (1989) provides a thorough introduction to techniques used to partition phenotypic variation into its components, to estimate the genetic variance and covariance underlying adaptive quantitative traits, and to estimate the intensities of genetic drift and selection on them. Wright (1968, 1969, 1977, 1978) describes many basic quantitative genetic procedures, including those used to estimate effective numbers of genes underlying quantitative traits and their relative effects. Mather and Jinks (1982) develop several statistical models of gene expression that can explain modes of population differentiation. More recently published papers provide many refinements to these quantitative genetic methods as well as to molecular techniques that can be used in concert with them.

Integration of Molecular and Quantitative Genetics in Conservation Biology

Quantitative genetics has provided, and continues to provide, important contributions to explaining patterns of resemblance among relatives, phenotypic changes resulting from changes in gene frequency, the combined expression of correlated traits, and genetic and phenotypic responses to selection. Molecular genetics is not necessary to understand these issues. However, the different strengths and limitations of molecular and quantitative genetics do not preclude the coordination of these approaches, and the possibilities afforded by their integration are now beginning to be recognized. For example, the use of molecular markers to facilitate directional selection for specific phenotypic goals (marker-assisted selection) is now a common technique to improve economic gain in the breeding of domesticated plants and animals (Lande and Thompson 1990). Quantitative genetics is necessary to explain how genetically based adaptations arise and are inherited, but the insights provided by this approach do not include a satisfactory understanding of the architectural details of the underlying genes. The integration of molecular and quantitative genetic approaches poses a leading challenge to geneticists because it offers the potential to better characterize the genetic architecture of quantitative variation and, therefore, identify genetic constraints to adaptive evolution.

One possible advantage of integrating molecular and quantitative genetic approaches in conservation biology is the potential to determine more precisely the number, arrangement, and effects of genes that constrain adaptive processes. For example, in combination with controlled breeding in a supplementation or captive brood program, a battery of markers used to "saturate" the genome could help to localize genes with major effects on fitness variation. The identity of these genes and their relative effects on the phenotype might aid propagation efforts if genetic management strategies appropriate for maintaining variation in a trait controlled by one or a few genes with detectable phenotypic effects can be feasibly implemented in a conservation program. Assuming that ample markers can be generated from the relevant species, determining the number of quantitative trait loci (QTLs; Hartl and Clark 1989) underlying quantitative traits and their location in the genome (Thoday 1961; Stam 1986; Paterson et al. 1988; Lander and Botstein 1989; Zeng 1994) might help to detect genetic problems that occur during captive propagation and thereby enhance the prospects of successfully reintroducing captive individuals to the wild.

However, the full range of applications of combined molecular and quantitative genetic analyses to problems in conservation genetics is far from clear. Furthermore, it should be recognized that the effectiveness with which molecular and quantitative genetic approaches can be integrated in naturally reproducing wild populations is likely to be far less than what has been observed in inbred lines of domesticated plants and animals because variation in the allele frequencies of QTLs in genetically variable populations can seriously bias estimates of their position and effect (see Cockerham and Weir 1983; Haley et al. 1994). Nevertheless, the rapidity with which research on this topic is proceeding warrants some optimism that coordination of molecular and quantitative genetic techniques can benefit conservation efforts in some instances.

Conclusions

An appropriate unit of intraspecific conservation has at least two desirable properties: discreteness, a property that reflects an ability to recognize the unit as separate from other such units, and distinctiveness, a quality that reflects the unit's conspicuous contribution to intraspecific diversity. Discreteness acknowledges intraspecific variation and distinctiveness provides a contextual framework for assessing this variation in terms of its ecological and evolutionary significance for the species as a whole. Assessing this significance indirectly with molecular genetic approaches is problematic. Therefore, where feasible, conservation geneticists should give greater emphasis to techniques designed to detect directly the adaptive variation within and among discrete and distinctive conservation units.

Conservation geneticists have long recommended the maintenance of genetic variability as part of the management of natural populations and the evolutionary lineages these populations represent. Regulation of effective population size and, where required, infusion of additional genetic material, have been cornerstones of this approach. However, this approach may be inadequate to protect populations in danger of imminent extinction because it does not evaluate their adaptive potential and may permit maladaptive genetic changes to occur undetected (see Hard 1995). Evaluation of genetic variability in life history, physiological, and behavioral traits is an essential component of a long-term strategy for genetic conservation, because this variability represents a population's capacity to respond to uncertain as well as predictable environmental change. This capacity is critical in populations that are declining or are already threatened or endangered.

Ignoring quantitative genetic variation in the con-

servation of genetic resources poses considerable risks to conservation units. These risks include an inability to detect a reduced capacity to avoid extinction in the first place and to rebound when conditions improve. Although many geneticists have acknowledged the importance of life history, physiology, and behavior in conservation, this suite of characteristics has been largely neglected by those assessing genetic variability in conserved populations. Estimation of genetic variation in these quantitative characteristics requires a substantial commitment of time and effort. Quantitative genetic analyses may even be impractical or undesirable under some circumstances involving threatened and endangered species, but in cases involving intensively manipulated populations such as captive broodstocks or marked wild populations, these analyses can be readily made under appropriate breeding designs. Clearly, the greatest application of quantitative genetics to conservation is before a population declines to this point, and thus it is best used as a proactive tool.

Quantitative genetic approaches have other important limitations that can weaken their applicability to conservation biology. Although determining whether a character simply has a genetic basis is relatively straightforward (Falconer 1989), the estimation and interpretation of genetic parameters can be difficult. Most of these estimates are specific to the population and environment in which they are measured. The estimates are sensitive to the influence of migration, mutation, selection, nonrandom mating, level of inbreeding, genotype-by-environment correlation or interaction, and ecological or social factors (Barker and Thomas 1987). In addition, the accuracy of these estimates depends on an assumption that the underlying genes are unlinked structural genes, not modifiers.

Nevertheless, at present, little evidence exists to support the general use of alternative (i.e., molecular) techniques to detect heritable variation in fitness or make inferences about future adaptation. Despite some logistical requirements that can limit its immediate utility, quantitative genetics provides proven techniques that are designed to analyze such variation directly. The integration of molecular and quantitative genetic approaches may increase the effectiveness of conservation efforts, but more research in this area is needed before coordinated approaches can be widely applied.

Uncertainty about the relationship between quantitative genetic and molecular genetic variability, the dynamics of quantitative genetic variation within and between populations, and the factors in nature that produce fitness variation all combine to indicate that the apparent simplicity of conserving intraspecific genetic diversity by the use of surrogate molecular markers is an illusion. Further work on the factors that maintain quantitative genetic variability within and among populations in traits that affect adaptation is necessary to diminish this uncertainty. No less important is the dialogue required between research and management to ensure that populations receive the most effective protection possible.

Acknowledgments

I thank Fred Allendorf, Peter Armbruster, William Bradshaw, Michael Lynch, Linda Park, Robin Waples, Gary Winans, and an anonymous reviewer for their insightful criticisms of the ideas presented here and their help in improving earlier versions of this paper.

References

Adams, M. W. 1977. An estimation of homogeneity in crop plants with special reference to genetic vulnerability in the dry bean, *Phaseolus vulgaris* L. Euphytica 26:655–679.

Allendorf, F. W. 1993. Delay of adaptation to captive breeding by equalizing family size. Conservation Biology 7:416–419.

Allendorf, F., N. Ryman, and F. Utter. 1987. Genetics and fishery management: past, present, and future. Pages 1–19 in N. Ryman and F. Utter, editors. Population genetics and fishery management. University of Washington Press, Seattle.

Antonovics, J. 1991. Genetically based measures of uniqueness. Pages 94–119 in G. H. Orians, G. M. Brown, Jr., W. E. Kunin, and J. E. Swierzbinski, editors. The preservation and valuation of biological resources. University of Washington Press, Seattle.

Arthur, A. E., J. S. Gale, and M. J. Lawrence. 1973. Variation in wild populations of *Papaver dubium*. VII. Germination time. Heredity 30:189–197.

Barker, J. S. F., and R. H. Thomas. 1987. A quantitative genetic perspective on adaptive evolution. Pages 3–23 in V. Loeschcke, editor. Genetic constraints on adaptive evolution. Springer-Verlag, Berlin.

Barton, N. H., and M. Turelli. 1989. Evolutionary quantitative genetics: how little do we know? Annual Review of Genetics 23:337–370.

Beacham, T. D., and R. E. Withler. 1985. Heterozygosity and morphological variability of pink salmon (*Oncorhynchus gorbuscha*) from southern British Columbia and Puget Sound. Canadian Journal of Genetics and Cytology 27:571–579.

Bonnell, M. L., and R. K. Selander. 1974. Elephant seals: genetic variation and near extinction. Science 184:908–909.

Bonner, J. T. 1965. Size and cycle. An essay on the structure of biology. Princeton University Press, Princeton, New Jersey.

Borlase, S. C., D. A. Loebel, R. Frankham, R. K. Nurthen, D. A. Briscoe, and G. E. Daggard. 1993. Modeling problems in conservation genetics using captive *Drosophila* populations: consequences of equalization of family sizes. Conservation Biology 7:122–131.

Bradshaw, A. D. 1965. Evolutionary significance of phenotypic plasticity in plants. Advances in Genetics 13:115–155.

Bradshaw, W. E. 1986. Pervasive themes in insect life cycle strategies. Pages 261–275 in F. Taylor and R. Karban, editors. The evolution of insect life cycles. Springer-Verlag, London.

Bradshaw, W. E., and C. M. Holzapfel. 1990. Evolution of phenology and demography in the pitcher-plant mosquito, *Wyeomyia smithii*. Pages 47–67 in F. Gilbert, editor. Insect life cycles: genetics, evolution, and co-ordination. Springer-Verlag, London.

Briscoe, D. A., and six coauthors. 1992. Rapid loss of genetic variation in large captive populations of *Drosophila* flies: implications for the genetic management of captive populations. Conservation Biology 6:416–425.

Bryant, E. H., and L. M. Meffert. 1992. The effect of serial founder-flush cycles on quantitative genetic variation in the housefly. Heredity 70:122–129.

Bryant, E. H., S. A. McCommas, and L. M. Combs. 1986. The effect of an experimental bottleneck upon quantitative genetic variation in the housefly. Genetics 114:1191–1211.

Bryant, E. H., L. M. Meffert, and S. A. McCommas. 1990. Fitness rebound in serially bottlenecked populations of the house fly. American Naturalist 136:542–549.

Bull, J. J. 1987. Evolution of phenotypic variance. Evolution 41:303–315.

Bulmer, M. G. 1971. The stability of equilibria under selection. Heredity 27:157–162.

Carson, H. L., and R. G. Wisotzkey. 1989. Increase in genetic variance following a population bottleneck. American Naturalist 134:668–673.

Carter, P. A., and W. B. Watt. 1988. Adaptation at specific loci. V. Metabolically adjacent enzyme loci may have very distinct experiences of selective pressures. Genetics 119:913–924.

Caughley, G. 1994. Directions in conservation biology. Journal of Animal Ecology 63:215–244.

Cavalli-Sforza, L. L., and W. F. Bodmer. 1971. The genetics of human populations. Freeman, New York.

Charlesworth, B. 1990. Optimization models, quantitative genetics, and mutation. Evolution 44:520–538.

Charlesworth, D., and B. Charlesworth. 1987. Inbreeding depression and its evolutionary consequences. Annual Review of Ecology and Systematics 18:237–268.

Cheverud, J., and six coauthors. 1994. Quantitative and molecular genetic variation in captive cotton-top tamarins (*Saguinus oedipus*). Conservation Biology 8:95–105.

Clark, A. G. 1987. Genetic correlations: the quantitative genetics of evolutionary constraints. Pages 25–45 in V. Loeschcke, editor. Genetic constraints on adaptive evolution. Springer-Verlag, Berlin.

Clegg, M. T., and B. K. Epperson. 1985. Recent developments in population genetics. Advances in Genetics 23:235–269.

Cockerham, C. C., and B. S. Weir. 1983. Linkage between a marker locus and a quantitative trait of sibs. American Journal of Human Genetics 35:263–273.

Comstock, R. E., and R. H. Moll. 1963. Genotype-environment interactions. Pages 164–196 in W. D. Hanson and H. F. Robinson, editors. Statistical genetics and plant breeding. National Academy of Sciences, National Research Council, Washington, DC.

Crow, J. F. 1986. Basic concepts in population, quantitative, and evolutionary genetics. Freeman, New York.

Crump, M. L. 1984. Intraclutch egg size variability in *Hyla crucifer* (Anura: Hylidae). Copeia 1984:302–308.

den Boer, P. J. 1968. Spreading the risk and stabilization of animal numbers. Acta Biotheoretica 18:165–194.

Dobzhansky, Th. 1948. Genetics of natural populations. XVI. Altitudinal and seasonal changes produced by natural selection in certain populations of *Drosophila pseudoobscura* and *Drosophila persimilis*. Genetics 33:158–176.

Dobzhansky, Th. 1970. Genetics of the evolutionary process. Columbia University Press, New York.

Dykhuizen, D., and D. L. Hartl. 1980. Selective neutrality of 6PGD allozymes in *E. coli* and the effect of genetic background. Genetics 96:801–817.

Eanes, W. F. 1987. Allozymes and fitness: evolution of a problem. Trends in Ecology & Evolution 2(2):44–48.

Ehrlich, P. R., and P. H. Raven. 1969. Differentiation of populations. Science 165:1228–1232.

Endler, J. A. 1977. Geographic variation, speciation, and clines. Princeton University Press, Princeton, New Jersey.

Ewens, W. J. 1979. Mathematical population genetics. Springer-Verlag, Berlin.

Falconer, D. S. 1989. Introduction to quantitative genetics, 3rd edition. Longman Group, Essex, UK.

Fisher, R. A. 1958. The genetical theory of natural selection. Dover, New York.

Foose, T. J., R. Lande, N. R. Flesness, G. Rabb, and B. Read. 1986. Propagation plans. Zoo Biology 5:139–146.

Frankham, R., B. H. Yoo, and B. L. Sheldon. 1988. Reproductive fitness and artificial selection in animal breeding: culling on fitness prevents a decline in reproductive fitness in lines of *Drosophila melanogaster* selected for increased inebriation time. Theoretical and Applied Genetics 76:909–914.

Frankham, R., G. J. Smith, and D. A. Briscoe. 1993. Effects on heterozygosity and reproductive fitness of inbreeding with and without selection on fitness in *Drosophila melanogaster*. Theoretical and Applied Genetics 86:1023–1027.

Gimelfarb, A. 1989. Genotypic variation for a quantitative trait maintained under stabilizing selection without mutations: epistasis. Genetics 123:217–227.

Gingerich, P. D. 1983. Rates of evolution: effects of time and temporal scaling. Science 222:159–161.

Goodnight, C. 1987. On the effect of founder events on the additive genetic variance. Evolution 41:80–91.

Goodnight, C. 1988. Epistasis and the effect of founder events on the additive genetic variance. Evolution 42:441–454.

Gupta, A. P., and R. C. Lewontin. 1982. A study of reaction norms in natural populations of *Drosophila pseudoobscura*. Evolution 36:934–948.

Gustafsson, L. 1986. Lifetime reproductive success and heritability: empirical support for Fisher's fundamental theorem. American Naturalist 128:761–764.

Haley, C. S., S. A. Knott, and J-M. Elsen. 1994. Mapping quantitative trait loci in crosses between outbred lines using least squares. Genetics 136:1195–1207.

Hard, J. J. 1995. Genetic monitoring of life-history characters in salmon supplementation: problems and opportunities. American Fisheries Society Symposium 15:212–225.

Hard, J. J., W. E. Bradshaw, and C. M. Holzapfel. 1992a. Epistasis and the genetic divergence of photoperiodism between populations of the pitcher-plant mosquito, *Wyeomyia smithii*. Genetics 131:389–396.

Hard, J. J., W. E. Bradshaw, and C. M. Holzapfel. 1993. The genetic basis of photoperiodism and its divergence among populations of the pitcher-plant mosquito, *Wyeomyia smithii*. American Naturalist 142:457–473.

Hard, J. J., R. P. Jones, Jr., M. R. Delarm, and R. S. Waples. 1992b. Pacific salmon and artificial propagation under the Endangered Species Act. NOAA (National Oceanic and Atmospheric Administration) Technical Memorandum NMFS (National Marine Fisheries Service) NWFSC-2, Northwest Fisheries Science Center, Seattle.

Hartl, D. L., and A. G. Clark. 1989. Principles of population genetics, 2nd edition. Sinauer, Sunderland, Massachusetts.

Hedrick, P. W. 1992. Genetic conservation in captive populations and endangered species. Pages 45–68 in S. K. Jain and L. W. Botsford, editors. Applied population biology. Kluwer Academic Publishers, Netherlands.

Hedrick, P. W., P. F. Brussard, F. W. Allendorf, J. A. Beardmore, and S. Orzack. 1986. Protein variation, fitness, and captive propagation. Zoo Biology 5:91–99.

Hedrick, P., S. Jain, and L. Holden. 1978. Multilocus systems in evolution. Evolutionary Biology 11:101–182.

Hedrick, P. W., and P. S. Miller. 1992. Conservation genetics: techniques and fundamentals. Ecological Applications 2:30–46.

Hegmann, J. P., and H. Dingle. 1982. Phenotypic and genotypic covariance structure in milkweed bug life history traits. Pages 177–188 in H. Dingle and J. P. Hegmann, editors. Evolution and genetics of life histories. Springer-Verlag, New York.

Hoelzel, A. R., and seven coauthors. 1993. Elephant seal genetic variation and the use of simulation models to investigate historical population bottlenecks. Journal of Heredity 84:443–449.

Holloway, G. J., S. R. Povey, and R. M. Sibly. 1990. The effect of new environment on adapted genetic architecture. Heredity 64:323–330.

Houle, D. 1989. Allozyme-associated heterosis in *Drosophila melanogaster*. Genetics 123:789–801.

Istock, C. A. 1983. The extent and consequences of heritable variation in fitness traits. Pages 61–96 *in* C. R. King and P. S. Dawson, editors. Population biology: retrospect and prospect. Columbia University Press, New York.

Istock, C. A., and W. G. Weisburg. 1987. Strong habitat selection and the development of population structure in a mosquito. Evolutionary Ecology 1:348–362.

Kaplan, R. H. 1987. Developmental plasticity and maternal effects of reproductive characteristics in the frog, *Bombina orientalis*. Oecologia (Berlin) 71:273–279.

Kaplan, R. H., and W. S. Cooper. 1984. The evolution of developmental plasticity in reproductive characteristics: an application of the "adaptive coin-flipping" principle. American Naturalist 123:393–410.

Kareiva, P. M., J. G. Kingsolver, and R. B. Huey, editors. 1993. Biotic interactions and global change. University of Washington Press, Seattle.

Kearsey, M. J., and B. W. Barnes. 1970. Variation for metrical characters in *Drosophila* populations. II. Natural selection. Heredity 25:11–21.

Keightley, P. D., and W. G. Hill. 1992. Quantitative genetic variation in body size of mice from new mutations. Genetics 131:693–700.

Kurtén, B. 1959. Rates of evolution in fossil mammals. Cold Spring Harbor Symposia on Quantitative Biology 24:205–215.

Lande, R. 1976. Natural selection and random genetic drift in phenotypic evolution. Evolution 30:314–334.

Lande, R. 1980. Genetic variation and phenotypic evolution during allopatric speciation. American Naturalist 116:463–479.

Lande, R. 1981. The minimum number of genes contributing to quantitative variation between and within populations. Genetics 99:541–553.

Lande, R. 1982. A quantitative genetic theory of life history evolution. Ecology 63:607–615.

Lande, R. 1983. The response to selection on major and minor mutations affecting a metrical trait. Heredity 50:47–65.

Lande, R. 1988a. Genetics and demography in biological conservation. Science 241:1455–1460.

Lande, R. 1988b. Quantitative genetics and evolutionary theory. Pages 71–84 *in* B. S. Weir, E. J. Eisen, M. M. Goodman, and G. Namkoong, editors. Proceedings of the second international conference on quantitative genetics. Sinauer, Sunderland, Massachusetts.

Lande, R. 1992. Neutral theory of quantitative genetic variance in an island model with local extinction and colonization. Evolution 46:381–389.

Lande, R. 1994. Risk of population extinction from fixation of new deleterious mutations. Evolution 48:1460–1469.

Lande, R., and S. J. Arnold. 1983. Measuring selection on correlated traits. Evolution 37:1210–1226.

Lande, R., and G. F. Barrowclough. 1987. Effective population size, genetic variation, and their use in population management. Pages 87–123 *in* M. E. Soulé, editor. Viable populations for conservation. Cambridge University Press, Cambridge, UK.

Lande, R., and R. Thompson. 1990. Efficiency of marker-assisted selection in the improvement of quantitative traits. Genetics 124:743–756.

Lander, E. S., and D. Botstein. 1989. Mapping Mendelian factors underlying quantitative traits using RFLP linkage maps. Genetics 121:185–199.

Leary, R. F., F. W. Allendorf, and K. L. Knudsen. 1992. Genetic, environmental, and developmental causes of meristic variation in rainbow trout. Acta Zoologica Fennica 191:79–95.

Lesica, P., R. F. Leary, F. W. Allendorf, and D. E. Bilderback. 1988. Lack of genic diversity within and among populations of an endangered plant, *Howellia aquatilis*. Conservation Biology 2:275–282.

Leibold, M., and A. J. Tessier. 1991. Contrasting patterns of body size for *Daphnia* species that segregate by habitat. Oecologia (Berlin) 86:342–348.

Lewontin, R. C. 1974. The genetic basis of evolutionary change. Columbia University Press, New York.

Lewontin, R. C. 1991. Twenty-five years ago in *Genetics*: electrophoresis in the development of evolutionary genetics: milestone or millstone? Genetics 128:657–662.

Lints, F. A., and M. Bourgois. 1982. A test of the genetic revolution hypothesis of speciation. Pages 423–436 *in* S. Lakovaara, editor. Advances in genetics, development and evolution of *Drosophila*. Plenum, New York.

Loebel, D. A., R. K. Nurthen, R. Frankham, D. A. Briscoe, and D. Craven. 1992. Modeling problems in conservation genetics using captive *Drosophila* populations: consequences of equalizing founder representation. Zoo Biology 1:319–332.

López-Fanjul, C., and A. Villaverde. 1989. Inbreeding increases genetic variance for viability in *Drosophila melanogaster*. Evolution 43:1800–1804.

Lush, J. L. 1945. Animal breeding plans. Collegiate Press, Ames, Iowa.

Lynch, M. 1984. The limits to life history evolution in *Daphnia*. Evolution 38:465–482.

Lynch, M. 1988. The rate of polygenic mutation. Genetical Research 51:137–148.

Lynch, M. 1991. The genetic interpretation of inbreeding depression and outbreeding depression. Evolution 45:622–629.

Lynch, M., S. Conery, and R. Bürger. In press. Mutational meltdowns in sexual populations. Evolution.

Lynch, M., and W. G. Hill. 1986. Phenotypic evolution by neutral mutation. Evolution 40:915–935.

Lynch, M., and R. Lande. 1993. Evolution and extinction in response to environmental change. Pages 234–250 *in* P. M. Kareiva, J. G. Kingsolver, and R. B. Huey, editors. Biotic interactions and global change. Sinauer, Sunderland, Massachusetts.

Lynch, M., and K. Spitze. 1994. Evolutionary genetics of *Daphnia*. Pages 109–128 *in* L. Real, editor. Ecological genetics. Princeton University Press, Princeton, New Jersey.

Lynch, M., K. Spitze, and T. Crease. 1989. The distribution of life-history variation in the *Daphnia pulex* complex. Evolution 43:1724–1736.

Mather, K., and J. L. Jinks. 1982. Biometrical genetics:

the study of continuous variation, 3rd edition. Chapman and Hall, New York.

Mayr, E. 1963. Animal species and evolution. Belknap Press, Cambridge, Massachusetts.

McCommas, S. A., and E. H. Bryant. 1990. Loss of electrophoretic variation in serially bottlenecked populations. Heredity 64:315–321.

Meffe, G. K. 1987. Embryo size variation in mosquito fish: optimality vs. plasticity in propagule size. Copeia 1987:762–768.

Meffe, G. K., and R. C. Vrijenhoek. 1988. Conservation genetics in the management of desert fishes. Conservation Biology 2:157–169.

Miller, P. S., and P. W. Hedrick. 1993. Inbreeding and fitness in captive populations: lessons from *Drosophila*. Zoo Biology 12:333–351.

Mitton, J. B. 1993. Theory and data pertinent to the relationship between heterozygosity and fitness. Pages 17–41 *in* N. W. Thornhill, editor. The natural history of inbreeding and outbreeding. Theoretical and empirical perspectives. University of Chicago Press, Chicago.

Mitton, J. B., and M. C. Grant. 1984. Associations among protein heterozygosity, growth rate, and developmental homeostasis. Annual Review of Ecology and Systematics 15:479–499.

Mousseau, T. A., and D. A. Roff. 1987. Natural selection and the heritability of fitness components. Heredity 59:181–197.

O'Brien, S. J., and six coauthors. 1987. East African cheetahs: evidence for two population bottlenecks. Proceedings of the National Academy of Sciences of the United States of America 84:508–511.

Orr, H. A., and J. A. Coyne. 1992. The genetics of adaptation: a reassessment. American Naturalist 140:725–742.

Orzack, S. H. 1985. Population dynamics in variable environments V. The genetics of homeostasis revisited. American Naturalist 125:550–572.

Paterson, A. H., E. S. Lander, J. D. Hewitt, S. Peterson, S. E. Lincoln, and S. D. Tanksley. 1988. Resolution of quantitative traits into Mendelian factors by using a complete linkage map of restriction fragment length polymorphisms. Nature 335:721–726.

Powell, J. R. 1978. The founder-flush speciation theory: an experimental approach. Evolution 32:465–474.

Price, T., and D. Schluter. 1991. On the low heritability of life-history traits. Evolution 45:853–861.

Prout, T., and J. S. F. Barker. 1993. F statistics in *Drosophila buzzatii*: selection, population size and inbreeding. Genetics 134:369–375.

Ralls, K., and J. Ballou. 1983. Extinction: lessons from zoos. Pages 164–184 *in* C. M. Schonewald-Cox, S. M. Chambers, B. MacBryde, and L. Thomas, editors. Genetics and conservation: a reference for managing wild animal and plant populations. Benjamin/Cummings, Menlo Park, California.

Robertson, A. 1968. The spectrum of genetic variation. Pages 5–16 *in* R. C. Lewontin, editor. Population biology and evolution. Syracuse University Press, Syracuse, New York.

Roff, D. A. 1990. Understanding the evolution of insect life cycles: the role of genetic analysis. Pages 5–27 *in* F. Gilbert, editor. Insect life cycles: genetics, evolution and co-ordination. Springer-Verlag, London.

Rose, M. R. 1982. Antagonistic pleiotropy, dominance, and genetic variation. Heredity 48:63–78.

Rose, M. R. 1984. Genetic covariation in *Drosophila* life history: untangling the data. American Naturalist 123:565–569.

Ryder, O. A. 1986. Species conservation and systematics: the dilemma of subspecies. Trends in Ecology & Evolution 1(1):9–10.

Schaal, B. A., W. J. Leverich, and S. H. Rogstad. 1991. A comparison of methods for assessing genetic variation in plant conservation biology. Pages 123–133 *in* D. A. Falk and K. E. Holsinger, editors. Genetics and conservation of rare plants. Oxford University Press, New York.

Schwaegerle, K. E., and B. Schaal. 1979. Genetic variability and founder effect in the pitcher plant, *Sarracenia purpurea* L. Evolution 33:1210–1218.

Selander, R. K. 1983. Evolutionary consequences of inbreeding. Pages 201–215 *in* C. M. Schonewald-Cox, S. M. Chambers, B. MacBryde, and W. L. Thomas, editors. Genetics and conservation. A reference for managing wild animal and plant populations. Benjamin/Cummings, Menlo Park, California.

Service, P. M., and M. R. Rose. 1985. Genetic covariation among life-history components: the effect of novel environments. Evolution 39:943–945.

Simmons, M. J., and J. F. Crow. 1977. Mutations affecting fitness in *Drosophila* populations. Annual Review of Genetics 11:49–78.

Soltis, P. S., D. E. Soltis, T. L. Tucker, and F. A. Lang. 1992. Allozyme variability is absent in the narrow endemic *Bensoniella oregona* (Saxifragaceae). Conservation Biology 6:131–134.

Soulé, M. E., editor. 1987. Viable populations for conservation. Cambridge University Press, Cambridge, UK.

Spitze, K. 1993. Population structure in *Daphnia obtusa*: quantitative genetic and allozymic variation. Genetics 135:367–374.

Stam, P. 1986. The use of marker loci in selection for quantitative traits. Pages 170–182 *in* C. Smith, J. W. B. King, and J. McKay, editors. Exploiting new technologies in livestock improvement: animal breeding. Oxford University Press, Oxford, UK.

Stearns, S. C. 1976. Life-history tactics: a review of the ideas. Quarterly Review of Biology 51:3–47.

Stearns, S. C. 1989. The evolutionary significance of phenotypic plasticity. BioScience 39:436–445.

Strauss, R. E. 1991. Correlations between heterozygosity and phenotypic variability in *Cottus* (Teleostei: Cottidae): character components. Evolution 49:1950–1956.

Strauss, S. H. 1986. Heterosis at allozyme loci under inbreeding and crossbreeding in *Pinus attenuata*. Genetics 113:115–134.

Templeton, A. R. 1986. Coadaptation and outbreeding depression. Pages 105–116 *in* M. E. Soulé, editor. Conservation biology. The science of scarcity and diversity. Sinauer, Sunderland, Massachusetts.

Tessier, A. J., A. Young, and M. Leibold. 1992. Population dynamics and body-size selection in *Daphnia*. Limnology and Oceanography 37:1–13.

Thoday, J. M. 1961. Location of polygenes. Nature 191: 368–370.

Travis, J. 1989. The role of optimizing selection in natural populations. Annual Review of Ecology and Systematics 20:279–296.

Turelli, M. 1984. Heritable genetic variation via mutation-selection balance: Lerch's zeta meets the abdominal bristle. Theoretical Population Biology 25:138–103.

Turelli, M., and L. Ginzburg. 1983. Should individual fitness increase with heterozygosity? Genetics 104:191–209.

Via, S. 1987. Genetic constraints on the evolution of phenotypic plasticity. Pages 47–71 *in* V. Loeschke, editor. Genetic constraints on adaptive evolution. Springer-Verlag, Berlin.

Via, S. 1991. Specialized host plant performance of pea aphid clones is not altered by experience. Ecology 72:1420–1427.

Via, S. 1993. Adaptive phenotypic plasticity: target or by-product of selection in a variable environment? American Naturalist 142:352–365.

Via, S. 1994. Population structure and local adaptation in a clonal herbivore. Pages 58–85 *in* L. Real, editor. Ecological genetics. Princeton University Press, Princeton, New Jersey.

Waller, D. M., D. M. O'Malley, and S. C. Gawler. 1987. Genetic variation in the extreme endemic *Pedicularis furbishiae* (Scrophulariaceae). Conservation Biology 1:335–340.

Waples, R. S. 1991. Pacific salmon, *Oncorhynchus* spp., and the definition of "species" under the Endangered Species Act. U.S. National Marine Fisheries Service Marine Fisheries Review 53(3):11–22.

Waples, R. S. 1995. Evolutionarily significant units and the conservation of biological diversity under the Endangered Species Act. American Fisheries Society Symposium 17:8–27.

Watt, W. B. 1994. Allozymes in evolutionary genetics: self-imposed burden or extraordinary tool? Genetics 136:11–16.

Williams, G. C. 1966. Adaptation and natural selection. A critique of some current evolutionary thought. Princeton University Press, Princeton, New Jersey.

Willis, J. H., and H. A. Orr. 1993. Increased heritable variation following population bottlenecks: the role of dominance. Evolution 47:949–957.

Wipking, W., and D. Neumann. 1986. Polymorphism in the larval hibernation strategy of the burnet moth *Zygaena trifolii*. Pages 125–134 *in* F. Taylor and R. Karban, editors. The evolution of insect life cycles. Springer-Verlag, London.

Wourms, J. P. 1972. The developmental biology of annual fishes. III. Pre-embryonic and embryonic diapause of variable duration in the eggs of annual fishes. Journal of Experimental Zoology 182:389–414.

Wright, S. 1951. The genetical structure of populations. Annals of Eugenics 15:323–354, London.

Wright, S. 1968. Evolution and the genetics of populations, volume 1. Genetic and biometric foundations. University of Chicago Press, Chicago.

Wright, S. 1969. Evolution and the genetics of populations, volume 2. The theory of gene frequencies. University of Chicago Press, Chicago.

Wright, S. 1977. Evolution and the genetics of populations, volume 3. Experimental results and evolutionary deductions. University of Chicago Press, Chicago.

Wright, S. 1978. Evolution and the genetics of populations, volume 4. Variability within and among natural populations. University of Chicago Press, Chicago.

Yamazaki, T., and six coauthors. 1983. Reexamination of diversifying selection of polymorphic allozyme genes by using population cages in *Drosophila melanogaster*. Proceedings of the National Academy of Sciences of the United States of America 80:5789–5792.

Yezerinac, S. M., S. C. Lougheed, and P. Handford. 1992. Morphological variability and enzyme heterozygosity: individual and population correlations. Evolution 46: 1959–1964.

Zeng, Z-B. 1994. Precision mapping of quantitative trait loci. Genetics 136:1457–1468.

Zink, R. M., M. F. Smith, and J. L. Patton. 1985. Association between heterozygosity and morphological variance. Journal of Heredity 76:415–420.

Zouros, E., and G. H. Pogson. 1994. The present status of the relationship between heterozygosity and heterosis. Pages 135–146 *in* A. R. Beaumont, editor. Genetics and evolution of aquatic organisms. Chapman and Hall, London.

PART FIVE

ECOSYSTEMS AND HABITAT

American Fisheries Society Symposium 17:329–333, 1995

SESSION OVERVIEW

Ecosystem and Habitat Conservation: More Than Just a Problem of Geography

PETER A. BISSON

*U.S. Forest Service, Olympia Forestry Sciences Laboratory
Olympia, Washington 98512, USA*

As more and more species become extinct or come under the protection of the U.S. Endangered Species Act (ESA; 16 U.S.C. §§ 1531 to 1544), public pressure mounts to stem the widespread loss of biological diversity. At the same time, implementation of recovery plans for threatened or endangered species may cause economic hardships that are translated into job losses and accompanying social problems. Perhaps nowhere have recent collisions of conservation with economic interests been more apparent than in the Pacific Northwest, where protection and management of the northern spotted owl *Strix occidentalis caurina* and Pacific salmon *Oncorhynchus* spp. have resulted in a scientific, economic, and political turmoil to which terms like "gridlock" and "train wreck" have been applied (FEMAT 1993), and which promises to be a major battleground relative to ESA reauthorization (Volkman and Lee 1994). Conditioned by this controversy, scientists from the region have questioned whether conventional use of natural resources is sustainable (Ludwig et al. 1993) and whether the ESA is the appropriate regulatory tool to protect biological diversity (Franklin 1993). Both of these topics have provoked lively debate in recent issues of *Ecological Applications* (1993, 3:547–589; 1994, 4:205–209), and the debate is likely to continue for some time.

The ESA requires critical habitat for threatened or endangered species to be identified and protected. For species with very small populations and limited geographic distribution, critical habitat may be measured in square meters. For anadromous salmonids, critical habitat is much more problematic. Chinook salmon *O. tshawytscha* spawning in Idaho, for example, may migrate thousands of kilometers from their natal spawning grounds to the Gulf of Alaska and back over the course of their life cycle. For most aquatic vertebrates, the geographic range of potentially critical habitat is somewhere between these two extremes. Critical habitat designations for species undertaking migrations usually include areas where important life history phases occur such as reproduction or overwintering.

It is useful to review some key concepts applied to ecosystem management. *Biotic integrity* is "the capability of supporting and maintaining a balanced, integrated, adaptive community of organisms having a species composition, diversity, and functional organization comparable to that of natural habitat of the region" (Karr and Dudley 1981). *Biological diversity* includes the variety and variability among living organisms and the ecological complexes in which they occur. It is divided into four levels (Temple 1991): genetic diversity (genetic variation within a species), phenotypic and morphological diversity (physical and life history variation within species), species diversity (species richness within an ecosystem), and community–ecosystem diversity (variability in habitats and ecological processes extending over a region). *Critical habitat*, according to the Endangered Species Act, includes areas that contain physical or biological features that are essential to the conservation of the species and that may require special management consideration or protection. These include areas important for population growth, food and water resources, shelter, and breeding and rearing sites, as well as areas representative of the species' historical distribution.

In practice, the protection of endangered species and the ecosystems upon which they depend has been based on identifying remaining patches of ecologically intact habitat and preventing these areas from being adversely altered by human activity, although some human use of the areas typically is allowed. Conservation of harvestable natural resources is, in effect, approached as a matter of zoning in space and time, the tools being spatial preserves and closed harvest seasons. For spotted owls, salmon, and other inhabitants of late-successional forests on federally owned forest lands in the Pacific Northwest, this exercise required the intervention of a U.S. President and the combined efforts of many scientists over many months to delineate a highly complex system of forest reserves and management areas (FEMAT 1993). Although the mobilization of scientific effort for the Forest Ecosystem Management Assessment Team (FEMAT) was unprecedented in conservation history, many questions remain about the project's ultimate success in preventing extinctions related to anthropogenic disturbances. For example, the forest management option preferred by FEMAT was believed to have only a 65% likelihood of achieving aquatic habitat of sufficient quality, distribution, and abundance to allow anadromous fishes to become well distributed across federal lands over the next century, even though it provided for wide buffer strips along all streams and designated key watersheds where conservation of salmonid habitat would receive top priority (FEMAT 1993). Part of the reason for this somewhat pessimistic assessment was that FEMAT authors realized protection of streams on federal lands would typically influence only parts of river basins inhabited by anadromous salmonids; other land owners might not be subject to such stringent environmental requirements.

This example illustrates a widespread problem in conserving evolutionarily significant units (ESUs): the scales at which conservation efforts are currently focused may not match the spatial, temporal, or geological scales appropriate for the protection of species with highly dynamic life cycles and complex metapopulation structures. In a wide variety of areas scattered across the landscape, aquatic habitat receives special protection and some degree of naturalness is retained (Figure 1). At very large landscape scales, often termed ecoregions because they possess characteristic biogeoclimatic conditions, large blocks of land (usually federally owned) such as national parks, wilderness areas, or (in the recent FEMAT example) key watersheds may encompass whole drainage systems and contain relatively pristine habitats. At the somewhat smaller landscape scale of a river basin, sizable blocks of land may occur in state parks, wildlife refuges, or extended reaches of main-stem rivers designated as wild and scenic. These patches tend to be smaller than major wilderness areas, but they can include habitats that are buffered against anthropogenic disturbance. The next smaller landscape division, the subbasin or watershed, contains a major tributary system within the river basin. Habitat conservation areas often found within subbasins include county parks, greenway belts, and privately owned reserves (such as those maintained by The Nature Conservancy). Finally, small habitat conservation areas may occur at

the scale of individual streams or standing water bodies, or even of reaches within a stream. Examples include municipal parks, conservation easements, and habitat restoration projects. The latter may consist of areas with previously degraded habitat or with new habitat created as mitigation for land or water development projects.

There is no question that preventing the degradation of remaining areas where biotic integrity is high must be a cornerstone of any program to conserve an ESU or a cluster of ESUs threatened with extinction (Reeves and Sedell 1992). These areas will serve as refuges and sources of colonists as environmental improvements are realized elsewhere. But the current system of habitat patches set aside in wilderness areas, parks, and various natural reserves may not provide all the habitat requirements of many fishes and other aquatic species. The majority of habitat reserves have boundaries dictated by land ownership patterns that rarely coincide with drainage divides, natural landscape units, and ecological boundaries. The size of habitat reserves may not encompass the migratory range of individuals within populations, and some patches may not be large enough to provide the full range of environmental conditions to which local populations are adapted and that are necessary to support all life history phases. The question, Which is better, many small refuges or a few large refuges? likely has different answers for different species. Habitat reserves may be managed for the benefit of one or a few species to the detriment of others, especially when management activities cause changes that depart from the natural range of conditions, leading to a decline in biological integrity. Furthermore, areas such as parks may be protected from the impacts of normal land uses but are often managed for human recreation, and this may simply mean the substitution of one type of anthropogenic disturbance for another.

There are additional, and in some ways even more daunting, problems with the current patchwork of large and small habitat refuges arrayed over the landscape. The arrangement of habitat refugia may not be favorable to the dispersal of organisms from one patch to another, which could have important genetic consequences for a population (Harrison 1991). Patch location and size may influence the vulnerability of patches to catastrophic disturbances, and their ability to recover from natural disturbances. Boundary conditions may facilitate the invasion of nonnative species or upset the structure and function of native assemblages. All these conditions, together or separately, can in-

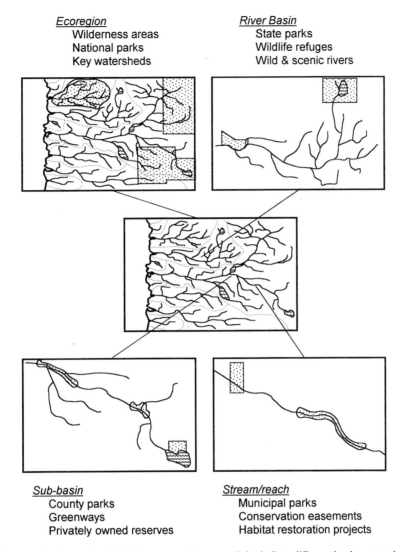

Ecoregion
  Wilderness areas
  National parks
  Key watersheds

River Basin
  State parks
  Wildlife refuges
  Wild & scenic rivers

Sub-basin
  County parks
  Greenways
  Privately owned reserves

Stream/reach
  Municipal parks
  Conservation easements
  Habitat restoration projects

FIGURE 1.—Examples of habitat conservation areas (stippled) at different landscape scales.

crease the risk that small, fragmented populations will become extinct (Sheldon 1987).

The development of geographic information system (GIS) technology has provided a powerful tool for landscape-level planning. Using a GIS, resource managers can examine the juxtaposition of landforms, plant and animal distributions, local climatic regimes, land uses, and other mappable factors in a way that facilitates efficient management decisions and long-range forecasting. Although GIS technology has revolutionized landscape planning, similar to the way in which DNA analysis has contributed to the understanding of phylogenetic and taxonomic relationships, it is still just a tool. The creation of

more patches, even with the aid of GIS, will not guarantee the conservation of biological diversity. There should be a sound conceptual and empirical basis for the location and arrangement of protected habitat areas in space and time, as well as a long-term strategy for the restoration of key ecological processes where habitat has been degraded (Karr 1991).

Two recent approaches to conserving aquatic ecosystems and biodiversity at a variety of landscape scales have been put forward by Sedell et al. (1994) and Moyle and Yoshiyama (1994). Both proposals encompass a hierarchical system of habitat conservation and management strategies for land-

TABLE 1.—Key elements of two recent proposals for landscape-based conservation of aquatic biodiversity: a habitat-oriented approach (Sedell et al. 1994) and a taxon-oriented approach (Moyle and Yoshiyama 1994).

Component	Description
Habitat orientation	
Riparian reserves	Portions of the landscape where riparian dependent and stream resources receive primary emphasis, including all permanently flowing streams, lakes, wetlands greater than 0.4 ha, and intermittent streams; riparian reserves protect all bodies of water, inner gorges, all riparian vegetation, 100-year floodplains, and landslide-prone areas
Key watersheds	May contain at-risk fish stocks or serve as sources of high-quality water; key watersheds receive no new roads in roadless areas and no net increase in roads elsewhere, and they receive the highest priority in restoration programs
Watershed analysis	Systematic procedure for characterizing watersheds, providing management prescriptions, and developing restoration strategies and monitoring programs
Watershed restoration	Procedures that restore watershed processes to recover degraded habitat; principal focus is on road removal and upgrading, silvicultural rehabilitation of riparian zones, and restoration of channel complexity with short-term use of in-channel structures
Taxon orientation	
Endangered species listing	Immediate listing under the Endangered Species Act of all species likely to be extirpated within the next 20 years
Management clusters	Implementation of management strategies for clusters of declining species that inhabit the same habitats or drainages, an assumption being that simultaneous protection of coexisting species will improve ecosystem health
Aquatic diversity management areas	Creation of a system of drainages and unique habitats that provides systematic regionwide protection of aquatic biodiversity; most of these water bodies, for which the first priority is to maintain local biodiversity, will be relatively small (<50 km^2) and managed by one governmental agency or landowner
Key watersheds	Representative watersheds more than 50 km^2 in area that are still dominated by native organisms and natural processes, or that have high potential to be restored to such a condition; the management goal for key watersheds is to ensure natural processes are allowed to continue with minimal human interference
Landscape management	Large-scale bioregional planning with the goal of protecting biodiversity and natural processes

scape units ranging in size from small to large (Table 1). Both proposals emphasize protection or restoration of ecosystem processes and natural assemblages of plants and animals rather than restoration measures directed at single species.

These two examples, based on principles of conservation biology, suggest new approaches that go beyond simply setting aside areas of unaltered habitat as preserves or refuges for endangered species. Investigations of physical and biological interactions controlling the structure and function of aquatic ecosystems have shown that drainage systems and the organisms within them are often highly dynamic and influenced by processes operating over several spatial and temporal scales (Naiman et al. 1992). These processes, which include disturbance–recovery cycles, synergistic interactions between environmental components, and biophysical linkages and feedback mechanisms, cause systems to be evolutionary. Such systems are not easily modeled, may rarely conform to steady-state assumptions (Gregory et al. 1991), and will never be fully understood (Holling 1993). Faced with inevitable uncertainty, natural resource managers have begun to explore alternative approaches to the protection of habitat and ecosystems and to the conservation of biodiversity (Angermeier and Williams

1993). New approaches place less emphasis on individual species and more emphasis on natural assemblages and evolutionary history, less emphasis on creating or maintaining certain habitat types in fixed locations and more emphasis on restoring ecological processes leading to the conditions created by natural disturbances, less emphasis on the total number or area of habitat preserves and more emphasis on the spatial and temporal relationships of preserves to one another.

It might be possible to undertake a risk analysis of particularly valued ecosystem resources, such as those we harvest or those at risk of extinction, by determining how well the current "zoning" scheme is working to protect different elements of the "space–time path" of the taxon(s) of interest. Are human uses of resource taxa or their ecosystems compatible with long-term conservation objectives? If not, can more careful selection of space–time windows for managing natural resources significantly reduce extinction risk? How can the overall risk of extinction be distributed more evenly throughout all elements of the space–time path so that all resource users share in conservation responsibility? These questions will be explored more fully in the papers that follow.

Acknowledgments

Participants in the session on ecosystem and habitat conservation at the 1994 symposium Evolution and the Aquatic Ecosystem in Monterey, California—Paul Angermeier, Kurt Fausch, Gary Grossman, Hiram Li, Gordon Reeves, Don Sada, and Isaac Schlosser—contributed freely of their ideas, and I am grateful for having had the opportunity to chair such a stimulating session. I also thank Gordon Reeves, Tom Backman, and Henry Regier for helpful comments on this paper.

References

Angermeier, P. L., and J. E. Williams. 1993. Conservation of imperiled species and reauthorization of the Endangered Species Act of 1973. Fisheries 18(7):34–38.

Franklin, J. F. 1993. Preserving biodiversity: species, ecosystems, or landscapes. Ecological Applications 3:202–205.

Gregory, S. V., F. J. Swanson, and W. A. McKee. 1991. An ecosystem perspective of riparian zones. BioScience 40:540–551.

Harrison, S. 1991. Local extinction in a metapopulation context: an empirical evaluation. Biological Journal of the Linnean Society 42:73–88.

Holling, C. S. 1993. Investing in research for sustainability. Ecological Applications 3:552–555.

Karr, J. R. 1991. Biological integrity: a long-neglected aspect of water resource management. Ecological Applications 1:66–84.

Karr, J. R., and D. R. Dudley. 1981. Ecological perspective on water quality goals. Environmental Management 5:55–68.

Ludwig, D., R. Hilborn, and C. Walters. 1993. Uncertainty, resource exploitation, and conservation: lessons from history. Science 260:17–36.

Moyle, P. B., and R. M. Yoshiyama. 1994. Protection of aquatic biodiversity in California: a five-tiered approach. Fisheries 19(2):6–18.

Naiman, R. J., and eight coauthors. 1992. Fundamental elements of ecologically healthy watersheds in the Pacific Northwest coastal ecoregion. Pages 127–188 in R. J. Naiman, editor. Watershed management: balancing sustainability and environmental change. Springer-Verlag, New York.

Reeves, G. H., and J. R. Sedell. 1992. An ecosystem approach to the conservation and management of freshwater habitat for anadromous salmonids in the Pacific Northwest. Transactions of the 57th North American Wildlife and Natural Resources Conference. 1992: 408–415.

Sedell, J. R., G. H. Reeves, and K. M. Burnett. 1994. Development and evaluation of aquatic conservation strategies. Journal of Forestry 92(4):28–31.

Sheldon, A. L. 1987. Rarity: patterns and consequences for stream fishes. Pages 203–209 in W. J. Matthews and D. C. Heins, editors. Community and evolutionary ecology of North American stream fishes. University of Oklahoma Press, Norman.

Temple, S. A. 1991. Conservation biology: new goals and new partners for managers of biological resources. Pages 45–54 in D. J. Becker, M. E. Kransy, G. R. Goff, C. R. Smith, and D. W. Gross, editors. Challenges in the conservation of biological resources: a practitioner's guide. Westview Press, Boulder, Colorado.

FEMAT (Forest Ecosystem Management Assessment Team). 1993. Forest ecosystem management: an ecological, economic, and social assessment. U.S. Forest Service, Portland, Oregon.

Volkman, J. M., and K. N. Lee. 1994. The owl and Minerva: ecosystem lessons from the Columbia. Journal of Forestry 92(4):48–52.

American Fisheries Society Symposium 17:334–349, 1995

A Disturbance-Based Ecosystem Approach to Maintaining and Restoring Freshwater Habitats of Evolutionarily Significant Units of Anadromous Salmonids in the Pacific Northwest

G. H. REEVES

U.S. Forest Service, Pacific Northwest Research Station,
3200 SW Jefferson Way, Corvallis, Oregon 97331, USA

L. E. BENDA

Department of Geological Sciences, University of Washington, Seattle, Washington 98195, USA

K. M. BURNETT

U.S. Forest Service, Pacific Northwest Research Station

P. A. BISSON

Technology Center, Weyerhaeuser Company, Tacoma, Washington 98477, USA

J. R. SEDELL

U.S. Forest Service, Pacific Northwest Research Station

Abstract.—To preserve and recover evolutionarily significant units (ESUs) of anadromous salmonids *Oncorhynchus* spp. in the Pacific Northwest, long-term and short-term ecological processes that create and maintain freshwater habitats must be restored and protected. Aquatic ecosystems throughout the region are dynamic in space and time, and lack of consideration of their dynamic aspects has limited the effectiveness of habitat restoration programs. Riverine–riparian ecosystems used by anadromous salmonids were naturally subjected to periodic catastrophic disturbances, after which they moved through a series of recovery states over periods of decades to centuries. Consequently the landscape was a mosaic of varying habitat conditions, some that were suitable for anadromous salmonids and some that were not. Life history adaptations of salmon, such as straying of adults, movement of juveniles, and high fecundity rates, allowed populations of anadromous salmonids to persist in this dynamic environment. Perspectives gained from natural cycles of disturbance and recovery of the aquatic environment must be incorporated into recovery plans for freshwater habitats. In general, we do not advocate returning to the natural disturbance regime, which may include large-scale catastrophic processes such as stand-replacing wildfires. This may be an impossibility given patterns of human development in the region. We believe that it is more prudent to modify human-imposed disturbance regimes to create and maintain the necessary range of habitat conditions in space (10^3 km) and time (10^1–10^2 years) within and among watersheds across the distributional range of an ESU. An additional component of any recovery plan, which is imperative in the short-term, is the establishment of watershed reserves that contain the best existing habitats and include the most ecologically intact watersheds.

Biodiversity is not a 'set-aside' issue that can be physically isolated in a few, or even many, reserves. . . . We must see the larger task—stewardship of all the species on all of the landscape with every activity we undertake as human beings—a task without spatial and temporal boundaries. (J. F. Franklin 1993)

Agencies responsible for the development of recovery plans for evolutionarily significant units (ESUs; Waples 1991) of anadromous salmonids *Oncorhynchus* spp. in the Pacific Northwest (PNW) of the United States face difficult tasks. First is the identification of ESUs. Second is the identification of factors that contribute to the decline of a particular ESU. A suite of factors, including habitat loss and degradation, overexploitation in sport and commercial fisheries, variable ocean conditions, and effects of hatchery practices, are responsible for the depressed status of these fish (Nehlsen et al. 1991). The relative importance of each in contributing to the decline of an ESU undoubtedly varies across the region. Any recovery program must address and incorporate consideration of all responsible factors to be successful.

The most common factor associated with declines of anadromous salmonids is habitat degradation, which includes destruction and modification of freshwater and estuarine habitats (Nehlsen et al. 1991; Frissell 1993). Stream and river systems throughout the PNW have been extensively altered by human activities such as agriculture, urbanization, and timber harvest (Bisson et al. 1992). Features of altered ecosystems include changes (generally reductions) in species diversity, changes in species distributions, and losses of habitat types or ecosystem states (Holling 1973; Rapport et al. 1985; Steedman and Regier 1987). Li et al. (1987), Bisson et al. (1992), and Reeves et al. (1993) noted that native salmonid assemblages are simplified in watersheds that have been impacted by various human activities. Native nonsalmonids or introduced species often dominate fish communities in altered ecosystems (Li et al. 1987; Bisson et al. 1992). Habitat degradation is widespread across the region as a result of past and present activities (Bisson et al. 1992; McIntosh et al. 1994). Degradation of terrestrial ecosystems in the PNW (Thomas et al. 1993) and elsewhere (e.g., Wilcove et al. 1986; Rolstad 1991) has resulted in similar changes in terrestrial species assemblages.

Past and many present approaches to management of freshwater habitats of anadromous salmonids have focused on mitigating losses rather than preventing them. This strategy has generally not been successful (Bisson et al. 1992) and habitat loss and degradation continue. Williams et al. (1989) also found that such a strategy failed to halt the decline of habitat quantity and quality for other freshwater fishes. Naturally variable ocean conditions increase the importance of freshwater habitats to anadromous salmonids (Thomas et al. 1993). As a result of this dependence on freshwater habitats and the extensive amount of habitat degradation that has occurred, protection and restoration of upslope and fluvial processes that create and maintain habitats must be an integral component of any recovery program.

Habitat losses may result from human activities that directly destroy habitats or change the long-term dynamics of ecosystems (Rapport et al. 1985; Webb and Thomas 1994). Recent proposals for restoring and protecting habitats of at-risk fishes (e.g., Reeves and Sedell 1992; Thomas et al. 1993; Moyle and Yoshiyama 1994) addressed habitat destruction, primarily through the establishment of watershed-level reserves in which human impacts would be minimized, as advocated by Sheldon (1988) and Williams et al. (1989). We are not aware of anyone who has explicitly addressed long-term ecosystem dynamics in the context of fish conservation. Williams et al. (1989) called for recovery efforts to restore and conserve ecosystems rather than simply habitat attributes, but they did not state how

to accomplish this. Williams et al. (1989) also noted that the failure to address this concern may be a major reason no fish species has ever been recovered after listing under the U.S. Endangered Species Act (ESA; 16 U.S.C. §§ 1531 to 1544).

The purpose of this paper is to examine components of strategies necessary to provide habitat for ESUs of anadromous salmonids in the PNW. Specifically, we will consider the role of natural disturbances in creating and maintaining habitats and how an understanding of this role might be incorporated into long-term recovery planning.

Ecosystem and Spatiotemporal Considerations

May (1994) noted that the most pressing challenge to conservation biology is the need to understand the responses of organisms over large temporal and spatial scales. Some relationships between habitat condition and individual salmonid response have been well established at the scales of habitat unit (e.g., Bisson et al. 1982; Nickelson et al. 1992), stream reach (e.g., Murphy et al. 1989), and (to a lesser extent) watershed (Schlosser 1991). But there is little understanding about how biological entities such as ESUs may respond to habitat patterns at large spatial scales. An initial hurdle in recovery planning for ESUs is identifying appropriate spatial and temporal scales on which to focus.

The ESA requires that ecosystems be considered in the development of recovery plans. The ESUs of anadromous salmonids generally encompass large geographic areas (e.g., Snake River basin in Idaho, upper Sacramento River and its tributaries in northern California). It is difficult to delineate the freshwater ecosystem of an ESU over such large areas. We believe that it is reasonable to consider the composite of individual watersheds within the geographic range of an ESU to be the "ecosystem" and to direct conservation and recovery efforts for freshwater habitats toward the populations that make up an ESU. Currens et al. (in press) suggest that appropriate temporal scales for populations are several decades to centuries and that spatial scales should begin at the watershed level (Figure 1). Although temporal considerations have not been addressed explicitly, recent proposals for restoring and conserving freshwater habitats of anadromous salmonids have emphasized watersheds (e.g., Reeves and Sedell 1992; Thomas et al. 1993; Moyle and Yoshiyama 1994). We concur with this direction and believe that for management and imple-

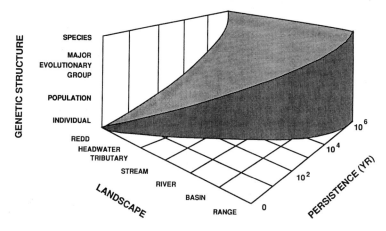

FIGURE 1.—A general hierarchial model of biological levels of organization for anadromous salmonids and the spatial and temporal scales that influence them (from Currens et al., in press).

mentation purposes, the individual watershed is the appropriate focus for recovery plans.

Within watersheds, recovery programs for ESUs must address not only root causes directly responsible for the immediate loss of habitat quantity and quality but also ecosystem processes that create and maintain habitats through time. In developing an ecosystem approach to the conservation and restoration of endangered organisms, it must be recognized that ecosystems are generally dynamic in space and time because of natural disturbances, particularly at large spaciotemporal scales (Botkin 1990).

A mosaic of conditions occurs within an ecosystem at any time as a consequence of disturbances (White and Pickett 1985). Any disturbed patch develops different habitat conditions or states over time. The assemblage of organisms in a particular patch changes with changing habitat conditions (Table 1; Huff and Raley 1991; Raphael 1991). Points along the trajectory of disturbance and recovery represent various states in the potential range of states that an ecosystem may exhibit. The locations of patches in particular states shift across the landscape due to the stochastic nature of most natural disturbances. In the PNW, terrestrial ecosystems are very dynamic in space and time as a result of natural disturbances such as fire and wind (Agee 1991, 1993). Holling (1973) noted that if resources are to be sustained, the dynamic nature of ecosystems and the need to maintain the diversity of ecosystem states must be recognized. Attempts to view and manage systems and resources in a static context may increase the rate of extinction of some organisms (Holling 1973).

Persistence in Dynamic Environments

It is unlikely that individual populations persist over long terms at the local scale in a dynamic environment (Hanski and Gilpin 1991; McCauley 1991; Mangel and Tier 1994). In dynamic environments, "... some patches are empty (but liable for colonization), while others are occupied (but liable to extinction). In such circumstances, the lights of individual patches wink on and off unpredictably, but the overall average level of illumination—the overall density of the metapopulation—may remain

TABLE 1.—Bird species found in different seral stages of Douglas-fir forests of Oregon and Washington (from Huff and Raley 1991).

Species	Seral stage		
	Early	Mid	Late
Chestnut-backed chickadee *Parus rufescens*	X	X	X
Hermit warbler *Dendroica occidentalis*	X	X	X
Western flycatcher *Empidonax difficilis*		X	X
Winter wren *Troglodytes troglodytes*		X	X
Red-breasted nuthatch *Sitta canadensis*	X		
Swainson's thrush *Catharus ustulatus*	X		
American robin *Turdus migratorius*	X		
Northern spotted owl *Strix occidentalis caurina*		X	
Pileated woodpecker *Dryocopus pileatus*		X	
Varied thrush *Ixoreus naevius*		X	

relatively steady" (May 1994). Metapopulations persist in dynamic environments through a suite of adaptations. Response to change varies with the level of biological organization (Karr and Freemark 1985; White and Pickett 1985). Physiological, morphological, and behavioral adaptations occur at the individual level. Life history patterns (Stearns 1977), reproductive rates, and modes of dispersal (Vrijenhoek 1985) are adaptations at the population level.

Several studies have documented the response of terrestrial populations to periodic catastrophic disturbances. Christensen (1985) cited examples of declines in small-mammal populations after fires in shrublands. Populations recovered after the vegetation did, and immigration from surrounding areas was a primary factor in the mammal recoveries. Colonizers of perturbed areas may be genetically predisposed to disperse (Sjorgen 1991) surplus to other populations (Hanski 1985; Pulliam 1988) or chance arrivals (Goodman 1987). Such adaptations increase the probability that metapopulations will persist through time.

The Dynamic Aquatic Environment

Aquatic ecologists and managers often do not have the long-term dynamic view of ecosystems held by terrestrial ecologists (White and Pickett 1985) and advocated by Holling (1973). Streams in the PNW (Resh et al. 1988) and elsewhere (Pringle et al. 1988; Reice 1994) are dynamic within relatively short time frames; typically a year to a decade, at the watershed scale, in response to floods or mass wasting (Swanston 1991). It is generally held that biological populations (some of them but not the entire assemblage) and physical features of these systems recover relatively quickly after such disturbances (e.g., Bisson et al. 1988; Lamberti et al. 1991; Pearson et al. 1992). Similar short-term responses of lotic fishes to disturbances have been noted in other areas (e.g., Hanson and Waters 1974; Matthews 1986). Over extended periods, habitat conditions in streams of similar size within a geomorphic region should be relatively uniform within and among watersheds (Vannote et al. 1980).

In contrast to terrestrial ecology, no theory predicts the mosaic of aquatic conditions or ecological states caused by disturbances and the corresponding responses of fish populations over extended periods. Minshall et al. (1989), Naiman et al. (1992), and Benda (1994) have proposed that aquatic ecosystems are dynamic in space and time at the watershed scale. The type, frequency, inten-

sity and effect of disturbance vary with channel size and location within the watershed (Benda 1994).

An Oregon Example

The natural disturbance regime in the central Oregon Coast Range includes infrequent stand-resetting wildfires and frequent intense winter rainstorms. Wildfires reduce the soil-binding capacity of roots. When intense rainstorms saturate soils during periods of low root strength, concentrated landsliding into channels and debris flows may result. Such naturally occurring disturbances in stream channels can have both immediate impacts on and long-term implications for anadromous salmonids. Immediate impacts include direct mortality, habitat destruction, elimination of access to spawning and rearing sites, and temporary reduction or elimination of food resources. Longer-term effects may be positive, however; landslides and debris flows introduce essential habitat elements, such as large wood and sediment, into channels and affect storage of these materials. The configuration of channel networks, the delivery, storage, and transport of sediment and wood, and the decomposition of woody debris interact to create, maintain, and distribute fish habitat over the long term.

Three streams in the central Oregon Coast Range were examined to explore some of the responses of salmonids and their habitats to the natural disturbance regime (G. H. Reeves, U.S. Forest Service, Pacific Northwest Research Station, unpublished data). The streams have gradients between 1 and 2.5% and drainage areas between 14 and 18 km^2. Benda (1994) examined these and other streams as part of a study to model watershed erosion and sedimentation. Summer habitats and assemblages of juvenile anadromous salmonids were inventoried in 1988 and 1989. The time since catastrophic wildfire and hillslope failure differed among streams.

The watershed of Harvey Creek was burned by an intense wildfire in the late 1800s, and the forest was principally 90–100-year-old Douglas fir *Pseudotsuga menziesii* at the time of the study. The channel contained a large volume of sediment in storage throughout the lower portion of the drainage network and thus was considered to be in an aggradational state (mean depth of deposits, 1.8 m). Evidence of burned wood in the channel indicated widespread landsliding followed the fire. Gravel was the dominant substrate (Figure 2). Larger substrate particles and large woody debris were buried in the gravel deposits. Deep pools (mean depth, 0.9 m),

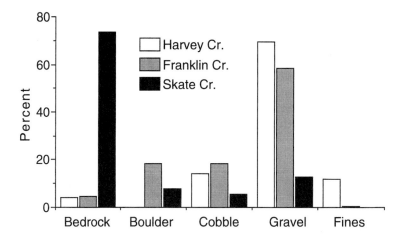

FIGURE 2.—Substrate composition in three streams of the central Oregon Coast Range that had differing histories of major natural disturbance. The time since the last major natural disturbance was 90–100 years for Harvey Creek, 160–180 years for Franklin Creek, and more than 330 years for Skate Creek. Cr = Creek (G. H. Reeves, unpublished data collected in July 1988 and 1989).

usually formed by scour around large wood, were the most common habitat units but were not hydraulically complex. Fewer pieces of large wood were observed in Harvey Creek than in the other study streams (Table 2), because wood deposited in the channel by the hillslope failure had been buried beneath sediment and little wood was being recruited from the relatively young surrounding forest. The juvenile salmonid assemblage was numerically dominated by age-0 coho salmon *Oncorhynchus kisutch*, but age-1 steelhead *O. mykiss* (about 1%) and cutthroat trout *O. clarki* (about 1%) were also present (Table 3).

The Skate Creek watershed was forested by trees more than 330 years of age, suggesting that the stream had not been subjected to a fire or hillslope failure for a long time. Habitat conditions in the stream were very simple. The substrate was predominantly bedrock and boulders with small, local-

ized patches of stored sediment (Figure 2). Riffles were thin sheets of water flowing over bedrock. Although large wood was more abundant than in the other streams examined (Table 2), the lack of a deformable gravel bed greatly limited the wood's effectiveness in forming pools. Therefore, pools were shallow (mean depth, 0.1 m) and often in bedrock depressions. Juvenile coho salmon were the only salmonids found in Skate Creek (Table 3).

Franklin Creek was intermediate in time since disturbance. Based on the present vegetation, we estimated that catastrophic wildfire and landsliding occurred 160–180 years ago in this watershed. Mean depth of sediment in the channel was 0.7 m, and there was a greater array and more even distribution of substrate types than in the other streams (Figure 2). Mean pool depth was 0.35 m, less than half the mean depth of pools in Harvey Creek. As a result of sediment transport from the channel that

TABLE 2.—Mean number of pieces of large wood (>0.3 m in mean diameter and >3 m long) per 100 m in three streams of the central Oregon Coast Range that had differing histories of major natural disturbance (G. H. Reeves, unpublished data collected in July 1988 and 1989).

Stream	Years since last major disturbance	Mean pieces of wood/100 m
Harvey Creek	90–100	7.9
Franklin Creek	160–180	12.3
Skate Creek	>330	23.5

TABLE 3.—Composition of the assemblage of juvenile anadromous salmonids in three streams of the central Oregon Coast Range that had differing histories of major natural disturbance (G. H. Reeves, unpublished data collected in July 1988 and 1989).

	Mean percent of estimated total numbers		
Stream	Age-0 coho salmon	Age-1 steelhead	Age-1 cutthroat trout
Harvey Creek	98.0	1.0	1.0
Franklin Creek	85.0	12.5	2.3
Skate Creek	100.0	0.0	0.0

partially excavated buried wood and of recruitment of wood from the surrounding riparian forest, Franklin Creek had more pieces of large wood than Harvey Creek, though fewer than Skate Creek (Table 2). The combination of these factors produced the most complex habitat conditions observed in the three streams. Coho salmon numerically dominated the juvenile salmonid assemblage, but steelhead and cutthroat trout were relatively more abundant than in Harvey Creek (Table 3). Botkin et al. (1995) found that the healthiest stocks of various anadromous salmonids in coastal Oregon and northern California occurred where riparian vegetation within 0.5 km of the stream was similar to that found along Franklin Creek.

These field observations and a simulation model developed by Benda (1994) indicate that under the natural disturbance regime, variation in the timing and location of erosion-triggering fires and storms results in episodic delivery of materials that cause stream channels to alternate between aggraded and degraded sediment states. This generates spatial and temporal variability in both habitat conditions (Figure 3) and components of the juvenile salmonid assemblage within and among watersheds. Benda's (1994) simulation model indicated that wildfires of a mean size of about 30 km² occurred in the central Oregon Coast Range over the past 3,000 years with a return interval of 200–300 years. The cumulative probability of wildfire increased with increasing watershed size; for a 200-km² drainage basin, the frequency of stand-resetting wildfires was once every 45 years.

At a coarse level of resolution, Benda's (1994) model predicted that channels in watersheds of similar drainage area have characteristic patterns of sediment delivery, storage, and transport that vary with position in the drainage network and drainage area. Under a natural fire regime, for example, streams in the upper drainage experience large sediment deposits (>1 m thick) infrequently (once every hundreds of years) because sources of mass failure are few and sediment bedload transport rates are low. Channels in the central part of the network (drainage area, 30–50 km²) have the highest probability of containing thick sediment deposits, partly due to relatively high cumulative probabilities of upstream mass wasting. These channels experience cycles of accumulation and flushing as sediment is transported in waves into and then out of them. Channels higher than sixth order with large drainage areas (>100 km²), are governed by lateral migrations rather than by cycles of filling and emptying. Sediment waves moving from tributaries into

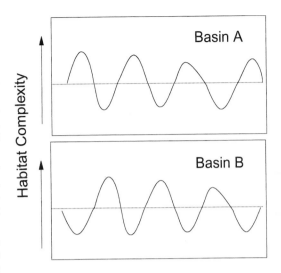

FIGURE 3.—Hypothetical historical conditions of fish habitat in different streams within and among watersheds in the central Oregon Coast Range (based on Benda 1994). The horizontal axis is time.

larger channels mix at tributary junctions. Although sediment waves occur once in 5–10 years, they probably are inconspicuous (depth, < 0.2m). Habitat conditions in unbraided channels in the lowest portion of the network likely are more uniform than in higher elevation channels. It is important to note that the occurrence of a particular state will be affected by local circumstances that influence sediment retention, such as the amount of large wood in the channel, but these were not modeled by Benda (1994).

In the model, stream channels draining watersheds similar in area to Harvey, Franklin, and Skate creeks oscillated over time between states of sediment aggradation and degradation (Benda 1994). For central Oregon Coast Range channels, the average period between the state characterized by sediment deposits of intermediate depth, as exhibited in Franklin Creek, and the sediment-poor state was estimated to be more than 100 years. The model also produced an average duration of gravel-rich conditions of 80 years (range, 50–300 years) in small basins. Harvey Creek has apparently been gravel-rich for 100 years, and may continue to be so for another 100 years, although gravel-rich areas will likely move downstream over time. Again, the duration of a particular condition would be affected by local circumstances that were not modeled by Benda (1994).

Juvenile salmonid assemblages are likely associated with each state predicted by the model. Benda's (1994) simulation indicated that sediment supply would be limited at any given location in these small streams a majority of the time. Based on field observations, coho salmon would have dominated such simplified habitats. When a channel segment was not in this degraded state, it would shift between states of aggradation and intermediate sediment supply. Two additional salmonid species, steelhead and cutthroat trout, are expected to occur in aggraded channels. The intermediate state is characterized by intermediate sediment depths and more complex habitat, which should support a juvenile salmonid assemblage containing greater proportions of trout. Benda (1994) has developed long-term average probabilities for the time a channel segment would have spent in each state. Applied to a population of channels (those with similar gradient, drainage area, etc.) for a particular time, these probabilities can be used to estimate the landscape-scale mosaic of habitat conditions or biodiversity. For example, in watersheds of an area similar to those in the field example (approximately 25 km^2), the frequency distribution developed by Benda (1994) indicates that a majority of channel segments in the central Oregon Coast Range should have limited sediment supplies at any particular time and thus should contain relatively simple habitats.

A natural mosaic of habitat conditions for anadromous salmonids has likely existed elsewhere in the PNW; the features and relative proportion of each channel state should vary with climate, vegetation, drainage pattern, and spatial scale. Meyer et al. (1992) found cycles of aggradation and degradation associated with wildfires and hillslope failures in a Wyoming stream like those just described for the central Oregon Coast Range. It seems reasonable to assume that channel conditions over time were similar to those observed in the Oregon streams we examined.

In summary, the natural disturbance regime of the central Oregon Coast Range is described by the frequency, size, and spatial distribution of wildfires and landslides, and this regime has been responsible for developing a range of channel conditions within and among watersheds. The structure and composition of the juvenile anadromous salmonid assemblage varies with channel conditions. A disturbance regime that resembles this natural regime must be incorporated into any recovery plan for freshwater habitats of ESUs of anadromous salmonids.

Adaptations of Anadromous Salmonids

Anadromous salmonid populations in the Pacific Northwest are well adapted to dynamic environments. Adaptations include straying by adults, high fecundity, and mobility of juveniles. Straying by adults is genetically controlled, directly or indirectly (Quinn 1984), and aids the reestablishment of populations in disturbed areas on large (Neave 1958) and local scales (Ricker 1989). Strays would be reproductively most successful where local populations have been reduced or extirpated (Tallman and Healey 1994), provided there are suitable spawning and rearing conditions. Individuals from more than one population may recolonize depopulated areas, increasing the genetic diversity of the new population.

Movements of juveniles from natal streams to other areas also facilitate the establishment of new populations. Some individuals may be genetically programmed to move; others may be displaced from high-density populations (Northcote 1992). Chapman (1962) suggested that juvenile salmonids that were unable to obtain territories and migrated downstream were less fit individuals. However, at least some may leave voluntarily if emigration improves survival. Tschaplinski and Hartman (1983) found that juvenile coho salmon moving downstream in a small British Columbia stream took up residence in unoccupied habitats and grew rapidly.

High fecundity contributes to the establishment and growth of a local population if conditions are favorable. Pacific salmon are relatively fecund for benthic-spawning fishes with large eggs. Pink salmon *Oncorhynchus gorbuscha*, the smallest species, typically possess 1,200–1,900 eggs per female (Heard 1991). Adult female chinook salmon *O. tshawytscha*, the largest species, may contain more than 17,000 eggs (Healey 1991). Both high fecundity and large eggs contribute to the reproductive success of species whose young have extended periods of intragravel residence. These traits also facilitate growth when conditions are suitable.

Human Alterations of Disturbance Regimes

Natural ecosystems generally have a large capacity to absorb change without being dramatically altered. Resilience of an ecosystem is the degree to which the system can be disturbed and still return to a domain of behavior in which processes and interactions function as before (Holling 1973). If a disturbance exceeds the resilience of the system, the domain may shift and the system will develop new conditions or states that had not previously been exhibited. Yount and Niemi (1990), modifying the disturbance definition of Bender et al. (1984), distinguished "pulse" disturbances from "press" disturbances. A pulse disturbance allows an ecosystem to remain within its normal bounds or domain and to recover the conditions that were present prior to disturbance. A press disturbance forces an ecosystem to a different domain or set of conditions. Yount and Niemi (1991) considered many anthropogenic disruptions, such as timber harvesting and urbanization, to be press disturbances. Gurtz and Wallace (1984) hypothesized that stream biota may not be able to recover from the effects of anthropogenic disturbances because these have no analogues in the natural disturbance regime, and organisms may not have evolved the appropriate breadth of habitat or reproductive requirements.

Modifications in the type of disturbance or in the frequency and magnitude of natural disturbances can alter the species composition, habitat features, and resilience of an ecosystem (White and Pickett 1985; Hobbs and Huenneke 1992). Alteration or loss of habitats as a result of changes in the disturbance regime can bring on extirpation of some species, increases in species favored by available habitats, and invasions of exotic organisms (Levin 1974; Harrison and Quinn 1989; Hansen and Urban 1992). We also believe that changes in the legacy of disturbance (the conditions that exist immediately following a disturbance) may be another important component of disturbance regimes that can be altered. Changes in the legacy can influence a system's resiliency by altering habitat conditions created immediately following a disturbance and by altering future conditions.

We do not mean to imply that every human action or activity causes a press disturbance; the impact of anthropogenic disruptions is minimal in some ecosystems (e.g., Attiwill 1994a, 1994b). However, we believe human activities that affect anadromous salmonids and their habitats, such as timber harvesting, urbanization, and agriculture, do generate press disturbances. These disturbances can result in the loss of habitats or ecosystem states necessary for various anadromous salmonids (Hicks et al. 1991; Bisson et al. 1992). Human activities in the PNW have altered the recovery potential of ecosystems, which may be as responsible for the decline of habitat as the direct impact of the activity itself.

A Disturbance-Based Ecosystem Approach to Freshwater Habitat Recovery

We believe that any long-term program for restoring and maintaining freshwater habitats for ESUs of anadromous salmonids must accommodate the dynamic nature of the PNW landscape. Given the dynamic nature of terrestrial ecosystems (Agee 1991, 1993), the links between terrestrial processes and aquatic ecosystems, the apparent adaptations of anadromous fish for persisting in a dynamic environment, and the limited available evidence (based on central Oregon Coast Range streams) of non-steady-state behavior of sedimentation and habitats, we believe a dynamic approach is advisable in any recovery program. In the following sections, we describe the components that should be included in this approach.

Watershed Scale Reserves: Short- and Long-term Considerations

Anadromous salmonids exhibit typical features of "patchy populations"; they exist in a dynamic environment and have good dispersal abilities (Harrison 1991, 1994). Conservation of patchy populations requires the conservation of numerous patches of suitable habitat and the potential for dispersal among them (Harrison 1991, 1994). Size and spacing of reserves should depend on the behavior and dispersal characteristics of the species of concern (Simberloff 1988). Rieman and McIntyre (1995) used logistic regression to investigate the influence of patch size, as well as stream width and gradient, on populations of bull trout *Salvelinus confluentus* at the reach, stream, and watershed scales. This approach could be helpful in identifying critical features of reserves for anadromous salmonids. In our current thinking on reserve planning for ESUs of anadromous salmonids, we consider patches to be watersheds, the size of which should depend on the species and geographic location. It is difficult to predict the exact number of patches required to sustain an organism (Lawton et al. 1994). Lande (1988) could do this for the northern spotted owl because data were available on essential life history variables. It is unlikely that predictions could be obtained for many other species, including ESUs of anadromous salmonids, because necessary life history data are often lacking (Lawton et al. 1994).

In the short term, reserves should be established in watersheds with good habitat conditions and functionally intact ecosystems to provide protection for these remaining areas. Reserves of this type are likely to be found in wildernesses and roadless areas on federal lands. Examples of watersheds that fulfill this requirement include some of the key watersheds identified by Reeves and Sedell (1992), the class I waters of Moyle and Yoshiyama (1994), and

FIGURE 4.—Distribution of tier 1 key watersheds identified by Thomas et al. (1993).

Systems should qualify based on the extent of habitat degradation and the degree to which their natural diversity and ecological processes are retained. Examples of such watersheds are some of the key watersheds identified by Reeves and Sedell (1992), some tier 1 key watersheds identified by Thomas et al. (1993), the class III waters of Moyle and Sato (1991), and the class III waters of Moyle and Yoshiyama's (1994) aquatic diversity management areas. Restoration programs implemented in these watersheds should be holistic in their approach. They should address instream habitat concerns, prevent further degradation, and restore ecological processes that create and maintain instream habitats.

It is imperative to recognize and acknowledge that identified reserves will experience natural and, often, anthropogenic disturbances. Thus, simply putting aside a fixed set of watersheds as reserves may not provide habitats of sufficient quantity and quality to ensure long-term persistence of ESUs. Conservation reserves have generally been established and managed without consideration of long-term disturbance dynamics and the biological and evolutionary processes that influence organisms contained within them (Western 1989). Consequently, their populations may have higher probabilities of extirpation in the long term than expected. Reasons for this include isolation of reserves from surrounding areas of suitable habitat resulting from habitat fragmentation (MacArthur and Wilson 1967; Diamond and May 1976); restriction or elimination of migration and dispersal (Elsenberg and Harris 1989; Harris and Elsenberg 1989); and boundary effects associated with surrounding areas, such as invasion of native and exotic competitors, disease, and pollution (Shonewald-Cox 1983; Wilcox 1990). Hales (1989) and White and Bratton (1980) noted that in dynamic landscapes, reserves may act as holding islands that persist only for relatively short ecological periods (100–200 years). Reserves should be large enough to allow operation of the natural disturbance regime and to support a mosaic of patches with different biological and physical attributes (Pickett and Thompson 1978).

Gotelli (1991) noted that reserve strategies such as those proposed by Harrison (1991, 1994) do not address the longevity of patches. This is a major concern in dynamic environments like those of the PNW. Modification of the strategy proposed by Harrison (1991, 1994) to accommodate a dynamic environment is a prudent approach in the development of a recovery strategy for anadromous salmo-

the tier 1 key watersheds of Thomas et al. (1993) (Figure 4). Ideally these reserves should be distributed across the range of an ESU and should contain subpopulations of it. Because of the critical importance of these watersheds in the short term, activities within them should be minimized or modified to protect the integrity of existing physical and ecological conditions.

Identification of watersheds that have the best potential for being restored should also be a short-term priority of any recovery strategy. These watersheds could serve as the next generation of reserves.

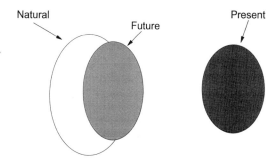

Natural   Future   Present

FIGURE 5.—Conceptual representation of the range of conditions experienced by aquatic ecosystems historically, currently, and under a new disturbance regime (modified from H. Regier, University of Toronto, personal communication).

nid habitats. Specifically, there is need for a shifting mosaic of reserves that change location in response to the ability of specific watersheds to provide suitable habitat conditions.

A New Human-Influenced Disturbance Regime

Under natural wildfire regimes of the PNW, the condition of freshwater habitats for anadromous salmonids was likely regulated by episodic delivery of sediment and wood to the channel. Given that human demands on ecosystems will only increase, we believe that returning the entire landscape to the natural wildfire regime will not be possible. Therefore, human activities will have to be molded into an analogous disturbance regime if habitats are to recover and persist. First must come an understanding of how the natural disturbance regime created and maintained habitats for anadromous salmonids through time and how it has been modified by human activity. Then it will be necessary to identify those human activities that can be altered to maintain desired ecological processes and leave the legacy that allows recovery and persistence of required freshwater habitats. In other words, the character of anthropogenic disruption must be shifted from a press to a pulse disturbance (Yount and Niemi 1990) (Figure 5). The following is an example of how we believe timber harvest and associated activities, as currently practiced on federal lands in the central Oregon Coast Range, have affected habitat and biodiversity of anadromous salmonids and how these could be adjusted to help create suitable conditions in space and time. We believe that timber management may offer more immediate opportunities than agricultural or urban processes for modifying practices to create a human-influenced disturbance regime that maintains components of the natural regime.

Disturbance caused by timber harvest differs from stand-resetting wildfires in the central Oregon Coast Range in several respects. One difference is the legacy of the disturbance. Wildfires left large amounts of standing and downed wood (Agee 1991), which was often delivered to channels along with sediment in storm-generated landslides (Benda 1994). This promoted development of high-quality habitats as sediment was transported from the system, leaving the wood behind (Benda 1994). Timber harvest, as it is generally practiced, reduces the amount of large wood available to streams (Hicks et al. 1991; Reeves et al. 1993; Ralph et al. 1994), so when harvest-related hillslope failures occur, sediment is the primary material delivered to the channel (Hicks et al. 1991). Because large wood is an integral component of aquatic habitats and a major influence on sediment transport and storage, the potential for developing complex habitats is much lower when small rather than large amounts of wood are in the channel. Consequently, channels may be simpler following timber harvest than they are after wildfires.

The interval between events also affects the conditions that develop after a disturbance (Hobbs and Huenneke 1992). Under the natural disturbance regime, variation in the timing and location of erosion-triggering fires and storms probably caused stream channels to alternate between aggraded and degraded sediment states, generating temporal variability in both fish habitats and assemblages of juvenile salmonids. Wildfires occurred on average about once every 300 years in the central Oregon Coast Range (Benda 1994). In watersheds smaller than 30 km^2, postfire development of the most diverse physical and biological stream conditions may have taken 150 years or more (see earlier discussion). Timber harvest generally occurs at intervals of 60–80 years on public lands and 40–50 years on private timberlands. This may not allow sufficient time for the development of conditions necessary to support the array of fishes found under natural disturbance regimes.

A third difference between timber harvest and a disturbance regime dominated by wildfire is the spatial distribution of each. Based on a fire frequency of once every 300 years, Benda (1994) estimated that on average, 15–25% of the forest in the central Oregon Coast Range would have been in early successional stages because of recent wildfires. In contrast, the area affected by timber harvest is much greater. For example, in the Mapleton District of the Siuslaw National Forest, which contains the watersheds studied by Benda (1994), approximately 35% of the forest is in early succession (J. Martin, Siuslaw National Forest, personal communication). If private lands were included, the percentage would be greater. The present forested landscape is more homogeneous with respect to seral stage than it was historically. Just as the distribution of terrestrial habitat has been altered by switching from a wildfire-driven to a harvest-driven disturbance regime, it is also possible that the distribution of aquatic habitats is different today than it was under the natural disturbance regime and thus less capable of supporting a diverse juvenile salmonid assemblage.

A fourth difference between the natural wildfire-driven and the current harvest-driven regime is the size of disturbance and the landscape pattern generated by the disturbance. Timber on federal lands has typically been managed by widely dispersed activities; approximately 174,000 km of roads exist across public lands in the range of the northern spotted owl (Thomas et al. 1993), and many millions of hectares have been affected by small harvests of approximately 16 ha. Wildfires, on the other hand, often generate a larger but more concentrated disturbance. When wildfires occurred in the central Oregon Coast Range, they tended to be large (mean, 3,000 ha), stand-resetting fires (Benda 1994). Consequently, the spatial pattern and amount of sediment delivered to channels would likely be different under these two disturbance regimes. In naturally burned areas, storms occurring during periods of low root strength would generate large volumes of sediment from nearly synchronous hillslope failures and channels would become aggraded. Subsequently, delivery would be reduced while source areas recharged. This, coupled with downstream flushing of stored sediments, would bring the channel to an intermediate level of sediment storage and a corresponding period of high-quality habitat. In unburned watersheds, sediment delivery rates would remain low. In contrast, timber harvest activities are dispersed; thus, we presume that mass wasting would be more widely distributed and would deliver sediment at elevated rates in most managed watersheds. Storm-generated landslides would be asynchronous, being governed through time by harvest schedules. Cycles of channel aggradation and degradation probably would not be apparent and sediment delivery, at a landscape scale, would likely be chronic rather than episodic. These factors would conspire to produce relatively low-quality habitats across the landscape and eliminate the potential for attaining the most complex habitat states.

In summary, the differences between present timber harvest disturbance regime and the natural disturbance regime have important implications for stream ecosystems and anadromous salmonids. Stream habitat, at a point in the channel, is less complex under the timber harvest regime (Hicks et al. 1991; Bisson et al. 1992) than under the natural regime, and the potential for achieving greater complexity is also reduced. This is primarily a result of the reduced legacy and shorter interval between disturbance events under the timber harvest regime. In addition, landscape-level habitat heterogeneity is reduced under the harvest regime because the disturbance is more dispersed and widespread.

The new disturbance regime created by timber harvest should address the concerns just listed. The legacy of hillslope failures associated with timber management activities needs to include more large wood. Benda (1990) identified the attributes of first- and second-order streams that favor the delivery of desirable material to fish-bearing channels. Increasing the extent of riparian protection along these streams, as proposed by Thomas et al. (1993), obviously increases the potential delivery of wood. Such a strategy may not result in wood loadings as large as occurred naturally because trees away from the riparian zone will have been removed. However, this strategy should increase wood loadings beyond what is currently possible and should allow channels to develop more complex habitats.

Longer intervals between harvest rotations could be another component of this new disturbance regime. In single basins in the central Oregon Coast Range, the desirable interval may be 150–200 years, although this is a first approximation. The exact interval would depend on the magnitude and areal extent of the natural disturbance regime and the time it takes for favorable habitat conditions to develop if adequate large wood and sediment are available. It will be different in other regions. Extending rotation time would also provide benefits to many terrestrial organisms.

Concentrating rather than dispersing management activities could be another element of the new disturbance regime. This would more closely resemble the pattern generated by natural disturbances than does the current practice of dispersing activity in small areas. For example, if a basin has four subwatersheds, it may be better to concentrate activities in one for an extended period (50–75 years) than to operate in 25% of each one at any time

A. Dispersion of Activity

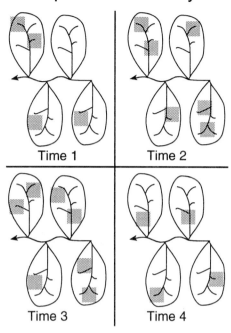

B. Concentration of Activity

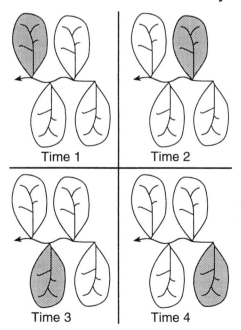

FIGURE 6.—Examples of patterns resulting from (**A**) dispersing and (**B**) concentrating land management activities in a watershed over time (modified from Grant 1990).

(Figure 6). Grant (1990) modeled such a scenario to determine its effects on patterns of peak flow and found that there was little difference between the two approaches. Franklin and Forman (1987) believed that dispersing activity (Figure 6A) increases habitat and landscape fragmentation and is more detrimental than concentrating activities (Figure 6B) to terrestrial organisms that require late-successional forests. We believe that concentrating activity would have similar benefits for the aquatic biota if the elements discussed previously are included. This approach could also be linked to planning future reserves and reducing risks in reserves, so it merits consideration in the development of habitat recovery efforts.

All of the elements discussed above must be included in the development of a new disturbance regime if the regime is to be successful at creating and maintaining habitats for anadromous salmonids. Exclusion of any element greatly reduces the potential for success. Our concept of designing a disturbance regime around human activities could complement parts of other strategies proposed for management of the central Oregon Coast Range (Noss 1993) and other parts of the PNW (e.g., Thomas et al. 1993). These call for reserves in which human activity is curtailed or eliminated. The proposed new disturbance regime could be applied to areas outside any such reserve system, particularly in the short term. It could also guide management strategies in reserves where limited human activity is allowed. The long-term goal of this effort would be to create refugia to replace and complement refugia in permanently designated reserves, such as wilderness areas and other withdrawn lands.

Conclusions

Plans directed at the freshwater habitat for ESUs of anadromous salmonids in the Pacific Northwest must be focused on restoring and maintaining ecosystem processes that create and maintain habitats through time. It is important to insure that as good habitats "wink out," either through anthropogenic or natural disturbances or through development into new ecological states, others "wink on." Designating the most intact remaining aquatic ecosystems as reserves is essential for meeting near-term requirements. In the long term, a static reserve system alone is unlikely to meet the requirements of these fish. Management must also be directed at developing the next generation of reserves. Strategies should be designed and implemented that treat land management activities as disturbance events to be manipulated so as to retain the ecological processes necessary to create and maintain freshwater habitat through time. Although necessary for anadromous salmonids, the approach of moving reserves and managing periodic disturbances may not be suitable for locally endemic or immobile biota. It is imperative to consider the needs of other organisms in the development of any habitat recovery program for ESUs of anadromous salmonids.

Many hurdles must be overcome to make our approach effective. First, biologists, managers, and planners need to think in longer time frames than they are generally accustomed to using. They need to acknowledge that ecosystems are dynamic in space and time over these longer periods. Simply designating reserves and expecting these to function as such for extended periods may be unrealistic; some benefits may accrue in the short term, but in the long run it is unlikely that habitats of sufficient quality and quantity will be available to sustain ESUs of anadromous salmonids. Expectations about habitat conditions in streams must change; a stream will not always have suitable habitats for anadromous salmonids, and all streams should not be expected to have suitable habitats at the same time. A consequence of a dynamic view is that, perspectives must be regional (Holling 1973). The percentage of the landscape that should contain suitable habitats must be identified and the temporal and spatial distributions of these habitats determined.

Finally, disturbance must be recognized as an integral component of any long-term strategy. This will be a difficult hurdle to overcome. It requires educating resource managers, scientists, administrators, politicians, and the public so they realize that periodic disturbance is not necessarily negative. To the contrary, disturbance may be necessary in order to have productive habitats for ESUs of anadromous salmonids in the PNW over long periods.

Acknowledgments

The development and refinement of the ideas within and context of this paper benefited from discussions and interactions with several people including Bernard Bormann, Ken Currens, Tom Dunne, Gordon Grant, Robert Gresswell, Brendan Hicks, Jim Hall, Hiram Li, Bruce Marcot, Tom Northcote, Scott Overton, Henry Regier, Bruce Rieman, Tom Spies, Fred Swanson, and Tommy Williams. Robert Gresswell and Reed Noss provided helpful reviews of early versions of this manuscript. Much of the research for this paper was done while the senior author was a visitor at Waikato University, Hamilton, New Zealand. We give special thanks to Jennifer Nielsen, who organized this symposium and was very patient with us as we wrote this paper.

References

Agee, J. K. 1991. Fire history of Douglas-fir forests in the Pacific Northwest. Pages 25–34 in L. F. Ruggiero, K. B. Aubry, A. B. Carey, and M. H. Huff, technical coordinators. Wildlife and vegetation of unmanaged Douglas-fir forests. U.S. Forest Service General Technical Report PNW-GTR-285.

Agee, J. K. 1993. Fire ecology of Pacific Northwest forests. Island Press, Washington, DC.

Attiwill, P. M. 1994a. The disturbance of forest ecosystems: the basis for conservative management. Forest Ecology and Management 63:247–300.

Attiwill, P. M. 1994b. Ecological disturbance and the conservative management of eucalypt forests in Australia. Forest Ecology and Management 63:301–346.

Benda, L. E. 1990. The influence of debris flows on channels and valley floors in the Oregon Coast Range, USA. Earth Surface Processes and Landforms 15: 457–466, Chichester, UK.

Benda, L. E. 1994. Stochastic geomorphology in a humid mountain landscape. Doctoral dissertation. University of Washington, Seattle.

Bender, E. A., T. J. Case, and M. E. Gilpin. 1984. Perturbation experiments in community ecology: theory and practice. Ecology 65:1–13.

Bisson, P. A., J. L. Nielsen, R. A. Palmason, and L. E. Grove. 1982. A system of naming habitat types in small streams, with examples of habitat utilization by salmonids during low streamflow. Pages 62–73 in N. B. Armantrout, editor. Aquisition and utilization of aquatic habitat inventory information. American Fisheries Society, Western Division, Bethesda, Maryland.

Bisson, P. A., J. L. Nielson, and J. W. Ward. 1988. Summer production of coho salmon stocked in Mt. St. Helens streams 3–6 years after the 1980 eruption. Transactions of the American Fisheries Society 117:322–335.

Bisson, P. A., T. P. Quinn, G. H. Reeves, and S. V. Gregory. 1992. Best management practices, cumulative effects, and long-term trends in fish abundance in Pacific Northwest river systems. Pages 189–232 in R. J. Naiman, editor. Watershed management: balancing sustainability and environmental change. Springer-Verlag, New York.

Botkin, D. B. 1990. Discordant harmonies: a new ecology for the twenty-first century. Oxford University Press, New York.

Botkin, D. B., K. Cummins, T. Dunne, H. Regier, M. Sobel, and L. Talbot. 1995. Status and future of salmon of western Oregon and northern California: findings and options. Center for the Study of the Environment, Report #8, Santa Barbara, California.

Chapman, D. W. 1962. Aggressive behavior in juvenile

coho salmon as a cause of emigration. Journal of the Fisheries Research Board of Canada 19:1047–1080.

Christensen, N. L. 1985. Shrubland fire regimes and their evolutionary consequences. Pages 86–100 *in* S. T. A. Pickett and P. S. White, editors. The ecology of natural disturbance and patch dynamics. Academic Press, Orlando, Florida.

Currens, K. P., and six coauthors. In press. A hierarchial approach to conservation genetics and production of anadromous salmonids in the Columbia River basin. U.S. National Marine Fisheries Service Fishery Bulletin.

Diamond, J. M., and R. M. May. 1976. Island biogeography and the design of nature reserves. Pages 163–186 *in* R. M. May, editor. Theoretical ecology. Saunders, Philadelphia.

Elsenberg, J. F., and L. D. Harris. 1989. Conservation: a consideration of evolution, population, and life-history. Pages 99–108 *in* D. Western and M. Pearl, editors. Conservation for the twenty-first century. Oxford University Press, New York.

Franklin, J. F. 1993. Preserving biodiversity: species, ecosystems, or landscapes? Ecological Applications 3:202–205.

Franklin, J. F., and R. T. T. Forman. 1987. Creating landscape patterns by forest cutting: ecological consequences and principles. Landscape Ecology 1:5–18.

Frissell, C. A. 1993. Topology of extinction and endangerment of native fishes in the Pacific Northwest and California (U.S.A.). Conservation Biology 7:342–354.

Goodman, D. 1987. Consideration of stochastic demography in the design and management of reserves. Natural Resources Modeling 1:205–234, Tempe, Arizona.

Gotelli, N. J. 1991. Metapopulation models: the rescue effect, the propagule rain, and the core-satellite hypothesis. American Naturalist 138:768–776.

Grant, G. E. 1990. Hydrologic, geomorphic, and aquatic habitat implications of old and new forestry. Pages 35–53 *in* A. F. Pearson and D. A. Challenger, editors. Forests—managed and wild: differences and consequences. University of British Columbia, Vancouver.

Gurtz, M. E., and J. B. Wallace. 1984. Substrate-mediated response of invertebrates to disturbance. Ecology 65:1556–1569.

Hales, D. 1989. Changing concepts of national parks. Pages 139–149 *in* D. Western and M. Pearl, editors. Conservation for the twenty-first century. Oxford University Press, New York.

Hansen, A. J., and D. L. Urban. 1992. Avian response to landscape patterns: the role of species life histories. Landscape Ecology 7:163–180.

Hanski, I. 1985. Single-species metapopulation dynamics: concepts, models, and observations. Biological Journal of the Linnean Society 42:17–38.

Hanski, I., and M. Gilpin. 1991. Metapopulation dynamics: brief history and conceptual domain. Biological Journal of the Linnean Society 42:3–16.

Hanson, D. L., and T. F. Waters. 1974. Recovery of standing crop and production rate of a brook trout population in a flood-damaged stream. Transactions of the American Fisheries Society 103:431–439.

Harris, L. D., and J. F. Elsenberg. 1989. Enhanced linkages: necessary steps for success in conservation of faunal diversity. Pages 168–181 *in* D. Western and M. Pearl, editiors. Conservation for the twenty-first century. Oxford University Press, New York.

Harrison, S. 1991. Local extirpation in a metapopulation context: an empirical evaluation. Biological Journal of the Linnean Society 42:73–88.

Harrison, S. 1994. Metapopulations and conservation. Pages 111–128 *in* R. J. Edwards, R. M. May, and N. R. Webb, editors. Large-scale ecology and conservation biology. Blackwell Scientific Publications, London.

Harrison, S., and J. F. Quinn. 1989. Correlated environments and the persistence of metapopulations. Oikos 56:293–298.

Healey, M. C. 1991. Life history of chinook salmon (*Oncorhynchus tshawytscha*). Pages 311–394 *in* C. Groot and L. Margolis, editors. Pacific salmon life histories. University of British Columbia Press, Vancouver.

Heard, W. R. 1991. Life history of pink salmon (*Oncorhynchus gorbuscha*). Pages 121–230 *in* C. Groot and L. Margolis, editors. Pacific salmon life histories. University of British Columbia Press, Vancouver.

Hicks, B. J., J. D. Hall, P. A. Bisson, and J. R. Sedell. 1991. Responses of salmonids to habitat changes. American Fisheries Society Special Publication 19: 483–518.

Hobbs, R. J., and L. F. Huenneke. 1992. Disturbance, diversity, and invasion: implications for conservation. Conservation Biology 6:324–337.

Holling, C. S. 1973. Resilience and stability of ecological systems. Annual Review of Ecology and Systematics 4:1–23.

Huff, M. H., and C. M. Raley. 1991. Regional patterns of diurnal breeding bird communities in Oregon and Washington. Pages 177–206 *in* L. F. Ruggiero, K. B. Aubry, A. B. Carey, and M. H. Huff, technical coordinators. Wildlife and vegetation of unmanaged Douglas-fir forests. U.S. Forest Service General Technical Report PNW-GTR-285.

Karr, J. R., and K. E. Freemark. 1985. Disturbance and vertebrates: an integrative perspective. Pages 153–168 *in* S. T. A. Pickett and P. S. White, editors. The ecology of natural disturbance and patch dynamics. Academic Press, Orlando, Florida.

Lamberti, G. A., S. V. Gregory, L. R. Ashkenas, R. C. Wildman, and K. M. S. Moore. 1991. Stream ecosystem recovery following a catastrophic debris flow. Canadian Journal of Fisheries and Aquatic Sciences 48:196–208.

Lande, R. 1988. Genetics and demography in biological conservation. Science 241:1455–1459.

Lawton, J. H., S. Nee, A. J. Letcher, and P. H. Harvey. 1994. Animal distributions: patterns and processes. Pages 41–58 *in* R. J. Edwards, R. M. May, and N. R. Webb, editors. Large-scale ecology and conservation biology. Blackwell Scientific Publications, London.

Levin, S. 1974. Dispersion and population interactions. American Naturalist 108:207–228.

Li, H. W., C. B. Schreck, C. E. Bond, and E. Rexstad. 1987. Factors influencing changes in fish assemblages of Pacific Northwest streams. Pages 193–202 *in* W. J. Matthews and D. C. Heins, editors. Community and evolutionary ecology of North American stream fishes. University of Oklahoma Press, Norman.

MacArthur, R. H., and E. O. Wilson. 1967. The theory of island biogeography. Princeton University Press, Princeton, New Jersey.

Mangel, M., and C. Tier. 1994. Four facts every conservation biologist should know about persistence. Ecology 75:607–614.

Matthews, W. J. 1986. Fish community structure in a temperate stream: stability, persistence, and a catastrophic flood. Copeia 1986:388–397.

May, R. M. 1994. The effects of spatial scale on ecological questions and answers. Pages 1–17 *in* R. J. Edwards, R. M. May, and N. R. Webb, editors. Large-scale ecology and conservation biology. Blackwell Scientific Publications, London.

McCauley, D. E. 1991. Genetic consequences of local population extinction and recolonization. Trends in Ecology & Evolution 6(1):5–8.

McIntosh, B. A., and six coauthors. 1994. Historical changes in fish habitat for select river basins of eastern Oregon and Washington. Northwest Science 68(Special Issue):36–53.

Meyer, G. A., S. G. Wells, R. C. Balling, Jr., and A. J. T. Jull. 1992. Response of alluvial systems to fire and climate change in Yellowstone National Park. Nature 357:147–150.

Minshall, G. W., J. T. Brock, and J. D. Varley. 1989. Wildfires and Yellowstone's stream ecosystems. BioScience 39:707–715.

Moyle, P. B., and G. M. Sato. 1991. On the design of preserves to protect native fishes. Pages 155–169 *in* W. L. Minckley and J. E. Deacon, editors. Battle against extinction: native fish management in the American west. University of Arizona Press, Tuscon.

Moyle, P. B., and R. M. Yoshiyama. 1994. Protection of aquatic biodiversity in California: a five-tiered approach. Fisheries 19(2):6–19.

Murphy, M. L., J. Heifetz, J. F. Thedinga, S. W. Johnson, and K. V. Koski. 1989. Habitat utilization by juvenile Pacific salmon (*Oncorhynchus*) in the glacial Taku River, southeast Alaska. Canadian Journal of Fisheries and Aquatic Sciences 46:1677–1685.

Naiman, R. J., and eight coauthors. 1992. Fundamental elements of ecologically healthy watersheds in the Pacific Northwest coastal ecoregion. Pages 127–188 *in* R. J. Naiman, editor. Watershed management: balancing sustainability and environmental change. Springer-Verlag, New York.

Neave, F. 1958. The origin and speciation of *Oncorhynchus*. Transactions of the Royal Society of Canada 52:25–39.

Nehlsen, W., J. E. Williams, and J. A. Lichatowich. 1991. Pacific salmon at the crossroads: stocks at risk from California, Oregon, Idaho, and Washington. Fisheries 16(2):4–21.

Nickelson, T. E., J. D. Rodgers, S. L. Johnson, and M. F. Solazzi. 1992. Seasonal changes in habitat use by juvenile coho salmon (*Oncorhynchus kisutch*) in Oregon coastal streams. Canadian Journal of Fisheries and Aquatic Sciences 49:783–789.

Northcote, T. G. 1992. Migration and residency in stream salmonids—some ecological considerations and evolutionary consequences. Nordic Journal of Freshwater Research 67:5–17.

Noss, R. F. 1993. A conservation plan for the Oregon coast range: some preliminary suggestions. Natural Areas Journal 13:276–290, Rockford, Illinois.

Pearson, T. N., H. W. Li, and G. A. Lamberti. 1992. Influence of habitat complexity on resistance to flooding and resilience of stream fish assemblages. Transactions of the American Fisheries Society 121:427–436.

Pickett, S. T. A., and J. N. Thompson. 1978. Patch dynamics and the design of nature reserves. Biological Conservation 13:27–37.

Pringle, C. M., and seven coauthors. 1988. Patch dynamics in lotic systems: the stream as a mosaic. Journal of the North American Benthological Society 7(4):503–524.

Pulliam, H. R. 1988. Sources, sinks, and population regulation. American Naturalist 132:652–661.

Quinn, T. F. 1984. Homing and straying in Pacific salmon. Pages 357–362 *in* J. D. McCleave, G. P. Arnold, J. J. Dodson, and W. H. Neil, editors. Mechanisms of migration in fish. Plenum, New York.

Ralph, S. C., G. C. Poole, L. L. Conquest, and R. J. Naiman. 1994. Stream channel morphology and woody debris in logged and unlogged basins of western Washington. Canadian Journal of Fisheries and Aquatic Sciences 51:37–51.

Raphael, M. G. 1991. Vertebrate species richness within and among seral stages of Douglas-fir/hardwood forests of northwestern California. Pages 415–424 *in* L. F. Ruggiero, K. B. Aubry, A. B. Carey, and M. H. Huff, technical coordinators. Wildlife and vegetation of unmanaged Douglas-fir forests. U.S. Forest Service General Technical Report PNW-GTR-285.

Rapport, D. J., H. A. Regier, and T. C. Hutchinson. 1985. Ecosystem behavior under stress. American Naturalist 125:617–640.

Reeves, G. H., and J. R. Sedell. 1992. An ecosystem approach to the conservation and management of freshwater habitat for anadromous salmonids in the Pacific Northwest. Transactions of the North American Wildlife and Natural Resources Conference 57:408–415.

Reeves, G. H., F. H. Everest, and J. R. Sedell. 1993. Diversity of juvenile anadromous salmonid assemblages in coastal Oregon basins with different levels of timber harvest. Transactions of the American Fisheries Society 122:309–317.

Reice, S. R. 1994. Nonequilibrium determinants of biological community structure. American Scientist 82:424–435.

Resh, V. H., and nine coauthors. 1988. The role of disturbance in stream ecology. Journal of the North American Benthological Society 7:433–455.

Ricker, W. E. 1989. History and present state of odd-year pink salmon runs of the Frasier River region. Canadian Technical Report of Fisheries and Aquatic Sciences 1702:1–37.

Rieman, B. E., and J. D. McIntyre. 1995. Occurrence of bull trout in naturally fragmented habitat patches of varied size. Transactions of the American Fisheries Society 124:285–296.

Rolstad, J. 1991. Consequences of forest fragmentation for the dynamics of bird populations: conceptual issues and evidence. Biological Journal of the Linnean Society 42:149–163.

Schlosser, I. J. 1991. Stream fish ecology: a landscape perspective. BioScience 41:704–712.

Shonewald-Cox, C. M. 1983. Guidelines to management: a beginning attempt. Pages 414–445 in C. M. Shonewald-Cox, S. M. Chambers, B. MacBryde, and L. Thomas, editors. Genetics and conservation: a reference for managing wild animal and plant populations. Benjamin Cummings Co., Menlo Park, California.

Sheldon, A. I. 1988. Conservation of stream fishes: patterns of diversity, rarity, and risk. Conservation Biology 2:149–156.

Simberloff, D. 1988. The contribution of population and community biology to conservation science. Annual Review of Ecology and Systematics 19:473–511.

Sjorgen, P. 1991. Extinction and isolation gradients in metapopulations: the case of the pool frog (Rana lessonae). Biological Journal of the Linnean Society 42:135–147.

Stearns, S. C. 1977. The evolution of life history traits: a critique of the theory and a review of the data. Annual Review of Ecology and Systematics 8:145–171.

Steedman, R. J., and H. J. Regier. 1987. Ecosystem science for the Great Lakes: perspectives on degradative and rehabilitative transformations. Canadian Journal of Fisheries and Aquatic Sciences 44(Supplement): 95–103.

Swanston, D. N. 1991. Natural processes. American Fisheries Society Special Publication 19:139–179.

Tallman, R. F., and M. C. Healey. 1994. Homing, straying, and gene flow among seasonally separated populations of chum salmon (Oncorhynchus keta). Canadian Journal of Fisheries and Aquatic Sciences 51: 577–588.

Thomas, J. W., and the Forest Ecosystem Management Assessment Team. 1993. Forest ecosystem management: an ecological, economic, and social assessment. Report of the Forest Ecosystem Management Assessment Team. United States Department of Agriculture, Forest Service, Portland, Oregon.

Tschaplinski, P. J., and G. F. Hartman. 1983. Winter distribution of juvenile coho salmon (Oncorhynchus kisutch) before and after logging in Carnation Creek, British Columbia, and some implications for overwinter survival. Canadian Journal of Fisheries and Aquatic Sciences 40:452–461.

Vannote, R. L., G. W. Minshall, K. W. Cummins, J. R. Sedell, and C. E. Cushing. 1980. The river continuum concept. Canadian Journal of Fisheries and Aquatic Sciences 37:130–137.

Vrijenhoek, R. C. 1985. Animal population genetics and disturbances: the effects of local extinction and recolonization on heterogeneity and fitness. Pages 265–285 in S. T. A. Pickett and P. S. White, editors. The ecology of natural disturbance and patch dynamics. Academic Press, Orlando, Florida.

Waples, R. S. 1991. Pacific salmon, Oncorhynchus spp., and the definition of "species" under the Endangered Species Act. U.S. National Marine Fisheries Service Marine Fisheries Review 53(3):11–22.

Webb, N. R., and J. A. Thomas. 1994. Conserving insect habitats in heathland biotypes: a question of scale. Pages 129–151 in R. J. Edwards, R. M. May, and N. R. Webb, editors. Large-scale ecology and conservation biology. Blackwell Scientific Publications, London.

Western, D. 1989. Conservation without parks: wildlife in rural landscapes. Pages 158–165 in D. Western and M. C. Pearl, editors. Conservation for the twenty-first century. Oxford University Press, New York.

White, P. S., and S. P. Bratton. 1980. After preservation: the philosophical and practical problems of change. Biological Conservation 18:241–255.

White, P. S., and S. T. A. Pickett. 1985. Natural disturbance and patch dynamics: an introduction. Pages 3–13 in S. T. A. Pickett and P. S. White, editors. The ecology of natural disturbance and patch dynamics. Academic Press, Orlando, Florida.

Wilcove, D. S., C. H. McLellan, and A. P. Dobson. 1986. Habitat fragmentation in the temperate zone. Pages 237–256 in M. E. Soule, editor. Conservation biology: the science of scarcity and diversity. Sinauer, Sunderland, Massachusetts.

Wilcox, B. A. 1990. In situ conservation of genetic resources. Pages 45–77 in G. H. Orions, G. M. Brown, Jr., W. E. Kunin, and J. E. Swierzbinski, editors. The preservation and valuation of biological resources. University of Washington Press, Seattle.

Williams, J. E., and seven coauthors. 1989. Fishes of North America endangered, threatened, or of special concern. Fisheries 14(6):2–20.

Yount, J. D., and G. J. Niemi. 1990. Recovery of lotic communities and ecosystems from disturbance—a narrative review of case studies. Environmental Management 14:547–570.

American Fisheries Society Symposium 17:350–359, 1995
© Copyright by the American Fisheries Society 1995

Desert Aquatic Ecosystems and the Genetic and Morphological Diversity of Death Valley System Speckled Dace

Donald W. Sada

Environmental Studies Program, University of Nevada–Las Vegas
2689 Highland Drive, Bishop, California 93514, USA

Hugh B. Britten and Peter F. Brussard

Biodiversity Research Center, Department of Biology, University of Nevada
Reno, Nevada 89557–0015, USA

Abstract.—The morphological and genetic diversities of fishes in North American deserts have been examined to estimate evolutionary rates, to create models of interbasin pluvial connectivity, and to justify protection of aquatic ecosystems throughout the region. Morphological and genetic studies comparing 13 populations of speckled dace *Rhinichthys osculus* from the Death Valley system, Lahontan basin, and lower Colorado River were conducted to quantify differences among populations. Differences in meristic and mensural characteristics among populations were highly significant, but differences in body shape were slight and best explained as representing two forms, one deep-bodied and short, the other elongate and slender. Starch gel electrophoretic assays of 23 loci showed isolated populations to be genetically unique. Fifty-nine taxa are identified as endemic to wetland and aquatic habitats in the Death Valley system: 16 forms of fish, 1 amphibian, 22 mollusks, 7 aquatic insects, 3 mammals, and 10 forms of flowering plants. Genetic and morphological differentiation of isolated speckled dace populations and the diversity and number of endemic forms associated with wetlands and aquatic habitats in the Death Valley system suggest that each desert wetland community functions as an evolutionarily significant unit.

Taxonomic descriptions of relict faunas occupying isolated aquatic habitats in the endorheic desert basins of western North America began with recognition of unique fishes during land surveys before the West was settled (see Minckley and Douglas 1991) and attention expanded quickly to the invertebrate fauna (Brues 1932). This fauna is depauperate compared to those in more mesic regions, but many studies over the last 50 years have shown that a wide diversity of endemic plants and animals is associated with regional wetlands that range in size from less than 0.1 ha to several thousand hectares. The small size of these habitats and the vulnerability of isolated populations to habitat degradation and invasion of nonnative species have been factors in the decline of many fish populations (Minckley and Deacon 1968; Williams et al. 1989; Rinne and Minckley 1991). Approximately 60% of the North American fish species currently listed as threatened or endangered by the U.S. Fish and Wildlife Service occupy the desert southwestern region of the United States (Williams and Sada 1985; Williams et. al 1989).

The Death Valley system (DVS) is an endorheic basin in southwestern Nevada and eastern California. It comprises the Owens River, Amargosa River, and Mojave River basins, which were tributary to Lake Manly (present day Death Valley) during

Pleistocene pluvial periods (Miller 1946; Hubbs and Miller 1948). The ecology, systematics, biogeography, and status of most fish species in the system have been actively studied (Miller 1948; Soltz and Naiman 1978; Echelle and Echelle 1993) with the exception of speckled dace *Rhinichthys osculus*, which occupies the Owens and Amargosa basins.

Concern for the declining status of DVS speckled dace caused us to conduct morphological, genetic, and status studies to determine characteristics of population variation and assist in the design of conservation programs to protect and enhance extant populations. In this paper we summarize results of these genetic and morphological studies and examine the utility of this analysis for identifying evolutionarily significant units. Then we consider other endemic organisms associated with DVS aquatic habitats and wetlands to determine if considering communities of these organisms provides stronger justification for regional wetland conservation than protection strategies based on either populations or single species. We believe that a wide diversity of endemic species associated with these habitats would suggest that conservation programs protecting endemic species could also protect desert aquatic and wetland ecosystems.

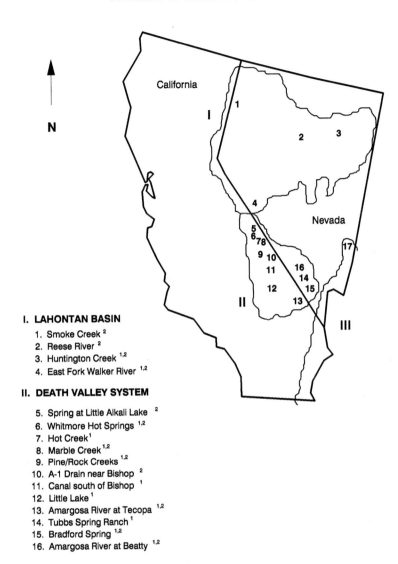

N

FIGURE 1.—Delineation of the Lahontan basin (I), Death Valley system (II), and Colorado River drainage (III), and locations (numbered 1–17) of *Rhinichthys osculus* populations examined in morphological and genetic studies. Superscript 1 on place names denotes museum collections examined in morphological studies; superscript 2 denotes populations sampled for genetic analysis.

The map includes the following legend:

I. LAHONTAN BASIN
1. Smoke Creek [2]
2. Reese River [2]
3. Huntington Creek [1,2]
4. East Fork Walker River [1,2]

II. DEATH VALLEY SYSTEM
5. Spring at Little Alkali Lake [2]
6. Whitmore Hot Springs [1,2]
7. Hot Creek [1]
8. Marble Creek [1,2]
9. Pine/Rock Creeks [1,2]
10. A-1 Drain near Bishop [2]
11. Canal south of Bishop [1]
12. Little Lake [1]
13. Amargosa River at Tecopa [1,2]
14. Tubbs Spring Ranch [1]
15. Bradford Spring [1,2]
16. Amargosa River at Beatty [1,2]

III. LOWER COLORADO RIVER DRAINAGE
17. Beaver Dam Wash [1,2]

Methods

Morphological and allozyme analyses were conducted on several speckled dace populations from the Lahontan basin, one population from the lower Colorado River drainage, and all known populations in the DVS (Figure 1). Of the 17 collections examined, 13 were used for the allozyme analysis (two proximate collections near Bishop, one of

them from Pine and Rock creeks and the other from A-1 Drain, were combined), and 14 collections from 13 localities were used for analysis of meristic and mensural characteristics.

Morphometrics.—Eighteen traditional truss measurements, fin lengths, and six meristic variables (lateral line scales and pores and rays of the dorsal, anal, pelvic, and pectoral fins, all counted by the

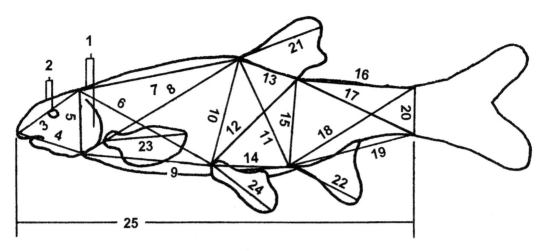

FIGURE 2.—Morphometric measurements taken on speckled dace from the Death Valley system: (1) head width; (2) least bony interorbital width; (3) snout tip to supratemporal canal; (4) snout tip to branchiostegal junction; (5) branchiostegal junction to supratemporal canal; (6) supratemporal canal to pelvic fin origin; (7) supratemporal canal to dorsal fin origin; (8) branchiostegal junction to dorsal fin origin; (9) branchiostegal junction to pelvic fin origin; (10) dorsal fin origin to pelvic fin origin; (11) dorsal fin origin to anal fin origin; (12) pelvic fin origin to base last dorsal ray; (13) dorsal fin base length; (14) pelvic fin origin to anal fin origin; (15) base last dorsal ray to anal fin origin; (16) base last dorsal ray to upper caudal peduncle; (17) base last dorsal ray to lower caudal peduncle; (18) anal fin origin to upper caudal peduncle; (19) anal fin origin to lower caudal peduncle; (20) vertical through junction of caudal vertebrae 5 and 6 anterior of hypural plate; (21) depressed dorsal fin length; (22) depressed anal fin length; (23) pectoral fin length; (24) pelvic fin length; (25) standard length. Measurements 1–20 were used in sheared principle components analysis.

methods of Hubbs et al. 1974) were recorded from 484 fish (Figure 2). Individuals were segregated by sex for analysis of mensural characters, and only fish longer than 25 mm standard length (SL) were examined. Truss measurements and two measures of body width were analyzed by principal components analysis with shearing to reduce the effects of size (Rohlf and Bookstein 1987). Components were sheared by locality and calculated from the covariance matrix of data transformed to $base_e$ logarithms. Proportional truss measurements (fractions of SL multiplied by 1,000), fin length measurements, and meristic counts were log-transformed ($base_e$) and tested by one-way analysis of variance (ANOVA) for differences among populations.

Protein electrophoresis.—The variability of isozymes coded by 23 gene loci (Table 1) was assayed for 311 speckled dace from 13 localities (Figure 1). Whole fish were stored at $-80°C$ until processed for electrophoresis. Approximately 1 g of muscle tissue was dissected from the caudal end of each fish. This was minced in an iced spot plate and combined with 0.75 mL of chilled 0.05 M tris-HCl extraction buffer (pH 7.1) in a centrifuge tube for further maceration. Electrophoretic and histochemical staining procedures followed those of May (1992).

Isozyme data analysis.—Estimates of polymorphism and heterozygosity and tests for conformance to Hardy–Weinberg expectations were made with the BIOSYS-1 program (Swofford and Selander 1981). A χ^2 test for heterogeneity was used to test the significance of allele frequency differences between populations, and fixation (F) statistics were calculated for all sampled populations. Differences among populations were assessed by principal components analysis of arcsine-transformed allele frequencies for polymorphic loci.

Distribution of endemic wetland species.—Endemic wetland species associated with DVS valley floor wetlands were identified from the literature and by consultation with regional scientists and land managers.

Results

Morphological Analysis

Meristic and proportional mensural characters were all within ranges documented for speckled dace (Hubbs et al. 1974). Highly significant differences among all populations for all meristic and mensural characters ($P < 0.001$, one-way ANOVA)

TABLE 1.—Enzymes and their abbreviations (Shaklee et al. 1990), enzyme numbers (IUBMBNC 1992), and buffers (May 1992) used in electrophoretic analysis of speckled dace.

Enzyme	Abbreviation[a]	Enzyme number	Number of loci	Buffer
Asparate aminotransferase	sAAT	2.6.1.1	1	R
Creatine kinase	CK	2.7.3.2	1	R
Esterase	EST	3.1.1.-	2	R
General (unidentified) protein	PROT		3	R
Glycerol-3-phosphate dehydrogenase	G3PDH	1.1.1.8	1	R
Glucose-6-phosphate 1-dehydrogenase	G6PHD	1.1.1.49	1	R
Glucose-6-phosphate isomerase	GPI	5.3.1.9	2	R
Isocitrate dehydrogenase (NADP$^+$)	IDH	1.1.1.42	1	C
L-Lactate dehydrogenase	LDH	1.1.1.27	2	4
Malic enzyme (NADP$^+$)	sMEP	1.1.1.40	1	4
Mannose-6-phosphate isomerase	MPI	5.3.1.8	1	4
Dipeptidase	PEPA	3.4.-.-	1	R
Tripeptide aminopeptidase	PEPB	3.4.-.-	1	R
Proline dipeptidase	PEPD	3.4.-.-	1	R
Phosphogluconate dehydrogenase	PGDH	1.1.1.44	1	4
Phosphoglucomutase	PGM	5.4.2.2	2	4
Superoxide dismutase	sSOD	1.15.1.1	1	R

[a]An "s" prefix specifies the cytosolic form of the enzyme.

showed the morphology of each population to be unique.

Plots of sheared principal components II and III indicated that the body shapes of both males and females segregated strongly in some populations but could not be differentiated in others. The most definitive patterns of body shape appeared when factor scores of strongly overlapping populations were clustered (Figure 3). These plots suggest that body shape is best explained by the existence of two forms, one that is short and deep and another that is long and slender.

Population Genetic Variability and Structure

Only Smoke Creek (*sAAT** locus), Bishop (*PGM-1**), and Huntington Creek (*PGD-1** and *PEPD**) populations were not in Hardy–Weinberg equilibrium at all assayed loci.

All populations were fixed for the same alleles at the remaining monomorphic loci except for those in the Owens basin at Whitmore Hot Springs and Little Alkali Lake. The alternative (**D*) allele of the *PEPA** locus segregated in these populations, providing the only instances in which populations were distinguished by a fixed allelic difference.

Polymorphism levels ranged from 4% in lower Marble Creek and Little Alkali Lake samples to 35% in the Huntington Creek population (Table 2). Mean observed heterozygosity ranged from 0.008 in the Little Alkali Lake population to 0.072 in the East Fork Walker River sample (Table 2).

Differences in allele frequencies among populations were significant for all polymorphic loci ($P <$ 0.05, χ^2 tests), indicating substantial genetic differentiation among populations. Furthermore, the mean fixation index, F_{ST}, among the 13 populations was 0.538 and the mean index for individuals, F_{IT}, was 0.621. A hierarchical analysis to elucidate the geographical scale at which panmixia breaks down showed significant allele frequency differences ($P <$ 0.05) among the four populations within the Owens River drainage at six of the eight polymorphic loci in the drainage (*PGM-1**, *PEPA**, *sAAT**, *EST-1**, and *GPI-1,2**). Likewise, all four polymorphic loci in the three Amargosa drainage populations (*PGDH**, *PGM-1**, *sAAT**, and *GPI-2**) differed significantly in allele frequencies ($P <$ 0.05). Significant differences in allele frequencies were detected between the two Smoke Creek populations at two of the four polymorphic loci (*PGM-1** and *GPI-1**; $P <$ 0.05). When the two Smoke Creek populations were tested for allele frequency heterogeneity with the two Humboldt populations, significant levels of heterogeneity occurred at eight of the nine polymorphic loci in these populations (*PGDH**, *PGM-1**, *MPI**, *PEPD**, *sAAT**, *CK**, *EST-1**, and *GPI-1**; $P <$ 0.05). There were no detected differences in allele frequencies between the Reese River and Huntington Creek populations (two Humboldt drainage populations from the Lahontan basin). This indicates that a higher level of gene flow may be occurring between the Humboldt drainage populations than among DVS populations. Finally, the population-to-drainage fixation index (F) was 0.525 and the population-to-total sample fixation index was 0.528, indicating that gene flow among these

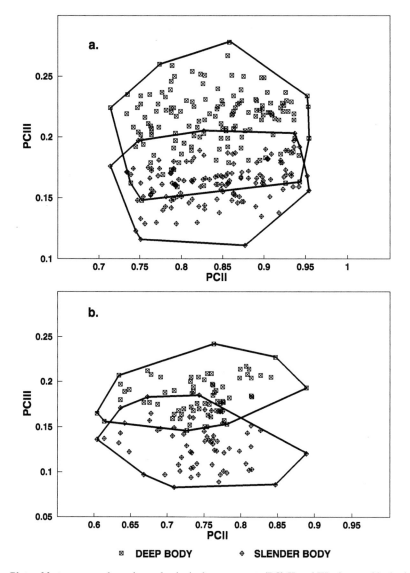

FIGURE 3.—Plots of factor scores along sheared principal components (PC) II and III, clustered by body type, for 484 Death Valley system, Lahontan basin, and lower Colorado River speckled dace *Rhinichthys osculus*. (**a**) females; (**b**) males.

populations is approximately the same within drainages as between drainages.

Factor score plots from analysis of principal components illustrate the genetic differences and relationships among the sampled populations (Figure 4). Death Valley system populations cluster separately from Lahontan basin populations. This indicates that ancestral DVS speckled dace invaded this basin from the Colorado River rather than from the Lahontan basin. Additional studies involving Colo-

rado River populations should be conducted to resolve questions about speckled dace diversity in this river system.

Other Taxa Endemic to Death Valley Wetlands

Fifty-nine plant and animal taxa are documented from wetland and aquatic habitats in the DVS (Table 3). This includes 10 plant taxa (Morefield and Knight 1992; Hickman 1993), 3 mammals (Hall and

TABLE 2.—Genetic variability estimates for 13 populations of speckled dace sampled in California and Nevada. Standard errors are in parentheses.

Population	Mean (SE) number of alleles per locus	Percentage of loci polymorphic	Mean (SE) heterozygosity	
			Direct count	Hardy–Weinberg expectation
Upper Smoke Creek	1.1 (0.1)	13.0	0.049 (0.028)	0.054 (0.030)
Lower Smoke Creek	1.3 (0.1)	17.4	0.054 (0.027)	0.071 (0.037)
Reese River	1.4 (0.1)	30.4	0.067 (0.026)	0.084 (0.035)
Huntington Creek	1.5 (0.2)	34.8	0.056 (0.020)	0.102 (0.037)
East Fork Walker River	1.4 (0.2)	30.4	0.072 (0.029)	0.078 (0.031)
Whitmore Hot Spring	1.1 (0.1)	8.7	0.029 (0.020)	0.028 (0.020)
Little Alkali Lake	1.0 (0.0)	4.3	0.008 (0.008)	0.008 (0.008)
Lower Marble Creek	1.0 (0.0)	4.3	0.017 (0.017)	0.020 (0.020)
Bishop, California	1.3 (0.1)	21.7	0.028 (0.015)	0.045 (0.024)
Tecopa, California	1.1 (0.1)	8.7	0.023 (0.016)	0.029 (0.021)
Ash Meadows	1.1 (0.1)	8.7	0.014 (0.013)	0.020 (0.015)
Beatty, Nevada	1.1 (0.1)	8.7	0.011 (0.009)	0.011 (0.009)
Beaver Dam, Washington	1.2 (0.1)	8.7	0.040 (0.028)	0.035 (0.024)

Kelson 1959), 7 aquatic insects (Polhemus 1979; Schmude 1992; Shepard 1992), 22 mollusks (Hershler and Sada 1987; Hershler 1989), 1 amphibian (Stebbins 1966), and 16 fish (Smith 1978; Soltz and Naiman 1978). Habitat occupied by these species is either aquatic (fish, mollusks, and insects), mesic alkali meadow (plants and mammals), or semiterrestrial (mollusks and amphibians). Although distributions of many of these taxa overlap with DVS

speckled dace, many inhabit wetland communities some distance from fish habitats.

Discussion

Several people have used isozyme data to estimate population structure and levels of gene flow between populations (e.g., Slatkin 1985; Allendorf et al. 1987). The isolation of desert wetlands sug-

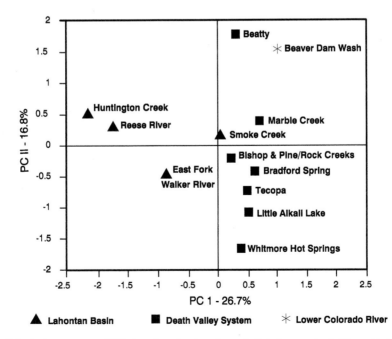

FIGURE 4.—Principal components (PC) plot of arcsine-transformed allele frequencies in 13 speckled dace populations assayed in the Death Valley system, Lahontan basin, and lower Colorado River.

TABLE 3.—Endemic taxa of valley floor wetlands in the Owens, Amargosa, and Mojave basins and Death Valley. Habitat abbreviations: ST = stream, SP = spring source, SPMAR = spring margin, MAM = mesic alkali meadow, MAF = mesic alkali flat.

Taxon	Habitat
Owens River Basin	
Plants	
Owens Valley checkerbloom *Sidalcea covellii*	MAM
Fish Slough milkvetch *Astragalus lentiginosus* var. *piscensis*	MAM
Owens Valley Mariposa lily *Calochortus excavatus*	MAM
Mollusks	
Owens springsnail *Pyrgulopsis owensensis*	SP
Fish Slough springsnail *Pyrgulopsis perturbata*	SP
Benton Valley springsnail *Pyrgulopsis aardhali*	SP
Fish	
Owens pupfish *Cyprinodon radiosus*	SP, ST
Owens tui chub *Gila bicolor snyderi*	ST
Long Valley speckled dace *Rhinichthys osculus* ssp.	SP, ST
Owens speckled dace *Rhinichthys osculus* ssp.	ST
Owens sucker *Catostomus fumeiventris*	ST
Mammals	
Owens Valley vole *Microtus californicus vallicola*	
Amargosa River Basin	
Plants	
Ash Meadows ivesia *Ivesia eremica*	MAF
Ash Meadows gumplant *Grindelia fraxino-pratensis*	SPMAR
Ash Meadows ladies tresses *Spiranthes infernalis*	SPMAR
Amargosa niterwort *Nitrophila mohavensis*	MAF
Spring-loving centaury *Centaurium namophilum*[a]	SPMAR
Tecopa birds beak *Cordylanthus tecopensis*	MAM
Insects	
Devils hole riffle beetle *Stenelemis calida calida*	SP
Amargosa naucorid bug *Ambrysus amargosus*	SP
Relict naucorid bug *Ambrysus* sp.	SP
Saratoga Springs bug *Belostoma saratogae*	SP
(No common name) *Microcylleopus similis*[a]	SP
Mollusks	
Sportinggoods tryonia *Tryonia anuglata*	SP
(No common name) *Tryonia variegata*	SP
(No common name) *Tryonia ericae*	SP
Point of Rocks tryonia *Tryonia elata*	SP
Fairbanks Spring springsnail *Pyrgulopsis fairbanksensis*	SP
Crystal Spring springsnail *Pyrgulopsis crystalis*	SP
Longstreet Spring springsnail *Pyrgulopsis* sp.	SP
Oasis Valley springsnail *Pyrgulopsis micrococcus*[b]	SP
Ash Meadows pebblesnail *Pyrgulopsis erythropoma*	SP
Distal-gland springsnail *Pyrgulopsis nanus*	SP
Median-gland springsnail *Pyrgulopsis pisteri*	SP
Elongate-gland springsnail *Pyrgulopsis isolatus*	SP
Amargosa springsnail *Pyrgulopsis amargosae*	SP
(No common name) *Assiminea* sp.	SPMAR
Fish	
Devils Hole pupfish *Cyprinodon diabolis*	SP
Ash Meadows Nevada pupfish *Cyprinodon nevadensis mionectes*	SP
Saratoga Springs pupfish *Cyprinodon nevadensis nevadensis*	SP
Warm Springs pupfish *Cyprinodon nevadensis pectoralis*	SP
Tecopa pupfish *Cyprinodon nevadensis tecopensis*	SP
Shoshone pupfish *Cyprinodon nevadensis shoshone*	SP
Nevada speckled dace *Rhinichthys osculus nevadensis*	SP, ST
Ash Meadows poolfish *Empetrichthys merriami*	SP
Mammals	
Ash Meadows vole *Microtus montanus nevadensis*	
Amargosa vole *Microtus californicus scirpensis*	
Amphibians	
Amargosa toad *Bufo nelsoni*	SPMAR

TABLE 3.—Continued.

Taxon	Habitat
Death Valley	
Plants	
Death Valley blue-eyed grass *Sisyrinchium funereum*	SPMAR
Insects	
(No common name) *Ambrysus funebris*	SP
(No common name) *Microcylleopus formicoideus*	SP
Mollusks	
Badwater snail *Assiminea infima*	SPMAR
Grapevine Springs elongate tryonia *Tryonia margae*	SP
Grapevine Springs squat tryonia *Tryonia rowlandsi*	SP
Robust tryonia *Tryonia robusta*	SP
Cottonball Marsh tryonia *Tryonia salina*	SP
Fish	
Salt Creek pupfish *Cyprinodon salinus salinus*	SP, ST
Cottonball Marsh pupfish *Cyprinodon salinus milleri*	SP
Mojave River Basin	
Fish	
Mojave tui chub *Gila bicolor mohavensis*	ST

[a]Also occurs in Death Valley.
[b]Also occurs in the Mohave River basin.

gests that genetic diversity within populations occupying these habitats should be low but that many populations should be genetically distinctive (Meffe and Vrijenhoek 1988). These assumptions have been confirmed by Turner (1974) and Echelle and Echelle (1993) in work with pupfish (genus *Cyprinodon*) throughout the southwestern United States, including the DVS.

Earlier workers recognized the distinctiveness of speckled dace populations within isolated Great basin drainages (Hubbs and Miller 1948; Hubbs et al. 1974). Data from our study indicate that speckled dace variation conforms more to relatively recent models (Slatkin 1985; Allendorf et al. 1987) that recognize the distinctiveness of individual populations than to earlier models that recognized distinctiveness only by focusing on similarities among populations in a basin. We used Slatkin's (1987) method with our data to calculate the genetically effective number of migrants between all populations for each generation (N_m) from F_{ST} values. The results indicate that approximately one genetically effective individual is exchanged per DVS population every five generations ($N_m = 0.21$). This level of gene flow is well below the level theoretically necessary to prevent populations from diverging in allele frequencies under a neutral model of selection (Hartl and Clark 1989). The significant differences in allele frequencies and the fixation of a unique allele at *PEPA** in the Whitmore Hot Springs and Little Alkali Lake populations further indicate a paucity of gene flow between DVS populations. Also, the relatively low heterozygosity es-

timates for most DVS speckled dace populations is consistent with the hypothesis that neutral or weakly selected genetic variability is lost due to genetic drift in small isolated populations. The mean heterozygosity over DVS populations (0.014) was substantially lower than heterozygosities in Lahontan basin and lower Colorado River populations, which were near the average documented for other cyprinid fishes (0.052; Buth et al. 1991). Even though the morphological distinctiveness of each DVS speckled dace population is clouded by the presence of only two body forms, differences in meristic and mensural characteristics among populations also suggest that each population is distinctive. Recognition of the taxonomic distinctiveness of many regional fishes has justified protection of many aquatic habitats throughout the desert Southwest (Williams and Sada 1985; Williams et al. 1985). However, no population of aquatic species in this region has been protected unless it had gained formal taxonomic recognition. Our genetic and morphological examination indicates that populations of even wide-ranging species may be unique and worthy of protection as evolutionarily significant units.

The number and diversity of species endemic to DVS wetlands suggest that factors causing the genetic and morphological differentiation of DVS speckled dace have also affected a wide diversity of plants and animals. Isolation and differentiation have permitted development of unique communities associated with each persistent water.

The biotic distinctiveness of these isolated waters

requires conservation programs to be designed around ecological requirements of the entire community. Williams et al. (1985) and Moyle and Yoshiyama (1994) have proposed protective strategies that embrace a broad variety of aquatic species. However, both these strategies focus attention on aquatic species; neither addresses the broader requirement of protecting communities that rely upon adjacent mesic soils as well as upon water. The ecology and distribution of species endemic to DVS wetlands suggests that conservation programs focusing upon only aquatic species may not be adequate. More definitive protective strategies should consider the entire ecosystem and be implemented to include the entire community of organisms reliant upon waters supporting the aquatic habitat. In this respect, tasks for these conservation programs should be expanded beyond those necessary to maintain aquatic ecosystems and include tasks necessary to maintain habitats in the broader context of wetlands.

The importance of recognizing the diverse taxa associated with aquatic habitats is supported by examining the ecology of several DVS endemic species. The Amargosa niterwort is restricted to mesic alkali flats in the Amargosa basin (Mozingo 1977), where most populations rely on soil moisture that is maintained by discharge from springs approximately 8 km away. This plant could be detrimentally affected by narrowly conceived conservation programs that inadvertently alter soil moisture and subsurface percolation in areas distantly associated with aquatic habitats. Although many endemic DVS taxa occupy terrestrial habitats close to flowing water, the population viability of most endemic plants, mammals, and mollusks (e.g., *Assiminea infima*) depends upon mesic soils moistened by water from sometimes distant springs and streams. Ecological diversity represented by the complete community of endemic DVS species suggests that programs designed to conserve these species would protect all aspects of wetland ecosystems in the region. We believe the uniqueness of these communities qualifies each as an evolutionarily significant unit.

Much of the knowledge about desert wetland ecosystems comes from comparatively recent studies of their species' ecology, genetics, and morphometry. The diversity of taxa dependent upon these aquatic resources is much greater than was appreciated before these studies began. Additional surveys and continued advances in taxonomy and ecology are necessary to define tasks that will conserve all members of these wetland communities.

The diversity of organisms that depend on desert wetlands suggests that programs conserving these species must be ecologically based on communities. This lesson can also be applied to conservation of wetland systems in more mesic regions. The challenge of conservation requires more than protecting a population in the isolation of its immediate environment. It also requires conserving populations as a part of their ecosystem so they can continue along their evolutionary pathways.

Acknowledgments

Genetics and morphology analyses were supported by funds from the California Department of Fish and Game's Endangered and Rare Fish, Wildlife, and Plant Species Conservation Account (contract FG0524), and assistance from the Nevada Biodiversity Initiative. Live material was collected under California Department of Fish and Game permit 279, Nevada Department of Wildlife permit 6923, and U.S. Fish and Wildlife Service Endangered Species Permit SADADW issued to D. Sada. Mary DeDecker and Anna Halford assisted with identifying endemic DVS plant taxa.

References

Allendorf, F. W., N. Ryman, and F. M. Utter. 1987. Genetics and fishery management: past, present, and future. Pages 1–19 *in* N. Ryman and F. Utter, editors. Population genetics and fishery management. University of Washington Press, Seattle.

Brues, C. T. 1932. Further studies on the fauna of North American hot springs. Proceedings of the American Academy of Arts and Sciences 67:184–303.

Buth, D. G., T. E. Dowling, and J. R. Gold. 1991. Molecular and cytological investigations. Pages 83–126 *in* I. J. Winfield and J. S. Nelson, editors. Cyprinid fishes systematics, biology, and exploitation. Chapman and Hall, London.

Echelle, A. A., and A. F. Echelle. 1993. Allozyme perspective on mitochondrial DNA variation and evolution of the Death Valley pupfishes (Cyprinodontidae: *Cyprinodon*). Copeia 1993:275–287.

Hall, E. R., and K. R. Kelson. 1959. Mammals of North America. Ronald Press, New York.

Hartl, D. L., and A. G. Clark. 1989. Principles of population genetics, 2nd edition. Sinauer, Sunderland, Massachusetts.

Hershler, R. 1989. Springsnails (Gastropoda: Hydrobiidae) of Owens and Amargosa River (exclusive of Ash Meadows) drainages, Death Valley system, California–Nevada. Proceedings of the Biological Society of Washington 102:176–248.

Hershler, R., and D. W. Sada. 1987. Springsnails (Gastropoda: Hydrobiidae) of Ash Meadows, Amargosa basin, California–Nevada. Proceedings of the Biological Society of Washington 100:776–843.

Hickman, J. C., editor. 1993. The Jepson manual. Higher

plants of California. University of California Press, Berkeley.

Hubbs, C. L., and R. R. Miller. 1948. The zoological evidence/correlation between fish distribution and hydrographic history in the desert basins of western United States. Bulletin of the University of Utah 38:17–166.

Hubbs, C. L., R. R. Miller, and L. C. Hubbs. 1974. Hydrographic history and relict fishes of the north-central Great Basin. California Academy of Sciences Memoirs, Volume VII.

IUBMBNC (International Union of Biochemistry and Molecular Biology, Nomenclature Committee). 1992. Enzyme nomenclature 1992. Academic Press, San Diego, California.

May, B. 1992. Starch gel electrophoresis of allozymes. Pages 1–27 in A. R. Hoelzel, editor. Molecular genetic analysis of populations: a practical approach. Oxford University Press, Oxford, UK.

Meffe, G. K., and R. C. Vrijenhoek. 1988. Conservation genetics in the management of desert fishes. Conservation Biology 2:157–169.

Miller, R. R. 1946. Correlation between fish distribution and Pleistocene hydrography in eastern California and southwestern Nevada, with a map of Pleistocene waters. Journal of Geology 54:43–53.

Miller, R. R. 1948. The Cyprinodont fishes of the Death Valley system of eastern California and southwestern Nevada. Miscellaneous Publications Museum of Zoology University of Michigan 68.

Minckley, W. L., and J. E. Deacon. 1968. Southwestern fishes and the enigma of 'endangered species'. Science 159:1424–1432.

Minckley, W. L., and M. E. Douglas. 1991. Discovery and extinction of western fishes: a blink of the eye in geological time. Pages 7–17 in W. L. Minckley and J. E. Deacon, editors. Battle against extinction. Native fish management in the American west. University of Arizona Press, Tucson.

Morefield, J. D., and T. A. Knight, editors. 1992. Endangered, threatened, and sensitive vascular plants of Nevada. U.S. Bureau of Land Management, Reno, Nevada.

Moyle, P. B., and R. M. Yoshiyama. 1994. Protection of aquatic biodiversity in California: a five-tiered approach. Fisheries 19(2):6–18.

Mozingo, H. N. 1977. Nitrophila mohavensis Munz and Roos. Mentzelia 3:24.

Polhemus, J. T. 1979. Family Naucoridae/creeping water bugs, saucer bugs. Pages 131–138 in A. S. Menke, editor. Semi-aquatic Hemiptera of California (Het-eroptera: Hemiptera). Bulletin of the California Insect Survey 21.

Rinne, J., and W. L. Minckley. 1991. Native fishes of arid lands: a dwindling resource of the desert southwest. U.S. Forest Service General Technical Report RM-206:1–45.

Rohlf, F. J., and F. L. Bookstein. 1987. A comment on shearing as a method for "size correction." Systematic Zoology 36:356–357.

Schmude, K. L. 1992. A revision of the riffle beetle genus Stenelemis (Coleoptera: Elmidae). Doctoral dissertation. University of Wisconsin, Madison.

Shaklee, J. B., F. W. Allendorf, D. C. Morizot, and G. S. Whitt. 1990. Gene nomenclature for protein-coding loci in fish. Transactions of the American Fisheries Society 119:2–15.

Shepard, W. D. 1992. Riffle beetles (Coleptera: Elmidae) of Death Valley National Monument, California. Great Basin Naturalist 52:378–381.

Slatkin, M. 1985. Gene flow in natural populations. Annual Review of Ecology and Systematics 16:393–430.

Slatkin, M. 1987. Gene flow and the geographic structure of natural populations. Science 236:787–792.

Smith, G. R. 1978. Biogeography of intermountain fishes. Great Basin Naturalist Memoirs 2:17–42.

Soltz, D. L., and R. J. Naiman. 1978. The natural history of native fishes in the Death Valley system. Natural History Museum of Los Angeles County Science Series 30:1–76.

Swofford, D. L., and R. B. Selander. 1981. BIOSYS-1: a FORTRAN program for the comprehensive analysis of electrophoretic data in population genetics and systematics. Journal of Heredity 72:281–283.

Stebbins, R. C. 1966. Field guide to western reptiles and amphibians. Houghton Mifflin, Boston.

Turner, B. J. 1974. Genetic divergence of Death Valley pupfish species: biochemical versus morphological evidence. Evolution 28:281–294.

Williams, J. E., and D. W. Sada. 1985. America's desert fishes: increasing their protection under the Endangered Species Act. Endangered Species Technical Bulletin 10:8–14. (U.S. Fish and Wildlife Service, Washington, DC.)

Williams, J. E., and six coauthors. 1985. Endangered aquatic ecosystems in North American deserts with a list of vanishing fishes of the region. Journal of the Arizona–Nevada Academy of Science 20.

Williams, J. E., and seven coauthors. 1989. Fishes of North America endangered, threatened, or of special concern: 1989. Fisheries 14(6):2–20.

American Fisheries Society Symposium 17:360–370, 1995

Evolutionarily Significant Units and Movement of Resident Stream Fishes: A Cautionary Tale

KURT D. FAUSCH

Department of Fishery and Wildlife Biology, Colorado State University
Fort Collins, Colorado 80523, USA

MICHAEL K. YOUNG

Rocky Mountain Forest and Range Experiment Station, U.S. Forest Service
222 South 22nd Street, Laramie, Wyoming 82070, USA

Abstract.—Many taxa of resident stream fishes are reported to be relatively sedentary throughout their lives. Such discrete populations would make identification and management of evolutionarily significant units (ESUs) straightforward. However, in contrast to this prevailing restricted movement paradigm, recent evidence indicates that even resident salmonids, such as interior stocks of cutthroat trout *Oncorhynchus clarki* living in headwater streams, move often, sometimes over relatively long distances. Resident stream fishes likely move in response to various ecological constraints, including the need to garner enough scarce resources or find a critical resource, or because the habitat they occupy becomes suboptimal or unsuitable. This emerging paradigm shift has important implications for defining and managing ESUs of resident stream fishes. For example, timing of sampling may affect which of several different "populations" mobile individuals are chosen to represent. Isolating small populations of native fishes above barriers to prevent invasion by exotic species may trade this risk for other environmental, demographic, or genetic risks caused by eliminating dispersal. Moreover, isolating small population fragments via natural or anthropogenic disturbances, or management actions, may create artificial ESUs. Biologists must understand not only the genetics and taxonomy, but also the spatial and temporal dynamics of component populations of species if they are to accurately identify and wisely manage ESUs.

Fish biologists often classify the diverse set of life history forms found in stream fish populations into distinct categories, such as resident, fluvial, adfluvial, and anadromous (Northcote 1984, 1992). Anadromous, adfluvial, and fluvial life history forms migrate to spend part of their life cycle in an ocean, large lake, or river, whereas resident stream fishes are thought to move little, except perhaps to find suitable spawning habitat (Allen 1951; Hesthagen 1988). Although these forms are best known for salmonids, other lotic fishes have been similarly classified (e.g., Funk 1955).

This classification has profound implications for how stream fishes are studied and managed (Gowan et al. 1994). Anadromous, adfluvial, and fluvial fish populations, though often reproductively isolated through spawning site fidelity, are mobile by definition, and therefore generally require different habitats in widely separated locations to accommodate different ontogenetic stages (McDowall 1988). Movement is acknowledged as part of their life history, and management is planned accordingly. In contrast, populations of resident fishes are presumed to form relatively discrete, independent, self-sustaining units controlled primarily by resources and disturbances that occur in situ (Chapman 1966; Grant and Kramer 1990). As a result, the reproduc-

tive isolation and the evolutionary distinctiveness used to define evolutionarily significant units (ESUs; sensu Waples 1991) may often be assumed, not demonstrated.

Despite this traditional view, recent work we have conducted indicates that the restricted movement of resident stream fishes (sensu Gerking 1959) is not as robust a concept as once thought. Here we review the evidence causing us to question this paradigm (Gowan et al. 1994) and include new information from nonsalmonids. We then discuss the ecological factors that influence movement, and we raise issues about the consequences such movement has for identifying and managing ESUs of resident stream fishes.

The Restricted Movement Paradigm

Adult resident stream fishes have been reported to remain within restricted home ranges (sensu Gerking 1953) for most or all of their lives (e.g., Miller 1957; Bachman 1984; Hill and Grossman 1987). Gowan et al. (1994) proposed that the theory of restricted movement, originally formalized by Gerking (1959) and supported by many studies since (e.g., Heggenes et al. 1991), has become a widely accepted paradigm, at least in the literature on resident stream salmonids.

The problem with most studies supporting this restricted movement paradigm (RMP, sensu Gowan et al. 1994) lies in the inferences drawn from the usual study design. Investigators generally captured fish from a set of defined reaches, marked and returned them to their "home reach," and at some time later (usually 7 to 365 d) recaptured fish from the same set of reaches. In a high proportion of the 40 studies listed by Gowan et al. (1994) that used this design, most fish that were recaptured were found in their home reach, leading investigators to conclude that the fish were sedentary. For example, in 30 of the 36 studies (83%) for which data were available, 50% or more of the recaptured fish were in their home reach, and in 14 (39%) of the studies, at least 80% of the recaptured fish were in the original release location (Figure 1). However, these recaptures usually represented only a relatively small proportion of the fish originally marked; the rest were assumed to have died or been missed in sampling. In fact, 50% or more of the marked fish were not recaptured in 26 (79%) of the 33 studies for which these data were available, and 80% or more were not recaptured in 12 (36%) of these studies. Gowan et al. (1994; see also Lindsey et al. 1959) pointed out the inherent circularity in concluding that stream salmonids are sedentary based on studies that focus recapture efforts primarily in reaches into which fish were released.

Although Gowan et al. (1994) focused on the validity of the RMP for stream salmonids, the paradigm originated from, and has been applied to, other fish taxa as well (see Gerking 1959). However, the same methods were used to assess residency. For example, Gerking (1950) concluded that a stream fish assemblage of catostomids and centrarchids was undisturbed by a large flood because 75% of fish recaptured afterwards were in their home reaches, yet intensive sampling captured only 25% of the 540 fish marked. Working with a similar fauna in a nearby stream, Gerking (1953) reported that 80–100% of individuals of four major species were recaptured in their home reaches after 34 d, but only 32% of 293 marked fish were recaptured. Hill and Grossman (1987) recaptured 77–92% of one cottid and two cyprinid species in their original sections and concluded that their annual home ranges were less than 20 m, but they did not relocate 67–81% of 365 marked fish.

The contention that nonsalmonid stream fishes are sedentary based on mark–recapture studies is also not consistent with other findings. For example,

the upstream movements of adults of over 20 fish species were significant enough to counter nutrient losses in the upper reach of a North Carolina stream (Hall 1972). Meffe and Sheldon (1990) reported that within 1 year, experimentally defaunated stream reaches had regained the original community structure and species' abundances of fishes, and they regarded this as evidence that stream fishes are part of an open system. As an alternative to the prevailing paradigm, Linfield (1985) proposed that cyprinid communities were "... totally mobile populations influenced predominantly by river flow characteristics and behavioural responses to flow and temperature ...."

We originally questioned the RMP based on results of several ongoing studies of trout populations in central Rocky Mountain streams (see Gowan et al. 1994 for a detailed discussion). In a long-term study of trout population response to habitat enhancement (log structures that form plunge pools) in six northern Colorado streams, Riley and Fausch (1995) found that populations of brook trout *Salvelinus fontinalis* and brown trout *Salmo trutta* increased significantly in sections where logs had been placed compared to adjacent control sections. However, the response could be explained primarily by immigration (Riley et al. 1992) rather than by increased overwinter survival, as had been proposed in reports of other long-term studies (e.g., Hunt 1971). The initial evidence for such immigration was large numbers of unmarked adult trout captured during each annual electrofishing effort, despite high capture efficiencies (Riley and Fausch 1992) that allowed permanently marking more than 90% of juvenile and adult fish each year. Alternative hypotheses that could explain the high percentage of unmarked fish, such as high mortality due to electrofishing followed by influx of unmarked fish from adjacent reaches, were rejected by experiments or other sampling (Riley et al. 1992).

Subsequent research, during which weirs were placed in two of the six streams, showed that in most years more trout moved across section boundaries during about 50–100 d from midsummer to fall than maximum-likelihood removal estimates indicated were present in the sections (Riley et al. 1992; Gowan 1995). Thus, immigration into sections was high enough to explain the rapid increase in trout abundance following habitat enhancement. However, despite high immigration, an average of 83% of marked trout that were recaptured had remained within their home reach between years (Riley et al. 1992), which clearly points out the dangers in mak-

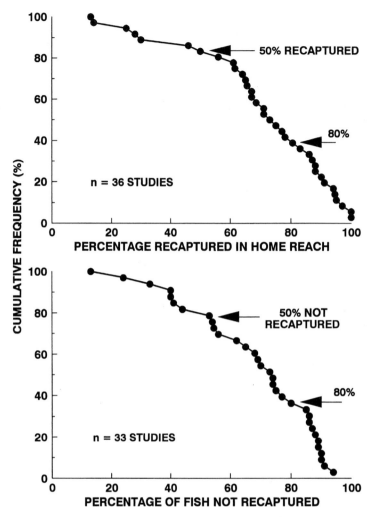

FIGURE 1.—Cumulative frequency of the percentage of recaptured fish found in their home reaches (**top:** N = 36 studies) and percentage of marked fish never recaptured (**bottom:** N = 33 studies), for studies of resident stream salmonids reviewed by Gowan et al. (1994). Percentages were cumulated from highest to lowest.

ing inferences about movement based only on re-captured fish.

Several investigators have used radiotelemetry to measure stream fish movements (e.g., Chisholm et al. 1987; Clapp et al. 1990; Tyus 1990), which allows continuous tracking of fishes far beyond the reaches where they were marked. In general, these studies have found that many fish move relatively long distances. For example, Todd and Rabeni (1989) found that smallmouth bass *Micropterus dolomieu* moved a minimum of 120–984 m/d between pools during all seasons; greater movement occurred at warmer water temperatures (range 4–27.5°C). Te-lemetry of 54 adult brown trout (250–530 mm total length, TL) in mountain watersheds of southern Wyoming showed that nearly 70% had annual home ranges longer than 50 m and 30% had home ranges exceeding 300 m of stream length (Young 1994). Some fish moved more than 1 km in 24 h, and three of the larger fish had home ranges of 23–96 km, which included downstream movements to a larger river. Clapp et al. (1990) and Meyers et al. (1992) found similar large home ranges (range 0.4–33 km) for adult brown trout in Michigan and Wisconsin streams, respectively. These results run counter to reports that resident trout generally remain within 20–50-m reaches their entire lives (Miller 1957; Bachman 1984).

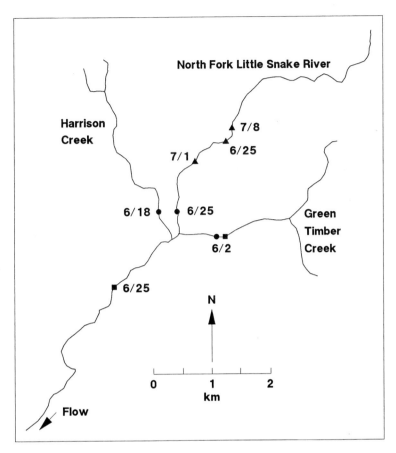

FIGURE 2.—Positions of three Colorado River cutthroat trout in early summer 1992 in the North Fork Little Snake River and two of its tributaries in south-central Wyoming. Each fish is shown as a different symbol placed at locations of capture on various dates (2 June is denoted 6/2; 8 July is 7/8).

Because neither brook trout nor brown trout are native to the Rocky Mountain region, one could argue that the high rates of movement we found might be due to the lack of adaptation by these species to the unique dynamics of habitats and flows in these mountain lotic systems. However, native stream salmonids are also known to display substantial movement. For example, adult Colorado River cutthroat trout *Oncorhynchus clarki pleuriticus* (170–237 mm TL) ranged over distances that often exceeded 500 m during June through early August in a southern Wyoming mountain watershed, based on radiotelemetry (M. K. Young, Rocky Mountain Forest and Range Experiment Station, U.S. Forest Service, unpublished data). They moved both upstream and downstream in a third-order stream after descending from tributaries where they had originally been captured (Figure 2). Bernard and Israelsen (1982) observed upstream and down-

stream movements of over 2 km by juvenile and adult cutthroat trout from April through November in a small tributary to a Utah river. The longest reported movements of native inland cutthroat trout exceeded 120 km (Bjornn and Mallet 1964).

Investigators have also measured turnover of marked individuals in defined reaches (Smith and Saunders 1958; Decker and Erman 1992), or measured recolonization into stream reaches depleted of fish, either intentionally or by natural disturbance (e.g., floods, drought, volcanism; see Hawkins and Sedell 1990). Almost always, they found rapid immigration of fishes (Shetter and Hazzard 1938; Gerking 1953; Matthews 1986; Meffe and Minckley 1987; Meffe and Sheldon 1990). For example, Peterson and Bayley (1993) found that fish communities in 18 experimentally depopulated reaches (46–113 m long) of Illinois streams quickly returned to predisturbance configurations. Colonization models indicated that fish assemblages would reach 70% of maximum similarity based on species composition within 60–140 h and 90% of original abundance within 100–270 h. Both marked slimy sculpins *Cottus cognatus* from adjacent reaches and unmarked individuals rapidly repopulated a repeatedly electrofished stream reach (Morgan and Ringler 1992). Of course, taxa differ in their colonization ability (Larimore et al. 1959; Schlosser 1987). For example, Fausch and Bramblett (1991) reported that certain fishes were capable of colonizing long distances upstream and persisting in intermittent tributaries of a western Great Plains river that were prone to intense flash floods, whereas other species colonized only short distances from the main river.

Some authors have proposed that individuals that move represent a mobile fraction of the population with different behavior than the larger portion that is sedentary (e.g., Funk 1955; Stott 1967; Solomon and Templeton 1976). However, others have found that individuals switch behavior (Harcup et al. 1984). Our work on movement of brook trout (Riley et al. 1992) and brown trout (Young 1994) indicates that individual fish in populations display a wide range of movement distances, suggesting the need for a model that is more complex than simply sedentary versus mobile fractions. In our view, future research should focus on the factors that cause fish to move under different environmental conditions (Gowan et al. 1994).

Why Stream Fishes Move

Theories of fish dispersal hold that movement is adaptive, conferring advantages to individuals in terms of their lifetime fitness (Northcote 1978; Gross 1987). In short, if the benefits of moving to one or more new habitats outweigh the energetic costs of movement and the risk of predation, life history types that move should be favored (Gross et al. 1988). Although such theories have most often been applied to diadromous fishes, resident stream fishes may also increase their fitness by moving to find habitat needed to complete certain life history stages, or to search for optimum habitat as present locations become unsuitable. Moreover, because streams are linear systems in which habitat is inherently patchy in space (Pringle et al. 1988; Scarsbrook and Townsend 1993) and dynamic through time (Resh et al. 1988; Poff and Ward 1989), it is clear that stream fishes often must move to find habitats needed to maximize fitness.

Current theories of population dynamics at landscape scales emphasize the need to consider both the physical arrangement of patches and the population dynamics within and among patches. Animals may need to move among patches to garner enough of scarce resources (i.e., habitat supplementation; Schlosser 1991, 1995a; Dunning et al. 1992) or to find a critical resource (i.e., habitat complementation), but in either case the physical arrangement and proximity of patches will affect the extent of fish movement required (Schlosser 1995b; Schlosser and Angermeier 1995, this volume). For example, large brown trout in two midwestern U.S. streams moved long distances to set up smaller home ranges for feeding during summer (i.e., habitat supplementation; Clapp et al. 1990) and to find overwinter habitat (habitat complementation; Meyers et al. 1992). Smaller brook trout and brown trout also have moved relatively long distances (200–400 m) to find suitable overwintering habitat (Cunjak and Power 1986; Chisholm et al. 1987).

In addition to moving to complement or supplement resources, stream fishes often must move because the habitat they occupy becomes either unsuitable or suboptimal. For example, both Clapp et al. (1990) and Meyers et al. (1992) found that habitats used by brown trout during winter were not suitable during summer due to high temperature. Similarly, juvenile coho salmon *Oncorhynchus kisutch* and coastal cutthroat trout *O. c. clarki* in a stream on Vancouver Island, British Columbia, used temporary floodplain habitat during winter that was dry during summer (Hartman and Brown 1987). In contrast, Näslund (1990; Näslund et al. 1993) found that within 8 years, mobile individuals of Arctic char *Salvelinus alpinus* began migration to two recently eutrophied lakes 5 km upstream from the oligotrophic lake they originally occupied, and these fish grew faster and matured earlier as a result. Similarly, both Riley and Fausch (1995) and Burgess and Bider (1980; Burgess 1985) found that trout populations in 100–250-m reaches increased rapidly after habitat enhancement, largely by immigration. In the six streams we studied, few of the immigrants came from adjacent 250-m control reaches (Riley and Fausch 1995), and in two of the streams marked fish were recaptured at eight distances up to 2 km upstream and downstream, suggesting that trout moved from many different locations throughout the watershed to encounter the added habitat (Gowan 1995). Thus, movement can also increase fitness by allowing individuals to rapidly exploit newly created resources (Northcote 1978, 1984).

Recent theories of animal population dynamics at landscape scales also hold that many species occur as loosely coupled subpopulations that form metapopulations (Hanski 1991). Models of metapopulation dynamics pit the probability of extinction in individual habitat patches against the probability of colonization from adjacent patches (Harrison and Quinn 1989; Harrison 1991). Given the patterns of extinction and recolonization observed among "resident" stream fishes, it is likely that many exist as metapopulations (e.g., Demarais et al. 1992; Propst et al. 1992; Schlosser and Angermeier 1995). This realization also highlights the importance of the movement required to drive metapopulation dynamics (Rieman and McIntyre 1993). Especially for small populations, such as those confined to headwater streams, even modest amounts of immigration may promote persistence. For example, Stacey and Taper (1992) found through demographic modeling based on observed population parameters that immigration of only five individuals per year into a small population of acorn woodpeckers *Melanerpes formicivorus* ($N = 52$ individuals) was required to counteract environmental variation and achieve persistence for more than the 70 years that the population had been known to exist. In contrast, 3 of 11 isolated headwater populations of the federally endangered Gila trout *O. gilae* in New Mexico were decimated or extirpated by stochastic environmental disturbances from floods, droughts, or fires in two successive years. This strongly suggests that refounding by adjacent source populations (sensu Pulliam 1988) would be required for long-term persistence throughout the species' range (Propst et al. 1992).

Consequences of Movement for Identifying Evolutionarily Significant Units

Waples (1991) defined ESUs based on two characteristics: reproductive isolation and evolutionary uniqueness (see also Behnke 1992). Here we address how movement may affect reproductive isolation, whether movement is heritable and thus useful for identifying ESUs, and how it might bias sampling for population genetics.

Because resident stream fishes have been perceived as relatively sedentary, reproductive isolation has often been assumed, after which genetic analysis may be conducted to define differences among populations. However, rarely has the spatial extent of any stream-dwelling fish population been examined. For example, if some proportion of fishes move to other waters to find critical habitat or reproduce, the range of a single population may include far more than the "type location." Despite

this, genetic analyses do often detect differences in allele frequencies in conspecifics from adjacent streams (Allendorf and Leary 1988), and have delineated others that are sympatric but apparently reproductively isolated (Allendorf et al. 1976; Ryman et al. 1979; Ferguson and Mason 1981; Carl and Healey 1984). If the number of immigrants to a population is low relative to population size, selection for locally adapted genes is strong (Waples 1991), or if immigrants have lower reproductive success than residents (Wiens 1976; Tallman and Healey 1994), the population may still differentiate genetically from its source (Wright 1969; Nelson and Soulé 1987). Thus, the extent to which populations may differ genetically, but still be linked reproductively, merits further investigation to address "genetically effective" movement rates.

By definition, movement is an important life history attribute for anadromous, adfluvial, and fluvial fishes. The evidence presented above suggests that it is also important for fishes with "resident" life histories, enabling them to find suitable habitat, food, and spawning areas in the linear hierarchies of stream systems where resources are patchily distributed across space (Frissell et al. 1986). However, despite its adaptive significance, the conditions under which migratory behavior is genetically determined are still under study, so identifying ESUs based on movement patterns alone appears to be unreliable. In some cases, migratory stocks have been shown to differ genetically from sympatric resident stocks (Foote et al. 1989; Verspoor and Cole 1989). Similarly, salmonids introduced into barren lakes have rapidly developed both inlet- and outlet-spawning populations with heritable differences in the direction that fry move to reach the lake (Lentsch 1985; Kaya 1989). However, Hindar et al. (1991) reported that life history polymorphism in salmonid populations need not be related to genetic differentiation. For example, fish displaying both migratory and resident life histories have originated from the same deme (Campbell 1977; Jonsson 1985) and the same parents (Nordeng 1983). Environmental influences, especially on juvenile development, appear to influence life history strategies such as the timing and amount of movement in several species of salmonids (Northcote 1969; Jonsson 1989; Metcalfe et al. 1989; Hindar and Jonsson 1993). Furthermore, the proportion of individuals opting for different strategies can vary annually depending on habitat conditions (Metcalfe et al. 1988).

Because many stream fishes are mobile, when fish are collected will influence the abundance and composition of a sample (Decker and Erman 1992). Ross et al. (1985) also concluded that ignorance of fish movements could lead to nonrandom samples of fish populations. Waples (1991) proposed that the timing of sampling could influence the genetic evaluation of populations, and Waples and Teel (1990) reported significant annual changes in allele frequencies. Given that seasonal and annual variability in population composition is likely, perhaps geneticists should consider the consequences of one-time sampling (e.g., Loudenslager and Gall 1980; Leary et al. 1993) for describing population genetic structure. For example, meristic and morphometric analyses were used to determine the genetic purity of Colorado River cutthroat trout from two tributaries and the main stem of the North Fork Little Snake River in southern Wyoming (Binns 1977). Fish in the main stem were reported to be genetically pure, fish from Harrison Creek were hybridized with rainbow trout *Oncorhynchus mykiss*, and fish from Green Timber Creek were assumed to be intermediate. However, in 1992 a single radio-tagged adult occupied all three locations within 3 weeks (Figure 2) and all telemetered fish originally captured in Harrison and Green Timber creeks eventually migrated to the North Fork Little Snake River (M. K. Young, unpublished data). Thus, depending on the timing of sampling these fish could have been chosen to represent populations in the main stem or one of the tributaries, all of which were presumed to be distinct.

Consequences of Movement for Preserving Evolutionarily Significant Units

Much current interest is focused on the possibility that many stream fishes make up metapopulations (e.g., Rieman and McIntyre 1993) consisting of subpopulations linked by immigration and emigration (Hanski and Gilpin 1991). Although models other than the classic metapopulation may be more realistic for stream fish populations (e.g., source–sink or patchy populations; Pulliam 1988; Schlosser and Angermeier 1995), the fundamental assumption is that individuals disperse. Therefore, if subpopulations are isolated and movement is prevented, their probability of extinction will likely increase. Extinction of small populations increases as population size is reduced, habitats become fragmented, and environmental variability increases (Shaffer 1987).

Fishery managers attempting to rehabilitate populations of rare native salmonids have often intentionally barricaded streams to prevent upstream movement of nonnative competitors (e.g., Stuber et al. 1988), which seemed appropriate under the restricted movement paradigm. However, if fish need to move during periods of harsh environmental conditions to find critical habitat, the result will be to trade the risk of extinction from invasion for the risk of extinction from environmental, demographic, or genetic factors (Rieman and McIntyre 1993). For example, the extirpation of several small cyprinids that inhabit plains streams in Oklahoma (Winston et al. 1991) and New Mexico (Bestgen and Platania 1991) after construction of mainstream dams was apparently due to such metapopulation fragmentation, followed by extinction with no possibility of recolonization. Moyle and Yoshiyama (1994) emphasized that reserves to support aquatic biodiversity must each contain all the critical resources necessary for survival of species (i.e., habitat complementation; Dunning et al. 1992) and must be large enough to provide the necessary resources from some portion of the watershed despite long-term temporal variation in the environment. Failure to include critical habitats above a barrier, or failure to realize that habitat will change due to natural (e.g., fire and succession) or anthropogenic (land management, potential climate change) processes, will result in the eventual extinction of fish populations (e.g., Propst et al. 1992).

A second problem that follows from management by isolation is the potential for creating genetically unique stocks by restricting movement. Whether migratory life histories are genetically determined is under debate (Skaala and Nævdal 1989; Hindar et al. 1991). However, if mobility is genetically influenced and mobile individuals are selected against by, for example, impassable barriers or hostile environments, their contribution to future generations will be eliminated and their portion of the genome lost. Such changes have occurred both naturally and due to human activities (e.g., Demarais et al. 1992). For example, Northcote (1981) found that heritable differences in rheotaxis by rainbow trout from populations upstream and downstream of a waterfall were genetically coded. As described above, others have also reported rapid evolution of genetically distinct inlet and outlet spawning forms of salmonids introduced to barren lakes (Lentsch 1985; Kaya 1989). Such relatively rapid changes can occur under strong selection in small isolated populations. For example, artificial selection and genetic drift reduced the proportion of polymorphic loci and average heterozygosity of hatchery westslope cutthroat trout *O. c. lewisi* in only 14 years (Allendorf and Phelps 1980). Thus, identifying "real" ESUs will require knowledge of the cause and time of population isolation.

The results of research to date indicate that resident stream fishes are generally more mobile than previously thought. This suggests that biologists must understand not only the genetics, taxonomy, and systematics of various units of a species, but also the spatial and temporal dynamics of its component populations in order to intelligently define and effectively manage evolutionarily significant units.

Acknowledgments

Funding for the research on trout population response to habitat enhancement in Colorado was generously provided by Federal Aid in Fish and Wildlife Restoration project F-88-R, issued through the Colorado Division of Wildlife to Kurt Fausch. We thank Charles Gowan and two anonymous reviewers for constructive criticism on draft manuscripts.

References

Allen, K. R. 1951. The Horokiwi Stream—a study of a trout population. New Zealand Marine Department Fisheries Bulletin 10.

Allendorf, F. W., and R. F. Leary. 1988. Conservation and distribution of genetic variation in a polytypic species, the cutthroat trout. Conservation Biology 2:170–184.

Allendorf, F. W., and S. R. Phelps. 1980. Loss of genetic variation in a hatchery stock of cutthroat trout. Transactions of the American Fisheries Society 109:537-543.

Allendorf, F. W., N. Ryman, A. Stennek, and G. Stahl. 1976. Genetic variation in Scandinavian brown trout (Salmo trutta L.): evidence of distinct sympatric populations. Hereditas 83:73–82.

Bachman, R. A. 1984. Foraging behavior of free-ranging wild and hatchery brown trout in a stream. Transactions of the American Fisheries Society 113:1–32.

Behnke, R. J. 1992. Native trout of western North America. American Fisheries Society Monograph 6.

Bernard, D. R., and E. K. Israelsen. 1982. Inter- and intrastream migration of cutthroat trout (Salmo clarki) in Spawn Creek, a tributary of the Logan River, Utah. Northwest Science 56:148–158.

Bjornn, T. C., and J. Mallet. 1964. Movements of planted and wild trout in an Idaho river system. Transactions of the American Fisheries Society 93:70–76.

Bestgen, K. R., and S. P. Platania. 1991. Status and conservation of the Rio Grande silvery minnow, Hybognathus amarus. Southwestern Naturalist 36:225–232.

Binns, N. A. 1977. Present status of indigenous populations of cutthroat trout, Salmo clarki, in southwest Wyoming. Wyoming Game and Fish Department, Fisheries Technical Bulletin 2, Cheyenne.

Burgess, S. A. 1985. Some effects of stream habitat improvement on the aquatic and riparian community of a small mountain stream. Pages 223–246 in J. A. Gore, editor. The restoration of rivers and streams. Butterworth, Boston.

Burgess, S. A., and J. R. Bider. 1980. Effects of stream habitat improvements on invertebrates, trout populations, and mink activity. Journal of Wildlife Management 44:871–880.

Campbell, J. S. 1977. Spawning characteristics of brown trout and sea trout Salmo trutta L. in Kirk Burn, River Tweed, Scotland. Journal of Fish Biology 11:217–229.

Carl, L. M., and M. C. Healey. 1984. Differences in enzyme frequency and body morphology among three juvenile life history types of chinook salmon (Oncorhynchus tshawytscha) in the Nanaimo River, British Columbia. Canadian Journal of Fisheries and Aquatic Sciences 41:1070–1077.

Chapman, D. L. 1966. Food and space as regulators of salmonid populations in streams. American Naturalist 100:345–357.

Chisholm, I. M., W. A. Hubert, and T. A. Wesche. 1987. Winter stream conditions and use of habitat by brook trout in high-elevation Wyoming streams. Transactions of the American Fisheries Society 116:176–184.

Clapp, D. F., R. D. Clark, Jr., and J. S. Diana. 1990. Range, activity, and habitat of large, free-ranging brown trout in a Michigan stream. Transactions of the American Fisheries Society 119:1022–1034.

Cunjak, R. A., and G. Power. 1986. Winter habitat utilization by stream resident brook trout (Salvelinus fontinalis) and brown trout (Salmo trutta). Canadian Journal of Fisheries and Aquatic Sciences 43:1970–1981.

Decker, L. M., and D. C. Erman. 1992. Short-term seasonal changes in composition and abundance of fish in Sagehen Creek, California. Transactions of the American Fisheries Society 121: 297–306.

Demarais, B. D., T. E. Dowling, and W. L. Minckley. 1992. Post-perturbation changes in populations of endangered Virgin River chubs. Conservation Biology 7:334–341.

Dunning, J. B., B. J. Danielson, and H. R. Pulliam. 1992. Ecological processes that affect populations in complex landscapes. Oikos 65:169–175.

Fausch, K. D., and R. G. Bramblett. 1991. Disturbance and fish communities in intermittent tributaries of a western Great Plains river. Copeia 1991:659–674.

Ferguson, A., and F. M. Mason. 1981. Allozyme evidence for reproductively isolated sympatric populations of brown trout Salmo trutta L. in Lough Melvin, Ireland. Journal of Fish Biology 18:629–642.

Foote, C. J., C. C. Wood, and R. E. Withler. 1989. Biochemical genetic comparison of sockeye salmon and kokanee, the anadromous and nonanadromous forms of Oncorhynchus nerka. Canadian Journal of Fisheries and Aquatic Sciences 46:149–158.

Frissell, C. A., W. J. Liss, C. E. Warren, and M. D. Hurley. 1986. A hierarchical framework for stream habitat classification: viewing streams in a watershed context. Environmental Management 10:199–214.

Funk, J. L. 1955. Movement of stream fishes in Missouri.

Transactions of the American Fisheries Society 85: 39–57.

Gerking, S. D. 1950. Stability of a stream fish population. Journal of Wildlife Management 14:193–202.

Gerking, S. D. 1953. Evidence for the concepts of home range and territory in stream fishes. Ecology 34:347–365.

Gerking, S. D. 1959. The restricted movement of fish populations. Biological Review 34:221–242.

Gowan, C. 1995. Trout responses to habitat manipulation in streams at individual and population scales. Doctoral dissertation. Colorado State University, Fort Collins.

Gowan, C., M. K. Young, K. D. Fausch, and S. C. Riley. 1994. Restricted movement in resident stream salmonids: a paradigm lost? Canadian Journal of Fisheries and Aquatic Sciences 51:2626–2637.

Grant, J. W. A., and D. L. Kramer. 1990. Territory size as a predictor of the upper limit to population density of juvenile salmonids in streams. Canadian Journal of Fisheries and Aquatic Sciences 47:1724–1737.

Gross, M. R. 1987. Evolution of diadromy in fishes. American Fisheries Society Symposium 1:14–25.

Gross, M. R., R. M. Coleman, and R. M. McDowall. 1988. Aquatic productivity and the evolution of diadromous fish migration. Science 239:1291–1293.

Hall, C. A. S. 1972. Migration and metabolism in a temperate stream ecosystem. Ecology 53:585–604.

Hanski, I. 1991. Single-species metapopulation dynamics: concepts, models and observations. Biological Journal of the Linnean Society 42:17–38.

Hanski, I., and M. Gilpin. 1991. Metapopulation dynamics: brief history and conceptual domain. Biological Journal of the Linnean Society 42:3–16.

Harcup, M. F., R. Williams, and D. M. Ellis. 1984. Movements of brown trout, Salmo trutta, L., in the River Gwyddon, South Wales. Journal of Fish Biology 24: 415–426.

Harrison, S. 1991. Local extinction in a metapopulation context: an empirical evaluation. Biological Journal of the Linnean Society 42:73–88.

Harrison, S., and J. F. Quinn. 1989. Correlated environments and the persistence of metapopulations. Oikos 56:293–298.

Hartman, G. F., and T. G. Brown. 1987. Use of small, temporary, floodplain tributaries by juvenile salmonids in a west coast rain-forest drainage basin, Carnation Creek, British Columbia. Canadian Journal of Fisheries and Aquatic Sciences 44:262–270.

Hawkins, C. P., and J. R. Sedell. 1990. The role of refugia in the recolonization of streams devastated by the 1980 eruption of Mount St. Helens. Northwest Science 64:271–274.

Heggenes, J., T. G. Northcote, and A. Peter. 1991. Spatial stability of cutthroat trout (Oncorhynchus clarki) in a small, coastal stream. Canadian Journal of Fisheries and Aquatic Sciences 48:757–762.

Hesthagen, T. 1988. Movements of brown trout, Salmo trutta, and juvenile Atlantic salmon, Salmo salar, in a coastal stream in northern Norway. Journal of Fish Biology 32:639–653.

Hill, J., and G. D. Grossman. 1987. Home range esti-

mates for three North American stream fishes. Copeia 1987:376–380.

Hindar, K., and B. Jonsson. 1993. Ecological polymorphism in Arctic charr. Biological Journal of the Linnean Society 48:63–74.

Hindar, K., B. Jonsson, N. Ryman, and G. Stahl. 1991. Genetic relationships among landlocked, resident, and anadromous brown trout, Salmo trutta L. Heredity 66:83–91.

Hunt, R. L. 1971. Responses of a brook trout population to habitat development in Lawrence Creek. Wisconsin Department of Natural Resources Technical Bulletin 48, Madison.

Jonsson, B. 1985. Life history patterns of freshwater resident and sea-run migrant brown trout in Norway. Transactions of the American Fisheries Society 114: 182–194.

Jonsson, B. 1989. Life history and habitat use of Norwegian brown trout (Salmo trutta). Freshwater Biology 21:71–86.

Kaya, C. M. 1989. Rheotaxis of young Arctic grayling from populations that spawn in inlet or outlet streams of a lake. Transactions of the American Fisheries Society 118:474–481.

Larimore, R. W., W. F. Childers, and C. Heckrotte. 1959. Destruction and re-establishment of stream fish and invertebrates affected by drought. Transactions of the American Fisheries Society 88:261–285.

Leary, R. F., F. W. Allendorf, and S. H. Forbes. 1993. Conservation genetics of bull trout in the Columbia and Klamath river drainages. Conservation Biology 7:856–865.

Lentsch, L. D. 1985. Evaluation of young-of-the-year production in a unique wild trout population. Master's thesis. Colorado State University, Fort Collins.

Lindsey, C. C., T. G. Northcote, and G. F. Hartman. 1959. Homing of rainbow trout to inlet and outlet spawning streams at Loon Lake, British Columbia. Journal of the Fisheries Research Board of Canada 16:695–719.

Linfield, R. S. J. 1985. An alternative concept to home range theory with respect to populations of cyprinids in major river systems. Journal of Fish Biology 27 (Supplement A):187–196.

Loudenslager, E. J., and G. A. E. Gall. 1980. Geographic patterns of protein variation and subspeciation in cutthroat trout, Salmo clarki. Systematic Zoology 29: 27–42.

Matthews, W. J. 1986. Fish faunal structure in an Ozark stream: stability, persistence and a catastrophic flood. Copeia 1986:388–397.

McDowall, R. M. 1988. Diadromy in fishes. Croom Helm, London.

Meffe, G. K., and W. L. Minckley. 1987. Persistence and stability of fish and invertebrate assemblages in a repeatedly disturbed Sonoran desert stream. American Midland Naturalist 117:177–191.

Meffe, G. K., and A. L. Sheldon. 1990. Post-defaunation recovery of fish assemblages in southeastern blackwater streams. Ecology 71:657–667.

Metcalfe, N. B., F. A. Huntingford, W. D. Graham, and J. E. Thorpe. 1989. Early social status and the development of life-history strategies in Atlantic salmon. Proceedings of the Royal Society of London B236:7–19.

Metcalfe, N. B., F. A. Huntingford, and J. E. Thorpe. 1988. Feeding intensity, growth rates, and the establishment of life-history patterns in juvenile Atlantic salmon. Journal of Animal Ecology 57:463–474.

Meyers, L. S., T. F. Thuemler, and G. W. Kornely. 1992. Seasonal movements of brown trout in northeast Wisconsin. North American Journal of Fisheries Management 12:433–441.

Miller, R. B. 1957. Permanence and size of home territory in stream-dwelling cutthroat trout. Journal of the Fisheries Research Board of Canada 14:687–691.

Morgan, C. R., and N. H. Ringler. 1992. Experimental manipulation of sculpin (Cottus cognatus) populations in a small stream. Journal of Freshwater Ecology 7:227–232.

Moyle, P. B., and R. M. Yoshiyama. 1994. Protection of aquatic biodiversity in California: a five-tiered approach. Fisheries 19(2):6–18.

Näslund, I. 1990. The development of regular seasonal habitat shifts in a landlocked Arctic charr, Salvelinus alpinus L., population. Journal of Fish Biology 36:401–414.

Näslund, I., G. Milbrink, L. O. Eriksson, and S. Holmgren. 1993. Importance of habitat productivity differences, competition and predation for the migratory behavior of Arctic charr. Oikos 66:538–546.

Nelson, K., and M. Soulé. 1987. Genetical conservation of exploited fishes. Pages 345–368 in N. Ryman and F. M. Utter, editors. Population genetics and fishery management. University of Washington Press, Seattle.

Nordeng, H. 1983. Solution to the "char problem" based on Arctic char (Salvelinus alpinus) in Norway. Canadian Journal of Fisheries and Aquatic Sciences 40:1372–1387.

Northcote, T. G. 1969. Patterns and mechanisms in the lakeward migratory behaviour of juvenile trout. Pages 183–203 in T. G. Northcote, editor. Symposium on salmon and trout in streams. H. R. MacMillan Lectures in Fisheries, University of British Columbia, Vancouver.

Northcote, T. G. 1978. Migratory strategies and production in freshwater fishes. Pages 326–359 in S. D. Gerking, editor. Ecology of freshwater fish production. Blackwell Scientific Publications, Oxford, UK.

Northcote, T. G. 1981. Juvenile current response, growth and maturity of above and below waterfall stocks of rainbow trout, Salmo gairdneri. Journal of Fish Biology 18:741–751.

Northcote, T. G. 1984. Mechanisms of fish migration in rivers. Pages 317–355 in J. D. McCleave, G. P. Arnold, J. J. Dodson, and W. H. Neill, editors. Mechanisms of migration in fishes. Plenum, New York.

Northcote, T. G. 1992. Migration and residency in stream salmonids: some ecological considerations and evolutionary consequences. Nordic Journal of Freshwater Research 67:5–17.

Peterson, J. T., and P. B. Bayley. 1993. Colonization rates of fishes in experimentally defaunated warmwater streams. Transactions of the American Fisheries Society 122:199–207.

Poff, N. L., and J. V. Ward. 1989. Implications of streamflow variability and predictability for lotic community structure: a regional analysis of streamflow patterns. Canadian Journal of Fisheries and Aquatic Sciences 46:1805–1818.

Pringle, C. M., and seven coauthors. 1988. Patch dynamics in lotic systems: the stream as a mosaic. Journal of the North American Benthological Society 7:503–524.

Propst, D. L., J. A. Stefferud, and P. R. Turner. 1992. Conservation and status of Gila trout, Oncorhynchus gilae. Southwestern Naturalist 37:117–125.

Pulliam, H. R. 1988. Sources, sinks, and population regulation. American Naturalist 132:652–661.

Resh, V. H., and nine coauthors. 1988. The role of disturbance in stream ecology. Journal of the North American Benthological Society 7:433–455.

Rieman, B. E., and J. D. McIntyre. 1993. Demographic and habitat requirements for conservation of bull trout. U.S. Forest Service General Technical Report INT-302, Ogden, Utah.

Riley, S. C., and K. D. Fausch. 1992. Underestimation of trout population size by maximum-likelihood removal estimates in small streams. North American Journal of Fisheries Management 12:768–776.

Riley, S. C., and K. D. Fausch. 1995. Trout population response to habitat enhancement in six northern Colorado streams. Canadian Journal of Fisheries and Aquatic Sciences 52:34–53.

Riley, S. C., K. D. Fausch, and C. Gowan. 1992. Movement of brook trout (Salvelinus fontinalis) in four small subalpine streams in northern Colorado. Ecology of Freshwater Fish 1:112–122.

Ross, S. T., W. J. Matthews, and A. A. Echelle. 1985. Persistence of stream fish assemblages: effects of environmental change. American Naturalist 126:24–40.

Ryman, N., F. W. Allendorf, and G. Stahl. 1979. Reproductive isolation with little genetic divergence in sympatric populations of brown trout (Salmo trutta). Genetics 92:247–262.

Scarsbrook, M. R., and C. R. Townsend. 1993. Stream community structure in relation to spatial and temporal variation: a habitat templet study of two contrasting New Zealand streams. Freshwater Biology 29:395–410.

Schlosser, I. J. 1987. A conceptual framework for fish communities in small warmwater streams. Pages 17–24 in W. J. Matthews and D. C. Heins, editors. Community and evolutionary ecology of North American stream fishes. University of Oklahoma Press, Norman.

Schlosser, I. J. 1991. Stream fish ecology: a landscape perspective. BioScience 41:704–712.

Schlosser, I. J. 1995a. Critical landscape attributes that influence fish population dynamics in headwater streams. Hydrobiologia 303:71–81.

Schlosser, I. J. 1995b. Dispersal, boundary processes, and trophic-level interactions in streams adjacent to beaver ponds. Ecology 76:908–925.

Schlosser, I. J., and P. L. Angermeier. 1995. Spatial variation in demographic processes of lotic fishes: conceptual models, empirical evidence, and implications for conservation. American Fisheries Society Symposium 17:392–401.

Shaffer, M. 1987. Minimum viable populations: coping with uncertainty. Pages 69–86 in M. E. Soulé, editor. Viable populations for conservation. Cambridge University Press, Cambridge, UK.

Shetter, D. S., and A. S. Hazzard. 1938. Species composition by age groups and stability of fish populations in sections of three Michigan trout streams during the summer of 1937. Transactions of the American Fisheries Society 68:281–302.

Skaala, Ø., and G. Nævdal. 1989. Genetic differentiation between freshwater resident and anadromous brown trout, Salmo trutta, within watercourses. Journal of Fish Biology 34:597–605.

Smith, M. W., and J. W. Saunders. 1958. Movements of brook trout, Salvelinus fontinalis (Mitchill) between and within fresh and salt water. Journal of the Fisheries Research Board of Canada 15:1403–1449.

Solomon, D. J., and R. G. Templeton. 1976. Movements of brown trout Salmo trutta L. in a chalk stream. Journal of Fish Biology 9:411–423.

Stacey, P. B., and M. Taper. 1992. Environmental variation and the persistence of small populations. Ecological Applications 2:18–29.

Stott, B. 1967. The movements and population densities of roach (Rutilus rutilus [L.]) and gudgeon (Gobio gobio [L.]) in the river Mole. Journal of Animal Ecology 36:407–423.

Stuber, R. J., B. D. Rosenlund, and J. R. Bennett. 1988. Greenback cutthroat trout recovery program: management overview. American Fisheries Society Symposium 4:71–74.

Tallman, R. F., and M. C. Healey. 1994. Homing, straying, and gene flow among seasonally separated populations of chum salmon (Oncorhynchus keta). Canadian Journal of Fisheries and Aquatic Sciences 51: 577–588.

Todd, B. L., and C. F. Rabeni. 1989. Movement and habitat use by stream-dwelling smallmouth bass. Transactions of the American Fisheries Society 118: 229-242.

Tyus, H. M. 1990. Potamodromy and reproduction of Colorado squawfish in the Green River basin, Colorado and Utah. Transactions of the American Fisheries Society 119:1035–1047.

Verspoor, E., and L. J. Cole. 1989. Genetically distinct populations of resident and anadromous Atlantic salmon, Salmo salar. Canadian Journal of Zoology 67:1453–1461.

Waples, R. S. 1991. Pacific salmon, Oncorhynchus spp., and the definition of "species" under the Endangered Species Act. U.S. National Marine Fisheries Service Marine Fisheries Review 53:11–22.

Waples, R. S., and D. J. Teel. 1990. Conservation genetics of Pacific salmon. I. Temporal changes in allelic frequency. Conservation Biology 4:144–156.

Wiens, J. A. 1976. Population responses to patchy environments. Annual Review of Ecology and Systematics 7:81–120.

Winston, M. R., C. M. Taylor, and J. Pigg. 1991. Upstream extirpation of four minnow species due to damming a prairie stream. Transactions of the American Fisheries Society 120:98–105.

Wright, S. 1969. The theory of gene frequencies. Evolution and genetics of populations, volume 2. Chicago University Press, Chicago.

Young, M. K. 1994. Mobility of brown trout in south-central Wyoming streams. Canadian Journal of Zoology 72:2078–2083.

American Fisheries Society Symposium 17:371–380, 1995

Safe Havens: Refuges and Evolutionarily Significant Units

HIRAM W. LI[1], KENNETH CURRENS[2], DANIEL BOTTOM[3], SHARON CLARKE[4],
JEFF DAMBACHER[3], CHRISTOPHER FRISSELL[5,6], PHILLIP HARRIS[5],
ROBERT M. HUGHES[7], DALE MCCULLOUGH[8], ALAN MCGIE[3],
KELLY MOORE[3], RICHARD NAWA[5], AND SANDY THIELE[7]

Watershed Classification Committee
Oregon Chapter of the American Fisheries Society

Abstract.—Use of a genetic refuge as a recolonization source to aid the recovery of evolutionarily significant units (ESUs) is an appealing concept. However, matching refuges with ESUs may prove problematical. Searching for "aquatic diversity areas" to protect Oregon's native fishes, we found that the best remaining habitats and populations are predominately at higher elevations. Relatively few populations inhabit refuge-quality habitats and these may be at the extremes of species' ranges. The ramification is this: are any of these marginal populations ESUs? Landscapes are so fragmented that what appears to be a metapopulation actually may comprise fragments of a "core-and-satellite" pattern. In the metapopulation model, each of the populations carries equal evolutionary weight; in the core-and-satellite model, the core is more evolutionarily important than satellite populations. Mistaking one model for the other will affect the designation of an ESU and strategies for designing a reserve system. We suggest two alternative approaches to estimation of evolutionary significance that may be helpful in minimizing these mistakes: (1) employ phylogenetic systematics to develop phylogenies of populations, or (2) search for congruent patterns between biogeography and phylogeny among members of the fish assemblage.

Our focus is on the criteria that we use to select habitats to conserve aquatic fauna. This may appear to differ from the aim of this symposium: to bring greater definition to "distinct population segment of any species," the smallest group that can be protected by the U.S. Endangered Species Act of 1973 (ESA; 16 U.S.C. §§ 1531 to 1544). The two objectives are related, however, because official designation of a species (or distinct population segment) as threatened or endangered triggers the need to conserve ecosystems; as stated in Section 2(b) of the ESA, "The purposes of this Act are to provide a means whereby the ecosystems upon which endangered species and threatened species depend may be conserved."

Clearly, the intent of the ESA is to protect and conserve ecosystems and endangered and threatened species are to be used as indicators of poor ecosystem health. From our perspective, identifying a "distinct population segment" that is threatened is only one criterion for recognizing ecosystems that need protection. Protection should also be based on identifying ecosystems most likely to sustain evolution of species. This requires detailed knowledge of the evolutionary dynamics involving landscapes, faunal assemblages, species, metapopulations, and distinct population segments, knowledge we seldom have. Using Oregon as an example, we focus on the problems that a fragmented landscape imposes on defining distinct population segments and delimiting refuges. In light of these problems, we offer alternative and complementary approaches to defining habitat needs for conservation.

Status of Oregon's Native Freshwater And Anadromous Fishes

Oregon's native freshwater fish fauna is imperiled. Forty-four percent of Oregon's native fishes are either endangered, threatened, or of special concern (Williams et al. 1989; Warren and Burr

[1]National Biological Service, Oregon Cooperative Fishery Research Unit, Department of Fisheries and Wildlife, 104 Nash Hall, Oregon State University, Corvallis, Oregon 97331–3803, USA.

[2]Northwest Indian Fisheries Commission, 6730 Martin Way, East, Olympia, Washington 98516-5540, USA.

[3] Oregon Department of Fish and Wildlife, 850 SW 15th Street, Corvallis, Oregon 97331, USA.

[4]U.S. Forest Service, Forest Science Laboratory, Oregon State University, Corvallis, Oregon 97331, USA.

[5]Department of Fisheries and Wildlife, 104 Nash Hall, Oregon State University, Corvallis, Oregon 97331–3803, USA.

[6]Flathead Biological Station, University of Montana, Polson, Montana 59860–9659, USA.

[7]ManTech Environmental Research Services Corp., 200 SW 35th Street, Corvallis, Oregon 97333, USA.

[8]Columbia River Inter-tribal Fish Commission, 729 NE Oregon, Suite 200, Portland, Oregon 97232-2107, USA.

1994). One hundred thirty-four anadromous salmo-nid stocks are at risk or of special concern (Nehlsen et al. 1991). In 18 years, Oregon coho salmon *On-corhynchus kisutch* went from supporting a peak in harvest to being petitioned for listing under the ESA (as have all coho salmon stocks from northern California to the Canadian border). Returns of spring-run chinook salmon *O. tshawytscha* are the lowest in recorded history—fewer than 100 fish returned to the Grande Ronde River in 1994 (Oregon Department of Fish and Wildlife, unpublished data). Populations of bull trout *Salvelinus confluentus* in Oregon are on the verge of being listed as threatened by the U.S. Fish and Wildlife Service. Pacific lampreys *Lampetra tri-dentata* returning to the Snake River in 1993 num-bered fewer than 20 (Washington Department of Fish and Game, unpublished data). Counts of Pacific lam-preys passing Winchester Dam on the Umpqua River, Oregon, fell from 37,000 in 1965 to 473 in 1993 (Or-egon Department of Fish and Wildlife, unpublished data).

Defining Reserves by Habitat Classification

Declines of fishes correspond with the deteriora-tion of catchment basins induced by poor land use practices (FEMAT 1993; Henjum et al. 1994; Li et al. 1994; McIntosh et al. 1994; Wissmar et al. 1994). Alarmed by this situation, the Oregon Chapter of the American Fisheries Society (AFS) gathered in 1990 to identify potential refuges for native fresh-water and anadromous fishes. The approach we used had to be fast and it had to compensate for the paucity of distributional and systematic information on the aquatic fauna. Speed was essential because timely work could influence land use policies that were being formulated, such as the President's For-est Management Plan (FEMAT 1993). Consequently, we chose to identify potential reserves or "aquatic diversity areas" (ADAs) using classification. Some of us have been influential in developing classifications for aquatic systems (e.g., Frissell et al. 1986; Hughes et al. 1986, 1987, 1990; Clarke et al. 1991) and we used that strength to our advantage. We presumed that preserving representative watershed basins in every ecoregion and in each zoogeographic province might preserve the evolutionary capacity of the fauna. Con-ceptually, we believed that watershed basins of the highest ecological integrity would be identified as ADAs. Depending on their distribution, ADAs could act as "seed sources" for natural gene flow and for recolonization of watersheds recovering from natural disturbances and human activity. This would be the

framework for the first classification, which would evolve as research provided new information.

Of necessity, the first classification was simple and based on expert opinion. We sent maps of ecoregions and watersheds, questionnaires, and sets of classification criteria to biologists in various state, federal, and academic institutions who were famil-iar with fishes and watershed conditions of ecore-gions and zoogeographical provinces of the state. They were asked to identify watersheds deserving protection under guidelines we devised (Table 1). They were required to fill out questionnaires to identify the criteria used in defining ADAs in their region and to identify the data bases used to sup-port their decisions. The maps and questionnaires were returned to our committee and sent to outside reviewers for comment. Any differences between the regional experts and the referees were arbi-trated by the Oregon Chapter, American Fisheries Society. Our classification criteria differ from those derived by Moyle and Yoshiyama (1994) in that ours place greater emphasis on fish taxa and less on comprehensive biodiversity; additionally, we were more explicit than Moyle and Yoshiyama (1994) in recognizing the importance of endemism as a crite-rion for ADA classification (Table 1). Once ADAs were identified, we determined the degree of protec-tion they gained by overlapping onto public lands.

Mapping showed that the distribution of ADAs is highly fragmented across Oregon (Figure 1). Two main points emerged. First, low- and midelevation ecoregions are underrepresented by ADAs and mountain regions are overrepresented (Table 2). Productive lowland areas were claimed early in the history of European-American settlement for agri-culture, cities, and logging (FEMAT 1993; Henjum et al. 1994), and few ADAs occur there. Most ADAs are on public lands that tend to be concen-trated at higher elevations. These areas are unsuit-able for agriculture and have less productive forests. Therefore, ADAs potentially protect mostly cold-water fishes inhabiting low-order streams at higher elevations, but not lowland species, subspecies, or populations. Second, many ADAs are isolated and without potential dispersal corridors (Figure 1). As a result, recolonization or gene flow from ADAs to disturbed habitats or fragmented populations will be difficult.

Habitat Fragmentation, Evolutionary Significance, and Metapopulations

Habitat fragmentation may influence how well we detect population segments that are evolutionarily

TABLE 1.—Comparison of two sets of criteria for evaluating potential refuges or preserves to protect aquatic biodiversity.

Criterion	Oregon Chapter, American Fisheries Society	Moyle and Yoshiyama (1994)
1	Supports a listed species or population sensitive to disturbance (e.g., the Metolius and upper Grande Ronde rivers support perhaps the healthiest populations of bull trout in Oregon)	Contains the resources and habitats necessary for the persistence of the species and communities it is designed to protect
2	Supports an endemic fish (e.g., the margined sculpin *Cottus marginatus* is endemic and confined to the Blue Mountains; McPhail and Lindsey 1986)	Contains the range and variability of environmental conditions necessary to maintain natural species diversity
3	Supports a rich, indigenous ichthyofauna, ideally without exotic fishes (e.g., all the fishes of the South Fork of the John Day River are indigenous)	Protected from edge and external threats
4	Serves a critical ecological function such as providing a dispersal corridor, conveying spawning gravels, or supplying high-quality water (e.g., several south-facing tributaries have riparian zones in good condition; they convey cold water to the Middle Fork of the John Day River, which is otherwise very warm, and provide coldwater refugia for spring chinook salmon)	Has interior redundancy of habitats to reduce the effects of localized species extinctions due to natural precesses
5	Associated with a long-term data set (e.g., the Andrews Forest site in the McKenzie River basin has served as a research site since 1948; it is a Long-term Ecological Research Site of the National Science Foundation and a monitoring site for both the international Man and the Biosphere program and the National Acid Precipitation Assessment Program)	Paired with another like preserve that contains most of the same species but is far enough distant that both are unlikely to be affected by a regional disaster
6		Supports populations of organisms large enough to have a low probability of extinction due to random demographic and genetic events

important and how well we can use ADAs as refuges. Fragmentation disrupts patterns of immigration and emigration that can result in gene flow among populations and recolonization of habitats. Most often, we only have genetic or ecological data from one point in time with which to assess longer-term interactions and the evolutionary significance of isolated populations in marginal habitats and ADAs.

Evolutionary Significance

Two basic quantitative approaches exist to examine relationships among taxa: phenetic analysis and phylogenetic analysis. In population genetics, phenetic analyses of geographical variation in neutral allelic frequencies are widely used to identify unique populations or groups of populations. Populations are grouped according to genetic similarity; a measure of genetic distance (e.g., Nei 1972) is used as a metric. This is a logical first step in identifying distinct population segments. Once distinctive patterns of genetic markers among populations are identified, the next step is to determine whether certain populations have distinctive physiological, life history, or ecological traits. The presumption is that these traits reflect potentially adaptive differences. Such classification of geographical aggregations of animals by similarity, rather than by ancestor–descendant relationships, however, may confound

effects of plesiomorphy, genetic drift, and parallel evolution. It also tends to put a premium on groups with unique traits, which may or may not be the most important traits for survival of a species.

In contrast, phylogenetic analyses are based on ancestor–descendant relationships and may be more appropriate for determining evolutionary significance. However, unlike phenetic analyses, phylogenetic analyses are not intended to analyze taxa below the species level. In most cases, identifying stable, conservative characteristics on which to base a phylogeny is difficult below the species level (Brooks and McLennan 1991). Differences among populations may be ephemeral as geographical barriers to reproductive isolation fall and populations introgress.

In some situations, though, phylogenetic analysis is useful in identifying persistent lineages, and it may aid in decisions to preserve the evolutionary capacity of the species. For example, because much of spawning habitat of steelhead (anadromous rainbow trout *Oncorhynchus mykiss*) and chinook salmon in the Deschutes River, Oregon (a Columbia River tributary), was destroyed by hydroelectric dams, biologists for federal and private power companies proposed expanding available habitat by removing ancient barriers to upstream migration on the White River, a large tributary of the Deschutes River. Phenetic analysis of native rainbow trout populations in

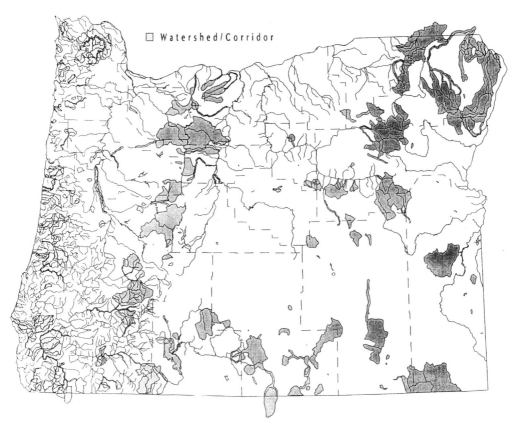

FIGURE 1.—Aquatic diversity areas in Oregon as defined by the Oregon Chapter of the American Fisheries Society.

the Deschutes River incorrectly suggested that highly differentiated White River populations may have been derived from undocumented introductions of nonnative hatchery strains. However, phylogenetic analyses showed that the White River populations were remnants of an ancestral fauna now known only from the Great Basin (Currens et al. 1990). Based on data from one point in time, the larger population of rainbow trout in the main stem of the Deschutes River may have seemed a logical source of the smaller tributary populations. In this case, however, artificial removal of

TABLE 2.—Distribution of aquatic diversity areas in relation to classes of available landscape.

Elevation class	Percentage of available landscape	Percentage of aquatic diversity areas in the landscape
Lowlands	11.1	4.6
Uplands	52.0	30.9
Mountains	36.9	64.5

ancient barriers would have led to loss of a different lineage of rainbow trout with possibly different disease resistance, greater tolerance of warm water, or other special ecological adaptations.

Metapopulation Dynamics

The long-term genetic and ecological interactions within species need to be considered in identifying populations for protection. The concept of an evolutionarily significant unit (ESU) as defined by Waples (1991) is explicitly linked to the concept of metapopulation—a population of populations. Genetically, the evolutionary significance of population structure and gene flow among breeding aggregations of a species was recognized by Wright (1931, 1940) and an evolutionary perspective has become fundamental for understanding geographical differences among isolated populations. More recently, Levins (1969) explored the consequences of metapopulations in terms of extinction and recolonization. In his model, metapopulations are composed of geographically isolated, genetic populations. Species persist because recolonization follows periodic extirpation of local populations. In a sense, then, each population of the metapopulation may be evolutionarily significant because persistence may depend on any or all populations of the species.

Metapopulations also can be described by other models (Harrison 1991), which may be more representative of what actually occurs in some species. For example, the core-and-satellite model describes a population complex whereby dispersal from a larger core population results in smaller satellite populations at and beyond its periphery. Satellite populations are more subject to extinction than the larger core (Harrison 1991). They may also be the source of evolutionary novelties (Scudder 1989). However, we believe the evolutionarily significant unit must include the core and the source of evolutionary material (Figure 2).

If the landscape is highly fragmented by human disturbance, as it is in Oregon, genetic or ecological data from one point in time could mislead interpretations of evolutionarily significant interactions, because what appears as several geographically isolated populations may once have been a large, contiguous, panmictic population. These fragments may be genetically distinct because of genetic drift in populations reduced to small numbers by human and natural disturbances. This is especially true when selectively neutral characters are used in the analysis. For example, Hatch (1990) suggested that genetic drift may be more common in small Oregon coastal watersheds than in larger ones because small watersheds have less capacity to buffer populations from the kinds of disturbances discussed by Reeves et al. (1995, this volume). Hatch (1990) found a distinct latitudinal cline in the frequency of the $sSOD-1*100$ allele among steelhead populations in large watersheds (larger than 350 km^2) along the Oregon Coast (Figure 3). No trend could be detected in small watersheds (smaller than 350 km^2).

The choice of the metapopulation model will also influence the design of genetic reserves, especially if financial resources are limited. Given that the ideal of protecting everything is probably unrealistic, one should design genetic reserves to protect evolutionary capacity and pattern. If a core-and-satellite pattern is detected, greater premium should be placed on protecting the core population. On the other hand, when the pattern suggests the classical metapopulation model of Levins (1969), all populations should be given equal weight initially; if choices have

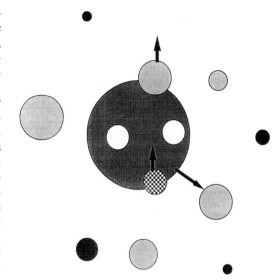

FIGURE 2.—Core-and-satellite model of metapopulations. The area of each circle represents the degree of genetic diversity. Arrows represent populations that are either diverging from or converging with the core population (largest circle). Spaces denote the degree of genetic isolation among populations. The black circles represent extinct populations. The checkered circle represents the ideal of hatchery supplementation programs (i.e., making the hatchery stock genetically indistinguishable from the wild stock). The open circles represent remnant genetic fragments of the core population, that might remain should severe landscape fragmentation occur. Should the core become fragmented, the distribution of populations might resemble a Levins (1969) metapopulation model.

to be made among them, there should be a tendency to protect unique populations.

The penalties for making a mistake are obvious. If we opt to protect a core population, but the classical metapopulation model is operating, only a fraction of the evolutionary potential will be protected. On the other hand, if a fragmented core-and-satellite pattern is mistaken for a classical metapopulation distribution and higher priority is given to protecting ecologically unique populations than more generalized ones, we risk leaving the core unprotected. A general complication is that populations at the margin of a species' range may be narrow specialists and unable to perform well under the species' more typical circumstances (Hoffmann and Blows 1994).

In order to reduce risk of making a wrong decision, hierarchical, temporal, and spatial scales must be considered. What we perceive on a particular temporal and spatial scale will influence the meta-

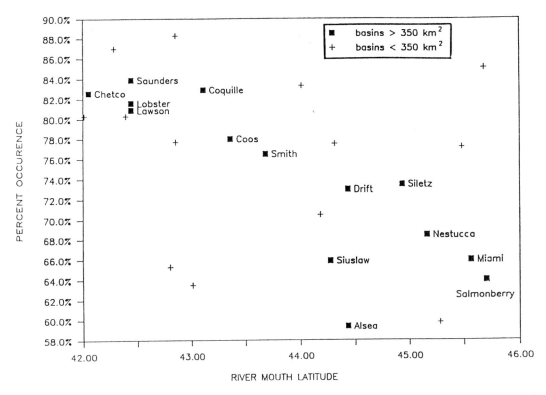

FIGURE 3.—Frequencies of the *sSOD-1*100* allele in coastal Oregon populations of wild steelhead versus latitudes of the mouths of natal rivers (from Hatch 1990). This allele codes a cytosolic (s) variant of the enzyme superoxide dismutase (SOD).

population pattern we identify. For example, the Levins (1969) model may accurately describe among-basin genetic and ecological relationships of fishes in isolated Pleistocene lake basins of the northern Great Basin accurately. But within basins, interactions may be better represented by the core-and-satellite model. Further complications are added when recurring disturbances create patchy landscapes at various stages of succession. The two basic metapopulation models must then be modified to consider the balance between local extinction, dispersal, and recolonization (Harrison 1991).

Biogeographic Analysis as a Tool for Conservation Biology

As described above, phenetic and phylogenetic analyses may not provide useful information in a fragmented landscape. Additionally, for ESA purposes, they only focus on single or very closely related taxa and do not examine other parts of the ecosystem. What then is the option? Brooks and McLennan (1991) argued that centers of endemism

have proven to be centers of high evolutionary activity and may serve that function in the future. If so, then centers of endemism should be protected.

New techniques in biogeographical analysis can help us address this problem because they allow us to examine relationships of historical faunal structure among areas of endemism (Mayden 1988; Brooks and McLennan 1991; Cracraft 1994). As outlined by Cracraft (1994), the analysis has two steps: (1) searching for congruent patterns of speciation and (2) examining the role of biotic dispersion. The work of Mayden (1988) and Cracraft (1991) exemplifies the power of the biogeographic approach.

We believe that the biogeographic approach will be an important factor in redefining our needs for ADAs in Oregon based upon fragmentary evidence. We cannot at this time completely regress the evolutionary history of the taxa onto the geological history of the area as suggested by Brooks and McLennan (1991). Our geological history is not yet well delineated and we do not have sufficient sys-

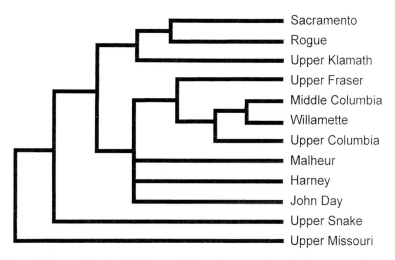

FIGURE 4.—Area cladogram of native fish faunas of selected western river basins. Faunal similarities decrease as connections (vertical lines) occur farther to the left.

tematic information to search for congruent patterns of divergence among our native fishes. Despite these difficulties, however, we gained insights into biogeographic relationships from a pilot study to explore the usefulness of historical ecology.

We developed an area cladogram using the method of Cracraft (1994) on raw distributions of fish taxa (scored as present [1] or absent [0]) among selected drainages of Oregon. Sacramento, upper Snake, and upper Missouri river faunas were outgroups. Data for the analysis came from species lists compiled by Bisson (1969), Moyle (1976), McPhail and Lindsey (1986), and Minckley et al. (1986) and from museum records of the Oregon State University Fish Collection. The analysis by itself did not separate effects of vicariance speciation patterns from effects of biotic dispersion. However, the consensus tree from the PAUP algorithm[9] showed that the John Day, Malheur, and Harney basin fish faunas have close affinities, branching off as a cluster from the Columbia River fauna (Figure 4). Bisson (1969) also concluded that the fish faunas of the John Day, Harney, and Malheur basins were very similar. Fishes of the South Fork basin of the John Day system above the barrier of Izee Falls, were more closely related to those of the Silvies River in the Harney basin than to the fauna of any other basin of the John Day River. Phenetic analyses of inland rainbow trout subspecies showed genetic di-

vergence of South Fork populations from those in the rest of John Day basin and greater similarity to those of the Harney basin (Figure 5; K. Currens, Oregon Cooperative Fishery Research Unit, unpublished data). We concluded that we will have gained a powerful analytical tool if we can refine our information about Oregon's fish fauna by examining phylogenies of individual species for congruence.

The use of biogeographical analyses to find "areas of evolutionary significance" may offer a useful complement to ESU designations and our efforts to find refuges. The ESA explicitly links the status of a species to the ecosystem upon which it depends. Identification of ESUs addresses unique populations or population groups, identification of "areas of evolutionary significance" more directly addresses evolutionary capacity of the biota.

Prospectus

Conserving biodiversity will depend on defining important units at different levels of biological organization. However, even if we are able to do so with fidelity, would the evolutionary capacity of the biota be retained? We think not. Native fishes will continue to be extirpated unless these efforts are coupled directly with policies to develop refuges and cease deleterious anthropogenic impacts. Extinction rates are accelerating (Wilson 1988). Even preserving areas of evolutionary significance requires us to assume that future environments will be similar to those past and that genetic instructions culled from the past will apply to environments of

[9]Swofford, D. L. 1985. PAUP: phylogenetic analysis using parsimony, version 2.4. Computer program distributed by the Illinois Natural History Survey, Champaign.

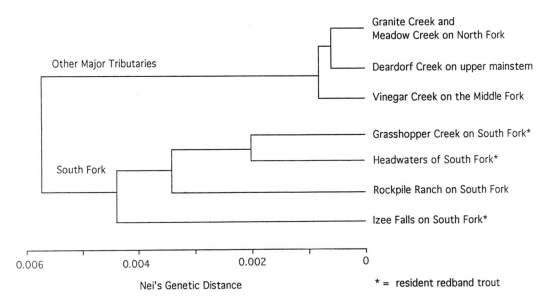

FIGURE 5.—Phenetic analysis of inland rainbow trout (including the redband subspecies) in the John Day River basin. Note that the population above Izee Falls is very different from other populations in the basin; its closest known affinities are with populations in the Silvies River of the Harney basin (not shown).

the future. In reality, however, the world is changing faster than the capacity of species to adapt. We must synchronize our efforts with the response times of biological systems: mean generation times of species; ecological succession of communities; evolutionary coadaptation of communities. For Pacific salmon, we need to find answers to management problems within two or three generations, approximately in the range of 9–21 years, or we will see massive extinctions. We need to set priorities because time is short.

To use Oregon as an example, we believe the first priority is to identify the status of the landscape and ADAs, however crudely. We must then ensure that the ADAs are protected. The next step is to identify potential habitats that can be rehabilitated quickly. At the same time, we need to identify the evolutionary relationships among populations in ADAs and those in unprotected areas and the historical biogeography of these areas. We need to add to our habitat reserve base quickly because random events can lead to extinction when population sizes are small (Pielou 1969). For instance, spawning and rearing areas for spring chinook salmon in the John Day basin are limited to a fraction of their estimated original range (Figure 6). When a truck accidentally spilled nearly 16,000 L of hydrochloric acid below the North Fork Wilderness Area of the John Day River in 1990, an estimated 98,000–145,000

fish were killed[10]. These included three species of fish sensitive to extinction: 9,500 Pacific lampreys, 1,700 spring chinook salmon, and 300 adult bull trout. Temporary loss of access to North Fork Wilderness Area habitats because of natural or anthropogenic disturbances is critical. The area is one of the last strongholds for wild chinook salmon and bull trout in the Columbia River basin. Like many other sensitive species in the region, spring chinook salmon are isolated in patches. Dispersal through lowland corridors to alternative upland sites is not possible, especially during the lower-flow period, because of high water temperatures (B. McIntosh and H. Li, Oregon State University, unpublished data).

We need additional reserves to lower the risk of extinctions. Research should be aimed at identifying areas that, with proper management, can become refuges. We need to determine how to connect dispersal corridors to patches of adequate habitat, and to determine how dispersal functions. Research on many issues is badly needed. However, speed is essential. If we concentrate on developing refuges (see Moyle and Yoshiyama 1994), we will be able to buy time for the other research issues

---

[10]Draft environmental assessment by the U.S. Fish and Wildlife Service, Oregon Department of Fish and Wildlife, and Confederated Tribes of the Umatilla Indian Reservation.

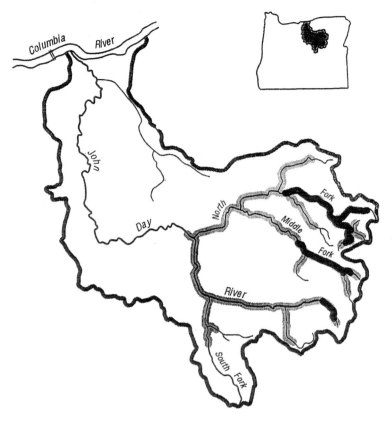

FIGURE 6.—Current (heavy black river reaches) and estimated historical (gray) holding and spawning distributions of spring chinook salmon in the John Day River basin, Oregon.

that are critical, such as phylogenetic and biogeographic analyses, that take time to develop.

Acknowledgments

We thank James D. Hall and two anonymous reviewers for their helpful comments, and we appreciate the patience and encouragement of the editor, Jennifer Nielsen. This is a contribution of the Oregon Chapter of the American Fisheries Society and paper 10,597 of the Agricultural Experiment Station of Oregon State University.

References

Bisson, P. A. 1969. Origin and distribution of the fish fauna of the Harney Basin, Oregon. Master's thesis. Oregon State University, Corvallis.

Brooks, D. R., and D. A. McLennan. 1991. Phylogeny, ecology, and behavior. The University of Chicago Press, Chicago.

Clarke, S. E., D. White, and A. L. Schaedel. 1991. Oregon, USA, ecological regions and subregions for water quality management. Environmental Management 15:847–856.

Cracraft, J. 1991. Patterns of diversification within continental biotas: hierarchical congruence among the areas of endemism of Australian vertebrates. Australian Systematic Botany 4:211–227.

Cracraft, J. 1994. Species diversity, biogeography, and the evolution of biotas. American Zoologist 34:33–47.

Currens, K. P., C. B. Schreck, and H. W. Li. 1990. Allozyme and morphological divergence of rainbow trout (*Oncorhynchus mykiss*) above and below waterfalls in the Deschutes River, Oregon. Copeia 1990: 730–746.

FEMAT (Forest Ecosystem Management Assessment Team). 1993. Forest ecosystem management: an ecological, economic, and social assessment. U.S. Forest Service, Portland, Oregon.

Frissell, C. A., W. J. Liss, C. E. Warren, and M. D. Hurley. 1986. A hierarchical framework for stream habitat classification: viewing streams in a watershed context. Environmental Management 10:199–214.

Harrison, S. 1991. Local extinction in a metapopulation context: an empirical evaluation. Biological Journal of the Linnean Society 42:73–88.

Hatch, K. M. 1990. Phenotypic comparison of thirty-eight steelhead (*Oncorhynchus mykiss*) populations from coastal Oregon. Master's thesis. Oregon State University, Corvallis.

Henjum, M. G., and seven coauthors. 1994. Interim protection for late-successional forests, fisheries, and watersheds: national forests east of the Cascade crest, Oregon and Washington. The Wildlife Society, Bethesda, Maryland.

Hoffmann, A. A., and M. W. Blows. 1994. Species borders: ecological and evolutionary perspectives. Trends in Ecology and Systematics 9:223–227.

Hughes, R. M., D. P. Larsen, and J. M. Omernik. 1986. Regional reference sites: a method for assessing stream potentials. Environmental Management 10: 629–635.

Hughes, R. M., E. Rexstad, and C. E. Bond. 1987. The relationship of aquatic ecoregions, river basins, and physiographic provinces to the ichthyogeographic regions of Oregon. Copeia 1987:423–432.

Hughes, R. M., T. R. Whittier, C. M. Rohm, and D. P. Larsen. 1990. A regional framework for establishing recovery criteria. Environmental Management 14:673–683.

Levins, R. 1969. Some demographic and genetic consequences of environmental heterogeneity for biological control. Bulletin of the Entomological Society of America 15:237–240.

Li, H. W., G. A. Lamberti, T. N. Pearsons, C. K. Tait, and J. L. Li. 1994. Cumulative impact of riparian disturbance in small streams of the John Day Basin, Oregon. Transactions of the American Fisheries Society 123(4):627–640.

Mayden, R. L. 1988. Biogeography, parsimony, and evolution in North American freshwater fishes. Systematic Zoology 37:329–355.

McIntosh, B. A., and six coauthors. 1994. Historical changes in fish habitat for select river basins of eastern Oregon and Washington. Northwest Science 68: 36–53.

McPhail, J. D., and C. C. Lindsey. 1986. Zoogeography of the freshwater fishes of Cascadia (the Columbia system and rivers north to the Stikine). Pages 622–698 *in* C. H. Hocutt and E. O. Wiley, editors. The zoogeography of North American freshwater fishes. Wiley-Interscience, New York.

Minckley, W. L., D. A. Hendrickson, and C. E. Bond. 1986. Geography of western North American freshwater fishes: description and relationships to intracontinental tectonism. Pages 521–613 *in* C. H. Hocutt and E. O. Wiley, editors. The zoogeography of North

American freshwater fishes. Wiley-Interscience, New York.

Moyle, P. B. 1976. Inland fishes of California. University of California Press, Berkeley.

Moyle, P. B., and R. M. Yoshiyama. 1994. Protection of aquatic biodiversity in California: a five-tiered approach. Fisheries 19(2):6–18.

Nei, M. 1972. Genetic distance between populations. American Naturalist 106:283–292.

Nehlsen, W., J. E. Williams, and J. A. Lichatowich. 1991. Pacific salmon at the crossroads: stocks at risk from California, Oregon, Idaho, and Washington. Fisheries 16(2):4–21.

Pielou, E. C. 1969. Mathematical ecology. Wiley-Interscience, New York.

Reeves, G. H., L. E. Benda, K. M. Burnett, P. A. Bisson, and J. R. Sedell. 1995. A disturbance-based ecosystem approach to maintaining and restoring freshwater habitats of evolutionarily significant units of anadromous salmonids in the Pacific northwest. American Fisheries Society Symposium 17:334–349.

Scudder, G. G. E. 1989. The adaptive significance of marginal populations: a general perspective. Pages 180–185 *in* C. D. Levings, L. B. Holtby, and M. A. Henderson, editors. Proceedings of the National Workshop on effects of habitat alteration on salmonid stocks. Canadian Special Publications in Fisheries and Aquatic Sciences.

Waples, R. S. 1991. Definition of "species" under the Endangered Species Act: application to Pacific salmon. NOAA (National Oceanic and Atmospheric Administration) Technical Memorandum NMFS (National Marine Fisheries Service) F/NWC-194, Northwest Fisheries Science Center, Seattle.

Warren, M. L., Jr., and B. M. Burr. 1994. Status of freshwater fishes of the United States: overview of an imperiled fauna. Fisheries 19(1):6–18.

Williams, J. E., and seven coauthors. 1989. Fishes of North America endangered, threatened, or of special concern: 1989. Fisheries 14(6):2–20.

Wilson, E. O. 1988. Biodiversity. National Academy Press, Washington, DC.

Wissmar, R. C., J. E. Smith, B. A. McIntosh, H. W. Li, G. H. Reeves, and J. R. Sedell. 1994. Ecological health of river basins in forested regions of eastern Washington and Oregon. Northwest Science 68:1–35.

Wright, S. 1931. Evolution in Mendelian populations. Genetics 16:97–159.

Wright, S. 1940. Breeding structure of populations in relation to speciation. American Naturalist 74:232–248.

American Fisheries Society Symposium 17:381–391, 1995

Observations on Habitat Structure, Population Regulation, and Habitat Use with Respect to Evolutionarily Significant Units: A Landscape Perspective for Lotic Systems

GARY D. GROSSMAN

Warnell School of Forest Resources
University of Georgia, Athens, Georgia 30602-2152, USA

JENNIFER HILL

Federal Energy Regulatory Commission, Office of Hydropower Licensing
825 N. Capitol Street, NE, Washington, DC 20426, USA

J. TODD PETTY

Warnell School of Forest Resources, University of Georgia

Abstract.—In this paper we attempt to synthesize a variety of developments in the fields of landscape and population ecology and apply these ideas to the physical and biological characteristics of lotic systems. First, most attempts to manage evolutionarily significant units (ESUs) are based on the notion that the physical characteristics of lotic habitats are stable. Yet, data from three permanent 100-m reaches in the Coweeta drainage of North Carolina indicated that these reaches possessed substantial annual variability with respect to both substratum composition and flow rates. In addition, substratum data demonstrated that these reaches were patchy environments, and that a landscape-based approach might facilitate the management of species in this system. Second, a simple landscape-driven difference equation model of population dynamics based on biological characteristics common to many fishes indicated that the critical habitat for population maintenance may not always be the area in which the species is most abundant. Finally, two tests for habitat selection by stream fishes indicated that more biologically realistic models (e.g., a landscape-based model that included prey abundance) and a model that included explicit tests for the mechanism of selection itself (e.g., energy gain) may greatly increase our ability to identify and manage habitats that are crucial for survival of ESUs.

At present we are at a crossroads with respect to preservation of the earth's genetic resources. Increasing population growth combined with questionable economic and management practices have driven us to a point at which the loss of biological diversity is occurring at an exponential rate, a rate previously unknown in modern times. Such losses dramatically increase our need for innovative management strategies for the conservation of evolutionarily significant units (ESUs). (We will use the term ESUs to refer to species, subspecies, or populations with extremely low abundances.) In this paper, we will attempt to synthesize a variety of developments in the fields of landscape and population ecology (Johnston and Naiman 1987; Pringle et al. 1988; Johnston et al. 1990; Schlosser 1991) and apply these ideas to the physical environment of lotic systems, as well as to both population dynamics and habitat selection by lotic animals. We focus our observations on the speciose stream fish assemblages of the eastern and midwestern United States, rather than on the depauperate western fauna.

Within the last decade, ecologists have recognized that most habitats are embedded within a larger landscape that may influence many of the ecological processes occurring in discrete habitat patches (Forman and Godron 1986; Pringle et al. 1988). For example, research on energy transformations in eastern woodland streams has shown that these systems depend upon the surrounding deciduous forest for most of their organic energy inputs (Minshall 1967). Studies of anthropogenic disturbances such as logging also demonstrate that activities occurring in habitats far from the streambank can have substantial deleterious effects on both fish and invertebrate populations in lotic systems (Moring 1975; Salo and Cundy 1987). Consequently, our understanding of rivers and streams from both a basic and applied perspective may be enhanced by the application of a landscape perspective to studies of these systems.

The field of landscape ecology is relatively recent; a text book on the subject was published in 1986 (Forman and Godron 1986) and the specialized journal *Landscape Ecology* began in 1987. Many of the conceptual underpinnings of this subdiscipline apparently have arisen from two disparate sources.

First, the general view that habitat types (i.e., woodlands, old fields) must be considered as part of a larger, interconnected landscape is an outgrowth of one of the canons of ecosystem ecology; one can not identify the true functional importance of subcomponents of ecological systems (i.e., patches of woodland, stream riffles) in isolation because everything is interconnected (Odum 1989). This approach highlights the problems inherent in the application of a strictly reductionistic approach to studies of ecological systems. The difference between landscape ecology and ecosystem ecology is that landscape ecology explicitly incorporates and focuses on the spatial components of ecological processes.

The second contribution to landscape ecology originates from studies of theoretical population genetics (Levins 1962, 1968). In these studies, populations or individuals were viewed as inhabiting a patchy landscape, each patch type having different fitness consequences for individuals. For example, individuals inhabiting patch A might have abundant food whereas those occupying patch B might encounter low food availability. All else being equal, the fitness of individuals in patch A would be substantially higher than those in patch B. The aggregate properties of the population are then obtained by summing the results for individuals in each patch type, and the fitness values for individuals are determined by the patches that they utilize. This perspective has been widely used in both theoretical population genetics and behavioral ecology to address the fitness consequences of life in a heterogeneous environment.

We will attempt to meld these two approaches in addressing three questions. First, can reaches in a North Carolina drainage be characterized as patchy environments with respect to their physical characteristics, and is this patchiness temporally stable? Second, can a fine-scale landscape approach increase our understanding of the population dynamics of ESUs and aid in the identification of habitats critical for the preservation of these units? Third, can mechanistic models of habitat selection aid us in identifying the critical factors affecting habitat use in lotic fishes?

Physical Environment of Lotic Systems

Traditionally, fisheries managers have viewed lotic environments as stable habitats comprising regularly ordered sequences of pools, runs, and riffles. However, since the work of Starrett (1951), it has become apparent that the physical environment of many streams is not particularly stable, especially with respect to flows (i.e., floods and droughts; Poff

and Ward 1989, 1990). Temporal and spatial variability in flow may influence many other physical attributes of stream reaches including temperature, substratum composition, and import and export of substrata. Many streams experience both floods and droughts with a frequency that is sufficient to affect individual fitness and population size as well as assemblage composition (Grossman et al. 1982, 1990; Schlosser 1985; Poff and Ward 1989, 1990).

What are the consequences of environmental variability for ESUs? First, ESUs are typically rare or have restricted distributions. It is possible that the persistence of ESUs in lotic systems over both ecological and evolutionary time is facilitated by environmental variability. This hypothesis, the basics of which were suggested by Andrewartha and Birch (1954) and further developed as the intermediate disturbance hypothesis of Connell (1978), is that moderate levels of environmental variability prevent species that are superior competitors from ever gaining sufficient advantage over inferior competitors to cause their extirpation. If this coexistence mechanism is operating in systems where ESUs are present, their survival may depend upon the maintenance of both temporal and spatial environmental variability (Grossman et al. 1982, 1990). This would call for the use of different management strategies by fisheries managers, such as plans that include preservation of the variances of the physical characteristics of reaches as well as average values.

Regardless of whether or not survival of ESUs is linked to environmental variation, an ability to quantify the habitat requirements of ESUs is essential for their successful management. Traditionally, fisheries managers studying lotic habitats focused their efforts on stream reaches. By a reach, we mean a given physiognomic unit of a stream (i.e., a pool, a riffle, a run) or a larger sequence of these units (e.g., a 100-m section of stream containing several habitat units). These physiognomic units are assumed to be relatively homogenous or, to paraphrase Shakespeare, a riffle by any other name would still flow as swift. Nonetheless, it is unclear whether this assumption is warranted, for relatively few studies have examined variation of physical characteristics (i.e., depth, velocity, substratum) within reaches themselves. For example, if riffles are relatively homogeneous, then a small number of random physical measurements may be sufficient to describe the physical habitat occupied by an ESU. However, if riffles are patchy habitats, then a much more intensive sampling program will be required to quantify an ESU's habitat. If units are patchy, the use of a landscape approach, with its focus on patchiness, might improve our understanding of these reaches. From a management perspective, fine-scale environmental variation of this type may be particularly important to small benthic fishes such as darters (Percidae), sculpins (Cottidae), madtoms (Ictaluridae), and some minnows (Cyprinidae) that may depend upon small patches of substratum for both foraging and reproduction (Page 1983; Moyle and Cech 1988). To test the assumption that the physical characteristics of stream reaches are stable, we made repeated physical measurements on permanent quadrats in three reaches of a southern Appalachian trout stream.

The Coweeta Creek drainage comprises a series of first- to fifth-order streams located on the U.S. Forest Service's Coweeta Hydrologic Station, Otto, North Carolina. The streams in this drainage are typical of many lotic habitats in the southern Appalachian Mountains. Our research team selected three 100-m reaches along a longitudinal gradient ranging from second to fifth order. Specifically, the study sites were located in a second-order (Upper Ball Creek, henceforth UBC), a fourth-order (Lower Ball Creek, LBC), and a fifth-order stream (Coweeta Creek, CC). In autumn 1991, we started at the lower end of each site and placed permanent benchmarks at 5-m intervals along both banks of each site. We then stretched a tape measure between benchmarks at the same position on opposite banks and located a series of permanent quadrats along this transect at intervals ranging from 0.5 to 1 m. We recorded the wetted width of each transect and obtained the following data from each quadrat: water depth, average velocity (at $0.6 \times$ depth from the surface; Bovee and Milhous 1978), and a visual estimate of the percent composition of bedrock, boulder, cobble, gravel, sand, silt, and organic debris in a 20-cm $\times$ 20-cm area directly under the quadrat mark. Substrata were classified on the basis of maximum diameter according to a modified Wentworth particle scale (Grossman and Freeman 1987). We have used similar techniques to quantify habitat availability in a variety of other studies (Grossman and Freeman 1987; Grossman et al. 1987; Grossman and Boule 1991; Grossman et al. 1995). We measured quadrats in autumn 1991, 1992, and 1993, and in spring 1993. Linear measurements were made with a straight edge or tape measure and velocity was measured with an electronic velocity meter.

Our first goal was to characterize the heterogeneity (patchiness) present within these reaches. For this analysis we have focused on the substratum due to its presumed importance to many species of benthic fishes. To assess whether a reach was spatially homogeneous or patchy, we first characterized a quadrat by its substratum. A quadrat was considered to be a member of a given substratum class if this class made up more than 50% of the total substratum. Substratum classes were: bedrock–boulder, cobble–gravel, sand–silt, and debris. An additional class, heterogeneous, was used for quadrats that were not dominated by the aforementioned classes. We then plotted these results on a graphical representation of the stream reach.

Besides determining whether or not these reaches were patchy, we also wanted to quantify the stability of these patches. Consequently, we compared the dominant substratum class of quadrats across all four sampling periods. We recorded the number of times the dominant substratum class of a quadrat changed between samples (maximum, three). In addition, if a quadrat went from submerged to exposed or vice versa, it was considered to have changed its dominant substratum class. We included these changes because if managers attempted to quantify the habitat requirements of an ESU in these reaches, their measurements would be affected by such shifts. In addition, we assessed variability in water flow by examining mean daily flow records from a gaging weir on an adjacent fourth-order stream (Swift and Cunningham 1986).

Our results demonstrate that stream reaches within the Coweeta drainage were, indeed, patchy environments (Figure 1). All five dominant substratum classes were present in the three reaches, often intermingled. The second-order reach (UBC) exhibited the greatest heterogeneity, whereas fourth- (LBC) and fifth- (CC) order reaches were considerably less heterogeneous and dominated by cobble–gravel riffles. Although we have presented data for only one season for each reach, similar substratum patchiness occurred in other seasons.

Each reach possessed areas that remained relatively constant over time as well as areas in which considerable transport of substrata occurred (Figure 2). As with substratum patchiness, the second-order reach exhibited the greatest amount of change in quadrats; the fifth-order site was the most stable. These changes also were reflected in the mean percentage composition of substrata within reaches (i.e., they changed from year to year: G. Grossman et al., University of Georgia, unpublished data). Flow data also exhibited substantial variability. Figure 3 shows that during the first 10 years of our study (1983–1992), annualized mean daily flows for the Coweeta drainage varied from some of the lowest to some of the highest in the 58-year period for which we have data.

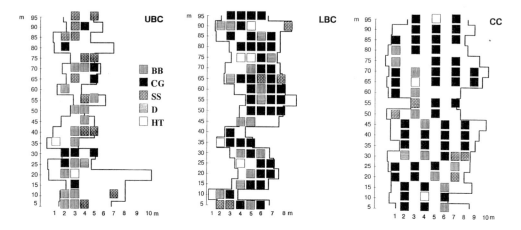

FIGURE 1.—Schematic representation of study sites in Upper Ball Creek (UBC, second order), Lower Ball Creek (LBC, fourth order), and Coweeta Creek (CC, fifth order) showing substratum composition in 20-cm × 20-cm permanent quadrats: BB = bedrock–boulder; CG = cobble–gravel; SS = sand–silt; D = debris; HT = heterogeneous. Blank quadrats were not sampled every year.

What are the implications of these findings for the management of ESUs? First, we believe that characterization of the habitat requirements of ESUs as well as habitat management in general, can be enhanced through the use of a landscape perspective. This perspective directly addresses the patchiness present in many lotic systems and deals with reaches as dynamic habitats rather than as static entities. Such a view may be of particular importance to the scientific management of benthic fishes such as darters, sculpins, madtoms, and some minnows. These species may exhibit specialization

for a given substratum (Winn 1958; Finger 1982; Page 1983) and small home ranges (Gerking 1959; Hill and Grossman 1987) that render the explicit identification of environmental patchiness essential to successful management.

Second, our results show that both mean daily flows and composition of the substratum were variable over the coarse of the study. Consequently, attempts to quantify the habitat requirements of ESUs should be made over several years to encompass the range of variation typically present in the habitat. This also suggests that studies of the habitat

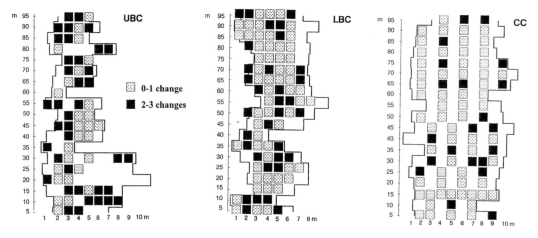

FIGURE 2.—Schematic representation of the permanent study sites in Upper Ball Creek (UBC), Lower Ball Creek (LBC), and Coweeta Creek (CC) showing the number of times the dominant substratum class changed in a quadrat between samples (maximum, three).

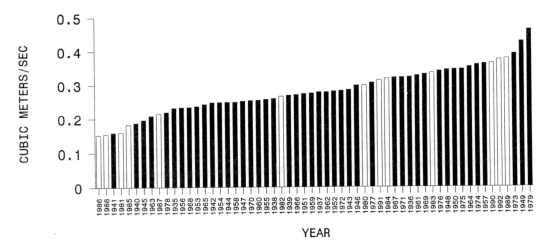

FIGURE 3.—Mean annual daily flows for Watershed 8, a fourth-order stream in the Coweeta drainage (data courtesy of the U.S. Forest Service's Coweeta Hydrologic Laboratory). The open histograms represent the first 10 years of our study.

needs of ESUs should be instituted before their populations decrease to critically low levels.

Landscape Approach to Population Dynamics

It seems worthwhile to explore the implications of patchy stream habitats for the population dynamics of ESUs. In a patchy environment, individuals may be distributed across a variety of patch types that differ in their effect on individual fitness (Levins 1962). Pulliam (1988) and Pulliam and Danielson (1991) explored the population-level consequences of this phenomena for birds and termed these models source–sink population models. Source–sink models lead to some counterintuitive results that may be of particular relevance to the management of ESUs in aquatic systems.

A common goal for the management of ESUs is identification of habitat patches that are favorable for the species. Fisheries managers commonly use habitat–abundance relationships to identify such areas (Fausch et al. 1988). Typically, researchers sample a variety of reaches occupied by the species, measuring both physical variables and abundance of the species of interest (Baltz 1990). The manager can then use correlation statistics to identify variables that are positively correlated with abundance, and they may infer that the physical attributes with the strongest correlation coefficients define the most favorable habitat for the ESU. However, it is not unusual for the predictive power of such models to range from moderate to low (i.e., R^2 values are

below 0.50; Layher and Maugham 1985; Fausch et al. 1988; Jowett 1992). This might be due to several reasons such as insufficient sampling, but it might also be caused by a logical flaw in this approach. Use of the landscape population models developed by Pulliam (1988) illustrates this shortcoming.

First, let us envision a hypothetical ESU with a given set of demographic characteristics (Tables 1 and 2). These characteristics are common to many fish species (Moyle and Cech 1988). Now imagine that this ESU occupies two habitat patches, A and B, that have differing fitness consequences for individuals (Table 2). Patch A is a small area of bedrock–boulder substratum surrounded by cobble–gravel riffle. Few individuals can inhabit patch A (it has a low carrying capacity), but fish within it experience low mortality rates and high fertility rates (Table 2). Conversely, patch B has a high carrying capacity, but fish within it experience higher mortality and zero fertility. We can now use a simple difference equation model to simulate changes in population size for the species (through summation of the output of this equation for both patches). This model is represented as

$$N_{t,X} = N_{0,X} + N_{0,X}(b_X - d_X) + (i_X - e_X);$$

$N_{t,X}$ = number of the ESU in patch X at time t,
$N_{0,X}$ = number of the ESU in patch X at time 0,
b_X = birth rate in patch X,
d_X = mortality rate in patch X,
i_X = immigration into patch X,
e_X = emigration from patch X.

TABLE 1.—Characteristics of a hypothetical evolutionarily significant unit (ESU). These traits are typical for many fish species.

1. Intraspecific competition is substantial among individuals
2. Nonreproductive habitat requirements are general
3. Habitat patches are occupied in relation to quality; if a patch is full, individuals enter the patch with the next highest quality value
4. The ESU has specialized habitat requirements for reproduction; hence, reproductive success varies among patches
5. Fecundity is high relative to density

From Table 1, we can assume that patches are occupied in proportion to their effect on individual fitness. This means that individuals will occupy patch A until carrying capacity is reached, at which point new individuals will begin to occupy patch B. In addition, once patch B reaches carrying capacity, all emigrants die. If we insert the parameters presented in Table 2 in the model and round all fractional values up to the nearest individual, densities of animals in patches A and B remain at the respective carrying capacities and population size of the ESU is stable (Table 2).

Now if an ESU is behaving in the manner portrayed by this landscape-based model (Pulliam 1988), managers charged with the preservation of this population may find their job quite difficult. For example, if traditional habitat–abundance models are used to identify the most favorable habitat for the ESU, the managers will mistakenly conclude that patch B, with its higher density, is the habitat that is crucial for preservation (Pulliam 1988). This would soon lead to extinction for the ESU, however, because the reproduction necessary for population maintenance occurs in patch A. Although our example is somewhat contrived, in that successful reproduction occurs only in patch A, one can vary the parameters in a more reasonable manner and still obtain the same result (i.e., the majority of recruits for the population may not originate in the portion of the habitat in which densities are highest). By treating the habitat as a patchy landscape, however, and focusing on patch-specific demography, managers may increase the probability that their efforts will have a positive effect on an ESU.

Models of Habitat Selection

Fisheries managers typically use two techniques to quantify habitat selection by lotic fishes. The first method is the correlational approach to habitat–abundance relationships described in the previous section. The second technique involves locating undisturbed specimens in the stream and recording a series of physical measurements (e.g., depth, velocity, substratum composition) at the position of the fish (Baltz and Moyle 1984; Baltz 1990). These physical measurements are then compared with habitat availability data obtained from random measurements within the reach. If this comparison yields significant differences, it is assumed that the ESU is exhibiting habitat selection or at least nonrandom microhabitat use. We refer to this type of study as a microhabitat use study. Although these two designs are probably the most frequently used by fisheries scientists attempting to quantify the habitat requirements of fishes, they both possess two shortcomings that may affect a manager's ability to identify and preserve critical habitat for an ESU.

First, the focus of both habitat–abundance and microhabitat use studies is on the physical habitat rather than on biological factors such as the abundances of prey, predators, or competitors—probably because physical variables are easier to quantify and manage than biological factors. Although there is no inherent reason why biological factors could not be included in habitat selection studies, such approaches appear to be uncommon (Orth 1987). It is also possible that exclusion of biological factors has contributed to the generally low predictive power of some habitat–abundance and microhabitat use studies.

An additional shortcoming of many habitat selection studies is that they neglect identification of the mechanism of selection itself. Thus, habitat selection typically is inferred from correlational analyses of physical data from reaches in which varying densities of fishes have been observed. Nonetheless, even when such approaches yield strong results (high r values), they still can not directly establish a causal mechanism for the relationship. In addition, investigators typically do not directly examine the stimuli upon which individuals are basing their "choice" nor do they frequently identify the fitness consequences of choosing a given microhabitat. Once again, these foibles may limit a manager's ability to correctly identify the habitat components

TABLE 2.—Parameters for a two-patch model, based on the characteristics in Table 1.

Parameter	Patch A	Patch B
Carrying capacity (number of animals)	25	5,000
Mortality rate (d)	0.001	0.50
Birth rate (b)	100	0

that are crucial for population maintenance or growth.

The two following approaches to the study of habitat selection by stream fishes should illustrate the advantages of including biological data with a landscape perspective and of the mechanistic approach to microhabitat use.

Landscape Approach to Microhabitat Selection

The mottled sculpin *Cottus bairdi* is the most abundant benthic fish in the Coweeta drainage (Freeman et al. 1988). We have studied microhabitat use by this species for the last 11 years (Grossman and Freeman 1987; Barrett 1989; G. Grossman et al., University of Georgia, unpublished data). Mottled sculpin have variable patterns of microhabitat use. Although these fish are always found on the bottom at low focal point velocities, they occur over wide ranges of average velocities and substrata (Grossman and Freeman 1987; Petty and Grossman, in press). Indeed, the use of physical microhabitat variables (average velocity, depth, composition of the substratum) by mottled sculpin typically does not differ significantly from values obtained by randomly sampling the environment (Grossman and Freeman 1987; Petty and Grossman, in press). In addition, physiological studies indicate that mottled sculpin are not metabolically restricted to low-velocity microhabitats (Facey and Grossman 1990, 1992). Finally, experimental and descriptive studies have shown that microhabitat use by mottled sculpin is not strongly influenced by predators (Grossman and Freeman 1987; Grossman et al. 1995) or the species' most common potential competitor (Barrett 1989).

These findings led us to reassess our approach to the study of microhabitat selection by this species. A dietary study of mottled sculpin in the Coweeta drainage (Stouder 1990) indicated they are generalized predators that feed primarily on benthic invertebrates. These results are supported by other dietary data for this species (Dineen 1951; Bailey 1952; Daiber 1956). Because the distribution of benthic macroinvertebrates in streams frequently is patchy (Resh and Rosenberg 1984; Downes et al. 1993), we hypothesized that mottled sculpin may be responding more to the patchy distribution of prey than to distinctive physical characteristics of the habitat. Specifically, we reasoned that if mottled sculpin are generalized predators without strong physiological constraints on microhabitat use, then a patch of sandy substratum with high prey abundance may be functionally equivalent to a patch of

cobble substratum with high prey abundance. Consequently, we tested the null hypothesis that prey abundance at patches occupied by sculpins was not significantly higher than prey abundance at randomly selected locations.

We tested this hypothesis by making microhabitat measurements on mottled sculpin in two sites in the Coweeta drainage using the methods of Grossman and Freeman (1987). At the location of each sculpin we also collected a benthic sample using a Hess sampler with a 0.1-m^2 capacity. In addition, we also selected a paired, random location for each specimen and collected an identical set of data at this location (Petty and Grossman, in press). Our data consisted of six samples: four from a 150-m-long site in autumn 1991, late spring 1992, summer 1992, and autumn 1992, and two samples (late spring 1992 and summer 1992) from a similar 150-m site upstream from the first site.

We compared microhabitat use by mottled sculpins with data from random measurements to quantify microhabitat selection within a season (Petty and Grossman, in press). We also tested the aforementioned null hypothesis regarding prey abundance with a Wilcoxon signed ranks test on paired patches (with sculpin and random) within a seasonal sample. In addition, we analyzed two restricted data sets. The first contained only prey that represented at least 5% of the diet of mottled sculpin in the Coweeta drainage on either a numerical or volumetric basis (Stouder 1990). The second deleted prey that we deemed to be too large to be consumed by sculpin (Petty and Grossman, in press). For each data set, values for numerical abundance and biomass were tested separately.

As with previous microhabitat studies, mottled sculpin did not exhibit microhabitat selection for physical factors in four of six seasonal samples (Petty and Grossman, in press). However, they did occupy patches with significantly higher numerical abundances of invertebrates in five of six seasonal samples (Petty and Grossman, in press). Similar results were obtained with biomass data and both numerical and biomass data for the two other data sets (Petty and Grossman, in press). Consequently, prey abundance appeared to elicit a stronger microhabitat selection response from mottled sculpin than physical habitat characteristics.

These results demonstrate an important point with respect to the management of ESUs. First, as previously mentioned, successful management of ESUs depends upon identification of the aspects of habitat that are essential for population maintenance. For a variety of reasons (e.g., ease of measurement) fisheries biologists generally focus on the physical habitat when attempting to identify critical resources for fishes. Nonetheless, the use of physical microhabitat data alone provided few insights into the factors influencing microhabitat selection by mottled sculpin. It was only when we included biological data (prey abundance) that we were able to detect a pattern in microhabitat use by the species. In addition, these findings have important consequences for management of ESUs. For example, fisheries managers tend to emphasize manipulation of physical habitat when they conceptualize a management strategy for an ESU. However, if they wanted to manage a habitat with an endangered mottled sculpin, it might be more effective to employ techniques that would directly increase the forage base for this species (nutrient additions, supplemental feeding, etc.) rather than to rely on manipulation of the physical habitat, which might or might not produce increases in prey abundance.

Energy-Based Model for Optimal Velocity Use

The previous example demonstrates the advantages of including biological variables in the assessment of the habitat requirements of ESUs. Nonetheless, it still does not provide a complete picture of the causal factors influencing habitat selection for mottled sculpin. Our next example describes a mechanistic, energy-based approach to habitat selection by rainbow trout *Oncorhynchus mykiss* and rosyside dace *Clinostomus funduloides*. These species are the numerically dominant water-column fishes in Coweeta Creek (Freeman et al. 1988).

Our model is based on the assumption that an individual will increase, and perhaps maximize, its fitness if it behaves in a manner that will maximize its net energy gain (Fausch 1984; Hill and Grossman 1993). For lotic fishes that forage in the water column, there is a clear physiological cost for maintaining position at a given velocity, as well as a quantifiable benefit to be gained by foraging at a given velocity. By measuring these costs and benefits, one can construct a simple cost-benefit model (Figure 4) to yield a prediction regarding the velocity that should maximize net energy gain for individuals of a given species. We developed models for two size-classes of trout (small, 65.7 ± 1.5 mm; medium, 89.9 ± 8.1 mm) and dace (medium, 48.3 ± 0.7 mm; large, 60.9 ± 0.7 mm) for spring, summer, autumn, and winter. Eight models (two size-classes × four seasons) were examined per species. Details of these models were presented by Hill and Grossman (1993), who used the respirometry data of

Facey and Grossman (1990) to estimate the energetic cost of maintaining position at a given velocity (i.e., to parameterize the cost curves). Hill and Grossman (1993) then derived benefit curves by measuring the amount of drift available (in calorific equilivents) at naturally occurring velocities in the stream, as well as by conducting foraging experiments to quantify the ability of rainbow trout and rosyside dace to capture prey at differing velocities. We then compared the cost and benefit curves to determine the velocity that yielded the maximum net energy gain for each size-class in each season (Figure 4). We tested the prediction of each model by comparing it with velocities occupied by the respective size-classes of trout and dace in Coweeta Creek during the four seasons (Hill and Grossman 1993).

To summarize our results, energy-based models of habitat selection yielded velocity predictions that were extremely close to those occupied by rainbow trout and rosyside dace in Coweeta Creek. The mean deviation between predicted and observed velocities was 2.5 cm/s for trout and 2.8 cm/s for dace. Velocities present in the stream during these seasons ranged from 0 to over 50 cm/s. The primary factor affecting the model was the relationship between velocity (independent variable) and the ability of the fish to capture prey (prey capture success). When an empirical model based on the third derivative of this curve (point of maximum rate of decline) was used as a predictor of optimal velocity, the mean absolute deviations between predicted and observed velocities decreased to 1.5 cm/s for trout and 1.9 cm/s for dace (Hill and Grossman 1993).

An energy-based mechanistic model of habitat selection for rainbow trout and rosyside dace proved to be an accurate predictor of the velocities occupied seasonally by these species in Coweeta Creek. These results have several implications for the management of ESUs. First, they provide an important datum for the scientific management of these species: changes in velocity may have strong impacts on trout and dace populations. If either species were an ESU, the optimal velocity data for Coweeta Creek would provide us with a starting point for the management of other streams inhabited by these fishes. For example, we could use such data to assess the potential impacts of water diversions, dams, or disturbances that may alter flow regimes. Or in a proactive sense, we could manipulate flows to increase the availability of optimal velocities, assuming that this would result in higher production of dace and rainbow trout.

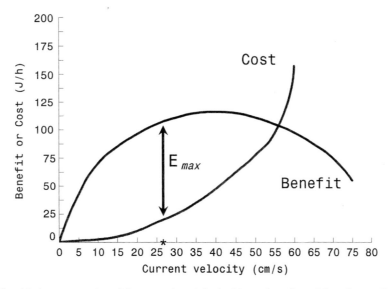

FIGURE 4.—Graphical representation of the energy-based (joules/h) cost-benefit model used to predict optimal water velocity use for rainbow trout and rosyside dace. The optimum velocity (asterisk) is predicted to be the one at which net energy gain by the fish is maximized (E_{max}).

Effective management of ESUs necessitates quantification of their habitat requirements. Nonetheless, most previous studies of the habitat requirements of lotic fishes have focused on the physical habitat. Our results for mottled sculpin, rainbow trout, and rosyside dace and those of Jowett (1992) for brown trout *Salmo trutta* demonstrate that habitat selection studies that include only physical factors may not provide sufficient information for the effective management of some lotic fishes. Consequently, the development of more biologically realistic approaches to the quantification of habitat requirements of lotic ESUs may increase our ability to prevent the extirpation of these fishes.

Through this application we have tried to demonstrate that stream reaches can be viewed as patchy habitats from both a temporal and spatial perspective, and that this patchiness may have profound consequences for assessing habitat requirements of ESUs as well as their population dynamics. In addition, although we should not ignore studies of the physical habitat requirements of ESUs, it is likely that the development of more biologically realistic, mechanistic models of habitat selection will facilitate the preservation and management of these fishes. Finally, we suspect that the application of the landscape perspective to entire streams as well as their reaches will enhance our ability to effectively manage aquatic ESUs.

Conclusions

The preservation of biological diversity requires an innovative approach to the management of natural resources. In this paper, we have attempted to apply some of the techniques of landscape ecology to lotic systems (Johnston and Naiman 1987; Pringle et al. 1988; Johnston et al. 1990; Schlosser 1991) by incorporating biologically relevant information that is often lacking from fisheries studies: quantification of the physical habitat of stream reaches, quantification of spatial and temporal variability within reaches, inclusion of the potential effects of patchiness on population regulation, and use of biologically based models of habitat selection.

Acknowledgments

We are grateful to the organizers of the symposium, Jennifer Nielsen, Peter Bisson, and the American Fisheries Society, for enabling us to participate. A host of people aided in various aspects of this research including Jeffrey Barrett, Maurice Crawford, Douglas Facey, Mary Freeman, Robert Ratajczak, Pedro Rincon, and Deanna Stouder. In addition, we thank Wayne Swank and the staff of the Coweeta Hydrologic Laboratory for providing help with all aspects of our research. The manuscript was reviewed by C. Jennings, C. Pringle, R. Ratajczak, P. Rincon, M. Van den Avyle, and B.

Wallace, and it also benefitted from the comments of D. Orth, D. Stauder, and an anonymous referee. Portions of this research were funded by the U.S. Forest Service McIntire–Stennis program (GEO-0047-MS), the National Science Foundation (BSR-9011661), and the Warnell School of Forest Resources.

References

Andrewartha, H. G., and L. C. Birch. 1954. The distribution and abundance of animals. University of Chicago Press, Chicago.

Bailey, J. E. 1952. Life history and ecology of the sculpin *Cottus bairdi punctulatus* in southwestern Montana. Copeia 1952:243–255.

Baltz, D. M. 1990. Autecology. Pages 585–607 *in* C. B. Schreck and P. B. Moyle, editors. Methods for fish biology. American Fisheries Society, Bethesda, Maryland.

Baltz, D. M., and P. B. Moyle. 1984. Segregation by species and size class of rainbow trout (*Salmo gairdneri*) and Sacramento sucker (*Catostomus occidentalis*) in three California streams. Environmental Biology of Fishes 10:101–110.

Barrett, J. C. 1989. The effects of competition and resource availability on the behavior, microhabitat use and diet of the mottled sculpin (*Cottus bairdi*). Doctoral dissertation. University of Georgia, Athens.

Bovee, K. D., and R. T. Milhous. 1978. Hydraulic simulation in instream flow studies: theory and technique. U.S. Fish and Wildlife Service Biological Service Program FWS/OBS 78/33.

Connell, J. H. 1978. Diversity in tropical rain forests and coral reefs. Science 199:1302–1310.

Daiber, F. C. 1956. A comparative analysis of the winter feeding habits of two benthic stream fishes. Copeia 1956:141–151.

Dineen, C. F. 1951. A comparative study of the food habits of *Cottus bairdi* and associated species of salmonidae. American Midland Naturalist 46:640–645.

Downes, B. J., P. S. Lake, and E. S. Schreiber. 1993. Spatial variation in the distribution of stream invertebrates: implications of patchiness for models of community organization. Freshwater Biology 30:119–132.

Facey, D. E., and G. D. Grossman. 1990. A comparative study of oxygen consumption by four stream fishes: the effects of season and velocity. Physiological Zoology 63:757–776.

Facey, D. E., and G. D. Grossman. 1992. The relationship between water velocity, energetic costs, and microhabitat use on four North American stream fishes. Hydrobiologia 239:1–6.

Fausch, K. D. 1984. Profitable stream positions for salmonids: relating specific growth rate to net energy gain. Canadian Journal of Zoology 62:441–451.

Fausch, K. D., C. L. Hawkes, and M. G. Parsons. 1988. Models that predict standing crop of stream fish from habitat variables 1950–1985. U.S. Forest Service General Technical Report PNW-GTR-213.

Finger, T. R. 1982. Interactive segregation among three species of sculpins (*Cottus*). Copeia 1982:680–694.

Forman, R. T., and M. Godron. 1986. Landscape ecology. Wiley, New York.

Freeman, M. C., and seven coauthors. 1988. Fish assemblage stability in a southern Appalachian stream. Canadian Journal of Fisheries and Aquatic Sciences 45:1949–1958.

Gerking, S. C. 1959. The restricted movement of fish populations. Biological Reviews of the Cambridge Philosophical Society 34:221–242.

Grossman, G. D., and M. C. Freeman. 1987. Microhabitat use in a stream fish assemblage. Journal of Zoology 212:151–176.

Grossman, G. D., and V. Boule. 1991. An experimental study of competition for space between rainbow trout (*Oncorhynchus mykiss*) and rosyside dace (*Clinostomus funduloides*). Canadian Journal of Fisheries and Aquatic Sciences 48:1235–1243.

Grossman, G. D., P. B. Moyle, and J. O. Whitaker, Jr. 1982. Stochasticity in structural and functional characteristics of an Indiana stream fish assemblage: a test of community theory. American Naturalist 120:423–454.

Grossman, G. D., A. DeSostoa, M. C. Freeman, and J. Lobon-Cervia. 1987. Microhabitat selection in a Mediterranean riverine fish assemblage. I. Fishes of the lower Matarraña. Oecologia 73:490–500.

Grossman, G. D., J. F. Dowd, and M. K. Crawford. 1990. Assemblage stability in stream fishes: a review. Environmental Management 14:661–671.

Grossman, G. D., R. E. Ratajczak, Jr., and M. K. Crawford. 1995. Do rock bass (*Ambloplites rupestris*) induce microhabitat shifts in mottled sculpin (*Cottus bairdi*)? Copeia 1995:343–353.

Hill, J., and G. D. Grossman. 1987. Home range estimates for three North American stream fishes. Copeia 1987:376–380.

Hill, J., and G. D. Grossman. 1993. An energetic model of microhabitat use for rainbow trout and rosyside dace. Ecology 74:685–698.

Johnston, C. A., and R. J. Naiman. 1987. Boundary dynamics at the aquatic-terrestrial interface: the influence of beaver and geomorphology. Landscape Ecology 1:47–57.

Johnston, C. A., N. E. Detenbeck, and R. J. Naiman. 1990. The cumulative effect of wetlands on stream water quality and quantity: a landscape approach. Biogeochemistry 10:105–141.

Jowett, I. G. 1992. Models of abundance of large brown trout in New Zealand rivers. North American Journal of Fisheries Management 12:417–432.

Layher, W. G., and O. E. Maugham. 1985. Relations between habitat variables and channel catfish populations in prairie streams. Transactions of the American Fisheries Society 114:771–781.

Levins, R. 1962. Theory of fitness in a heterogeneous environment, I. The fitness set and adaptive function. American Naturalist 96:361–378.

Levins, R. 1968. Evolution in changing environments. Princeton University Press, Princeton, New Jersey.

Minshall G. W. 1967. Role of allochthonous detritus in

the trophic structure of a woodland spring brook community. Ecology 48:139–149.

Moring, J. R. 1975. The Alsea watershed study: effects of logging on the aquatic resources of three headwater streams of the Alsea River, Oregon. Parts 1–3. Oregon Department of Fish and Wildlife, Fisheries Research Report 9, Corvallis.

Moyle, P. B., and J. J. Cech. 1988. Fishes: an introduction to ichthyology. Prentice Hall, New York.

Odum, E. 1989. Ecology and our endangered life support system. Sinauer, Sunderland, Massachusetts.

Orth, D. J. 1987. Ecological considerations in the development and application of instream flow-habitat models. Regulated River Research and Management 1:171–181.

Page, L. 1983. Handbook of darters. Tropical Fish Hobbyist Publications, Inc., Neptune, New Jersey.

Petty, J. T., and G. D. Grossman. In press. Patch selection by the mottled sculpin (Pisces:Cottidae) in a southern Appalachian stream. Freshwater Biology.

Poff, N. L., and J. V. Ward. 1989. Implications of stream variability and predictability for lotic community structure: a regional analysis of streamflow patterns. Canadian Journal of Fisheries and Aquatic Sciences 46:1805–1818.

Poff, N. L., and J. V. Ward. 1990. Physical habitat template of lotic systems: recovery in the context of historical pattern of spatiotemporal heterogeneity. Environmental Management 14:629–645.

Pringle, C. M., and seven coauthors. 1988. Patch dynamics in lotic systems: the stream as a mosaic. Journal of the North American Benthological Society 7:503–524.

Pulliam, H. R. 1988. Sources, sinks and population regulation. American Naturalist 132:652–661.

Pulliam, H. R., and B. Danielson. 1991. Sources, sinks, and habitat selection: a landscape perspective on population dynamics. American Naturalist 137: S50–S66.

Resh, V. H., and D. M. Rosenberg, editors. 1984. The ecology of aquatic insects. Praeger, New York.

Salo, E. O., and T. W. Cundy, editors. 1987. Streamside management: forestry and fishery interactions. University of Washington Institute of Forest Resources Contribution #57.

Schlosser, I. J. 1985. Flow regime, juvenile abundance, and the assemblage structure of stream fishes. Ecology 66:1484–1490.

Schlosser, I. J. 1991. Stream fish ecology: a landscape perspective. BioScience 41:704–712.

Starrett, W. C. 1951. Some factors affecting the abundance of minnows in the Des Moines River, Iowa. Ecology 32:13–27.

Stouder, D. J. 1990. Dietary fluctuations in stream fishes and the effects of benthic species interactions. Doctoral dissertation. University of Georgia, Athens.

Swift, L. W. Jr., and G. B. Cunningham. 1986. Routines for collecting and summarizing hydrometeorological data at Coweeta Hydrologic Laboratory. Pages 301–320 in W. K. Michener, editor. Research data management in the ecological sciences. University of South Carolina Press, Columbia.

Winn, H. E. 1958. Comparative reproductive behavior and ecology of fourteen species of darters (Pisces-Percidae). Ecological Monographs 28:155–191.

American Fisheries Society Symposium 17:392–401, 1995

Spatial Variation in Demographic Processes of Lotic Fishes: Conceptual Models, Empirical Evidence, and Implications for Conservation

ISAAC J. SCHLOSSER

Department of Biology, University of North Dakota
Box 9019 University Station, Grand Forks, North Dakota 58202-9019, USA

PAUL L. ANGERMEIER

National Biological Service, Virginia Cooperative Fish and Wildlife Research Unit[1]
Virginia Polytechnic Institute and State University
Blacksburg, Virginia 24061-0321, USA

Abstract.—Identification, protection, and restoration of endangered fish taxa require an understanding of spatial variation in the demographic attributes of fish populations. Five "metapopulation" models have been proposed to describe spatial variation in demographic processes. The models differ in the nature of spatial variation in reproduction, extinction, and colonization and include (1) classic metapopulation models; (2) mainland–island or source–sink models; (3) patchy population models; (4) hybrid models, which combine attributes of (2) and (3); and (5) nonequilibrium metapopulation models. Few studies of natural fish populations have yielded the data needed to effectively discriminate the most appropriate model(s) for lotic fishes. However, two published studies, one of an anadromous salmonid in a coastal Oregon stream and the other of cyprinids in a northern Minnesota stream, support the hybrid model of spatial variation in demographic processes, combining patchy population processes at smaller spatial scales with mainland–island or source–sink interactions at larger spatial scales. If other studies support the importance of the hybrid model, regulatory agencies will need to emphasize the identification, protection, and restoration of ecosystem processes that generate and maintain natural patch dynamics in mainland or source areas on the landscape.

Effective conservation of fish species requires integration of appropriate scientific principles across a range of conceptual levels of organization ranging from molecular systematics to population, community, and ecosystem ecology. A sound understanding of spatial variation in demographic factors within and between populations, and of the influences of assemblage and ecosystem attributes on those factors will be particularly important in conservation efforts (Lande 1988; Simberloff 1988). Regulatory agencies will need a multilevel understanding so they can prioritize the population, habitat, or ecosystem attributes to be protected in order to assure maximum conservation of fishery resources with minimal economic costs. Our objectives in this paper are to (1) briefly examine the nature of spatial and temporal variation in physical attributes of stream ecosystems and life history characteristics of stream fishes; (2) review basic conceptual models describing spatial variation in demographic processes in populations and the type of data needed to effectively test these models; (3) review two studies of fish, one from the Pacific Northwest and one from the upper Midwest of the United States, that provide data useful for assessing the appropriateness of these models for describing variation in demographic processes of lotic fishes at the spatial scale of a drainage basin; and (4) explore the implications of these models for future research and regulations aimed at effective conservation of fishery resources. We focus on fish in lotic environments because of our familiarity with those ecosystems.

Hierarchical Lotic Habitats

Lotic ecosystems can be best viewed as hierarchically organized physical environments; at successively higher levels of organization, they incorporate microhabitat, habitat, stream reach, subbasin, and drainage basin units (Frissell et al. 1986). Development and persistence of each level in the physical hierarchy is controlled by processes functioning at particular temporal and spatial scales. At drainage basin and stream segment levels, factors controlling spatial heterogeneity act over large spatial scales but at a variety of temporal scales; they include wildfires (Reeves et al. 1995, this volume), large

---

[1]The Unit is jointly sponsored by the National Biological Service, Virginia Polytechnic Institute and State University, and the Virginia Department of Game and Inland Fisheries.

storms (e.g., hurricanes), tectonic uplift, glaciation, climate change, and alluvial or colluvial valley filling. These factors generate inherent spatial heterogeneity between drainage basins or stream segments in slope, mineral substrate, presence or absence of geological barriers, and nature of the terrestrial vegetation (Frissell et al. 1986). These factors also potentially generate considerable temporal variation in large-scale spatial heterogeneity. Short-term asynchronous disturbances, including floods, fires, and unusual weather can generate successional processes at large spatial scales so that basins, subbasins, and stream segments with identical geological histories may exhibit mosaics of varying ecological features (Reeves et al. 1995).

Factors acting at levels from reach to microhabitat operate over smaller spatial and shorter temporal scales and include the expenditure of kinetic energy associated with the downhill transport of water and sediment, obstructions in the channel, extent of channel meandering, and animal activity (Keller and Swanson 1979; Richards 1982; Naiman et al. 1988). Interactions between the energy flux associated with water and sediment transport, channel obstructions, and channel meandering create substantial longitudinal heterogeneity (differences in depth, substrate, and current velocity related to pool–riffle development) and lateral heterogeneity (stream margins, backwaters, and isolated pools) (Richards 1982; Moore and Gregory 1988; Junk et al. 1989; Schiemer and Spindler 1989). In smaller streams, beavers *Castor canadensis* create well-defined longitudinal habitat variation between their ponds and upstream or downstream areas and lateral habitat variation between their ponds and the riparian zone (Johnston and Naiman 1987; Naiman et al. 1988). Furthermore, the profound alterations in stream habitat induced by beaver dams are highly variable in their persistence and reoccurrence. Thus, beaver activity adds not only spatial but also temporal variability to the landscape (Naiman et al. 1988).

The habitat heterogeneity created by these physical and biological factors is spatially and temporally nested: physical units at any given level form physical units at the next higher level of organization (Frissell et al. 1986). Consequently, habitat heterogeneity can be defined at a variety of spatial scales, ranging from microhabitat to drainage basin to ecoregion, and at a variety of temporal scales, ranging from season to year to geological time periods (Frissell et al. 1986). In an ecological context, however, the appropriate scale for defining heterogeneity will be determined by the ecological process being considered. In the case of spatial and temporal variation in demographic processes of stream fishes, the appropriate scale will be strongly influenced by the life history characteristics of the particular taxonomic unit being examined, especially the migratory movements of the fish and the temporal persistence of habitats required by particular life stages.

Life Cycles and Movements

Associated with the spatial and temporal heterogeneity of physical habitats in lotic ecosystems, fish exhibit complex life cycles and habitat use patterns (Figure 1). Life begins at spawning, after which an incubation period lasts anywhere from a few days to several months. Hatched fish usually move to feeding habitats, where most growth and development occurs. These feeding habitats normally constitute a mosaic (Figure 1) of several habitat types (pools, riffles, stream margins, etc.), all of which can potentially be used by fish during the growing season and which serve as nursery and rearing habitats for larval and juvenile stages. In the case of anadromous salmonids, the mosaic of feeding habitats used during the transition from larval to juvenile and eventually to adult stages can extend over thousands of kilometers, spanning small streams, rivers, estuaries, and open ocean.

Depending on the life cycle of the species and local climatic conditions, the fish then go through one to several seasons favorable and unfavorable to growth until sexual maturity is reached (Figure 1). In north temperate areas, where great differences in summer and winter climate occur, favorable and unfavorable periods engender movement primarily between summer feeding habitats and winter refugia (Peterson 1982; Schlosser 1987; Cunjak 1988). Extreme variations in stream discharge, however, may induce movements between feeding habitats available during moderate to high stream flows and deeper refugia during reduced flows (Larimore et al. 1959). Once sexual maturity is reached, the fish move to appropriate spawning sites for egg deposition and reinitiation of the cycle.

Although the depiction of fish life cycles in Figure 1 is basic and simple, it has three important implications for understanding the impact of habitat heterogeneity on stream fish populations. First, in any given region of the landscape, fish population dynamics are potentially influenced by the four rate-dependent demographic processes of immigration, emigration, natality, and mortality, with mortality being strongly influenced by growth during

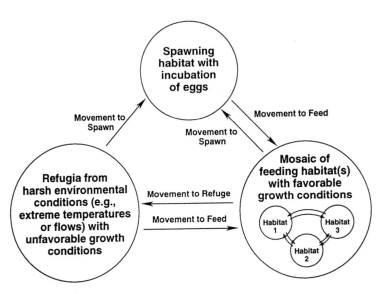

FIGURE 1.—Basic life cycle of stream fish with emphasis on patterns of habitat use and movement (based on Jones 1968; Northcote 1978; and Schlosser 1991).

early life (Werner and Gilliam 1984). Second, because various life history stages of fish require different physical habitats, the spatial and temporal distribution of habitat types, and the connection of these habitats via dispersal pathways, will be critical for completion of fish life cycles (Bisson et al. 1982; Schiemer and Spindler 1989). Third, depending on the life history of the particular fish taxon, the habitat heterogeneity required to complete a life cycle can extend over a large range of spatial and temporal scales. This latter point suggests that fish ecologists need to develop models of how habitat templates (Frissell et al. 1986) constrain life history patterns (Poff and Ward 1990; Winemiller and Rose 1992) and of how habitat and life history patterns influence demographic processes.

Models of Large-Scale Spatial Variation in Demographic Processes

Historically, most ecologists, and fish biologists in particular, have focused on the "local spatial scale" and the "local population" to understand how environmental factors interact to influence population size and population dynamics (Matthews and Heins 1987 and references therein). The local spatial scale can be defined as the scale at which individuals routinely interact with each other, and the local population can be defined as a set of individuals of the same species interacting with each other with a high degree of probability (Hanski and Gilpin 1991).

More recently, as ecologists have become aware of the fundamental importance of large-scale spatial and temporal environmental heterogeneity (Forman and Godron 1986) and the critical role dispersal processes play in the life history of most vertebrate taxa (Hansson 1991), they have increasingly focused their attention on what Levins (1969) termed the "metapopulation." The spatial scale of the metapopulation can be defined as the scale at which individuals infrequently move from one place (population) to another, typically across unsuitable habitats and with substantial risk of failing to locate another suitable habitat (Hanski and Gilpin 1991). The biological scale of the metapopulation can be defined as a set of local populations that interact with each other via individuals that move among the populations (Hanski and Gilpin 1991). The key aspect of the metapopulation concept, which differentiates it from population-level processes, is the focus on the co-occurrence of extinction and colonization in different populations within a region. This framework for modeling populations is particularly interesting with respect to fishes, because critical avenues of fish dispersal are typically linear and easily obstructed (e.g., stream channels).

Several metapopulation models have been proposed that differ in the nature of their spatial variation in reproduction, extinction, and colonization (Harrison 1991). The classic metapopulation model of Levins (1969) is represented by several similar-sized populations, all of which are subject to extinc-

A. CLASSIC METAPOPULATION

B. MAINLAND/ISLAND
or SOURCE/SINK

C. PATCHY POPULATION

D. HYBRID

E. NONEQUILIBRIUM
METAPOPULATION

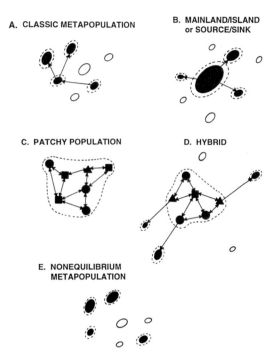

FIGURE 2.—Five models of spatial variation in population processes (adapted from Harrison 1991). Symbols represent habitat patches that are occupied (solid) or vacant (open). Arrows indicate migration (colonization) and dashes indicate the boundaries of "populations."

tion with equal and temporally independent probabilities (Figure 2A). These populations persist regionally because of recolonization from adjacent populations. Movement among adjacent populations in the classic metapopulation model is, however, sufficiently low that individual populations do not function as subcomponents of one patchy population (see below).

The mainland–island or source–sink metapopulation models differ from the classic model in that one or more of the populations has a greater potential for providing emigrants than the other populations (Figure 2B). In the mainland–island relationship, a large mainland population normally is the origin of immigrants to outlying island populations (Harrison 1991). Sources and sinks are characterized by qualitative demographic differences, sinks being relatively unsuitable in some way for survival and reproduction (Pulliam 1988; Dunning et al. 1992). The higher reproduction and survival in sources makes them the primary origins of immigrants to adjacent sink populations. Because local recruitment in sinks is not sufficient to maintain the

populations there, immigration from source areas is the primary rate-dependent process determining population size in the sinks. Furthermore, mainland or source populations and island or sink populations are likely to differ in their responses to temporal environmental variability. Populations in island or sink areas may retract to mainland or source areas during unfavorable conditions. In contrast, mainland or source areas tend to always be occupied (Harrison 1991).

The patchy population model is superficially similar to the classic metapopulation model in that it depicts units of local populations fluctuating somewhat independently. Units in the patchy population model, however, are coupled by very high dispersal between different types of habitat patches (Taylor 1988; Harrison 1991). In particular, dispersal takes place at a faster rate than local extinctions so that individuals in the different habitat patches act as a single population rather than as a metapopulation (Figure 2C). "Patchy population" dynamics are likely to characterize highly mobile taxa in spatially heterogeneous, temporally variable habitats, where use of several complementary habitats is needed for persistence of the population (Harrison 1991; Dunning et al. 1992; Schlosser 1995a).

The hybrid population model combines characteristics of the patchy population and mainland–island or source–sink models (Figure 2D). Occurrence of this type of model depends on the spatial arrangement of the habitat patches (Harrison 1991). High dispersal takes place among one group of habitat patches. This group acts as a mainland or source area on the landscape, but dispersal might be low enough to cause extinctions in peripheral habitat patches and hence cause the peripheral populations to behave as part of a classic metapopulation (Harrison 1991).

Finally, nonequilibrium models emphasize the importance of many local extinctions and infrequent recolonization (Figure 2E). This pattern is usually associated with the regional decline of a species as the species' habitat undergoes long-term fragmentation, reduction, or deterioration (Harrison 1991). In particular, fragmentation reduces recolonization from either adjacent metapopulations or mainland or source areas.

Empirical Examples for Lotic Fishes

Are these models of variation in demographic processes appropriate for lotic fishes? Few empirical data have been gathered to specifically test these conceptual models for fish or, for that matter, most

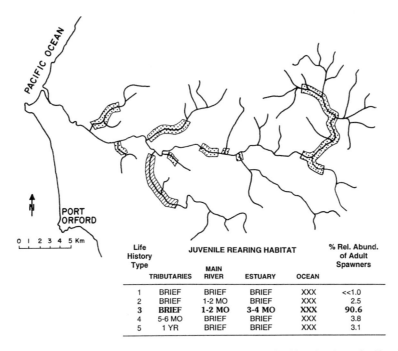

FIGURE 3.—Distribution of spawning (stippled or hatched areas) by fall chinook salmon in Sixes River, Oregon (adapted from Reimers 1973). Spawning occurred in all of the areas marked, but 60–70% of all spawning occurred in the 5-km section of stream indicated with cross-hatching. The inset table reflects the distribution of five juvenile life history types in freshwater and marine habitats (MO = month), along with their percentage contributions to returning adult spawners (based on data in Reimers 1973).

other taxonomic groups (Doak and Mills 1994). It is very difficult to document, over large geographic scales, such highly dynamic relationships as variation in reproductive success, the rate and distance of dispersal by juveniles and adults, and the temporal persistence of dispersers in various regions of the landscape. We are not aware of any study of lotic fish that has fully documented all of these factors. Two examples from the literature, however, give some insight into the conceptual model that is likely to be most appropriate.

In an early and insightful study, Reimers (1973) investigated spatial variation in demographic attributes of fall chinook salmon *Oncorhynchus tshawytscha* in Sixes River, Oregon. He examined spatial distribution of spawning effort, variation in the use of freshwater, estuary, and ocean environments by juvenile life history types, and subsequent recruitment of juvenile life history types into the adult spawning population. Most of the spawning in the drainage basin occurred in smaller tributary streams, but there was considerable spatial variation in the distribution of spawning activity; 60–70% of all spawning occurred in one 4–5-km stretch in

one tributary (Figure 3). Based on growth characteristics reflected in the formation of the scales of the fish, Reimers (1973) discriminated five life history types among the juveniles, the types varying in length of time spent in freshwater tributaries, the main river, or the estuary before they moved to the ocean (Figure 3). These life history types also differed dramatically in their relative contributions to the group of spawners returning to the stream. Individuals that spend 3–4 months in the estuary accounted for more than 90% of returning spawners, whereas individuals that spent 5 months to a year in freshwater tributaries but little time in the estuary accounted for only 6–7% of return spawners (Figure 3).

Reimers (1973) did not address several factors needed to determine the exact nature of the population structure for these salmon. In particular, the fidelity between spawning location and life history type would have to be known to assess whether straying from more productive tributaries was important in maintaining subpopulations in less productive areas. Furthermore, an examination of long-term variability in spawning and recruitment

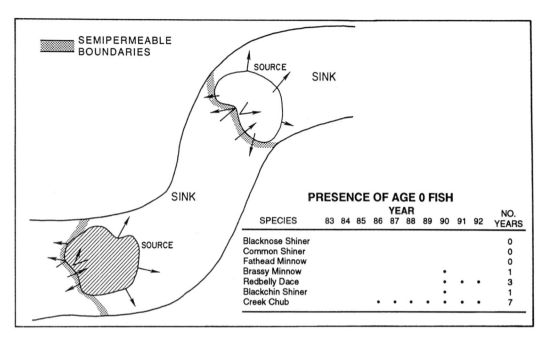

FIGURE 4.—Conceptual framework for viewing exchanges between beaver pond and stream environments at the fish trophic level. Direction of stream flow is from right to left. Arrows indicate dispersal or attempts at dispersal. Hatched and unhatched enclosed areas represent beaver ponds of differing depth and volume. Beaver dams are represented as boundaries that are only semipermeable to fish movement. The inset table indicates the occurrence by year of age-0 cyprinids of species known to be represented by individuals older than age 0 (adapted from Schlosser 1995b). Species are blacknose shiner *Notropis heterolepis*, common shiner *Luxilus cornutus*, fathead minnow *Pimephales promelas*, brassy minnow *Hybognathus hankinsoni*, northern redbelly dace *Phoxinus eos*, blackshin shiner *Notropis heterodon*, and creek chub *Semotilus atromaculatus*.

would be useful in establishing the influence of temporal environmental variability on these demographic relationships. Reimers (1973) suggested, for instance, that annual variability in stream temperature has a strong influence on the use of the river as a rearing habitat. Finally, the relationship between life history type and genetics needs to be more clearly established for these fish. This is particularly critical because current emphasis in the definition of evolutionarily significant units (ESUs; Waples 1995, this volume) relies primarily on genetic distance, as defined by biochemical tests, rather than on ecological distance, as defined by the type of life history variation observed by Reimers (1973).

More recently, Schlosser (1995b) used a long-term descriptive study in a headwater stream in northern Minnesota to assess the effect of annual variation in stream discharge and spatial proximity of beaver ponds on fish abundance and recruitment in adjacent lotic ecosystems. Considerable annual variation in fish density occurred in the stream over

the 10-year period. Increased fish density was associated with elevated stream discharge and creation of downstream beaver ponds. Weir traps used to monitor directional fish movement indicated that annual changes in lotic fish density were associated with the amount of fish dispersal occurring along the stream segment. Downstream movement out of an upstream pond was largely restricted to periods of elevated stream discharge, when the dam was more permeable to fish passage (Figure 4). Upstream movement out of a downstream pond occurred over a broader range of discharges because the boundary at the upstream side of the pond was generally permeable.

Individual species varied considerably in their tendencies to move up- or downstream because the differing sizes of the upstream and downstream beaver ponds affected the composition of fish dispersing from these source areas. Most fish movement occurred over relatively brief time periods, suggesting life history and developmental processes were critical in influencing the timing of dispersal. Size structure of fishes captured in the stream indicated that predominantly fish older than age 1 were dispersing along the stream. But the occurrence of age-0 individuals (Figure 4) indicated that only one species, the creek chub, routinely reproduced in the stream, implying that the stream was a reproductive sink for most of the species. The extent to which the stream ultimately acts as a sink will depend on the flexibility in life history characteristics of the taxa and the large-scale (basin-level) spatial distribution and accessibility of other beaver ponds (Schlosser 1995b).

In summary, both studies (Reimers 1973; Schlosser 1995b) revealed strong spatial variation in juvenile recruitment and fish production within drainage basins. In Oregon, production of juvenile chinook salmon was concentrated in one tributary of the main channel (egg stage) and the associated estuary (juvenile stage). Production of juvenile fish in the Minnesota drainage basin appeared to be concentrated in beaver ponds. Smaller peripheral populations associated with the core area of fish production were maintained in both drainages. These smaller populations were in upstream tributaries in Oregon and adjacent lotic environments in Minnesota.

These results tend to support either the mainland–island or source–sink models of spatial variation in demographic processes (Figure 2B). Other studies conducted at even finer spatial scales within estuaries or beaver ponds, however, have revealed considerable habitat patchiness (Johnston and Naiman 1987; Livingston 1992) within which most fish used several types of habitats during their development (Bisson et al. 1982; Schiemer and Spindler 1989; Schlosser 1991; Livingston 1992; Rahel 1994). This suggests that incorporation of patchy population processes into either mainland or source areas, as represented by the hybrid model (Figure 2D), is the most appropriate representation of spatial variation in demographic processes for these fishes (see also Schoener 1991).

Implications for Fish Conservation

In its current state, metapopulation theory might best be considered a tool for thought and an identifier of useful data to be gathered in studies of natural demographic processes, regardless of which model is ultimately found to be most useful (Harrison 1994). If, however, other empirical studies support the patchy population and mainland–island or patchy population and source–sink models of spatial demographic variation, such hybrid models

will have several important implications for future efforts at fish conservation.

First, a hybrid model paradigm suggests we need to enhance our understanding of how mainland or source areas are organized and distributed on the landscape. We must focus in particular on conceptually and empirically linking the hierarchical model of habitat heterogeneity, as proposed by Frissell et al. (1986), with hybrid models of spatial variation in demographic processes. It will be especially important to identify the spatial and temporal scales over which mainland or source areas are organized. For example, an entire drainage may act as a mainland or source for adjacent drainages, while dynamics among habitat patches within the drainage may be those of a patchy population. Furthermore, the nature of these spatial relationships may vary over time (Carline et al. 1991; Reeves et al. 1995). It will also be essential to identify the mixture and arrangement of habitat patches that make up mainland or source areas, along with the key ecosystem processes generating and maintaining those patches. In all likelihood, the degree of habitat complementation and supplementation (Dunning et al. 1992; Schlosser 1995a), the availability of refugia during times of harsh physical conditions (Schlosser 1987; Sedell et al. 1990), energy supplies and predator–prey interactions (Gregory et al. 1991; Schlosser 1991), and long-term disturbance regimes in the drainage basin (Reeves et al. 1995) will all fundamentally influence the spatial distribution and temporal persistence of mainland or source areas (Schlosser 1991, 1995a).

Second, if hybrid models effectively represent demographic processes for lotic fishes, we need to focus increased attention on documenting the nature of physical linkages between peripheral populations and mainland or source areas. In particular, we need to refine our understanding of the distances fish will travel from mainland or source areas to peripheral habitats (Echelle et al. 1976; Fausch and Young 1995, this volume) and the role habitat boundaries and barriers to dispersal play in colonization of peripheral areas (Baumgartner 1986; Gilliam et al. 1993; Schlosser 1995a, 1995b). Notably, fragmentation of populations into islands over large geographic areas because of reduced colonization has frequently been implicated in fish endangerment and extirpation (Brown 1986; Sheldon 1987; Winston et al. 1991; Frissell 1993; Angermeier 1995). According to hybrid metapopulation models, reduced colonization might be due either to collapse of the mainland or source population or to creation of migratory barriers.

Third, we need to increase our understanding of the biological processes linking peripheral habitats with mainland or source areas. In particular, we need to evaluate in more detail the relative roles of life history attributes, ecological factors (e.g., food and habitat availability), and genetics in controlling intraspecific variation in the tendency of fish to disperse into peripheral habitats. Numerous studies in streams indicate that dispersal from other habitats is an important determinant of fish assemblage structure (Gorman 1986; Angermeier and Schlosser 1989; Osborne and Wiley 1992). Depending on the role of genetics in the dispersal process, small peripheral populations may be critical in maintaining overall genetic diversity within the drainage basin (Scudder 1989; Riddell 1993). We know very little about the relationship between the genetic markers currently being used to define ESUs and dispersal into peripheral areas.

Fourth, we need to better understand how mainland or source areas influence levels of biological organization above single populations. In particular, we need to assess if mainland and source areas, or islands and sinks, are shared by several taxonomic groups because of commonalities in life history or dispersal processes (Taylor 1991; Winemiller and Rose 1992) and how such sharing affects the structure of fish assemblages and the attributes of ecosystems. Winemiller and Rose (1992) provided a conceptual framework for grouping taxa on the basis of life history characteristics; these groups might be expected to share sources and sinks on the landscape. Efforts to interface population processes with assemblage structure will be increasingly important as regulatory agencies shift their conservation efforts away from species and populations toward assemblage and ecosystem organization (Angermeier and Schlosser 1995, this volume).

Finally, identification, protection, and restoration of mainland or source areas, along with the ecosystem processes generating them, will be essential if conservation efforts by regulatory agencies are to succeed. The loss of mainland or source areas is likely to be a primary reason why species shift towards nonequilibrium metapopulation dynamics and subsequently lose isolated or peripheral populations (Sheldon 1987; Dennis et al. 1991; Rieman et al. 1993). Thus, although fish conservation efforts need a multitiered approach, encompassing several levels of conceptual organization (Moyle and Yoshiyama 1994), the fundamental long-term problem faced by regulatory agencies is maintaining or reestablishing the core mainland or source areas on the landscape of fish reproduction and survival.

Acknowledgments

Final preparation of this manuscript was supported by a National Science Foundation grant (STIA RII-EPSCoR) to the state of North Dakota. P. Bisson, W. Ensign, and two anonymous reviewers made numerous suggestions that greatly improved the manuscript. S. Harrison and P. Reimers graciously granted us permission to use previously published information.

References

Angermeier, P. L. 1995. Ecological attributes of extinction-prone species: loss of freshwater fishes of Virginia. Conservation Biology 9:143–158.

Angermeier, P. L., and I. J. Schlosser. 1989. Species-area relationships for stream fishes. Ecology 70:1450–1462.

Angermeier, P. L., and I. J. Schlosser. 1995. Conserving aquatic biodiversity: beyond species and populations. American Fisheries Society Symposium 17:402–414.

Baumgartner, J. V. 1986. The genetics of differentiation in a stream population of the threespine stickleback, Gasterosteus aculeatus. Heredity 57:199–208.

Bisson, P. A., J. L. Nielsen, R. A. Palmason, and L. E. Grove. 1982. A system of naming habitat types in small streams, with examples of habitat utilization by salmonids during low stream flow. Pages 62–73 in N. B. Armantrout, editor. Acquisition and utilization of aquatic habitat inventory information. American Fisheries Society, Western Division, Bethesda, Maryland.

Brown, K. L. 1986. Population demographic and genetic structure of plains killifish from the Kansas and Arkansas River basins in Kansas. Transactions of the American Fisheries Society 115:568–576.

Carline, R. F., T. Beard, and B. A. Hollender. 1991. Response of wild brown trout to elimination of stocking and to no-harvest regulations. North American Journal of Fisheries Management 11:253–266.

Cunjak, R. A. 1988. Physiological consequences of overwintering in streams: the cost of acclimatization. Canadian Journal of Fisheries and Aquatic Sciences 45:443–452.

Dennis, B., P. L. Munholland, and J. M. Scott. 1991. Estimation of growth and extinction parameters for endangered species. Ecological Monographs 61:115–143.

Doak, D. F., and L. S. Mills. 1994. A useful role for theory in conservation biology. Ecology 75:615–626.

Dunning, J. B., B. J. Danielson, and H. R. Pulliam. 1992. Ecological processes that affect populations in complex landscapes. Oikos 65:169–175.

Echelle, A. A., A. F. Echelle, and B. A. Taber. 1976. Biochemical evidence for congeneric competition as a factor restricting gene flow between populations of darter (Percidae: Etheostoma). Systematic Zoology 25:228–235.

Fausch, K. D., and M. K. Young. 1995. Evolutionarily significant units and movement of resident stream fishes: a cautionary tale. American Fisheries Society Symposium 17:360–370.

Forman, R. T., and M. Godron. 1986. Landscape ecology. Wiley, New York.

Frissell, C. A., W. J. Liss, C. E. Warren, and M. D. Hurley. 1986. A hierarchical framework for stream habitat classification; viewing streams in a watershed context. Environmental Management 10:199–214.

Frissell, C. A. 1993. Topology of extinction and endangerment of native fishes in the Pacific Northwest and California (U.S.A.). Conservation Biology 7:342–354.

Gilliam, J. F., D. F. Fraser, and M. Alkins-Koo. 1993. Structure of a tropical fish community: a role for biotic interactions. Ecology 74:1856–1870.

Gorman, O. T. 1986. Assemblage organization in stream fishes: the effects of rivers on adventitious streams. American Naturalist 128:611–616.

Gregory, S. V., F. J. Swanson, W. A. McKee, and K. W. Cummins. 1991. An ecosystem perspective of riparian zones. BioScience 41:540–551.

Hanski, I., and M. Gilpin. 1991. Metapopulation dynamics: a brief history and conceptual domain. Biological Journal of the Linnean Society 42:3–16.

Hansson, L. 1991. Dispersal and connectivity in metapopulations. Biological Journal of the Linnean Society 42:89–103.

Harrison, S. 1991. Local extinction in a metapopulation context: an empirical evaluation. Biological Journal of the Linnean Society 42:73–88.

Harrison, S. 1994. Metapopulations and conservation. Pages 111–128 in R. J. Edwards, R. May, and N. R. Webb, editors. Large-scale ecology and conservation biology. Blackwell Scientific Publications, Oxford, UK.

Johnston, C. A., and R. J. Naiman. 1987. Boundary dynamics at the aquatic-terrestrial interface: the influence of beaver and geomorphology. Landscape Ecology 1:47–57.

Junk, W. J., P. B. Bayley, and R. E. Sparks. 1989. The flood pulse concept in river-floodplain systems. Pages 110–127 in D. P. Dodge, editor. Proceedings of the international large river symposium. Canadian Special Publication in Fisheries and Aquatic Sciences 106.

Jones, F. R. 1968. Fish migration. Edward Arnold, London.

Keller, E. A., and F. J. Swanson. 1979. Effects of large organic material on channel form and fluvial processes. Earth Surface Processes 4:361–380.

Lande, R. 1988. Genetics and demography in biological conservation. Science 241:1455–1460.

Larimore, R. W., W. F. Childers, and C. Heckrote. 1959. Destruction and reestablishment of stream fish and invertebrates effected by droughts. Transactions of the American Fisheries Society 88:261–285.

Levins, R. 1969. Some demographic and genetic consequences of environmental heterogeneity for biological control. Bulletin of the Entomological Society of America 15:237–240.

Livingston, R. J. 1992. Medium-sized rivers of the Gulf coastal plain. Pages 351–358 in C. T. Hackney, S. M. Adams, and W. H. Martin, editors. Biodiversity of the southeastern United States: aquatic communities. Wiley, New York.

Matthews, W. J., and D. C. Heins, editors. 1987. Ecology and evolution of North American stream fishes. University of Oklahoma Press, Norman.

Moore, K. M. S., and S. V. Gregory. 1988. Response of young-of-the-year cutthroat trout to manipulation of habitat structure in a small stream. Transactions of the American Fisheries Society 117:162–170.

Moyle, P. B., and R. M. Yoshiyama. 1994. Protection of aquatic biodiversity in California: a five tiered approach. Fisheries 19:6–18.

Naiman, R. J., C. A. Johnston, and J. C. Kelley. 1988. Alteration of North American streams by beaver. BioScience 38:753–762.

Northcote, T. G. 1978. Migratory strategies and production in freshwater fishes. Pages 326–359 in S. D. Gerking, editor. Ecology of freshwater fish production. Blackwell Scientific Publications, Oxford, UK.

Osborne, L. L., and M. J. Wiley. 1992. Influence of tributary spatial position on the structure of warm water fish communities. Canadian Journal of Fisheries and Aquatic Sciences 49:671–681.

Peterson, N. P. 1982. Immigration of juvenile coho salmon (Oncorhynchus kisutch) into riverine ponds. Canadian Journal of Fisheries and Aquatic Sciences 39:1308–1310.

Poff, N. L., and J. V. Ward. 1990. Physical habitat template of lotic systems: recovery in the context of historical pattern of spatiotemporal heterogeneity. Environmental Management 14:629–645.

Pulliam, H. R. 1988. Sources, sinks, and population regulation. American Naturalist 132:652–661.

Rahel, F. 1994. Foraging in a lethal environment: fish predation in hypoxic waters of a stratified lake. Ecology 75:1246–1253.

Reeves, G. H., L. E. Benda, K. M. Burnett, P. A. Bisson, and J. R. Sedell. 1995. A disturbance-based ecosystem approach to maintaining and restoring freshwater habitats of evolutionarily significant units of anadromous salmonids in the Pacific northwest. American Fisheries Society Symposium 17:334–349.

Reimers, P. E. 1973. The length of residence of juvenile fall chinook salmon in Sixes River, Oregon. Research Reports of the Fish Commission of Oregon, Volume 4 Number 2, Portland.

Richards, K. 1982. Rivers: form and process in alluvial channels. Methuen, New York.

Riddell, B. E. 1993. Spatial organization of Pacific salmon: what to conserve? Pages 23–41 in J. G. Cloud and G. H. Thorgaard, editors. Genetic conservation of salmonid fishes. Plenum, New York.

Rieman, B., D. Lee, J. McIntyre, K. Overton, and R. Thurow. 1993. Consideration of extinction risks for salmonids. U.S. Forest Service Fish Habitat Relationships Technical Bulletin No. 14, Eureka, California.

Schiemer, F., and T. Spindler. 1989. Endangered fish species of the Danube River in Austria. Regulated Rivers Research and Management 4:397–407.

Schlosser, I. J. 1987. A conceptual framework for fish communities in small warmwater streams. Pages

17–24 *in* W. J. Matthews and D. C. Heins, editors. Ecology and evolution of North American stream fishes. University of Oklahoma Press, Norman.

Schlosser, I. J. 1991. Stream fish ecology: a landscape perspective. BioScience 41:704–712.

Schlosser, I. J. 1995a. Critical landscape attributes that influence fish population dynamics in headwater streams. Hydrobiologia 303:71–81.

Schlosser, I. J. 1995b. Dispersal, boundary processes, and trophic level interactions in streams adjacent to beaver ponds. Ecology 76:908–925.

Schoener, T. W. 1991. Extinction and the nature of the metapopulation: a case study. Acta Oecologica 12: 53–75.

Scudder, G. G. E. 1989. The adaptive significance of marginal populations: a general perspective. Pages 180–185 *in* C. D. Levings, L. B. Holtby, and M. A. Henderson, editors. Proceedings of the national workshop on effects of habitat alteration on salmonid stocks. Canadian Special Publication of Fisheries and Aquatic Sciences 105.

Sedell, J. R., G. R. Reeves, F. R. Hauer, J. A. Stanford, and C. P. Hawkins. 1990. Role of refugia in recovery from disturbances: modern fragmented and disconnected river systems. Environmental Management 14: 711–724.

Sheldon, A. L. 1987. Rarity: patterns and consequences for stream fishes. Pages 203–209 *in* W. J. Matthews and D. C. Heins, editors. Ecology and evolution of North American stream fishes. University of Oklahoma Press, Norman.

Simberloff, D. 1988. The contribution of population and community biology to conservation science. Annual Review of Ecology and Systematics 19:473–511.

Taylor, A. D. 1988. Large-scale spatial structure and population dynamics in arthropod predator-prey systems. Annals Zoologici Fennici 25:63–74.

Taylor, B. 1991. Investigating species incidence over habitat fragments of different areas—a look at error estimation. Biological Journal of the Linnean Society 42:177–191.

Waples, R. S. 1995. Evolutionarily significant units and the conservation of biological diversity under the Endangered Species Act. American Fisheries Society Symposium 17:8–27.

Werner, E. E., and J. F. Gilliam. 1984. The ontogenetic niche and species interactions in size structured populations. Annual Review of Ecology and Systematics 15:395–425.

Winemiller, K. O., and K. A. Rose. 1992. Patterns of life-history diversification in North American fishes: implications for population regulation. Canadian Journal of Fisheries and Aquatic Sciences 49:2196–2218

Winston, M. R., C. M. Taylor, and J. Pigg. 1991. Upstream extirpation of four minnow species due to damming of a prairie stream. Transactions of the American Fisheries Society 120:98–105.

American Fisheries Society Symposium 17:402–414, 1995

Conserving Aquatic Biodiversity: Beyond Species and Populations

PAUL L. ANGERMEIER

National Biological Service, Virginia Cooperative Fish and Wildlife Research Unit[1]
Virginia Polytechnic Institute and State University, Blacksburg, Virginia 24061-0321, USA

ISAAC J. SCHLOSSER

Department of Biology, University of North Dakota, Grand Forks, North Dakota 58202, USA

Abstract.—Although biodiversity encompasses diversity at many organizational levels, the vast majority of conservation efforts has been directed at particular species and populations. This approach has not been effective at protecting aquatic populations or biodiversity in general. Biological conservation can be more effective by making policy more preventive (less reactive), and by broadening its focus to include large-scale biotic elements (e.g., landscapes) and key ecological and evolutionary processes. Species-specific approaches should be complemented by efforts to protect distinctive assemblages (defined to include guilds, communities, landscapes, and biomes), which provide the ecological and evolutionary context for populations. Aquatic assemblages vary considerably with respect to composition and organization, but no widely accepted framework reflects that diversity. Before policy to protect assemblages can be established, biologists must develop a framework for taking stock of assemblage diversity. Such a framework would enable managers to collect data on distribution, status, and trends, and to build compelling cases for conservation. We develop a general taxonomy of aquatic assemblages after reviewing current approaches to ecological classification. The framework consists of four major hierarchical levels: geographic regions, regional landscapes, primary communities, and secondary communities. At each level, assemblages are distinguished directly on the basis of biotic factors and indirectly on the basis of physical factors. The resulting classification is based on the best available knowledge, but it can be adapted to incorporate new information. We envision that the framework can be used by state or provincial resource managers to develop detailed classifications of aquatic assemblages, which should help conservation biologists to articulate the diversity and value of those assemblages. Such classifications may be especially useful for prioritizing efforts in areas where biodiversity loss is not already severe. However, comprehensive conservation strategies should focus on protecting ecological integrity rather than on accurately monitoring biodiversity loss.

The primary goal of biological conservation is to retain indefinitely as much of Earth's biotic diversity (biodiversity) as possible, with emphasis on those biotic elements most vulnerable to human impacts. Evaluating progress toward that goal requires a clear understanding of what biodiversity is and what ought to be conserved. Widespread usage of the term "biodiversity" by scientists, managers, policy makers, and the public has generated many (often disparate) conceptions of biodiversity (Angermeier and Karr 1994). Perhaps the most common misconception is that biodiversity is synonymous with species diversity. Yet, all thorough treatments (e.g., Office of Technology Assessment [OTA] 1987; Reid and Miller 1989; Noss 1990a) clearly characterize biodiversity as consisting of diversity from multiple organizational levels (e.g., species, genes, ecosystems). Biodiversity is much more than a catalog of

species and genes; the ecological and evolutionary contexts of biotic elements are as important as their identities.

A comprehensive view of biodiversity encompasses elements from any of several organizational levels within each of three distinct hierarchies: taxonomic, genetic, and ecological (Table 1). Because native elements from all hierarchies and levels are inherently valuable, all are (more or less) similarly appropriate targets of conservation. However, the ability of biologists to distinguish elements varies considerably among organizational levels. For example, protocols to distinguish species are more sophisticated than those to distinguish communities. Nevertheless, focusing on a single hierarchy (e.g., taxonomy) or organizational level (e.g., species) for purposes of measuring or conserving biodiversity is an extremely arbitrary and narrow approach that ignores the existence and value of most biodiversity.

It is beyond the scope of this essay to discuss conservation approaches appropriate for all elements of biodiversity. Herein, we focus on ecologi-

[1]The Unit is jointly sponsored by the National Biological Service, Virginia Polytechnic Institute and State University, and the Virginia Department of Game and Inland Fisheries.

TABLE 1.—Representative nested levels of organization in three hierarchies used to characterize biodiversity (adopted from Angermeier and Karr 1994). An element at any level (e.g., a species, a gene, an ecosystem) may be an appropriate target of conservation. Comprehensive views of biodiversity integrate diversity from all levels and hierarchies.

Taxonomic hierarchy	Genetic hierarchy	Ecological hierarchy
Biota	Genome	Biosphere
Kingdom	Chromosome set	Biome
Division or phylum	Chromosome	Landscape
Class	Gene complex	Ecosystem or community
Order	Gene	Guild
Family	Allele	Population
Genus		
Species		

cal elements (Table 1) because we are most familiar with the processes relevant to their conservation and because we believe that conservation of ecological elements at all organizational levels is necessary to conserve elements in other biotic hierarchies. Furthermore, current conservation policy is inadequate for protecting ecological elements, especially those associated with large spatial scales (e.g., landscapes).

Conservation of a particular ecological element requires an understanding of the element's evolutionary and ecological context. Elements are generated and maintained through many complex processes that link higher and lower organizational levels. Hierarchy theory (O'Neill et al. 1989) dictates that the behavior of a hierarchical system (e.g., community) is constrained by the behavior of its component systems (e.g., guilds, populations) and the behavior of the higher-level system (e.g., landscape, biome) of which it is a component. For example, persistence of a community is tied directly to processes that regulate dynamics of component populations over entire landscapes (Schlosser and Angermeier 1995, this volume). Thus, effective conservation of any ecological element requires attention to the key organizational processes responsible for its generation and maintenance. Failure to recognize or mitigate impairment of those processes (over a wide range of spatiotemporal scales) is likely to result in eventual loss of the target element.

Our primary objectives in this essay are to assess the adequacy of current policy in the United States for conserving aquatic biodiversity, to review rationales for conserving aquatic assemblages (broadly defined), and to formulate a general framework for classifying aquatic assemblages for the purpose of

conservation. Finally, we discuss application of the framework in the context of comprehensive conservation strategies. Although our discussion applies to all aquatic systems, our perspective is admittedly biased by our experience with freshwater stream fishes. Nevertheless, our ultimate goal is to provide guidance for changes in conservation policy that will more effectively protect all elements of aquatic biodiversity.

Conservation Policy in the United States

Biological conservation in the United States is founded on the Endangered Species Act of 1973 (ESA; 16 U.S.C. §§ 1531 to 1544) and various state-level analogs. The ESA is one of the most powerful and far-reaching environmental laws in the world, but it was not designed to protect biodiversity in general. Rather, its purpose is to function as a "safety net" to prevent species extinction. Although conservation of "ecosystems" is an explicit goal of the ESA, narrow implementation has limited protection mostly to individual organisms (of threatened or endangered species) and to their "critical habitat." Neither the ESA nor the numerous other federal laws related to conserving biodiversity (see Table 9.1 in OTA 1987) explicitly protect most elements of biodiversity (Table 1) or the key processes responsible for maintaining those elements.

Thus, despite the great value of other elements to society, the vast majority of conservation efforts are expended on a few selected species or populations, and usually on those that have been driven nearly to extinction. Increasingly, genetic elements are targets of conservation (exemplified by several papers in this volume), but conservation approaches rarely focus on ecological elements above the level of population. This strategy seems misguided for two reasons. First, higher-level elements (e.g., communities, landscapes) form the ecological and evolutionary context for populations and species (Pickett et al. 1992); preserving the integrity of that context is critical to species persistence. Second, higher-level elements are inherently valuable to society because they form the infrastructure of our irreplaceable life-support systems (Ehrlich and Mooney 1983).

The current model for biological conservation can be summarized by the following sequence. First, a region's ecosystems are exploited until the most sensitive species experience widespread declines. Eventually, species drop below the threshold of imminent extinction; they attain "protected" status. Efforts to protect remaining fragments of popula-

ECOLOGICAL DIVERSITY POLICY

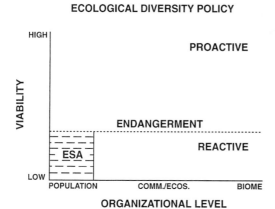

FIGURE 1.—Conceptual diagram of the relationship between conservation policy, represented by the Endangered Species Act of 1973 (ESA), and ecological diversity, represented in two dimensions (organizational level and viability). The axis for organizational level is adapted from the ecological hierarchy in Table 1. The dashed line labeled "endangerment" represents the threshold below which an element of biodiversity is in danger of extirpation. Policy implemented below this threshold (e.g., the ESA) is reactive, whereas policy implemented above the threshold is proactive (preventive). Conservation policy is dominated by two features: reactive implementation and narrow focus on populations. Biotic elements outside the "ESA box" are largely unprotected by current conservation policy.

tions and suitable habitat are begun, including elimination or mitigation of causes of decline. Key historical habitats are restored and populations recolonize or are stocked. When recovery is sufficient to ensure long-term persistence of imperiled species, protective status is relaxed, and the ecosystems are again fully exploitable.

The two dominant features of conservation policy in the United States—reactive implementation and focus on populations—are illustrated in Figure 1. Conservation mechanisms do not activate until populations decline to an "endangerment" threshold, whereupon the ESA acts as a safety net to prevent extinction. However, without mechanisms to prevent declines, the safety net can become overloaded, thereby limiting its effectiveness. Preventing endangerment is likely to be more cost-effective than emergency efforts to pull elements back from the brink of extinction.

Mechanisms to explicitly protect higher-level ecological elements are virtually nonexistent in current conservation policy (Figure 1). Thus, even if the ESA performed perfectly in its limited intended

capacity, the potential for loss of most elements of biodiversity remains high without additional safety net mechanisms. In most cases, ecologists have only superficial knowledge of how much diversity exists at organizational levels above populations. Information on rates of diversity loss for higher-level elements or the intensity of human impact that they can tolerate is largely unavailable. Nevertheless, we contend that just as species should not be driven to extinction, neither should distinctive communities or landscapes be eliminated by short-sighted human uses of natural resources. Conservation biologists need better tools to help them articulate the diversity and value of higher-level ecological elements.

A potential policy mechanism to prevent endangerment of elements of aquatic biodiversity in the United States stems from the Clean Water Act (CWA; 33 U.S.C. §§ 1251 to 1387). Over the last two decades, implementation of the CWA has increasingly emphasized the goal of maintaining biological integrity (Karr 1991). Important advantages of conservation policy based on the CWA over ESA-based policy are that the CWA applies to all aquatic systems, not just those with endangered species, and that the integrity concept applies to all ecological elements, not just populations (Angermeier and Karr 1994). Fuller implementation and enforcement of both the ESA and the CWA should make conservation of aquatic biodiversity more effective.

The current, narrowly focused policy for conservation of biodiversity is not especially effective. Although biodiversity loss undoubtedly would be greater with no policy, current annual expenditures of over US$100 million on endangered species are not preventing species loss in the United States, let alone loss of other elements of biodiversity. Several recent status assessments of North America's aquatic fauna (Miller et al. 1989; Williams et al. 1989; Master 1990; Williams et al. 1993) indicate that species extinction and endangerment is increasing but recovery is not keeping pace. Of the federally protected aquatic species with official recovery plans, the status of only 4% has improved (Williams and Neves 1992). Future rates of fish extinction are likely to be greatest in high-endemism areas of the southeastern United States, where many states have large backlogs of imperiled species that are not yet federally protected (Warren and Burr 1994). The extent of decline and loss of other types of elements of aquatic biodiversity is largely unknown.

Significant improvement in the effectiveness of conservation policy will require fundamental shifts in approach, including greater emphasis on higher-level ecological elements (e.g., landscapes), and on maintaining biological integrity of systems (Angermeier and Karr 1994). In particular, the primary goal of conservation should be to prevent endangerment of biotic elements through protection of ecological and evolutionary processes that generate and maintain them rather than to ward off extinction through heroic recovery efforts. Protecting key processes will require an expansion of the spatiotemporal scales over which conservation approaches are typically applied. Because human impacts are applied at large spatiotemporal scales (e.g., those associated with landscapes), conservation approaches should also focus on large-scale elements and processes (Angermeier and Karr 1994).

Conservation of Aquatic Assemblages

Definitions

Although ecologists agree on the general meaning of the terms "assemblage," "community," and "ecosystem," specific uses are highly variable. For clarity, we briefly describe our use of these terms herein. First, we do not view community and ecosystem as distinct hierarchical levels (Table 1), but rather as complementary perspectives on the same level (King 1993). The community perspective focuses on dynamics of organism distribution and abundance, whereas the ecosystem perspective focuses on dynamics of energy and materials moving through and around organisms. Second, we use assemblage more inclusively than community. For example, a community is a relatively local interactive assemblage of populations (Drake 1991), but we also use assemblage to refer to higher-level ecological elements (e.g., landscapes, biomes) with some populations that are relatively remote or noninteractive. Thus, we view all levels above population in the ecological hierarchy (i.e., guild through biosphere in Table 1) as assemblages.

The validity of assemblages as ecological elements is arguable. There is a long-standing debate among ecologists over whether assemblages are meaningful entities or merely convenient constructs of co-occurring populations. Opposing perspectives reflect biases toward particular taxa and organizational processes. For example, some forest ecologists (e.g., Davis 1986; Hunter et al. 1988) contend that assemblages are not integrated products of coevolution, but transitory artifacts of species-specific distributions that track (and lag behind) continually changing climatic conditions. Such assemblages may be useful as descriptive tools, but have little heuristic or predictive value. In contrast, some avian ecologists (e.g., Diamond 1975; Grant and Schluter 1984) argue that assemblages often are highly structured, especially by competition, and that co-occurring species are selected from the available species pool by complex deterministic processes. The structure of such assemblages may be predicted from knowledge of resource distribution in the environment, resource requirements by potential assemblage members, and interspecific interactions. The full range of perspectives, from highly integrated (Werner 1984) to largely unstructured (Sale 1984) entities, has been applied to fish assemblages. However, ecologists' understanding of assemblage organization typically is too superficial to allow reliable distinctions between deterministically and stochastically structured assemblages (Drake 1991). Nevertheless, most aquatic assemblages are probably organized by complex combinations of stochastic and deterministic processes (e.g., Strange et al. 1993). Notably, most studies of assemblage organization focus on local assemblages (i.e., communities) rather than the higher-level (larger-scale) elements that we include in our use of the term.

In a conservation context, it may not matter which view of assemblage organization is more "correct"; viability of *all* types of natural assemblages (at all hierarchical levels) should be preserved. We support Noss's (1990b) view that assemblages ought to be conserved even if they are dynamic entities with many interchangeable components. Despite loose integration, assemblages are not random sets of populations, and they can be defined operationally on the basis of ecological attributes or societal benefits. Because these features are measurable and predictable over time frames relevant to society (decades rather than millennia), assemblages are appropriate targets of conservation. Thus, even if current assemblages will not persist for another 1,000 years because of inherent climatic change, their immediate value justifies their conservation.

Why Conserve Assemblages?

Broadening the scope of conservation policy to protect assemblages provides several advantages over the current species-specific focus. First, it is not cost-effective to expend limited resources on species-specific efforts. The sheer number of species likely to become endangered over the next few decades will quickly exhaust society's time, money, and patience for conservation of biodiversity (Franklin 1993). Current rates of endangerment and recovery suggest that resources for conservation may already be overextended. Furthermore, most of the species likely to be extirpated are probably unknowable because of their small size and inconspicuous habits (e.g., invertebrates). Assemblage-level approaches offer the only hope for conserving relatively large portions of our poorly known biota.

Assemblage-level conservation is more likely to balance conflicting needs of various components, and thereby protect ecological integrity, than is species-specific conservation. For example, a species-specific strategy to protect winter-run chinook salmon *Oncorhynchus tshawytscha* in California could allow managers to regulate river flows to enhance the winter run at the expense of other declining runs, or to enhance salmon runs at the expense of estuarine biota (Moyle and Yoshiyama 1994). Such "options" would be less tenable in a strategy to protect the entire assemblage. Species-specific approaches often encourage simplistic, short-term means of achieving narrow conservation goals, such as supplemental stocking to maintain population size. Programs designed to conserve assemblages are more likely to espouse maintenance of complex ecological processes (e.g., flow fluctuation, predation) to meet their broader goals. Finally, because assemblages form the ecological and evolutionary context for populations (Pickett et al. 1992), protecting the integrity of assemblages will enhance the maintenance of component populations. Indeed, some important population processes *require* intact higher-level elements such as landscapes in which to operate (Schlosser and Angermeier 1995).

How should assemblages be prioritized for conservation? As for species, both utilitarian and nonutilitarian rationales can be developed for conserving assemblages. Notably, such rationales are based in value, not in science. Utilitarian value of assemblages might stem from economic goods (e.g., fishery) and services (water purification) or from less tangible benefits (aesthetic appeal, cultural significance). Because market-based mechanisms often support the maintenance of assemblages with high utilitarian value, legal mandates for their conservation may not be necessary. Nonutilitarian value of an assemblage might stem from the value of component populations or from attributes of the assemblage itself. For example, an assemblage might warrant conservation because it includes rare species. This rationale is an extension of species-specific approaches and does not require substantial revision of current policy. In contrast, if assemblage value is a function of the particular combination of populations or particular organizational processes, current approaches are inadequate for prioritizing conservation efforts. In particular, if conservation goals include protection of assemblages with unusual combinations of species or unusual dynamics, there is no widely accepted protocol for deciding which assemblages meet such criteria. Yet, unusual assemblages are the most likely to be lost and to need immediate conservation.

Sometimes assemblages are ranked with regard to conservation value, usually in association with choosing sites for reserves. Assemblage uniqueness is one criterion of conservation value, but species diversity is the most widely used criterion (Rabe and Savage 1979; Margules and Usher 1981; Maitland 1985). However, diversity is only one of many ecological axes along which assemblages may be ordered (Schoener 1986). Given that the primary goal of conservation is to preserve all biodiversity, the use of species diversity as the primary criterion for assessing conservation value of assemblages seems unjustified. For example, many species-poor assemblages may be valuable because of their unusual organization, or assemblages of several regionally rare species may be more valuable than assemblages of many common species (Winston and Angermeier, in press). Furthermore, species-rich assemblages often do not include those species most in need of protection (Prendergrast et al. 1993).

Assemblage Classification

Need for Classification

Establishing safety net policies to prevent loss of valuable assemblages is likely to require legal mandates analogous to the ESA. Demonstrating a need for such policies is likely to require a systematic accounting (especially of declines) of various assemblages. A major obstacle to conducting such an accounting is the lack of a widely accepted protocol for identifying and delineating (classifying) biotic assemblages (Orians 1993). Before a compelling case for conservation can be developed, fundamental questions must be addressed. How many types of assemblages exist? How many of each type remain? Where are they? Which ones are most imperiled? The continued inability of ecologists to answer these questions will relegate the conservation of assemblages to haphazard preservation of fragments of disintegrating systems. A widely accepted assemblage classification would have the heuristic value of raising awareness of biodiversity and the practical value of providing a framework within which to assess status and trends of valuable biotic resources.

The lack of a widely accepted taxonomy of assemblages reflects the complexity of the problem and the relative newness of efforts to solve it. Although formal classifications of biotic assemblages have existed since the 1830s, such attempts are centuries younger than the classification of species, which is still a dynamic and contentious area of research (Mayden and Wood 1995, this volume). Given that many types of assemblages are already imperiled (e.g., old-growth forests, mature prairies, large rivers, arid region springs, coastal wetlands), conservation biologists cannot wait for the emergence of a perfect assemblage taxonomy based on scientific unanimity. A crude but adaptable classification scheme would be better than none at all.

Several requirements of an assemblage classification appropriate for conservation are already identifiable. For example, a scheme must be based on features relevant to the conservation value of assemblages and be applicable over a wide range of spatial scales (local to regional). Furthermore, such a classification must focus on native elements, thereby providing a basis for monitoring biodiversity loss due to additions of artificial diversity (Angermeier 1994). In addition, Orians (1993) suggested that an appropriate classification scheme should (1) clearly identify recurring assemblages, (2) predict relationships between biotic and physical features, (3) correspond to species' distributions, and (4) be hierarchical. A few classification schemes that meet most of these criteria already have been developed for aquatic systems of individual states (Edwards et al. 1989; Moyle and Ellison 1991), but no standard scheme is applied consistently among authors or across regions.

Current Classification Approaches

Ecological systems have been classified on the basis of many factors and for many purposes. We do not attempt an exhaustive review of ecological classification, especially of the huge literature in government agency documents. Rather, we summarize a selected (but representative) portion of the literature on classification of aquatic systems, emphasizing literature relating to fish.

Both physical and biotic factors may be used in a classification scheme. Physical factors include climate, geology, topography, soils, and hydrology at spatial scales large enough to encompass both aquatic and terrestrial components of the biosphere. At smaller spatial scales, important physical factors include such water properties as temperature, depth, current velocity, and spatial complexity and such temporal dynamics as tidal cycle, water permanence, and predictability. Physical factors are used to classify the places where organisms live rather than assemblages themselves, but the primary purpose of many such classifications is to provide predictions about the assemblages living in those places.

Biotic factors can be used directly to classify assemblages in several ways. First, portions of the biota may form the structural matrix in which other organisms live. Physiognomy, growth forms, or other descriptors of vegetation structure are frequently used to classify assemblages, especially in terrestrial systems such as forests or prairies. A few sedentary aquatic animals such as corals are prominent structural features in some aquatic systems. Second, assemblages may be classified on the basis of their composition. Typically, taxonomic composition is used (e.g., species or family), but functional attributes (reflecting food and habitat use) or life history attributes (reflecting reproductive strategy or behavior) may provide more ecologically meaningful classifications (Corkum and Ciborowski 1988). Finally, assemblages might be classified on the basis of key organizational processes, which may be regulated by physical factors (e.g., severe temperatures, scouring flows) or biotic ones (e.g., food availability, disease). This approach requires synthesis of extensive ecological information and is the least well developed (but see Schoener 1986).

In a conservation context, both physical and biotic classification factors have advantages. Biotic factors, which directly distinguish assemblages (the targets of conservation), are more closely related to the conservation values of society. Physical factors are only indirect indicators of assemblages but are easier to measure precisely. Given how little is known about most assemblages, classification approaches incorporating both types of factors are likely to be the most practical.

Schemes for classifying aquatic systems can be ordered along two major axes: (1) a spectrum of physical and biotic factors, and (2) a spatial scale (Figure 2). Representative schemes from the ecological literature form three primary groups. At the left side of Figure 2 are schemes that classify habitat units entirely on the basis of physical (geomorphic and hydraulic) features. These approaches are especially well developed for streams (Bisson et al. 1982; Frissell et al. 1986; Hawkins et al. 1993), but

ECOLOGICAL TAXONOMIES

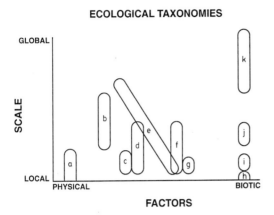

FIGURE 2.—Conceptual ordination of representative schemes for classifying aquatic systems ecologically. Horizontal position reflects the relative importance of physical and biotic classification factors in each scheme. Vertical position reflects the spatial scale of units identified in each scheme. Scales smaller than those appropriate for mapping assemblages are not shown. Each ellipse represents one or more published schemes: (a) Hutchinson (1957), Bisson et al. (1982), Frissell et al. (1986), and Hawkins et al. (1993); (b) Omernik (1987); (c) Sly and Busch (1992); (d) Edwards et al. (1989); (e) Cowardin et al. (1979); (f) Moyle and Ellison (1991); (g) Savage and Rabe (1991); (h) Schlosser (1982), Angermeier and Karr (1983), and Berkman and Rabeni (1987); (i) Tonn and Magnuson (1982) and Rahel and Hubert (1991); (j) Rahel (1984), Hughes et al. (1987), Matthews and Robison (1988), and Keller and Crisman (1990); (k) Moyle and Cech (1988) and Banarescu (1990).

are also used for lakes (Hutchinson 1957; Busch and Sly 1992). At the right side of Figure 2 are schemes that directly classify assemblages on the basis of their taxonomic compositions. Some of these distinguish large-scale assemblages such as zoogeographic regions (Moyle and Cech 1988; Banarescu 1990), whereas others distinguish assemblages at smaller regional scales (Rahel 1984; Hughes et al. 1987; Matthews and Robison 1988; Keller and Crisman 1990) or among nearby localities (Tonn and Magnuson 1982; Rahel and Hubert 1991). Efforts to use functional or life history attributes to classify assemblages have focused on guilds within local communities (Schlosser 1982; Angermeier and Karr 1983; Berkman and Rabeni 1987; see review by Austen et al. 1994). No classification scheme has been based on organizational processes. No scheme based entirely on physical or biotic factors has been applied over a wide range of spatial scales (Figure 2).

Schemes in the center of Figure 2 incorporate

some mix of physical and biotic classification factors. Cowardin et al. (1979), represented by the diagonal ellipse (e) in Figure 2, used more biotic factors as the spatial scale decreased, and they adopted Bailey's (1978) ecoregions at large spatial scales; consequently, their scheme incorporated a greater variety of biotic factors and a greater range of spatial scales than any other. Although narrower in spatial scope, the scheme developed by Sly and Busch (1992; ellipse c) distinguished units by factors similar to those in the scheme of Cowardin et al. (1979). The only biotic factor in the left-most scheme of this group (ellipse b; Omernik 1987) was terrestrial vegetation growth form. Schemes in the center group were devised for various purposes. Cowardin et al. (1979) identified distinctive types of aquatic habitat. Omernik (1987) investigated inherent properties of regional water quality. Savage and Rabe (1979), Edwards et al. (1989), and Moyle and Ellison (1991) assessed sites for conservation value. None of these schemes used ecological attributes or (as previously mentioned) organizational processes as explicit classification factors.

Information on how assemblages are organized is far too incomplete to support a meaningful classification of most assemblages on the basis of organizational processes. Classification schemes for current use must be based largely on physical factors and assemblage composition attributes. In schemes developed for use within limited geographic regions, discriminatory axes based on physical and biotic factors and axes based on taxonomic and ecological attributes may indicate similar arrays of assemblages. Phylogenetically similar taxa are often ecologically similar and associated with similar habitats. Furthermore, physical factors such as landform features and aquatic habitat structure impose major constraints on assemblage composition and organization (Schlosser 1987; Swanson et al. 1988; Poff and Ward 1990). Thus, within a region, physical factors or predominant taxa may be adequate indicators of assemblage type. However, relationships between physical and taxonomic attributes may change markedly among regions.

Comparisons of fish assemblages from different geographic regions illustrate the variability in relationships between habitat, taxonomy, and ecology. For example, small streams in Poland and Ontario supported fish assemblages that differed markedly in size distribution and benthic association despite similarities in habitat (Mahon 1984). In contrast, Moyle and Herbold (1987) concluded that assemblages in cold headwater streams of North America and Europe were ecologically similar, but assemblages in warmer streams of eastern North America were distinct from those in western North America and Europe. Finally, Tonn et al. (1990) found ecologically similar fish assemblages in small lakes of Finland and Wisconsin, despite taxonomic dissimilarities between the regions.

Ideally, assemblages should be classified primarily on the basis of biotic factors, but where biological information is scarce, physical factors may have to be used (cautiously) to distinguish assemblages. In general, we expect the assemblage diversity indicated by biologically based classifications to exceed that indicated by physically based classifications. Schemes based on physical factors are especially likely to "overlook" rare assemblages (Kirkpatrick and Brown 1994).

Development of a comprehensive protocol for classifying assemblages will require ecologists to think more pluralistically about assemblage organization and pattern. Pluralism recognizes ecological differences among assemblages; pluralistic theory attempts to explain those differences on the basis of interacting organizational processes (Schoener 1986). For example, organization of fish communities along gradients of stream size may be viewed as the integration of effects of changes in habitat complexity and stability, fish size, predation risk, and population stability (Schlosser 1982, 1987). Schoener (1986) identified 10 ecological axes along which assemblages can be ordered, including the relative importance of physical and biotic processes, number of species, species rarity, temporal variation in population sizes, and importance of history. Despite gaps in current knowledge of assemblage organization, many assemblages can be accurately characterized by suites of biotic and physical attributes that are closely associated with organizational processes (Schoener 1986). Understanding the distribution of assemblages along and across these axes will require innovative analyses of a wide range of physical and biotic factors, but such analyses may ultimately shed light on the number of assemblage types that exist, how to best distinguish them, and which ones are most imperiled. The typology of assemblage organization and pattern is a pressing research area in which ecology can become more relevant to conservation biology (Doak and Mills 1994).

Framework for an Assemblage Taxonomy

Many tools are already available for comprehensive classifications of aquatic assemblages that will be relevant to conservation. We offer several guidelines for framing such schemes. First, there is no "best" classification scheme. Conservation is a manifestation of science and social value, both of which are dynamic. A framework for assemblage classification should incorporate the best available knowledge, but also be adaptable and able to accommodate new information. Second, a globally applicable classification scheme is desirable but not essential. Most assemblage conservation is likely to be implemented within states, provinces, or regions; this is the spatial scale at which ecologically meaningful and legally defensible classification is most critical. However, legislative support for assemblage conservation, especially at federal levels, will require some geographic standardization. To be feasible, federal standards should apply only to coarse levels in the classification framework; finer levels should be determined by states and provinces. Third, assemblage classification should integrate biotic and physical factors. Biotic factors should be emphasized because the biota is the ultimate target of conservation, but limited knowledge may necessitate the use of physical factors to distinguish some assemblages. Finally, assemblage classification should focus on taxa that are well known, ecologically important, and highly valued. In general, vertebrates are the best taxa on which to base aquatic assemblage classifications because they are relatively well-studied, important to assemblage organization, and considered valuable by society.

We propose a general hierarchical taxonomy of aquatic assemblages for use in conservation contexts (Table 2). It includes four major levels, each with two or three nested sublevels that distinguish assemblages at progressively smaller spatial scales. A suite of biotic and physical factors is used to distinguish assemblages at each level. The number of levels appropriate for a given region depends on biotic complexity and current understanding of how assemblages are organized. The first two major levels refer to large-scale units of the biosphere that contain both aquatic and terrestrial components. The last two major levels refer specifically to aquatic systems, including the adjoining riparian components.

Level I distinguishes regional assemblages on the basis of biogeographic patterns such as distribution of higher taxa (e.g., families) and endemism, and on the basis of major differences in climate and geologic history. Assemblages at this level (and sublevels) are delineated by the zoogeographic regions, subregions, and provinces of Moyle and Cech (1988) or Banarescu (1990), which reflect large-scale (i.e., subcontinental) interactions between

TABLE 2.—Framework for a general taxonomy of aquatic assemblages for use in biodiversity conservation. Biotic and physical factors that might be used to classify assemblages at each of four major hierarchical levels are listed (in no particular order). Where feasible, biotic factors should be more heavily weighted than physical factors. Levels I and II refer to large-scale units that contain both aquatic and terrestrial components. Levels III and IV refer specifically to aquatic systems, including their adjoining riparian components. Each major level in the hierarchy may comprise several sublevels; more sublevels (i.e., more assemblage types) are expected at (numerically) higher levels. Hypothetical examples of assemblage types that might be distinguished in Virginia are listed for each hierarchical level.

| Level | Potential classification factors | | Hypothetical types in Virginia |
	Biotic	Physical	
I. Geographic region	Occurrence of higher taxa (e.g., families)	Climate; geologic history	Mississippi versus Atlantic drainages
II. Regional landscape	Terrestrial vegetation growth form; occurrence of lower taxa (e.g., species)	Soils; topography; hydrology; system connectivity; boundary attributes	Mountains versus coastal plain
III. Primary community	Occurrence of higher taxa (e.g., classes, families)	Salinity; flow; water body size; temperature	Streams with versus without fish; headwater streams versus mainstem rivers; rivers versus lakes
IV. Secondary community	Occurrence of lower taxa (e.g., species); occurrence of trophic or life history guilds; predominant organizing processes (e.g., predation); invasion resistance	Nutrient availability; habitat complexity; stability; predictability; substrate composition; acid-neutralizing capacity	Systems with versus without piscivores; systems regulated by competition versus disturbance; systems predominated by natives versus exotics; oligotrophic versus eutrophic; acid-sensitive versus alkaline

geomorphic features (Swanson et al. 1988) and fish dispersal and evolution (Smith 1981).

Level II distinguishes landscape assemblages on the basis of terrestrial vegetation growth forms, soils, topography, geology, and hydrology. Geomorphic features (Swanson et al. 1988) that determine permeability of habitat boundaries, spatial connectivity of habitat types, and availability of dispersal corridors strongly affect classification within this level because of their importance to metapopulation dynamics (Schlosser and Angermeier 1995). Assemblage delineation is similar to that for physiographic provinces (Fenneman 1946) or ecoregions (Bailey 1978; Omernik 1987) and reflects landscape-derived constraints on aquatic systems (Ross 1963; Lotspeich and Platts 1982). The ability of regional landscape units derived from terrestrial attributes (e.g., ecoregions) to reflect differences in aquatic biota is highly variable (Hughes et al. 1994). However, we suggest that such units are reasonable starting points in delineating aquatic landscape-level assemblages. Their utility may be improved by incorporating aquatic attributes such as hydrologic regime (e.g., Biggs et al. 1990) and spatiotemporal relationships among habitat types required by component populations (Schlosser and Angermeier 1995).

Level III distinguishes widely distributed types of communities on the basis of occurrence of certain higher taxa (e.g., fish versus no fish) and a suite of physical variables such as salinity (e.g., freshwater versus marine), flow (e.g., lotic versus lentic), and water body size (e.g., stream versus river). Assemblages at this level (and sublevels) are delineated as units analogous to those identified by Cowardin et al. (1979). Less widely distributed types of communities (Level IV) are distinguished on the basis of biotic composition (ecological or taxonomic) or organizational processes, and on the basis of physical factors such as water chemistry, water permanence, habitat complexity, and substrate composition. Assemblage delineation follows guidelines such as those used by Frissell et al. (1986), Edwards et al. (1989), and Moyle and Ellison (1991).

Concluding Remarks

The proposed framework provides an objective basis for recognizing and taking stock of aquatic biodiversity at organizational levels above populations. Answering the question "How many types of assemblages exist?" is one of the first and most difficult obstacles to developing a comprehensive strategy for biodiversity conservation. After assemblages are identified and delineated, state and provincial management agencies can gather data on geographic distribution, status, and trends for each assemblage type. Ultimately, such data provide the basis for cost-effectively establishing conservation priorities, siting preserves, or selecting systems to restore. Our framework should enable agencies to identify a manageable number of assemblage types within their areas of interest. A similar framework developed by Edwards et al. (1989) distinguished 48 types of systems in Texas. Specific assemblage categories and classification criteria must be determined by experts familiar with the biota within particular areas. However, we believe the framework will allow data from different area-specific schemes to be comparable, especially at large spatial scales. National or regional assemblage taxonomies might be developed by aggregating state or provincial taxonomies.

The proposed framework is admittedly crude, but imperfection is no excuse for inaction. Conservation biology is a crisis discipline (Soule 1985); crises demand new approaches. Ecologists already are being asked to decide which assemblages they *most* want to conserve. Such decisions should be based on appropriate ecological information rather than being made haphazardly or on the basis of economic forces. Our framework provides managers with an adaptable tool to aid in such decisions while choices still exist. To be effective, conservation must begin while representative assemblages at all spatial scales are still viable. The utility of our framework is diminished if most native assemblages are already eliminated or badly degraded. For example, it may be of limited use in California, where so few native assemblages remain that they all warrant preservation (Moyle and Yoshiyama 1994).

A classification scheme must accommodate new information to retain its heuristic and practical value. Users of our framework should ensure that classification categories and criteria are updated to reflect current knowledge. Changes in scheme structure might be justified by new information on geographic distributions of key taxa or by fundamental changes in assemblage composition or organization, such as that expected following global climate change. However, the greatest potential for changes in assemblage classification is associated with progress in understanding assemblage organization. Aquatic ecologists are just beginning to recognize the patterns associated with various mechanisms of assemblage formation and maintenance (Drake 1991), and to distinguish between inherently different assemblages and those merely at different successional stages (Bourgeron 1988, Reeves et al. 1995, this volume). As understanding of assemblage organization advances, assemblages that seem similar (or different) today may turn out to be very different (or similar).

Finally, we offer an important caution. Assemblage classifications can be valuable accounting tools for ecologists and resource managers, but keeping accurate tallies of lost or imperiled assemblages is not a compelling long-term goal for conservation biologists. Escalating loss of biodiversity is an incriminating testimony to society's mismanagement of the biosphere. Conservation biologists must work to get society out of the business of creating inviable biotic relics and into the business of protecting ecological integrity, especially at large spatiotemporal scales. Important roles for conservation biologists in this endeavor include raising public awareness of biodiversity, articulating the value of biodiversity, and demonstrating the dependency of economic stability on ecological integrity (Angermeier and Karr 1994). The primary challenges for society include limiting human population growth and resource consumption. Without significant success in these areas, biodiversity accountants will evolve into biological historians.

Acknowledgments

J. R. Karr, P. B. Moyle, and M. R. Winston kindly reviewed an earlier draft and helped us clarify many of the ideas presented here.

References

Angermeier, P. L. 1994. Does biodiversity include artificial diversity? Conservation Biology 8:600–602.

Angermeier, P. L., and J. R. Karr. 1983. Fish communities along environmental gradients in a system of tropical streams. Environmental Biology of Fishes 9:117–135.

Angermeier, P. L., and J. R. Karr. 1994. Biological integrity versus biological diversity as policy directives: protecting biotic resources. BioScience 4:690–697.

Austen, D. J., P. B. Bayley, and B. W. Menzel. 1994. Importance of the guild concept to fisheries research and management. Fisheries 19(6):12–20.

Bailey, R. G. 1978. Description of the ecoregions of the United States. U.S. Forest Service, Intermountain Region, Ogden, Utah.

Banarescu, P. 1990. Zoogeography of freshwaters. Volume 2: distribution and dispersal of freshwater animals in North America and Eurasia. AULA-Verlag, Wiesbaden, Germany.

Berkman, H. E., and C. F. Rabeni. 1987. Effect of siltation on stream fish communities. Environmental Biology of Fishes 18:285–294.

Biggs, B. J. F., and six coauthors. 1990. Ecological characterisation, classification, and modelling of New Zealand rivers: an introduction and synthesis. New Zealand Journal of Marine and Freshwater Research 24:277–304.

Bisson, P. A., J. L. Nielsen, R. A. Palmason, and L. E. Grove. 1982. A system of naming habitat types in small streams, with examples of habitat utilization by salmonids during low stream flow. Pages 62–73 in N. B. Armantrout, editor. Acquisition and utilization of aquatic habitat inventory information. American Fisheries Society, Western Division, Bethesda, Maryland.

Bourgeron, P. S. 1988. Advantages and limitations of ecological classification for the protection of ecosystems. Conservation Biology 2:218–220.

Busch, W.-D. N., and P. G. Sly, editors. 1992. The development of an aquatic habitat classification system for lakes. CRC Press, Boca Raton, Florida.

Corkum, L. D., and J. J. H. Ciborowski. 1988. Use of alternative classifications in studying broad-scale distributional patterns of lotic invertebrates. Journal of the North American Benthological Society 7:167–179.

Cowardin, L. M., V. Carter, F. C. Golet, and E. T. LaRoe. 1979. Classification of wetlands and deepwater habitats of the United States. U.S. Fish and Wildlife Service, FWS/OBS-79/31.

Davis, M. B. 1986. Climatic instability, time lags, and community disequilibrium. Pages 269–284 in J. Diamond and T. J. Case, editors. Community ecology. Harper and Row, New York.

Diamond, J. 1975. Assembly of species communities. Pages 342–444 in J. M. Diamond and M. L. Cody, editors. Ecology and evolution of communities. Harvard University Press, Cambridge, Massachusetts.

Doak, D. F., and L. S. Mills. 1994. A useful role for theory in conservation biology. Ecology 75:615–626.

Drake, J. A. 1991. Community-assembly mechanics and the structure of an experimental species ensemble. American Naturalist 137:1–26.

Edwards, R. J., G. Longley, R. Moss, J. Ward, R. Matthews, and B. Stewart. 1989. A classification of Texas aquatic communities with special consideration toward the conservation of endangered and threatened taxa. Texas Journal of Science 41:231–240.

Ehrlich, P. R., and H. A. Mooney. 1983. Extinction, substitution, and ecosystem services. BioScience 33:248–254.

Fenneman, N. M. 1946. Physical divisions of the United States. Map (scale 1:7,000,000). U.S. Geological Survey, Reston, Virginia.

Franklin, J. F. 1993. Preserving biodiversity: species, ecosystems, or landscapes? Ecological Applications 3:202–205.

Frissell, C. A., W. J. Liss, C. E. Warren, and M. D. Hurley. 1986. A hierarchical framework for stream habitat classification; viewing streams in a watershed context. Environmental Management 10:199–214.

Grant, P., and D. Schluter. 1984. Interspecific competition inferred from patterns of guild structure. Pages 201–233 in D. R. Strong, Jr., D. Simberloff, L. G. Abele, and A. B. Thistle, editors. Ecological communities: conceptual issues and the evidence. Princeton University Press, Princeton, New Jersey.

Hawkins, C. P., and ten coauthors. 1993. A hierarchical approach to classifying stream habitat features. Fisheries 18(6):3–12.

Hughes, R. M., S. A. Heiskary, W. J. Matthews, and C. O. Yoder. 1994. Use of ecoregions in biological monitoring. Pages 125–151 in S. L. Loeb and A. Spacie, editors. Biological monitoring of aquatic systems. Lewis Publishers, Boca Raton, Florida.

Hughes, R. M., E. Rexstad, and C. E. Bond. 1987. The relationship of aquatic ecoregions, river basins, and physiographic provinces to the ichthyogeographic regions of Oregon. Copeia 1987:423–432.

Hunter, M. L., G. L. Jacobson, and T. Webb. 1988. Paleoecology and the coarse-filter approach to maintaining biological diversity. Conservation Biology 2:375–385.

Hutchinson, G. E. 1957. Treatise on limnology, volume 1. Wiley, New York.

Karr, J. R. 1991. Biological integrity: a long neglected aspect of water resource management. Ecological Applications 1:66–84.

Keller, A. E., and T. L. Crisman. 1990. Factors influencing fish assemblages and species richness in subtropical Florida lakes and a comparison with temperate lakes. Canadian Journal of Fisheries and Aquatic Sciences 47:2137–2146.

King, A. W. 1993. Considerations of scale and hierarchy. Pages 19–45 in S. Woodley, J. Kay, and G. Francis, editors. Ecological integrity and the management of ecosystems. St. Lucie Press, Ottawa.

Kirkpatrick, J. B., and M. J. Brown. 1994. A comparison of direct and environmental domain approaches to planning reservation of forest higher plant communities and species in Tasmania. Conservation Biology 8:217–224.

Lotspeich, F. B., and W. S. Platts. 1982. An integrated land-aquatic classification system. North American Journal of Fisheries Management 2:138–149.

Mahon, R. 1984. Divergent structure in fish taxocenes of north temperate streams. Canadian Journal of Fisheries and Aquatic Sciences 41:330–350.

Maitland, P. S. 1985. Criteria for the selection of important sites for freshwater fish in the British Isles. Biological Conservation 31:335–353.

Margules, C., and M. B. Usher. 1981. Criteria used in assessing wildlife conservation potential: a review. Biological Conservation 21:79–109.

Master, L. 1990. The imperiled status of North American aquatic animals. Biodiversity Network News 3(3):1–2, 7–8, The Nature Conservancy, Arlington, Virginia.

Matthews, W. J., and H. W. Robison. 1988. The distribution of the fishes of Arkansas: a multivariate analysis. Copeia 1988:358–374.

Mayden, R. L., and R. M. Wood. 1995. Systematics, species concepts, and the evolutionarily significant unit in biodiversity and conservation biology. American Fisheries Society Symposium 17:58–113.

Miller, R. R., J. D. Williams, and J. E. Williams. 1989. Extinctions of North American fishes during the past century. Fisheries 14(6):22–38.

Moyle, P. B., and J. J. Cech, Jr. 1988. Fishes: an introduction to ichthyology. Prentice-Hall, Englewood Cliffs, New Jersey.

Moyle, P. B., and J. P. Ellison. 1991. A conservation-oriented classification system for the inland waters of California. California Fish and Game 77:161–180.

Moyle, P. B., and B. Herbold. 1987. Life-history patterns and community structure in stream fishes of western North America: comparisons with eastern North America and Europe. Pages 25–32 *in* W. J. Matthews and D. C. Heins, editors. Community and evolutionary ecology of North American stream fishes. University of Oklahoma Press, Norman.

Moyle, P. B., and R. M. Yoshiyama. 1994. Protection of aquatic biodiversity in California: a five-tiered approach. Fisheries 19(2):6–18.

Noss, R. F. 1990a. Indicators for monitoring biodiversity: a hierarchical approach. Conservation Biology 4:355–364.

Noss, R. F. 1990b. Can we maintain biological and ecological integrity? Conservation Biology 4:241–243.

Office of Technology Assessment. 1987. Technologies to maintain biological diversity. OTA (Office of Technology Assessment)-F-330, Washington, DC.

Omernik, J. M. 1987. Ecoregions of the conterminous United States. Supplementation to the Annals of the Association of American Geographers 77:118–125, Washington, DC.

O'Neill, R. V., A. R. Johnson, and A. W. King. 1989. A hierarchical framework for the analysis of scale. Landscape Ecology 3:193–205.

Orians, G. H. 1993. Endangered at what level? Ecological Applications 3:206–208.

Pickett, S. T. A., V. Parker, and P. L. Fiedler. 1992. The new paradigm in ecology: implications for conservation biology above the species level. Pages 65–88 *in* P. L. Fielder and S. K. Jain, editors. Conservation biology: the theory and practice of nature conservation, preservation, and management. Chapman and Hall, New York.

Poff, N. L., and J. V. Ward. 1990. Physical habitat template of lotic systems: recovery in the context of historical pattern of spatiotemporal heterogeneity. Environmental Management 14:629–645.

Prendergrast, J. R., R. M. Quinn, J. H. Lawton, B. C. Eversham, and D. W. Gibbons. 1993. Rare species, the coincidence of diversity hotspots and conservation strategies. Nature 365:335–337.

Rabe, F. W., and N. L. Savage. 1979. A methodology for the selection of aquatic natural areas. Biological Conservation 15:291–300.

Rahel, F. J. 1984. Factors structuring fish assemblages along a bog lake successional gradient. Ecology 65:1276–1289.

Rahel, F. J., and W. A. Hubert. 1991. Fish assemblages and habitat gradients in a Rocky Mountain–Great Plains stream: biotic zonation and additive patterns of community change. Transactions of the American Fisheries Society 120:319–332.

Reeves, G. H., L. E. Benda, K. M. Burnett, P. A. Bisson, and J. R. Sedell. 1995. A disturbance-based ecosystem approach to maintaining and restoring freshwater habitats of evolutionarily significant units of anadromous salmonids in the Pacific northwest. American Fisheries Society Symposium 17:334–349.

Reid, W. V., and K. R. Miller. 1989. Keeping options alive: the scientific basis for conserving biodiversity. World Resources Institute, Washington, DC.

Ross, H. H. 1963. Stream communities and terrestrial biomes. Archiv fuer Hydrobiologie 59:235–242.

Sale, P. F. 1984. The structure of communities of fish on coral reefs and the merit of a hypothesis-testing, manipulative approach to ecology. Pages 478–490 *in* D. R. Strong, Jr., D. Simberloff, L. G. Abele, and A. B. Thistle, editors. Ecological communities: conceptual issues and the evidence. Princeton University Press, Princeton, New Jersey.

Savage, N. L., and F. W. Rabe. 1979. Stream types in Idaho: an approach to classification of streams in natural areas. Biological Conservation 15:301–315.

Schlosser, I. J. 1982. Fish community structure and function along two habitat gradients in a headwater stream. Ecological Monographs 52:395–414.

Schlosser, I. J. 1987. A conceptual framework for fish communities in small warmwater streams. Pages 17–24 *in* W. J. Matthews and D. C. Heins, editors. Ecology and evolution of North American stream fishes. University of Oklahoma Press, Norman.

Schlosser, I. J., and P. L. Angermeier. 1995. Spatial variation in demographic processes of lotic fishes: conceptual models, empirical evidence, and implications for conservation. American Fisheries Society Symposium 17:392–401.

Schoener, T. W. 1986. Overview: kinds of ecological communities—ecology becomes pluralistic. Pages 467–479 *in* J. Diamond and T. J. Case, editors. Community ecology. Harper and Row, New York.

Sly, P. G., and W.-D. N. Busch. 1992. A system for aquatic habitat classification of lakes. Pages 15–26 *in* W.-D. N. Busch and P. G. Sly, editors. The development of an aquatic habitat classification system for lakes. CRC Press, Boca Raton, Florida.

Smith, G. R. 1981. Late Cenozoic freshwater fishes of North America. Annual Review of Ecology and Systematics 12:61–93.

Soule, M. E. 1985. What is conservation biology. BioScience 35:724–734.

Strange, E. M., P. B. Moyle, and T. C. Foin. 1993. Interactions between stochastic and deterministic processes in stream fish community assembly. Environmental Biology of Fishes 36:1–15.

Swanson, F. J., T. K. Kratz, N. Caine, and R. G. Woodmansee. 1988. Landform effects on ecosystem patterns and processes: geomorphic features of the earth's surface regulate the distribution of organisms and processes. BioScience 38:92–98.

Tonn, W. M., and J. J. Magnuson. 1982. Patterns in the species composition and richness of fish assemblages in northern Wisconsin lakes. Ecology 63:1149–1166.

Tonn, W. M., J. J. Magnuson, M. Rask, and J. Toivonen. 1990. Intercontinental comparison of small-lake fish assemblages: the balance between local and regional processes. American Naturalist (136)345–375.

Warren, M. L., Jr., and B. M. Burr. 1994. Status of freshwater fishes of the United States: overview of an imperiled fauna. Fisheries 19(1):6–18.

Werner, E. E. 1984. The mechanisms of species interactions and community organization in fish. Pages 360–382 *in* D. R. Strong, Jr., D. Simberloff, L. G. Abele, and A. B. Thistle, editors. Ecological communities: conceptual issues and the evidence. Princeton University Press, Princeton, New Jersey.

Williams, J. E., and seven coauthors. 1989. Fishes of North America endangered, threatened or of special concern: 1989. Fisheries 14(6):2–20.

Williams, J. E., and R. J. Neves. 1992. Introducing the elements of biological diversity in the aquatic en-

vironment. Transactions of the North American Wildlife and Natural Resources Conference 57: 345–354.

Williams, J. D., M. L. Warren, Jr., K. S. Cummings, J. L. Harris, and R. J. Neves. 1993. Conservation status of freshwater mussels of the United States and Canada. Fisheries 18(9):6–22.

Winston, M. R., and P. L. Angermeier. In press. Assessing conservation value using centers of population density. Conservation Biology.

PART SIX

PANEL DISCUSSION

American Fisheries Society Symposium 17:417–418, 1995

Session Overview
Results of Facilitated Discussion of Issues

Dale P. Burkett

U.S. Fish and Wildlife Service, Great Lakes Coordination Office
1405 S. Harrison Road, Room 308
East Lansing, Michigan 48823, USA

Issues for Resource Agencies to Address

Prior to formal presentations by resource agency representatives, a facilitated session was conducted in which the audience was encouraged to identify issues for the represented resource agencies to address. Issues identified ranged from questions to suggested changes in policy and broke out into two major categories: conceptual challenges and recommendations.

Conceptual challenges for resource agencies to address:

• Define the evolutionarily significant unit (ESU) concept and clarify it with respect to the metapopulation concept.
• Use the ESU concept or develop another suitable concept.
• Make the ESU concept sufficiently flexible to apply to all organisms, if it is accepted.
• Determine and manage specific U.S. Endangered Species Act (ESA; 16 U.S.C. §§ 1531 to 1544) criteria that might be affected by application of the ESU concept.
• Delineate effects of increasing long-term harvest on species and take into account implications of these effects upon ESU guidelines.

Recommendations to the resource agencies broke down into the following categories: prevention and proactive response; regulatory and policy; implementation; collaborative efforts; and education and information needs.

Prevention and proactive response recommendations:

• Develop sideboards governing methods to work within the ESA to enhance or supplement endangered or threatened ESUs.
• Develop appropriate technologies to store eggs and sperm in gene banks.
• Explore the need for new law(s) to address prelisting issues.

Regulatory and policy recommendations:

• Implement existing mandates by providing adequate funding and staff.
• Recognize all aquatic organisms in the final ESU concept.
• Move ESU concept development closer to science and further from politics.
• Integrate biological facts into policy.
• Identify and fill data gaps based upon agreed ESU delineation requirements.
• Make biological assessments of the regulatory environment subject to peer review.
• Analyze existing laws with respect to interaction effects and make any needed modifications to ensure that intended purposes of existing laws are achieved in concert with ESU objectives.
• Identify and resolve interjurisdictional issues.

Implementation recommendations:

• Address the issue of application of ESUs on privately owned lands and waters.
• Translate ESU conservation guidelines into management actions.
• Evaluate species and stocks and proactively list them instead of reacting to petitions.
• Prioritize efforts and focus upon greatest and most immediate need.
• Review agency funding and staffing and make adjustments to support effective implementation.

Collaborative effort recommendation:

• Have state and federal agencies and entities work together in voluntary collaboration to implement ESUs effectively.

Education and information recommendation:

• Develop and implement appropriately targeted ESU education and information programs.

Issues for the Academic Community and the Public to Address

After formal presentations by resource agency representatives, a facilitated session was conducted in which panelists were encouraged to identify issues for the academic community and the public to address. These needs broke out into the following major categories: working concept, policy, application, information and education, and application constraints.

Working concept needs:

- Conduct proactive analyses of various techniques that can be used to determine ESUs.
- Evaluate and communicate, in clear, concise, and easily understood terms, strengths, weaknesses, and uses of various tools that can be used to determine ESUs.
- Shape ESU concepts to bridge gaps in understanding.

Policy needs:

- Increase academic involvement in ESU policy development.
- Have nongovernmental organizations bring adversarial groups to the table to ensure their participation in ESU policy development and to encourage their ownership of products.

Application needs:

- Allow scientific community input during the draft phase and institute a peer review process for biological opinions.

Information and education category needs:

- Encourage scientists to convey the importance of ESUs more effectively in popular journals and through vehicles such as letters to editors.
- Develop methods to educate specific public audiences with targeted messages to improve appropriate understanding of ESU concepts.

Funding and staffing constraints present serious challenges in addressing ESUs. Finer levels of ESU resolution will require significantly greater investments of funds and staff resources. It will also require cooperative multientity planning efforts and restructuring current programs.

American Fisheries Society Symposium 17:419–422, 1995

Conservation Guidelines on Significant Population Units: Responsibilities of the National Marine Fisheries Service

WILLIAM W. FOX, JR. AND MARTA F. NAMMACK[1]

National Marine Fisheries Service, Office of Protected Resources
1335 East West Highway, Silver Spring, Maryland 20910, USA

In this paper we cover our agency's concerns related to the topic of this conference, defining unique units in population conservation, from the perspectives of three areas: regulatory, administrative, and programmatic. We begin by explaining the principal statutory obligations of the National Marine Fisheries Service (NMFS) as they relate to the topic of this conference and then very briefly mention NMFS' relationships with other federal agencies in meeting those statutory obligations.

The NMFS has three principal laws that define our mandate for the conservation, management, protection, and recovery of biological species, their subunits and their habitat. These are the U.S. Endangered Species Act (ESA; 16 U.S.C. §§ 1531 to 1544), the Marine Mammal Protection Act (MMPA; 16 U.S.C. §§ 1361 to 1407), and the Magnuson Fishery Conservation and Management Act of 1976 (MFCMA; 16 U.S.C. §§ 1801 et seq.). The NMFS also has other important mandates under such laws as the Fish and Wildlife Coordination Act (16 U.S.C. §§ 661 to 666c), Comprehensive Environmental Response Compensation and Liability Act of 1980, or "Superfund," (42 U.S.C. §§ 9601 to 9675), the Clean Water Act of 1990 (33 U.S.C. §§ 1251 to 1387), and the Oil Pollution Act of 1990 (33 U.S.C. §§ 2701 to 2761). Our sister agency in the National Oceanic and Atmospheric Administration (NOAA), the National Ocean Service (NOS), with whom we work closely, administers the Marine Protection, Research and Sanctuaries Act of 1972 (33 U.S.C. §§ 1441 to 1445) and the Coastal Zone Management Act of 1972 (16 U.S.C. §§ 1451 to 1464). There are many other lesser authorities that relate to NMFS mission; however, we concentrate our comments on just the three principal laws.

The essential first step in executing any conservation mandate is, of course, to define the basic biological unit around which we will conduct a program of research to understand its dynamics and status and ultimately establish a management program, whether the management program is simply regulating harvest or an emergency effort to recover a badly diminished unit. Each of our principal laws embodies this step, but each does so somewhat differently. It is our hope that this conference will provide information and advice on ways to improve our use of biological units below the species level for the conservation of species.

The Endangered Species Act

The main purpose of the ESA is to provide for the recovery of endangered and threatened species. The definition of species in the ESA, includes "any subspecies of fish or wildlife or plants, and any distinct population segment of any species of vertebrate fish or wildlife which interbreeds when mature."

The salient question about this definition is what does the word "distinct" in "distinct population segment" really mean? The legislative history is replete with concern over how this provision could be abused—the example discussed during the ESA's reauthorization in the late 1970s is the potential listing of the population of common squirrels in Central Park in New York City. From this discussion, it would appear that a population segment with mere short-term geographic isolation alone was not distinct under the ESA and, therefore, not intended to receive its protection. In other words, distinct should be interpreted to include the sense of "distinction," that is, factors measuring the significance or value of the population segment to the species as a whole.

The NMFS was faced with this problem, perhaps in its ultimate complexity, when the agency was petitioned to list five runs or sets of runs of Snake and Columbia river salmon *Oncorhynchus* spp. Furthermore, a report from the American Fisheries Society (AFS) indicated that over 200 runs of salmon on the Pacific Coast of the United States may require protection under the ESA.

Well, as much of the general public knows, species of Pacific salmon have the characteristic of returning to the stream where they began life to

---

[1]The views expressed herein are those of the authors and the Office of Protected Resources but do not necessarily represent those of the National Marine Fisheries Service.

spawn once and then die; that is, the species are subdivided into runs that are geographically distinct and that interbreed when mature. If that were not enough, Pacific salmon spawn in year-class cycles that, depending on the species, are centered 2–4 years apart, and thus Pacific salmon exhibit temporal as well as geographic separation. Of course, there is straying to various degrees both in space and time, which prevents complete reproductive isolation. However, is every geographic run to be treated as a species under the ESA? Moreover, is every year-class of every geographic run to be treated likewise?

These questions are made more difficult to answer because there are still millions of individuals of each of the five North American Pacific salmon species throughout their range. Furthermore, because Pacific salmon exhibit high adaptability and ease to culture, humans have been introducing cultured fish for decades to create new runs or to strengthen existing runs to mitigate the effects of the hydropower system and other sources of mortality. These introductions have significantly adulterated endemic fish populations and, therefore, questions are raised about the very existence of endemic fish in many locations.

The NMFS developed a science-based policy that, after determining that an endemic run existed and that it was substantially reproductively isolated, assessed the value or significance of a particular run to the species as a whole. The scientific basis for the policy received considerable public and scientific scrutiny, including an open presentation and debate at the AFS annual meeting in 1990. The scientific basis was published subsequently in the peer-reviewed literature (Waples 1991).

The NMFS policy on the definition of species under the ESA for Pacific salmon was published in the *Federal Register* on 20 November 1991. Under this policy, a run of Pacific salmon is considered a distinct population segment, and thus, a species under the ESA, if it represents an evolutionarily significant unit (ESU) of a biological species. A stock must satisfy two criteria to be considered an ESU:

1. it must be substantially reproductively isolated from other conspecific population units; and
2. it must represent an important component in the evolutionary legacy of the species.

Once a population segment is considered to be distinct, and thus a species under the ESA, NMFS evaluates its status to determine if it warrants protection as threatened or endangered.

Because the criteria for determining whether a Pacific salmon population is an ESU are subjective, it becomes important to evaluate the appropriate grouping level. Identifiable ESUs may even be listed together (i.e., as a taxonomic species) for the sake of convenience but should be managed as separate units. In general, ESUs should correspond to more comprehensive units unless there is clear evidence that evolutionarily important differences exist between smaller population segments. The NMFS policy on the definition of species for Pacific salmon does not necessarily result in the splitting of Pacific salmon runs—in fact, it may more often result in lumping of runs. For example, NMFS listed spring and summer runs of Snake River chinook salmon *O. tshawytscha* as one ESU because of their genetic and phenotypic similarities. Also, NMFS denied the petition for Illinois River winter steelhead *O. mykiss* because the population was not considered to be an ESU; rather, it was considered to be part of a much larger ESU. The NMFS, therefore, initiated a broader status review. This is consistent with the view expressed by the U.S. Congress that the authority to list distinct vertebrate populations should be used sparingly.

The NMFS and the Fish and Wildlife Service (FWS) share jurisdiction under the ESA. We have joint administrative regulations and an understanding, though perhaps overly vague, of our respective jurisdictions over species. However, we have not always had identical policies and practices with regard to some important details not explicitly covered by the regulations. One of those is the topic of this conference, that is, defining the unit that receives protection under the ESA.

In the years before NMFS received the five Pacific salmon petitions, NMFS and FWS had been working together on a joint species definition policy. Those petitions caused NMFS to charge ahead separately on the species policy. However, last fall, NMFS and FWS formed a working group to review and develop joint policies on all outstanding issues. We are nearing the completion of several joint policies. One of these is a joint policy on the recognition of distinct vertebrate population segments under the ESA. We purposefully waited for this conference before completing this policy—Michael Spear of the FWS will outline the salient features of our current draft in his presentation.

Section 7(a) (2) of the ESA requires all federal agencies to consult with either NMFS or FWS if any actions of those federal agencies may affect a species listed under the ESA. Through this consultation process federal agencies must mitigate the effects of their actions and may be bound by a limit on the take of listed species imposed by NMFS or FWS. The controlling document is the biological opinion issued by NMFS or FWS.

For other than federal agencies, Section 10 of the ESA allows NMFS and FWS to issue permits for incidental takes and for research or enhancement that benefits the listed species. Section 6 allows NMFS or FWS to delegate portions of their responsibilities to the states.

Finally, the ESA requires the designation of critical habitat for the listed species. Critical habitat may not be adversely affected. For species under NMFS jurisdiction, however, a critical habitat designation provides little additional protection except potentially in areas where the listed species does not presently exist but did at one time and will after recovery.

The NMFS Endangered Species Program is small, on the order of US$8 million out of a NMFS appropriation of $234 million for the current fiscal year. However, this is up nearly 50% from just 5 years ago, and an additional $1–2 million is in the works from reprogramming this fiscal year. The president's budget for the 1995 fiscal year requests another $8 million for an overall doubling of the program. We remain hopeful that the U.S. Congress will provide us with the requested resources.

Proper administration of the ESA is complex. Listing the three "species" of Snake River salmon taxed our intellectual, personnel, and fiscal resources. While we have been able to address the harvest sector because of our responsibilities under another law, which I will address in a moment, we have been challenged with the need to understand the complex effects of the federal Columbia River power system, fish hatcheries, and timber harvesting in order to produce biological opinions and Section 10 permits and then to defend them in the ensuing litigation.

Let me now turn briefly to our other two principal laws and contrast their need for defining and identifying unique units for population conservation.

Marine Mammal Protection Act

The main purpose of the MMPA is to protect marine mammals from takings through a general moratorium, with a few exceptions, so that they may remain at or return to their optimum sustainable population (OSP). Similarly, under the MMPA, we are not concerned with only taxonomic species but also with the smaller units of populations—in this case stocks rather than subspecies and distinct population segments. As defined in the MMPA a population stock or stock means a group of marine mammals of the same species or smaller taxa in a common spatial arrangement and that interbreed when mature. The goal of the MMPA is to maintain all population stocks or stocks of marine mammals at OSP.

Like the ESA, the underlying principle is protection; the policies formulated under this principle are to be risk averse. With regard to population stocks, the effect of a risk-averse policy is to favor splitting. In other words, if differences among groups of marine mammals can be detected or are even reasonably suspected based on scientific information, rather than scientifically proven, these groups are treated as population stocks under the law. Each must be at OSP, each is subject to the increased protection of a depletion finding (i.e., no takes may be allowed), and each is subject to a determination of whether the excepted takings to the general moratorium disadvantage the population stock.

The NMFS and FWS also jointly administer the MMPA. However, in this case the greatest burden is placed on NMFS. The FWS has responsibility for just five species: the polar bear *Ursus maritimus*, sea otter *Enhydra lutris*, walrus *Odobenus rosmarus*, manatee *Trichechus* spp., and dugong *Dugong dugong*. The remainder, all seals, sea lions, dolphins, and whales, are NMFS responsibility. The MMPA establishes a total federal preemption of marine mammal management, but it also provides for a return of management to the states and, most recently, the ability to enter into cooperative agreements with Alaska natives comanaging subsistence harvests.

The MMPA was just reauthorized with amendments that require the identification of all marine mammal population stocks and a determination of their status. This is an immense job. However, once again the president's 1995 fiscal year budget requests an additional $12 million for NMFS' responsibilities under the MMPA, nearly doubling the current program.

Magnuson Fishery Conservation and Management Act of 1976

The main purpose of the MFCMA is to conserve and manage the harvest of marine fish and shellfish so as to produce optimum yield from each fishery. Although the focus of the MFCMA is on fisheries, "the term 'fishery' means one or more stocks of fish which can be treated as a unit for purposes of conservation and management and which are identified on the basis of geographical, scientific, technical, recreational, and economic characteristics, or any fishing of such stocks. . . . The term 'stock of fish' means a species, subspecies, geographical grouping, or other category of fish capable of management as a unit." It is important to note that the definition of the smallest management unit under the MFCMA does not have to be strictly biological as it does under the ESA and MMPA.

The MFCMA has seven national standards for developing conservation and management plans or regulations. One of those (number 3) is directly relevant to the issue of population conservation units: "To the extent practicable, an individual stock of fish shall be managed throughout its range, and interrelated stocks of fish shall be managed as a unit or in close coordination." Unlike the ESA and the MMPA, the provisions of the MFCMA are not required to be afforded to the smallest biological unit. Presumably, this is because the MFCMA is to maintain an optimum harvest rather than protect or recover populations and because, at least in theory, corrective measures may be taken to rectify any problem before the extraordinary assistance of the ESA is required.

The jurisdictions of NMFS, the FWS, and the states are interlocking under the MFCMA. Fishery management plans are developed by regional fishery management councils. The states are full voting members on each council and the FWS participates as a nonvoting member. The president's 1995 fiscal year budget requests an increase of $42 million for NMFS responsibilities under the MFCMA. Perhaps this budget request can be considered a pre-ESA listing effort to avoid the need for the ESA in marine fisheries, because over 40% of the marine fish stocks that can be assessed are overexploited.

Summary

In summary, each of our conservation laws provides for the management of unique population units. The NMFS has developed and implemented a policy for defining distinct population segments of Northwest Pacific salmon, the ESU. This conference was convened to examine that policy, not only for Pacific salmon but also as a concept for all vertebrate species. The NMFS and FWS are presently developing a joint policy on defining distinct vertebrate population segments. We look forward to the information and advice from this conference to assist us in putting the finishing touches on our draft joint policy before it is circulated for public comment.

References

Waples, R. S. 1991. Pacific salmon, *Oncorhynchus* spp., and the definition of "species" under the Endangered Species Act. U.S. National Marine Fisheries Service Marine Fisheries Review 53(3):11–22.

American Fisheries Society Symposium 17:423–424, 1995

Considerations in Defining the Concept of a Distinct Population Segment of Any Species of Vertebrate Fish or Wildlife

MICHAEL SPEAR

U.S. Fish and Wildlife Service
911 Northeast 11th Avenue, Portland, Oregon 97232, USA

The U.S. Endangered Species Act (ESA; 16 U.S.C. §§ 1531 to 1544) defines species, in part, as "any distinct population segment of any species of vertebrate fish or wildlife which interbreeds when mature." The authority to list a species as endangered or threatened thus extends to distinct population segments of vertebrates.

The Fish and Wildlife Service (FWS) and National Marine Fisheries Service (NMFS) need to agree on an interpretation of the cited phrase for the purposes of adding species to or removing them from the lists or reclassifying them under the ESA. This policy interpretation must be comprehensive in scope because it would govern the interpretation of distinct population segments for all species' listings under U.S. law, including listings of foreign species. At the same time, the agencies must recognize the limits of their resources and ensure that the ESA is administered so that it retains its focus on conserving unique and irreplaceable biological entities.

Listings of populations under the ESA have been controversial because of a perceived lack of consistency and adherence to general guiding principles. Several high-profile species are listed under the ESA as populations (e.g., bald eagle *Haliaeetus leucocephalus*, grizzly bear *Ursus arctos horribilis*, gray wolf *Canis lupis*, and marbled murrelet *Brachyramphus marmoratum*).

It is important in light of the ESA's general principle of following the best available scientific information in determining the status of a species that this interpretation follow sound biological principles. Any interpretation adopted should also be aimed at carrying out the purposes of the ESA. That is, "to provide a means whereby the ecosystems upon which endangered species and threatened species depend may be conserved, to provide a program for the conservation of such endangered species and threatened species, and to take such steps as may be appropriate to achieve the purposes of . . . treaties and conventions," including the Convention on International Trade in Endangered Species of Wild Fauna and Flora (CITES).

The U.S. Congress has instructed us to exercise the authority to list distinct population segments "sparingly" (96th Congress, 1st session, Senate Report 151). Of 300 native vertebrate species listed under the ESA, only about 20 are given separate status as distinct population segments.

The FWS has been attempting for several years to develop consensus on the proper application of its authority to treat certain vertebrate populations as species under the ESA. The current draft is the outcome of several rounds of circulation and review of documents among the FWS's regions and with NMFS.

The most significant difference between the present draft and previous attempts to interpret this authority involves the stepwise application of two sets of standards, first to consider the distinctness of the population under consideration and then to assess its significance. A population that satisfies the tests for distinctness and significance would be treated as a species for purposes of status evaluation and could be added to or removed from the endangered or threatened lists or reclassified from one status to the other independently.

The Assessment Process

Distinctness would be the first quality addressed in the evaluation; a population segment could be considered distinct if it satisfies one of the following criteria.

1. The population segment is markedly separated from other populations of the same taxon as a consequence of physical, physiological, ecological, or behavioral factors. Genetic or morphological discontinuity may provide evidence of this separation.

This first criterion is intended to identify biologically isolated population segments that are capable of being evaluated independently of one another. Examples of populations that have been recognized and listed consistent with this criterion include the Mohave population of desert tortoise *Gopherus agassizi* and the coastal population of the western snowy plover *Charadrius alexandrinus*. The two populations of the desert tortoise in the Mohave and

Sonoran deserts are geographically and reproductively separate, behaviorally different in selection of habitat, and presumably adapted to the different climatic regimes in the two areas. The coastal population of the snowy plover is also geographically and reproductively separated from other populations of the same subspecies. In the latter case, the boundary of 80 km from the coast has been used to delimit the listed population because it coincides with the natural discontinuity between biological units. Also consistent with this criterion was our rejection of a petition to list a "southwestern population" of the northern goshawk *Accipiter gentilis* that was not separated by any biological discontinuity from neighboring "populations" of the same species and would have included members of more than one recognized subspecies.

> 2. The population segment is delimited by international governmental boundaries within which there are differences in control of exploitation, management of habitat, conservation status, or regulatory mechanisms.

The second criterion recognizes the importance of international boundaries and the significance of national environmental laws and regulations in contributing to the well being of native species. There is ample precedent under both the ESA and CITES for assigning different status to species at the level of national populations. Among the populations now protected in this way are the populations of the gray wolf and grizzly bear in the lower 48 states.

The next level of examination would address significance; if a population segment satisfies one or more of the above criteria for distinctness, its biological and ecological significance would be considered in light of congressional guidance that the authority to list distinct population segments be used sparingly, while encouraging the conservation of genetic diversity. In carrying out this examination, the FWS would consider available scientific evidence of the distinct population segment's importance to the taxon to which it belongs. This consideration might include, but would not be limited to, the following:

1. persistence in an unusual or unique ecological setting;

2. evidence that its loss would leave a significant gap in the range of a species;
3. evidence that it represents the only surviving natural occurrence of a species; or
4. evidence that it differs markedly from other populations of the species in its genetic characteristics.

These are only examples of the attributes that might be considered in determining whether a population segment was significant. Because precise circumstances are likely to vary considerably from case to case, it is not possible to describe prospectively all the classes of information that might bear on the biological and ecological importance of a distinct population segment. It is entirely possible that there may be cases of populations that are significant in ways that are not now anticipated. The examples cited above as distinct listed populations would also satisfy this criterion. The gray wolf and grizzly bear are both small remnants of species that once ranged widely south of Canada and whose near extirpation in the conterminous states creates a large gap within these species' historic ranges. The listed population of the snowy plover ranges from Mexico north to Washington State in a distinctive ecological setting. The Mohave population of the desert tortoise is one of two regionally distributed components whose disappearance would significantly alter the range of this species.

The petition that was denied for the goshawk population also presents an interesting case, in that the three-state area delineated is comparable in extent to populations that have been recognized as significant but failed the test of being distinct and thus capable of being independently evaluated. In fact, we currently are reviewing this species continentwide, as well as throughout the range of one of its subspecies, for possible listing under the ESA.

Finally, the biological status of a population segment that satisfied the tests for distinctness and significance would be considered in relation to the ESA's standards for listing: would the population segment, treated as if it were a species, fit the ESA's definition of endangered or threatened?

American Fisheries Society Symposium 17:425–429, 1995

A National Biological Service Perspective on Defining Unique Units in Population Conservation

TIM L. KING AND J. LARRY LUDKE

National Biological Service, Leetown Science Center,
1700 Leetown Road, Kearneysville, West Virginia 25430, USA

The intent of this conference is to establish a forum in which the best science available can be applied to discuss ways in which the scientific and management communities can define subunits of fish and other aquatic species for conservation purposes. We applaud the leadership that the American Fisheries Society has demonstrated by organizing and coordinating this conference. The conference is timely and significant as federal and state managers, scientists, conservation interests, lawmakers, business and development interests, and the general public all grapple with the complex issues of conserving this nation's biological heritage while moving forward with sustained development.

Our purpose for participating in the conservation guidelines session of this forum is to provide a National Biological Service (NBS) perspective on administrative and programmatic application of the concepts of "evolutionarily significant" and "irreplaceable" units in aquatic biological diversity. Additionally, we will attempt to summarize some of the practical concerns that we have as research managers in a nonadvocacy federal bureau. Certainly the collective knowledge and wisdom reside within this body to take us a step closer to understanding how to conserve aquatic biological diversity better.

Because the National Biological Service is a new bureau within the Department of the Interior (DOI), it may be helpful to acquaint you briefly with the NBS. Shortly after assuming office as Secretary of the Interior, Bruce Babbitt announced his intention to consolidate biological research activities, then scattered among various DOI bureaus, into a single organizational entity. In the wake of the northern spotted owl debacle, Secretary Babbitt embarked on a bold attempt to get DOI's house in order. His vision was, and is, to instill a strong ethic supporting natural resource conservation throughout DOI; to build partnerships to sustain economic development without sacrificing natural resources; and, aggressively, to seek to facilitate the creation of an information-rich environment in which decision makers can formulate enlightened management policies, regulations, and programs.

The NBS is intended to be a focal point for catalyzing the generation of and enhancing the availability of data needed by resource managers to make informed decisions (National Research Council 1993). More specifically, the NBS

- serves as the biological research arm of the DOI;
- is an independent, scientific, nonadvocacy agency; and
- will be a leader in forging the national and international partnerships required to provide timely and quality information to natural resource managers.

The National Biological Service

Authorities and Regulatory Responsibilities

To appreciate the position of the NBS with regard to application of evolutionary units of biological diversity in the regulatory arena, it is necessary to understand the status and intended relationship of the NBS to the management process.

1. The NBS exists as a result of administrative fiat through the Secretary of the Interior's authority and is funded by annual appropriation from the U.S. Congress. The NBS does not have independent authorizing legislation. An authorization bill has passed the U.S. House of Representatives, though with considerable constraints regarding research and survey activities on private lands and the use of volunteers.

2. The NBS conducts research, inventories, and surveys under numerous authorities of the Secretary of the Interior, among which are the Fish and Wildlife Act of 1956 (16 U.S.C. § 742); Migratory Bird Treaty Act of 1918 (16 U.S.C. §§ 703 to 708, 709a, 710, 711) and the Migratory Bird Conservation Act (16 U.S.C. §§ 715 to 715r); Fish and Wildlife Coordination Act (16 U.S.C. §§ 661 to 666c); Nonindigenous Aquatic Nuisance Prevention and Control Act of 1990 (16 U.S.C. §§ 4701 to 4751); and Endangered Species Act of 1973 (ESA; 16 U.S.C. §§ 1531 to 1544).

3. The NBS is an independent, scientific organization and as such has a nonadvocacy role. It is

not regulatory but serves as a primary information provider to agencies that do have regulatory responsibilities.

A fundamental concern of the NBS as part of the scientific community, and as a source of information to managers, is the accuracy with which regulations and policy reflect biological fact. To the scientist, truth is based on the best accumulation of information at a given point in time. In light of new knowledge, or better information, acceptance of what is true may change. Today, however, we live in the information age—in the midst of an information explosion. Our legal and social institutions cannot evolve as rapidly as knowledge accumulates. This may result in greater variance and widening gaps separating current state of knowledge (representing truth to the scientist) from the knowledge base that was operational when a particular law or regulation was formulated. This relationship between society's rules affecting natural resources and the underpinning state of scientific knowledge is the linkage that provides the brush strokes of veracity and relevance to natural resource managers in the application of their art.

As scientists, we are mindful of the need for a strong connection between the knowledge base and its application in the form of laws and regulations. Where the knowledge base is undergoing rapid development, such as it currently is in molecular genetics, scientists have a responsibility to reconcile emerging technology and its results with the knowledge base of the past. Scientists must assist managers by helping them understand the application and meaning of emerging biological information in the regulatory setting.

The NBS concern is to provide valid, useful, proactive information to the decision maker. It is especially important to understand the relationship of the knowledge base to the formulation and application of regulations—where the linkages are strong and where they are weak and when they are obsolete. Such understanding paves the way for new knowledge to provide improved understanding, better tools, and a firm biological foundation for consideration in the development and application of laws and regulations.

Mission and Partnerships

We have chosen to discuss NBS administrative interests and concerns relative to aquatic biological diversity in terms of the bureau's mission. We emphasize partnerships and teamwork as opposed to competition because we believe there is a vast family of federal, state, academic, and private interest stakeholders in the societal application of the concept of evolutionarily significant units (ESUs) of biological diversity.

We are experiencing an intense period of change in this nation. We are simultaneously reinventing, downsizing, streamlining, and revitalizing government. We are attempting to initiate a new way of managing, with emphasis in more holistic ecosystems approaches. Land-managing agencies are energetically devising ecosystem templates upon which they will base new strategies to manage at the landscape scale. National standards have not been established, so uniformity in approach does not exist among the various bureaucracies in their endeavors to implement this new vision of order. The bad news is that this revolution fosters temporary inefficiency, lack of uniformity and standardization, and confusion. The good news is that the circumstances of change breed innovation, creativity, and, ultimately, progress and new understanding.

We have been unkind to the land and water upon which plants and animals (including ourselves) depend. With a greater understanding of ecosystems, we may be better able to determine which activities are, and those that are not compatible with different ecosystems. Which are our vulnerable or threatened habitats and species in greatest need of protection? Who are the partners with whom we need to cooperate and the managers for whom we need to provide information to better conserve biological diversity? Finally, with whom must we establish dialogue and seek consensus to reverse the adversarial trend of the past and create an environment of cooperation for solving our mutual problems?

Program Activities

Several of the NBS program activities in research, inventory, monitoring, and information transfer provide an infrastructure that supports a better understanding of aquatic biological diversity and ESUs. Currently, the available resources are insufficient to accomplish the task of inventorying the nation's biological resources, determining trends, and conducting the vast amount of research needed to determine causes and effects of population and species declines; to develop better tools and technologies needed to assess status and trends; and to disseminate and share available knowledge. The NBS mission was developed with the realization that we have insufficient information about this nation's biological resources and that it will take a collective effort with all stakeholders working together to provide a better understanding of the status of natural resources, of real and potential threats to their survival, and of the current needs of management to conserve them.

Peter Raven, who chaired the National Research Council Committee on the Formation of the NBS, has said "A national biological survey should do more than just catalog species and their ranges. It must also investigate such issues as how to manage biological resources in a sustainable manner." The earth and its biota are being challenged at an ever-increasing rate. It is difficult in many instances to discern what is "natural" any longer. At a time when we need, but lack, the necessary understanding of biological processes, biological diversity is being challenged with accelerated change emanating from every direction. How can species and populations that have evolved over millennia adapt to drastic modification in habitat occurring over decades or years? It is in this context that we struggle with the very pragmatic problem of determining priorities for biological conservation.

The NBS is committed to working in partnership with others in a quest to understand and identify where our conservation priorities must lie. Secretary Babbitt has established partnership and ecosystem initiatives as programmatic priorities for the NBS. The NBS is cooperating with the Smithsonian Institution in support of a National Biological Diversity Center. A conceptual framework has been developed to initiate a National Biological Information Infrastructure, a national information network making biological information more universally available to managers and scientists. Specific research initiatives include increased emphasis on genetics and systematics, ecosystems management, nonindigenous species introductions and their effects, and species at risk, to mention a few.

An ambitious effort has begun to publish the first of a planned biennial status and trends report on the condition of the nation's biological resources. The first report will focus on the status and trends of well-studied taxa, ecological components, and communities. Authors have been solicited from government and private sectors, and their reports will be synthesized into a state-of-the-knowledge summary, targeted for completion in August 1994.

Biological Diversity

The Problem

- The nation's biological diversity is in decline, and there are many unanswered questions about how it should be managed to support the sustainability

of all the goods and services upon which we depend.
- In the United States alone, over 760 species of fishes, crayfish, and freshwater mussels are considered at risk. Only 70 of the 297 species of freshwater mussels in the United States are considered to have stable populations. In some areas biologists are recommending that populations of mussels be brought into captive refugia because of the real threat that exotic zebra mussels *Dreissena polymorpha* pose to their survival.
- Nationally, 775 species of plants and animals have been listed as threatened or endangered, and there is an enormous backlog of candidate species that have been nominated for listing.
- Recovery programs have been developed for only half the listed species and far fewer have been implemented.

Underlying these alarming trends is the reality that we do not know how human alteration and degradation of ecosystems affect those ecosystems' ability to provide in a sustainable way the goods and services upon which society depends. Declines in the quantity and quality of the nation's biological resources are due in part to a lack of basic knowledge of the biota, ignorance of trends, and inefficient use of existing information. Such a lack of information and understanding is critical because we continually make important, and often irreversible, decisions concerning living resources.

Typical Symptoms of the Problem

The following are examples of manifestations of problems posed by threatened populations of species (Ralls et al. 1992).

- Research on threatened populations is neglected until a taxon is endangered. Similarly, research on husbandry and other techniques critical to implementing sound recovery strategies is postponed until the last few individuals remain.
- The resources we provide for research on threatened taxa are so meager that only the most economically important (e.g., spotted owl *Strix occidentalis*) or high-profile (e.g., California condor *Gymnogyps californianus*, Atlantic salmon *Salmo salar*) species can be studied. Moreover, economic conflicts over rapidly diminishing habitat prompt population viability analyses of threatened populations in the absence of adequate data.
- Decisions concerning the listing of a species are often made in the absence of genetic information

for comparative evaluation of the species' validity and relationships.

Tough Questions for Management Agencies

A major shortcoming of present fisheries knowledge is the lack of detailed and precise information relating life history of fish stocks to local genetic adaptation. Decisions on managing stocks (herein defined as an intraspecific group of randomly mating individuals that has temporal and spatial integrity), and whether to list populations and species, may be based on partial information derived from only a portion of the technologies available. Certainly, an ESU could best be defined by simultaneously examining three distinct levels of organization (Clayton 1981), including (1) the molecular level (allozymes, mitochondrial DNA restriction fragment length polymorphisms and sequences, nuclear variable number of tandem repeats, and randomly amplified polymorphic DNA); (2) the organismic level (ecophenotypical, physiological, and behavioral traits); and (3) the ecological level (environmental factors and species interactions).

The following are examples of "evolutionary unit" issues of current practical concern.

Genotypic and ecophenotypic contradiction.—Sturgeon abundance and distribution have decreased dramatically over the last century. This decline has been attributed primarily to loss of habitat and to overharvesting. At present, two-thirds of the North American sturgeon are considered threatened, endangered, or of special concern (Williams et al. 1989). Recently, questions have arisen concerning the systematics of sturgeon in the genus *Scaphirhynchus*. The genus consists of three putative species: shovelnose sturgeon *S. platorynchus*, pallid sturgeon *S. albus* (federally listed as endangered), and Alabama sturgeon *S. suttkusi*. The Alabama sturgeon is currently being reviewed for endangered species status by the U.S. Fish and Wildlife Service (FWS).

Meristic and morphometric differences were used to distinguish the three *Scaphirhynchus* species. However, two sources of genetic data appear to cloud the systematics. Congruence of allozyme data (Phelps and Allendorf 1983) with the mitochondrial cytochrome *b* sequence data (W. B. Schill, National Biological Service-Leetown Science Center, personal communication) suggests the three species of *Scaphirhynchus* are possibly phenotypic variants of the same species. This presents a management dilemma that would best be resolved by further research into population genetics (e.g., estimating

gene exchange), systematics, and life history characteristics (i.e., identifying physiological, behavioral, or ecological isolating mechanisms) of these important fishery resources.

Genotypic contradictions and ecophenotypic data of no value.—The eastern oyster *Crassostrea virginica* ranges from St. Lawrence Bay, Nova Scotia, through the Gulf of Mexico to the Yucatan Peninsula, Mexico, into the West Indies, and may extend to Brazil. Throughout this vast range, genetic differentiation and nonheritable variation in morphology and physiology have been reported. Shell morphology has proven susceptible to environmental influences but exhibits no predictive variation among geographic populations. Population genetic surveys have indicated eastern oysters comprise at least four geographic races; however, allozyme and mitochondrial DNA analyses have produced different results in two portions of the species range.

An extensive allozyme survey suggested no allele frequency differentiation between Atlantic Coast and Gulf of Mexico eastern oysters (Buroker 1983); the only differences observed were between one Laguna Madre, Texas, population and all northern Gulf of Mexico and Atlantic Coast eastern oysters. Further analyses found discontinuous allele frequencies between Laguna Madre populations and all other northern Gulf of Mexico populations (King et al. 1994). The discontinuity in genetic variability occurred between two populations separated by 26 nautical kilometers. In contrast, a mitochondrial DNA study reported major differences between Atlantic and Gulf of Mexico populations but found no differences between Gulf of Mexico populations and Laguna Madre populations (Reeb and Avise 1990). Results of these studies underscore the need for caution in inferring population genetic structure and estimates of gene exchange from any single class of genetic markers.

How different does different have to be?—Atlantic salmon once served as a major food source for precolonial New Englanders. However, beginning in the early nineteenth century, Atlantic salmon populations and their associated habitat were altered by dam construction, logging, pollution, and excessive harvesting. A combination of these factors is believed to have caused the precipitous decline in Atlantic salmon populations. As a result of this marked decrease in abundance (reflected as low and declining Atlantic salmon runs and low juvenile densities), Atlantic salmon in five Maine rivers were designated as "category 2" candidates for listing under the ESA in 1991. In October 1993, all anadromous U.S. Atlantic salmon were included in a petition to region 5 of the FWS for a rule to list the species under the ESA.

To provide the most informed response to this petition and for planning and implementing biologically sound management programs knowledge of intraspecific genetic structure and a thorough understanding of the evolutionary relationships among naturally reproducing populations of Atlantic salmon are essential. However, guidelines are needed to determine levels of differentiation needed to be considered "evolutionarily significant" or "irreplaceable." Stated differently, how different (e.g., genetically, ecophoenotypically, or behaviorally) must two geographic populations be to be considered distinct management units?

Intuitively, the best approach is that which combines multiple classes of data and includes all levels of consideration. We have a real need for taxonomists, systematists, geneticists, ethologists, and ecologists to work together to devise an approach the manager can apply. We are making great progress in developing more tools with greater sophistication to assist us in understanding relatedness of animals and plants. Application of these tools may result in greater likelihood for error in the sense that we may show differences where they do not exist. Likewise, instances have occurred where no differences were identified when they actually existed. Barraged with this cacophony of information, the resource manager must make a decision, often from contradictory data. The manager is more comfortable when faced with information indicating differences because the likelihood of failing to protect that which should be protected is minimized.

In summary, the NBS functions as a nonadvocacy, fact-finding research entity within the DOI. Consistent with Secretary Babbitt's vision that the NBS provide information to managers so that they can make timely decisions in an information-rich environment is our commitment to work with other government scientists, academics, and private sector interests to understand better, and manage to conserve, biological diversity. A heightened understanding of evolutionarily significant or irreplaceable units should lead to greater certainty in management and the potential for reduced risk to resources in formulating policy, regulations, and law.

References

Buroker, N. E. 1983. Population genetics of the American oyster *Crassostrea virginica* along the Atlantic coast and the Gulf of Mexico. Marine Biology 75:99–112.

Clayton, J. W. 1981. The stock concept and the uncoupling of organismal and molecular evolution. Canadian Journal of Fisheries and Aquatic Sciences 38:1515–1522.

King, T. L., R. Ward, and E. G. Zimmerman. 1994. Population structure of Eastern oysters (*Crassostrea virginica*) inhabiting the Laguna Madre, Texas and adjacent bay systems. Canadian Journal of Fisheries and Aquatic Sciences 51 (Supplement 1):215–222.

National Research Council. 1993. A biological survey for the nation. National Academy Press, Washington, DC.

Phelps, S. R., and F. Allendorf. 1983. Genetic identity of pallid and shovelnose sturgeon (*Scaphirhynchus albus* and *S. platorynchus*). Copeia 1983:696–700.

Ralls, K., R. A. Garrott, D. B. Siniff, and A. M. Starfield. 1992. Research on threatened populations. *In* D. R. McCullough and R. H. Barrett, editors. Proceedings of Wildlife 2001: populations. Elsevier Science Publishers, New York.

Reeb, C. A., and J. C. Avise. 1990. A genetic discontinuity in a continuously distributed species: mitochondrial DNA in the American oyster, *Crassostrea virginica*. Genetics 124:397–406.

Williams J. E., and seven coauthors. 1989. Fishes of North America endangered, threatened, or of special concern: 1989. Fisheries 14(6):2–20.

American Fisheries Society Symposium 17:430–433, 1995

Roles, Responsibilities, and Opportunities for the Bureau of Land Management in Aquatic Conservation

MICHAEL P. DOMBECK AND JACK E. WILLIAMS[1]

Bureau of Land Management, 1849 C Street, NW, Washington, DC 20240, USA

The Department of the Interior's Bureau of Land Management (BLM) administers more land—nearly 109 million hectares—than any other federal agency. Objectives for managing these public lands are rooted in the legal mandates of multiple use and sustained yield.

During the past decade, it has become increasingly clear that our traditional approach to managing the land has been inadequate. One only has to witness the loss of biodiversity, the spread of exotic species, or the reduction in land productivity to see that the health of the land is diminishing and that long-term sustainability of our landscape is in jeopardy. As Aldo Leopold (1941) wrote in his essay *Wilderness as a Land Laboratory*,

> The effort to control the health of land has not been very successful. It is now generally understood that when soil loses fertility, or washes away faster than it forms, and when water systems exhibit abnormal floods and shortages, the land is sick.

Numerous reports by the American Fisheries Society have served as repeated wake-up calls for all land managers. For example, a 1989 report (Williams et al. 1989) documented a 45% increase in the number of freshwater fish taxa in North America designated as rare, and a 1991 report (Nehlsen et al. 1991) listed 214 anadromous salmonid stocks in the West at risk of extinction. Although aquatic habitat conditions on BLM lands are less than desired, these lands include some of the best remaining habitats, including habitat for at least 109 of the 214 anadromous fish stocks. It is particularly disturbing, however, that large increases in aquatic at-risk taxa occur in the West, exactly where most federal lands are located, including almost all lands administered by the BLM. These endangered and threatened species are indicators of the declining health of the land. As Stewart Udall (1991) wrote in his *Foreword to Battle Against Extinction: Native Fish Management in the American West*,

> It is unfortunate that we must deal at the level of individual species. This forces us to focus attention on

single parts of ecosystems, while ecosystems themselves should be the subjects of our efforts. Endangered species are nonetheless the messengers of change, and we must heed their messages.

In the past, multiple-use management has more closely resembled a series of dominant single uses, each partitioned in separate parcels across the landscape. In one area, perhaps recreation was the dominant theme; in another, timber production. Multiple-use lands tend to have been either "locked up" through wilderness designation or "put up" for use. The middle ground of use in a sustainable manner has been hard to find. The BLM planning, budget, and program structures encouraged competition among BLM programs and among our interested publics. One of our most challenging administrative tasks is how to eliminate our functional program by program approach but at the same time encourage professional identity and resource accountability.

Today, we see BLM's role in ecosystem management as one of finding this middle ground between preservation and exploitation. As a multiple-use agency, BLM must allow use of the land, but land use cannot proceed at a pace that diminishes biodiversity or causes the need to list additional species or populations pursuant to the U.S. Endangered Species Act (16 U.S.C. §§ 1531 to 1544). We must strive for sustainability of ecological and economic systems. In this regard, agencies like the BLM and Forest Service should exemplify stewardship responsibilities.

During our past tug-of-war to allocate use of the land, the overall health of the landscape has diminished. Even the long-term viability of our wilderness areas now is being questioned because of invasions of exotic species and other problems originating outside wilderness boundaries. Fewer and fewer islands seem to exist. As John Muir[2] was fond of pointing out, "When we try to pick out anything by itself, we find it hitched to everything else in the universe." Noxious weeds are now judged to be spreading an incredible 14% annually, which equates to 932 more hectares each day affected on

---

[1]Present address: Intermountain Research Station, 316 E. Myrtle, Boise, Idaho 83702, USA.

[2]Muir, J. 1869. The unpublished journals of John Muir. [27 July 1869].

BLM lands. In northern California, yellow star thistle has spread 60% annually for each of the last 6
years. This spread of exotic species indicates declining land health and demonstrates a long-term, negative interaction between land use and the disruption of natural disturbance cycles. We all know of
similar invasions of exotic species in aquatic ecosystems, although they are hidden from our view and
therefore less obvious (Courtenay and Moyle 1992).

The emerging concepts of ecosystem management, if implemented with appropriate goals, the
best available science, and broad public involvement, can provide a suitable answer to such problems. We would like to explore BLM's emerging
concepts of ecosystem management, how these
emerging concepts relate to our existing mandates,
and how they relate to defining units of conservation in the aquatic environment.

Ecosystem Management

Ecosystem management is the integration of ecological, economic, and social components in order
to manage lands in a manner that safeguards resource sustainability, biological diversity, and productivity. In January 1994, the BLM articulated the
following principles of ecosystem management as
described in *Ecosystem Management in the BLM:
from Concept to Commitment*.

1. Sustain the productivity and diversity of ecological systems—the health of the land should be
   our primary goal.
2. Employ the best available science—science
   should provide the cornerstone for resource al-
   locations and establishment of land-use goals.
3. Involve all the players—all interested segments
   of the public should be involved in assessment,
   planning, and management-plan development.
4. Determine desired future ecosystem conditions
   based on historic, ecological, economic, and social considerations.
5. Minimize and repair impacts to the land—minimize our waste products.
6. Adopt an interdisciplinary approach to problem
   solving rather than continuing the functional,
   programmatic approach of the past.
7. Maintain a long-term perspective—for example,
   base planning on long-term horizons and goals.
8. Reconnect isolated parts of the landscape—for
   example, reconnect rivers to their floodplains.
9. Practice adaptive management—monitor our actions and make adjustments to management programs when necessary.

Director of the BLM Mike Dombeck, Assistant
Secretary of the Interior Bob Armstrong, and Secretary of the Interior Bruce Babbitt recently addressed these themes at the first-ever gathering of
all BLM managers—the BLM Summit—held in
April 1994 at Lake Tahoe. The summit focused on
five themes for BLM managers: ecosystem-based
management, collaborative leadership, improvement of the way we do business, service to our
publics, and diversification of our workforce. Ecosystem management is not simply a new strategy or
initiative. Rather, it is a fundamental change in the
way the BLM will administer nearly 109 million
hectares of public lands. Our mangers were told to
define ecosystem management "on the ground." As
Director Dombeck said at the summit, "If the land
isn't healthy, we have failed as managers." Again
from Aldo Leopold (*The Ecological Conscience*
1947), "No important change in human conduct is
ever accomplished without an internal change in
our intellectual emphases, our loyalties, our affec-
tions, and our convictions." At the BLM Summit we
strove to make this kind of internal change. Only
time and the resulting health of the land will determine our success.

The need for healthy land has never been greater.
As the health of our lands diminishes, our management options become progressively more and more
restricted. Healthy landscapes will provide habitat
complexity and diversity, which help maintain biological diversity. By providing healthy watersheds,
we maintain evolutionarily significant units (ESUs),
we conserve candidate species, and we ensure the
survival of the many smaller, lesser known species,
that, in the words of Edward O. Wilson (1987), "are
the little things that run the world"—the soil micro-
organisms and aquatic invertebrates that constitute
the unheralded ESUs.

We sometimes need to be reminded of the relationships among healthy ecosystems, biological diversity, and productivity. A recent report by Tilman
and Downing (1994) of the University of Minnesota
published in the 27 January 1994 issue of *Nature*
clearly described this critical linkage. Tilman and
Downing measured drought resistance of grasslands
composed of plots containing various levels of plant
species richness. They found that those grasslands
with the highest levels of plant diversity were more
drought resistant; that is, the most diverse plots
provided more productivity during drought because
the species rich plots contained more drought-tol-
erant species. It sounds deceptively simple—higher
diversity results in stability, resiliency, and higher
productivity—and it is a concept equally at home in

aquatic systems. Maintaining complex, diverse habitat structure, or rather, maintaining the processes that create the habitat complexity, is the key.

The management of BLM public lands has largely been influenced by two legal mandates: multiple use and sustained yield. The Federal Land Policy and Management Act of 1976 (FLPMA; 43 U.S.C. §§ 1701 to 1784), which is BLM's authorizing legislation, defines multiple use as "harmonious and coordinated management of the various resources without permanent impairment of the productivity of the land and the quality of the environment with consideration being given to the relative values of the resources and not necessarily to the combination of uses that will give the greatest economic return or the greatest unit output."

Past interpretations of multiple use have pitted development interests against preservation interests. Yet the definition of multiple use in FLPMA appears robust enough to encompass, or perhaps even encourage, a more conservative approach to land management.

Now, let us look at the emerging models for ecosystem management. Each example has its own strengths and weaknesses. In many respects, we have not found the perfect model of ecosystem management. For that matter, with all the local variation, complexities, and issues, finding a single model of ecosystem management may not be in our best interests.

In the Pacific Northwest, the Forest Service and BLM have developed a watershed management strategy, known as PACFISH (interim strategy for the management of anadromous-fish-producing watersheds on federal lands in eastern Oregon and Washington, Idaho, and portions of California), for aquatic and riparian systems. This strategy constitutes the aquatic and riparian components of the final supplemental environmental impact statement (EIS) within the range of the northern spotted owl *Strix occidentalis*. The strategy also will be included in the "Eastside" EIS process now underway for eastern Oregon and Washington and also will appear in the upper Columbia River basin EIS (outside the range of the northern spotted owl). It is through the EIS process that the strategy will be modified to include regional concerns and be amended into the land-use plans of both agencies. The strategy also may be applied to Alaska in the future.

The PACFISH strategy consists of the following elements: (1) riparian goals, (2) quantified riparian management objectives, (3) riparian habitat conservation areas (RHCAs), (4) standards and guidelines that guide all land management activities in RHCAs, (5) key watersheds, (6) watershed analysis, (7) restoration, and (8) monitoring. The goal of the strategy is to restore proper ecological functions to riparian areas and thereby provide the habitat complexity necessary in aquatic systems to support natural diversity and productivity of invertebrate and fish communities. Among the key features of this strategy are (1) a greatly expanded view of what constitutes riparian areas, including all floodplain and wetland areas, intermittent tributaries, and unstable zones, such as landslide-prone areas that can contribute substantial woody debris and silt into streams; (2) quantified management objectives for riparian and aquatic habitats; and (3) watershed analysis.

Successful implementation of this strategy will depend upon a variety of factors. First, success will depend upon involvement of the public, states, and others through the National Environmental Policy Act (NEPA; 42 U.S.C. §§ 4321, 4331 to 4335, 4341 to 4347) review process, which refines the strategy and institutionalizes it into the land-use plans of the BLM and Forest Service. Second, once the NEPA process is completed, there are a variety of needs and opportunities for partners during implementation. For example, watershed analysis requires input from many landowners and other parties to identify cumulative effects and restoration needs. There is a watershed analysis handbook to identify accepted procedures. The BLM and Forest Service are considering a certification process that would include public participation in ensuring that sound and scientifically acceptable procedures are followed. Although the goal of the strategy is to restore the health of riparian and aquatic systems and thereby meet the needs of entire communities rather than individual species, there is a key watershed process by which certain watersheds can be given priority attention, increased protection to address the needs of unique populations or other conservation units, or both. Here again, we need your assistance in determining which habitats and populations need special attention through key watershed designation.

Numerous other examples of watershed-level conservation strategies are emerging. Along the Henry's Fork River in Idaho, for example, ranchers, scientists, environmentalists, and agency representatives have formed a coalition to stop degradation of this entire watershed and begin restoration activities. Many of these initiatives are laudable but will require close scrutiny to ensure their scientific credibility. Applying the best available science to ecosystem management is a particular concern for the BLM, which has virtually no research capabilities of its own. The BLM will rely even more heavily upon professional societies, academia, and other federal agencies for support in this area.

In terms of regulatory needs, the existing authorities of FLPMA appear to be broad enough to encompass and encourage the concepts of ecosystem management and sustainability. Not all BLM lands, however, are managed under FLPMA. Unlike the majority of public lands managed under FLPMA, the 850,500 ha of BLM lands in western Oregon are managed pursuant to the Oregon and California Act of 1937 (O&C Act; 43 U.S.C. § 1181a). The O&C Act directs that these western Oregon lands be managed for "permanent forest production, and the timber thereon shall be sold, cut, and removed in conformity with the principal of sustained yield for the purpose of providing a permanent source of timber supply, protecting watersheds, regulating stream flow, and contributing to the economic stability of local communities." Some forest products groups have questioned the inclusion of these lands within the ecosystem management framework of the Northwest forest plan (the EIS within the range of northern spotted owl), in part because of O&C Act language.

One last item relative to our legal mandates—the BLM does not have a specific "viability" requirement, such as the one the Forest Service uses to ensure that all native populations are maintained at viable levels. This requirement has proved useful in recognizing the significance of ESUs and other population units before they reach federal candidate species status. In general, because of our adjacent land base in many states and overall similarity of missions, it is important that the BLM and Forest Service have compatible planning, budget, and regulatory frameworks.

In summary, this is a time of great change for federal agencies like the BLM. Reorganization of planning, budget structures, and office staffs is occurring at a rapid pace. There are opportunities to shape the structure, function, and even the mission of BLM. We welcome and encourage your assistance in this process.

In observing the magnitude of conservation problems facing us today, Edward O. Wilson (1992) described in his book *The Diversity of Life,*

> The solution will require cooperation among professions long separated by academic and practical tradition. Biology, anthropology, economics, agriculture, government, and law will have to find a common voice.

No less than the welfare of our natural heritage is at stake.

References

Courtenay, W. R., Jr., and P. B. Moyle. 1992. Crimes against biodiversity: the lasting legacy of fish introductions. Transactions 57th North American Wildlife and Natural Resources Conference 57:365–372.

Leopold, A. 1941. Wilderness as a land laboratory. The Living Wilderness 6:3.

Leopold, A. 1947. The ecological conscience. Bulletin of the Garden Club of America September 1947.

Nehlsen, W., J. E. Williams, and J. A. Lichatowich. 1991. Pacific salmon at the crossroads: stocks at risk from California, Oregon, Idaho, and Washington. Fisheries 16(2):4–21.

Tilman, D., and J. A. Downing. 1994. Biodiversity and stability in grasslands. Nature 367:363–365.

Udall, S. L. 1991. Foreword. Pages ix–xi *in* W. L. Minckley and J. E. Deacon, editors. Battle against extinction: native fish management in the American West. University of Arizona Press, Tucson.

Williams, J. E., and seven coauthors. 1989. Fishes of North America endangered, threatened, or of special concern: 1989. Fisheries 14(6):2–20.

Wilson, E. O. 1987. The little things that run the world (the importance and conservation of invertebrates). Conservation Biology 1(4):344–346.

Wilson, E. O. 1992. The diversity of life. Harvard University Press, Cambridge, Massachusetts.

American Fisheries Society Symposium 17:434–435, 1995

A Forest Service Perspective on Defining Unique Units in Population Conservation

GORDON HAUGEN

U. S. Forest Service
333 Southwest First, Portland, Oregon 97204, USA

The U.S. Department of Agriculture Forest Service's perspective on understanding the regulatory, administrative, and programmatic concerns of responding to and managing evolutionarily significant or irreplaceable units is presented here. The terms "evolutionarily significant" and "irreplaceable" units are not Forest Service terms, yet we work with them on a daily basis through our consultation efforts with the National Marine Fisheries Service (NMFS). The Forest Service deals with federally listed and proposed threatened and endangered species and sensitive species. Sensitive species are those plant and animal species identified by a regional forester for which population viability is a concern as evidenced by (1) a significant current or predicted downward trend in population numbers or density or (2) a significant current or predicted downward trend in habitat capability that would reduce a species' existing distribution.

The Forest Service has 13 laws and Executive Orders that provide the authority to manage wildlife, fish, and plant resources on national forests and grasslands. For this discussion, there are two primary statutes of interest: The National Forest Management Act of 1976 (16 U.S.C. §§ 1600 to 1614) and the Endangered Species Act (ESA; 16 U.S.C. §§ 1531 to 1544). These statutes, along with implementing regulations, form the basis for the Forest Service's fish and wildlife policy. This policy directs the Forest Service in the day-to-day management of habitats for plant and animal species on the national forests. The National Forest Management Act calls for the management of plant and animal communities throughout planning areas (a planning area equates to a national forest) and further requires that the management of habitats maintains viable populations of existing native and desired nonnative vertebrate species in the planning areas. A viable population is one that has the estimated number and distribution of reproductive individuals to insure that its continued existence is well distributed in the planning area. The Endangered Species Act mandates the Secretary of Agriculture, with respect to the national forest system, to conserve fish, wildlife, and plants, including those that are listed as threatened and endangered, and mandates that Forest Service actions not only avoid jeopardizing a listed species, but also lead toward recovery of the listed species. The combined mandates of these two statutes form the basis for the Secretary of Agriculture's fish and wildlife policy and the Forest Service's sensitive species policy. Both of these policies direct the Forest Service to ensure that Forest Service actions do not cause a loss of species viability nor cause a trend toward the federal listing of a species.

The Forest Service has a self-imposed policy to conduct and prepare a biological evaluation similar to the biological assessment required by the ESA for all actions that are funded, permitted, or carried out by its various units. The biological evaluation is the process that determines if an action would lead to the loss of species viability or cause a trend toward the federal listing of a species on the regional forester's sensitive species list. It is also the process that is used to determine if an action would result in a "may affect" determination for a listed or proposed species. A "may affect" determination would require the Forest Service to enter into a so-called "formal" consultation with the NMFS or the U.S. Fish and Wildlife Service.

With the above in mind, what effect does a newly listed species have on the Forest Service's continued implementation of ongoing actions and on the implementation of new proposals?

The Snake River Basin Consultation Effort

When the Snake River basin Pacific salmon *Oncorhynchus* spp. were listed, the Forest Service conducted an initial assessment of some 10,800 ongoing actions to determine if those actions constituted a "may affect" situation to the listed fishes. This analysis was conducted on 10 national forests in three Forest Service regions. The initial analysis indicated that approximately 4,800 ongoing actions had the potential to affect the listed Pacific salmon. The area covered by the Snake River basin listing was divided into 52 sections in seven watersheds for the purpose of developing comprehensive biological assessments. This effort to comply with the ESA regulations put a tremendous burden on Forest Service personnel and budgets as well as on the consultation staff of NMFS.

Cooperation between NMFS, the Bureau of Land Management (BLM), and the Forest Service has been good in this effort. The agencies are consistently in communication with each other at all levels. In some cases, the Forest Service and the BLM are preparing joint biological assessments through the designation of a lead agency.

Funding associated, directly or indirectly, with the consultation effort is a real concern. With limited dollars, choices need to be made between habitat improvement projects and consultations, or development of range allotment management plans, permit administration, and consultations. The multiple lawsuits that followed the listings have also placed a high demand on personnel and funds. It is very difficult and frustrating for a biologist to have a priority of preparing a biological assessment and an administrative record and providing on-the-ground project monitoring. Due to these complications, multiple listings of new species can bring projects and programs, even beneficial ones, to a complete standstill.

From a programmatic standpoint, the Forest Service and the BLM put forth the PACFISH initiative (interim strategy for the management of anadromous-fish-producing watersheds on federal lands in eastern Oregon and Washington, Idaho, and portions of California) to provide improved management of the watersheds supporting anadromous fish. An environmental assessment for PACFISH was completed. Upon adoption of the proposed interim direction, Forest Service plans were amended to incorporate new management standards. These standards remain in effect until ecosystem-based environmental impact statements that provide long-term conservation strategies can be completed. The Forest Service is working with NMFS and the BLM to develop long-term monitoring strategies to determine the effectiveness of such future management strategies.

Summary

The Forest Service is, and has been, proactive in the management of anadromous fish habitat. One would hope that being proactive would be viewed as a positive position to take; however, being proactive has raised questions by many of the public who have been, and still are, consumptive users of resources managed on national forest system lands. This has lead to copious amounts of planning, time-consuming responses, and either perceived or actual litigation that has used limited staffing and funding that would otherwise be available to get work done which will benefit the fish and wildlife habitat resources. In closing, I would ask the scientific community to help shape the process behind determining the evolutionarily significant unit (ESU), and, more importantly, when doing so to include both the nonconsumptive as well as the consumptive users of national forest resources so that there is broad acceptance of the science used in shaping the ESU concept. This will go a long way in helping to make policy that can be developed, stand the test of time, and defuse litigation.

DATE DUE

APR 2 8 '87			
GAYLORD			PRINTED IN U.S.A